Encyclopedia of Cloud Computing

Encyclopedia of Cloud Computing

Editors

SAN MURUGESAN
BRITE Professional Services and
Western Sydney University, Australia

IRENA BOJANOVA
National Institute of Standards and Technology (NIST)
USA

WILEY

Library of Congress Cataloging-in-Publication Data

Names: Murugesan, San, editor. | Bojanova, Irena, 1963– editor.
Title: Encyclopedia of cloud computing / editors, San Murugesan, Irena Bojanova.
Description: Chichester, West Sussex, United Kingdom ; Hoboken, NJ : Wiley, [2015] |
    Includes bibliographical references and index.
Identifiers: LCCN 2015040295 (print) | LCCN 2015051450 (ebook) |
    ISBN 9781118821978 (cloth) | ISBN 9781118821954 (Adobe PDF) |
    ISBN 9781118821961 (ePub)
Subjects: LCSH: Cloud computing–Encyclopedias.
Classification: LCC QA76.585 .E556 2016 (print) | LCC QA76.585 (ebook) |
    DDC 004.67/8203–dc23
LC record available at http://lccn.loc.gov/2015040295

A catalogue record for this book is available from the British Library.

Set in 10/12pt Times by SPi Global, Pondicherry, India

1   2016

Editorial Advisory Board

Contents

About the Editors

San Murugesan is editor-in-chief of *IT Professional,* director of BRITE Professional Services, and adjunct professor at the Western Sydney University, Australia. He is a corporate trainer, a consultant, and an author. He is a former senior research fellow of the US National Research Council at NASA Ames Research Center, California, and served in various positions at ISRO Satellite Centre, Bangalore. His expertise and interests span a range of areas: cloud computing, green IT, IT for emerging regions, Internet of Things (IoT), smart systems, and mobile applications.

He is a co-editor of *Harnessing Green IT: Principles and Practices* (Wiley and IEEE Computer Society, 2012), *Understanding and Implementing Green IT* (IEEE Computer Society, 2011), *Handbook of Research on Web 2.0, 3.0, and X.0: Technologies, Business, and Social Applications* (Information Science Reference, 2009), and *Web Engineering* (Springer, 2001). He serves as editor of *Computer* and edits and contributes to its "cloud cover" column. He also serves as associate editor of *IEEE Transaction on Cloud Computing.*
He is standing chair of the COMPSAC Symposium on IT in Practice (ITiP). Dr. Murugesan is a Fellow of the Australian Computer Society, a Fellow of IETE, and a senior member of the IEEE Computer Society (IEEE CS). For further information, visit his web site at www.bitly.com/sanprofile (accessed November 22, 2015).

Irena Bojanova, PhD, is a computer scientist at the National Institute of Standards and Technology (NIST). She managed academic programs at University of Maryland University College (UMUC), Johns Hopkins University (JHU) and PIsoft Ltd., and co-started OBS Ltd. (now CSC Bulgaria). She received her PhD in computer science / mathematics from Bulgarian Academy of Sciences and her MS and BS degrees in mathematics from Sofia University, Bulgaria.

Dr. Bojanova was the founding chair of IEEE CS Cloud Computing Special Technical Community and acting editor-in-chief of *IEEE Transaction on Cloud Computing.* She is a co-chair of the IEEE Reliability Society IoT Technical Committee and a founding member of the IEEE Technical Sub-Committee on Big Data. Dr. Bojanova is the Integrity Chair, IEEE CS Publications Board, an associate editor-in-chief and editor of IT

Trends Department of *IT Professional* and an associate editor of *International Journal of Big Data Intelligence* (IJBDI). She writes cloud and IoT blogs for IEEE CS Computing Now (www.computer.org, accessed February 13, 2015). Dr. Bojanova is a senior member of IEEE CS and can be reached at irena.bojanova@computer.org.

About the Authors

Saeid Abolfazli is a research lead and big data analyst at Xchanging and YTL Communications, Malaysia. His research interests include mobile cloud computing, parallel and distributed systems, and big data analytics. Saeid completed his PhD at the University of Malaya in 2014 where he served as a research member and part-time lecturer. His PhD was fully funded by the University of Malaya as a high-impact research project. Saeid was also a lecturer for the Ministry of Education and Khorasan Technical and Vocational Organization between 2000 and 2006. He is a member of the IEEE Society and IEEE CS Cloud Computing special technical community (STC). He served as Publicity Chair of ISSRE in 2015 and reviewer for several international conferences and journals on computer science. For further details, visit his Web page at www.mobilecloudfamily. com/saeid (accessed November 21, 2015) and contact him at abolfazli@ieee.org.

Lailani L. Alcantara is an associate professor at the College of International Management and Graduate School of Management in Ritsumeikan Asia Pacific University (APU). Her research interests include innovation, entrepreneurship, international expansion, and social networks. Her current research projects include ubiquitous computing business and innovation diffusion. She has published articles in the *Journal of International Management, Management Decision, Japanese Journal of Administrative Science, Management Research Review, Journal of Transnational Management*, and *Asian Business and Management*. She has received research grants from the Ministry of Education, Culture, Sports, Science and Technology in Japan, the Highly Commended Paper Award from Emerald Publishing, and outstanding research and teaching awards from APU. For further details, visit her Web page at https://sites.google.com/site/lailanipark/home (accessed November 22, 2015).

Peter Altevogt is a performance architect at IBM Germany Research and Development GmbH in Boeblingen in Germany. His interests include hardware architectures, information management, cloud computing, high-performance computing systems and performance analysis using discrete-event simulations and queueing modeling. He has built up and led performance teams for IBM BladeCenter systems, IBM information management software products, and cloud management software. He is currently working on performance analysis of next-generation processor systems. He has authored numerous publications and holds various patents. For further details, contact him at peter.altevogt@de.ibm.com.

Frederico Alvares is a teaching and research associate at Ecole des Mines de Nantes, France. He received his PhD in Computer Science from University of Nantes and his interests include cloud computing, autonomic computing, and self-adaptive component-based software development. His major contributions include a mechanism for the coordination and synchronization of autonomic cloud services and the application of discrete control techniques to build correct self-adaptive component-based applications. For further details contact him at frederico.alvares@inria.fr.

Alaka Ananth is currently working as a lecturer in the Department of Computer Science and Engineering at the National Institute of Engineering, Mysore, India. She had completed her postgraduate program at the National Institute of Technology Karnataka, Surathkal Mangalore, India. Her research interests include cloud computing and algorithms. She has five papers to her credit. For further details contact her at alaka.bhoomi@gmail.com.

Alexandru-Florian Antonescu is a research associate in the Department of Products and Innovation at SAP Switzerland. He received his PhD from the University of Bern (Switzerland) in 2015. He obtained his Master degree in the management of information technology, and his diploma in computer science from University "Politehnica" of Bucharest (Romania). His research interests include distributed computing, scalability of cloud systems, large-scale statistical data analysis, and mobile computing. For his PhD he investigated the use of service-level agreements in cloud environments for scaling distributed infrastructures. For further details contact him at antonescu@iam.unibe.ch.

Mehmet N. Aydin is an associate professor of management information systems (MIS) in the Faculty of Engineering and Natural Sciences at KADİR HAS University. He holds a PhD degree in MIS from the University of Twente in the Netherlands. His interests include cloud computing, agile software development, and business applications of social network analysis. He has published over 40 articles as journal papers (e.g., in the *Journal of Database Management, Information Frontiers*, and the *Journal of Enterprise Management*), book chapters (e.g., in the Springer series on Lecture Notes in Computer Science), and conference proceedings (IFIP 8.1, CAISE). His contribution to the research field of method engineering and agile methods has been cited in various studies. He can be reached at mehmet.aydin@khas.edu.tr.

Kapil Bakshi is a distinguished systems engineer for Cisco Systems Inc. He has over 19 years of experience in the industry in various roles at Oracle, Hewlett-Packard, and other salient high-tech companies. His current areas of focus are cloud architectures, big-data analytics, datacenter applications, and software-defined networks. He holds leadership positions in several industry forums and bodies. Kapil is a prolific author and industry contributor, with several publications, papers, and books. He also holds patents in data analytics and service provider domains. He holds a BSEE, a BSCS and an MBA from University of Maryland, College Park, and an MS in computer engineering from Johns Hopkins University. He can be reached at kabakshi@cisco.com, Twitter: @kapil_bakshi and https://www.linkedin.com/in/kbakshi (accessed November 22, 2015).

Diana Barreto is a research assistant at the Cloud Computing and Distributed Systems Laboratory (CLOUDS Lab) at the University of Melbourne, Australia. Her interests include distributed systems, cloud computing (especially platform as a service) and multicloud environments. She received her BSc in 2006 in systems engineering at the Pontificia Universidad Javeriana, Colombia. In 2008 she received a graduate diploma in software construction at Universidad de los Andes, Colombia, and in 2013 she finished her MSc degree in information technology at the University of Melbourne. She worked for more than 6 years, designing and developing distributed applications for telecommunications and financial companies. For further details contact her at dianibar@gmail.com.

Salman A. Baset is a research staff member at IBM T. J. Watson Research Center, Yorktown Heights, NY. His research interests include cloud computing, DevOps and configuration management, telco and IPTV analytics, and Internet protocols. He has co-designed the RELOAD protocol for building peer-to-peer communication systems, which is now an Internet RFC. Currently, he serves as the vice-chair of the SPEC OSG cloud subcommittee on cloud benchmark standardization; he is a core reviewer on the OpenStack Chef-based deployment project (https://github.com/stackforge/openstack-chef-repo, accessed November 22, 2015), and a chair of the distributed and fault-tolerant computing (DFTC) professional interest community at IBM. He is a recipient of the Young Scholars Award from the Marconi Society in 2008 and a best paper award at IPTCOMM in 2010.

Pierfrancesco Bellini is a professor of programming methods for electronic calculators, School of Engineering, at the University of Florence. His interests include ontology design, formal methods, temporal logics, distributed systems, and software engineering. He received a PhD in electronic and informatics engineering from the University of Florence. He has been involved in European Commission projects such as ECLAP, AXMEDIS, VARIAZIONI, IMAESTRO, WEDELMUSIC, and he is currently involved in industrial projects such as TRACE-IT, RAISSS, and ICARO CLOUD. He has been the program co-chair of the WEDELMUSIC, ICECCS, and AXMEDIS conferences. He has been co-editor of the ISO MPEG SMR standard in MPEG 4. For further details, visit his Web page at http://www.disit.dinfo.unifi.it/bellini (accessed November 22, 2015) or contact him at pierfrancesco.bellini@unifi.it.

Larry Beser is currently a master technology consultant with HP Enterprise Services. His interests include demonstrating how applying enterprise architecture principles to extreme complexity delivers unique value to his global clients. Larry's roles include chief technologist and chief architect with EDS as well as HP. He was recognized for his work by being named a Distinguished SE in the program's inaugural group at EDS. While at Cisco, he invented new tools and methods for creating cloud-adoption strategies for global clients and developed 3D datacenter modeling for visualizing problem and solution complexity. His publications span datacenter energy management, cloud strategy, and enterprise architecture. For further details, visit his Web page at www.LarryBeser.com (accessed November 22, 2015), or contact him at Larry@LarryBeser.com.

Kashif Bilal is an assistant professor at COMSATS Institute of Information Technology, Abbottabad, Pakistan. His research interests include energy-efficient high-speed networks, green computing, and robustness in datacenters. Currently, he is focusing on exploration of network traffic patterns in real datacenters and development of the datacenter network workload generator. He received his PhD in electrical and computer engineering from North Dakota State University, Fargo, United States in 2014. He also received the College of Engineering (CoE) Graduate Student Researcher of the year 2014 award at NDSU. For further details, contact him at kashifbilal@ciit.net.pk.

Thomas Michael Bohnert holds a PhD from the University of Coimbra and a diploma degree in computer engineering from the University of Applied Sciences Mannheim, for which he also worked as research associate and lecturer at the Institute for Software Engineering and Communications. After academic tenures, he joined SIEMENS Corporate Technology, the company's corporate research department, responsible for defining and driving a future Internet strategy. In 2008 he joined SAP AG, working at the SAP Research Labs Zurich, Switzerland, first as senior researcher and later as technical director. In mid-2009 he was appointed chief future Internet strategist. From 2012 onwards he was with Zurich University of Applied Sciences teaching service engineering and continuing future Internet research as head of the ICC Lab.

Irena Bojanova. See her biography in the "About the Editors" section.

Travis Breaux is an assistant professor of computer science at Carnegie Mellon University (CMU). His interests include software engineering, privacy and security, and legal compliance. Dr. Breaux is the director of the Requirements Engineering Lab, which has been funded by the US Department of Homeland Security, the National Science Foundation, and the National Security Agency. He is a co-founder of the IEEE Workshop on Requirements Engineering and Law, Chair of the USACM Committee on Privacy Security, and the designer of the engineering privacy course that is part of CMU's Masters in Privacy program. He is a senior member of the IEEE and ACM. For further details, visit his Web page http://www.cs.cmu.edu/~breaux/ (accessed November 22, 2015) or contact him at breaux@cs.cmu.edu.

V. K. Cody Bumgardner is the chief technology architect at the University of Kentucky, Lexington, KY. His interests include distributed "cloud" computing optimization, computational economics, and research-focused computing. Bumgardner is the author of the book *OpenStack in Action* and focuses on researching, implementing, and speaking about cloud computing and computational economics. For further details, visit his Web page http://codybum.com (accessed November 22, 2015) or contact him at cody@uky.edu.

Rajkumar Buyya is professor and director of the Cloud Computing and Distributed Systems (CLOUDS) Laboratory at the University of Melbourne, Australia. He is also serving as the founding CEO of Manjrasoft, a spinoff company of the university, commercializing its innovations in cloud computing. He has authored over 500 publications and four text books including *Mastering Cloud Computing* published by McGraw-Hill and Elsevier / Morgan Kaufmann (2013) for the Indian and international markets respectively. A scientometric analysis of cloud computing literature by German scientists ranked Dr. Buyya as the world's top-cited author and the world's most productive author in cloud computing. He served as the foundation editor-in-chief (EiC) of *IEEE Transactions on Cloud Computing* and is co-EiC of *Software: Practice and Experience.* For further information on Dr. Buyya, please visit his cyberhome at www.buyya.com (accessed November 22, 2015).

Wei Cai is a PhD candidate at The University of British Columbia (UBC). His interests include cloud gaming, mobile cloud computing, and cognitive software systems. He received an MSc and a BEng from Seoul National University and Xiamen University, and has completed research visits to Academia Sinica, the Hong Kong Polytechnic University and National Institute of Informatics, Japan. Wei has published more than 20 technical papers and his current ongoing work is a cognitive platform for ubiquitous cloud-based gaming. He has received awards of best paper from SmartCom2014 and CloudComp2013, a UBC doctoral four-year fellowship, a Brain Korea 21 scholarship, and an excellent student scholarship from the Bank of China. For further details visit his Web page at http://ece.ubc.ca/~weicai/ (accessed November 22, 2015), or contact him at weicai@ece.ubc.ca.

Rodrigo N. Calheiros is a research fellow at the Department of Computing and Information Systems, the University of Melbourne, Australia. His interests include cloud computing, big data analytics, and large-scale distributed systems. Since 2008, he has been involved in the design and development of the CloudSim toolkit, which has been used by many universities and companies worldwide. He is also serving as associate editor for the journal *Computers and Electrical Engineering.* For further details, contact him at rnc@unimelb.edu.au.

Eric Carlson is a systems engineer / program manager at TechTrend, Inc. His interests include cloud computing, systems engineering, investment analysis, and acquisition support. He has been instrumental in establishing the foundations for cloud computing within the Federal Aviation Administration (FAA), was one of the key authors of the FAA Cloud Computing Strategy, and worked across the agency to help bring the technology to fruition. He holds a Master of Science in systems engineering from George Washington University, a Bachelor of Business Administration and Minor in computer information systems from James Madison University, and is a certified project management professional (PMP). For further details, please contact him at carlsonem@gmail.com.

Mark A. Carlson, principal engineer, industry standards at Toshiba, has more than 35 years of experience with networking and storage development and more than 18 years' experience with Java technology. Mark was one of the authors of the CDMI cloud storage standard. He has spoken at numerous industry forums and events. He is the co-chair of the SNIA cloud storage and object drive technical working groups, and serves as vice-chair on the SNIA Technical Council.

Daniele Cenni is a research fellow and holds a PhD in engineering from the Department of Information Engineering, University of Florence, Italy. His interests include sharing systems (P2P), social networks, information retrieval, and cloud computing. He has participated in European research and development projects like ECLAP and AXMEDIS, and in regional research projects such as ICARO. He was professor for the course operating systems (Master of Science in Engineering, Faculty of Engineering, University of Florence). He has published technical articles in international journals and conferences on these subjects, and has received a research grant from University of Florence since 2007. For further details, visit his Web page http://www.disit.dinfo.unifi.it/cenni (accessed November 22, 2015) or contact him at daniele.cenni@unifi.it.

Saurav Kanti Chandra is a senior architect at Infosys Limited. His interests include technology evangelism in the areas of networking, the future of the Internet, network function virtualization, Internet of Things and software-defined networking. He has authored papers in IEEE conferences and has been the speaker in international conferences. For further details, contact him at kantichandra@infosys.com.

K. Chandrasekaran is a professor in the Department of Computer Science and Engineering at the National Institute of Technology Karnataka, Surathkal, Mangalore, India. His interests include distributed systems, cloud computing, big data, and green computing. He serves as associate editor and one of the editorial members of *IEEE Transactions on Cloud Computing*. He is also a member of the IEEE Computer Society's cloud computing STC. He serves as a member of various professional societies including the IEEE (senior member), the ACM (senior member), and the Computer Society of India (life member). He has more than 180 research papers published by various international journals, and conference papers. He has coordinated many sponsored projects supported by various organizations, which include government and industrial bodies besides some consultancy projects sponsored by industry. For further details, please contact him at his e-mail address: kchnitk@ieee.org.

Jack C. P. Cheng is an assistant professor at the Hong Kong University of Science and Technology (HKUST). His interests include construction management, building information modeling, construction ICT, and sustainable built environment. He is a professional member of the Hong Kong Institute of Building Information Modeling (HKIBIM), a member of the Autodesk Industry Advisory Board (AIAB), and a certified Carbon Auditor Professional (CAP). He has received the Champion Prize 2012 Environmental Paper Award organized by the Hong Kong Institution of Engineers (HKIE) in Hong Kong and three best paper awards in different international conferences. In 2012, he also received the School of Engineering Distinguished Teaching Award at HKUST. For further details, visit his Web page: http://www.ce.ust.hk/Web/FacultyStaffDetail.aspx?FacultyStaffId=70 (accessed November 22, 2015) or contact him at cejcheng@ust.hk.

Fangyuan Chi is a graduate student (MASc) at The University of British Columbia. Her interests include cloud gaming, distributed systems, and data mining. She is currently working on cloud gaming, and has published a conference paper in this field. For further details, contact her at fangchi@ece.ubc.ca.

Giuseppina Cretella received her PhD in Computer and Electronic Engineering from the Second University of Naples. She is involved in research activities dealing with Semantic Web and Semantic Web Services, Knowledge Discovery, Reverse Engineering and Cloud Computing. She participated in research projects supported by international and national organizations, such as: mOSAIC Cloud FP7 project, CoSSMic Smart Cities FP7 and Cloud@Home.

Yong Cui is a full professor at Tsinghua University, and co-chair of IETF IPv6 Transition WG Software. His interests include computer network architecture and mobile cloud computing. Having published more than 100 papers in refereed journals and conferences, he received the National Award for Technological Invention in 2013, the Influential Invention Award of the China Information Industry in both 2012 and 2004. He co-authored five Internet standard documents, including IETF RFC 7283 and RFC 7040, for his proposal on IPv6 transition technologies. He serves on the editorial board on both IEEE TPDS and IEEE TCC. For further details, visit his Web page http://www.4over6.edu.cn/cuiyong/ (accessed November 22, 2015) or contact him at cuiyong@tsinghua.edu.cn.

William Culhane is a computer science graduate student at Purdue University. His interests include distributed systems, especially with regard to data management and processing algorithms. His recent work with the LOOM system finds optimal aggregation overlays based on mathematical modelling. He has a Bachelor's degree from the Ohio State University and a Master's degree from Purdue University. For further details contact him at wculhane@purdue.edu.

Nitin Dangwal is a technology architect at Infosys Limited. His interests include cloud computing, access management, and identity federation. He has 11 years of intensive experience in IAM across product development, support, and implementation sides, especially in identity federation. He has previously published papers on security testing in cloud and federation in cloud. He can be contacted at nitin_dangwal@infosys.com.

Wolfgang Denzel is a research staff member at the IBM Zurich Research Laboratory in Switzerland. His interests include the architectural design and performance analysis of datacenters, high-performance computing systems, server-interconnect fabrics, and switches. He contributed to the design of various IBM products in the field of switching and high-performance computing systems and participated in numerous European and US government projects in this field. He has produced more than 40 publications and patents. For further details, contact him at wde@zurich.ibm.com.

Neha Mehrotra Dewan is a technology architect at Infosys Limited. Her interests include cloud computing, identity and access management, identity governance, and QA services. She has 11 years of intensive experience in JEE-based product development, sustenance, support, and deployment primarily in the IdAM domain. She has published papers in cloud-computing areas like "Cloud testing versus testing cloud", "Security testing in cloud" and "Federation in cloud." She is a certified IBM cloud computing advisor and can be contacted at neha_mehrotra@infosys.com.

Beniamino Di Martino is full professor of information systems and vice-director of the Department of Industrial and Information Engineering at the Second University of Naples (Italy). His interests include semantic Web and semantic Web services, cloud computing, big-data intelligence, high-performance computing and algorithmic patterns recognition. He is project coordinator of the EU-funded FP7-ICT-2010-256910 project "mOSAIC-Open-Source API and Platform for Multiple Clouds" and has been participating in several national and international research projects, including: EU-SMARTCITIES CoSSmiC, EU-ARTEMIS Crystal, EU-IST OntoWeb and APART, MUR PRIN "Cloud@Home," "Mosaico," and "Iside," FAR – Laboratori Pubblico-Privati – "LC3," CNR PF and Agenda 2000. He is chair of the nomination committee for the "Award of Excellence in Scalable Computing," and active member of: the IEEE Working Group for the IEEE P3203 Standard on Cloud Interoperability; the IEEE Intercloud Testbed Initiative; the Cloud Standards Customer Council, and the Cloud Computing Experts' Group of European Commission. For further details contact him at beniamino.dimartino@unina.it.

Vladimir Dimitrov is a full-time professor at the University of Sofia, Bulgaria. His interests include grid and cloud computing, database systems, and the formalization and verification of distributed systems. Currently, he is chief of the Computer Informatics Department at the Faculty of Mathematics and Informatics, University of Sofia, and director of the Master degree program on information systems in the same faculty. Vladimir Dimitrov is one of the key initiators for the Bulgarian segment of the European Grid. His recent research and publications are on the use of Z-notation and CSP for the formal specification and verification of distributed systems, such as distributed business processes and MapReduce framework. Vladimir Dimitrov is a member of the editorial board of *IT Professional*. For further details, contact him at cht@fmi.uni-sofia.bg.

Dan Dunn has been an IT consultant to the private and public sectors since the late 1980s and has spent time as a nonappropriated and Federal employee. Starting as a desktop technician and systems administrator, since the mid-1990s he has specialized in the planning and execution of large-scale infrastructure and datacenter transformations. Notably, he has planned and managed worldwide network refreshes, multiyear datacenter migrations, and prepared a Cabinet-level US agency for Y2K, which required a full refresh of the infrastructure. His experience in the private sector has provided valuable insight for his work now as a leader in Capgemini Government Solutions technology and cloud services. He can be reached at Daniel.Dunn@capgemini-gs.com.

Andy Edmonds is a senior researcher at the Zurich University of Applied Sciences. His interests include distributed systems, service-oriented architectures and cloud computing. He currently co-chairs the Open Grid Forum's Open Cloud Computing Interface working group. For further details, visit his Web page http://blog.zhaw.ch/icclab (accessed November 22, 2015), or his linkedin page http://ch.linkedin.com/in/andyedmonds (accessed November 22, 2015) or contact him at edmo@zhaw.ch.

Antonio Esposito is a PhD student at the Department of Industrial and Information Engineering at the Second University of Naples. His interests include semantic Web and semantic Web services, cloud computing, design and cloud patterns, and reverse engineering. He participates in the EU-SMARTCITIES CoSSmiC Smart Cities FP7 project and the EEE Intercloud Testbed Initiative. He has received the MS degree (magna cum laude) in computer and electronic engineering from the Second University of Naples. For further details contact him at antonio.esposito@unina2.it.

Patrick Eugster is an associate professor of computer science at Purdue University, currently on leave at the Technical University of Darmstadt. His interests include distributed systems, programming languages, and distributed programming. Patrick worked on a variety of subjects related to distributed systems, including event-based and reactive programming, wireless sensor networks, and cloud-based big data processing. Recognition for his research includes a CAREER award from the US National Science Foundation (2007), an induction into the DARPA Computer Science Study Panel (2010), and a Consolidator award from the European Research Council (2013). His research has also been funded by several companies including Google, NetApp, Cisco, and Northrop Grumman. For further details, visit his Web page http://www.cs.purdue.edu/homes/peugster/ (accessed November 22, 2015) or contact him at p@cs.purdue.edu.

Felix Freitag is associate professor at the Department of Computer Architecture of the Universitat Politècnica de Catalunya (UPC). His current interests include performance evaluation of distributed systems with a focus on cloud computing. He has published extensively in academic conferences and journals. Currently he leads the EU-funded research project, Clommunity. He received his PhD in telecommunications engineering in 1998. For further details contact him at felix@ac.upc.edu.

Doyle Friskney EDD is the AVP and chief technology officer at the University of Kentucky, Lexington, KY. His interests include the application of technology in research and instruction. His focus is on the development of collaborative learning spaces. Friskney has extensive understanding of technology in higher education research environments. In collaboration with the Teaching and Academic Support Center and libraries, he developed a new faculty and student support model that provides comprehensive support across organizational units. He has been appointed by the Governor to the Communications Advisory Council for the state of Kentucky. Recently, he was involved in efforts to establish a statewide higher education research and education network. He also presented at a number of higher education conferences. For further details contact him at doyle@uky.edu.

Renate Fruchter is the Founding Director of the Project Based Learning Laboratory (PBL Lab) in the Civil and Environmental Engineering Department at Stanford University. Her current research interests include collaboration technologies in support of cross-disciplinary, geographically distributed teamwork, e-Learning, cloud computing, and quantified-self data analysis. She has received the John L. Tishman Distinguished Lecture in Construction Engineering and Management (2014), the ASCE Computing in Civil Engineering award (2013); the Learning!100 award for innovation (2011), the US Distance Learning Association (USDLA) Twenty-First Century Award for Best Practices in Distance Learning Distinction (2010), and doctor honoris causa for ground-breaking work in PBL from AAU Denmark. For further details, visit her Web page http://pbl.stanford.edu (accessed November 22, 2015), or contact her at fruchter@stanford.edu.

G. R. Gangadharan is an assistant professor in the Institute for Development and Research in Banking Technology (IDRBT), Hyderabad, India. His research interests focus on the interface between technological and business perspectives and include Internet technologies and green IT. He received his PhD in information and communication technology from the University of Trento, Italy and the European University Association. He is a senior member of IEEE and ACM. Contact him at geeyaar@gmail.com.

Abdullah Gani is an associate professor at the Faculty of Computer Science and Information Technology, at the University of Malaya where he is director of the Center for Mobile Cloud Computing Research and dean of the faculty. He obtained his Bachelor and Master degrees from the University of Hull, United Kingdom, and his PhD from the University of Sheffield. He was the primary investigator for a high-impact research project from 2011 to 2016. Abdullah is an IEEE senior member. For further details visit his Web page at http://web.fsktm.um.edu.my/~abdullah/ (accessed November 22, 2015) or write to him at Abdullah@um.edu.my.

David G. Gordon works as a compliance engineer at Aetna in Denver, Colorado. Prior to this, he received an MBA from the University at Buffalo and a PhD in engineering and public policy from Carnegie Mellon University, where his research with Dr. Travis Breaux focused on the intersection between legal requirements and software engineering. Aside from his professional interests, he is an active singer and actor, having performed with the Buffalo Philharmonic, the Pittsburgh Symphony, the Pittsburgh Opera, and various theater companies. For more information or to contact him, please visit http://davegordonltd.com.

Mohammad Hadi Sanaei is with the Department of Information Technology at Shahsavand Co. He obtained his MSc in information technology management from the Ferdowsi University of Mashad. Hadi completed a BE in software engineering in 2010. His interests include mobile commerce, cloud-based mobile commerce systems, wireless communication and VOIP systems. For further details, contact him at mh.sanaei@gmail.com.

Piyush Harsh is a researcher at the InIT Cloud Computing Lab at the Zurich University of Applied Sciences. He has been researching ways to improve datacenter automation, and OSS/BSS solutions for clouds. In the past he co-chaired the OpenNebula interoperability working group. His research interests include distributed algorithms, security and authentication solutions, communication protocols, and online privacy. Find more about him at http://piyush-harsh.info (accessed November 22, 2015) or contact him at harh@zhaw.ch.

Ragib Hasan is an assistant professor in the Department of Computer and Information Sciences at the University of Alabama at Birmingham. His primary research interests are computer security, cloud security, and digital forensics. His research has been funded by a National Science Foundation CAREER award, as well as by grants from the Department of Homeland Security, the Office of Naval Research, Google, and Amazon. He was a recipient of the 2013 Google RISE award, the 2014 Best of Blogs and Online Activism award from Deutsche-Welle, and the 2013 Information Society Innovation Fund award for his Shikkhok.com online education platform for children in South Asia. For further details, visit his Web page http://www.ragibhasan.com (accessed November 22, 2015), or contact him at ragib@cis.uab.edu.

Carol M. Hayes is a research associate at the University of Illinois College of Law. Her interests include Internet policy, intellectual property, cybersecurity, and online privacy. She is licensed to practice law in Washington State and has written several academic articles focusing on topics like the intersection of cybersecurity and privacy online. She served as a Christine Mirzayan Science and Technology Policy Graduate Fellow at the National Academy of Sciences in Washington, DC, and as an Invited Foreign Scholar at the Institute of Intellectual Property in Tokyo, Japan. For further details, please contact her at carol.mullins@gmail.com.

Alex Heneveld is the co-founder and chief technology officer at Cloudsoft Corporation, and the founder and a committer on Apache Brooklyn, the leading application blueprinting software. His interests include application architecture, autonomic management, and the representation of semistructured information. Alex holds a PhD (informatics) and an MSc (cognitive science) from the University of Edinburgh, where he was a Marshall Scholar, and an AB (mathematics) from Princeton University. For further details, visit http://www.cloudsoftcorp.com/company/people/alex-heneveld/ (accessed November 22, 2015).

Craig Hill is a distinguished systems engineer in the US Federal area and has been with Cisco Systems for 19 years. His focus covers a broad range of both current and evolving technologies, protocols, and large-scale architectures in IP/MPLS, Campus, WAN, datacenter, SDN, and security, while covering a broad customer set, including DoD/IC, and large enterprises. He also serves as a senior field advisor on future product direction in Cisco's chief development office. He is a 19-year certified CCIE (#1628), has numerous white-paper publications spanning broad topics, and holds an electronics engineering degree from Capitol College. He can be reached at crhill@cisco.com, on Twitter: @netwrkr95 or on Linked-In: http://www.linkedin.com/in/crhill/ (accessed November 22, 2015).

Thorsten Humberg is a research assistant at Fraunhofer Institute for Software and Systems Engineering ISST. His interests and research topics include information security and compliance management, for example in the context of cloud computing. He received his computer science degree in 2010 from the

University of Dortmund, Germany. For further details, visit the ISST Web page or contact him at Thorsten. Humberg@isst.fraunhofer.de.

Shigeru Imai is a PhD student at Rensselaer Polytechnic Institute. His interests include adaptive middleware for cloud computing and fault-tolerant data-streaming systems. He is currently working on cost-efficient distributed data processing over multiple cloud environments. He is a recipient of the Yamada Corporation Fellowship for 2012–13. For further details, visit his web site at http://www.cs.rpi.edu/~imais/ (accessed November 22, 2015).

Chamikara Jayalath received his PhD, on geodistributed big-data processing, from Purdue University. Chamikara has published several peer reviewed papers on big-data processing. He has also made significant contributions to the Apache Web Services project where he was a project management committee member. For further details contact him at chamikaramj@gmail.com.

Jan Jürjens is professor of software engineering at Technical University Dortmund (Germany), Scientific Coordinator at Fraunhofer Institute for Software and Systems Engineering ISST, head of the Compliance Innovation Lab at the Fraunhofer Innovation Center for Logistics and IT, and senior member of Robinson College (University of Cambridge, United Kingdom). Previous positions include a Royal Society Industrial Fellowship at Microsoft Research Cambridge, a nonstipendiary Research Fellowship at Robinson College, and a postdoc position at TU München. Jan holds a Doctor of Philosophy in computing from the University of Oxford and is author of *Secure Systems Development with UML* (Springer, 2005; Chinese translation 2009) and other publications, mostly on software engineering and IT security. For more information see http://jan.jurjens.de (accessed November 23, 2015).

Sowmya Karunakaran is a Ph.D. student at IIT-Madras. Her research interests include dynamic pricing of online services, cloud computing economics, and modeling online user behavior. She has received various best paper awards and excellence certifications for her work. Her PhD thesis is on developing dynamic pricing algorithms based on bidder behavior for cloud computing services. Her research includes designing information systems that adapt to the user's behavioral biases using techniques such as game theory and machine learning. Sowmya has a Master's degree in management from IIM-Bangalore, and she has worked as a software consultant in the IT services industry for more than 6 years. For further details, contact her at sowmya.karu@gmail.com.

Philip Kershaw is technical manager for Earth Observation at the Centre for Environmental Data Archival, STFC Rutherford Appleton Laboratory (http://www.ceda.ac.uk, accessed November 22, 2015). His interests include federated identity management and access control, the application of cloud computing for the environmental sciences, and application development for environmental informatics. Philip is currently technical lead for the JASMIN Cloud, a project to deploy a private and hybrid cloud customized for the science community. JASMIN is a UK-based big-data processing and analysis facility for the climate science, Earth observation and environmental sciences communities funded by the UK Natural Environment Research Council. He is chair of a UK cloud-computing working group, formed to support and coordinate activities around the development of cloud computing technologies for the UK research community. He has contributed to the development of federated identity management systems for the environmental sciences including the Earth System Grid Federation and for the EU Framework 7 research project, Contrail, focused on the development of federated cloud capabilities. For further details, visit his blog at http://philipkershaw.blogspot.co.uk/ (accessed November 22, 2015) or follow at @PhilipJKershaw.

Jay P. Kesan is a professor at the University of Illinois College of Law. His interests include patent law, cyberlaw, and entrepreneurship. At the University of Illinois, Professor Kesan is also employed in the Institute of Genomic Biology, the Department of Electrical and Computer Engineering, the Information Trust Institute, the Coordinated Science Laboratory, the College of Business, and the Department of Agricultural and Consumer Economics. He has served as a Thomas A. Edison Scholar at the US Patent and Trademark Office (USPTO). For further details, please visit his faculty information page at http://www.law.illinois.edu/faculty/profile/jaykesan or contact him at kesan@illinois.edu (accessed November 22, 2015).

Osman Khalid is an assistant professor at COMSATS Institute of Information Technology, Abbottabad, Pakistan. His research interests include opportunistic networks, recommendation systems, and trust and reputation networks. He received his PhD in electrical and computer engineering from North Dakota State University, Fargo, ND, United States, and his Master degree in computer engineering from CASE, Pakistan. For further details, contact him at osman@ciit.net.pk.

Amin M. Khan is currently working towards his PhD degree at the Department of Computer Architecture of the Universitat Politècnica de Catalunya (UPC). His current interests include distributed systems, cloud computing and machine learning. He is a member of the IEEE and the ACM. He received his Master degree in informatics from University of Edinburgh and University of Trento in 2007. For further details, visit his Web page http://aminmkhan.com (accessed November 22, 2015), or contact him at mkhan@ac.upc.edu.

Samee U. Khan is an associate professor at the North Dakota State University, Fargo, ND, United States. Prof. Khan's research interests include optimization, robustness, and security of: cloud, grid, and big-data computing. Prof. Khan received a PhD in 2007 from the University of Texas, Arlington, TX, United States. His work has appeared in over 250 publications. He is on the editorial boards of number of leading journals, such as *IEEE Transactions on Computers, IEEE Access*, and *IEEE Cloud Computing Magazine.* He is a Fellow of the Institution of Engineering and Technology (IET, formerly IEE), and a Fellow of the British Computer Society (BCS). He is a senior member of the IEEE. For further details, visit his Web page: http://sameekhan.org/ (accessed November 22, 2015).

Tibor Kiss is a software engineer at IBM Storage Department in Hungary. His interests include parallel and distributed systems and quantitative finance. He started working for IBM in 2005 on hardware systems development with a focus on performance engineering and modeling. Recently he has joined the IBM's storage department where he is working on parallel filesystem development for cloud. He received a BSc in computer engineering from College of Dunaujvaros. For further details, contact him at tibor.kiss@gmail.com.

Kirill Kogan is a research assistant professor at IMDEA Networks. His current research interests are in design, analysis, and implementation of networked systems, broadly defined (in particular network processors, switch fabrics, packet classification, network management, service architecture, cloud, and fog computing). He received his PhD from Ben-Gurion University in 2012. During 2012–14 he was a postdoctoral fellow at the University of Waterloo and Purdue University. He is a former technical leader at Cisco Systems where he worked on design and architecture of C12000 and ASR1000 routers during 2000–12. For further details contact him at kirill.kogan@gmail.com.

Yousri Kouki is a cloud architect and researcher at Linagora. His interests include cloud computing, service-oriented computing, autonomic computing and green IT. His professional contributions include a solution for cloud service-level management. His present activities aim to design and implement an open PaaS solution that supports the lifecycle of cloud applications across multiple IaaS. For further details, contact him at yousri.kouki@gmail.com.

Nir Kshetri is a professor at the University of North Carolina-Greensboro and a research fellow at Kobe University. Nir holds a PhD from University of Rhode Island. His current research areas include global cybercrime, cybersecurity, and cloud economy. He is the author of four books including *Cybercrime and Cybersecurity in the Global South* (Palgrave, 2013), and *The Global Cyber-Crime Industry* (Springer-Verlag, 2010). Nir has published over 80 journal articles and has given lectures or presented research papers in 45 countries. He received the Emerald Literati Network Award for Excellence in 2010 and 2013. Nir participated as lead discussant at the peer review meeting of the UNCTAD's Information Economy Report 2013. Contact him at nbkshetr@uncg.edu.

Olga Kulikova is an information security advisor at KPMG IT Advisory, the Netherlands. She has experience with engagements related to information risk management, cloud security, identity and access management and cyber defense. Olga's most recent interest is in advising on data protection, cloud assurance and third-party contract arrangements. Olga received her Bachelor of Science degree in electrical engineering from the Bauman Moscow State Technical University and Master of Science degree in information management from the Technical University Delft. Olga has a Cloud Computing Security Knowledge certificate from the CSA. For further details, contact her at kulikova.olga@kpmg.nl.

Anand Kumar is a professor in the Electrical and Electronics Engineering Department and is the Associate Dean, Academic Resource Planning at BITS Pilani, Dubai Campus. He obtained his Master's degree and doctorate from Rice University, Houston, TX, United States. His research interests include cloud computing, VLSI design, MEMS and telecom. He has spent 7 years at Motorola and Ericsson developing telecom software. He has spent over 13 years in academia in a variety of industry-based interdisciplinary projects and research. He has over 25 publications. He served as the program co-chair for the ICCCTAM-12, 2012. He has won the Outstanding Teaching Award in 2004. For further details, visit his Web page: http://universe. bits-pilani.ac.in/dubai/akumar/profile (accessed November 22, 2015) or contact him at akumar@dubai. bits-pilani.ac.in.

Kincho H. Law is currently professor of civil and environmental engineering at Stanford University. His research interests include engineering ontology and information interoperability, legal informatics, e-government services, enterprise integration, Web services and cloud computing. His works also deal with various aspects of structural health monitoring and control systems, smart manufacturing, computational social science and cyberphysical systems. He serves on several editorial boards including the ASME *Journal of Computing Information and Science in Engineering* and the ASCE *Journal of Computing in Civil Engineering.* He has authored and co-authored over 400 articles in journals and conference proceedings. Prof. Law is the recipient of the 2011 ASCE Computing in Civil Engineering Award. For further details, visit his Web page http://eil. stanford.edu (accessed November 23, 2015) or contact him at law@stanford.edu.

Thomas Ledoux is a senior assistant professor at Mines Nantes and member of the INRIA Ascola team. His interests include software engineering, autonomic computing, green computing and / or cloud computing. He held a PhD from University of Nantes in 1998 in which his main contribution was the design of a reflective middleware named OpenCorba. He heads several national projects for the Ascola team and serves on a number of conference program committees. Currently, he investigates self-adaptive component-based applications for large-scale distributed systems. For further details, visit his Web page, www.emn.fr/ledoux (accessed November 23, 2015), or contact him at thomas.ledoux@mines-nantes.fr.

Victor C. M. Leung is a professor of electrical and computer engineering and holder of the TELUS Mobility Research Chair at The University of British Columbia. His research interests are in the broad areas of wireless networks and mobile systems. He has contributed more than 850 technical papers, some of which have won best paper awards. He was a winner of the 2012 UBC Killam Research Prize and the IEEE Vancouver Section Centennial Award. He was a distinguished lecturer of the IEEE Communications Society and he is a Fellow of IEEE, the Royal Society of Canada, the Canadian Academy of Engineering, and the Engineering Institute of Canada. For further details, visit his Web page, http://ece.ubc.ca/~vleung/ (accessed November 23, 2015), or contact him at vleung@ece.ubc.ca.

Zheng Li is a PhD candidate at the School of Computer Science at the Australian National University (ANU), and a graduate researcher with the Software Systems Research Group at National ICT Australia (NICTA). His research interests include cloud computing, Web services, software cost / effort estimation, and empirical software engineering. Before starting his PhD, he received a Masters by Research degree from the University of New South Wales (UNSW). He worked for 4 years in industry as a software test engineer before coming from overseas for further graduate degrees. He is the first author of 20+ peer-reviewed journal and conference publications. For further details, visit his Web page, https://sites.google.com/site/zhenglihomepage (accessed November 23, 2015) or contact him at imlizheng@gmail.com.

Jyhjong Lin is a full professor and the chair of the Department of Information Management at the Ming Chuan University in Taiwan. His research interests include software engineering, system / business architecture and management, and Web / cloud applications. Prof. Lin received his PhD degree in 1995 from the Computer Science Engineering Department at the University of Texas at Arlington in the United States.

Simon Liu is the Director of the US National Agricultural Library. His research interests include cyber security, knowledge management, artificial intelligence, scientific computing, big data, public access, and open science. Liu has published one book and more than 80 book chapters, journal articles, and conference papers. He has served as a speaker at more than 50 conferences, workshops, and seminars. He has also served as the editor-in-chief of *IT Professional* magazine and editor of two international journals. Liu has received awards from the HHS Secretary, NIH Director, IEEE Computer Society, and the Computer Science Corporation. Contact him at simonliu@gwu.edu.

Saif Ur Rehman Malik is an assistant professor at COMSATS Institute of Information Technology, Islamabad, Pakistan. His research interests include formal methods (verification, modeling, and analysis) and their applications in large-scale computing systems. He received his PhD in electrical and computer engineering from North Dakota State University, Fargo, United States in 2014. Besides his PhD, he completed his Master degree in computer science (MSCS) from COMSATS Institute of Information Technology, Islamabad, Pakistan in 2009. For further details, contact him at saif_ur_rehman@comsats.edu.pk.

Victor Marek is a professor of computer science at the University of Kentucky, Lexington, KY. His interests include computer science logic, constraint solving, databases, and distributed computation. Marek contributed to a variety of areas, including nonmonotonic logic, answer set programming, and logic programming theory. He has received various awards including the Sierpinski Prize from the Polish Mathematical Society. For further details, visit his Web page at http://www.cs.uky.edu/~marek (accessed November 23, 2015) or contact him at marek@cs.uky.edu.

C. Marimuthu is a PhD student at the Department of Computer Science and Engineering, National Institute of Technology Karnataka, Surathkal, Mangalore, India. His research interests include engineering software as a service, green software engineering, and software architecture. His current research work involves providing tools support to develop green and sustainable software. For further details, contact him at muthucwc.seopro@gmail.com.

Dan C. Marinescu joined the computer science department at the University of Central Florida in August 2001 as Provost Professor of Computer Science. He has been an associate and then full professor in the Department of Computer Science at Purdue University, in West Lafayette, Indiana, since 1984 and an adjunct professor in the School of Electrical Engineering at Purdue. His contributions span several areas: (i) scheduling and workflow management in large-scale distributed systems, (ii) parallel algorithms and performance evaluation of parallel and distributed systems, (iii) parallel algorithms and systems for the study of the {\tt 3D} structure of biological macromolecules and viral assemblies, and (iv) quantum computing and quantum information theory. He has published more than 220 papers in these areas, in refereed journals, and in conference proceedings. He has also published several books.

Fabrizio Marozzo is a research technician at the University of Calabria, Italy. His interests include parallel computing, distributed systems, data mining and cloud computing. He received a Ph.D. in systems and computer engineering in 2013. In 2011–12 he visited the Barcelona SuperComputing Center (BSC) for a research internship. He coauthored several papers published in conference proceedings, edited volumes and international journals. He has been a member of the program committee of scientific conferences and edited volumes. He was the recipient of two national awards for best Master thesis in the ICT area: the Javaday award (2010) and the AICA / Confindustria thesis award (2010). For further details, visit his Web page at http://www.fabriziomarozzo.it (accessed November 23, 2015) or contact him at fmarozzo@dimes.unical.it.

Thijs Metsch is a senior researcher at Intel Labs Europe. His interests include orchestration of cloud, grid, and HPC systems based on data science. He co-chairs the Open Cloud Computing Interface working group, which established one of the first standards in cloud computing. For further details, visit his LinkedIn profile at http://de.linkedin.com/in/thijsmetsch (accessed November 23, 2015), or contact him at thijs.metsch@intel.com.

Shakti Mishra is an assistant professor in the Institute for Development and Research in Banking Technology (IDRBT), Hyderabad, India. Her research interests include distributed systems and formal methods. She holds a PhD from the National Institute of Technology, Allahabad, India. Contact her at mishra.mahi@gmail.com.

R. K. Mittal is a senior professor and former director, special projects at Birla Institute of Technology and Science, Pilani (BITS Pilani), India. His research interests include cloud computing, software reliability, robotics, MEMS, nanotechnology, and e-waste. He holds a Master's degree in mechanical engineering and a doctorate in software engineering from BITS Pilani. He has published over 80 papers in international and national peer-reviewed journals and conferences, has guided numerous Masters' dissertations and five doctoral theses, while three are ongoing. He has co-authored two textbooks: *Robotics and Control* (New Delhi, India: McGraw-Hill, 2003) and *Elements of Manufacturing Processes* (New Delhi, India: Prentice-Hall, 2003), several in-house course notes, lab manuals and monographs. For further details, visit his Web page at http://www.bits-pilani.ac.in/pilani/rkm/Profile (accessed November 23, 2015), or contact him at rkm.bits@gmail.com.

San Murugesan. See his biography in the 'About the Editors' section.

Leandro Navarro is associate professor at the Department of Computer Architecture of the Universitat Politècnica de Catalunya (UPC). His current interests include distributed systems with a focus on community computer networks. He has published extensively in academic conferences and journals. Currently he leads the EU-funded research project, Confine, and the Erasmus Mundus Joint Doctorate in Distributed Computing. He received his PhD in telecommunication engineering in 1992. For further details contact him at leandro@ac.upc.edu.

Ganesh Neelakanta Iyer is a QA Architect at Progress Software Development, Hyderabad. He is also a visiting faculty at IIIT, Hyderabad, and adjunct professor at BVRIT, Narsapur, India, in the Department of Computer Science and Engineering. He has several years of industry experience at Progress Software, Hyderabad, Sasken Communication Technologies and NXP semiconductors, Bangalore. His research interests include cloud computing, game theory, software testing with cloud and mobile, and the IoT. He has several international publications including book chapters, journals and conferences. He serves as the chair for several international conferences. He has also delivered many workshops / seminars / invited talks in various research and industry forums as well as in academic institutions both in India and abroad. For further details, visit his Web page at http://ganeshniyer.com/home.html (accessed November 23, 2015).

Paolo Nesi is a full professor of distributed systems at the University of Florence. His interests include distributed systems, smart cloud, knowledge modeling, and data mining. He has been a member of many international conference committees and editor of international publications and journals. Paolo Nesi published more than 230 articles in international journals and congresses and has been chair and / or program chair of a number of international conferences of IEEE, KSI, EC, and program committee member of a number of major conferences. He has been coordinator of a number of large research and development projects of the European Commission, and has worked for ISO MPEG. For further details, visit his Web page at http://www.disit.dinfo.unifi.it/nesi (accessed November 23, 2015) or contact him at paolo.nesi@unifi.it.

Liam O'Brien is an enterprise solution architect with Geoscience Australia. His interests include enterprise, systems, and software architecture, SOA, cloud computing, and software modernization and reuse. He has over 25 years' experience in research and development in software engineering. He has published 65+ peer-reviewed research papers in international journals, conferences, and workshops. He received the best paper award at ICIW 2010 and several SEI Acquisition Support Program team-excellence awards. He received BSc and PhD degrees from the University of Limerick, Ireland. He is a member of the IEEE, the Australian Computer Society, and the Australian Service Science Society. For further details you can contact him at liamob99@hotmail.com.

Bahadir Odevci is the founder of imonacloud.com, which is a cloud-based technology startup company. He holds an MBA degree (2007) from University of Amsterdam Business School, and a BSc degree (2000) in computer engineering from the Bogazici University, Istanbul. He worked in Amsterdam, Moscow, and Istanbul, providing technical expertise and leadership for large-scale banking transformations. He worked at IBM Netherlands as lead IT architect for 3 years. Bahadir has published papers in the *Journal of Enterprise Architecture*, and has given numerous presentations on the business rules approach, SOA, and enterprise architecture for Open Group IT Practitioners, Business Rules Community and Marcus Evans EA conferences across Europe and internally at IBM. He can be reached at bahadir.odevci@imona.com.

Pratik Patel is a graduate student at Rensselaer Polytechnic Institute. His interests include mobile cloud computing, distributed computing, operating systems, and network programming. His latest work includes leveraging the actor model in mobile cloud computing. For further details, contact him at prpatel05@gmail.com.

Siani Pearson is a principal research scientist at Hewlett Packard Labs. Her interests include privacy, accountability, and security. Siani has recently held the posts of scientific coordinator of a major European research project on accountability for the cloud (A4Cloud) and Vice-President of the UK chapter of the Cloud Security Alliance. She is a member of several boards including the HP Privacy and Data Protection Board; *IEEE Transactions on Cloud Computing* editorial board, and the advisory boards of several universities and EU projects. She has received a fellowship of the British Computer Society and the Ron Knode service award (2013). For further details, visit her Web page at http://www.labs.hpe.com/people/siani_pearson/ (accessed November 23, 2015) or contact her at Siani.Pearson@hpe.com.

Rajiv Ranjan is a senior research scientist, Julius Fellow, and project leader with CSIRO Computational Informatics, Canberra, where he is working on projects related to cloud and service computing. Previously, he was a senior research associate (lecturer level B) at the School of Computer Science and Engineering, University of New South Wales (UNSW). He has a PhD (March 2009) in computer science and software engineering from the University of Melbourne. He is broadly interested in the emerging areas of cloud, grid, and service computing, with 80+ scientific publications. Most of the publications are A*/A ranked, according to the ARC's Excellence in Research for Australia (ERA). For further details contact him at rranjans@gmail.com.

Sean Rhody is an experienced information technology executive with a strong background in solution architecture, application development and cloud computing. He is the former CTO of Capgemini Government Solutions and has previously been a partner at CSC Consulting. Sean has been active in the technology industry for many years and has been an influential contributor to the industry – founding and managing *Java Developers Journal*, coauthoring multiple books on technology, and providing frequent speaking engagements. Sean is a certified global software engineer and has led IT strategy, technology implementation and cloud / datacenter migration programs over a career spanning 23 years in IT. He can be reached at seanrhody@gmail.com.

Alexis Richardson is CEO at Weaveworks, makers of Weave. He maintains a bio at https://www.linkedin.com/pub/alexis-richardson/0/2/b34 (accessed November 23, 2015) and may be e-mailed at alexis.richardson@gmail.com.

Sonal Sachdeva is a senior technology architect at Infosys Limited. His interests include access management and continuous delivery. He has 13 years of intensive experience in architecture design, development, sustenance, and QA services. He has provided IDAM domain consultancy and architected solutions and has expertise in advanced authentication, password management services, single sign on, federation, custom authentication and authorization, and all other major IDAM features He is a certified CSQA. He can be contacted at sonal_sachdeva@infosys.com.

Sherif Sakr is a senior researcher in the software systems research group at National ICT Australia (NICTA), ATP lab, Sydney, Australia. He is also a conjoint senior lecturer in the School of Computer Science and Engineering (CSE) at the University of New South Wales (UNSW). Dr. Sakr's research interest is data and information management in general, particularly in areas of indexing techniques, query processing and optimization techniques, graph data management, social networks, and data management in cloud computing. Dr. Sakr has published more than 70 refereed research publications in international journals and conferences such as: *Proceedings of the VLDB Endowment* (PVLDB), *IEEE Transactions on Service Computing* (IEEE TSC), *ACM Computing Survey* (ACM CSUR), the *Journal of Computer, Systems and Science* (JCSS), *Information Systems*, the *Journal of Computer Science and Technology* (JCST), the *Journal of Database Management* (JDM), *IEEE Communications Surveys and Tutorials* (IEEE COMST), *Scientometrics*, VLDB, SIGMOD, ICDE, WWW, CIKM, ISWC, BPM, ER, ICWS, ICSOC, IEEE SCC, IEEE Cloud, TPCTC, DASFAA, ICPE, JCDL, and DEXA. Sherif is an IEEE senior member. For further details, visit his Web page (http://www.cse.unsw.edu.au/~ssakr/, accessed November 23, 2015), or contact him at ssakr@cse.unsw.edu.au.

Pierangela Samarati is a professor at Università degli Studi di Milano, Italy. Her interests include data protection, security, and privacy. She has been a computer scientist at SRI International, California, and a visiting researcher at Stanford University, California, and George Mason University, Virginia. She has participated in several research projects and published more than 240 papers. She is the chair of the IEEE Systems Council TC-SPCIS, and of the Steering Committees of ESORICS and of ACM WPES. She is a member of several steering committees. She has served as General Chair, Program Chair, and PC member of several conferences. She is ACM Distinguished Scientist and IEEE Fellow. For further details, visit her Web page at http://www.di.unimi.it/samarati (accessed November 23, 2015), or contact her at pierangela.samarati@unimi.it.

Zohreh Sanaei is a research lead and big-data analyst at Xchanging and YTL Communications, Malaysia. Her interests include mobile cloud computing, resource scheduling in distributed computing and service-oriented computing. She has completed her PhD with distinction from University of Malaya in 2014, where she served as a full-time research assistant. She completed her MSc and BE in software engineering and information systems in 2001 and 2008 respectively. Her doctorate education was fully funded by the University of Malaya as a high-impact research project. She received best paper award at PGRES'13, on-time PhD graduation prize, and a distinction award for her PhD thesis. Zohreh is a reviewer for several international conference and computer science journals. For further details, visit her Web page at http://mobilecloudfamily.com/zohreh (accessed November 23, 2015) and contact her at sanaei@ieee.org.

Lalit Sanagavarapu is a senior technology manager in the strategic business unit, Institute for Development and Research in Banking Technology (IDRBT), Hyderabad, India. His areas of research include cloud computing, information security, and analytics. Contact him at slmohan@idrbt.ac.in.

Cameron Seay is an assistant professor of information technology at North Carolina Agricultural and Technical State University in Greensboro, North Carolina. His interests include cloud computing, enterprise computing, big-data analytics, and high-performance computing. Cameron has been instrumental in expanding enterprise systems computing in undergraduate curricula, and is a founding member of the enterprise computing community at Marist College in Poughkeepsie, New York. In addition, he has created several K-12 STEM projects in Eastern NC. He has received numerous industry awards, including five IBM Faculty Awards, and was named Outstanding Faculty from Georgia State's College of Education and Distinguished Faculty from the Georgia State's Department of Educational Psychology. For further details contact Dr. Seay at cwseay@ncat.edu.

Krishnananda Shenoy is principal architect at Infosys Limited. His focus includes technology-led business development. His interest areas are communication technologies (wireless communication covering 3G, LTE, WiFi and Bluetooth), cloud services for the IoT, engineering analytics in IoT for predictive maintenance and predictive diagnostics. He can be contacted at e-mail krshenoy@infosys.com.

Mohammad Shojafar is currently a PhD candidate in information and communication engineering at the DIET Department of La Sapienza University of Rome. His research interests include wireless communications, distributed computing and mathematical / AI optimization. He is an author / coauthor of more than 30 peer-reviewed publications in well known IEEE conferences (e.g., PIMRC, INFOCOM, ICC, ISDA) and major CS journals. From 2013, he was a member of the IEEE SMC Society Technical Committee on Soft Computing in Europe. Recently, he served as a reviewer and TPC member of NaBIC2014, SoCPaR 2014, and WICT 2014. He has received a fellowship award for foreign nationals educated abroad. For further details, visit his Web page at www.mshojafar.com (accessed November 23, 2015) and contact him at shojafar@diet.uniroma1.it.

Alan Sill directs the US National Science Foundation Center for Cloud and Autonomic Computing at Texas Tech University, where he is also a senior scientist at the High-Performance Computing Center and adjunct professor of physics. He serves as Vice-President of standards for the Open Grid Forum, and co-chairs the US National Institute of Standards and Technology's "Standards Acceleration to Jumpstart Adoption of Cloud Computing" working group. Sill holds a PhD in particle physics from the American University. He is an active member of the Distributed Management Task Force, IEEE, TM Forum, and other cloud-standards working groups, and has served either directly or as liaison for the Open Grid Forum on several national and international standards roadmap committees. For further details, visit http://cac.ttu.edu (accessed November 23, 2015) or contact him at alan.sill@ttu.edu.

Mark Smiley, PhD, is a principal architect at the MITRE Corporation. His interests include cloud computing, big data, embedded applications and scientific visualization. He is an experienced software architect, software development manager, designer, and developer. He built an award-winning software organization (Crossroads A-List, 2002), brought six new software product lines to market, and wrote the software for a renal perfusion monitor. He currently advises organizations on cloud computing and software architecture. In the past few years he won two MIP grants to develop cloud technology. He has earned a Special Recognition for Commitment to the Public Interest, a Director's Award for his work on cloud computing, and several Spot Awards. You may contact him at CloudArchitect2000@gmail.com or at msmiley@mitre.org.

Jungmin Son is a PhD candidate at the Cloud Computing and Distributed Systems (CLOUDS) Laboratory, Department of Computing and Information Systems at the University of Melbourne, Australia. His interests include cloud resource provisioning, datacenter network optimization, and distributed computing. He has received the Australian Postgraduate Award (APA) funded by the Australian Government for his exceptional research potential. For further details, contact him at jungmins@student.unimelb.edu.au.

Ram D. Sriram is currently the chief of the Software and Systems Division, Information Technology Laboratory, at the National Institute of Standards and Technology. His research interests include computer-aided design and engineering, large-scale software development, health and bio informatics, and cyber physical social systems. Sriram has coauthored or authored more than 250 publications, including several books. Sriram is a recipient of several awards, including an NSF Presidential Young Investigator Award (1989), ASME Design Automation Award (2011), and ASME CIE Distinguished Service Award (2014). Sriram is a Fellow of ASME and AAAS, a life member of ACM, a senior member of the IEEE, and a life member of AAAI. For further details visit his web site at http://www.nist.gov/itl/ssd/rsriram.cfm (accessed November 23, 2015).

Julian Stephen is a computer science graduate student at Purdue University. His research interests include assured data analysis in distributed settings and developing programming abstractions for it. He is currently working on confidentiality-preserving data analysis for cloud-based platforms. Before starting his PhD, Julian worked at Oracle. For further details, visit his Web page https://www.cs.purdue.edu/homes/stephe22/ (accessed November 23, 2015) or contact him at stephe22@purdue.edu.

Ivan Stojmenovic was a full professor at the University of Ottawa, Canada. His contributions spanned several areas including computer networks, cloud computing and wireless communications. He received the 2012 Distinguished Service Award from IEEE ComSoc Communications Software TC. He was cited over 13 000 times and his h-index is 56. Google Scholar lists him as the top researcher in parallel and distributed systems by citations, and among the top ten in wireless networks and algorithms. He served as editor in chief of several journals, including *IEEE Transactions on Parallel and Distributed Systems*. He was a Fellow of the IEEE and Canadian Academy of Engineering, and Member of the Academia Europaea. Sadly, he passed away in November 2014 in a road accident and he is missed by the professional community.

Edwin Sturrus is an information security advisor at KPMG IT Advisory, the Netherlands. He has experience with engagements related to information risk management, cloud security, identity and access management, and cybercrime. His most recent interest is in advising, publishing, and lecturing on cloud risk management, cloud assurance, and access governance. Edwin received his Master of Science degree in Economics and Informatics from the Erasmus University of Rotterdam and has a Cloud Computing Security Knowledge certificate from the Cloud Security Alliance. For further details, contact him at sturrus.edwin@kpmg.nl.

Jim Sweeney is a cloud consulting architect specializing in helping customers large and small make the leap to private and public clouds. His interests include bicycling, rock climbing, and spending time at the local Maker's club, building 3D printers or teaching Arduino programming classes. Previously Jim served as president and CTO of a desktop virtualization startup and prior to that he was the chief technology officer for GTSI Corporation where he led GTSI's offerings in virtualization and cloud computing. His previous publication, *Get Your Head in the Cloud: Unraveling the Mystery for Public Sector*, is available on Amazon, Barnes & Noble, and iTunes. For further details, visit his Web page at www.jimsweeney.info (accessed November 23, 2015), or contact him at dcjims@gmail.com.

Domenico Talia is a full professor of computer engineering at the University of Calabria and co-founder of DtoK Lab. His research interests include cloud computing, distributed knowledge discovery, parallel and distributed data-mining algorithms, and mobile computing. Talia published seven books and more than 300 papers in international journals such as CACM, *Computer,* IEEE TKDE, IEEE TSE, IEEE TSMC-B, IEEE Micro, ACM Computing Surveys, FGCS, *Parallel Computing, Internet Computing* and conference proceedings. He is a member of the editorial boards of several scientific journals, among them *IEEE Transactions on Computers* and *IEEE Transactions on Cloud Computing.* For further details, contact him at talia@dimes.unical.it.

Manoj V. Thomas is currently working on a PhD in the Department of Computer Science and Engineering, National Institute of Technology Karnataka, Surathkal, Mangalore, India. He obtained his Bachelor of Technology from RIT, Kottayam, Kerala and Master of Technology from NITK, Surathkal with First Rank and a gold medal. He has more than 10 years of teaching experience and his areas of interests include computer networks, cloud computing and cloud security. He is a life member of the Computer Society of India. For further details, please contact him at his e-mail address: manojkurissinkal@gmail.com.

Srimanyu Timmaraju is working on a Master's degree in information technology from the University of Hyderabad, India. His areas of research include cloud computing and data mining. Contact him at tnsmanyu@gmail.com.

Deepnarayan Tiwari is a research fellow in the Institute for Development and Research in Banking Technology (IDRBT), Hyderabad, India, and is currently working on his PhD from the University of Hyderabad, India. His areas of research include cloud computing and cryptography. Contact him at dtiwari@idrbt.ac.in.

Paolo Trunfio is an associate professor at the University of Calabria, Italy. His interests include cloud computing, distributed data mining, parallel computing and peer-to-peer networks. In 2007, he was a visiting researcher at the Swedish Institute of Computer Science (SICS) in Stockholm, with a fellowship of the CoreGRID Researcher Exchange Programme. Previously, he was a research collaborator at the Institute of Systems and Computer Science of the Italian National Research Council (ISI-CNR). He is a member of the editorial board of the International Scholarly Research Notices journal, and has served in the program committee of several conferences. He is the author of about 100 scientific papers published in international journals, conference proceedings and edited volumes. For further details, contact him at trunfio@dimes.unical.it.

Piotr Tysowski is a management consultant in the digital business technology practice of Strategy& (formerly Booz & Company). He received a PhD in Electrical and Computer Engineering at the University of Waterloo, where he also received Bachelor and Master degrees. He was the recipient of the Alexander Graham Bell Canada Graduate Scholarship. He holds an MBA from Wilfrid Laurier University. Previously, he worked at Research In Motion (now known as BlackBerry) as a senior software engineer and product manager. He is an inventor of over 30 US patents. His interests include cloud computing security and privacy, mobile computing, software engineering, and technology product strategy. For further details, visit his Web page at www.tysowski.com (accessed November 23, 2015) or contact him at pktysowski@uwaterloo.ca.

Muhammad Usman Shahid Khan is a PhD student at North Dakota State University, Fargo, United States. His research interests include data mining, recommender systems, network security, and cloud computing. He received his Master degree in information security from the National University of Science and Technology, Pakistan, in 2008. For further details, visit his Web page: http://ndsu.edu/pubweb/~muhkhan (accessed November 23, 2015).

Carlos A. Varela is an associate professor at Rensselaer Polytechnic Institute. His interests include Web-based and Internet-based computing, middleware for adaptive distributed systems, concurrent programming models and languages, and software verification. He is an associate editor of the ACM *Computing Surveys* journal and *IEEE Transactions on Cloud Computing*. He has received several research grants, including the NSF CAREER award. For further details, visit his web site at http://www.cs.rpi.edu/~cvarela/ (accessed November 24, 2015).

Bharadwaj Veeravalli is an associate professor at National University of Singapore. His mainstream research interests include cloud/grid/cluster computing, scheduling in parallel and distributed systems, bioinformatics and computational biology, and multimedia computing. He is one of the earliest researchers in the field of divisible load theory (DLT). He had successfully secured several externally funded projects. He has published over 100 papers in high-quality international journals and conferences. He has coauthored three research monographs in the areas of PDS, distributed databases, and networked multimedia systems. He is currently serving on the editorial board of *IEEE Transactions on Computers, IEEE Transactions on SMC-A, Multimedia Tools and Applications* (MTAP) and *Cluster Computing*, as an associate editor. For further details, visit his Web page http://cnl-ece.nus.edu.sg/elebv/ (accessed November 24, 2015).

B. Vijayakumar is a professor and head of the Department of Computer Science at BITS Pilani, Dubai Campus. His research interests include multimedia systems, Web data mining, component-based software development and distributed database systems. He has 30+ publications in refereed journals and conferences. He has served as technical program chair for the International Conference on Cloud Computing Technologies, Applications and Management-ICCCTAM-2012 held at BITS Pilani, Dubai Campus in December 2012. For further details, visit his Web page, http://universe.bits-pilani.ac.in/dubai/vijay/profile (accessed November 24, 2015) or contact him at vijay@dubai.bits-pilani.ac.in.

Sabrina De Capitani di Vimercati is a professor at Università degli Studi di Milano, Italy. Her interests include data protection, security, and privacy. She has been a visiting researcher at SRI International, California and at George Mason University, Virginia, in the United States. She has published more than 200 papers and chapters. She is member of the steering committees of the European Symposium on Research in

Computer Security and of the ACM Workshop on Privacy in the Electronic Society. She is the chair of the IFIP WG 11.3 on Data and Application Security and Privacy. She is co-recipient of the ACM-PODS'99 Best Newcomer Paper Award, and is a senior member of the IEEE. For further details, visit her Web page at http://www.di.unimi.it/decapita (accessed November 23, 2015) or contact her at sabrina.decapitani@unimi.it.

Yang Wang is currently a Canadian Fulbright visiting research chair at the Illinois Institute of Technology. His research interests include cloud computing, virtualization technology, big data analytics, and Java Virtual Machine on multicores. He has been a research fellow at the National University of Singapore (2011–12) and an Alberta Industry R&D associate (2009–11).

Montressa Washington is a certified business transformation consultant with over 16 years of experience including leadership and technical responsibilities with IBM and Accenture. She has expertise in change management, business transformation, and technology assimilation, and has guided government and commercial clients in the implementation of collaborative, cloud, and social media technologies. She received an MBA from the Johns Hopkins University Carey Business School, a BA in English from the University of Maryland College Park, and is presently completing a PhD in management at the Weatherhead School of Management, Case Western Reserve University. As a Management Design Fellow she contributes to diverse investigative themes, encompassing the various aspects of technology integration. Ms. Washington can be reached at mlw41@case.edu.

Rudy J. Watson serves as Associate Vice Dean of the Information and Technology Systems Department and Program Chair, Technology Management in the Graduate School at the University of Maryland University College. He is a PMI-certified project management professional with broad and diverse experience in information technology, including 33 years with IBM. Dr. Watson was the Executive Project Manager for the Mission Oriented Cloud Architecture (MOCA) project. MOCA was a US Air Force-directed research and development project investigating solutions to seven functional areas (cloud computing foundation, resilience, compliance, analytics, deep-packet inspection, multitenancy, and secure collaboration). He holds a PhD in logistics, technology and project management from the George Washington University. Dr. Watson can be contacted at rudy.watson@umuc.edu.

Abraham Williams is a technology manager in the Institute for Development and Research in Banking Technology (IDRBT), Hyderabad, India. His areas of research include cloud computing, networks, and security. Contact him at akwilliams@idrbt.ac.in.

Shams Zawoad is a PhD student at the Computer and Information Sciences Department of the University of Alabama at Birmingham (UAB) and a graduate research assistant at the Secure and Trustworthy Computing Lab (SECRETLab) at UAB. His research interests include cloud forensics, antiphishing, mobile malware, and secure provenance. He received his BSc in computer science and engineering from the Bangladesh University of Engineering and Technology (BUET) in January 2008. Prior to joining UAB in 2012, he had been working in the software industry for 4 years. For further details, visit his Web page at http://students.cis.uab.edu/zawoad/ (accessed November 23, 2015), or contact him at zawoad@cis.uab.edu.

Lingfang Zeng is an associate professor at Huazhong University of Science and Technology. He was Research Fellow for 4 years in the Department of Electrical and Computer Engineering, National University of Singapore (NUS), Singapore, during 2007–8 and 2010–13, supervised by Dr. Bharadwaj Veeravalli. He has published over 40 papers in major journals and conferences, including ACM Transactions on Storage, IEEE Transactions on Magnetics, Journal of Parallel and Distributed Computing, Journal of Network and

Computer Applications, FAST, SC, IPDPS, MSST and CLUSTER, and serves for multiple international journals and conferences. He is a member of IEEE. His recent work focuses on datacenter technology, in-memory computing, and nonvolatile systems. For further details, visit his Web page at http://stlab.wnlo.hust. edu.cn/cszeng/index.html (accessed November 23, 2015).

Wolfgang Ziegler is a senior researcher at the Fraunhofer Institute SCAI. His interests include distributed computing, service-level agreements, data protection and data security, software licensing technology. He is Area Director Applications of the Open Grid Forum (OGF) and Chair of OGF's working group, which has developed a recommendation for service-level agreements. His current focus is on standards for cloud computing, data protection, and software licensing technology for cloud environments. For further details, contact him at wolfgang.ziegler@scai.fraunhofer.de.

Nazim Ziya Perdahci is an assistant professor of IT at the Department of Informatics at Mimar Sinan Fine Arts University. He holds a PhD degree (2002) in physics from the Bogazici University, Istanbul, Turkey. His interests include network science, social network analysis of information exchange platforms, and complex systems. He worked as an Assistant Professor of IT at the Department of Information Technologies of the Isik University, Istanbul, Turkey. Nazim has published eight articles as journal and conference papers. He also has two patents. He has opened a part-time elective course about cloud computing recently at the Kadir Has University. He can be reached at nz.perdahci@msgsu.edu.tr.

Albert Y. Zomaya is currently the Chair Professor of High Performance Computing and Networking in the School of Information Technologies, University of Sydney. He is also the Director of the Centre for Distributed and High Performance Computing. Dr. Zomaya published more than 500 scientific papers and articles and is author, coauthor or editor of more than 20 books. He is the editor-in-chief of the *IEEE Transactions on Computers* and Springer's *Scalable Computing*, and serves as an associate editor for 22 leading journals. Dr. Zomaya is the recipient of many awards, such as the IEEE TCPP Outstanding Service Award and the IEEE TCSC Medal for Excellence in Scalable Computing. He is a chartered engineer, a Fellow of AAAS, IEEE, and IET. For further details, visit his Web page: http://sydney.edu.au/engineering/it/~zomaya/ (accessed November 23, 2015).

Reviewers

Foreword

With technology, perfect timing is the key to success. A technology could be superior, and there could be a need for it, but if the timing is not right success usually does not take place. There are many examples. James Gosling, inventor of Java, developed the Network-Extensive Window System (NeWS) and a few other technologies, but only Java caught on widely. When he invented Java there were multiple operating systems and Java was a common language across many platforms. This enabled a "write once, run anywhere" paradigm. Even more important – it was the client-server era, when Java's concepts of virtual machines and object orientation thrived.

The same is true for the *Encyclopedia of Cloud Computing*. This is a book for *this time*. A couple of years earlier, it would have been premature; a few years later, it may be old news. Cloud computing has matured over the past few years and now it is part of the offering of many major IT companies, not just technology leaders and early adopters. It is being used by many industries and in different ways. It is delivered as public and private cloud and anything in between. It is accepted by industry, academia, and even governments. While Amazon Web Services are still the leaders, there are also many other cloud service providers (such as Azure, Google, HP, and IBM) as well as numerous cloud enablers who supply the necessary cloud software tools and hardware. The cloud computing field is hard to navigate. Wikipedia entries, and numerous technical and business documents satisfy some of the needs. However, a comprehensive compendium of materials on cloud computing did not exist until now. This is the offering by San Murugesan and Irena Bojanova. Irena and San undertook a challenging task to spearhead editing a massive volume focusing on 56 topics in the area of cloud computing. Defining the encyclopedia content, choosing experts to author the chapters, coordinating with them, and delivering a coherent book, is a monumental effort.

The record of Murugesan and Bojanova in the area of cloud computing makes them ideal editors for the encyclopedia. They both were up to the challenge. Irena was the founding chair of the IEEE Special Technical Community (STC) on cloud computing, one of the first established STCs, where she helped define the STC agenda and execute its first deliverables. San is the editor-in-chief of *IEEE Professional*, a unique publication targeting computing professionals, and edits the "Cloud cover" column in *IEEE Computer.* They both also guest edited theme issues on cloud computing at *IEEE Computing Now* and elsewhere. Irena was editor-in chief of *IEEE Transaction on Cloud Computing.* I have personally witnessed these efforts as president of the IEEE Computer Society.

This is a remarkable book. It can be read selectively (certain parts or chapters) or from cover to cover. It offers readers a comprehensive and detailed view of cloud computing, covering horizontal technologies as well as vertical solutions and applications delivered in the cloud. It meets current needs and I strongly recommend it to technical and business people alike. It will also have lasting practical and historical value, covering the foundations of cloud computing for generations to come.

Dejan S. Milojicic, Senior Researcher and Manager
Hewlett Packard Laboratories, Palo Alto
IEEE Computer Society President, 2014
Managing Director of Open Cirrus Cloud Computing Testbed, 2009–13

Preface*

Several converging and complementary factors are driving the rise of cloud computing. The increasing maturity of cloud technologies and cloud service offerings coupled with users' greater awareness of the cloud's benefits (and limitations) is accelerating the cloud's adoption. Better Internet connectivity, intense competition among cloud service providers (CSPs), and digitalization of enterprises, particularly micro-, small-, and medium-sized businesses, are increasing the cloud's use. Cloud computing is not just an IT paradigm change, as some perceive. It is redefining not only the information and communication technology (ICT) industry but also enterprise IT in all industry and business sectors. It is also helping to close the digital (information) divide, driving innovations by small enterprises and facilitating deployment of new applications that would otherwise be infeasible.

Cloud computing is becoming ubiquitous and the new normal. To better understand and exploit the potential of the cloud – and to advance the cloud further – practitioners, IT professionals, educators, researchers, and students need an authoritative knowledge source that comprehensively and holistically cover all aspects of cloud computing. Several books on cloud computing are now available but none of them cover all key aspects of cloud computing comprehensively. To gain a holistic view of the cloud, one has to refer to a few different books, which is neither convenient nor practicable. There is not one reference book on the market that comprehensively covers cloud computing and meets the information needs of IT professionals, academics, researchers, and undergraduate and postgraduate students.

This encyclopedia serves this need and is the first publication of this kind. It targets computing and IT professionals, academics, researchers, university students (senior undergraduate and graduate students), and senior IT and business executives. This publication contains a wealth of information for those interested in understanding, using, or providing cloud computing services; for developers and researchers who are interested in advancing cloud computing and businesses, and for individuals interested in embracing and capitalizing on the cloud. This encyclopedia is a convenient ready-reference book with lots of relevant and helpful information and insights.

About the Encyclopedia

The *Encyclopedia of Cloud Computing* is a comprehensive compendium of cloud computing knowledge and covers concepts, principles, architecture, technology, security and privacy, regulatory compliance, applications, and social and legal aspects of cloud computing. Featuring contributions from a number of subject experts in industry and academia, this unique publication outlines and discusses technological trends and

---

* This work was completed by Irena Bojanova and accepted for publication prior to her joining NIST.

developments, research opportunities, best practices, standards, cloud adoption and other topics of interest in the context of the development, operation, management, and use of clouds. It also examines cloud computing's impact, now and in the future.

The book presents 56 chapters on a wide array of topics organized in ten parts. After gaining an overview of cloud computing in Chapter 1, readers can study the rest of the chapters in sequence or hop to a chapter that interests them. We present a brief preview of the encyclopedia below.

Book Preview

Part I: Introduction to Cloud Computing

In this part, we present an overview of cloud computing concepts, cloud services, cloud-hosting models, and applications. We also outline the benefits and limitations of cloud computing, identify potential risks of the cloud, and discuss its prospects and implications for business and IT professionals. This introductory chapter should enable readers gain an overall, holistic view of cloud computing and help them comprehend the rest of the chapters.

Part II: Cloud Services

Cloud vendors offer an array of cloud services that enterprises, individuals and IT personnel can use. In this part, we present five chapters that outline a range of generic and specialized services. To help readers familiarize themselves with available cloud services and make appropriate choices that meet their requirements, in the "Cloud Services and Cloud Service Providers" chapter we outline cloud services and tools offered by different vendors. In the "Mobile Cloud Computing" chapter, we present a comprehensive overview of mobile cloud computing (MCC) – its definition, motivation, building blocks, and architectures – and outline challenges in MCC that deserve future research. Next, in the "Community Clouds" chapter, we identify the design goals of community clouds and discuss cloud application scenarios that are specific to a few different communities of users. In the "Government Clouds" chapter, we discuss primary considerations for government agencies in adopting the cloud and cloud service offerings in the federal marketplace, and offer some insight into the progression towards the adoption of clouds by government organizations worldwide. In the last chapter of this part, "Cloud-Based Development Environments (PaaS)," we discuss the fundamentals of platform as a service (PaaS) and focus on a specific cloud-computing layer, called "application Platform as a Service" (aPaaS). Then we articulate basic approaches to aPaaS and provide a concise comparison of leading PaaS solutions.

Part III: Cloud Frameworks and Technologies

Clouds adopt special frameworks for their implementation and employ several concepts and technologies, virtualization being a key technology supporting the cloud. In this part we present seven chapters covering a few different popular cloud reference frameworks, different types of virtualization, and datacenter networks. In the "Cloud Reference Frameworks" chapter, we review major cloud reference frameworks – the NIST Cloud Reference Architecture, the IETF Cloud Reference Framework, the CSA Cloud Reference Model, and others – and provide the context for their application in real-world scenarios. In the next chapter, "Virtualization: An Overview," we introduce the concepts of virtualization, including server, storage, and network virtualization, and discuss the salient features of virtualization that make it the foundation of cloud computing. Then we describe advanced concepts such as virtualization for disaster recovery and the business continuity. In the "Network and I/O Virtualization" chapter, we discuss how network and computer input-output virtualization techniques can be leveraged as key underpinnings of a cloud network, and examine their use and benefits.

Next, in the "Cloud Networks" chapter, we provide a comprehensive overview of the characteristics, categories, and architecture of cloud networks, which connect compute, storage and management resources of cloud infrastructure and provide network services to all tenants of the cloud computing environment. In the "Wireless Datacenter Networks" chapter, we analyze the challenges of traditional datacenter networks (DCNs), introduce the most popular and efficient wireless technology, 60 GHz RF technology, and several classic architectures for wireless DCNs, and outline recent research on developing high performance in wireless DCNs. Next, as open-source software and tools for building clouds are gaining popularity and acceptance, in the "Open-Source Cloud Software Solutions" chapter, we highlight and compare leading open-source cloud software solutions for IaaS, PaaS, and SaaS, and discuss the features of open-source cloud infrastructure automation tools. Lastly, in the "Developing Software for Cloud: Opportunities and Challenges for Developers" chapter, we discuss challenges in developing SaaS clouds, the popular SaaS development platforms available for the public cloud and private cloud, and best practices to transform traditional Web applications to cloud-based multitenant SaaS applications.

Part IV: Cloud Integration and Standards

In this part, we present three chapters covering topics such as cloud portability, integration, federation, and standards. In "Cloud Portability and Interoperability," we discuss the problem of cloud interoperability and portability and outline relevant methodologies, research projects, proposals for standards, and initiatives aimed at resolving interoperability and portability issues from different viewpoints. Next, in "Cloud Federation and Geo-Distribution," we outline motivation for cloud federation and discuss the challenges in the location and access of data stored and shared between datacenters, computation on such distributed data, and communication of data across datacenters. Lastly, in "Cloud Standards," we discuss key unique features of cloud-computing standards, map cloud standards categories to cloud service layers, and detail the currently available set of cloud standards with respect to these categories.

Part V: Cloud Security, Privacy, and Compliance

Cloud security, privacy, and compliance continue to be key barriers to wider adoption of the cloud. In this part we present six chapters focusing on these issues and potential solutions to address them. First, in "Cloud Security: Issues and Concerns," we present an overview of key security issues and concerns arising in the cloud scenario, in particular with respect to data storage, management, and processing. Next, in "Securing the Clouds: Methodologies and Practices," we provide an overview of security issues posed by SaaS, IaaS, and PaaS cloud service models and discuss methodologies, best practices, practical approaches, and pragmatic techniques to address these issues. In the "Cloud Forensics" chapter, we discuss the challenges of cloud forensics, and techniques that support reliable forensics, and explore key open problems in this area. Then, in "Privacy, Law, and Cloud Services," we examine legal issues surrounding digital privacy, including application of the Fourth Amendment and the Stored Communications Act. Having an understanding of these issues helps consumers alter their service choices to protect their digital privacy when migrating data to the cloud. Next, in "Ensuring Privacy in Clouds," we examine the philosophical foundations of privacy, review privacy risks particular to the cloud, and discuss IT best practices for addressing those risks and emerging research in privacy relevant to cloud-based systems. Lastly, in "Compliance in Clouds," we provide an overview of compliance and security challenges in adopting clouds and discuss information security certificates as a means to evaluate service providers with respect to their compliance.

Part VI: Cloud Performance, Reliability, and Availability

In this part, which consists of seven chapters, we discuss other key factors that influence cloud adoption such as performance, reliability, and availability of the cloud. In "Cloud Capacity Planning and Management," we present a comprehensive overview of capacity planning and management of cloud infrastructure: first, we state the problem of capacity management in the context of cloud computing from the viewpoint of service providers; next, we provide a brief discussion on when capacity planning should take place, and, finally, we survey a number of methods proposed for capacity planning and management. Next, in "Fault Tolerance in the Cloud," we discuss fault tolerance in the cloud and illustrate various fault tolerance strategies and provide taxonomy of fault-tolerance approaches.

In the "Cloud Energy Consumption" chapter, we discuss the cloud energy consumption and its relationship with cloud resource management, and how cloud computing infrastructures could realize their potential of reducing the energy consumption for computing and data storage, thus shrinking their carbon footprint. Next, in "Cloud Modeling and Simulation," we provide an introduction to cloud modeling and simulation technologies with focus on performance and scalability and describe modeling of contention, data segmentation, and workload generation. Then we illustrate various stages of cloud modeling using the simulation of OpenStack image deployment on the cloud as an example. In "Cloud Testing: An Overview," we describe various cloud test dimensions such as elasticity and scalability testing, security testing, performance testing, compatibility testing, API integration testing, live upgrade testing, disaster recovery testing, and multitenancy testing, and introduce approaches for automation of cloud integration testing. In "Testing the Cloud and Testing as a Service," we discuss testing of cloud infrastructure and applications and outline how cloud services could be leveraged for testing and quality assurance of software, hardware, Web apps and information systems. Lastly, in "Cloud Services Evaluation," we discuss objectives of evaluation of cloud services from different vendors, the features that are commonly evaluated, the *de facto* benchmarks, and a practical methodology for cloud service evaluation.

Part VII: Cloud Migration and Management

To migrate applications to the cloud successfully, several technical and nontechnical aspects, such as access control, service-level agreement, legal aspects, and compliance requirements have to be addressed. In this part, we present ten chapters that focus on these considerations. In the first chapter, "Enterprise Cloud Computing Strategy and Policy," we discuss how a well crafted cloud-computing strategy that uses a structured engineering approach to balance requirements, schedule, cost, and risk, would help in the successful application of cloud computing. Next, in "Cloud Brokers," we discuss different cloud broker mechanisms and their properties, and describe a typical cloud broker architecture. In "Migrating Applications to Clouds," we present a methodology for cloud migration that encompasses an application-description process, a cloud-identification process, and an application-deployment process, and we illustrate the method with an example. Next, in "Identity and Access Management," we elaborate on the essence of identity and access management (IAM) and why effective IAM is an important requirement for ensuring security, privacy, and trust in the cloud computing environment. In the chapter "OAuth Standard for User Authorization of Cloud Services," we describe the IETF OAuth specification – an open Web standard that enables secure authorization for applications running on various kinds of platforms. In "Distributed Access Control in Cloud Computing Systems," we highlight the issue of distributed access control, access control policies and models, distributed access control architecture for multicloud environments, and trust and identity management in the cloud computing environment.

Next, in "Cloud Service Level Agreement (SLA)," we describe the typical elements of a cloud SLA with examples, give an overview of SLAs of well known public cloud service providers, and discuss future directions in SLAs of cloud-based services. In "Automatic Provisioning of Intercloud Resources Driven by Nonfunctional Requirements of Applications," we present an automatic system that performs provisioning of resources on public clouds based on the nonfunctional requirements of applications by translating the high-level nonfunctional requirements from administrators into VM resource parameters. The system also selects the most appropriate type of VM and its provider, and allocates actual VMs from the selected provider. Next, in "Legal Aspects of Cloud Computing," we discuss the legal landscape of the cloud, providing a high-level overview of relevant laws and regulations that govern it, including how countries have addressed the problem of transborder dataflow; and describe the increasingly important role played by contracts between cloud service providers and their clients. Lastly, in "Cloud Economics," we discuss economic considerations of cloud computing, such as cloud pricing, supply chain, market models, stakeholders, and network effects.

Part VIII: Cloud Applications and Case Studies

Clouds are used in many different applications in several domains. In this part, we present four chapters focusing on cloud applications in four different domains. First, in "Engineering Applications of the Cloud," we outline two applications to illustrate the potential of the cloud service environment for supporting information interoperability and collaborative design – one in product information sharing and the other in distributed collaboration. Next, in "Educational Applications of the Cloud," we present a broad overview of cloud computing in education for activities such as instruction, front-office interaction, and back-office operations, and discuss the potential of cloud computing in academic research. Next, in "Personal Applications of Clouds," we discuss personal cloud applications associated with collaboration, social networking, personal computing, storage of personal data, and some forms of entertainment access. Finally, in "Cloud Gaming," we outline platforms that provide cloud-gaming services and classify the services into four models and the corresponding platforms into four architectural frameworks. We also examine their advantages and disadvantages, discuss their associated challenges, and identify potential future developments in cloud gaming.

Part IX: Big Data and Analytics in Clouds

Clouds are excellent platforms for big data storage, management and analytics. In this part, in nine chapters, we cover various aspects of the big data-cloud nexus. In the first chapter, "An Introduction to Big Data," we provide an overview of big data and big data analytics and their applications, and outline key big-data technologies. In "Big Data in a Cloud," we provide an overview of NoSQL databases, discuss an analytic cloud and how it contrasts with a utility cloud, and focus on running big data analytics in a virtualized environment. Next, in the chapter "Cloud-Hosted Databases," we outline popular technologies for hosting the database tier of software applications in cloud environments and discuss their strengths and weaknesses. In "Cloud Data Management," we present recent advances in research and the development of cloud data management, including practical cases of current and anticipated data challenges, the state-of-the-art data-management technologies for extremely large datasets in the cloud, and lessons learned on building data-management solutions in the cloud.

Apache Hadoop – an open-source Java-based framework – is the driving force behind big data. In "Large-Scale Analytics in Clouds (Hadoop)," we provide an overview of Hadoop and its variants and alternatives. The MapReduce framework is a simple paradigm for programming large clusters of hundreds and thousands of servers that store many terabytes and petabytes of information. In the next chapter, "Cloud Programming Models (Map Reduce)," we present an overview of the MapReduce programming model and its variants and implementation. In "Developing Elastic Software for the Cloud," we describe the Google App Engine as an example of a PaaS programming framework, and MapReduce and Simple Actor Language System and Architecture (SALSA) as examples of distributed computing frameworks on PaaS and IaaS. In the "Cloud Service for

Distributed Knowledge Discovery" chapter, we discuss the functional requirements of a generic distributed knowledge system, and how these requirements can be fulfilled by a cloud. As a case study, we describe how a cloud platform can be used to design and develop a framework for distributed execution of knowledge discovery in database (KDD) workflows. Lastly, in "Cloud Knowledge Modeling and Management," we present and discuss the main aspects of cloud knowledge modeling focusing on the state of the art solutions.

Part X: Cloud Prospects

In the last part, comprising four chapters, we examine the prospects of the cloud. First, in "Impact of the Cloud on IT Professionals and the IT Industry," we discuss the impact of cloud computing on IT, changes to existing IT roles, and newly created IT roles. Next, in "Cloud Computing in Emerging Markets," we analyze the current status of cloud computing in emerging markets and examine the fundamental forces driving its adoption and major constraints these countries face in using and deploying it. In "Research Topics in Cloud Computing," by aggregating information from several sources (publications in journals, conferences and workshops; white papers from major industry players; objectives of major cloud computing laboratories in universities; reports on government and industry research funding; and the major cloud research and development projects in several countries), we present cloud computing research trends in industry and academia. Finally, in our concluding chapter, "Cloud Computing Outlook: The Future of the Cloud," we discuss the future of the cloud and outline major trends that are poised to make the cloud ubiquitous and the new normal.

We believe that this encyclopedia, which covers a range of key topics on cloud computing in a single volume, should be helpful to a spectrum of readers in gaining an informed understanding of the promise and potential of the cloud. We welcome your comments and suggestions at CloudComputingEncyclopedia@gmail.com.

Acknowledgments

Publication of this unique and comprehensive book, the *Encyclopedia of Cloud Computing,* which covers most aspects of cloud computing, would not have been possible without the contribution, support, and cooperation of several people, whom we acknowledge here.

First, we would like to thank the chapter authors for contributing enthusiastically to the book, and sharing their expertise, experiences and insights with the readers. We gratefully acknowledge their contribution and cooperation.

Next, we extend our gratitude to the reviewers, (see page xxxvi) who have reviewed the chapter manuscripts, provided valuable feedback to authors, and assisted us in making final decisions. We thank Dejan S. Milojicic, senior researcher and manager at HP Laboratories, Palo Alto, and IEEE Computer Society president 2014, for writing a foreword for the book. We also thank our advisory board members (see page v) for their valuable support and suggestions.

The editorial team at Wiley deserves our appreciation for its key role in publishing this volume and in ensuring its quality. In particular, we would like to thank Anna Smart, former commissioning editor; Sandra Grayson, associate editor; and Karthika Sridharan, production editor for their enthusiastic support and cooperation. Next, we would like to thank Sandeep Kumar, project manager, SPi Global and David Michael, copy editor, for their excellent production management and copyediting. We highly commend their professionalism and commitment.

Finally, we would like to thank our respective family members for their encouragement, support, and cooperation, which enabled us to make this venture a reality.

San Murugesan
Irena Bojanova

Part I
Introduction to Cloud Computing

1

Cloud Computing: An Overview

San Murugesan[1] and Irena Bojanova[2]*

[1] *BRITE Professional Services and Western Sydney University, Australia*
[2] *National Institute of Standards and Technology (NIST), USA*

1.1 Introduction

Cloud computing is receiving keen interest and is being widely adopted. It offers clients applications, data, computing resources, and information technology (IT) management functions as a service through the Internet or a dedicated network. Several converging and complementary factors have led to cloud computing's emergence as a popular IT service-delivery model that appeals to all stakeholders. Considered as paradigm change in IT, it is being adopted for a variety of applications – personal, academic, business, government, and more – not only for cost savings and expediency but also to meet strategic IT and business goals. It is transforming every sector of society and is having a profound impact, especially on the IT industry and on IT professionals – application developers, enterprise IT administrators, and IT executives. Driven by advances in cloud technology, the proliferation of mobile devices such as smartphones and tablets, and use of a variety of applications supported by ubiquitous broadband Internet access, the computing landscape is continuing to change. There is an accompanying paradigm shift in the way we deliver and use IT.

Cloud computing is a radical new IT delivery and business model. Users can use cloud services when and where they need them and in the quantity that they need, and pay for only the resources they use. It also offers huge computing power, on-demand scalability, and utility-like availability at low cost.

Cloud computing is no longer hype. Individuals are using cloud-based applications, such as Web mail and Web-based calendar or photo-sharing Web sites (e.g., Flickr, Picasa) and online data storage. Small- and medium-sized enterprises are using cloud-based applications for accounting, payroll processing, customer

---

*This work was completed by Irena Bojanova and accepted for publication prior to her joining NIST.

---

relationship management (CRM), business intelligence, and data mining. Large enterprises use cloud services for business functions, such as supply-chain management, data storage, big data analytics, business process management, CRM, modeling and simulation, and application development. Research studies reveal that users give convenience, flexibility, the ability to share information, and data safety as major reasons for engaging in cloud computing activities.

As cloud computing is moving towards mainstream adoption, there is considerable excitement and optimism, as well as concerns and criticism. Many people have incomplete information or are confused about cloud computing's real benefits and key risks, which matter to them. Given its transformational potential and significance, it is important that students, IT professionals, business managers and government leaders have an informed, holistic understanding of cloud computing and how they can embrace it.

In this chapter, we present an overview of cloud computing concepts, cloud services, cloud-hosting models, and applications. We also outline the benefits and limitations of cloud computing, identify its potential risks, and discuss the prospects for the cloud and what businesses and individuals can do to embrace cloud computing successfully. Finally, we discuss the prospects and implications of cloud computing for businesses, the IT industry, and IT professionals.

1.2   Cloud Computing

In its evolution since the mid-1970s, computing has passed through several stages – from mainframe computers to minicomputers to personal computers to network computing, client-server computing, and distributed computing. Now, coming full circle, computing is migrating outward to the clouds, to distant computing resources reached through the Internet.

Depending on how you view cloud computing, it can be described in different ways. There are several definitions, but the National Institute of Standards and Technology (NIST) offers a classic definition that encompasses the key elements and characteristics of cloud computing (Mell and Grance, 2011):

> Cloud computing is a model for enabling ubiquitous, convenient, on-demand network access to a shared pool of configurable computing resources (e.g., networks, servers, storage, applications, and services) that can be rapidly provisioned and released with minimal management effort or service provider interaction.

The International Organization for Standardization (ISO) provides a similar definition, choosing to call cloud computing an "evolving paradigm": "Cloud computing is a paradigm for enabling network access to a scalable and elastic pool of shareable physical or virtual resources with self-service provisioning and administration on-demand" (ISO/IEC DIS 17789:2014, 2014).

Gartner defines cloud computing in simplistic terms as "A style of computing where scalable and elastic IT-enabled capabilities are provided as a service to multiple customers using Internet technologies" (http://www.gartner.com/it-glossary/cloud-computing, accessed November 25, 2015).

Another definition encompasses several key characteristics of cloud computing and presents a broader and practical view of it (Vaquero *et al.*, 2009):

> Clouds [are] a large pool of easily usable and accessible virtualized resources such as hardware, development platforms and/or services. These resources can be dynamically reconfigured to adjust to a variable load (scale), allowing also for an optimum resource utilization. This pool of resources is typically exploited by a pay-per-use model in which guarantees are offered by the Infrastructure Provider by means of customized SLAs [service-level agreements].

Table 1.1   *Cloud characteristics*

Cloud characteristic	Description
On-demand self-service	Computing capabilities (e.g. server time and network storage) can be unilaterally automatically provisioned as needed).
Broad network access	Capabilities are accessible through heterogeneous thin or thick client platforms (e.g., mobile phones, tablets, laptops, and workstations).
Resource pooling	Computing resources (e.g. storage, processing, memory, and bandwidth) are pooled to serve multiple consumers, and are dynamically assigned and reassigned according to demand. Customers have no control over the exact location of resources, but may be able to specify location (e.g., country, state, or datacenter).
Rapid elasticity	Capabilities can be elastically provisioned and released commensurate with demand. Available capabilities often appear to be unlimited.
Measured service	Resource use is automatically controlled and optimized through metering capabilities, appropriate to type of service (e.g., storage, processing, bandwidth, and active user accounts).
Multitenancy	Cloud computing is a shared resource that draws on resource pooling as an important feature. It implies use of same resources by multiple consumers, called tenants.

1.2.1   Key Cloud Characteristics

Cloud computing has the following key distinguishing characteristics:

- on-demand self-service;
- broad network access;
- resource pooling;
- rapid elasticity and scalability;
- measured service;
- multitenancy.

These characteristics, briefly outlined in Table 1.1, differentiate cloud computing from other forms of traditional computing.

The cloud draws on some of the older foundations of IT such as centralized, shared resource pooling, utility computing, and virtualization, and incorporates new mechanisms for resource provisioning and dynamic scaling. It adopts new business and revenue models and incorporates monitoring provisions for charging for the resources used. Cloud computing became more widely available only with the adoption of broadband Internet access and advances in virtualization and datacenter design and operation. Philosophical and attitude changes by IT vendors and users were also drivers for cloud's popularity.

1.2.2   Cloud computing attributes

Computing clouds have several distinguishing attributes. They:

- have massive resources at their disposal and support several users simultaneously;
- support on-demand scalability of users' computational needs;

- offer ubiquitous access – stored data and applications are accessible by authorized users anywhere, anytime;
- facilitate data sharing, enterprise-wide data analysis, and collaboration;
- are generally self-healing, and can self-reconfigure providing continuous availability in case of failure of their computing resources;
- offer enhanced user experience via a simplified Web-browser user interface.

1.3   Cloud Service Models

A computational or network resource, an application or any other kind of IT service offered to a user by a cloud is called a cloud service. Cloud services range from simple applications such as e-mail, calendar, word processing, and photo sharing to various types of complex enterprise applications and computing resources offered as services by major providers. For comprehensive information on cloud offerings currently available from several vendors, see the Cloud Computing Directory (http://www.cloudbook.net/directories/product-services/cloud-computing-directory, accessed November 25, 2015) and also refer to Chapter 2.

Depending on the type of services offered, cloud services can be classified into three major categories (see Table 1.2): software as a service (SaaS), platform as a service (PaaS), and infrastructure as a service (IaaS). In addition to these foundational services, several cloud support services, such as security as a service and identity and access management as a service, are on offer. Each service category can be used independently or used in combination with others.

1.3.1   Software as a Service

"Software as a service" clouds are also called *software clouds*. In the SaaS model, an application is hosted by a cloud vendor and delivered as a service to users, primarily via the Internet or a dedicated network. It eliminates the need to install and run the application locally, on a user's computer, and thereby also relieves the users from the burden of hardware and software maintenance and upgrades. The software license is not

Table 1.2   *Cloud service models*

Service model	Capability offered to the user	Controllability by users
Software as a service (SaaS)	Use of applications that run on the cloud.	Limited application configuration settings, but no control over underlying cloud infrastructure – network, servers, operating systems, storage, or individual application capabilities.
Platform as a service (PaaS)	Deployment of applications on the cloud infrastructure; may use supported programming languages, libraries, services, and tools.	The user has control of deployed applications and their environment settings, but no control of cloud infrastructure – network, servers, operating systems, or storage.
Infrastructure as a service (IaaS)	Provisioning of processing, storage, networks, etc.; may deploy and run operating systems, applications, etc.	The user has control of operating systems, storage, and deployed applications running on virtualized resources assigned to the user, but no control over underlying cloud infrastructure.

owned by the user. Users are billed for the service(s) used, depending on their usage. Hence, costs to use a service become a continuous expense rather than a huge up-front capital expense at the time of purchase. Examples of SaaS include Webmail, Google Apps, Force.com CRM, Quicken online accounting, NetSuite's Business Software Suite, Sun Java Communications Suite, and Paychex payroll management system.

1.3.2   Platform as a Service

In the PaaS model, the platform and tools for application development and middleware systems are hosted by a vendor and offered to application developers, allowing them simply to code and deploy without directly interacting with the underlying infrastructure. The platform provides most of the tools and facilities required for building and delivering applications and services such as workflow facilities for application design, development, testing, deployment, and hosting, as well as application services such as Web service integration, database integration, security, storage, application versioning, and team communication and collaboration. Examples of PaaS include Google App Engine, Microsoft Azure, Amazon's Web services and Sun Microsystems NetBeans IDE. The PaaS cloud is also called *platform cloud or cloudware*.

1.3.3   Infrastructure as a Service

In an IaaS cloud, raw computer infrastructure, such as servers, CPU, storage, network equipment, and datacenter facilities, are delivered as a service on demand. Rather than purchasing these resources, clients get them as a fully outsourced service for the duration that they need them. The service is billed according to the resources consumed. Amazon Elastic Compute Cloud (EC2), GoGrid, and FlexiScale are some of the examples of IaaS clouds. This type of cloud is also called a *utility cloud* or *infrastructure cloud*.

An IaaS cloud exhibits the following characteristics:

- availability of a huge volume of computational resources such as servers, network equipment, memory, CPU, disk space and datacenter facilities on demand;
- use of enterprise-grade infrastructure at reduced cost (pay for the use), allowing small and midsize enterprises to benefit from the aggregate compute resource pools;
- dynamic scalability of infrastructure; on-demand capacity can be easily scaled up and down based on resource requirements.

1.3.4   Cloud Support Services

In order to embrace the promise of clouds fully and successfully, adopters must use one or more of the three foundational cloud services – software as a service (SaaS), infrastructure as a service (IaaS), and platform as a service (PaaS). But they must also address several other related factors, such as security, privacy, user access management, compliance requirements, and business continuity. Furthermore, would-be adopters may have to use services from more than one service provider, aggregate those services, and integrate them with each other and with the organization's legacy applications / systems. Thus they need to create a cloud-based system to meet their specific requirements. To assist them in this, and to facilitate transition to the cloud, a cloud ecosystem is emerging that aims to offer a spectrum of new cloud support services that augment, complement, or assist the popular SaaS, IaaS, and PaaS offerings. Examples of such cloud support services are data storage as a service (DSaaS), analytics as service (AaaS), desktop as a service (DAAS), security as a service (SecaaS), identity and access management as a service (IAMaaS), and monitoring as a service (MaaS).

1.3.4.1   Data Storage as a Service (DSaaS)

With cloud storage, data is stored in multiple third-party servers, rather than on dedicated servers used in traditional networked storage, and users access a virtual storage. The actual storage location may change as the cloud dynamically manages available storage space; however, the users see a static location for their data. Key advantages of cloud storage are reduced cost and better data safety and availability. Virtual resources in the cloud are typically cheaper than dedicated physical resources connected to a PC or the network. Data stored in a cloud is generally safe against accidental erasure or hard-drive failures as Cloud Service Providers (CSPs) keep multiple copies of data across multiple physical machines continually. If one machine crashes, the data that was on that machine can be retrieved from other machine(s) in the cloud. Cloud vendors generally offer better security measures than a small business could afford. Enterprise data storage in clouds, however, raises some concerns, which are discussed later.

1.3.4.2   Analytics as a Service (AaaS)

Analytics as a service (AaaS), also known as data analytics as a service (DAaaS), refers to the provision of analytics platforms – software and tools – on a cloud for analysis and mining of large volumes of data (big data). Several vendors, such as IBM, Amazon, Alpine Data Labs, and Kontagent, offer such services. Customers can feed their data into the platform and get back useful analytic insights. It lets clients use particular analytic software for as long as it is needed and they pay only for the resources used. As a general analytic solution, AaaS has potential use cases in a range of areas and offers businesses an alternative to developing costly in-house high-performance systems for business analytics. An AaaS platform is extensible and scalable and can handle various potential use cases. It lets businesses get their data analytics initiatives up and running quickly.

1.3.4.3   Desktop as a Service (DaaS)

Desktop as a Service (DaaS) is a cloud service in which the back-end of a virtual desktop infrastructure (VDI) is hosted by a cloud service provider. It provides users with the ability to build, configure, manage, store, execute, and deliver their desktop functions remotely. Examples of such service are VMware Horizon Air, Amazon WorkSpaces and Citrix XenDesktop. Clients can purchase DaaS on a subscription basis and the service provider manages the back-end responsibilities of data storage, backup, security and upgrades. DaaS is well suited for a small or mid-size businesses that want to provide their users with a virtual desktop infrastructure (VDI), but find that deploying a VDI in-house is not feasible due to cost, implementation, staffing and other constraints.

1.3.4.4   Security as a Service (SecaaS)

Security as a Service (SecaaS) refers to the provision of security applications and services via the cloud, either to cloud-based infrastructure and software or from the cloud to the customers' on-premises systems. This enables enterprises to make use of security services in new ways, or in ways that would not be cost effective if provisioned locally. The services provided include authentication, virus detection, antimalware / spyware, intrusion detection, encryption, e-mail security, Web security, and security event management.

1.3.4.5   Identity and Access Management as Service (IAMaaS)

Identity and access management as a service (IAMaaS) offers cloud-based IAM services to clients and requires minimal or no on-premises presence of hardware or software. Services include user provisioning, authentication, authorization, self-service, password management, and deprovisioning.

1.3.4.6 *Monitoring as a Service (MaaS)*

Monitoring-as-a-service (MaaS) facilitates the deployment of monitoring functionalities for various other services and applications within the cloud. Monitoring focuses on how services are performing. The common application for MaaS is online state monitoring, which continuously tracks certain states of applications, networks, systems, instances or any element that may be deployable within the cloud.

1.4 Cloud Computing Deployment Models

Based on where the cloud is deployed and by whom, who owns and manages it, and who its primary users are, clouds are classified into five categories: public cloud, private cloud, virtual private cloud, community cloud, and hybrid cloud.

1.4.1 Public Cloud

The public cloud is the most common and widely known form of cloud, and is open for anyone – business, industry, government, nonprofit organizations and individuals – to use. The cloud infrastructure is, however, owned and managed by the cloud service provider – the organization that offers the cloud services. Public cloud services are offered on a pay-per-usage model; however, some applications on public clouds are accessible for free.

1.4.2 Private Cloud

A private cloud is deployed, provided, and controlled by an enterprise behind its firewall for its own use. Unwilling to head into public clouds because of concerns surrounding them and compliance requirements, some enterprises deploy their own cloud computing environments for their own (and their business partners') exclusive use. Thus, by having their own cloud, they gain operational efficiencies, effectively use their existing resources, if any, and have full control over the cloud, the applications, and data on the cloud.

1.4.3 Virtual Private Cloud

A virtual private cloud (VPC) is a segment of a public cloud, designated for a user with additional provisions and features for meeting that user's specific security and compliance requirements. Virtual private clouds provide users with more control over the resources they use than a pure public cloud does. An example of this type of cloud is Amazon's VPC.

1.4.4 Community Cloud

A community cloud is known as an industry cloud or vertical cloud. It is optimized and specially deployed for use by a particular industry sector or a group of users so that it meets specific requirements to address issues that are crucial to them. AcademyOne's Navigator Suite (aimed at academics and students) and Asite Solutions (specifically designed for construction industry) are examples of these types of clouds.

1.4.5 Hybrid Clouds

A hybrid cloud is a combination of two or more of the above cloud models. In this model, an enterprise makes use of both public and private clouds – deploying its less critical, low-risk services on a public cloud and business-critical core applications on its internal private cloud. A hybrid model allows for selective implementation addressing concerns about security, compliance, and loss of control, as well as enabling adoption of public clouds that offer cost benefits and more application options.

1.5   Benefits, Limitations, and Concerns associated with Cloud Computing

Cloud computing offers several substantial benefits to its users – individuals and enterprises. But it also has limitations and poses some risks, the effects of which depend on the application type and liabilities involved. In embracing cloud computing, therefore, users must understand, acknowledge, and address its limitations and risks.

1.5.1   Benefits of Cloud Computing

The key benefits of embracing a cloud include reduced capital and operational cost, improved flexibility, on-demand scalability, easier and quicker application deployment, ease of use, and availability of vast cloud resources for every kind of application or use. Many applications, including e-mail, office document creation, and much data storage continue to move into the clouds to reap the benefits of this new paradigm in IT.

Cloud computing frees users and businesses from the limitations of local computing resources and allows them to access the vast computational resources and computation power out in the cloud. For users to make use of cloud resources from anywhere in the world at any time, all that is needed is an Internet connection and a Web browser. The cloud lets the users run even computationally intensive or storage-intensive applications, as all of their computing and storage needs are sourced from the cloud.

Public clouds eliminate significant capital expenses for hardware and upfront license fees for software, as well as the headaches of hardware and software maintenance and upgrade by users. Cloud applications can be deployed instantly and simultaneously to thousands of users in different locations around the world, and can be regularly updated easily. Further, as clouds provide improved business continuity and data safety, they are particularly attractive to small- and medium-size enterprises, as well as enterprises in disaster-prone areas. Startups and application developers can use computing clouds to try their ideas without having to invest in their own infrastructure.

Other benefits of using a cloud are:

- lower operational and service cost to users – they pay for what they use;
- on-demand scalability to meet peak and uncertain computing demands;
- shared access to data / application-supporting collaboration and teamwork;
- greater data safety than most businesses can provide and manage in their own on-premises IT systems;
- ease of, and quicker, application deployment;
- freedom to use a vast array of computational resources on the cloud.

1.5.2   Limitations of Cloud Computing

There are a few limitations that users must consider before moving to the cloud. The key limitations of the cloud are:

- need for a reliable, always-available high-speed network access to connect to clouds;
- possibility of slow response at times due to increased traffic or uncertainties on the network, or higher load on computers in the cloud;
- additional vulnerabilities to security of data and processes on clouds;
- risk of unauthorized access to users' data;

- loss of data due to cloud failure (despite replication across multiple machines);
- reliability and continued availability of services offered by cloud service providers.

1.5.3   Cloud Concerns

Despite its promises, cloud computing's mainstream adoption is constrained by perceived and real barriers and concerns. Security and privacy of data and applications on the cloud are two of the top concerns of users in moving into clouds followed by reliability and availability of cloud services, as well as adherence to compliance requirements, where applicable. External clouds raise additional concerns about loss of control and sharing data outside the enterprise firewall.

Many people think that because they don't know where their data is stored remotely, and because the applications are accessed over the Internet, cloud services are insecure. They believe that if data and applications were physically housed in computers under their control, they would protect them better. But this is not necessarily the case as economies of scale allow a CSP to offer more sophisticated security, disaster recovery, and service reliability than an individual institution (particularly a small enterprise) can afford to deploy on its own.

Cloud computing security concerns and requirements can differ considerably among the stakeholders – end-user service consumers, cloud service providers and cloud infrastructure providers – and are determined by the specific services they provide or consume. The Cloud Security Alliance (CSA) has identified seven top cloud security threats and outlined impact of those threats as well as remediation for them (Cloud Security Alliance, 2009, 2010). They are:

1. Abuse and nefarious use of cloud computing.
2. Insecure application programming interfaces.
3. Malicious insiders.
4. Shared technology vulnerabilities.
5. Data loss/leakage.
6. Account, service & traffic hijacking.
7. Unknown risk profile.

Based on a 2013 survey, CSA has also identified nine critical threats to data security in the order of severity:

1. Data breaches.
2. Data loss.
3. Account hijacking.
4. Insecure APIs.
5. Denial of service.
6. Malicious insiders.
7. Abuse of cloud services.
8. Insufficient due diligence.
9. Shared technology issues. (Cloud Security Alliance, 2013)

Many enterprise computing applications must meet compliance requirements, which depend on the type of business and customer base. To better ensure the desired level of service delivery and to limit liabilities, service level agreements (SLAs) with the cloud vendors are highly recommended when consuming cloud services. A cloud SLA specifies terms and conditions as well as expectations and obligations of the cloud service provider and the user.

By careful planning and incorporating the user's requirements into cloud service offerings, both the cloud vendors and users can reduce risk and reap the rewards of cloud-based hosted services.

1.6   Migrating to Clouds

A new mindset is needed to embrace cloud computing. To use and benefit from clouds successfully, an enterprise must prepare itself strategically, culturally and organizationally, and take a holistic view of cloud computing. It must develop its strategic plan and follow a phased, pragmatic, step-by-step approach that provides a business context for its cloud adoption. It must choose a cloud option that is appropriate for the application, considering and managing the risks of migrating to clouds by applying safeguards. Moving into clouds is not just about technology; the cloud migration should also factor in the role of people, processes, and services, and the change-management process. Migration to clouds will also demand a new kind of IT management and governance framework.

1.6.1   Choosing your Cloud

A major decision that IT managers and enterprises have to make is the type of cloud – public clouds, private clouds, or variations of them – that is well suited for their application. To arrive at a better decision, they have to understand the differences between these deployments, and understand the risks associated with each in the context of the characteristics and requirements of their applications. They also have to consider:

- performance requirements, security requirements, and cloud service availability and continuity;
- amount of data transfer between the user and the clouds and/or between the clouds;
- sensitive nature of the applications;
- control of their application and data;
- total costs involved;
- whether the external cloud providers are trusted;
- terms and conditions imposed by the external cloud providers; and
- in-house technical capabilities. (Claybrook, 2010)

1.7   Cloud Prospects and Implications

Computing clouds are powerful change-agents and enablers. Soon the core competency for most enterprises would be using IT services and infrastructure that cloud computing offers as hosted services, not building their own IT infrastructure. Cloud computing will profoundly change the way people and enterprises use computers and their work practices, as well as how companies and governments deploy their computer applications. It is transforming the way we think about computing environments and will drastically improve access to information for all, as well as cutting IT costs. Ongoing developments – the increasing maturity of clouds, the introduction of new cloud computing platforms and applications, the growth in adoption of cloud computing services, and the emergence of open standards for cloud computing – will boost cloud computing's appeal to both cloud providers and users.

Clouds will enable open-source and freelance developers to deploy their applications in the clouds and profit from their developments. As a result, more open-source software will be published in the cloud. Clouds will also help close the digital divide prevalent in emerging and underdeveloped economies and may help save our planet by providing a greener computing environment.

Major stumbling blocks for enterprises moving their applications into the cloud in a big way are reliability, performance, bandwidth requirements, trust, and security issues. However, these barriers are gradually being lowered or removed. Government regulations and other compliance requirements lag behind market developments and demand, and these aspects need to be addressed swiftly.

Driven by economic imperatives and the promise of flexibility and convenience, cloud computing will gain wider acceptance. Like the Internet, cloud computing is a transformational technology. It will mature rapidly, as vendors and enterprises come to grip with the opportunities and challenges that it presents.

Cloud computing creates new possibilities for businesses – IT and non-IT – and there will be new investments. Researchers will be better able to run experiments quickly on clouds, share their data globally, and perform complex analysis and simulations. Universities and training institutions will offer new courses and programs, focused on cloud computing.

Some IT professionals, particularly those who work with on-premises IT systems, might be afraid of losing their jobs because of the ongoing adoption of cloud computing. The truth is that while some might lose their current job, they might be absorbed in other roles. So, they should be prepared to learn new skills and evolve in these new roles. They might need to learn how to deploy and manage applications in the cloud and minimize risks, as well as how to work with cloud providers. There will be a need for professionals to develop new kinds of cloud applications and to design, deploy, and maintain computing clouds.

Cloud service providers, the IT industry, professional and industry associations, governments, and IT professionals all have a role to play in shaping, fostering, and harnessing the full potential of the cloud ecosystem.

1.8   Conclusions

The cloud ecosystem is evolving to provide a vast array of services that support and aid deployment of cloud-based solutions for a variety of applications across many different domains. Further, new types of cloud deployments, new models that deliver value-added services, and new costing and business models are on the horizon. Besides cloud service providers and users, many new players that perform niche roles are getting into the cloud arena. Cloud-based applications are being adopted widely by individuals and businesses in developed countries, and even more so in developing economies such as India, South Africa, and China. Governments in many countries are promoting adoption of clouds by businesses – particularly micro, small, and medium enterprises, as well as individuals. As a result, a new bigger cloud ecosystem is emerging.

References

Claybrook, B. (2010) Cloud vs. in-house: Where to run that app? *Computer World*, http://www.computerworld.com/article/2520140/networking/cloud-vs--in-house--where-to-run-that-app-.html (accessed November 25, 2015).

Cloud Security Alliance (2009) *Security Guidance for Critical Areas of Focus in Cloud Computing V2.1*, https://cloudsecurityalliance.org/csaguide.pdf (accessed November 25, 2015).

Cloud Security Alliance (2010) *Top Threats to Cloud Computing V1.0*, http://cloudsecurityalliance.org/topthreats/csathreats.v1.0.pdf (accessed November 25, 2015).

Cloud Security Alliance (2013) The Notorious Nine: Cloud Computing Top Threats in 2013, Cloud Security Alliance, https://downloads.cloudsecurityalliance.org/initiatives/top_threats/The_Notorious_Nine_Cloud_Computing_Top_Threats_in_2013.pdf (accessed November 25, 2015).

ISO/IEC DIS 17789:2014 (2014) *Information Technology – Cloud Computing – Reference Architecture*, International Organization for Standardization, Geneva.

Mell, P. M., and Grance, T. (2011) *The NIST Definition of Cloud Computing.* Special Publication 800-145. NIST, Gaithersburg, MD, http://www.nist.gov/customcf/get_pdf.cfm?pub_id=909616 (accessed November 25, 2015).

Vaquero, L. M, Rodino-Merino, L., Caceres, J., and Lindner, M. (2009) A break in the clouds: Towards a cloud definition. *ACM SIGCOMM Computer Communication Review* **39**(1), 50–55.

Cloud Vocabulary

The following are some of the key terms commonly used in cloud computing:

Cloudburst. The term is used in a positive and a negative sense. Cloudburst (positive) refers to the dynamic deployment of a software application that runs on an enterprise's in-house computing resources to a public cloud to address a spike in demand. But cloudburst (negative) conveys the failure of a cloud computing environment due to its inability to handle a spike in demand.

Cloudstorming. This term refers to the act of connecting multiple cloud computing environments.

Cloudware. This is a general term referring to a variety of software, typically at the infrastructure level, which enables building, deploying, running, or managing applications in a cloud computing environment.

Cloud provider. A cloud provider is an organization that makes a cloud computing environment available to others, such as an external or public cloud.

Cloud enabler. This term refers to an organization or a vendor that is not a cloud provider per se but makes technology and services available, such as cloudware, which enables cloud computing.

Cloud portability. This term refers to the ability to move applications (and often their associated data) across cloud computing environments from different cloud providers, as well as across private or internal cloud and public or external clouds.

Cloud interoperability. This term refers to the ability of two or more systems or applications to exchange information and to use the information that has been exchanged together.

Cloud sourcing. This term refers to leveraging services in the network cloud – raw computing, storage, messaging, or more structured capabilities, such as vertical and horizontal business applications, even community – to provide external computing capabilities, often to replace more expensive local IT capabilities. While it might provide significant economic benefits, there are some attendant tradeoffs, such as security and performance. These services are delivered over the network but generally behave as if they are local.

Part II
Cloud Services

2

Cloud Services and Service Providers

K. Chandrasekaran and Alaka Ananth

National Institute of Technology Karnataka, India

2.1 Introduction

Cloud computing has become popular. It provides resources to consumers in the form of different services, such as software, infrastructure, platforms, and security. The services are made available to users on demand via the Internet from a cloud computing provider's servers, as opposed to being provided from a company's own on-premises servers. Cloud services are designed to provide easy, scalable access to applications, resources and services, and are managed by cloud service providers. A cloud service can be dynamically scaled to meet the needs of its users. Examples of cloud services include online data storage and backup solutions, Web-based e-mail services, hosted office suites and document collaboration services, database processing, managed technical support services, and more.

Cloud services can be broadly classified into three major types: software as a service (SaaS), platform as a service (PaaS), and infrastructure as a service (IaaS). Many other services, such as security as a service (SeaaS), knowledge as a service, and analytics as a service (AaaS) are emerging.

Many companies have come forward to adopt a cloud environment and ensure that both the users and the companies benefit from this. Amazon, Microsoft, Google, EMC, Salesforce, IBM, and many more companies provide various tools and services to give cloud support for their customers. In the subsequent sections we provide an overview of different tools and services offered by various cloud service providers.

2.2 Providers of Infrastructure as a Service

Infrastructure as a service (IaaS) is a provision model in which an organization offers storage, hardware, servers, and networking components to users for their use on demand. The client typically pays on a per-use basis.

Encyclopedia of Cloud Computing, First Edition. Edited by San Murugesan and Irena Bojanova.
© 2016 John Wiley & Sons, Ltd. Published 2016 by John Wiley & Sons, Ltd.

The service provider owns the infrastructure and is responsible for housing, running, and maintaining it. This subsection provides information about a few major IaaS providers such as Amazon and Rackspace.

2.2.1   Amazon Elastic Compute Cloud

Amazon Elastic Compute Cloud (EC2) is an IaaS offering by Amazon Web Services, a leading provider of IaaS. Powered by the huge infrastructure that the company has built to run its retail business, Amazon EC2 provides a true virtual computing environment. By providing a variety of virtual machine or instance types, operating systems, and software packages to choose from, Amazon EC2 enables the user to instantiate virtual machines of his choice through a Web-service interface. The user can change the capacity and characteristics of the virtual machine by using the Web-service interface, so the virtual machines have therefore been termed "elastic." Computing capacity is provided in the form of virtual machines or server instances by booting Amazon Machine Images (AMI), which can be instantiated by the user. An AMI contains all the necessary information needed to create an instance. The primary graphical user interface (GUI) is the AWS Management Console (Point and Click) and a Web-service application program interface (API), which supports both simple object access protocol (SOAP) and query requests. The API provides programming libraries and resources for Java, PHP, Python, Ruby, Windows, and .Net. The infrastructure is virtualized by using Xen hypervisor and different instance types:

- standard instances – suitable for most applications;
- microinstances – suitable for low-throughput applications;
- high-memory instances – suitable for high-throughput applications;
- high-CPU instances – suitable for compute-intensive applications;
- cluster compute instances – suitable for high-performance computing (HPC) applications.

An instance can be obtained on demand on an hourly basis, thus eliminating the need for estimating computing needs in advance. Instances can be reserved beforehand and a discounted rate is charged for such instances. Users can also bid on unused Amazon EC2 computing capacity and obtain instances. Such instances are called spot instances. Those bids that exceed the current spot price are provided with the instance, which allows the user to reduce costs. The spot price varies and is decided by the company.

Instances can be placed in multiple locations. These locations are defined by regions and availability zones. Availability zones are distinct locations that are engineered to be insulated from failures and provide inexpensive, low-latency network connectivity. Thus, by placing the instances in multiple availability zones, one can achieve fault tolerance and fail-over reliability. The Amazon EC2 instances can be monitored and controlled by the AWS Management Console and the Web-service API. However, AWS provides Amazon Cloud Watch, a Web service that provides monitoring for AWS cloud resources, starting with Amazon EC2. It enables customers to observe resource utilization, operational performance, and overall demand patterns – including metrics such as CPU utilization, disk reads and writes, and network traffic.

Elastic load balancing (ELB) enables the user to distribute and balance the incoming application's traffic automatically among the running instances based on metrics such as request count and request latency. Fault tolerance and automatic scaling can be performed by configuring the ELB as per specific needs. ELB monitors the health of the instances running and routes traffic away from a failing instance.

An instance is stored as long as it is operational and is removed on termination. Persistent storage can be enabled by using either elastic block storage (EBS) or Amazon Simple Storage Service (S3). EBS provides a highly reliable and secure storage while Amazon S3 provides a highly durable storage infrastructure designed for mission-critical and primary data storage. Storage is based on units called objects, whose size can vary from 1 byte to 5 gigabytes of data. These objects are stored in a bucket and retrieved via a unique, developer-assigned key. Storage is accessible through a Web-service interface and provides authentication procedures to protect against unauthorized access.

2.2.2   Rackspace Cloud

Rackspace Cloud offers IaaS to clients. It offers three cloud computing solutions – cloud servers, cloud files, and cloud sites. Cloud servers provide computational power on demand in minutes. Cloud files are for elastic online file storage and content delivery; and cloud sites are for robust and scalable Web hosting.

Cloud servers is an implementation of IaaS where computing capacity is provided as virtual machines, which run in the cloud servers' systems. The virtual machine instances are configured with different amounts of capacity. The instances come in different flavors and images. A flavor is an available hardware configuration for a server. Each flavor has a unique combination of disk space, memory capacity, and priority for CPU time. Virtual machines are instantiated using images. An image is a collection of files used to create or rebuild a server. A variety of prebuilt operating system images is provided by Rackspace Cloud – 64-bit Linux Distributions (Ubuntu, Debian, Gentoo, CentOS, Fedora, Arch and Red Hat Enterprise Linux) or Windows Images (Windows Server 2003 and Windows Server 2008). These images can be customized to the user's choice.

The cloud servers' systems are virtualized using the Xen Hypervisor for Linux and Xen Server for Windows. The virtual machines generated come in different sizes, measured based on the amount of physical memory reserved. Currently, the physical memory can vary from 256 Mb to 15.5 GB. In the event of availability of extra CPU power, Rackspace Cloud claims to provide extra processing power to the running workloads, free of cost.

Cloud servers can be run through the Rackspace Cloud Control Panel (GUI) or programmatically via the Cloud Server API using a RESTful interface. The control panel provides billing and reporting functions and provides access to support materials including developer resources, a knowledge base, forums, and live chat. The cloud servers' API was open sourced under the Creative Commons Attribution 3.0 license. Language bindings via high-level languages like C++, Java, Python, or Ruby that adhere to the Rackspace specification will be considered as Rackspace approved bindings.

Cloud servers scale automatically to balance load. This process is automated and initiated from either the Rackspace Cloud Control Panel or the Cloud Server API. The amount to scale is specified, the cloud server is momentarily taken offline, the RAM, disk space and CPU allotment are adjusted, and the server is restarted. A cloud server can be made to act as a load balancer using simple, readily available packages from any of the distribution repositories. Rackspace Cloud is working on beta version of Cloud Load Balancing product, which provides a complete load-balancing solution.

Cloud servers are provided with persistent storage through RAID10 disk storage, thus data persistency is enabled, leading to better functioning.

2.3   Providers of Platform as a Service

Platform as a service (PaaS) is a category of cloud computing services that provides a computing platform for development of applications. Its offerings facilitate the deployment of applications without the cost and complexity of buying and managing the underlying hardware and software and provisioning hosting capabilities. It may also include facilities for application design, development, testing, and deployment. It also offers services such as Web-service integration, security, database integration, storage, and so forth. This section gives an overview of the few companies that offer PaaS.

2.3.1   Google Cloud Platform

Google Cloud Platform enables developers to build, test and deploy applications on Google's highly scalable and reliable infrastructure. Google has one of the largest and most advanced networks across the globe and offers software infrastructure such as MapReduce, BigTable, and Dremel.

Google Cloud Platform includes virtual machines, block storage, NoSQL datastore, big data Analytics, and so forth. It provides a range of storage services that allows easy maintenance and quick access to users' data. Cloud Platform offers a fully managed platform as well as flexible virtual machines allowing users to choose according to their requirements. Google also provides easy integration of the user's application within Cloud Platform.

Applications hosted on Cloud Platform can automatically scale up to handle the most demanding workloads and scale down when traffic subsides. Cloud Platform is designed to scale like Google's own products, even when there is a huge traffic spike. Managed services such as App Engine or Cloud Datastore provide autoscaling, which enables the application to grow with the users. The user has to pay only for what he uses.

2.3.2   Google App Engine

Google App Engine lets the user run Web applications on Google's infrastructure. App Engine applications are easy to build, easy to maintain, and easy to scale as traffic and data storage needs grow. There are no servers to maintain App Engine.

App can be served from the user's own domain name (such as http://www.example.com/) using Google Apps. Otherwise it can be served using a free name on the appspot.com domain. An application can be shared with the world, or it can limit access to members of organization.

Google App Engine supports apps written in several programming languages. Apps can be built using standard Java technologies, including the JVM, Java servlets, and the Java programming language – or any other language using a JVM-based interpreter or compiler, such as JavaScript or Ruby. App Engine also features a Python runtime environment, which includes a fast Python interpreter and the Python standard library. It features a PHP runtime, with native support for Google Cloud SQL and Google Cloud Storage, which works just like using a local mySQL instance and doing local file writes. Finally, App Engine provides a Go runtime environment, which runs natively compiled Go code. This runtime environment is built to ensure that the application runs quickly, securely, and without interference from other apps on the system.

App Engine costs nothing to start. All applications can use up to 1 GB of storage and enough CPU and bandwidth to support an efficient app serving around 5 million page views a month, absolutely free. When billing is enabled for the application, free limits are raised, and one has to only pay for resources one uses above the free levels.

2.3.3   SAP HANA Cloud Platform

The SAP HANA Cloud Platform is an open-standard, Eclipse-based, modular platform as a service. SAP HANA Cloud Platform applications are deployed via command-line tools to the cloud as Web application archive (WAR) files or Open Services Gateway initiative (OSGi) bundles. These bundles are normal jar components with extra manifest headers. The applications run within the Java-based SAP HANA Cloud Platform runtime environment powered by SAP HANA, and can be maintained using Web-based management tools.

The main features of the SAP HANA Cloud Platform are as follows:

- an enterprise platform built for developers;
- native integration with SAP and non-SAP software;
- in-memory persistence;

- a secure data platform;
- a lightweight, modular runtime container for applications.

The SAP HANA Cloud Platform lets the users quickly build and deploy business and consumer applications that deliver critical new functionality to meet emerging business needs. It also helps connecting users with customers in more engaging experiences. It provides connectivity based on the cloud connectivity service. As a result, the platform streamlines the integration of new applications at the lowest possible total cost of ownership. Support for open programming standards provides a low barrier to entry for developers. This makes them productive from the start in building enterprise applications that can integrate with any SAP or non-SAP solution.

2.3.4   Captiva Cloud Toolkit

A tool called Captiva Cloud Toolkit to assist software development is offered by EMC. The EMC Captiva Cloud Toolkit is a software developer kit (SDK) comprised of modules that help Web application developers to add scanning and imaging functionality quickly and directly to their Web-based business applications. It is ideal for document-capture vendors, commercial software developers, and enterprises that want to create custom Web-based applications that are fully scan enabled, complementing their business solution offerings.

Using the Captiva Cloud Toolkit, developers can quickly create a working scan-enabled Web-based business application. As a result, time to market is shortened and development, testing, and support costs are greatly reduced. The enterprise's return on investment is also achieved quickly and its ability to compete in an increasingly competitive distributed document capture market is accelerated.

There are a few modules that are commonly used in most of the process development. These are basic modules that import images from various sources like Fax, e-mail, scanners or from any repository. Major modules are:

- *Scan.* Basically scanning happens at page level to bring images page by page into Captiva. Scanning is the entry point to Captiva where one can import any kind of document like pdf, tiff, or jpg.
- *MDW.* Multi Directory Watch is another entry point to Captiva. It can be pointed to any folder/repository from where Captiva could import documents directly. It is very useful if business is receiving documents in the form of softcopy – for example as attached files sent by e-mail. It also acts as a scan module.
- *IE.* Image enhancement is a kind of filter or repairing tool for images that are not clear. It enhances image quality so it can be processed easily. One can configure IE in accordance with business requirements and images being received. The main functions of IE include deskewing and noise removal.
- *Index.* Indexing is data-capturing activity in Captiva. One can capture key data from various fields. For example, if a bank form is being processed, the account number and sort code could be the indexing field. Indexing could be added as required by the business. A validation field could be added to avoid unwanted data entry while indexing any document.
- *Export.* Export is the exit point of Captiva where images / data are sent to various repositories like File, Net, Document or a data repository. The exported data is used for business requirement of various business divisions. For example, if we are capturing an account number and sort code for bank applications, this could be mapped to any department where it is needed.
- *Multi.* Multi is the last process in Captiva. It deletes batches that have gone through all the modules and have been exported successfully. Multi could be configured according to the needs of the business. In cases when it is required to take a backup of batches, this module could be avoided.

These modules are the very basic modules of Captiva for indexing and exporting. But for more flexibility and automation, Dispatcher is used, which is more accurate in capturing the data.

2.4   Providers of Software as a Service

Software as a service (SaaS) is a software distribution model in which applications are hosted by a vendor or service provider and made available to customers over a network. SaaS is becoming an increasingly common delivery model as underlying technologies that support Web services and service-oriented architecture (SOA) mature and new developmental approaches, such as Ajax, become popular. This section briefly outlines a few companies offering SaaS.

2.4.1   Google Cloud Connect

Google Cloud Connect is a feature provided by Google Cloud by integrating cloud and API for Microsoft Office. After installing a plug-in for the Microsoft Office suite of programs, one can save files to the cloud. The cloud copy of the file becomes the master document that everyone uses. Google Cloud Connect assigns each file a unique URL, which can be shared to let others view the document.

If changes are made to the document, those changes will show up for everyone else viewing it. When multiple people make changes to the same section of a document, Cloud Connect gives the user a chance to choose which set of changes to keep.

When the user uploads a document to Google Cloud Connect, the service inserts some metadata (information about other information) into the file. In this case, the metadata identifies the file so that changes will track across all copies. The back end is similar to the Google File System and relies on the Google Docs infrastructure. As the documents sync to the master file, Google Cloud Connect sends the updated data out to all downloaded copies of the document using the metadata to guide updates to the right files.

2.4.2   Google Cloud Print

Google Cloud Print is a service that extends the printer's function to any device that can connect to the Internet. To use Google Cloud Print, the user needs to have a free Google profile, an app program or web site that incorporates the Google Cloud Print feature, a cloud-ready printer, or printer connected to a computer logged onto the Internet.

When Google Cloud Print is used through an app or Web site, the print request goes through the Google servers. Google routes the request to the appropriate printer associated with the user's Google account. Assuming the respective printer is on and has an active Internet connection, paper and ink, the print job should execute on the machine. The printer can be shared with other people for receiving documents through Google Cloud Print.

Google Cloud Print is an extension built into the Google Chrome Browser but it should be enabled explicitly. Once enabled, the service activates a small piece of code called a connector. The connector's job is to interface between the printer and the outside world. The connector uses the user's computer's printer software to send commands to the printer.

If one has a cloud-ready printer, one can connect the printer to the Internet directly without the need for a dedicated computer. The cloud printer has to be registered with Google Cloud Print to take advantage of its capabilities.

2.4.3   Microsoft SharePoint

Microsoft offers its own online collaboration tool called SharePoint. Microsoft SharePoint is a Web application platform that comprises a multipurpose set of Web technologies backed by a common technical infrastructure. By default, SharePoint has a Microsoft Office-like interface, and it is closely integrated with the

Office suite. The Web tools are designed to be usable by nontechnical users. SharePoint can be used to provide intranet portals, document and file management, collaboration, social networks, extranets, web sites, enterprise search, and business intelligence. It also has system integration, process integration, and workflow automation capabilities. Unlike Google Cloud Connect, Microsoft SharePoint is not a free tool but it has additional features that cannot be matched by Google or any other company.

2.4.4   Sales Cloud

Salesforce.com is a cloud computing and social enterprise SaaS provider. Of its cloud platforms and applications, the company is best known for its Salesforce customer relationship management (CRM) product, which is composed of Sales Cloud, Service Cloud, Marketing Cloud, Force.com, Chatter and Work.com.

Sales Cloud refers to the "sales" module in salesforce.com. It includes Leads, Accounts, Contacts, Contracts, Opportunities, Products, Pricebooks, Quotes, and Campaigns. It includes features such as Web-to-lead to support online lead capture, with auto-response rules. It is designed to be a start-to-end setup for the entire sales process. Sales Cloud manages contact information and integrates social media and real-time customer collaboration through Chatter. The Sales Cloud gives a platform to connect with customers from complete, up-to-date account information to social insights, all in one place and available anytime, anywhere. Everything occurs automatically in real time, from contact information to deal updates and discount approvals.

Salesforce.com created the Sales Cloud to be as easy to use as a consumer web site like Amazon and built it in the cloud to eliminate the risk and expense associated with traditional software. With its open architecture and automatic updates, the Sales Cloud does away with the hidden costs and drawn-out implementation of traditional CRM software. By continuing to innovate and embrace technologies like mobile, collaboration, and social intelligence, the Sales Cloud has continued to pull ahead of the competition.

2.5   Providers of Data Storage as a Service

Data Storage as a service (DSaaS) is another important service in the field of cloud computing. One need not buy large-sized hard drives. Instead, one can use this service and pay rent to the company, which provides this service. This section gives a brief overview of a few major storage providers.

2.5.1   Amazon Storage as a Service

Amazon Storage as a Service known as Amazon S3 is storage for the Internet. It is designed to make Web-scale computing easier for developers. Amazon S3 provides a simple Web-services interface that can be used to store and retrieve any amount of data, at any time, from anywhere on the Web. It gives any developer access to the same highly scalable, reliable, secure, fast, inexpensive infrastructure that Amazon uses to run its own global network of web sites. The service aims to maximize benefits of scale and to pass those benefits on to developers.

Amazon S3 was built to meet users' requirements and expectations. Along with its simplicity, it also takes care of other features like security, scalability, reliability, performance, and cost. Thus, Amazon S3 is a highly scalable, reliable, inexpensive, fast, and also easy to use service, which meets design requirements.

Amazon S3 provides a highly durable and available store for a variety of content, ranging from Web applications to media files. It allows users to offload storage where they can take advantage of scalability and pay-as–you-go pricing. For sharing content that is either easily reproduced or where one needs to store an original copy elsewhere, Amazon S3's Reduced Redundancy Storage (RRS) feature provides a compelling solution. It also provides a better solution with Storage for Data Analytics. Amazon S3 is an ideal solution for storing pharmaceutical data for analysis, financial data for computation, and images for resizing. Later this content can be sent to Amazon EC2 for computation, resizing, or other large-scale analytics without incurring any data transfer charges for moving the data between the services.

Amazon S3 offers a scalable, secure, and highly durable solution for backing up and archiving critical data. For data of significant size, the AWS Import / Export feature can be used to move large amounts of data into and out of AWS with physical storage devices. This is ideal for moving large quantities of data for periodic backups, or quickly retrieving data for disaster recovery scenarios. Another feature offered by Amazon S3 is its Static Website Hosting. Amazon S3's web site hosting solution is ideal for web sites with static content, including html files, images, videos, and client-side scripts such as JavaScript.

2.5.2   Google Cloud Storage

Google Cloud Storage is an Internet service to store data in Google's cloud. It enables application developers to store their data on Google's infrastructure with very high reliability, performance, and availability. The service combines the performance and scalability of Google's cloud with advanced security and sharing capabilities. It is safe and secure. Data is protected through redundant storage at multiple physical locations. It is different from Google Drive as Google Drive is for users whereas Google Cloud Storage is for developers.

Cloud Storage provides many features like high capacity and scalability, consistency in data storage, and RESTful programming interface. It uses authentication and authorization to interact with APIs.

Tools for Google Cloud Storage are:

- *Google Developers Console.* This is a Web application where one can perform simple storage management tasks on the Google Cloud Storage system.

Gsutil is a Python application that lets the user to access Google Cloud Storage from the command line.

2.6   Other Services

Some companies offer various services under a single name. In this subsection, we discuss a few such providers. We also include other services like Knowledge as a Service (KaaS) by Salesforce.com and Queuing by Amazon Simple Queue Service (SQS).

2.6.1   IBM SmartCloud

Various cloud services are also offered to consumers by IBM. IBM cloud computing consists of cloud computing solutions for enterprises as offered by the global information technology company, IBM. All offerings are designed for business use and marketed under the name IBM SmartCloud. IBM cloud includes IaaS, SaaS, and PaaS offered through different deployment models.

IBM SmartCloud is a branded ecosystem of cloud computing products and solutions from IBM. It includes IaaS, PaaS, and SaaS offered through public, private, and hybrid cloud delivery models. IBM places these offerings under three umbrellas: SmartCloud Foundation, SmartCloud Services, and SmartCloud Solutions. Figure 2.1 illustrates the architecture of IBM SmartCloud.

SmartCloud Foundation consists of the infrastructure, hardware, provisioning, management, integration, and security that serve as the underpinnings of a private or hybrid cloud. Built using those foundational components, PaaS, IaaS, and backup services make up SmartCloud Services. Running on this cloud platform and infrastructure, SmartCloud Solutions consist of a number of collaboration, analytics and marketing SaaS applications.

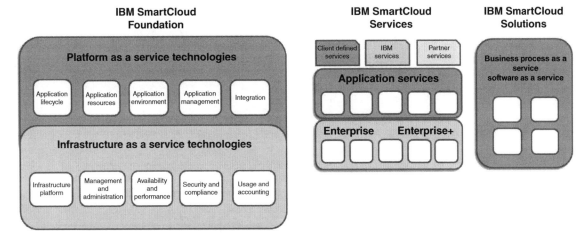

Figure 2.1   *IBM SmartCloud architecture. Source: Adapted from IBM (2012)*

Along with IaaS, PaaS, and SaaS, IBM also offer business process as a service (BPaaS). Infrastructure cloud services provide consumers with processing, storage, networks, and other fundamental computing resources where the consumer is able to deploy and run arbitrary software, which can include operating systems and applications. In Platform cloud services consumers can deploy consumer-created or acquired applications onto the cloud infrastructure created using programming languages and tools supported by the provider. Application cloud services allow consumers to use the provider's applications, running on a cloud infrastructure. The applications are accessible from various client devices through a thin client interface such as a Web browser (e.g., Web-based e-mail). Business process cloud services are any business process (horizontal or vertical) delivered through the cloud service model (multitenant, self-service provisioning, elastic scaling, and usage metering or pricing) via the Internet with access via Web-centric interfaces and exploiting Web-oriented cloud architecture. The BPaaS provider is responsible for the related business functions.

2.6.2   EMC IT

EMC is one of the leading global enterprises that require dynamic scalability and infrastructure agility in order to meet changing applications as well as business needs. In order to reduce the complexity and optimize the infrastructure, EMC chose cloud computing as the ideal solution to address the challenges. Offering IT as a service (ITaaS) reduces energy consumption through resource sharing.

By virtualizing the infrastructure, allocation of the resources on demand is possible. This also helps to increase efficiency and resource utilization. EMC IT provides its business process units with IaaS, PaaS, and SaaS services. Figure 2.2 gives an overview of the services offered by EMC, which are explained below.

IaaS offers EMC business units the ability to provision infrastructure components such as network, storage, compute, and operating systems, individually or as integrated services.

Platform as a service provides secure application and information frameworks on top of application servers, Web servers, databases, unstructured content management, and security components as a service to business units. EMC IT offers database platforms and application platforms for the purpose of development.

SaaS provides applications and tools in a services model for businesses. EMC IT offers widely used applications to business units like BI, CRM, ERP, and master data management. EMC IT brought together several

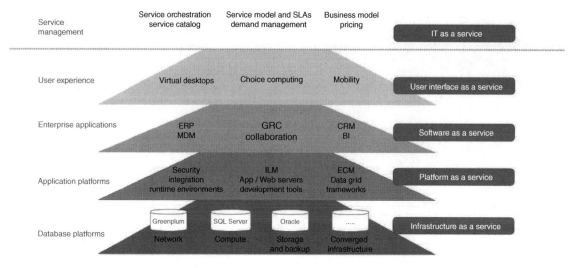

Figure 2.2 *Cloud services by EMC. Source: Adapted from EMC (2011)*

existing business solutions under unified architecture called Business Intelligence as a Service. It also offers ERP and CRM as a service.

User interface as a service (UIaaS) provides an interface rather than the actual device used.

2.6.3   Microsoft Windows Azure

Cloud computing is providing a new way to imagine IT at Microsoft, called Microsoft IT (MIST). Cloud computing is now the preferred and default environment for new and migrated applications at Microsoft. Windows Azure is a cloud-computing platform and infrastructure created by Microsoft. It provides both PaaS and IaaS services.

Windows Azure Cloud Services (Web and Worker Roles/PaaS) allows developers to deploy and manage application services easily. It delegates the management of underlying role instances and operating system to the Windows Azure platform.

The migration assessment tool (MAT) for Windows Azure encapsulates all the information to be aware of before attempting the application migration to Windows Azure. Based on responses to a series of simple binary questions, the tool generates a report that outlines the amount of development effort involved to migrate the application, or the architecture considerations for a new application.

The Windows Azure pricing calculator analyzes an application's potential public cloud requirements against the cost of the application's existing infrastructure. This tool can help to compare current operational costs for an application, against what the operating costs would be on Windows Azure and SQL Azure.

Azure Virtual Machines provide IaaS to the clients who request virtual machines (VMs). It provides the ability to create VMs on demand from a standard image or the one given by customer. One has to pay for the VM only when it is running. To create a VM, one has to specify a virtual hard drive (VHD) and the size of the VM. Azure Virtual Machines offer a gallery of stock VHDs. These include Microsoft-provided options, such as Windows Server 2008 R2, Windows Server 2012, and Windows Server 2008 R2 with SQL Server, along with Linux images provided by Microsoft partners. One can also create VM from one's own VHD and add it to the gallery.

2.6.4   Salesforce Service Cloud: Knowledge as a Service

Salesforce.com provides another service known as knowledge as a service (KaaS) in the form of Service Cloud. It includes Accounts, Contacts, Cases, and Solutions. It also encompasses features such as the Public Knowledge Base, Web-to-case, Call Center, and the Self-Service Portal, as well as customer service automation. Service Cloud includes a call-center-like case-tracking feature and a social networking plug-in for conversation and analytics.

Service Cloud delivers the world's first enterprise-grade knowledge base to run entirely on an advanced, multitenant cloud platform. That means that one can obtain all cloud-computing benefits that salesforce.com is known for delivering without expensive datacenters or software. Powerful knowledge management is provided, without the hassle of on-premises software. Unlike stand-alone applications, this knowledge base is fully integrated with everything else. Service Cloud offers all the tools one needs to run the entire service operation. When the consumer's knowledge base is a core part of the CRM solution, knowledge as a process can be managed. One can continually create, review, deliver, analyze, and improve the knowledge, and because it is delivered by the Service Cloud, the user's knowledge is available wherever other customers need it. Agents have the right answers at their fingertips to communicate over the phone, send out in an e-mail, or share via a chat client. The same knowledge base serves up answers to the service Web site as a part of company's public site. If one wants to take advantage of social channels like Twitter or Facebook, one can easily share knowledge that is tapped into the wisdom of the crowd to capture new ideas or answers. All this is done securely.

2.6.5   Amazon Simple Queue Service (SQS)

Amazon Web Services offers a queuing service called Amazon Simple Queue Service (SQS). It is a fast, reliable, scalable, fully managed message queuing service. SQS makes it simple and cost-effective to decouple the components of a cloud application. It can be used to transmit any volume of data, at any level of throughput, without losing messages or requiring other services to be always available.

SQS is a distributed queue system that enables Web service applications to queue messages quickly and reliably that one component in the application generates to be consumed by another component. A queue is a temporary repository for messages that are waiting to be processed.

SQS offers various features like allowing multiple readers and writers at the same time, providing access control facilities, and guaranteeing high availability of sending and retrieving messages due to redundant infrastructure. It also provides for variable length messages as well as configurable settings for each queue.

2.7   Nonmajor providers

Some of the nonmajor providers include Dell, VMware, Manjrasoft, and Akamai. Dell provides IaaS and also supports datacenter virtualization and client virtualization. VMware supports cloud through vCloud, vCloud Hybrid Services. Manjrasoft provides Aneka Platform to work in a .NET framework. Aneka provides a set of services for cloud construction and development of applications. Akamai provides cloud services as well as cloud security. Akamai's Global Traffic Manager ensures global load balancing.

2.8   Conclusion

In this chapter, we have discussed various companies that support cloud computing by providing tools and technologies to adapt to cloud environment. Each section briefly described a particular service like IaaS, PaaS, SaaS, etc. Each subsection gave an overview of a particular company tools which offer the services.

Table 2.1   *Details of cloud service providers*

Provider name	Service model	Deployment model	Server operating system
Amazon Web Services	IaaS	Public	Widows, Linux
Google App Engine	PaaS	Public	Windows
Windows Azure	IaaS	Public	Widows, Linux
IBM Cloud	IaaS	Private, Hybrid	Widows, Linux
Salesforce Platform	PaaS	Public	Widows, Linux
Rackspace	IaaS	Public, Private, Hybrid	Widows, Linux
SAP HANA Cloud	PaaS	Public	Linux

Table 2.2   *List of tools / services offered by companies*

Company name	Tools / services
EMC	Captiva Cloud toolkit, EMC IT
Google	Google App Engine, Google docs, Google Cloud Print, Google Cloud Connect
Amazon	Amazon EC2, Amazon S3, Amazon SQS
Microsoft	Windows Azure
IBM	IBM Smart Cloud
Salesforce	Sales Cloud, Service Cloud
SAP LABS	SAP HANA Cloud
Rackspace	Rackspace Cloud
VMware	vCloud
Manjrasoft	Aneka Platform
Redhat	OpenShift Origin, OpenShift Enterprise
Gigaspaces	Cloudify

Table 2.1 gives some details of the service model, deployment model, and server operating systems of a few cloud service providers.

An attempt has been made to list the tools / services offered by each company in Table 2.2. A brief explanation for major providers has been given in the previous sections. Although there are a larger number of companies that offer cloud services, we have chosen to present those that have progressed the most.

References

EMC (2011) *IT's Journey to the Cloud: A Practitioner's Guide*. White Paper, https://www.emc.com/collateral/software/white-papers/h7298-it-journey-private-cloud-wp.pdf (accessed November 26, 2015).

IBM (2012) *Transitioning to IBM Smart Cloud Notes*. Smart Cloud White Paper, https://www-304.ibm.com/connections/files/basic/anonymous/api/library/cb2c6245-fd7d-4a55-ae69-82b19553d026/document/3cfd72c6-e99c-4ada-8ff3-c11ed7eaed7d/media (accessed November 26, 2015).

3

Mobile Cloud Computing

Saeid Abolfazli,[1] Zohreh Sanaei,[1] Mohammad Hadi Sanaei,[1] Mohammad Shojafar,[2] and Abdullah Gani[1]

[1] University of Malaya, Malaysia
[2] La Sapienza University of Rome, Italy

3.1 Introduction

Mobile devices, especially smartphones, are creating dependency. People do not leave home without them. However, smartphones' small size, lightness, and mobility impose severe limitations on their processing capabilities, battery life, storage capacity, and screen size and capability, impeding execution of resource-intensive computation and bulky data storage on smartphones (Abolfazli *et al.,* 2013). Resource-intensive mobile applications are mobile applications whose execution requires a large central processing unit (CPU) capable of dealing with many transactions per second, a lot of random access memory (RAM) to load the code and data, extensive disk storage to store contents, and long-lasting batteries, which are not available in today's mobile devices. Enterprise systems, three-dimensional games, and speech recognition software are examples of such resource-intensive applications.

To address the shortcomings of mobile devices, researchers have proposed frameworks to perform resource-intensive computations outside mobile devices, inside cloud-based resources. This has led to mobile cloud computing (MCC) (Sanaei *et al.,* 2013a). Mobile cloud computing infrastructures include a multitude of mobile devices, cloud-based resources (individual/corporate computing devices that inherit cloud computing technologies and principles), and networking infrastructures that are managed via software systems known as cloud-based mobile augmentation (CMA) (Abolfazli *et al.,* 2014a). The augmented mobile device can execute intensive computations that would not be done otherwise. Thus, the mobile application programmers do not consider the deficiencies of mobile devices while programming applications and users will not experience device limitations when employing intensive applications.

Cloud-based mobile augmentation can overcome the resource deficiencies of mobile devices and execute "three main categories of applications, namely (i) computing-intensive software such as speech recognition

---

Encyclopedia of Cloud Computing, First Edition. Edited by San Murugesan and Irena Bojanova.
© 2016 John Wiley & Sons, Ltd. Published 2016 by John Wiley & Sons, Ltd.

and natural language processing, (ii) data-intensive programs such as enterprise applications, and (iii) communication-intensive applications such as online video streaming applications" (Abolfazli *et al.,* 2014a). To fulfill the diverse computational and quality of service (QoS) requirements of numerous different mobile applications and end users, several CMA solutions (reviewed in Abolfazli *et al.,* 2012a, 2014a) have been undertaken that suggest four major architectures for MCC. The major differences in these MCC architectures derive from various cloud-based resources with different features, namely multiplicity, elasticity (defined later in this chapter), mobility, and proximity to the mobile users. Multiplicity refers to the abundance and volume of cloud-based resources and mobility is unrestricted movement of the computing device while its wireless communication is maintained, uninterrupted. Resources employed can be classified into four types: distant immobile clouds (DIC), proximate immobile clouds (PIC), proximate mobile clouds (PMC), and hybrid (H). Consequently, efforts can be classified under four architectures, namely MDICC, MPICC, MPMCC, and HMCC (the prefix "M" represents mobile devices and the suffix "C" indicates computing action).

Deploying various cloud-based resources in CMA solutions is not straightforward and the diversity of existing resources complicates system management and maintenance. Several ongoing challenges, such as high augmentation overheads, application dependency on the underlying platform (known as portability), lack of interoperation among various mobile devices and cloud-based resources, absence of standardization, and mobility management, require future work before MCC solutions can be successfully adopted.

The remainder of this chapter is as follows. Section 3.2 presents an overview of MCC including its motivation, definition, and major building blocks. Section 3.3 briefly reviews four fundamental architectures for MCC. Section 3.4 provides a brief discussion on open challenges and section 3.5 concludes the chapter.

3.2   Mobile Cloud Computing

In this section, we present the motivation for augmenting mobile devices and define mobile cloud computing (MCC). We also devise a taxonomy of MCC building blocks, and explain them. Major differences between cloud computing and MCC are also explained.

3.2.1   Motivation

The motivation for MCC lies in the intrinsic deficiencies of mobile devices and the ever-increasing computing requirements of mobile users (Abolfazli *et al.,* 2013, 2014a). The miniature nature and mobility requirement of mobile devices impose significant constraints on their CPU, RAM, storage, and batteries. Mobile device manufacturers are endeavoring to enhance the computing capabilities of mobile devices by employing energy-efficient multicore processors, large fast RAM, massive low-overhead storages, and high charge density (long-life) batteries. However, technological limits, financial and time deployment costs, weight and size, and user safety regulations hinder mobile device empowerment.

Alternatively, researchers have used the concept of cloud computing to address the limitations of mobile devices and fulfill users' demands, which has led to the state-of-the-art MCC (Abolfazli *et al.,* 2014a). Mobile cloud computing researchers envision enhancing the computational capabilities of contemporary mobile devices to enable users to perform unrestricted computing, anywhere, anytime, from any device.

3.2.2   Definition

Mobile cloud computing (MCC) "is a rich mobile computing technology that leverages unified elastic resources of varied clouds and network technologies toward unrestricted functionality, storage, and mobility to serve a multitude of mobile devices anywhere, anytime through the channel of Ethernet or Internet regardless of heterogeneous environments and platforms based on the pay-as-you-use principle" (Sanaei *et al.*, 2013a).

Computing resource richness in MCC is realized by exploiting the computational power of computing entities, including giant clouds, desktop computers in public places, and resource-rich mobile devices that inherit cloud computing technologies and principles, which are named cloud-based resources. Resource elasticity enables automatic on-demand provisioning and deprovisioning of computing resources. Elasticity allows service consumers to use an amount of computing resources that matches their requirements. The resources can be acquired instantaneously when necessary and can be released when not in use with minimum effort. Hence, mobile users pay (depending on the service delivery model) only for the resources consumed.

Moreover, accessing varied cloud-based resources in MCC does not require communication through the Internet, whereas in stationary computing, cloud services are mainly delivered via this risky channel. In MCC, services can more effectively be delivered using a local network via WLAN, regardless of networking technologies and standards.

It is noteworthy that MCC involves execution of only those applications that require extensive computational resources beyond native mobile devices. If a user starts an application on a mobile device and connects to the cloud to monitor resource utilization or VM status inside the cloud, it is not MCC. Similarly, when a mobile user uses an application such as Facebook that is hosted on a cloud server, there is no MCC.

3.2.3 Building Blocks

In this section, the main building blocks of MCC are studied from two aspects: hardware and software. In every MCC system, hardware building blocks provide a rich mobile computation platform that can be employed by varied software programs. Our taxonomy is illustrated in Figure 3.1 and explained as follows.

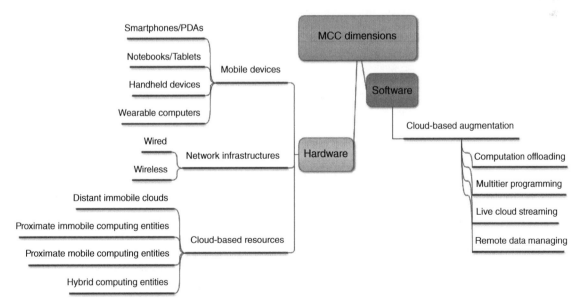

Figure 3.1 *Taxonomy of mobile cloud computing building blocks*

3.2.3.1   *Hardware*

Hardware infrastructures, including heterogeneous resource-constraint mobile devices, cloud-based resources, and networking infrastructures are tangible building blocks of MCC. Heterogeneity in MCC is inherited from mobile and cloud computing technologies and is intensified by the existence of a multitude of dissimilar devices, infrastructures, technologies, and features (Sanaei *et al.,* 2012a).

- *Mobile devices:* MCC is complicated by a multitude of heterogeneous battery-operated wireless-connected mobile devices (e.g., smartphones, tablets, and wearable computers) that feature varied limited computing capabilities.
- *Cloud-based resources:* computing entities that are built based on cloud computing technologies and principles (e.g., elasticity and pay-as-you-use) are called cloud-based resources. Four types of cloud-based resources are identified in Abolfazli *et al.* (2014a) as distant immobile clouds (DIC), proximate immobile clouds (PIC), proximate mobile clouds (PMC), and hybrid (H). These are discussed in section 3.3.
- *Networking infrastructures:* efficient, reliable, and high-performance networking in MCC necessitates the deployment of both wired and wireless networking technologies and infrastructures. Although mobile devices perform only wireless communications, immobile cloud-based resources require wired communication to transmit digital contents to different computing devices in a reliable and high-speed medium.

3.2.3.2   *Software*

The software building block of MCC comprises augmentation protocols and solutions to efficiently leverage cloud-based resources to mitigate the shortcomings of mobile devices.

Cloud-based mobile augmentation (CMA) "is the-state-of-the-art mobile augmentation model that leverages cloud computing technologies and principles to increase, enhance, and optimize computing capabilities of mobile devices by executing resource-intensive mobile application components in the resource-rich cloud-based resources" (Abolfazli *et al.,* 2014a). Major CMA approaches consist of computation offloading, live cloud streaming, multitier programming, and remote-data managing.

- *Computation offloading* is the process of identifying, partitioning, and migrating resource-intensive components of mobile applications to cloud-based resources. Identifying intensive components and partitioning can take place in three different ways: static, dynamic, and hybrid. Static partitioning is a one-time process of identifying and partitioning the resource-intensive components of a mobile application at design and development time. The benefit of static partitioning is that it does not impose runtime overheads on a mobile device and once the application is partitioned the same partitions can be used for an infinite number of executions. However, static partitioning is not adaptable to environment changes. On the other hand, in dynamic partitioning the identification and partitioning take place at runtime to better meet changes in the MCC environment. The challenge in dynamic partitioning is the excessive overhead of identifying intensive tasks, monitoring the environment, partitioning the application, and offloading the components. The third approach is to use a hybrid model where part of the application is partitioned at design time and part during runtime to mitigate the partitioning overhead and adapt to the environmental changes. Despite significant efforts in offloading (efforts are reviewed in our previous work (Abolfazli *et al.,* 2013, 2014a), offloading performance is degraded due to the overhead of partitioning and content offloading (Shiraz *et al.,* 2012).
- *Multitier programming* is used (Sanaei *et al.,* 2012b) to alleviate the overheads of code partitioning and offloading by building loosely coupled applications that perform resource-intensive computations

(often Web services) in the remote resources and minimize mobile-side computations. Resource-intensive computations are always available in the remote servers to be called for execution. Thus, the overhead of identifying, partitioning, and migrating tasks from mobile device to remote resources is omitted. Upon successful execution of the intensive tasks, the results are synchronized with the native components in the mobile device. In this model, only data is transmitted to the remote resources and codes are not migrated from the mobile device. Thus, the transmission overhead is significantly reduced. At application runtime, when the execution reaches the resource-intensive components, it pauses local execution and transmits application stack memory and raw data to the remote resources for execution. Upon completion of the execution, results are integrated and execution is resumed. However, application functionality in these solutions depends on the remote functions and services whose failure affects the application's execution. For instance, the speech-recognition component in navigation applications is a resource-intensive task whose execution with acceptable accuracy is impossible inside the mobile device.

- *Live cloud streaming* is another approach that aims to augment mobile devices by performing all of the computations outside the mobile device. Results are delivered to users as precompiled screenshots, streaming live to the mobile device. This approach requires low latency, high throughput, and a reliable wireless network, which is challenging to establish using current technologies.
- *Remote data-managing solutions* such as Dropbox virtually expand mobile storage by storing users' digital contents in the cloud-based resources. Parallel to the growth in computing, digital data is increasing sharply, demanding a huge amount of space on mobile devices, which further hinders mobile device adoption and usability. Although cloud storage enhances storage in mobile devices and improves data safety, trust and data security and privacy concerns prevent some users from leveraging remote storage.

3.2.4   Mobile Cloud Computing versus Cloud Computing

Although MCC inherits cloud-computing traits, significant fundamental differences exist between these two technologies. These are summarized in Table 3.1.

Cloud computing aims to provide rich elastic computing resources for desktop clients, whereas MCC envisions serving mobile users and realizing unrestricted functionality on the go. Service providers also differ in cloud computing and MCC. Resources in cloud computing consist of one or more unified computing entities,

Table 3.1   *Comparison of major cloud computing and mobile cloud computing characteristics*

Characteristics	Cloud computing	Mobile cloud computing
Service consumers	Desktop users	Mobile users
Service providers	Giant datacenters	Cloud-based resources
Network carrier	Wired	Wired/wireless
Objectives	Elasticity, pay-as-you-use	Mobile augmentation
Energy solutions	Conserve energy and emit less CO_2 on the server side	Conserve client's battery
Mobility	Neither client nor server	Both client and server
End-users' major considerations	Monetary ownership and maintenance costs of proprietary resources	Temporal, energy, and communication overhead

known as cloud datacenters, working in a parallel and distributed manner under corporate ownership located on the vendor's premises. Resources in MCC can be any computing device inheriting cloud technologies and principles capable of mitigating the resource deficiencies of mobile devices – these are referred to as cloud-based resources (Abolfazli *et al.,* 2014a).

Cloud computing leverages only wired communications, whereas MCC uses both wired and wireless devices. Although wireless is the dominant communication mode in MCC, immobile cloud-based resources leverage wired networks to enhance the computational experience of the end users. Wired network is beneficial in areas such as live virtual machine (VM) migration (Clark *et al.,* 2005) which is an emerging phenomenon that aims to mitigate the impacts of user mobility in augmentation process. Live VM migration in MCC allows immobile computing entities to transfer the running computational tasks over the wired network to a computing device, which is proximate to the new location of the nomadic user.

Another major difference between cloud computing and MCC is in their objectives. The former aims to reduce the ownership and maintenance costs of running private datacenters by introducing the concepts of resource elasticity and pay as you use. Cloud computing promises on-demand elastic resources by which desktop users can automatically provision computing resources and pay accordingly. Researchers in cloud computing endeavor to improve resource utilization rates, minimize the energy cost of intensive computing, and reduce negative impacts on the environment. However, MCC envisions augmenting computational capabilities of mobile devices by enabling long-time execution of resource-intensive mobile computing tasks. Mobility is not provisioned for cloud datacenters or desktop service consumers, whereas it is necessary for service consumers and feasible/beneficial for service providers. Users in cloud computing are concerned about high resource availability and saving monetary costs of executing resource-intensive computations on demand, whereas mobile service consumers use cloud-based services to enhance application execution time, reduce energy consumption of the mobile device, and reduce the wireless communication cost (in the absence of monthly flat communication plans).

3.3 Mobile Cloud Computing Architectures

Numerous MCC proposals have been investigated over the last few years, developing four major architectures for MCC, which are briefly discussed in this section. Table 3.2 presents major characteristics of four MCC architectures. It is noteworthy that these architectures are applicable to all cloud-based augmentation models described in Figure 3.1.

Table 3.2 *Major characteristics of varied mobile cloud computing architectures*

Characteristics	MDICC	MPICC	MPMCC	HMCC
Architecture	*Client-server*		*Client-server/peer-to-peer*	
Heterogeneity	High	High	Low	Medium
WAN latency	High	Medium	Low	Medium
Resource elasticity	High	Medium	Low	High
Resource multiplicity	Low	Medium	High	High
Resource availability	High	Low	Medium	High
Mobility implication	*Medium (client-side mobility)*		*High since both can move*	
Utilization cost	Low	Medium	High	Medium
Security and privacy	High	Medium	Low	High
Trust	High	Medium	Low	High

3.3.1 Mobile Distant-Immobile-Cloud Computing (MDICC)

A general abstract architecture for a typical MCC system consists of a mobile device that consumes the computing resources of a computing entity using a typical offloading framework like MAUI (Cuervo *et al.,* 2010) via a bidirectional wireless link. The first proposed architecture for MCC is depicted in Figure 3.2 (a) where the mobile user is consuming computational resources of public clouds using the Internet. Computational tasks in this model are executed inside the DIC resources (i.e., public cloud service providers such as Amazon EC2) and the results are sent back to the mobile client. The main advantages of DIC are high computational capabilities, resource elasticity, and relatively high security. High computing capabilities and elasticity of DIC minimize remote computational time and conserve the mobile battery. The utilization cost of DIC datacenters is the least possible cost given the ultimate goal of cloud computing, which is to reduce the computing costs.

However, existing architecture, hardware, and platform heterogeneities between DIC resources (x86 architecture) and mobile devices (ARM-based) complicate code and data portability and interoperability among mobile and cloud computers (Sanaei *et al.,* 2013a). Heterogeneity also imposes excess overheads by employing handling techniques such as virtualization, semantic technology, and middleware systems. Moreover, DIC are coarse location granular resources (meaning they are few in number and located far away from the

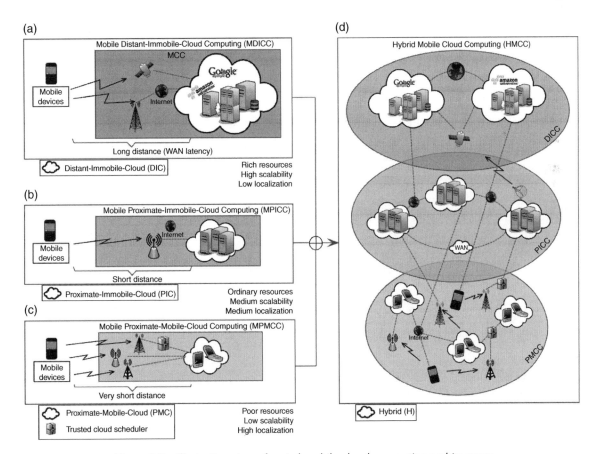

Figure 3.2 *Illustrative view of varied mobile cloud computing architectures*

majority of mobile service consumers) and are intrinsically immobile computing resources. Thus, exploiting DIC to augment mobile devices leads to long WAN latency due to many intermediate hops and high data communication overheads in the intermittent wireless networks. Long WAN latency degrades the application's execution performance and wastes the limited mobile battery. Moreover, service consumers' mobility on one hand and lack of cloud mobility on the other hand intensify WAN latency, and degrade effectiveness and efficiency of MCC solutions.

In fact WAN latency will likely remain in wireless communications for a long time, despite significant improvements in data transmission speed and network bandwidth (Satyanarayanan *et al.,* 2009). The main delaying factor in WAN latency is the processing delay at each intermediate hop to perform tasks such as decompression, decryption, security checking, virus scanning, and routing for each packet (Abolfazli *et al.,* 2013). Thus, the greater the number of hops, the longer is the WAN latency.

3.3.2 Mobile Proximate-Immobile-Cloud Computing (MPICC)

To mitigate the effects of long WAN latency, researchers (Satyanarayanan *et al.,* 2009) have endeavored to access computing resources with the least number of intermediate hops and have proposed alternative architecture for MCC, depicted in Figure 3.2 (b). Hence, mobile devices utilize computing resources of desktop computers in nearby public places such as coffee shops and shopping malls. Instead of travelling through numerous hops to performing intensive computations in DIC, tasks are executed inside the one-hop distance public computers in the vicinity (called PIC). These are medium location-granular (compared to the coarse grain resources, PICs are more numerous and are located nearer to mobile service consumers) and feature moderate computational power that provides less scalability and elasticity. Employing the computing resources of PICs holds several implications that require future research, particularly with regard to service providers' and consumers' security and privacy, isolating a computer's host OS from a guest mobile OS, incentivizing computer owners and encouraging them to share resources with nearby mobile devices, lack of on-demand availability (shops are open at certain hours and on certain days), and lack of PIC mobility.

3.3.3 Mobile Proximate-Mobile-Cloud Computing (MPMCC)

The third MCC architecture, depicted in Figure 3.2 (c), has been proposed recently (Abolfazli *et al.,* 2012b; Marinelli 2009) to employ a cloud of nearby resource-rich mobile devices (i.e., PMCs) that are willing to share resources with proximate resource-constraint mobile devices. Rapidly increasing popularity and ever increasing numbers of contemporary mobile devices, especially smartphones and tablets, are enabling the vision of building PMC to be realized. Two different computing models are feasible: peer-to-peer and client-server. In peer-to-peer, service consumers and providers can communicate with each other directly to negotiate and initiate the augmentation. However, the service consumer needs to perform an energy-consuming node discovery task to find an appropriate mobile service provider in vicinity. Moreover, peer-to-peer systems are likely vulnerable to fraudulent service providers that can attack the service consumer device and violate its privacy. The alternative MPMCC communication model is arbitrated client-server in which mobile client communicates with a trusted arbitrator and requests the most reliable and appropriate proximate node. The arbitrator can keep track of different service providers and can perform security monitoring. The crucial advantages of exploiting such resources are resource pervasiveness, and short WAN latency. In client-server and peer-to-peer models, the number of intermediate hops is small due to the service provider's proximity to the consumer. The resource pervasiveness of PMCs enables execution of resource-hungry tasks anytime anywhere (either in an *ad hoc* ecosystem or an infrastructure environment where mobile network operators (MNO) like Verizon can manage the process). Therefore, short WAN latency, and the ubiquity of mobile service providers leverage such resources in high latency-sensitive, low security-sensitive mobile computational tasks.

In this architecture, scalability and elasticity are limited due to the constrained computing power of individual mobile devices. Mobility management is another challenging feature of this architecture. Unrestricted mobility of mobile service providers and mobile service consumers in this architecture significantly complicates seamless connectivity and mobility, and noticeably degrades the efficiency of augmentation solutions. When a user (either service consumer or provider) starts moving across heterogeneous wireless networks with dissimilar bandwidths, jitters, and latencies, the varying network throughputs, increasing mobile-cloud distance, and frequent network disconnections increase WAN latency and directly affect application response time and energy efficiency. However, the utilization cost of PMCs is likely to be higher than the cost of other immobile resources due to their ubiquity and negligible latency, although they feature finite computing resources. Other shortcomings of this model are security, privacy, and data safety in mobile devices. Mobile devices are not safe to store user data because they are susceptible to physical damage, robbery, hardware malfunction, and risk of loss.

3.3.4   Hybrid Mobile Cloud Computing (HMCC)

Each of the three architectures features advantages and disadvantages that hinder optimal exploitation for efficient mobile computation augmentation. Sanaei *et al.* (2013b) demonstrated the feasibility of consolidating various resources to build a HMCC model (see Figure 3.2 (d)) that addresses these deficiencies, to optimize CMA. SAMI (Sanaei *et al.*, 2012b) is a multitier infrastructure that convergences public clouds, MNOs, and MNOs' trusted dealers to optimize resource-constrained mobile devices. In the core of the multitier infrastructure, a resource scheduler program evaluates each computational task and allocates appropriate resources to meet computational needs and fulfill user's QoS requirements (like cost, security, and latency) optimally. However, addressing the challenges – especially seamless mobility, significant MCC heterogeneity, wireless network intermittency, and current wireless networking – and developing an optimal generic scheduler is not a trivial task. Moreover, increasing number of mobile service consumers and hybrid cloud-based resources increase system complexity and complicate management and maintenance. Lightweight resource discovery and scheduling algorithms are essential for this MCC architecture.

3.4   Outstanding Challenges

Although advancements in MCC research are very impressive, several outstanding challenges require further work.

3.4.1   Lightweight Techniques

Resource-poverty in mobile computing is the major factor that necessitates the development of lightweight techniques for mobile consumers (e.g., offloading techniques), cloud service providers (e.g., light resource scheduling methods), and network providers (lightweight signal handoff). Native CPU, RAM, storage, and batteries are major resources to be conserved when performing intensive computation. To realize lightweight approaches in MCC, shrinking application dependency on underlying platforms (towards portability) is significantly beneficial. Moreover, exploiting lightweight communication technology (i.e., wireless local area networks compared to cellular), omitting excess/redundant native computation outsourcing protocols, using light data compression algorithms, and leveraging nearby high performance cloud-based resources are feasible and beneficial solutions.

3.4.2   Portability

Many and varied mobile and cloud operating systems, programming languages, application programming interfaces (APIs), and data structures fragment the MCC domain and hinder the portability of content among various computing entities. Portability in MCC is illustrated in Figure 3.3. It refers "to the ability of (i) migrating cloud components from one cloud to other clouds, (ii) migrating mobile components from one smartphone to other smartphones, and (iii) migrating data across heterogeneous clouds" (Sanaei *et al.,* 2013a) with little/no modification or configuration. These are non-trivial tasks in the absence of standards, technologies, and solutions to handle the heterogeneity in MCC. To achieve portability in MCC, automatic code convertor solutions, such as PhoneGap (http://phonegap.com/, accessed November 29, 2015), which regenerate codes for different platforms, together with early standardization, and lightweight heterogeneity handling techniques such as service oriented architecture that enable developing loosely coupled mobile-cloud applications, are promising.

3.4.3   Interoperability

Interoperability in MCC, shown in Figure 3.4, refers to the collaboration of interclouds, mobile clouds, and intermobiles with heterogeneous APIs and data structures. This is a challenging task. Lack of interoperability in MCC breeds vendor lock-in problems – user data and applications are locked inside certain clouds. When

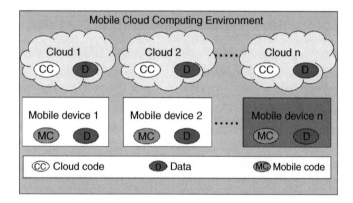

Figure 3.3   *Portability in mobile cloud computing*

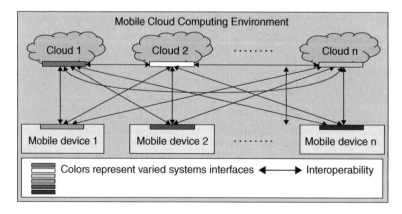

Figure 3.4   *Interoperability in mobile cloud computing*

users change service providers (mainly due to quality and cost issues), migrating content originates high monetary and time costs of moving codes and data from one format to another and transmitting them from the old provider to the new cloud provider. Risks of code and data corruption during conversion and transmission still threaten cloud consumers in MCC. Addressing interoperability in MCC demands standard protocols and common APIs such as the Open Cloud Computing Interface (OCCI) (http://occi-wg.org, accessed November 29, 2015).

3.4.4 Seamless Connectivity

Mobility in MCC is an inseparable property of service consumers and mobile service providers that requires seamless connectivity. Establishing and maintaining continuous, consistent wireless sessions between moving service consumers and other computing entities (e.g., mobile devices and clouds) in the presence of heterogeneous networking technologies requires future research and development. Lack of seamless connectivity increases application execution time and mobile energy consumption due to frequent disconnections and interruptions, which substantially degrade user experience. Seamless connectivity across heterogeneous wireless ecosystems requires solutions such as next-generation all IP-based infrastructures wireless networks.

3.4.5 Live VM Migration

Executing resource-intensive mobile applications via VM migration-based application offloading frameworks, involves the encapsulation of an application and migrating it to the remote VM. These are nontrivial tasks due to additional overheads of deploying and managing a VM on mobile devices. Live VM migration between distributed cloud-based resources (especially for DICs) is vital in executing resource-intensive applications, considering wireless bandwidth, network intermittency, and mobility constraints. When a roaming user increases distance from the offloaded contents (code or data), the increased distance prolongs access latency and degrades user-observed application performance. Thus, mobilizing the running VM and migrating it to resources nearer to the user without perceivable service interruption becomes essential to avoid user experience degradation. Therefore, optimal solutions such as reactive and proactive migration of VM instances (a proactive model requires predicting the new user destination) to a place closer to the mobile user without service interruption, and solutions alike VMware vMotion (http://www.vmware.com/files/pdf/VMware-VMotion-DS-EN.pdf, accessed November 29, 2015) are vital to smooth live migration of VM and to avoid user experience degradation. Consider, for example, a scenario that reactively initiates the migration of a VM to a server in Detroit as soon as you depart from a New York airport and ends migration before you reach Detroit.

3.5   Conclusions

Mobile Cloud Computing (MCC) is the state-of-the-art distributed mobile computing technology that aims to alleviate resource deficiency of a multitude of heterogeneous resource-constrained mobile devices by performing resource-intensive computations inside cloud-based resources. Mobile augmentation solutions are influenced to a large extent by the cloud-based resources employed. Granularity (resource multiplicity and proximity), computational capability, mobility, and heterogeneity between mobile devices and resources are major resource properties that affect augmentation performance and need consideration in design and development of imminent augmentation solutions. Despite impressive MCC findings, several challenges, such as lightweight low-latency architecture, live VM migration, efficient resource scheduling (automatically allocate resources to intensive mobile tasks), and seamless mobility, require further efforts to enable MCC in real scenarios.

References

Abolfazli, S., Sanaei, Z., Ahmed, E., *et al.* (2014a) Cloud-based augmentation for mobile devices: Motivation, taxonomies, and open challenges. *IEEE Communications Surveys and Tutorials* **16**(1), 337–368.

Abolfazli, S., Sanaei, Z., Alizadeh, M., *et al.* (2014b) An experimental analysis on cloud-based mobile augmentation in mobile cloud computing. *IEEE Transactions on Consumer Electronics (February)*, 1–9.

Abolfazli, S., Sanaei, Z. and Gani, A. (2012a.) *MobileCloud Computing: A Review on Smartphone Augmentation Approaches*. Proceedings of the IEEE Mobile Cloud Computing Workshop, Singapore. IEEE, pp. 199-204.

Abolfazli, S., Sanaei, Z., Gani, A. *et al.* (2013) Rich mobile application: Genesis, taxonomy, and open issues. *Journal of Network and Computer Applications* **40**, 345–362.

Abolfazli, S., Sanaei, Z., Shiraz, M., *et al.* (2012b) *MOMCC: Market-Oriented Architecture for Mobile Cloud Computing based on Service Oriented Architecture*. Proceedings of the IEEE Mobile Cloud Computing Workshop, Beijing. IEEE, pp. 8– 13.

Clark, C., Fraser, K., Hand, S. *et al.* (2005) Live migration of virtual machines. Proceedings of the Symposium on Networked Systems Design and Implementation, Boston, pp. 273–286.

Cuervo, E., Balasubramanian, A., Cho, D. K., *et al.* (2010) *MAUI: Making Smartphones Last Longer with Code Offload*. Proceedings of the ACM Conference Mobile Systems, Applications, Services, San Francisco. ACM, pp. 49–62.

Marinelli, E. E. (2009) *Hyrax: Cloud Computing on Mobile Devices using MapReduce*. Computer Science Department, Carnegie Mellon University, Pittsburgh, PA.

Sanaei, Z., Abolfazli, S., Gani, A., *et al.* (2012a) Tripod of requirements in horizontal heterogeneous mobile cloud computing. Proceedings of Computing, Information Systems and Communications, Singapore, pp. 217–222.

Sanaei, Z., Abolfazli, S., Gani, A., *et al.* (2013a) Heterogeneity in mobile cloud computing: Taxonomy and open challenges. *IEEE Communications Surveys and Tutorials* **16**(1), 369–392.

Sanaei, Z., Abolfazli, S., Khodadadi, T., *et al.* (2013b) Hybrid pervasive mobile cloud computing: Toward enhancing invisibility. *Information – An International Interdisciplinary Journal* **16**(11), 1–12.

Sanaei, Z., Abolfazli, S., Shiraz, M., *et al.* (2012b) *SAMI: Service-Based Arbitrated Multi-Tier Infrastructure Model for Mobile Cloud Computing*. Proceedings of IEEE Mobile Cloud Computing, Beijing. IEEE, pp. 14–19.

Satyanarayanan, M., Bahl, P., Caceres, R., and Davies, N (2009) The case for VM-based cloudlets in mobile computing. *IEEE Pervasive Computing* **8**(4), 14–23.

Shiraz, M., Abolfazli, S., Sanaei, Z., and Gani, A. (2012) A study on virtual machine deployment for application outsourcing in mobile cloud computing. *Journal of Supercomputing* **63**(3), 946–964.

4

Community Clouds

Amin M. Khan, Felix Freitag, and Leandro Navarro

Universitat Politècnica de Catalunya, Spain

4.1 Introduction

Cloud computing infrastructures have emerged as a cost-effective, elastic, and scalable way to build and support Internet applications, but issues like privacy, security, control over data and applications, performance, reliability, availability, and access to specific cloud services have led to different cloud deployment models. Among these deployment models, the public cloud offers services of generic interest over the Internet, available to anybody who signs in. On the other hand, the private cloud model aims to provide cloud services only to a specific user group, such as a company, and the cloud infrastructure is isolated by firewalls, avoiding public access. When a private cloud is combined with the public cloud – for instance when some functionality of the cloud is provided by the public cloud and some remains in the private cloud – then this cloud model is called a hybrid cloud. The community cloud bridges different aspects of the gap between the public cloud, the general purpose cloud available to all users, and the private cloud, available to only one cloud user with user-specific services. The concept of community cloud computing has been described in its generic form in Mell and Grance (2011) as a cloud deployment model in which a cloud infrastructure is built and provisioned for exclusive use by a specific community of consumers with shared concerns and interests, owned and managed by the community, or by a third party, or a combination of both. The community cloud model assumes that cloud users can be classified into communities, where each community of users has specific needs. A specific cloud, the community cloud, addresses the cloud requirements of such a group of users.

We discuss community clouds in the rest of this chapter as follows. In section 4.2 we explain the concept of community cloud computing in detail. In section 4.3 we give examples of cloud user communities for which the community cloud model could be applied. In section 4.4 we derive the requirements of community clouds in relation to user communities. Section 4.5 presents the potential and advantages of community

Encyclopedia of Cloud Computing, First Edition. Edited by San Murugesan and Irena Bojanova.
© 2016 John Wiley & Sons, Ltd. Published 2016 by John Wiley & Sons, Ltd.

clouds, and in section 4.6 we examine how existing community cloud solutions fulfill the challenges for their communities. Finally in section 4.7 we conclude our study on community clouds and outline some future scenarios based on current trends.

4.2   The Concept of Community Clouds

The wide availability of commercial cloud solutions has led to widespread adoption of cloud use by all kinds of stakeholders (Buyya *et al.*, 2011). It is a natural evolution from this growing number of cloud users that, among these users, certain clusters or communities of users arise, where each community is characterized by shared interests and common concerns. Cloud communities can extend over several users and they are not necessarily limited geographically. A community, however, is characterized by having common interests and concerns. For a community cloud to be tailored to a group of users, this group will need to have some specific requirement from the community cloud, different from those of the other communities. While communities may be of different sizes, some kind of critical mass related to this community will be needed to make it worth the effort of developing a tailored community cloud. The community cloud solution may have an advantage over the private cloud of sharing the cost of the cloud development and maintenance among its community members.

General-purpose cloud solutions provided by the public cloud do not optimally fit the specific needs of the different user communities because, for instance, certain security concerns or performance requirements regarding clouds are insufficiently addressed by such generic cloud solutions. This leads to increasing interest in building community cloud solutions that fit specific communities of users (Marinos and Briscoe, 2009). A community cloud offers features that are tailored to the needs of a specific community. Each community cloud is specialized to provide particular features – the ones that are needed by its community. The difference between one community cloud and another for their users is that the provision of different features, such as performance, security, and ease of use of the cloud, is emphasized.

The specific requirements of a community of its particular cloud will influence the design of the community cloud. While the concept of addressing the needs of a particular community is common to community clouds, its requirements may lead to very different cloud designs between one community cloud and another. The tailored solutions for each community cloud may affect the hardware needed to build the cloud, the cloud management platforms, the services available in the cloud, and the applications provided to the user community.

The advantage of a community cloud lies in being able to offer optimized cloud solutions to specific user communities. They can see the drawbacks of using off-the-shelf cloud solutions, motivating them to look at community clouds, because when a community cloud is used, certain user requirements can be better satisfied. An important condition, however, is that there is enough commonality between the requirements of the community, for example in terms of cloud services that the community uses or cloud performance needs, in order to allow the finding of a specific but common cloud solution.

Community clouds are implemented using different designs depending upon requirements. Figure 4.1 depicts some of these possible architectures. One common approach is that a public cloud provider sets up separate infrastructure and develops services specifically for a community to provide a vertically integrated solution for that market. Similarly, a third-party service provider can focus on a particular community and only specialize in building tailor-made solutions for that community. Another option is that community members that already have expertise in cloud infrastructures come together to federate their private clouds and collectively provision cloud services for the community (Buyya *et al.*, 2010). Another radical model involves building community cloud services using resources contributed by individual users by either solely relying on user machines or using them to augment existing cloud infrastructures (Marinos and Briscoe, 2009).

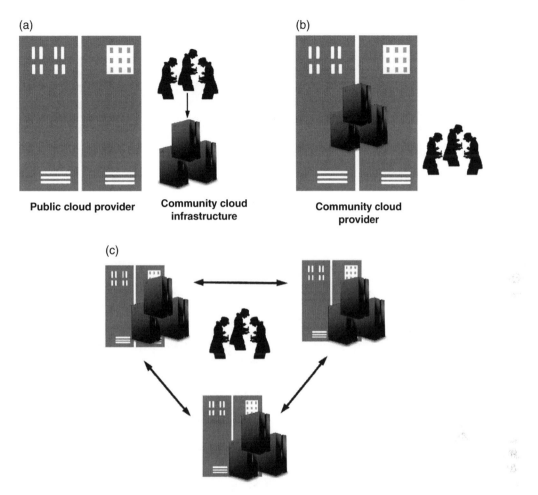

Figure 4.1 *Different architectures for community cloud. (a) Public cloud provider's separate infrastructure for a community. (b) Purpose-built cloud infrastructure for a community. (c) Multiple private clouds combined for community*

4.3 Cloud User Communities

Community cloud systems are designed according to the requirements of the community of users they are to serve. Several community clouds have recently been reported for enterprise communities. Enterprises belonging to the same industrial sector often use similar but independent cloud solutions, and, comparing these solutions, it can be seen that these clouds are optimized in similar aspects, which allow these enterprises to gain advantages for reaching common goals. Instead of these private clouds, building a community cloud for such enterprises shares the cost of the cloud solutions among them, and may also offer opportunities for collaboration for mutual benefit, even among competitors. Enterprise use of community cloud computing can also extend to the following two situations, beside the common scenario of consuming the services provided by a third-party cloud provider. One is where a number of organizations or departments within an organization sign peering agreements to share cloud resources to provide services, giving rise to federated cloud systems (Moreno-Vozmediano *et al.*, 2012). The other is where an organization can make use of the resources

Table 4.1   *Community cloud success stories*

Community	Solutions
Government agencies	UK's G-Cloud, EU's Cloud for Europe, Amazon's GovCloud, IBM's Federal Community Cloud, Lockheed Martin's SolaS Secure Community Cloud
Healthcare industry	Optum's Health Care Cloud
Financial services industry	NYSE Technologies' Capital Markets Community Platform, CFN Services' Global Financial Services Cloud
Media industry	Siemens' Media Community Cloud, IGT Cloud
Aviation industry	SITA's ATI Cloud
Higher education	Oxford's Online Research Database Service
Telecom industry	Ericsson Cloud System

of its customers to meet peaks in demand or to deliver a better experience to them. Table 4.1 lists some of the popular community clouds. In the following we identify some user communities that adopt community cloud solutions.

4.3.1   Government Departments and Agencies

Government departments and contractors need to work with sensitive data that cannot be stored or shared in a public cloud environment. This is because of the nature of the data and the requirement for privacy, which is also enforced by government regulations and laws. This requires cloud infrastructure to be set up separately for government departments and the private contractors working with them, which allows the data to be handled in a confident and secure manner. Major public cloud vendors offer solutions specifically tailored for this, which adhere to government legislation and standards, and are mostly designed for particular countries as requirements differ for different regions. There are many cloud solutions in the United States – for example, GovCloud from Amazon, Federal Community Cloud from IBM, and SolaS Secure Community Cloud from Lockheed Martin. Similarly, there is the G-Cloud program in the United Kingdom and the Cloud for Europe program in European Union, and other countries are also looking to develop similar cloud systems.

4.3.2   Health Service Providers

Stakeholders in health services include patients, physicians, health IT administrators, hospitals, pharmacies, insurance companies and government entities. Public and private organizations working in health services deal with private data related to patients and health professionals, and they are legally obliged to keep such data secure. Coordination among IT health systems through a community cloud better satisfies the common requirements of this user community for improved and secure access to information to support health decisions about patients. Community cloud solutions allow the secure exchange of this data between organizations and pave the way for designing innovative and useful applications for health services that are not possible with public cloud infrastructure. Examples include the Health Care Cloud by Optum.

4.3.3   Financial Services Providers

Enterprises operating in the financial market share the need for cloud platforms that offer low latency to access real-time information and the highest security standards for data privacy. These requirements are not met by the public clouds and private clouds require too large an investment to make it feasible for most organizations.

For the financial community, community cloud platforms have been developed with optimization and services specifically designed for financial services providers. Examples include Capital Markets Community Platform from NYSE Technologies and the Global Financial Services Cloud from CFN Services.

4.3.4 Media Industry

Media production is a collaborative task, often carried out within an ecosystem of partners. For content production, media companies need to apply low-cost and agile solutions to be efficient. For instance, computer-game development companies make use of cloud infrastructures to transfer huge files efficiently. Sharing of content between partners allows collaborative and faster decision making. A community cloud allows fast exchange of digital media content and deploys services for specific B2B workflow executions to simplify media content production. Examples of such clouds include IGT Cloud from IGT and CA Technologies and Media Community Cloud from Siemens.

4.3.5 Aviation Industry

In the air transport industry, many different players need to interact. Business applications are often shared across stakeholders in order to assure that critical functions are satisfied. A high level of adoption of common infrastructure is already taking place to ease the orchestration of the operational processes. A community cloud framework to better manage air transport unifies the different cloud systems that are used by this community. Examples of such clouds include the ATI cloud from SITA.

4.3.6 Higher Education Institutions

Departments and schools in universities share the use of IT for teaching, learning, and research. From an administrative perspective, these units are often independent while needing similar services. Through a community cloud, management and configuration of servers can be shared and become more cost-effective, while the accounting service of the community cloud allows the consumed services to be chargeable to each member of the community. Examples of community clouds in higher education include the Online Research Database Service (ORDS) offered by University of Oxford to institutions in the United Kingdom.

4.3.7 Legal Services Providers

Judicial services and law firms need access to historical and current data from legal proceedings and other government legislation. The ongoing work of these organizations deals with sensitive data. Storing this data on a public cloud can pose a risk or violate government regulations on data protection. A community cloud in this situation allows the data to be exchanged in a secure manner and also makes it possible to design applications that are able to aggregate data from multiple organizations that are working together and provide useful insights to support decision making.

4.3.8 Engineering and Manufacturing Industry

Engineering companies involved in computer-aided design and manufacturing require access to computing and storage resources. The data and processes form the intellectual property of these organizations and the security and privacy of this information is very important. The applications are very compute and data intensive. Transferring data to and from public clouds and running the processes in public clouds may not meet performance requirements and in some cases may not be the most cost-efficient solution either. Community clouds allow these companies to address their requirements in an effective manner, sharing the costs of setting up and running the infrastructure among themselves.

4.3.9 Energy Industry

Energy companies working in oil and gas exploration need to exchange data between the partners and the subcontractors. The main requirements in this case are that data can be very large in size and must be available in timely fashion to locations all over the globe. A community cloud set up by the partners working in energy sector addresses these requirement and different members contribute their infrastructure and facilities to make this possible.

4.3.10 Scientific Research Organizations

Scientific research is also benefiting from bringing together computing resources across the organizations. This is seen as the extension of earlier grid computing systems (Foster and Kesselman, 2003), where organizations share their clusters and private clouds with others and reap the benefits of having greater resource availability on demand. Another approach also followed in some projects is the use of voluntary computing systems to form a community cloud that makes use of resources contributed by users connected to the Internet to solve research problems (Cappos *et al.*, 2009).

4.3.11 ICT Services Providers

Community clouds have been used by Internet, mobile, and telecom services providers. Examples include clouds for service providers taking advantage of customer-premises equipment (CPE) or network edge devices to improve delivery of services, or better use of provider's facilities. Cases were reported from Telefónica pooling together Wi-Fi of different customers to provide higher bandwidth – an idea that can be extended to providing other services like video-on-demand. This model also helps for content distribution networks (CDN) in that this can allow content to be distributed efficiently for users. For the latter, Ericsson is looking at placing cloud nodes at the base stations, which provides the flexibility of cloud model for networking services. This idea builds on the concepts of software-defined networking (SDN) and network virtualization. These examples show that interesting cloud-based solutions can be developed in the future that make use of a federation of a large number of smaller cloud installations set up either at the service providers' or the consumers' end.

4.3.12 Online Social Network Users

Online social networks, like Facebook, Twitter and Google Plus among many others, connect users that already know each other in real life and the basic premise is that such users are more likely to collaborate among themselves and share resources. Compared to other user communities, a specific feature of this community is that mutual trust and resource-sharing agreements can be easily established. The context of this community imposes the requirement on the cloud system that it should be closely integrated with the API provided by commercial online social networks. For example, social cloud projects integrated the Facebook API to provide a distributed storage service by which a user can store files on friends' machines (Chard *et al.*, 2012). The particular needs of this scenario match the community-cloud model, where a user community has identical underlying issues and challenges. The integration of social networks brings some additional complexity to the implementation of the community cloud system, but also makes issues like resource discovery and user management easier to tackle.

4.3.13   Mobile Device Users

Mobile devices increasingly provide better computing capacity, memory, and storage, even though battery power and bandwidth are still limitations. Interesting cloud applications are being envisaged that either employ the mobile devices both as providers and consumers of cloud services or use resources on mobile devices to augment the traditional cloud infrastructure. Issues like high mobility and limited capacity of the devices, however, pose constraints for providing applications in community clouds consisting of or augmented by mobile devices.

4.3.14   Clouds in Community Networks

Community wireless mesh networks are already based on the principle of reciprocal resource sharing. The social and community aspects make it easy to extend this sharing of bandwidth to other computing resources. Particular challenges for community clouds in such networks exist, however. For example, these networks are limited in resources, are often unstable, have variable network bandwidth between links, and have a high churn of nodes. A generic cloud solution is therefore not an option for this user community, but a community cloud focused on community networks is required. The strong sense of community and technical knowledge of participants of such networks are positive factors that can be taken into account in the design of such a community cloud tailored to the local needs and built on infrastructure provided by community members.

4.4   Requirements and Challenges

This section considers requirements for community clouds emphasized by different user communities. These requirements provide the foundation for the design of the community cloud system, and need to be satisfied for it to be deployed and adopted successfully by the community.

4.4.1   Security

There are many security challenges that need to be addressed to ensure users' trust in the system. With multiple independent cloud providers from the community, security becomes even more important in a community cloud. The data and applications running on different cloud systems have to be protected from unauthorized access, damage, and tampering. With the rise in sophisticated security threats and increase in data breaches, this is one of the most critical factors impacting any community cloud system.

4.4.2   Autonomy

Community cloud systems may be formed based on individual cloud systems that are set up and managed independently by different owners. The main requirement for a cloud owner for participating in such a community cloud is that the local cloud setup should adhere to the common API provided by the community cloud. The cloud owner should also contribute some set of mutually agreed resources to the community.

4.4.3   Self-Management

Depending on the type of community, self-management capabilities may be an important requirement of the community cloud system. Community clouds should manage themselves and continue providing services without disruption, even when part of the infrastructure runs into issues. Self-management should also help in the coordination between different cloud owners that become part of a federated community cloud.

4.4.4   Utility

For the community cloud to be accepted it should provide applications that are valuable for the community. Usage strengthens the value of the community cloud, motivating its maintenance and update.

4.4.5   Incentives for Contribution

Some types of community cloud may be built upon collective efforts. Such a community cloud builds on the contribution of the volunteers in terms of computing, storage, network resources, time, and knowledge. For these community clouds to be sustainable, incentive mechanisms are needed to encourage users to contribute actively towards the system.

4.4.6   Support for Heterogeneity

The hardware and software used by members in a community cloud can have varying characteristics, which can result from the use of different hardware, operating systems, cloud platforms, or application software. The community cloud system should handle this heterogeneity seamlessly.

4.4.7   Standard API

The cloud system should make it straightforward for the application programmers to design their applications in a transparent manner for the underlying heterogeneous cloud infrastructure. The API should provide the appearance of middleware that obviates the need to customize the applications specific to each cloud architecture. This is essential for community clouds that result from the federation of many independently managed clouds. Each such cloud may be using a different cloud management platform that may provide a different API. Providing a standard API for the community cloud ensures that applications written for one community cloud can also be deployed for another community cloud in the future, and as they are integrated into the community cloud they can be easily deployed on new cloud architectures.

4.4.8   QoS and SLA Guarantees

The community cloud system needs mechanisms for ensuring quality of service (QoS) and enforcing service-level agreements (SLA). With business processes increasingly reliant on the cloud services and applications, businesses need strong guarantees before transitioning to a community cloud system.

4.5   Potential of Community Clouds

This section states the potential of community clouds, highlighting their key advantages.

4.5.1   Security, Control and Privacy

In some cases, the security and privacy of data are so important that giving access and control of this information to a public cloud provider is not feasible. Private clouds are also not an option because of the huge investment needed to set up and maintain such infrastructure. Community clouds address both of these concerns, provided that the members of the community have existing trust relationships among themselves and require similar applications.

4.5.2 Elasticity, Resilience and Robustness

Community clouds bring cooperating entities together to pool their infrastructures. In contrast to private clouds, this provides redundancy and robustness, and results in a more resilient system than those provided by public clouds, which can act as a point of failure as evident from the outages, even though rare, at many public providers in recent years. In the situation of peaks in demand and outages at the location of one member, resources from other members can help to alleviate these problems.

4.5.3 Avoiding Lock-In

Services from a public cloud can suffer from vendor lock-in problems. In the absence of standardization in clouds, moving data and applications from one cloud vendor to another is not feasible. A community cloud allows users to consume resources either from a variety of vendors or from infrastructure tailored to the community, and so provides a protection against vendor lock-in.

4.5.4 Cost Effectiveness

Private and hybrid clouds require a huge capital investment for building and maintaining the infrastructure in-house. Community clouds allow this cost to be distributed among the members of the community.

4.5.5 Enhanced Requirement Satisfaction

In some cases, public clouds may not be able to meet the performance and functionality demands of the community. For example, the applications may be time critical and need better network speed than a public cloud service provider can commit.

4.6 Achievements of Community Clouds

We now look at the different aspects of existing community cloud solutions and study how they address specific requirements of the community they are designed to serve.

4.6.1 Community Clouds for Governmental Organizations

The prime motivation for community clouds for organizations and departments working with government services is the issue of the privacy of citizens' data, which cannot be stored or processed in public clouds. There are many regulations for working with private data, for example the US International Traffic in Arms Regulations (ITAR), the Federal Risk and Authorization Management Program (FedRAMP), the Federal Information Processing Standard (FIPS), and the Federal Information Security Management Act (FISMA) in the United States, the Data Protection Act in the United Kingdom, and the Data Protection Directive in European Union. To address these concerns, Amazon in the United States has set up an isolated cloud infrastructure for government agencies and their customers, which is separate from public cloud offerings from Amazon, which is accredited by FedRAMP for dealing with private and sensitive data. SolaS Secure Community Cloud also places a strong emphasis on security using a perimeter networking demilitarized zone (DMZ) and distinct virtual datacenters (VDCs) using virtualization firewall capability to offer complete logical separation for data and applications. IBM has similarly set up dedicated federal datacenters (FDC) that can ensure certified computing capabilities for government organizations.

4.6.2   Community Clouds for Financial Industry

For financial industry applications, security and privacy of data are important, as in any other cloud-based solutions, but the driving factors behind the community cloud for this market are challenging demands on the service's performance. For instance, public clouds are too slow to support the applications for high-performance electronic trading. NYSE Technologies has addressed this problem by setting up customized infrastructure, highly secure datacenters and a high-speed network to meet high performance and security demands of the applications for the financial industry. This secure and high performance infrastructure allows the development and deployment of sophisticated applications for financial industry, which are not feasible within public clouds.

4.6.3   Community Clouds for the Health Sector

The prime issue for organization in the health sector, preventing it from using public clouds, is the privacy of sensitive data of patients and related information. The storage and usage of this data are governed by strict government regulations like the Health Insurance Portability and Accountability Act (HIPAA) in the United States, the Privacy Act 1988 in Australia, and similar data-protection regulations in the United Kingdom, the European Union, and many other countries. Optum, in the United States, provides cloud solutions for the health sector and ensures compliance with HIPAA and other federal data-protection laws. Another advantage resulting from the community cloud is that Optum can offer data analytics from historic and real-time data to support decision making and other application for improving patients' healthcare.

4.6.4   Community Clouds for Aviation Industry

The main need of the aviation industry is to have customized, flexible and on-demand applications and software that can assist with industry and business process optimization and enable operational agility for growth and cost savings. SITA, with its strong background of working in the air transport industry, has developed the ATI cloud for the aviation industry, addressing its needs with specifically tailored cloud applications, high-performance architecture and strong SLA guarantees.

4.7   Conclusion

A community cloud is a cloud deployment model in which a cloud infrastructure is built and provisioned for exclusive use by a specific community of consumers with shared concerns and interests, owned and managed by the community or by a third party, or a combination of both. The wide offer of commercial cloud solutions has led to the widespread adoption of cloud usage by all kinds of stakeholders. It is a natural evolution from this growing number of cloud users that certain clusters or communities of users arise, where each community is characterized by shared interests and common concerns. Community clouds have become a reality today with community cloud solutions deployed in many important economic sectors and for different stakeholders, like manufacturing industry, telecommunication providers, financial service providers, health services, government agencies, and education. Among the advantages of community clouds is that specific cloud-user requirements are satisfied and in a more cost-effective way. High security standards and high cloud performance can be achieved with community clouds at a cost that is shared among the community. The opportunity of community cloud lies in being able to offer optimized cloud solutions to specific user communities. An important condition, however, is that the requirements of the community are shared, in order to be able to determine a specific cloud solution that is worth the effort of developing a community cloud.

References

Buyya, R., Broberg, J., and Goscinski, A. (2011) *Cloud Computing: Principles and Paradigms*. John Wiley & Sons, Inc., New York.

Buyya, R., Ranjan, R., and Calheiros, R. N. (2010). InterCloud: Utility-oriented federation of cloud computing environments for scaling of application services. Tenth International Conference on Algorithms and Architectures for Parallel Processing (ICA3PP '10), Busan, Korea, pp. 13–31.

Cappos, J., Beschastnikh, I., Krishnamurthy, A., and Anderson, T. (2009) Seattle: a platform for educational cloud computing. Fortieth ACM Technical Symposium on Computer Science Education (SIGCSE '09). Chattanooga, TN, pp. 111–115.

Chard, K., Bubendorfer, K., Caton, S., and Rana, O. (2012) Social cloud computing: A vision for socially motivated resource sharing. *IEEE Transactions on Services Computing* **5**(4), 551–563.

Foster, I. and Kesselman, C. (2003) *The Grid 2: Blueprint for a New Computing Infrastructure*. Elsevier, Dordrecht.

Marinos, A. and Briscoe, G. (2009) Community cloud computing. First International Conference on Cloud Computing (CloudCom '09), Beijing, China, pp. 472–484.

Mell, P. M., and Grance, T. (2011) *The NIST Definition of Cloud Computing*. Special Publication 800-145. NIST, Gaithersburg, MD.

Moreno-Vozmediano, R., Montero, R. S., and Llorente, I. M. (2012) IaaS cloud architecture: From virtualized datacenters to federated cloud infrastructures. *Computer* **45**(12), 65–72.

5

Government Clouds

Sean Rhody and Dan Dunn

Capgemini Government Solutions, USA

5.1 Introduction

There is a great need for modernization and transition to cloud based technologies within government agencies. The US Federal government is moving to cloud, based on various directives and legislative actions. While the government agencies have much in common with commercial institutions in terms of need, they differ significantly in how they can make use of cloud computing, and have many restrictions relating to security and privacy that impact how they can use cloud technology. This chapter will explore in detail the impact of cloud computing on government agencies, and identify the most significant constraints that government agencies face.

5.2 The Federal Government's Journey to the Cloud

The US Federal Government is composed of three branches – executive, legislative and judicial – supported by numerous administrative agencies. According to the US Office of Personnel Management, there are over 2.7 million federal employees, excluding those in agencies with secret clearances. There are 456 federal agencies, with the majority of employees and agencies supporting the executive branch of the government. The government estimates that it spent $79.5 billion on information technology (IT) in 2012, whereas private estimates that take into account quasi-government agencies like the US Postal Service and Fannie Mae place the spend at closer to $120 billion. One way to get acquainted with these numbers is to look at the Federal IT Dashboard (IT Dashboard, 2014).

The federal landscape for IT spending is complex and influenced by factors such as the economy, political leadership, legislation, and policy. For the past several years, the struggling economy has pressured the various IT organizations to pursue economies. The lack of agreement in Congress on a budget, and the resultant

Encyclopedia of Cloud Computing, First Edition. Edited by San Murugesan and Irena Bojanova.
© 2016 John Wiley & Sons, Ltd. Published 2016 by John Wiley & Sons, Ltd.

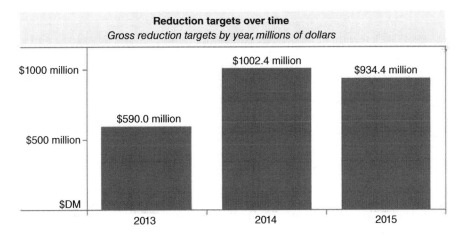

Portfolio stat commodity IT reduction targets

Government - wide gross reduction targets
Total cost savings and avoidance targets, 2013–15

$2.53 billion

Consolidation areas – *gross reduction targets, milions of dollars* (2013–15)

IT infrastructure	Mainframes and servers	738.6
	Mobile	387.8
	Telecommunications	298.3
	Desktop computers	122.9
	Other IT infrastructure	8.2
Enterprise IT systems	Other enterprise IT	290.7
	Email	183.8
	Web infrastructure	106.1
	Collaboration tools	27.0
	Identity and access management	16.4
	Security	0.3
Business systems	Finance	201.6
	Other Administrative functions	139.2
	Human resources	5.9

0.0 200.0 400.0 600.0

Reduction targets over time
Gross reduction targets by year, millions of dollars

$1000 million — $1002.4 million (2014), $934.4 million (2015)
$590.0 million (2013)
$500 million
$DM
2013   2014   2015

Data reported by agencies as of october 16, 2012.
Consolidation areas based on OMB memorandum M–11–29

Figure 5.1 *PortfolioStat – cost reduction targets. Source: IT Dashboard (2014)*

Continuing Resolutions that have effectively frozen IT spending for years, have also contributed to a climate where continued cost savings are vital to meeting the mission needs of the various agencies.

One of the ways in which the government has begun to address this need is through a program called PortfolioStat. This program is an annual budgeting and review process that was instituted by the Office of

Management and Budget (OMB). According to www.itdashboard.gov/portfolio_stat (accessed December 4, 2015), PortfolioStat required the various federal agencies to establish 3-year IT spending plans "to consolidate or eliminate duplicative, low-value, or wasteful IT spending and improve the Portfolio management processes with the organization." Each agency was charged with cost reductions, and an overall plan for consolidation was established (see Figure 5.1) for a 3-year total reduction of over $2.5 billion. The breakdown of savings is targeted heavily at IT infrastructure and enterprise systems, including computing equipment, mobile devices, and commonly used collaboration tools.

The adoption of sequestration has also affected the IT community. Faced with further budget reductions, the Office of the Chief Information Officer (OCIO) has had to increase the cost reductions in its yearly plans in order to meet the further reduced IT budgets. *A major pressure on the federal IT community is the need to reduce capital expenditures.*

A second factor affecting the federal government IT community is the large number of independent datacenters. In 2010 the White House launched the Federal Data Center Consolidation Initiative (FDCCI) with the goal of eliminating at least 800 datacenters by 2015. This effort acknowledges the fact that there is a great deal of redundancy and excess capacity in the real estate and support systems that house and power the IT capacity of the government. Additionally, through analysis of agency datacenter consolidation plans, the utilization of computer resources has been shown to be under 30% of capacity for many agencies (Kundra, 2011). Another major pressure on government IT is the overcapacity and underutilization of IT infrastructure.

One facet of the government IT process that exacerbates this pressure is the need for certification and accreditation of facilities and computer systems. While some forms of this process may be performed in the private sector, in the federal IT landscape this is a necessary and fairly extensive process that takes time and dollars to complete. Given the redundancy in the number of datacenters and systems, the magnitude of this cost is increased significantly, adding to the burden of the IT organization. *The federal IT community needs to reduce or eliminate redundant systems, and use the computing resources it employs more effectively to meet the mission needs of the various agencies.*

Until recently, the IT community in the government did not have a satisfactory method for acquiring IT services that account for scalability and reliability requirements without acquiring significant excess capacity. In many agencies, computer resource use is periodic in nature. As an example consider the flow of information to the Internal Revenue Service (IRS). There are monthly and quarterly peaks, but there is a very significant peak in resource utilization that occurs near the annual tax deadline of April 15. Under such conditions, and given the prevailing architectural approaches for resilience that required multiple instances of a solution to avoid single points of failure, in a pre-cloud environment where IT resources are neither virtualized nor tied to a utility/consumption model, it is not surprising that dramatic utilization differences occur, as systems had to be sized for peak usage, rather than average usage. *The federal IT community needs a consumption-based model that can adjust to dramatic swings in resource consumption (i.e. elastic capacity).*

As in the private sector, the US federal IT community is faced with several disruptive technology movements. Mobility, and the advent of the bring-your-own-device (BYOD) model, in which the federal government is asked to allow access to protected resources by third-party devices, is a new and disruptive IT approach that is gaining traction due to the demand by knowledge workers for the ability to use the device they desire to achieve their daily work requirements.

Device form factors, and the growth of the post-PC movement to tablets, phablets, phones, and other devices, is also pressing government agencies to accommodate security models that allow modes of access that were not possible in the past. Government telecommuting initiatives have also contributed to a growing need in the IT community to grant access to data outside of the existing internal network.

The rapid growth and maturation of virtualization and cloud technologies has also affected the federal IT point of view. While the federal government is seldom a first mover in the IT space, one of the key concepts of the commercial cloud computing model – a utility or consumption pricing model and dynamic, elastic capacity – has started an inevitable movement in the government. This is the concept of paying for IT services not on a license-based or server-based model but rather on the basis of dynamic, changeable usage, similar to the consumption of electrical power from a utility. Datacenter managers and agency CIOs have noted the low utilization of their current equipment footprint, and have realized that virtualization can, to a limited extent, provide them some of the elastic capacity that they need within their current inventory of equipment. There are still limitations to this approach including power, network, and the physical construction of the existing datacenters, and the ultimate capacity is limited to what is available in one agency datacenter.

The first White House CIO, Vivek Kundra, launched an ongoing federal initiative to reform information management in the federal government (Kundra, 2010), which included a shift to "cloud-first" thinking. This initiative included tasking agency CIOs with creating a cloud strategy for their agencies, shortlisting several "must-move" services to begin a migration to cloud-based computing, and creating contract vehicles that would allow agencies to purchase cloud services (such as the General Services Administration (GSA) infrastructure as a service blanket purchase agreement (BPA)). This and other initiatives are starting the process for the government to move to cloud computing, and to make a fundamental procurement shift from capital expense to operating expense.

A basic frustration felt throughout federal government IT organizations is the difficulty faced in procuring computing resources in a timely manner. Based on traditional contracting methods it can take more than 6 months to acquire additional equipment. There has not been a mechanism to allow the government to acquire for "as long as necessary but no longer" – they have traditionally purchased and maintained equipment. In many cases, agencies are seeking agility and flexibility in IT space, looking to be able to trial new, innovative solutions quickly, and move rapidly from proof of concept to production, and, when the solution becomes obsolete, to be able to decommission and migrate rapidly to a different solution. Agencies have been frustrated in this desire, as contracts lock them into a vendor for many years, make it difficult to modify the services (either to decrease or increase capacity), and in general do not lend themselves to the freedom of movement many commercial IT organizations enjoy.

A frequently cited wish of the government IT specialist is the basic ability to buy computing services on an operating-expense basis, rather than on a capital-expense basis. This would represent a major change and improvement to the mechanism by which the government IT does business. Amazon Web Services (AWS) is often cited as a proper model but in reality many nuances of services would suffice for the government. The main needs are:

- Dynamic capacity without extensive contractual negotiations or procurements. The ability to change the level of consumption in response to changes in usage (such as from peak periods or a temporary surge in usage due to some temporary event) is a definite requirement for future government IT.
- Avoiding vendor lock-in is also a desirable feature. Although moving from one vendor to another is a painful event, at times there is a need to move from one supplier to another. The government needs the ability to move when a need exists, without penalty for shifting services.
- Preintegrated services. One major effect that mobile devices have had on computing is the concept of "apps", and the corollary which implies that services can build on one another, such as when an app for finding local restaurants leverages GPS, maps and crowd sourced ratings to help a user find the best restaurant to meet his needs. In practice this need is much more complex, but software as a service (SaaS) built on top of cloud infrastructure cannot maintain the same silo approach as in the past – the need to share information between services is too great to be ignored.

In summary, the federal government realizes that migration to cloud computing is a necessary step in the continued evolution of IT in the government. Economic pressures have reduced budgets and forced IT departments to make difficult choices, but they have also hastened the recognition of cloud computing as a potential source of cost reduction. The government has also recognized that it has a tremendous amount of redundancy of IT solutions, a problem compounded by low utilization of computer resources on many solutions. The 25-point plan launched by the White House CIO outlines an approach to improve the landscape of government IT by addressing these issues and transforming the IT organization of the government.

5.3   The Pre-Flight Check – Primary Considerations for Government agencies

In one sense, the IT needs of the federal government are similar to the needs of businesses in the private sector. There is a basic need for sufficient infrastructure capacity, enough computer power to accomplish tasks, sufficient network bandwidth for all users to work without extensive delays, adequate storage to maintain information, a secure infrastructure that prevents unauthorized access to information, and the ability to respond in sufficient time (as defined by the situation) to new requests. These needs are little different between government and private sector, although there may be increased emphasis on some aspects in the government. But in another sense there are significant differences between what may be considered an acceptable IT risk in the private sector and what is acceptable from a government perspective. The next several sections describe some key considerations for cloud computing that have significant differences from the private sector.

5.3.1   Legal, Policy and Regulatory Requirements

In addition to the normal needs that a government IT organization has, there are legal, policy and regulatory requirements that go far beyond average private sector needs. From a cloud-computing perspective, there are a few significant areas that any government agency must cope with.

The Federal Data Center Consolidation Initiative (FDCCI) and the Cloud First initiative of the White House's 25-point plan, and corresponding legislative requirements are forcing many agencies to migrate to cloud solutions, or face budget penalties that will further affect the already reduced spending currently in effect.

There is also direction from the Office of Management and Budget (OMB), in the form of Circular A-11, in particular Exhibit 300s and Exhibit 53.

> Exhibit 300s are companions to an agency's Exhibit 53. Exhibit 300s and the Exhibit 53 together with the agency's Enterprise Architecture program, define how to manage the IT Capital Planning and Control Process. Exhibit 53A is a tool for reporting the funding of the portfolio of all IT investments within a Department while Exhibit 300A is a tool for detailed justification of Major IT Investments. Exhibit 300B is for the management of the execution of those investments through their project life cycle and into their useful life in production. (http://www.whitehouse.gov/sites/default/files/omb/assets/egov_docs/fy13_guidance_for_exhibit_300_a-b_20110715.pdf, accessed November 28, 2015)

The process of creating and approving these documents is significant, and they affect the funding of IT investments over multiple years. The amount of planning, justification and oversight exceeds the normal IT process within a private sector corporation.

5.3.2   Security and Privacy

Security and privacy in the Federal government are subject to a host of regulations, but there are three sets that are commonly discussed, the 3 Fs – FIPS, Federal Information Security Management Act (FISMA) and the Federal Risk Management Authorization Management Program (FedRAMP). Each of these has some impact on cloud providers and the way in which they provide services.

FIPS 140 is a series of publications regarding security from NIST (the National Institute of Standards and Technology). There are four levels of FIPS 140 security (FIPS 140-2, levels 1–4), each of which is a higher order of security than the previous level. FIPS 140 relates to cryptographic controls and modules for securing electronic communications. A new version, FIPS 140-3 is under development.

The Federal Information Security Management Act, is a set of policies, procedures and responsibilities that federal agencies (and by extension third party cloud service providers) must adhere to in order to receive certification and authorization to operate (ATO). Providers are required to have an inventory of systems and their interfaces to other systems, a categorization of information and information systems according to level of risk (as defined by FIPS 199), a set of security controls (as defined by FIPS 200), to implement security controls, and assess those controls, to authorize operation, and then continually monitor the security controls. Detailed information regarding FISMA is available from NIST at http://csrc.nist.gov/groups/SMA/fisma/index.html (accessed November 28, 2015).

The Federal Risk Management Authorization Management Program is a program aimed at simplifying the process of certifying "aaS" solutions throughout the federal government. It is designed to be a "certify once, operate everywhere" process. Without FedRAMP, providers would be required to recertify solution offerings with every new customer purchase. With FedRAMP, cloud solution providers (CSPs) can simplify their path to implementation. To achieve a FedRAMP certification, a CSP must implement the FedRAMP requirements, which are FISMA compliant, and have a third-party assessment organization (3PAO) perform an independent assessment to audit the cloud systems under review. The FedRAMP Joint Authorization Board (JAB) reviews the assessment and grants authorization. Detailed information regarding FedRAMP can be obtained at http://www.fedramp.gov (accessed November 28, 2015).

5.3.3 Data Location Requirements

There are several government requirements regarding the storage and location of data, and access to it by personnel. In general, data must be stored in the continental United States (or CONUS), unless specifically excepted. This requirement can have significant impact on cloud providers. In order to operate as a provider for the federal government, cloud providers must offer storage options that are only located in CONUS. Data cannot be stored outside CONUS, even if the provider normally uses offshore locations as nodes for additional redundancy or data replication.

An additional requirement is the need for dedicated support personnel who are at minimum US residents, and usually US citizens who have government clearances. Many cloud solution providers use offshore resources to provide round-the-clock support as well as to balance the costs of labor. To be a provider to the federal government, these resources must be US residents or, better, located in a US facility, that has a separate, dedicated capacity for US-only processing. No foreign nationals, or resources without clearance are allowed to access the dedicated capacity, and physical security as well as electronic safeguards must be in place.

An additional concern is compliance with export restrictions such as the International Traffic in Arms Regulations (ITAR). These restrictions apply to the management of data and often require additional audit trails for compliance purposes. An example of a cloud provider service that meets all of these regulations is AWS GovCloud (http://aws.amazon.com/govcloud-us/faqs/, accessed November 28, 2015).

To summarize, Federal IT faces many of the same challenges as private-sector commercial organizations when it comes to cloud computing. The federal sector also has some fairly significant requirements from legal, regulatory, security, and data perspectives that increase the effort necessary for CSPs to deliver solutions to the federal government.

5.4   The Federal Market Place – Primary Offerings

Governments across the globe are facing challenges very similar to those that the private sector is facing but in some respects they are actually outpacing their civilian counterparts in adopting, or at minimum experimenting with, cloud technologies. Also in step with much of the private sector, governments are still in the process of consolidating infrastructure and services to high-density capable datacenters, implementing virtualization and positioning themselves for future moves to the cloud.

One of the primary concerns and to some extent a limiting factor in the adoption of cloud services within the United States federal government and other governments as well is around security concerns and laws and regulations prohibiting the use of commercial vendors that are not certified by the government. The limited number and relatively recent entry of providers authorized to operate in this space has required agencies to build their own services or wait in anticipation as vendors made it through the process. It is projected that the adoption of the services described below will increase over the next several years as facilities and IT enterprises are transformed and the commercial industry provides more choices to the Federal governments.

- *Infrastructure as a Service (IaaS)*. The ongoing movement to new, high-density powered and cooling datacenter facilities is providing the opportunity to establish the infrastructure components necessary to provide IaaS within the government IT landscape. Like the private sector, some government departments and agencies are seeing delays in datacenter transitions due to obligations arising from long-term contracts with communications and datacenter providers and are currently only moving services that reap the greatest benefits and leaving other services behind (for now) to avoid cancellation penalties and contract buyout costs. One such benefit on the radar is data storage/archival and retrieval as a service to be hosted within the private cloud and available to multiple agencies. As with the private sector, the maturity of IaaS is such that most agencies do not yet have the capability to provide on-demand, catalogue-based services for infrastructure rapid provisioning.
- *Platform as a Service (PaaS)*. The implementation of PaaS has not become prevalent within Federal markets due to the maturity levels needed to provide the combined infrastructure, database, storage, and other components in a self-service catalogue. The private sector has the ability to procure PaaS through external vendors; however, Federal security requirements often prohibit the use of noncompliant providers.
- *Software as a Service (SaaS)*. The value of SaaS is recognized within federal governments and is being adopted for specific applications. The parallels with the private sector remain consistent, there are internal systems suitable for SaaS as well as external (customer-facing) systems that are being considered and implemented. Such enterprise applications as internal (corporate) e-mail are predominant in early adoption with client-facing (public) access for such areas as case management and other services becoming more common.

5.4.1   Summary of Commercial Offerings to Federal Government

The landscape of Cloud Services being offered to federal government is growing exponentially as companies large and small find their niche and compete for a piece of the market. The traditional behemoths like IBM, Hewlett Packard, AT&T, and a plethora of others have well established datacenters and are providing a wide variety of options for both physical hosting of private clouds (IaaS) and for the SaaS and PaaS needs as well. Other companies, small and large, are finding their niche in providing specific solution sets that provide a specific service.

5.5 The Next Generation of Cloud Adopters

Governments around the globe have recognized the value of cloud services for several years and many have specific policies that encourage the use of the cloud and virtually all major powers are actively moving towards the cloud. The following provides some insight in to the progression towards cloud adoption by government organizations worldwide.

5.5.1 Japanese Government

In 2009, the Japanese Government introduced a comprehensive cloud services initiative called the Kasumigaseki Cloud as part of its Digital Japan Creation Project (ICT Hatoyama Plan). The Kasumigaseki Cloud initiative goal is to (i) establish a cloud-based infrastructure that will meet the present and future needs of the government's IT systems, and (ii) enable a more efficient service organization by providing a centralized pool of resources. The expected outcome of the plan is to eliminate redundancies in services among the various ministries and the need for separate ministries to maintain autonomous IT systems.

Additional goals have also been established as part of the project: (i) establish a National Digital Archive digitize government documents and other information; (ii) define and introduce standardized formats and metadata that will improve public access. The standardization of document formats across ministries is intended to reduce the number of documents required by the public to file and subsequently provide efficiencies within the government and improve the public access experience.

As with all Government initiatives in the cloud space, the initiative is expected to provide cost savings through the consolidation of the IT Infrastructure, improve IT operational efficiencies and – a concern to governments across the globe – provide tangible benefits to the environment.

5.5.2 US Government

In 2009 the Cloud Computing Mall for government agencies was designed and deployed to provide a portal for government agencies to learn, use, acquire, manage, and secure cloud services.

In 2010, Recovery.com Launched. It was established by the 2009 Recovery Act and required the Recovery Board to create and manage a web site "to foster greater accountability and transparency in the use of funds made available in this Act." The site displays for the American public the distribution of all Recovery funds by federal agencies and how the recipients are spending those funds. The site also offers the public the ability to report suspected fraud, waste, or abuse related to Recovery funding.

In 2011, *Federal Cloud Computing Strategy (Cloud First),* a publication from the Office of the White House, US Chief Information Officer, outlined the strategy for cloud computing and outlined high level benefits and guidance. The strategy was designed to:

- articulate the benefits, considerations, and tradeoffs of cloud computing;
- provide a decision framework and case examples to support agencies in migrating towards cloud computing;
- highlight cloud-computing implementation resources;
- identify federal government activities and roles and responsibilities for catalyzing cloud adoption.

The US Federal Budget for 2011 incorporated cloud computing as a major part of its strategy to achieve efficiency and reduce costs. It stated that all agencies should evaluate cloud computing alternatives as part of their budget submissions for all major IT investments, where relevant. Specifically:

- By September 2011, all newly planned or performing major IT investments acquisitions had to complete an alternatives analysis that included a cloud-computing-based alternative as part of their budget submissions.
- By September 2012, all IT investments making enhancements to an existing investment had to complete an alternatives analysis that included a cloud-computing-based alternative as part of their budget submissions.
- By September 2013, all IT investments in steady state had to complete an alternatives analysis that included a cloud-computing-based alternative as part of their budget submissions.
- In 2010, the US Government established FedRAMP to streamline the process of determining whether cloud services met federal security requirements. As of May 2013, nine vendors had met the requirements and were granted "authority to operate" all the IaaS category.
- In 2013, the General Services Administration (GSA) was exploring the concept of cloud brokerage as a service. This initiative was intended to address agencies challenges in managing multiple service providers and the legacy infrastructure and services. In late 2013, GSA put out to bid for a pilot to evaluate and learn more about the potential benefits.

5.5.3   UK Government

In 2010 the United Kingdom implemented its G-Cloud strategy to bring economic efficiencies and sustainability to the government's information and communication technology (ICT) operations. The initiative was committed to the adoption of cloud computing and delivering computing resources. Specific mention was made that the initiative would make "fundamental changes" in how the public sector procured and operated ICT. The G-Cloud strategy detailed how the government will achieve this as follows:

- achieve large, crossgovernment economies of scale;
- deliver ICT systems that are flexible and responsive to demand in order to support government policies and strategies;
- take advantage of new technologies in order to deliver faster business benefits and reduce cost;
- meet environmental and sustainability targets;
- allow government to procure in a way that encourages a dynamic and responsive supplier marketplace and supports emerging suppliers.

5.6   The Foreseeable Future – Current Initiatives

5.6.1   Green IT

Governments are very aware of the widespread concerns regarding the environment and our collective use and potential waste of resources. Whether deliberately, as in the case of Japans Kasumigaseki Cloud initiative, or as an unplanned cause and effect, the movement of government computing to the cloud is providing a real positive effect on the environment.

The combined effect of consolidating IT services into new high-density, energy-efficient datacenters and thereby closing inefficient facilities can reduce the carbon footprint considerably. The datacenter builds during the 1900s to accommodate the .com boom are now obsolete as virtually every energy-consuming component of datacenters has undergone changes and they are now more efficient and scalable. The prevalence of hardware virtualization and the ability to consolidate 20 servers, serving 20 different applications into one server and 20 applications has pushed the datacenters to upgrade power, and chilling, and to rearrange or build new to meet the new, smaller, and highly condensed infrastructure footprint. Hardware and virtualization software vendors are including the "Green IT" benefits of their products in their literature and conversations, and some will even calculate the potential reduction in the customer's carbon footprint based on customized transformation requirements.

As the datacenters of old are retired or upgraded and the worldwide effort to consolidate and improve services and efficiencies through cloud technologies comes to fruition, the positive environmental impact will be of real value.

5.6.2   Cloud Service Management (Brokerage)

As agencies further adopt cloud services of all varieties they are often met with the same challenges they have been experiencing for decades but with less direct control. Although the cloud concept promises less complexity in both providing services and management of those services, the reality is that IT managers still find themselves in the historically difficult position of managing their legacy IT and cloud-service providers, The challenge of problem identification and resolution is further complicated because of the nature of cloud services. The US General Services Administration is leading an effort to alleviate this management problem and is reviewing the concept of cloud brokerage services.

The cloud broker provides a set of services that fills the gaps many agencies have found when trying to establish, monitor and manage cloud services and their providers. The following are the primary roles for the cloud broker and the value that is expected:

- *End-to-end service visibility* through the use of a robust tool set and state-of-the art monitoring capabilities the cloud broker will enable the agency to be proactively notified of service issues and to monitor the end-to-end service. The traditional problem of multiple vendors to provide a complete service (e.g. SaaS, PaaS and communications links) are aggregated and treated as a single service by the cloud broker.
- *Single-point-of contact (the cloud broker)* for the end-to-end service and bound by specific SLAs to measure performance and use metrics. The problem of vendors pointing fingers and not working together is no longer the user's issue.
- *Provides a set of proven end-to-end service options* from the GSA service catalog, which, as more agencies participate, will reduce time to deliver, provide a wider selection of E2E services, and result in potential cost savings.

In summary, cloud computing presents a new approach for governments in managing the cost, utilization, redundancy, and complexity of their IT projects. Due to the unique nature of government computing, there are additional constraints and requirements beyond those of normal private-sector organizations but the need and ability to include cloud computing in the portfolio of IT solutions is clear and compelling.

References

IT Dashboard (2014), www.itdashboard.gov/portfolio_stat (accessed November 28, 2015).

Kundra, V. (2011) *Federal Cloud Computing Strategy, 2011*, http://www.whitehouse.gov/sites/default/files/omb/assets/egov_docs/federal-cloud-computing-strategy.pdf (accessed November 28, 2015).

Kundra, V. (2010) *Twenty-Five Point Implementation Plan to Reform Federal IT*. Department of Homeland Security, http://www.dhs.gov/sites/default/files/publications/digital-strategy/25-point-implementation-plan-to-reform-federal-it.pdf (accessed November 28, 2015).

6

Cloud-Based Development Environments: PaaS

Mehmet N. Aydin,[1] Nazim Ziya Perdahci,[2] and Bahadir Odevci[3]

[1] Kadir Has University, Turkey
[2] Mimar Sinan Fine Arts University, Turkey
[3] Imonacloud.com, Turkey

6.1   Introduction

Cloud computing has been promoted as a panacea for long-standing problems concerning the use, development, and management of information technologies (IT). One might argue that it is yet another "catchy term," and has evolved from other unfilled promises such as application service provisioning, IT outsourcing, and service-oriented architecture. The premise of these attempts, including cloud computing, is that the paradigm underpinning such IT aspects as development, management, distribution, use, and pricing, is subject to change. We have already witnessed such changes in the market along with alternative ecosystems in which new players are taking off or established firms are about to change (IDC, 2010; Gartner, 2011; Rymer and Ried, 2011). In this chapter, we focus on a specific cloud-computing layer, which is called platform as a service (PaaS). The emergence of this layer raises several questions regarding the existence of traditional development environments and tools, as well as the viability of alternative ecosystems along with new positions in the market. It is this questioning that shows the need for making sense of what underpins the very idea of PaaS and how the idea is manifested in the market, and finally what implications can be drawn for academics and practitioners.

Cloud computing essentially refers to a model for enabling ubiquitous, convenient, on-demand network access to a shared pool of configurable computing resources (e.g., networks, servers, storage, applications, and services), which can be provisioned and released rapidly with minimal management effort or service provider interaction (Mell and Grance, 2011). As such, cloud-computing-based services reflect salient features of cloud computing including metered billing ("pay as you go"), resource pooling, virtualization, automation, and elasticity.

Given the evolving conceptual ground of cloud computing, three abstraction layers (infrastructure, platform, and software as services, abbreviated Iaas, PaaS, and SaaS respectively) have been proposed and to a large extent agreed upon to frame out cloud-computing endeavors. Each layer has its own characteristics and requires in-depth analysis of its essential elements with exemplary solutions. As discussed in later sections, PaaS mediates between SaaS and IaaS conceptually and practically, and appears to be a challenging and promising area for academics and practitioners. Thus, in the next section, we elaborate the fundamentals of PaaS and outline core components of a typical platform as a service. Given the proliferation of such platforms, appropriate categories are needed, which are provided in the third and fourth sections. Further attention is given to aPaaS as a particular category in the fourth section. We then articulate basic approaches to aPaaS and discuss metadata aPaaS by utilizing our industry experience as well as a review of recent studies on this subject. The last section concludes by considering implications of the present study for academics and practitioners.

6.2   Fundamentals of the PaaS Model

Cloud computing offerings and deployments can be described in two different ways – by the service architecture and by the network deployment architecture. Network deployment architecture shows how the resources are physically deployed, and generally falls into two categories: private clouds and public clouds. The service architecture defines the set of resources that a cloud provider is offering. The most common model in use today conceptualizes three service models in terms of a layered aspect of hardware and software stack. The three service layers comprise software as a service (SaaS), platform as a service (PaaS), and infrastructure as a service (IaaS) layers. At the bottom of the stack is the IaaS, which comprises the facilities, hardware, abstraction, core connectivity, delivery, and API layers. The addition of integration and middleware services forms the PaaS module.

Platform as a service acts as a bridge between IaaS and SaaS – it is the runtime environment, the middleware of the cloud service architecture. In a cloud-computing model, what is meant by a platform is a "standard for the (virtualized) hardware of a computer system, determining what kinds of software it can run." In principle, a platform would not remain visible to cloud consumers, as it is the layer that provides integration and middleware services for Web-based applications. However, cloud-computing platforms are very different from traditional platforms, with many additional components exclusively built for SaaS. Such components, unique to SaaS applications, include, for example, SaaS-to-SaaS and SaaS-to-on-premises integration tools, billing and metering applications, and infrastructure provided over the cloud. Moreover, a SaaS platform has to be multitenant. Nearly all of these new components, which were unnecessary in traditional platforms, have come to be provided as a service – i.e. on-demand, and over the cloud. Thus PaaS evolved from the SaaS concept, which emerged as an independent cloud-service model, combining an application platform with managed cloud infrastructure services. Platform-as-a-service cloud providers have become additional actors in the cloud ecosystem, offering new products to the software market.

The focus of PaaS has widened its scope in recent years, touching upon not only the runtime environment of a SaaS application but also the tools and methods of developing apps in the cloud. Platform as a service, in this context, can be defined as a "complete application platform as a service" that offers independent software vendors (ISVs) as well as individuals the means to produce "multi-tenant SaaS solutions or various application integration solutions" in a fast and efficient way (Beimborn *et al.*, 2011). Thus, some of the PaaS offerings comprise a predetermined development environment with all tools for development, testing, and deployment cycles of an application.

Emerging PaaS solution providers take the responsibility for managing the computing infrastructure, including runtime execution stack, database, middleware, security, performance, availability, reliability, and scalability. They provide middleware technologies as a service, such as applications servers, enterprise service buses (ESBs), business process modeling (BPM) tools, and message brokers in order for SaaS solutions to launch consistently. In addition to providing a complete set of technologies required in developing SaaS applications, there is a new trend in platform providers, which is to offer marketing and/or selling opportunities.

It is generally agreed that the core components of PaaS offerings should at least comprise the following (Beimborn *et al.*, 2011; Teixeira *et al.*, 2014):

* application runtime environments;
* database system and middleware.

Supplementary components, which differentiate contemporary PaaS offerings, can be listed as follows:

* facilitated development environment (cloud IDE);
* marketplace of applications deployed into the corresponding PaaS.

One of the main components of a PaaS computing platform is its runtime environment, which must satisfy such features as scalability, security, and portability. Run-time environments support multitenancy architectures: all tenants share a common codebase and instances of an application.

Another important component of a PaaS, which focuses particularly on fast and efficient development of SaaS applications, is the Web-integrated development environment (WIDE) (also known as a Cloud IDE). Often, WIDEs support the use of multiple programming languages and offer a wide range of libraries and tools for modeling, implementing, testing, and version control. Depending on the application domain, various database management systems (DBMs) are also included. Integration of discrete data from services is usually done by representational state transfer (RESTful) (Richardson and Ruby, 2007).

Platform as a service offerings can be analyzed further with respect to application development capabilities. On the one hand, there are those platforms providing a core application as their own, letting ISVs develop extensions (or add ons) for it. In this case, any software makes sense only as an add-on to the provider's core application. Other platforms let users to develop their own stand-alone SaaS style applications. Forrester's study (see Figure 6.1) is an attempt to clarify how the ongoing interaction between ISVs and cloud service providers create an ecosystem that has ushered in migrating their on-premise software solutions to PaaS solutions.

In addition to the core components of PaaS, many cloud providers enrich their platforms by providing an online marketplace, which serves two purposes: giving support to ISVs' sales activities, and providing a software repository for application developers.

Software development on a PaaS is a new paradigm. It has caused a shift from the traditional waterfall model (or its modified flavors) of sequential software development processes to agile software development methodologies consisting of multiple processes at every phase. At the end of the day, developers want to focus on coding for the business, so that they solve business problems. They do not want to configure operating systems (OSs), storage blocks, or IP addresses on virtual machines. Platform as a service, in this context, should also focus on presenting developers with agile, self-service provisioning of a full application platform, and tools to use for agile development of applications. In that sense, PaaS lowers the economic barrier for any project ensuring developers can access computing resources they could not otherwise afford.

Basically, two major categories of PaaS offerings have come into prominence: integration platform as a service (iPaaS) and application platform as a service (aPaaS).

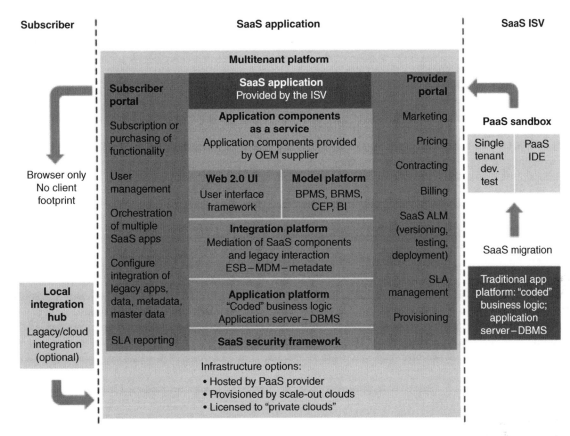

Figure 6.1 Forrester's PaaS reference architecture. Source: Rymer and Ried, 2009. Reproduced with permission of Forrester Research, Inc.

6.3 Integration Platforms as a Service (iPaaS)

The shift to the cloud is one of the biggest shifts in IT ever. The number of SaaS solution providers is growing by leaps and bounds. In addition to pure SaaS players, legacy software vendors have started to convert their product ranges from traditional single-tenant on-premise application models to multitenant SaaS model cloud offerings (Hai and Sakoda, 2009). Business users are purchasing independent SaaS applications in ever increasing numbers. It is estimated that there are over a thousand cloud computing companies specializing in SaaS around the world (Gartner, 2011).

One of the challenges for this shift is SaaS integration. Integration services in the cloud play a critical role for the success of any application within an organization or across organizations (Marian, 2012). In this regard, cloud integration can take place as cloud-to-cloud integration or cloud-to-on-premise integration. One of the fundamental prerequisites of integration is a bridge among SaaS applications. The way for the SaaS model to handle integration has been to leverage a set of Web application programming interfaces (APIs). There are about 5000 APIs published by SaaS solution providers. The number is doubling every 6 months. With this rapid growth of cloud services and APIs, the question is how to harness all that technology and how to make it work seamlessly.

Standardization efforts to achieve seamless integration are promising, but require an ecosystem that is mature enough in terms of united and strong collaborations among industry partners. For instance, as an open community leading cloud standard, Open Cloud Computing Interface (OCCI) was initiated to create a protocol (RESTful) for all kinds of API management. The current release of OCCI aims to support other models including IaaS and SaaS. Yet another standardization effort, Cloud Application Management for Platforms (CAMP), submitted to OASIS for nonassertion mode use, is backed by Oracle Corp., CloudBees Inc., Cloudsoft Corp., Software AG, Huawei Technologies Co., Rackspace Inc., Redhat Inc.

Organizations may be tempted by the ease of use of Web service integration tools available in most development environments but in the long term this would raise serious issues by creating connections liable to break by software upgrades, or possibly by additional integrations.

Some SaaS solution providers supply integration as part of their service offering but, in general, companies are not willing to solve these integration problems because it would pull them away from their core business.

For example, in order to link a shipping application and a billing application companies need to write some custom point-to-point integration code. Over time, with ever increasing SaaS incorporation into business, they would need to write hundreds of such applications. And if anything changes, they would need to update every connection.

As more companies move their business into SaaS, there have been new emerging partners called integration platform as a service (iPaaS), which are cloud integration platforms that provide all the benefits and architecture of SaaS. A typical iPaaS can be used for cloud-to-cloud, cloud-to-on-premise, or even on-premise-to-on-premise integration including legacy applications. To achieve this, they provide specialized connectors for SaaS applications that help the sites to make connections using different protocols. In most of the cases these connectors can significantly reduce the integration effort. In such component architecture the traditional role of a programmer is replaced by a component assembler who constructs middleware from prefabricated connectors, which are displayed as icons in a graphical user interface (GUI).

6.4   Application Platform as a Service (aPaaS)

Among the PaaS, a major category is those products that offer a combination of some form of computing infrastructure, which is made accessible over public clouds.

What they offer is a set of development tools and services, much like traditional integrated development environments (IDEs), which allow them to create applications and have them deployed and executed over the PaaS infrastructure. Such platforms have come to be referred to as application-Platform-as-a-Service (aPaaS). What distinguishes an aPaaS from a basic PaaS offering is the inclusion of two more core components: a Web 2.0 user interface (UI) and a model platform. The UI is a what-you-see-is-what-you get (WYSIWYG)-style software development editor. The model platform helps implementing the business logic. In this respect aPaaS is what can be called an extended application server. There are three different approaches to application development on aPaaS. Analyst reports (e.g., Gartner, 2011) and industry players (e.g. OrangeScape) tend to agree on three service layers, which comprise metadata application platform as a service (metadata aPaaS), framework application Platform as a service (framework aPaaS), and Instance application Platform as a Service (Instance aPaaS) layers.

6.4.1   Instance aPaaS

The Instance aPaaS layer lies at the bottom of aPaaS stack, which is closer to the IaaS layer of the cloud-computing architecture. A cloud instance refers to a virtual server instance (a virtual machine) from an IaaS solution provider. In this service model, Web-based applications are developed on and deployed to dedicated cloud instances. Instance aPaaS solution providers take responsibility for managing development and deployment components, including on-demand instance creation, Web application and database servers, monitoring and process management, databases and a number of application server and framework choices for server-side application development.

Independent software vendors, on the other hand, are responsible for the planning of instance resources such as adding or removing instances, adding or removing capacity from instances, or deploying new application versions. Thus, Instance aPaaS tenants are experienced software developers and systems programmers who get direct command line access to the operating system but have to take care of all the details of actually deploying the code.

Platform consumers are billed according to platform support options, the number and types of cloud instances, or the bandwidth allocated to them on a pay-as-you-go basis.

6.4.2 Framework aPaaS

In framework aPaaS, developers deploy code conforming to framework conventions and let the platform worry about deploying code onto compute nodes. Framework aPaaS presents a platform stack chosen by the provider; it is often tightly coupled to its infrastructure. Third-party libraries, which satisfy platform restrictions, can be included. For example, Google App Engine abstracts Java and Python application server instances using the Servlet request model and common gateway interface (CGI). VMforce provides an external environment for deploying abstractions that are structured around the Spring Framework. Salesforce.com's Heroku is another example of framework aPaaS.

This approach to aPaaS takes into account instances aPaaS with a frame consisting of interoperable elements. Such framing is taken for granted by the users (i.e., developers needing some or all of these elements). One way of framing can be based on key phases of a software development process, which are development, distribution, and operations phases (Giessmann and Stanoevska-Slabeva, 2013). In this regard, one can identify specific services for each element of the frame. For instance, the framework aPaaS may provide additional features such as monitoring, community features for the operation phase with integration with other elements of the framework. The advantage of this integration is that developers are facilitated with additional features without any configuration problem.

6.4.3 Metadata aPaaS

In this section, we will investigate fundamental metadata aPaaS features as well as provide a basic comparison amongst metadata aPaaS solutions in the market to date; Force.com, ImonaCloud.com, Mendix.com and OrangeScape.com.

At a high abstraction level, metadata aPaaS offers visual tools to customize data models, application logic, workflow, and user interfaces. It also enables users (including nondevelopers) to customize applications according to their own business needs.

An architectural concern for a typical aPaaS is the application metadata development environment, which is one of the most distinguishing architectural characteristics of aPaaS, it targets novice coders (even "noncoders") with a cloud computing application development framework for building standalone and/or integrated applications. Thus cloud service providers are responsible for providing and maintaining hardware resources, operating systems, development frameworks, and control structures. Users (clients), who can use these tools to build and host applications in the cloud, are responsible for installing and managing the applications, they want to use.

From a customer (tenant) perspective, multitenancy as an architecture principle adopted in aPaaS (Bezemer *et al.*, 2010) enables customers to share the same hardware resources, by offering them one shared application and database instance, while allowing them to configure the application to fit their needs as if it runs on a dedicated environment.

Basically, a typical metadata aPaaS is a metadata-driven application development environment that targets novice coders with a cloud computing application development framework for building standalone and / or integrated applications. With Force.com and ImonaCloud.com, developers using Web-based visual programming tools describe an application's data model, application logic and UI components. In this approach, deployment of the application metadata to the platform infrastructure is instant.

6.5   A Brief Comparison of Leading PaaS Solution Providers

Given the dynamic nature of the PaaS market, a comprehensive analysis of the PaaS market in terms of providers and customers is not viable. It is, however, important to use examples to demonstrate salient characteristics of PaaS.

OrangeScape.com and Mendix.com use a standalone IDE that is downloaded to a developer's personal computer (PC). This way, while the development of an app's metadata takes place in a richer user interface, after the development is over, deployment up to the cloud infrastructure takes place through just a few clicks.

With metadata aPaaS, the developer gains access to a metadata-driven development environment. Classically, metadata is defined as "data about data." As a programming language concept, a less ambiguous definition would be "data about the containers of data." A major obstacle in developing metadata aPaaS systems based on metamodeling is the need for an underlying problem-solving algorithm. For this reason metadata aPaaS environments tend to be special-purpose in nature, designed for use in particular applications. Most of the metada aPaaS options target business analysts.

In such an environment the task of a business analyst becomes that of developing the business dictionary, the business logic, presentation layer artefacts rather than of hardcoding an entire application program.

Developers configure custom data objects, code, and presentation elements; they create metadata interpreted by the platform at runtime. The decoupling of application metadata from the application runtime is a very strong proposition to solve cloud computing's vendor lock-in problem. Considering that an PaaS is deployed on multiple IaaS offerings, one can expect that application metadata can easily be transferred across multiple PaaS runtimes deployed across major IaaS vendors, thus giving a chance to end users to opt for an infrastructure offering of their choice.

End users would prefer to have a chance to choose among IaaS vendors according to their availability, quality of service (QoS) and pricing strategies. This would ultimately encourage competition amongst IaaS vendors, solving the lock-in problem. OrangeScape offers private and public cloud options to accommodate its metadata aPaaS offering according to business needs. ImonaCloud is already deployed into major IaaS and PaaS offerings; applications developed and used on ImonaCloud can seamlessly be transferred across the majority of IaaS vendors, whereas apps developed on Force.com platform lock in to its vendor (Armbrust *et al.*, 2010).

Another innovative aspect in PaaS architectures is the marketplace offering complementary to the application development platform. Marketplaces are becoming an inevitable component of PaaS offerings, so that the platform attracts customers and thus developers. While Azure, as an instance aPaaS, is offering applications in its Azure Marketplace, Google App Engine, as a framework aPaaS, is offering Google Apps Marketplace, Force.com as a metadata aPaaS, is offering applications in various vertical sectors, as well as applications extending SalesForce's CRM functionality in its AppExchange.com Marketplace. Integration at the metadata level of applications offered in the marketplace is a critical consideration in terms of attracting more customers. The ImonaCloud.com application development platform offers its developers the possibility of extending the functionality of any application placed in its marketplace.

6.6   Conclusion

The customer's choices among PaaS models and offerings differ according to needs. Providers of aPaaS choose the technology frameworks and tools on behalf of their customers to improve productivity. One should bear in mind that PaaS technology stack choices are often optimized for their platforms. If keeping technology stack choices open is a priority – if you want to run your application using a specific library of your choice – using IaaS offerings may be more valuable than the services provided by a PaaS provider.

Each different aPaaS type forces developers to change their programming models to suit the platform style. Instance aPaaS requires the least significant changes, framework aPaaS requires the most significant changes, and metadata aPaaS forces teams to abandon most of their existing models entirely.

Let us use an analogy to make it clearer. Instance aPaaS is like shopping and cooking for yourself. Framework aPaaS is going out to a restaurant, where cooking is taken care of for you, but you have to make it to the restaurant. Finally metadata aPaaS is ordering online, you continue your daily routine and only focus on eating. Choices made differ according to how you feel that day, your budget, or diet.

As a result, one needs to understand the competencies, restrictions, and business model for each of the PaaS models before adopting them.

References

Armbrust, M., Fox, A., Griffith, R. *et al.* (2010). A view of cloud computing. *Communications of the ACM* **53**(4), 50–58.

Beimborn, D., Miletzki, T., and Wenzel, S. (2011) Platform as a service (PaaS). *Business and Information Systems Engineering* **3**(6), 381–384.

Bezemer, C. P., Zaidman, A., Platzbeecker, B. *et al.* (2010). *Enabling Multi-Tenancy: An Industrial Experience Report.* Proceedings of the 2010 IEEE International Conference on Software Maintenance, September 2010. IEEE, pp. 1–8.

Diomidis, S. (2014) Developing in the cloud. *IEEE Software* **31**(2), 41–43.

Gartner (2011) *PaaS Road Map: A Continent Emerging,* Gartner, Stamford, CT, http://www.gxs.co.uk/wp-content/uploads/wp_gartner_paas_road_map.pdf (accessed November 28, 2015).

Giessmann, A. and Stanoevska-Slabeva, K. (2013) *What are Developers' Preferences on Platform as a Service? An Empirical Investigation.* Forty-Sixth Hawaii International Conference on System Sciences, January 2013, IEEE, pp. 1035–1044.

Hai, H. and Sakoda, S. (2009) SaaS and Integration best practices. *Fujitsu Science Technology Journal* **45**(3), 257.

IDC (2010) *Market Analysis – Worldwide Public Application Development and Deployment as a Service: 2010–2014 Forecast,* IDC, Framingham, MA.

Marian, M. (2012) iPaaS: Different Ways of Thinking. *Procedia Economics and Finance* **3**, 1093–1098.

Mell, P. M., and Grance, T. (2011) *The NIST Definition of Cloud Computing.* Special Publication 800-145. NIST, Gaithersburg, MD, http://www.nist.gov/customcf/get_pdf.cfm?pub_id=909616 (accessed November 25, 2015).

Richardson, L. and Ruby, S. (2008) *RESTful Web Services*, O'Reilly Media, Inc., Sebastopol, CA.

Rymer, J. R. and Ried, S. (2011) *The Forrester Wave™: Platform-As-A-Service for App Dev and Delivery Professionals, Q2 2011,* https://www.salesforce.com/assets/pdf/misc/wave_platform-as-a-service_for_app_dev_and_delivery.pdf (accessed November 28, 2015).

Teixeira, C., Pinto, J. S., Azevedo, R. *et al.* (2014) The building blocks of a PaaS. *Journal of Network and Systems Management* **22**(1), 75–99.

Part III
Cloud Frameworks and Technologies

7

Cloud Reference Frameworks

Kapil Bakshi[1] and Larry Beser[2]

[1] Cisco Systems Inc.
[2] HP Enterprise Services

7.1 Introduction

Cloud reference frameworks are important tools for enabling meaningful dialogs when comparing cloud technical architectures, for informing business stakeholders evaluating cloud services, and providing for a common cloud taxonomy across organizational boundaries. Reference frameworks are composed of a variety of reference architectures and reference models, which collectively describe all of the relevant aspects in a context that can then be applied to particular stakeholder viewpoints and interests.

The majority of this chapter describes commonly discussed cloud and related reference frameworks. These frameworks most commonly arise from standards bodies, consortiums, and forums where the need for common ground generally overrides proprietary interests. As a result, any particular reference framework embodies the perspectives and interests of the organization from which it emerged, such as a security or any other architectural view.

Systemic interoperability, being a critical factor for market success in general, is enabled through the development and adoption of reference frameworks. By leveraging each framework's architectural view in the context of any particular initiative, business outcomes are more clearly mapped to enabling technologies, which lowers risk while enhancing investment return.

Encyclopedia of Cloud Computing, First Edition. Edited by San Murugesan and Irena Bojanova.
© 2016 John Wiley & Sons, Ltd. Published 2016 by John Wiley & Sons, Ltd.

7.2 Review of Common Cloud Reference Frameworks

7.2.1 NIST Cloud Reference Framework

The National Institute of Standards and Technology (NIST) promotes the US economy and public welfare by providing technical leadership for the measurement and standards infrastructure. The Institute has published a reference architecture document for cloud computing to foster adoption of cloud computing, and its implementation depends upon a variety of technical and nontechnical factors. This document was published in the form of Special Publication 500-292 (Liu *et al.*, 2011). Titled *NIST Cloud Computing Reference Architecture (RA) and Taxonomy,* this document explains the components and offerings of cloud computing. The NIST Cloud RA is a vendor-neutral architecture and provides flexibility for innovation within the framework. The NIST Cloud RA is presented in two parts:

- a complete overview of the actors and their roles;
- the necessary architectural components for managing and providing cloud services such as service deployment, service orchestration, cloud service management, security, and privacy.

The NIST cloud-computing reference architecture defines five major actors:

- A cloud consumer represents a person or organization that uses the service from a cloud provider via a business relationship. A cloud consumer uses the appropriate service based on a service catalog and a service contract with the cloud provider.
- A cloud provider acquires and manages the computing infrastructure and cloud software that provides the services and makes arrangements to deliver the cloud services to the cloud consumers through network access. The types of services can be infrastructure as a service (IaaS), software as a service (SaaS), and platform as a service (PaaS).
- A cloud carrier provides network connectivity and transport of cloud services from cloud providers to cloud consumers.
- A cloud auditor conducts independent assessments of cloud services, information system operations, performance, and security of the cloud implementation.
- A cloud broker manages the usage, performance, and delivery of cloud services, and negotiates relationships between cloud providers and cloud consumers. They can provide intermediation, aggregation, and arbitrage functions.

A cloud provider's activities can be described in five major areas, as shown in Figure 7.1: service deployment, service orchestration, cloud service management, security, and privacy. Service deployment models include public cloud, private cloud, community cloud, or hybrid cloud. The differences between these service models are based on how cloud resources are provided to a cloud consumer.

Service orchestration is the process of composing system components, like hardware and software resources, in an abstracted fashion, to support the cloud providers for the creation of a service.

Cloud service management includes service-related functions (business support, provisioning/configuration, and portability/interoperability) for services consumed by cloud consumers. Business support can include customer care, contractual issues, inventory management, accounting/billing, reporting/auditing, and pricing management. Provisioning/configuration consists of resource provisioning, monitoring and reporting, SLA, and metering management. Cloud service management may also include mechanisms to support data portability, service interoperability, and system portability.

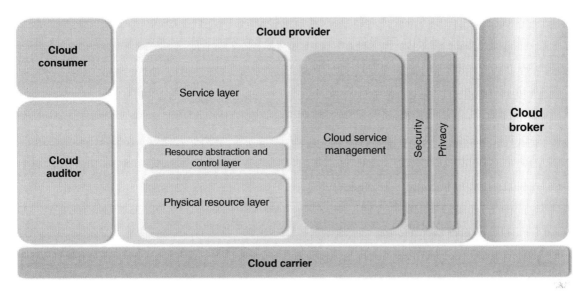

Figure 7.1 *NIST cloud reference architecture. Source: Liu et al. (2011)*

The NIST Cloud RA makes security a concern of cloud providers, consumers, and other parties. Cloud-based systems address security requirements such as authentication, authorization, availability, confidentiality, identity management, integrity, audit, security monitoring, incident response, and security policy management. With regard to privacy aspects of RA, personal information (PI) and personally identifiable information (PII) should be protected and processed appropriately by the cloud providers.

Key Takeaway

This framework enables use case and scenario planning methods by providing major actor and service constructs for cloud.

7.2.2 IETF (Draft) Cloud Reference Framework

This section discusses the Cloud Reference Framework submitted to Internet Engineering Task Force (IETF) as a draft. The IETF is an open international community of network designers, operators, vendors, and researchers concerned with the evolution of the Internet architecture and operations. The IETF-proposed draft cloud reference framework describes different layers for interoperability, integration, and operations of virtualized applications (Figure 7.2). The IETF draft reference framework provides standardization of cloud functional elements and the interfaces between the functions.

The cloud reference framework consists of the following horizontal layers:

- user/customer side services/functions and resources layer (USL);
- access delivery layer (ADL);
- cloud service layer (CSL);
- resource control (composition and orchestration) layer (RCL);

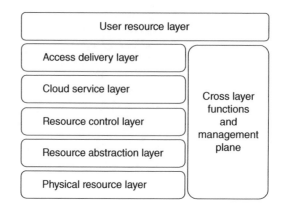

Figure 7.2   *IEFT cloud reference architecture. Source: Khasnabish et al., 2013*

- resource abstraction and virtualization layer (RAVL);
- physical resource layer (PRL).

The vertical cross-layer cloud management functions perform the following:

- configuration management;
- services registry and discovery;
- monitoring with logging, accounting, and auditing;
- service-level agreement (SLA) management;
- security services and infrastructure management.

The cloud reference framework also describes Intercloud. Intercloud provides a capability to enable portability and interoperability of independent cloud domains, and cloud provisioning and operation services.

The USL functions include access to information and identity tasks in the cloud, including visualization and administrative management functions.

The ADL hosts infrastructure components to deliver cloud-based services to customers and their access by end users. The ADL may include endpoint functions such as user portal and service gateways, like distributed cache and content delivery network (CDN) gateways. The Intercloud functions of the ADL include Intercloud infrastructure to support cloud federation, federated identity, cloud services registry/discovery, and cloud brokering functions.

The cloud service layer provides functionality for the three cloud services models, namely, IaaS, PaaS, and SaaS. The CSL develops these services based on the basic technical resources of CPU, memory, hard-disk space, and bandwidth.

The resource control layer manages and integrates the virtual resources to the upper layers and provides the ability to create efficient, secure, and reliable services. Additionally, the RCL layer has the following responsibilities:

- resources composition and orchestration;
- resource schedule control;
- Intercloud resource control;
- resource availability control;
- resource security management;
- services lifecycle management.

The RAVL provides the abstraction of the physical resources to the higher layers. The abstracted physical resources are abstracted first, next they are composed by the cloud management software (at composition and abstraction layers), and finally they are deployed as virtual resources on the virtualized physical resources. The function of the RAVL is to convert physical resources into a virtual resources pool. As part of the RAVL, the networking (resources) layer converts network capabilities and capacities (such as bandwidth, ports, etc.) into a set of resource pools, which can be leveraged by the upper layers. The resource pools include virtual switch, virtual router, virtual firewall, virtual network interface, virtual network link, and virtual private network (VPN) resources.

The PRL consists of resources like CPU, memory, hard disk, network interface card (NIC), and network ports.

As noted, in addition to the above-mentioned layers, this framework also provides vertical functions that run across all of those layers. The functions provided by the vertical layer are cloud management plane, cloud configuration management, cloud service registry / repository, cloud monitoring, accounting, and audit management, cloud SLA management, and cloud security services management.

Key Takeaway

This framework provides the essential architecture building blocks for engaging cloud networking and associated communications services.

7.2.3 Cloud Security Alliance: Cloud Reference Model

The Cloud Security Alliance (CSA) is a nonprofit organization that promotes research into best practices for securing cloud computing and the ability of cloud technologies to secure other forms of computing. The CSA has developed a cloud reference model, coupled with a security and compliance model (Figure 7.3).

The CSA's Cloud Reference Model has IaaS as the foundation of all cloud services, with PaaS building upon IaaS, and SaaS in turn building upon PaaS. As a result, in addition to capability inheritance, information security issues and risk are also inherited.

Infrastructure as a service includes the infrastructure resource stack, including facilities, hardware platforms, abstraction, and connectivity functions. It provides the capability to abstract resources, as well as to deliver physical and logical connectivity to those resources. IaaS provides a set of Application Programming Interfaces (APIs), which allows management and other interaction with the infrastructure.

Platform as a service provides a layer of integration with application development frameworks, middleware capabilities, and functions such as database, messaging, and queuing. These services allow developers to build applications on the platform with programming languages and tools.

Software as a service builds upon the underlying IaaS and PaaS stacks and provides a self-contained operating environment that is used to deliver the entire software user experience, including the content, its presentation, the application, and management capabilities. Based on IaaS, SaaS, or PaaS, consumers can leverage their content and metadata to develop and deploy applications.

The security and compliance models portray cloud service mapping, which can be compared against a catalog of security controls to determine which controls exist and which do not — as provided by the consumer, the cloud service provider, or a third party (Figure 7.3). The security stack addresses several layers including physical, compute and storage, trusted computing, network, management, information, and applications. The security stack can in turn be compared to a compliance framework or set of requirements such as PCI, DSS, HIPPA, or FedRAMP.

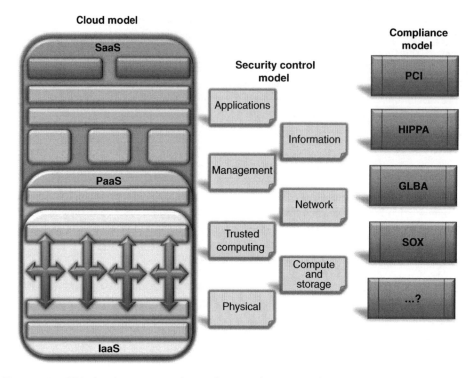

Figure 7.3   *CSA cloud, security, and compliance reference model. Source: Archer et al. (2011)*

The CSA also has 13 other domains, which highlight strategic and tactical security within a cloud environment and can be applied to any combination of cloud service and deployment models. The domains are divided into two broad categories: governance and operations. The governance domains are broad and address strategic and policy issues, while the operational domains focus on more tactical security concerns and implementation within the architecture.

The governance domain consists of:

- governance and enterprise risk management;
- legal issues, including contracts and electronic discovery;
- compliance and audit;
- information management and data security;
- portability and interoperability.

The operational domain focuses on:

- business continuity and disaster recovery;
- datacenter operations;
- incident response, notification, and remediation;
- application security;
- encryption and key management
- Identity and access management;
- virtualization;
- security as a service.

Key Takeaway

This reference model provides the essential architectural building blocks for engaging security and compliance for cloud initiatives.

7.2.4 Distributed Management Task Force Common Information Model

Founded in 1992, the Distributed Management Task Force, Inc. (DMTF) focuses on collaboration and the development of systems management standards, as well as their validation, promotion, and adoption in IT organizations. Its standards provide common management infrastructure components for instrumentation, control, and communication in a platform-independent and technology-neutral way.

Initially developed in 1997, their Common Information Model (CIM) provides a common definition of management information for systems, networks, applications, and services, and allows for vendor extensions. The CIM's common definitions enable vendors to exchange semantically rich management information between systems throughout the network. As a conceptual information model not bound to a particular implementation, CIM allows for the interchange of management information between management systems and applications in a vendor-neutral fashion. This can be either "agent to manager" or "manager to manager" communications that provides for distributed system management. There are two parts to CIM: the CIM specification and the CIM schema.

The CIM specification describes the language, naming, and Meta schema, and mapping techniques to other management models such as simple network management protocol (SNMP), management information bases (MIBs), distributed management task force (DMTF), and management information formats (MIFs). The Meta schema is a formal definition of the model. It defines the terms used to express the model and their usage and semantics. The elements of the Meta schema are classes, properties, and methods. The Meta schema also supports indications and associations as types of classes and references as types of properties.

The CIM schema provides the actual model descriptions. The CIM schema supplies a set of classes with properties and associations that provide a well-understood conceptual framework within which it is possible to organize the available information about the managed environment (DMTF Architecture Working Group, 2012). The CIM schema itself is structured into three distinct layers:

- The Core schema is an information model that captures notions that are applicable to all areas of management.
- Common schemas are information models that capture notions that are common to particular management areas, but independent of a particular technology or implementation. The common areas are systems, devices, networks, applications, metrics, databases, the physical environment, event definition and handling, management of a CIM infrastructure (the interoperability model), users and security, policy and trouble ticketing/knowledge exchange (the support model). These models define classes addressing each of the management areas in a vendor-neutral manner.
- Extension schemas represent organizational or vendor-specific extensions of the common schema. These schemas can be specific to environments, such as operating systems (for example, UNIX or Microsoft Windows). Extension schema fall into two categories, technology-specific areas such UNIX98 or product-specific areas that are unique to a particular product such as Windows.

The formal definition of the CIM schema is expressed in a managed object file (MOF) which is an ASCII or UNICODE file that can be used as input into an MOF editor, parser, or compiler for use in an application. The unified modeling language (UML) is used to visually portray the structure of the CIM schema.

Key Takeaway

The DMTF CIM provides essential architecture building blocks for engaging cloud management systems and related interoperability concerns

7.2.5 ISO/IEC Distributed Application Platforms and Services (DAPS)

The International Organization for Standardization (ISO) is the world's largest developer of voluntary international standards. Founded in 1947, they have published more than 19 500 International Standards covering almost all aspects of technology and business. JTC 1/SC38 consists of three working groups: Web services, service oriented architecture (SOA), and cloud computing. JTC 1/SC38 also includes a study group on cloud computing whose goals include providing a taxonomy, terminology, and value proposition for cloud computing.

The working group on cloud computing was established in February 2012. It is committed to usage scenarios and use cases as an analysis tool to identify specific characteristics and requirements of cloud computing. The relationship between usage scenarios and use cases is illustrated in Figure 7.4. Standard templates are provided as well to facilitate the methodology.

Among the major scenarios covered are high level scenarios (including provisioning methods such as IaaS, PaaS, and "generic"), cloud delivery scenarios (such as PaaS-based CRM), business support, migration, portability, interoperability, mobility, and cloud computing for the public sector.

The working group's collection of use case scenarios and use cases provide a real-life method to identify where standards are or can be applied to cloud reference architectures and help quantify what gaps may exist. This method also enables relevant stakeholders to be identified and their collaboration to occur in context.

As of 2013, ISO/IEC JTC 1/SC 38 has eight published standards and five standards are under development in DAPS. The cloud computing standards focused on by the working group include standards that define cloud computing vocabulary and reference architecture, including general concepts and characteristics of

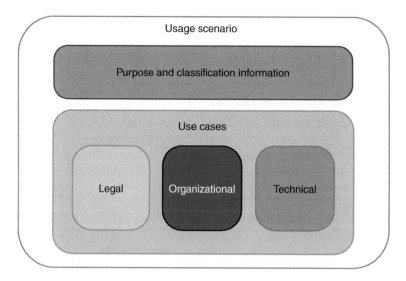

Figure 7.4 *Usage scenarios and use cases. Source: ISO/IEC JTC 1/SC 38/WG 3 (2013)*

cloud computing, types of cloud computing, components of cloud computing, and cloud computing roles and actors (ISO/IEC JTC 1/SC 38/WG 3, 2013). Primary goals of the working group standards activity include interoperability and enabling future standards work.

Key Takeaway

This standard provides reference examples and exposes methods and techniques for decomposing business requirements in use case and scenario planning efforts, primarily toward application and systems interoperability.

7.2.6   Open Grid Forum Open Cloud Computing Interface (OCCI)

The Open Grid Forum (OGF) is a community of users, developers, and vendors leading the global standardization effort for distributed computing (including clusters, grids, and clouds). The OGF community consists of thousands of individuals in industry and research, representing over 400 organizations in more than 50 countries. It is an open community committed to driving the rapid evolution and adoption of applied distributed computing.

The purpose of the Open Cloud Computing Interface Working Group (OCCI-WG) is the creation of practical solutions that interface with cloud infrastructures offered as a service. They focused initially on solutions that covered the provisioning, monitoring, and definition of Cloud Infrastructure Services (IaaS). The current release of the Open Cloud Computing Interface is suitable to serve many other models in addition to IaaS, including PaaS, and SaaS.

The OCCI goals are interoperability, portability, and integration in a vendor-neutral context with minimal cost. The current specification consists of three documents. This specification describes version 1.1 of OCCI. Future releases of OCCI may include additional rendering and extension specifications. The documents of the current OCCI specification suite are:

- OCCI Core: describes the formal definition of the OCCI core model.
- OCCI HTTP Rendering: defines how to interact with the OCCI core model using the RESTful OCCI API. The document defines how the OCCI core model can be communicated and thus serialized using the HTTP protocol.
- OCCI Infrastructure: contains the definition of the OCCI Infrastructure extension for the IaaS domain. The document defines additional resource types, their attributes, and the actions that can be taken on each resource type.

By focusing on the delivery of API specifications for the remote management of cloud infrastructures, the work enables the development of vendor-neutral interoperable tools. Their scope is the high-level functionality required for lifecycle management of virtual machines or workloads running on virtualization technology supporting service elasticity. The API work is supported by use cases that provide context and applicability of the API in lifecycle management. Reference implementations are specifically excluded, as are details relating to supporting infrastructure design (such as storage and network hardware configuration).

Key Takeaway

The OGF OCCI provides essential architecture building blocks for engaging cloud infrastructure analysis and design from an interoperability perspective.

7.2.7   Open Security Architecture (OSA) Secure Architecture Models

Open Security Architecture (OSA) is a not-for-profit-organization supported by volunteers for the benefit of the security community. The OSA is divided into three categories: the Control Catalog, the Pattern Landscape, and the Threat Catalog. The OSA provides a single, consistent, clearly defined control catalog intended to simplify requirements from numerous standards, governance frameworks, legislation, and regulations.

Patterns show the best practice set of controls that should be specified for a given situation, consisting of security architectures that address specific security problems. Applying OSA patterns in your work gives you a fast start, improves the quality of the solution you deploy, and reduces overall effort (Figure 7.5).

The Control Catalog, based on NIST 800-53 (2006), provides details for all controls required to create a security solution. Controls are mapped against other standards, regulations, legislation, and governance standards. To ensure consistency between patterns and application of controls, the OSA has defined "actors" for the use cases. OSA actors are prototypical business roles which can be used singly or in combination, depending on the intent of the use case.

The OSA Threat Catalog is a list of generic risks that need to be taken into account when rating the applicability of a control to a given pattern. For the classification of top-level threats, the OSA proposes to categorize the threat space into sub-spaces according to a model of three orthogonal dimensions labeled motivation, localization, and agent. The threat agent is the actor that imposes the threat on a specific asset. Threat agents can be human, technological, or *force majeure* (environmental).

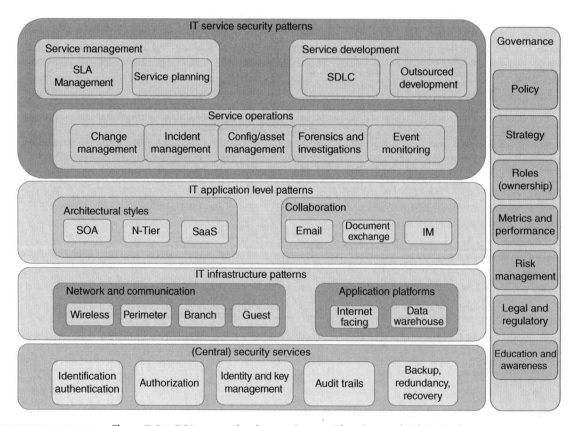

Figure 7.5   *OSA pattern landscape. Source: Phaedrus and Tobias (n.d.)*

The Cloud Computing Pattern published by the OSA illustrates the application of controls, actors, and threats in the construction of a cloud security reference pattern. The pattern enables the evaluation of a given cloud solution according to its function and capabilities in a vendor-agnostic fashion.

Key Takeaway

The OSA secure architecture model provides essential architecture building blocks for engaging cloud security and compliance. It also offers guidance in the application of the architecture building blocks to use case and scenario planning efforts for cloud security and compliance perspectives.

7.2.8   Organization for the Advancement of Structured Information Standards

The Organization for the Advancement of Structured Information Standards (OASIS) is a nonprofit consortium that drives the development, convergence, and adoption of open standards for the global information society. OASIS promotes industry consensus and produces worldwide standards for security, cloud computing, SOA, Web services, the smart grid, electronic publishing, emergency management, and other areas.
OASIS cloud-related standards include:

- AMQP: advanced message queuing protocol offers organizations an easier, more secure approach to passing real-time data streams and business transactions. By enabling a commoditized, multi-vendor ecosystem, AMQP creates opportunities to transform the way business is done over the Internet and in the cloud.
- IDCloud: identity in the cloud identifies gaps in existing identity management standards for the cloud and the need for profiles to achieve interoperability within current standards. IDCloud performs risk and threat analyses on collected use cases and produces guidelines for mitigating vulnerabilities.
- OData: Open Data is a REST-based protocol that simplifies the sharing of data across applications for reuse in the enterprise, cloud, and mobile devices. OData enables information to be accessed from a variety of sources including relational databases, file systems, content management systems, and traditional web sites.
- SAML: security assertion markup language provides a framework for communicating user authentication, entitlement, and attribute data between online partners.
- SOA-RM: SOA reference model defines the foundation upon which specific SOA concrete architectures can be built.
- TOSCA: topology and orchestration specification for cloud applications enhances the portability of cloud applications and the IT services that comprise them. TOSCA enables the interoperable description of application and infrastructure cloud services, the relationships between parts of the service, and the operational behavior of these services, independent of the supplier that creates the service, the particular cloud provider, or hosting technology. TOSCA facilitates higher levels of cloud service and solution portability without lock in.

OASIS perspectives on cloud generally are in data/messaging or security contexts. They are included in the category of "open standard" bodies promoting interoperability and ease of management in heterogeneous environments. The OASIS Cloud Application Management for Platforms (CAMP) technical committee advances an interoperable protocol that cloud implementers can use to package and deploy their applications. CAMP defines interfaces for self-service provisioning, monitoring, and control. Common CAMP-use cases include moving premise applications to the cloud (private or public), and redeploying applications across cloud platforms from multiple vendors.

Key Takeaway

The OASIS cloud-related standards provides essential architecture building blocks for engaging cloud service design and cloud service interoperability initiatives

7.2.9   SNIA Cloud Data Management Interface Standard

The Storage Networking Industry Association (SNIA) is an association of producers and consumers of storage networking specifications and standards. It works towards its goal "to promote acceptance, deployment, and confidence in storage-related architectures, systems, services, and technologies, across IT and business communities" by forming and sponsoring technical work groups for storage networking standards and specifications.

The Cloud Data Management Interface (CDMI) is a SNIA standard that specifies a protocol for self-provisioning, administering, and accessing cloud storage. With help of the CDMI, the clients can discover cloud storage capabilities and leverage the CDMI to manage data and containers. In addition, metadata can be set on containers and their contained data elements through CDMI. The CDMI can be used for administrative and management applications to manage containers, accounts, security access, monitoring/billing information, and storage that is accessible by other protocols. The CDMI exposes the capabilities of the underlying storage and data services to potential clients. Figure 7.6 portrays a cloud reference model.

The cloud storage reference model includes functions and layers that enable clients to discover the capabilities available in the cloud storage, manage containers and the data that is placed in them; and allow metadata to be associated with containers and the objects they contain. This is portrayed in above three layers of the model. CDMI defines RESTful HTTP operations for the above functions.

The CDMI defines the functions to manage the data and as a way to store and retrieve the data. The means by which the storage and retrieval of data is achieved is termed a data path. Hence, CDMI specifies both a data path and a control path interface.

The container metadata layer is used to configure and expose the data requirements of the storage provided through the storage protocol (e.g., block protocol or file protocol). For example, for the underlying file system for a block protocol (e.g., iSCSI), the CDMI container provides a useful abstraction for representing the data system metadata.

The SNIA's CDMI is based on an object model with categorized data, container, domain, queue, and capability objects. The CDMI defines two namespaces that can be used to access stored objects: a flat-object ID namespace and a hierarchical path-based namespace. Objects are created by ID by performing HTTP commands against a special URI.

The CDMI uses many different types of metadata: HTTP metadata, data system metadata, user metadata, and storage system metadata. HTTP metadata is metadata that is related to the use of the HTTP protocol (e.g., content length, content type). The CDMI data system metadata, user metadata, and storage system metadata are defined in the form of name-value pairs. Data-system metadata are metadata that are specified by a CDMI client and are a component of objects. Data-system metadata abstractly specify the data requirements associated with data services that are deployed in the cloud storage system.

Key Takeaway

This cloud interface standard provides essential architecture building blocks for engaging cloud storage design and infrastructure integration.

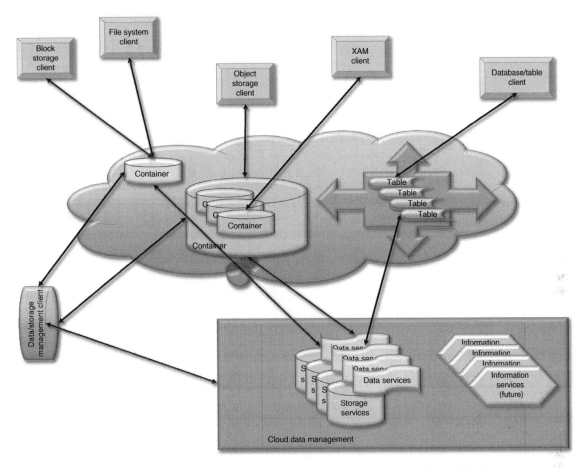

Figure 7.6 *Cloud storage reference model. Source: Bairavasundaram (2012)*

7.2.10 The European Telecommunications Standards Institute Cloud Standard

The European Telecommunications Standards Institute (ETSI) produces globally applicable standards for information and communications technologies (ICT), including fixed, mobile, radio, converged, broadcast, and Internet technologies. Their work in cloud is an extension of an earlier focus on grid computing. The ETSI has four published standards for cloud focused on standardization, interoperability testing, and service-level agreements. The ETSI working programs include security, interoperability, connected things, wireless systems, fixed networks, content delivery, and public safety.

The latest published cloud standard from ETSI (ETSI TS 103 142 Test Descriptions for Cloud Interoperability) describes the testing requirements and scenarios for interoperability as defined by another standards group, the OCCI from the Open Grid Forum. This illustrates the interlocking interests of the many open standards groups and how they collaborate to drive agnostic interoperable solutions.

In order to unleash the potential of cloud computing, the European Commission (EC) published a Communication on Cloud Computing, released on September 27, 2012, identifying cutting through the jungle of standards as one of the key actions to foster mass adoption of cloud computing. The ETSI was requested by the EC to coordinate with stakeholders in the cloud standards ecosystems and devise standards and roadmaps in support of EU policy in critical areas such as security, interoperability, data portability, and reversibility. The first meetings were held in Cannes in December 2012.

Related to cloud standards projects at ETSI are those addressing "connected things" (ETSI, 2013). An ever increasing number of everyday machines and objects are now embedded with sensors or actuators and have the ability to communicate over the Internet. These "smart" objects can sense and even influence the real world. Collectively, they make up what is known as the "Internet of Things" (IoT). The IoT draws together various technologies including radio frequency identification (RFID), wireless sensor networks (WSNs), and machine-to-machine (M2M) service platforms. The ETSI is addressing the issues raised by connecting potentially billions of these "smart objects" into a communications network by developing the standards for data security, data management, data transport, and data processing. This will ensure interoperable and cost-effective solutions, open up opportunities in new areas such as e-health and smart metering, and allow the market to reach its full potential.

Key Takeaway

The ETSI cloud standards outline architecture building blocks for engaging cloud networking and communications interoperability.

7.2.11   The Open Group Cloud Model

The Open Group (TOG) is a vendor and technology-neutral industry consortium that provides innovation and research, standards and certification development on topics of IT Architecture. It is currently developing several related cloud-reference models, namely Cloud Computing Reference Architecture, the Cloud Ecosystem Reference Model, and the Distributed Computing Reference Model. This section will discuss details of the published Distributed Computing Reference Model (DCRM) (Figure 7.7).

The DCRM contains several components that are focused on the interfaces between the components. The DCRM is leveraged in the context of portability and interoperability of cloud reference architecture. The management systems and marketplaces are particular kinds of components, shown separately because of their particular relationships to platforms and infrastructure.

The application data, applications, platforms, and infrastructure stack can be applied to enterprise systems, cloud systems, and user devices. In cloud systems, applications may be exposed as software as a service (SaaS), platforms may be exposed as platform as a service (PaaS), and infrastructure may be exposed as infrastructure as a service (IaaS).

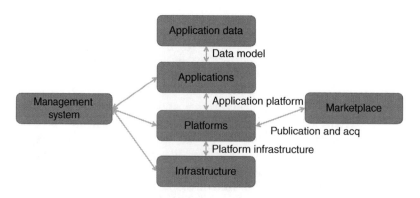

Figure 7.7   *Distributed computing reference model. Source: Bakshi and Skilton (2013)*

In the DCRM, an application can have several facets – a cloud SaaS service, enterprise application service, composition of cloud and enterprise application services, an application program running on a server, and / or application mobile device. They can also be SOA style – applications consisting of collections of services. These may be explicitly programmed as service compositions; for instance, using the OASIS standard Web services business process execution language (WS-BPEL). Applications interfaces include application data through data models, through application-application interfaces, and through application-platform interfaces (APIs).

An application platform consists of the hardware and software components that provide the services used to support applications. It exposes an application platform interface to applications. In the DCRM, a platform may be a PaaS service, where interfaces could be application-platform interfaces, platform-platform interfaces, infrastructure through platform-infrastructure interfaces, management systems through platform management interfaces, or marketplaces through publication and acquisition interfaces. Platforms are used by developers to develop applications and other programs and by systems managers to run and configure applications, either through management systems or directly.

In the DCRM, infrastructure includes cloud IaaS services, hardware in servers, PCs and mobile devices, and virtualized hardware resources in enterprise systems. A cloud infrastructure service makes hardware components available as cloud resources, generally through a virtualization layer. The functional interface in this case supports the loading and execution of machine images. The management interface enables the hardware resources to be provisioned and configured, and enables machine images to be deployed on them. Infrastructure interfaces to platforms through platform-infrastructure interfaces; management systems through infrastructure management interfaces.

Management systems are components to manage cloud and enterprise IT resources. Management systems interface with applications through APIs, platforms through platform management interfaces, and infrastructure through infrastructure management interfaces. Management systems are used by systems managers to manage applications, platforms, and infrastructure.

A cloud marketplace enables cloud providers to make their products available and enables cloud consumers to choose and obtain the products that meet their requirements. The products may be services, machine images, applications, or other cloud-related products. They can have associated descriptions, prices, terms of use, and so forth, so that consumers can select them and contract for their use. Marketplaces interface with platforms through product publication and acquisition interfaces.

Key Takeaway

This reference model provides essential architecture building blocks for engaging cloud portability and cloud interoperability perspectives and offers guidance in the application of the architecture's building blocks to use case and scenario planning efforts for cloud portability and cloud interoperability initiatives.

7.3 Conclusion

Cloud reference frameworks are a critical tool for architecting, engineering, and standard setting in any major cloud initiative. Today's most influential groups are promoting "open" standards as a rule, which recognize the various requirements for heterogeneous ecosystems. They provide the mechanism for introducing transformational change, while still supporting legacy systems.

Collaboration across the groups is a part of the "Open" movement, wherein each group's interest / focus is applied in the context of one or several other reference frameworks using scenarios and use cases. In this fashion, reference frameworks constitute the common taxonomy and architectural landscape for cloud and its evolution.

Essential to the success of a cloud effort is the appropriate application of a framework to one's own business environment, business requirements, and technical capabilities. Each framework presented in this chapter has a particular focus and use as described in its key takeaway. By aligning a framework with the actual work at hand, the architecture and engineering efforts become better informed and better integrated. This reduces overall risk and works to ensure anticipated business outcomes.

References

Archer, J., Boehme, A., Cullinane, A., *et al.*(2011) *Security Guidance for Critical Areas of Focus in Cloud Computing. Cloud Security Alliance (CSA)*, https://cloudsecurityalliance.org/download/security-guidance-for-critical-areas-of-focus-in-cloud-computing-v3/ (accessed November 29, 2015).

Bairavasundaram, L., Baker, S., Carlson. M., *et al.* (2012) *SNIA Cloud Data Management Interface v1.0.2, Technical Position*, http://snia.org/sites/default/files/CDMI%20v1.0.2.pdf (accessed November 29, 2015).

Bakshi, K. and Skilton, M. (2013) *Guide: Cloud Computing Portability and Interoperability*, Berkshire, United Kingdom, http://www.opengroup.org/cloud/cloud/cloud_iop/index.htm (accessed December 4, 2015).

DMTF Architecture Working Group (2012) *Common Information Model (CIM) Metamodel Version 3.0.0* (Document Number: DSP0004, December 13).

ETSI (2013) *Building The Future, ETSI Work Programme 2013–2014*, http://www.etsi.org/images/files/WorkProgramme/etsi-work-programme-2013-2014.pdf (accessed November 29, 2015).

Khasnabish, B., Ma, S., So, N., *et al.* (2013) IETF Cloud Reference Framework. IETF Fremont, California. http://datatracker.ietf.org/doc/draft-khasnabish-cloud-reference-framework/?include_text=1.

Liu, F., Tong, J., Mao, J., *et al.* (2011) *NIST Cloud Computing Reference Architecture*. Special Publication 500-295, ITL NIST, Gaithersburg, MD, http://www.nist.gov/customcf/get_pdf.cfm?pub_id=909505 (accessed November 29, 2015).

ISO/IEC JTC 1/SC 38/WG 3 (2013) *Methodology and Guidelines for Cloud Computing Usage Scenario and Use Case Analysis*, International Standards Organization, Geneva.

Phaedrus, R. and Tobias, S. (n.d.) *Open Security Architecture. Release 08.02*,

http://www.opensecurityarchitecture.org/cms/library/patternlandscape/251-pattern-cloud-computing (accessed December 4, 2015).

8

Virtualization: An Overview

Jim Sweeney

Cloud Consultant, USA

8.1 Introduction

Used since the 1960s in computing, virtualization is a broad term that refers to the abstraction of computing resources, a technique for hiding the physical characteristics of computing resources (CPU, memory, disk and network interfaces) from the way in which other systems, applications, or end users interact with those resources. Virtualization creates an external interface that hides an underlying implementation by combining resources at different physical locations or by simplifying a control system.

Another key concept is *encapsulation*, which means that all of the files that are associated with the virtualized operating system (OS), application, and support software are saved as one big file, or virtual disk. Through encapsulation, the state of a virtual machine can be saved to disk and then the virtual machine can be restarted in the time that it takes to reload the data from the disk. The recent development of new virtualization platforms for VMware, Citrix and others has refocused attention on this mature concept and taken it to the X86 world and off of mainframes.

Virtualization software such as vSphere from VMware, XEN from Citrix, and Hyper-V from Microsoft can transform or virtualize the hardware resources of an x86-based computer – including the CPU, RAM, hard disk, and network controller – to create a fully functional virtual machine that runs its own operating system and applications just like a "real" computer. Note that virtual machines can cover most x86 operating systems (i.e. Windows, Linux, or Solaris x86). Multiple virtual machines share hardware resources without interference so that a single computer can run several operating systems and applications, commonly referred to as workloads, at the same time. See Figure 8.1.

In general, virtualization happens by placing a thin layer of software directly on the computer hardware or on a host operating system. This software layer contains a virtual machine monitor or *hypervisor*, which

Encyclopedia of Cloud Computing, First Edition. Edited by San Murugesan and Irena Bojanova.
© 2016 John Wiley & Sons, Ltd. Published 2016 by John Wiley & Sons, Ltd.

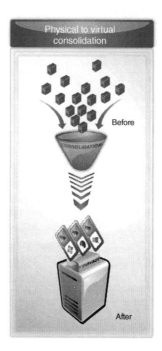

Figure 8.1 *Virtualization allows many workloads to run on one physical server. Source: Used with permission from UNICOM Global*

allocates hardware resources dynamically and transparently so that multiple operating systems, each contained within its own virtual machine, run concurrently on a single physical computer. In the case of VMware, the hypervisor is referred to as the ESX server. (See Figure 8.2.)

It is important to understand that a virtual machine can be just about any X86 operating system with its associated applications. For example, a Microsoft Hyper-V physical host could be running three separate virtual machines; The first virtual machine could be running the Solaris X86 operating system and a database, the second virtual machines could be a Windows Server 2012 and MS Exchange, and the third a Linux operating system with the LAMP stack being used by developers!

8.2   Origins of Virtualization

The concepts in virtualization can be traced back to the days of mainframes in 1960s. At the time there was a major effort by IBM to improve the concept of time sharing. A time-sharing system that would let multiple users access the same computer simultaneously was not an easy problem to solve. While many others were working on improving time-sharing and adding cool interfaces in order to make it easier to submit batch jobs, adding these features added complexity to the mainframe OS.

Meanwhile a small IBM engineering team in Cambridge, Massachusetts, came up with a novel approach that gave each user a virtual machine with its own operating system. With a virtual machine the operating system does not have to be complex because it only has to support one user. In 1972, IBM released a version of its CP/CMS product to the general mainframe-using public. The "CP" was the part that created the virtual machine and stood for "control program"; CMS stood for "console monitor system."

In those days most people were focusing on time sharing. In time sharing you divide up the CPU, memory and other system resources between users. The concept of time sharing did not just go away. MultiCS, an

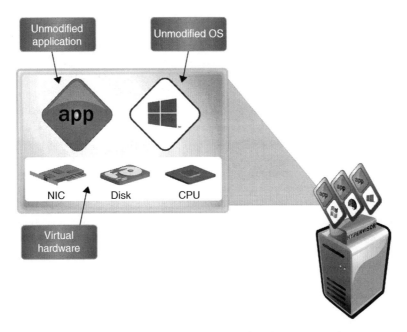

Figure 8.2 *A look inside a virtual machine. Source: Used with permission from UNICOM Global*

example of a time sharing OS from that period developed by MIT eventually went on to become the first UNIX operating system. UNIX is a very good example of a multiuser system that uses the time-sharing technique to virtualize applications rather than entire virtual machines.

Building on this work, VMware introduced virtualization on the X86 platform in 2001. However, it was not until much more work was done and Virtualization Infrastructure 3 (VI3) was released in June of 2006 that server virtualization really took off. Following closely on its heels, Xen was released in 2003.

Xen originated as a research project at the University of Cambridge, led by Ian Pratt, senior lecturer at Cambridge. Ian co-founded XenSource, Inc. with another Cambridge University alumnus, Simon Crosby. Zen was acquired by Citrix in October 2007 and has been continually updated and maintained by them since that time.

Shortly thereafter, Microsoft introduced its first real release of Hyper-V, its server virtualization offering. While there was a beta release that was included in Windows Server 2008, the first official release did not occur until June, 2008.

8.3   Types of Virtualization

The three major types of virtualization are hardware virtualization, storage virtualization and network virtualization.

8.3.1   Hardware Virtualization

When most people think of virtualization today, they are normally thinking of just *hardware virtualization*. Hardware virtualization is the sharing of the physical system resources (CPU, memory, network, and local storage) to enable multiple virtual machines to run on the same physical server. There are three major types of hardware or server virtualization: full virtualization, para virtualization, and hardware-assisted virtualization.

8.3.1.1   Full Virtualization

Full virtualization is almost complete simulation of the actual underlying hardware to allow a virtual machine consisting of a guest operating system and applications to run unmodified. Today's modern virtualization vendors can virtualize any x86 operating system using a combination of binary translation and direct execution techniques. Binary translation translates kernel code to replace nonvirtualizable instructions with new sequences of instructions that have the intended effect on the virtual hardware. Meanwhile, as many CPU requests as possible are directly executed on the host processor for higher performance. In this manner, each virtual machine has all the services of the physical system, including a virtual BIOS, virtual devices, and virtualized memory management. It is important to note that almost 75% of the CPU requests that are made by the virtual machine are passed directly to the host CPU for execution, which is what enables the speed of today's modern virtualized systems.

This combination of binary translation and direct execution provides full virtualization as the guest OS is fully abstracted (completely decoupled) from the underlying hardware by the virtualization layer. The guest OS is not aware it is being virtualized and therefore requires no modification. Full virtualization is the only option that requires no hardware assist or operating system assist to virtualize sensitive and privileged instructions. The hypervisor translates all operating system instructions on the fly and caches the results for future use, while user level instructions run unmodified at native speed.

8.3.1.2   Paravirtualization

Paravirtualization is different from full virtualization. Here the designers were trying to achieve greater speed at the cost of adding direct communication between the virtual machine (VM) operating system and the host operating system. The goal in paravirtualization is lower virtualization overheads but the performance advantage of paravirtualization over full virtualization can vary greatly depending on the workload. It also means that significant kernel-level modifications to the virtual machine OS are required in order for it to work. As paravirtualization cannot support unmodified operating systems, its compatibility and portability is sometimes poor.

8.3.1.3   Hardware-Assisted Virtualization

In the early 2000s, hardware manufacturers, specifically CPU vendors, started to embrace virtualization rapidly. They developed and released new features into their CPUs to simplify virtualization techniques and provide greater performance for the virtual machines. Both virtualization technology (VT-x) from INTEL Corporation and AMD-V from AMD target privileged instructions with a new feature called CPU execution mode. With these new CPU features, certain CPU calls are automatically recognized and handled directly by the CPU, removing the need for either binary translation or paravirtualization.

In addition to hardware or server virtualization there are others types of virtualization such as *storage virtualization* and *network virtualization* that are used in cloud computing.

8.3.2   Storage Virtualization

In general, storage virtualization refers to the abstraction of the physical resources of the storage (disks, memory, controllers, etc) from the user. Most of the storage vendors these days provide some level of storage virtualization. Storage virtualization means that all storage platforms are presented in a singular manner to the virtualized servers, meaning that the virtualized servers can easily and quickly access the storage while allowing the storage device to perform the potentially complicated translation to the hardware.

The advantages of this technology are many. First, it is more efficient for the requestor of the data to know only that it resides on the disk array itself and let the disk manager automatically move blocks of data around the array to gain efficiencies in both amount of physical resources used and performance.

The primary advantage with storage virtualization is that storage devices from many different vendors can be managed as if they were one heterogeneous storage platform. Now one can migrate data (and virtual machines, but we will cover that shortly) between storage platforms. This is one of the amazing benefits that virtualization provides and it is one of the many features that make virtualization a cornerstone of cloud computing. Again, we will get there shortly, so please continue on!

8.3.3   Network Virtualization

In keeping with the theme of this chapter, network virtualization presents logical networking devices and services – logical ports, switches, routers, firewalls, load balancers, VPNs and more – to workloads irrespective of the underlying physical hardware. In all other respects, virtual machines run on the virtual network exactly the same as if they were running on a physical network.

By reproducing the physical network in software one can have better flexibility and agility when deploying virtual server workloads. A virtualized network has the same features and guarantees of a physical network, but it delivers the operational benefits and hardware independence of virtualization.

Remember the old days of computing? Getting a computer from the server folks, begging for some GBs or TBs of storage from the storage admins and, finally, getting network attributes assigned and configured by the network admins? With network virtualization this is not necessary. All of the attributes of the physical network are mapped logically and presented to the virtual machines. The advantage is that provisioning is fast and nondisruptive to the organization. Another advantage is that a virtualized network allows for automated maintenance as remapping of physical to logical devices can occur without interruption. Physical resources can then be maintained, replaced, or upgraded and the mapping restored all without interruption to the workloads running on them!

In summary, all of these virtualization technologies, server, storage, and networking are now included in what is commonly referred to as the *software-defined datacenter* (SDDC). An SDDC is a new vision for IT infrastructure. It extends the previous three virtualization concepts using the tools of abstraction, pooling, and automation, to all of the datacenter's resources and services in order to achieve IT as a service (a concept that is crucial to the cloud!).

8.4   Advantages of Virtualization

There are several compelling forces driving companies and government agencies to adopt a strategy for virtualization in their datacenters. The recognition that there is a finite budget for operations and a finite amount of energy for power and cooling has made organizations aware that virtualization is a practical way to optimize their datacenters, reduce costs, consume less energy, and implement sustainable energy conservation measures to serve the public good.

8.4.1   The Return on Investment of Virtualization

Most nonvirtualized servers operate at 5–15% of capacity, which is highly inefficient. Unfortunately, this problem scales. Since there are many more physical servers than there need to be, today's datacenters are several times bigger and correspondingly more inefficient than they need to be. With server virtualization, operational efficiencies of 70–75% are commonplace and efficiencies of 85% are not unheard of. But don't forget that this advantage of virtualization is actually a double-edged sword! Given that most of the power

Table 8.1   *The growth of datacenters in power and heat*

	Power required/ rack (kW)	Heat output/ sq. ft. (W)
Datacenters of the recent past	2	40
2007 datacenters (average)	10	200
Datacenters of the near future	25	500

Source: Used with permission from UNICOM Global

Table 8.2   *IT equipment and the power it uses*

Component	IT equipment usage (%)	Percentage of total power
Servers	56	33.6
Storage	27	16.2
Networking	17	10.2

Source: Used with permission from UNICOM Global

input to a physical server is output back out of the back of the server as heat, we need to cool that heat. For every watt of power not expelled as heat, another watt less is needed to cool that heat. So virtualization really provides us with a *return on investment* (ROI) that is out of this world.

8.4.1.1   *An Example*

Recent aggregate data show that datacenters' power usage today has greatly exceeded past levels, and estimates indicate that this trend will continue (see Table 8.1.)

8.4.2   The Datacenter

Companies have reached a critical point where they recognize that they can no longer maintain the status quo for how their datacenters use power. In Table 8.1, the projection that a rack could consume an estimated 25 kW of power is based on very dense systems, such as the HP C-class blade system, which provides significant computing and performance advantages. Racks configured for high density; however, can have unanticipated costs. For example, the C-class system is a 10U chassis with up to 16 half-height blades, each with up to dual socket, 12 core CPUs, for a total of 384 cores per chassis. With four chassis per 42U rack, there are a total of 1536 cores per rack. This is the good news, from a density standpoint. But it also means that each chassis requires approximately 5.5 kW or more, and therefore more than 20 kW will be needed at the rack.

High-density computing means increased power density, causing much faster thermal rise than in older, lower-density systems. A datacenter with an average heat load of 40 W per square foot can cause a thermal rise of 25 °F in 10 minutes, while an average heat load of 300 W per square foot can cause the same rise in less than a minute. In this scenario, there is not only an increase in the amount of money spent on electricity – the datacenter infrastructure is also at greater risk in case of a disaster, where a 25 °F increase in 1 minute can potentially cause sensitive electronic components to literally fry in 3 to 4 minutes.

8.4.3   IT Equipment

Recent estimates indicate that 60–70% of a datacenter's electricity is used by the IT equipment, with the rest dedicated to cooling, power conditioning and lights. Table 8.2 gives a breakdown of the three major types of IT equipment and the total percentage of power they use in the datacenter.

Do you see now why server virtualization has such a high ROI? In one particular case, my previous employer helped a government agency remove almost 750 servers from its datacenter. They started small with virtualization but quickly became experts at it and, after a lot of work on their part, ended up with a consolidation ratio of 37.5 average virtual machines per physical host! That's 750 physical hosts that are not running at 5% efficiency, and not belching most of their power input back out as heat that we have to cool.

We haven't even mentioned space savings yet. Is there a datacenter out there today that, having virtualized its server workloads, is bursting at the seams? Removing 70–80% of the servers in a datacenter would certainly help with that problem, wouldn't it? So that's why server virtualization has such a high return on investment. I helped write the server virtualization TCO calculator for my former employer (Unicom Government Systems). While it is a little dated, just by going there and plugging in a few numbers you start to get an idea of why server virtualization is such a treat for the datacenter managers as well as the finance team! You can find the calculator at http://unicomgov.com/ideas-innovations/virtualization-zone/calculators/ (accessed November 29, 2015).

8.4.4 Virtualized Server Infrastructure

The act of virtualizing a single physical computer is just the beginning. While VMware's estimated market share for server virtualization is in the 70% range, almost all of the vendors offer a robust virtualization platform that can scale from one physical machine to hundreds of interconnected physical computers and storage devices to form an entire virtual infrastructure. What makes VMware different, and the reason we will focus on features from them that will enable cloud computing is other vendors have stayed with server virtualization and have not added features that move their offerings "up-the-stack," so to speak, to Infrastructure-as-a-Service cloud platforms.

Traditional server virtualization when thought of in light of todays cloud computing, really comprises a combination of server, storage and network virtualization. There are several components of virtualization which form the backbone for a traditional infrastructure-as-a-service cloud.

8.5 VMware® VI3 and vSphere

VMware® and their Virtual Infrastructure 3 (VI3) was the first generation of server virtualization to gain industry-wide acceptance. Because of its reliability, and the number of built-in features, Virtual Infrastructure 3 quickly became the gold standard for virtualizing servers. It was the first product that reliably and consistently allowed multiple virtual servers (VMs) and their applications to run unmodified on off-the-shelf physical hardware. While there are several non-x86 vendors offering cloud, most people are looking at x86 workloads, so we will focus on that. In addition, as VMware currently is the vendor with the biggest market share in private clouds, we will consider features from them. Many other vendors have similar features in their offerings.

VI3, as it was commonly called, had a number of important building blocks that made it so popular:

- *vMotion* – The ability to manually migrate (or move) a running virtual machine from one physical server to another without disrupting the users connected to that virtual machine. This assumes, of course, that the virtual machine is sitting on some kind of shared storage, but just about everyone architected their virtual environments that way! Figure 8.3 shows vMotion technology.
- Dynamic Resource Schedule (DRS) – An automatic (or manual) feature that allows the VMware hypervisor to move virtual machines around a cluster of physical machines as needed to gain maximum use of resources across all the virtual machines.
- High Availability (HA) – A feature that allowed VI3 to restart automatically virtual machines that are no longer running due to failure of an underlying physical machine. Again, shared storage was necessary here, but that is the recommended architecture for virtualization.

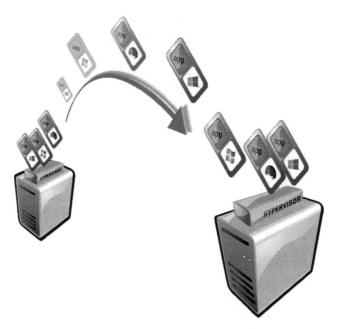

Figure 8.3 *VMware and vMotion. Source: Used with permission from UNICOM Global*

These are some of the features that form the absolute backbone of Cloud Computing.

8.5.1   Recent Feature Additions

Their current offering, *vSphere*, is not billed as a virtualization platform, but rather a "cloud computing" platform. Regardless, there are a number of impressive new features in vSphere that are worth considering.

8.5.2   Scale Up

The resource limits for both virtual and physical machines have been greatly expanded. Virtual machines can now have up to 32 virtual CPUs (increased from eight) and up to 1 TB of RAM (previously 255 GB). The underlying physical hardware can scale to 2 TB of RAM and 128 total cores! This means you can now run many more virtual machines on fewer, but larger, physical machines, making datacenter management a bit easier. It also means you can now run workloads that you may have thought previously were too large for a virtualized environment.

In addition, it is now possible to "hot add" both CPU and memory to your virtual machines. You cannot, however, "hot delete" these same features. Taking away memory or CPUs while an application is running can have disastrous results and is not permitted. You can, however, "hot add" and "hot remove" both storage and network devices. These features are a wonderful addition to the product line. As of vSphere 5.5 you can even hot add Hot-pluggable SSD PCIe devices.

8.5.3   vStorage Thin Provisioning

When you create your virtual machines and specify the hard disk size there is an option for thin provisioning. When setting up the vSphere cluster in our lab, the engineers tried this feature and it truly works as advertised.

We set up a test virtual machine and gave it a 60 GB disk drive but checked the option to thin provision it. After installing the OS, there was an 8 GB hard disk there! As we add additional software to that virtual machine, the size increases as we need it. This feature is independent of any thin provisioning that you may do at the LUN level of your storage. One word of caution: as with any thin provision technology, overprovisioning means that you will need to pay closer attention to your storage to avoid running out! It is a *very* bad idea to use thin provisioning at the server layer and also use thin provisioning at the storage layer. This is double trouble just waiting to happen.

8.5.4 Fault Tolerance (FT)

In VI3, we had high availability (HA), the ability to restart virtual machines that are no longer running when the physical machines fail. High availability is a good feature but it has one big disadvantage: there is a short outage while the virtual machines from the failed host are restarted. What if you cannot afford even a short outage? Enter fault tolerance. With fault tolerance, an identical virtual machine runs in lock-step with the original but is running on a separate host. To the external world, they appear as one instance (one IP address, one MAC address, one application), but in reality, they are fully redundant instances of virtual machines. If an unexpected hardware failure occurs that causes the active (i.e., primary) virtual machine to fail, a second, (i.e., passive) virtual machine immediately picks up and continues to run, uninterrupted, without any loss of network connections or transactions. This technology works at a layer underneath the operating system and application, so it can be used on any OS and on any application that is supported by VMware without modification.

Now VMware has even gone further with something called vSphere App HA. This new feature works in conjunction with vSphere HA host monitoring and virtual machine monitoring to improve the application uptime. vSphere App HA can be configured to restart an application service when an issue is detected with that application. It is possible to protect several commonly used, off-the-shelf applications. vSphere HA can also reset the virtual machine should the application fail to restart.

8.5.5 Virtual Distributed Switch (VDS)

Perhaps the greatest feature of vSphere is the virtual distributed switch (VDS). In VI3, each host had its own series of virtual switches containing all of the network characteristics of all of the virtual machines on that host. It was – and is – a very powerful feature that offers very granular control over the networking characteristics of the cluster. Unfortunately, that was also the downside of this feature. Network engineers hated the fact that certain aspects of the network were now in the hands, and under the control, of the server/storage managers. Horror stories abound about network problems caused by improper configuration of the virtual switches. vSphere still has this capability but has added the capability to support a virtual distributed switch. This allows for even more granular control (for example, monitoring traffic between virtual machines on the same physical server), while giving network administrators control over the network and the server administrators control over the servers.

Of course, that is not all. VMware then moved to virtualize the entire network with the software defined datacenter that we have already discussed.

8.6 Virtualization and the Private Cloud

Some virtualization vendors have improved and expanded their offerings to allow their customers to build private clouds on top of their virtualized infrastructure. In other cases, the cloud vendors themselves have layered significant capabilities on top of the virtualization layer to facilitate cloud computing. This is the case with Amazon Web Services (AWS), which uses an open source version of Xen on which to offer a cloud computing service to their customers.

However, no one virtualization vendor has added more capabilities that have moved them up the stack to cloud computing than VMware. For that reason, the cloud-ready features that we talk about in this chapter will reference VMware products. Other vendors have some of these pieces, but VMware leads the pack with a robust set of features and functionality.

8.6.1 Cloud-Ready Features

So if we already have a VMware-virtualized environment within a software defined datacenter, what would be needed to take a virtualized infrastructure and make it an IaaS cloud infrastructure? Assuming that vSphere provides the "ability to dynamically scale and provision computing power in a cost efficient way using a virtualized environment within a common redundant physical infrastructure," what more do we need? The first step is creating some kind of self-service portal.

8.6.2 vCloud Director

The vCloud Director has three very important and far-ranging jobs to do:

* Act as the overall cloud coordinator talking to vSphere and (through vCenter) controlling the creation of virtual machines.
* Act as the portal where users come to request new virtual machines. The user can choose characteristics of their virtual machine: number of processors, amount of RAM, size of disk, which virtual networks to use, which prebuilt template to deploy, other software to add to the template, and so forth.
* Function as the administrator portal where cloud managers set up templates, load software packages, develop approval workflows (using an add-on product from VMware called vCloud Request Manager), create and manage user groups, allowing them to have access to virtual machines of various sizes and lengths of times with certain templates, and so forth.
* Function as the "glue" tying discrete locations (vCloud cells) together into one cloud.

VMware vCloud Director also has built-in logical constructs to facilitate a secure multi-tenant environment, if users require it.

Other vendors have similar offerings that are very good in their own right. Computer Associates has a product called CA Automation Suite for Cloud™. It performs the same functions as vCloud Director, but has the ability to provision both virtual and bare metal physical machines as well (Cisco UCS blades, for example). BMC is also a strong competitor in this section of the marketplace; and server OEMs (Cisco, HP, IBM, and Oracle) are getting into the game too. So there are several vendors with strong offerings from which to choose.

8.6.3 vCenter Chargeback

What about charging the users for the services they require and use? VMware provides a tool called vCenter Chargeback which meters usage and provides reporting to allow for individual chargeback or showback to users. Administrators can set up charges for:

* Product licensing – Fixed cost charges for each Microsoft OS license used on the virtual machine.
* Usage – Actual CPU, memory, network, and storage usage can be separately monitored and charged.
* Administrative fees – A per-virtual machine fee can be set up for administrative overhead.

While we reference the VMware product, companies like Computer Associates, BMC and others have built chargeback into their cloud provisioning and management software, which run on top of many of the offerings from other virtualization vendors. There are also smaller companies that specialize in the chargeback/ showback aspect of cloud, giving you several options from which to choose.

We now have created a flexible infrastructure with a portal for users and we can charge them correctly for the resources they use. It's time to add security to the mix!

8.6.4   VMware vShield

vCloud Director is designed for secure multitenancy so that multiple organizations do not affect one another. For securing user access, the product ships with the built-in LDAP/Active-Directory integration. In addition, user access and privileges within vCloud Director are controlled through role-based access control (RBAC). There are additional steps that can be taken to harden the environment.

vShield provides all of the important pieces of network security, including:

- vShield Firewall – used when configuring access to the virtual machines;
- vShield Edge Virtual Appliances – built-in devices deployed automatically by vCloud to facilitate routed network connections using three different methods, including MAC encapsulation.

In addition, vShield features a built-in management console called vShield Manager.

8.6.5   vCloud API

Finally, vCloud API is a set of programming tools, application programmer interfaces and constructs around which users can make their applications "cloud aware." Developers can now build "cloud awareness" into their applications, making it easier to design, deploy and manage applications within a cloud environment.

8.7   Limitations of Virtualization

From its humble beginnings, virtualization has grown from servers to storage and networks and with some of the newly added features forms the basis for many of today's cloud offerings. But does that mean that server virtualization is right for everyone? Is it right for every workload? There are many advantages that virtualization brings to a data center but it is not a panacea. There are use cases where virtualization is not a good fit. Here are some examples:

- In one case, a small department was evaluating virtualization for its environment. It had a total of five servers in its group. It also had a total of five workloads, an e-mail server, a database server, some test and development servers, and a small NFS storage device. Given how virtualization and DR/HA works, it is typical to architect an environment with no less than three physical hosts. That way, if a physical server fails there are still two physical hosts remaining over which to spread the workload and still have some redundancy. As you can imagine, the ROI for this project just would not have come close to some of the bigger projects that I have implemented. After listening intently, I calmly went against everything in this chapter and told them they should keep their current systems!
- In a second case, the customers had almost no experience with server virtualization. Yet they were adamant that they wanted to virtualize their Microsoft Exchange environment. Without experimenting with small, more expendable workloads like file and print servers, or Web servers, the customers wanted to

implement Microsoft Exchange immediately in a virtualized environment. The folly here is not that the customers wanted to run Microsoft Exchange in a virtualized environment. Can Microsoft Exchange run well in a virtualized environment? Absolutely. But the phrase "walk before you run" is appropriate here. Microsoft Exchange is a complex environment that requires more skill and knowledge in a virtualized environment. Certainly more skill than basic Web servers, or file and print servers. They didn't start walking before they tried to run and the project was a disaster!

• The last example is one of pure religion. I had some customers who only had about 50 servers to virtualize. The customers understood the benefits of virtualization but half of the team was Microsoft bigots and half were on the side of VMware. (There are plenty of other vendors in there too, but in this example it was just these two. We don't want to forget Citrix and several other prominent vendors. There are at least a dozen at this point!) They already had a site license from Microsoft that allowed them to implement Microsoft Hyper-V. Finally they just wanted a very basic environment. They weren't going to use all of the fancy features of VMware and they were very price conscience. Microsoft Hyper-V is really the right solution for this customer. The moral of the story? Understand your requirements and where you want to go with virtualization. There are a lot of options out there for you!

8.8   Disaster Recovery and Business Continuity

Before we move from virtualization to cloud computing, there are two concepts that really helped virtualization make the leap from datacenter to cloud:

• business continuity; and
• disaster recovery.

First they are different, although many people mistakenly believe that they are the same. First, these two terms are often married together under the acronym BC/DR. That may lead people to think them as one in the same but they are very different. Disaster recovery is the process by which an entity resumes business after a disruptive event. The event might be something huge, like a natural disaster (say earthquake or tornado) or the terrorist attacks on the World Trade Center, or the event could be something small, like software malfunctions caused by computer viruses or even a datacenter power outage.

Business continuity is a much more comprehensive approach to planning how the entity can continue to make money after the disastrous event until its normal facilities are restored.

But how does virtualization help companies with DR/BC? A lot of planning that needs to occur to clearly delineate roles and procedures in cases of a natural disaster. But virtualization can assist. To see how, we need to remember a key concept of virtualization – encapsulation. In a virtual environment the OS, the applications, and everything on the physical hard drive of a physical server is encapsulated into a file. It didn't seem so important earlier, but it is a lynchpin of virtualization. Now, thanks to encapsulation, workloads are just big files and there is a lot of flexibility in them. They can be copied, for example, and become a backup somewhere else. And since my storage is virtualized and is presented to the virtualized servers the same way no matter what the underlying hardware, those workloads can reside on storage in another datacenter that is not necessarily from the same manufacturer. Since virtualization "normalizes" the underlying physical server resources, these workloads can even be copied from one brand of server to another without a problem! VMware even has a product to assist called Site Recovery Manager (SRM). Site Recovery Manager assists in the planning for DR/BC in three ways:

Site Recovery Manager allows the administrators to document and tweak a disaster recovery plan. This is not as trivial. Suppose the primary datacenter is wiped out due to a natural disaster. Suppose also that everything has been replicated and copied to the secondary datacenter. What workloads have to come up immediately?

In what order? E-mail is certainly high on the priority list. But there are probably things that need to come up before it, like maybe DNS or Active Directory or a dozen other workloads that would be forgotten in an emergency. SRM replaces erroneous run books that are often outdated as soon as they are written.

Site Recovery Manager allows administrators to test their DR plan. This is important as many public companies, like banks for example, are required to test their DR/BC plans on a quarterly basis. Without SRM, this is an enormous undertaking, which is not only complex but expensive. Site Recovery Manager automates the testing process, so that you can test and retest as necessary without disruption to the normal business activities.

Finally, in a disaster situation, SRM allows an administrator to hit one button and have all necessary workloads come up, in order, intact and ready to go. In short the failover and the failback can be started with one click of the mouse.

8.9   Conclusion

Hopefully you now have a good understanding of the concepts of virtualization, and how they can benefit your organization. Virtualization for X86 platforms has been around since the late 1980s and is a mature and valuable platform, significantly reducing cost and improving agility and reliability in today's modern data-centers. The virtualization technologies of today also include features that make cloud computing possible such as encapsulation and the ability to migrate workloads across disparate platforms and locations. Finally, also included are some of the very lynchpins of cloud computing such as chargeback and rapid provisioning.

Before cloud computing, virtualization technologies proved that disaster recovery and even business continuity was easier, more reliable and faster with a virtualized environment.

Its time now to start looking at cloud computing. With the basics of virtualization understood you will quickly see how it is the lynchpin of a modern cloud architecture!

Additional Resources

Everything VM (2010) *The History of Virtualization*, http://www.everythingvm.com/content/history-virtualization (accessed November 29, 2015).

Smoot, S. R. and Tan, N. K. (2011) *Private Cloud Computing: Consolidation, Virtualization, and Service-Oriented Infrastructure*, Morgan Kaufmann, Burlington, MA.

Stewart, V., Slisinger, M., Malesh, V., and Herrod, S. (2012) *Virtualization Changes Everything: Storage Strategies for VMware vSphere and Cloud Computing*, CreateSpace Independent Publishing Platform.

Sweeney, J. (2011) *Get Your Head in the Cloud: Unraveling the Mystery for Public Sector*, http://www.amazon.com/Get-Your-Head-Cloud-Unlocking/dp/110564720X/ref=sr_1_3?ie=UTF8&qid=1396798304&sr=8-3&keywords=Get+Your+Head+in+the+Cloud (accessed November 29, 2015).

9

Cloud Network and I/O Virtualization

Kapil Bakshi and Craig Hill

Cisco Systems Inc., USA

9.1 Introduction

An essential characteristic of a cloud environment is its ability to offer elastic, on-demand and measured access to pooled resources. Virtualization is a key trait to achieve these cloud characteristics, whereby one abstracts runtime from the underlying infrastructure. This is especially true for cloud networks and computer input output (I/O) systems. This chapter focuses on the virtualization techniques for the data-in-motion aspect of cloud networks and I/O subsystems. It discusses these techniques and the benefits, applications, and use cases of cloud environments. The chapter addresses key virtualization techniques for network infrastructure and for the connection of virtually pooled resources in multiple clouds. It also focuses on computer I/O subsystem virtualization techniques, which connect virtual networks to virtual machines in a typical cloud environment.

The network virtualization part is discussed in several related contexts. This chapter assumes a basic understanding of Open Systems Interconnection (OSI) network layers and leverages the layer constructs. First, the traditional concepts of Open Systems Interconnection (OSI) layer 2 virtualization, like virtual local area network (VLAN), are presented. Following this section, network function virtualization (NFV) is discussed. This is the virtualization of networking functions on industry-standard high-volume servers and hypervisors. This is a rapidly emerging form of network virtualization. Next, the chapter discusses, the notion of network device partition for cloud resource pooling, where the physical network devices can be logically partitioned, for delivering pooled cloud network services. This is a common leverage technique in cloud environments. As network traffic patterns for cloud workloads evolve, the cloud datacenter networks need to be revisited for these new requirements. The chapter starts to focus on the network techniques to connect

Encyclopedia of Cloud Computing, First Edition. Edited by San Murugesan and Irena Bojanova.
© 2016 John Wiley & Sons, Ltd. Published 2016 by John Wiley & Sons, Ltd.

clouds for key use cases like virtual machine mobility, including cloud datacenter interconnect and Internet protocol (IP) localization. The second major section of the chapter focuses on the computer I/O subsystem virtualization techniques. This includes details of network interface virtualization, virtualizing, and pooling of common storage protocols like fiber channel (FC) on Ethernet (FCoE), followed by a review of the virtualization of I/O subsystems for virtual and physical workloads to access them in a pooled fashion. These are a sample of varied I/O virtualization techniques deployed in a cloud environment. All virtualization techniques in this chapter are reviewed with benefits and use cases in mind.

9.2   Network Virtualization

9.2.1   Network Segmentation

Network segmentation is the most fundamental and commonly used virtualized technique in networks. There are several benefits of network segmentation, as it pools network resources and simplifies network operations by extending and maintaining the virtualization performed in the devices (both physical and virtual) and between these devices, securely and over a common physical network infrastructure. Network segmentation references two key components in networks, which map to the Open Systems Interconnection (OSI) network layers:

- segmentation in the device (router, switch);
- techniques to maintain the segmentation as the virtualized content is moved over the network for local area networks (LAN), wide area networks (WAN), and cloud datacenters (DC).

Network segmentation techniques are normally categorized in OSI layer 2 (Ethernet layer), or OSI layer 3 (IP layer). The virtual local area network (VLAN) is the most common technology within the networking industry used for network segmentation at layer 2.

The VLANs target virtualizing the "bridge" function in a layer 2 device (i.e., network switch), which includes Ethernet MAC addresses, the port that the MAC address is associated with, as well as a VLAN trunk for communication between switching devices. Each VLAN is identified over the VLAN trunk via a "tag." This notion follows the IEEE 802.1Q standards (Donahue, 2011), which identify the VLAN association of a particular MAC address when the traffic is sent over the trunk. Virtual local area networks have been a *de facto* standard for more than 20 years in both campuses and datacenters for segmenting user groups within a layer 2 switch. Virtual local area networks have become even more critical as the cloud world has evolved; as virtual machines (VM) are created, in most cases, a VLAN and IP address is also associated with that VM, offering IP connectivity and VLANs for virtual segmentation.

As VLANs have evolved into a vital component in cloud networks, limitations for VLANs are beginning to arise, prompting efforts at innovation. The two key areas in which VLANs are exhausting their capabilities are scale (number of VLANs) and the complexity that basic bridge networks introduce. To address this complexity and these limitations, a key innovation is being adopted in the cloud network environments, namely, virtual extensible LAN (VxLAN) (Mahalingam *et al.*, 2012). This is one of the leading protocols, which has emerged to overcome VLAN limitations. VxLAN is an emerging standards-based technique, which is clearly targeting the cloud computing designs that leverage VLAN-like capabilities; however, it offers much larger VLAN scale. By expanding the VLAN tag information from 4000 to 16 million (tag in VxLAN is referred to as a Segment ID), VxLAN offers a new level of scale in segmentation allocation.

9.2.2 Use Cases for Network Segmentation

There are several use cases for network segmentation, as it is commonly leveraged in cloud and enterprise networks (see Figure 9.1 for an example). Network segmentation provides a means to group several network end points logically, with a common set of appointments. For example, you can give all member of a project access to their respective project files, by virtue of segment membership. Hence, one can also leverage in company departmental classification, which can be represented in a cloud context.

Companies are typically subject to several regulatory requirements, which oblige them to segment their data, organization, and processes. Network segmentation can help achieve compliance with such regulations by restricting access to application services. Hence data flowing between finance and outside departments could be regulated and protected by firewalls if desired.

Another potential use case for network segmentation could be to isolate malware-infected workloads from approved workloads. One can implement a scheme where, after a check on a cloud workload, it can be placed into a functional segment as opposed to a quarantined network segment.

9.2.3 Network Function Virtualization

Network function virtualization (NFV), illustrated in Figure 9.2, is the virtualization of functions onto consolidated industry-standard servers and hypervisors, which could be located in cloud datacenters, network nodes, and in the end-user premises (Chiosi *et al.,* 2012). Virtualizing network functions could potentially offer many benefits and use cases including:

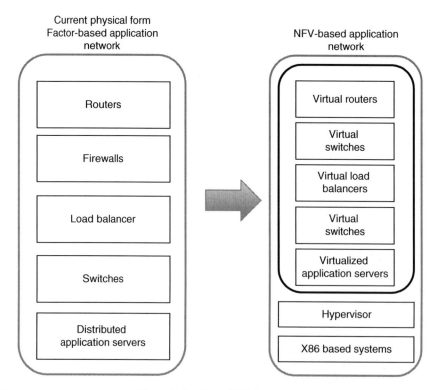

Figure 9.1   *Virtual LAN use case*

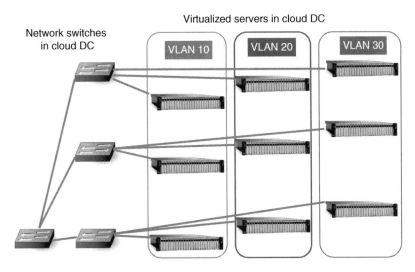

Figure 9.2   *Network function virtualization*

- consolidation for reduced equipment costs and power consumption;
- agility for increased speed of service and market creation;
- enabling multitenancy, which allows use of a single platform for different applications, users, and tenants.

This section covers implementation of the NFV instance, namely virtualized switches, routers, and virtual firewall as part of layer 4 through layer 7 network services (Figure 9.1).

With virtual networking layer 2 and layer 3 devices, virtual machines in a cloud environment can be connected in the same manner as physical machines (Gray and Nadeau, 2013), allowing the building of complex virtual networks for cloud virtual connectivity. Virtual layer 2 switches allow virtual machines on the same server host to communicate with each other using the same protocols that would be used over physical switches without the need for additional networking hardware. Server virtual switches support common capabilities like VLANs that are compatible with standard VLAN implementations. A virtual machine can be configured with one or more virtual Ethernet adapters, each of which has its own IP address and MAC address. As a result, virtual machines have the same properties as physical machines from a networking standpoint.

A virtual switch works in much the same way as a modern Ethernet switch. It maintains a MAC:port forwarding table and performs the functions of frame MAC lookup and forwarding to ports. A virtual layer 2 switch resides either in the hypervisor as a kernel module or in use space. It provides capabilities, like the layer 2 forwarding engine, VLAN tagging and manipulation, layer 2 security, and segmentation.

Similarly, as shown in Figure 9.3, virtual routers provide layer 3 functions and capabilities for cloud use cases.

A virtual router runs typical router operating systems and hosted capabilities like virtual private network (VPN); routing protocols; network address translation (NAT); dynamic host configuration protocol (DHCP); ACLs, and authentication, authorization, and accounting functions.

Just as network layer 2 and layer 3 functions can be virtualized, similarly, several layer 4 through layer 7 functions, such as virtual firewall, virtual load balancer, virtual application accelerator, can also be virtualized. Let us explore some of these virtual network functions for virtual firewalls.

A virtual firewall (VF) is a firewall service running entirely within a virtualized form factor on a hypervisor, providing the usual packet filtering and monitoring that a physical firewall provides. The virtual firewall

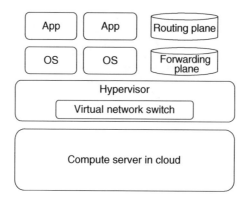

Figure 9.3   *Virtual routing function*

can be implemented in several ways; it can be a software firewall on a guest VM, a purpose-built virtual security appliance, or a virtual switch with additional security capabilities. A virtual firewall can be used for a common layer 3 gateway firewall or a transparent layer 2 firewall.

9.2.4   Use Cases for Network Function Virtualization

This creates interesting use cases for deployment of a virtual router in cloud. The software router can be deployed by an enterprise or a cloud provider as a virtual machine (VM) in a provider-hosted cloud. A virtual router can be deployed as a single-tenant router in virtual form for WAN gateway functionality for multitenant provider-hosted clouds. The virtual router serves primarily as a router per tenant, where each tenant gets its own routing instance; hence, its own VPN connections, firewall policies, QoS rules, access control, and so on.

A common use case for a virtual router is a virtual private network (VPN) service to the tenant edge in the cloud, which enables an enterprise to connect distributed sites directly to its cloud deployment. A virtual router can serve as a multiprotocol label-switching (MPLS) router enabling managed connectivity from an enterprise site to the cloud deployment. There are many other use cases for a software virtual router in the cloud.

A virtual firewall addresses use cases like the invisibility of packets moving inside a hypervisor, between virtual machine (VM-to-VM) traffic. The virtual network traffic may never leave the physical host hardware, therefore security administrators cannot observe VM-to-VM traffic. The virtual firewall also provides multitenancy function in a cloud and can also provide functions for intertenant firewalls for application segregation.

9.2.5   Network Device Partitioning

Following the notion of hypervisor and virtualization of x86 hardware, the networking industry has taken a similar approach in virtualizing the physical aspects of a router and/or a switch. This notion provides the ability to run feature-rich routing and switching software virtually with multiple instances on the physical network device natively.

In the earlier section, we discussed how VLANs and VRFs could be used to virtualize layer 2 and layer 3 tables in routers and switches. The ability to virtualize the entire switch or router also exists and has proven quite useful in operational networks today.

As shown in Figure 9.4, some network devices offer virtual device context (VDC), where the operator can partition the entire operating system into multiple virtual OSs, giving the perception of multiple logical devices within a single physical hardware.

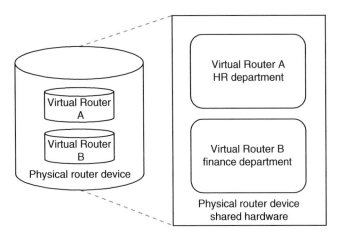

Figure 9.4   *Network device partitioning*

9.2.6   Use Cases for Network Device Partitioning

The network partition approach for virtualization has proven useful in operational environments that need the control and features of the switch. The key benefit is that the capability exists to logically partition the network device, while paying for a single physical device. The cost savings are obviously a key factor here, as is the reduction in power, cooling, and physical hardware maintenance. This capability highlights true capital expenditure (CAPEX) savings for the end customers.

The use of network partition opens up a number of use cases that can provide added benefits:

* offering a secure network partition between different user departments traffic;
* provides empowered departments the ability to administer and maintain their own configurations;
* provide a device context for testing new configuration or connectivity options without impacting production systems;
* consolidation of multiple departments switch platforms into a single physical platform while still offering independence from the OS, administration and traffic perspective;
* use of a device context for network administrator and operator training purposes.

9.2.7   Wide Area Networks and Cloud Datacenter Interconnect Virtualization

The challenge in cloud datacenter network design is to incorporate scale, high availability, as well as a flexible solution, to expand resources outside of the physical location. Cloud datacenter interconnect (DCI), offers the ability to transparently extend connectivity between multiple cloud datacenters, offering the same look and feel as if there was a cloud single datacenter, while gaining the benefits that multiple cloud datacenters offer, such as:

* high availability;
* increased uptime;
* security;
* the ability to expand beyond the physical building while appearing as a single datacenter to the applications, as well as users accessing those applications and/or data.

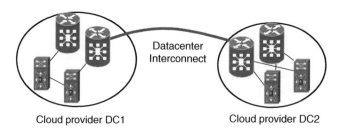

Figure 9.5   *Sample cloud DCI scenario*

The choices for DCI technologies are typically made based on the existing infrastructure, or transport options available between the cloud datacenters – specifically, optical, dark fiber, or Ethernet, multiprotocol label-switching (MPLS), as well as the option of IP. These choices are dictated by the service provider offering the transport between two or more cloud datacenters.

Let us consider one technology as an example of DCI. In an MPLS network, data packets are assigned labels and packet-forwarding decisions are made solely on the contents of these labels, without the need to examine the packet itself. This allows one to create end-to-end circuits across any type of transport medium, using any protocol (Rosen *et al.*, 2001). Looking at MPLS offerings for layer 2 extensions for a cloud datacenter, MPLS allows for point-to-point Ethernet "psuedowire" extensions (also akin to Ethernet over MPLS), which is ideal when three or more cloud datacenters target full interconnection among them. Ethernet over MPLS is a very popular and widely deployed technology, used for DCI.

As shown in Figure 9.5, there are also several deployed techniques for IP transport options for DCI, where layer 2 MPLS and other techniques can be tunneled over IP.

9.2.8   Cloud Datacenter Interconnect Use Cases

Cloud DCI technologies have generally targeted several use cases accommodating server communications, requiring extending VLAN layer 2 domains between cloud datacenters, offering those applications the functionality of being co-located, while still located on the same VLAN.

This increases the ease of workload and virtual machine mobility, and offers the server operator (and hypervisor vendors) the ability to move active VMs between geographically disbursed datacenters or to relocate applications from an extremely busy server to a server that is less utilized (but which is located in a remote datacenter). This is one of several use cases, and the DCI technologies allow this functionality, either in a "hot" (active VM) or "cold" (VM/workloads in standby) mode.

9.2.9   Network Localization for Virtual Machine Mobility

Virtual workload mobility is quickly becoming an important use case to load balance virtual resources across cloud datacenters, not only to provide high availability of access to the cloud applications and data, but leverage computing resources that are used less, regardless of whether the resources are geographically dispersed.

The challenge with leveraging virtual workload mobility is making the consumers of the applications (e.g. the end user) aware that the host (physical or logical) has moved. For example, consider the case where a user in Chicago is accessing a cloud application that is located in Washington DC. Because of a failure of the VM hosting the application, the cloud datacenter operator relocates the application to a less utilized datacenter in San Francisco. Other than a brief disruption of service, the end user is not aware of this move. The traffic pattern for the user in Chicago must traverse to Washington DC, then across the country to San Francisco, creating a "hair-pin" traffic pattern that is inefficient, and adds significant delay and response time for the end user of the application. This example highlights the primary challenge with virtual host mobility integrated with efficient routing.

IP localization methods, specifically locator ID separation protocol (LISP), aim at solving this routing inefficiency by efficiently populating the IP routing network with the optimum path (Chiappa, 2014). Prior to LISP, these methods were not trivial, and other than injecting route into the customer's IP backbone, which can be very inefficient and does not scale if a large number of host routes exist within the enterprise, the solutions to solve this were minimal. LISP addresses this key requirement, which is becoming increasing common in cloud environments. LISP is based on Request for Comments (RFC) 6830, and offers a host IP and location IP address split, offering a flexible way for enterprise backbone networks to recognize host (physical or logical) mobility, and detect the move very rapidly (Farinacci *et al.*, 2013).

The key benefit reference to building cloud computing networks with LISP is the ability for LISP to handle the mobility aspect if a workload is relocated from one datacenter to another. The host mobility use case is one of several target use cases that LISP addresses and is the optimum routing method for host mobility in the industry today, and, because of this, is ideal when building cloud computing environments for stations, hosts, or mobile devices accessing the applications in the datacenter.

9.3   I/O Virtualization

Input/output virtualization (I/O virtualization or IOV) is a mechanism to abstract the upper layer network and I/O protocols from the underlying physical transport. This is typically discussed in the context of server virtualization and adjacent switches. The growing trend of server virtualization and its use in the cloud puts a strain on server I/O capacity. As a result, network traffic, storage traffic, and interprocess communications traffic cause bottlenecks on the I/O subsystems of a compute node.

Input/output virtualization addresses performance bottlenecks by consolidating I/O to a single connection whose bandwidth ideally exceeds the I/O capacity of the server itself, thereby ensuring that the I/O link itself is not a bottleneck. That bandwidth is then dynamically allocated in real time across multiple virtual connections to both storage and network resources, thereby increasing both VM performance and the potential number of VMs per server. The following section discusses several I/O virtualization techniques, namely fiber channel over Ethernet (FCoE), converged network adapter (CNA), single-root I/O virtualization (SR-IOV), and multiroot I/O virtualization (MR-IOV). There are several other techniques for I/O virtualization; however, these are some of the more common techniques being employed.

9.3.1   Network Interface Virtualization

Network interface virtualization (Figure 9.6) takes a single physical adapter and presents multiple virtual adapters as if they were physical adapters to the server and network. This technique enables leveraging virtual adapters to virtual machines as if they are network adapters. With interface virtualization you can also connect to a virtual switch in the hypervisor or directly to the virtual machines. Each interface creates a virtual cable between the network and virtual workloads.

9.3.2   Fiber Channel over Ethernet

Fiber channel over Ethernet (FCoE) is a storage protocol that runs the fiber channel (FC) directly over Ethernet. FCoE makes it possible to communicate fiber channel traffic across existing 10 Gb Ethernet-switching infrastructure and converges storage and IP protocols onto a single cable transport and interface. Typically, Ethernet-based networks are for TCP/IP networks and fiber channel networks are for storage networks. Fiber

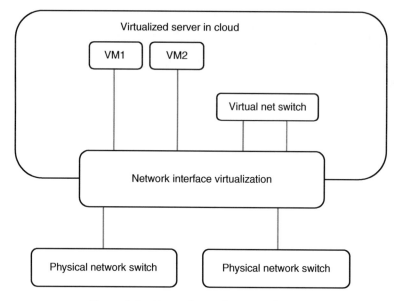

Figure 9.6   *Network interface virtualization*

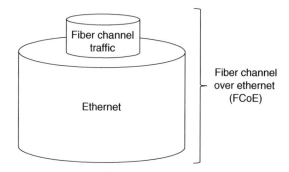

Figure 9.7   *Encapsulation of FC traffic in Ethernet*

channel supports high-speed (1/2/8/16 GB) data connections between computing systems and storage devices. FCoE shares fiber channel and Ethernet traffic on the same physical cable or lets organizations separate fiber channel and Ethernet traffic on the same hardware (Figure 9.7). The goal of FCoE is to consolidate I/O and reduce cabling and switching complexity and reduce interface card counts in a server.

Fiber channel over Ethernet uses a lossless Ethernet fabric and its own frame-encapsulated format. It retains fiber channel's character tics and leveraged Ethernet links for fiber channel for communications. It works with standard Ethernet cards, cables, and switches to handle fiber channel traffic at the data link layer, using Ethernet frames to encapsulate, route, and transport FC frames across an Ethernet network from one switch with fiber channel ports and attached devices to another, similarly equipped switch.

9.3.3   Converged Network Adapter

A converged network adapter (CNA) is a network-interface card (NIC) that contains a fiber channel (FC), host bus adapter (HBA), and an Ethernet NIC. It connects servers to FC-based storage area

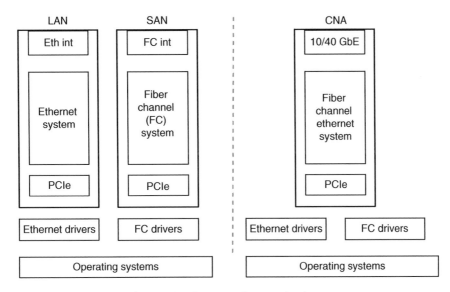

Figure 9.8   *Converged network adapter*

networks and Ethernet-based networks. In networks without CNAs, servers have to have at least two adapters, namely FC HBA to connect the server to the storage network, and Ethernet NIC to connect the server to the LAN.

As shown in Figure 9.8, a single CNA to connect servers to storage and networks reduces costs by requiring fewer adapter cards, cables, switch ports, and peripheral component interconnect express (PCIe) slots. Converged network adaptors also reduce the complexity of administration because there is only one connection and cable to manage. A CNA connects to the server via a PCIe to Ethernet and FC networks. The server sends both FC SAN and LAN and traffic to an Ethernet port on a converged FCoE switch using the fiber channel over Ethernet (FCoE) protocol for the FC SAN data and the Ethernet protocol for LAN data. The converged switch can de-encapsulate FC from FCoE and the Ethernet traffic is then sent directly to the LAN.

9.3.4   Single-Root I/O Virtualization

The PCI Special Interest Group (PCI-SIG) has developed several I/O virtualization specifications to help standardize I/O virtualization implementations, and single-root I/O virtualization (SR-IOV) is one of the specifications. As shown in Figure 9.9, SR-IOV is a specification that allows a PCIe device to appear to be multiple separate physical PCIe devices (Intel Corp, 2011). The single-root (SR-IOV) standard specifies how multiple guests, physical or virtual machines (VM) on a single server with a single PCIe controller or root can share I/O devices without requiring a hypervisor on the main data path. SR-IOV works by introducing the idea of physical functions (PFs) and virtual functions (VFs). Physical functions are full-featured PCIe functions; virtual functions are smaller functions that lack configuration resources.

Support is needed for SR-IOV in the BIOS as well as in the operating system instance or hypervisor that is running on the hardware. The hypervisor must work in conjunction with the PF drivers and the VF drivers to provide the necessary functionality, such as presenting PCI configuration space to VMs and allowing a VM to perform PCI resets on a VF.

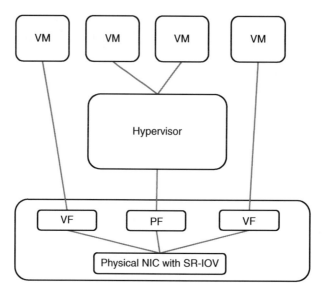

Figure 9.9 *Single root virtualization for multiple virtual hosts*

9.3.5 Multiroot I/O Virtualization

Multiple root I/O virtualization (MR-IOV) extends the concept to allow multiple independent systems with separate PCIe roots to connect to I/O devices through a switch. Both SR-IOV and MR-IOV require specific support from I/O cards. PCI-SIG is standardizing the mechanisms that enable PCIe devices to be directly shared, with no runtime overheads. Multiroot IOV provides a means to share directly between operating systems in a multiple systems environment.

Hence, MR-IOV provides methods for exporting and managing multiple views of the same device in a PCIe fabric that can be connected to more than one root port. From each root port's perspective, the fabric appears to be a traditional PCIe-based fabric and there is no change to the root complex. In addition, there is no change to non-multiroot (MR) endpoints. However, there need to be changes to MR-aware switches and endpoints to support multiple virtual hierarchies.

A common example for MR-IOV implementation is a blade server with a PCIe "backplane," where a new PCIe hierarchy is required. Essentially, a fabric is constructed which logically partitions the PCIe fabric into multiple virtual hierarchies (VHs) all sharing the same physical hierarchy.

9.3.6 I/O Virtualization Use Cases

The I/O virtualization techniques described in the above sections can be implemented in several cloud infrastructure virtualization use cases. These use cases provide benefits ranging from cost-effectiveness, to reducing complexity in cloud infrastructure implementation, to streamlining operational elements of a cloud network and reducing power and cooling requirements. Some sample use cases for I/O virtualization are:

- Input/output consolidation for discrete servers: the converged network adapters (CNA) in discrete severs in rack mount servers, consolidate access Ethernet and fiber channel switches into a fiber channel over Ethernet (FCoE) switches (Gai and DeSanti, 2009). This consolidation not only reduces the infrastructure to manage, power, cool and implement, but also reduces cost of only maintaining a single access network.

- Blade computing efficiencies. A blade server is an enclosure that provides power, cooling and interconnectivity to a set of server blades. The I/O virtualization techniques bring several efficiencies, reducing the number of cables, adapters, access switches and power requirements.
- Input/output assembly efficiencies. Single and multiroot I/O virtualization techniques allow the PCIe device to present multiple instances of itself up to the operating systems instance or hypervisor. This reduces cost, cabling, infrastructure, and operational processes.

9.4   Conclusion: End-to-End Network and I/O Virtualization Design

In conclusion, network virtualization can be seen in traditional areas of network segmentation, network services, and intercloud connectivity. The I/O virtualization is focused on the interface between the compute and I/O subsystems. These focus on the abstraction of protocols, networking functions, network devices, and I/O access. Let us review a select few future considerations on the topic of network and I/O virtualization. First, network and I/O virtualization can be adopted pervasively for cloud networks, where the actual integration of virtualized environment is conducted by orchestration and programmatic functions. This is a key point to note as cloud networks evolve to accommodate scale, complexity, and cost functions via agile provisioning of cloud services. Second, the key focus in future of virtualization techniques in cloud will to enable cloud software applications for agile, easy and cost-effective deployment and operations. Hence, any application-centric cloud infrastructure should be a key consideration in building any cloud network and I/O systems. The network and I/O virtualization function, coupled with orchestration and application enablement, are the key future traits for cloud networks.

Abbreviations

Cloud datacenters (CDC)
Converged network adapter (CNA)
Datacenter interconnect (DCI)
Fiber channel over Ethernet (FCoE)
Forwarding plane (FP)
Input output (I/O)
Institute of Electrical and Electronics Engineers (IEEE)
Internet protocol(IP)
Local-area network (LAN)
Media access control (MAC)
Metro-area network (MAN)
Multiprotocol label switching (MPLS)
Multiroot I/O virtualization (MR-IOV).
Network address translation (NAT)
Network functions virtualization (NFV)
Open systems interconnection (OSI)
Peripheral component interconnect express (PCIe)
Physical functions (PFs)
Quality of service (QoS)
Routing plane (RP)

Single-root I/O virtualization (SR-IOV)
Virtual extensible LAN (VxLAN)
Virtual functions (VFs)
Virtual local area network (VLAN)
Virtual machines (VM)
Virtual private network (VPN)
Virtual route forwarding (VRF)
Wide area network (WAN)

References

Chiappa, J. (2014) *An Architectural Introduction to the LISP (Location-Identity Separation System)*, http://tools.ietf.org/html/draft-ietf-lisp-introduction-03 (accessed November 30, 2015).

Chiosi, M., Clarke, D., Feger, J., *et al.* (2012) *Network Functions Virtualisation: An Introduction, Benefits, Enablers, Challenges and Call for Action.* SDN and OpenFlow World Congress, Darmstadt-Germany, https://portal.etsi.org/NFV/NFV_White_Paper.pdf (accessed November 30, 2015).

Donahue, G. (2011) *Network Warrior*, O'Reilly Media, Inc., Sebastopol, CA.

Farinacci, D., Fuller, V., Meyer, D., and Lewis, D. (2013) *The Locator/ID Separation Protocol (LISP)*. IETF Draft. http://tools.ietf.org/html/rfc6830 (accessed November 30, 2015).

Gai, S. and DeSanti, C. (2009) *Data Center Networks and Fibre Channel over Ethernet (FCoE)*, Cisco Press, Indianapolis, IN.

Gray, K. and Nadeau, T. (2013) *SDN: Software Defined Networks*, O'Reilly Media, Inc., Sebastopol, CA.

Intel Corp. (2011) *PCI-SIG SR-IOV Primer*, http://www.intel.com/content/dam/doc/application-note/pci-sig-sr-iov-primer-sr-iov-technology-paper.pdf (accessed November 30, 2015).

Mahalingam, M., Dutt, D., Duda, K., *et al.* (2012) *VXLAN: A Framework for Overlaying Virtualized Layer 2 Networks over Layer 3 Networks*, http://tools.ietf.org/html/draft-mahalingam-dutt-dcops-vxlan-00 (accessed November 30, 2015).

Rosen, E., Viswanathan, A., and Callon, R. (2001) RFC 3031. Multiprotocol Label Switching Architecture, http://datatracker.ietf.org/doc/rfc3031/ (accessed November 30, 2015).

10

Cloud Networks

Saurav Kanti Chandra and Krishnananda Shenoy

Infosys Limited, India

10.1 Introduction

Cloud networks are intended to provide network services in a similar fashion to the way in which compute and storage services are provided in a cloud-computing environment. The primary purpose of cloud networks is to interconnect the physical and virtualized infrastructure resources, connect them to Internet, and expose the network as a service. The usage of cloud networks in a popular commercial model of cloud computing called infrastructure as a service (IaaS) can be taken as an example to underline the importance of cloud networks. Infrastructure as a service provides on-demand infrastructure consisting of storage, servers, and network over the Internet. Subscribers configure a network and manage an infrastructure in order to run applications. Amazon's EC2 is an example of such a service. Cloud networks play a pivotal role in IaaS by performing following tasks:

- They connect storage and virtual servers optimally.
- They help in creating network isolation or separate network segments for applications.
- They allow the configuration of network security with firewall, authentication, and authorization.
- They help in load balancing amongst the application server instances.
- They help in steering the traffic as desired.
- They allow the creation of secure connectivity with the enterprise's on-premises network. They let the data traffic from users' on-premises networks travel to the cloud infrastructure in a secure way over the Internet. Typically, a virtual private network (VPN) is used for this purpose. The VPN uses security protocols and encryption to secure private data communication over the Internet.

Encyclopedia of Cloud Computing, First Edition. Edited by San Murugesan and Irena Bojanova.
© 2016 John Wiley & Sons, Ltd. Published 2016 by John Wiley & Sons, Ltd.

- They are extendable to keep the scaled-out virtual instances connected.
- The network configuration is exposed as a service in the form of application programming interfaces (APIs) to the user.

10.2   Characteristics of Cloud Networks

Key characteristics of cloud networks are elasticity, autonomic networking, geodistribution and high availability, and programming interface.

10.2.1   Elasticity

Elasticity of cloud networks can be defined as the property by which a network is grown or shrunk seamlessly in response to demand from users. Shrinking or growing a network helps extend the network to connect additional virtual resources and to resize the network after virtual resources are decommissioned. Shrinking and growing are done in compliance with the network security requirements and quality of service requirements of traffic. In a nutshell, elasticity of cloud networks is the ability to reconfigure network as and when cloud services require it. Virtualization enables elasticity in cloud infrastructure because the virtual instances can be instantiated and decommissioned faster than the physical configurations. Network is also virtualized to enable the network elasticity.

10.2.2   Autonomic Networking

Autonomic networking is the ability of cloud networks to manage themselves by responding to events happening in the network and correcting themselves. In a cloud infrastructure, a network grows and becomes complex over time. The management of the network also becomes complex. Due to the lack of scalable management capability, the growth of the network stops beyond a particular size. Autonomic networking capability keeps the network flexible to grow to a greater scale. Autonomic networking is executed by monitor and control functions. Monitoring is done proactively. Discovery, awareness, and analysis is carried out perpetually to diagnose the state and events of a network. It creates the knowledge base to validate the events and respond to that. Response is typically a control action regulated by policies. The typical control actions are a reconfiguration of network resources, authorization, and authentication.

10.2.3   Geodistribution and High Availability

Cloud infrastructure is distributed across locations. This is needed for various reasons, such as compliance with local laws and regulations, disaster recovery, backup, and for having the infrastructure permanently available. Cloud networks support the geodistribution of infrastructure. They support the movement of virtual instances from one location to another location. Cloud networks extend the network domain to multiple sites by network tunnels. Applications residing on virtual machines, located in different locations, use the extended network domain for faster communication to each other.

10.2.4   Programming Interface

The "*x* as a service" model of cloud computing uses application programming interfaces (APIs) extensively. Cloud networks provide APIs for configuration and controllability. Cloud network APIs are used to configure network interface, access control lists, routing, virtual private network, network firewalls, IP address design, and so forth. In the case of infrastructure as a service, the cloud networks' APIs are given

Table 10.1   *Cloud networks for different cloud computing hosting models*

Type of cloud computing	Features of cloud networks
Public cloud	Internet connectivity of the infrastructure
	Network isolation across users
Private cloud	Interconnection among resources
	Secure connectivity over internet to enterprise's network
Hybrid cloud	Internet connectivity of the infrastructure
	Interconnection among resources

to the users of the cloud service. In platform as a service, the cloud networks APIs are given to users on need basis. In software as a service, the cloud networks' APIs are not used directly by the user. Cloud service providers use the cloud networks' APIs internally to manage the cloud infrastructure.

Consumption of the APIs triggers an automated network configuration within the cloud networks. The configuration requests are often mediated by an orchestration layer. The orchestration layer configures the virtual and physical network elements by using management and control protocols. In some cases it uses network operating system APIs, where the network elements provide a programmability option from the network operating system. It can also use the granular controls provided by the controller of software defined networking (SDN).

10.3   Types of Cloud Networks

The cloud networks depend on the cloud type, size and service models, as outlined below.

10.3.1   Public, Private, and Hybrid Clouds

The cloud networks vary depending on the type of the cloud. Table 10.1 provides characteristics of cloud networks in different types of cloud computing infrastructure:

10.3.2   Size Based – Small, Medium, and Large

The cloud networks depend on the size of the cloud infrastructure. A small-sized private cloud by a small enterprise has a simple network configuration. A large and distributed cloud infrastructure has a complex network requirement for securely connecting hundreds of physical and virtual resources.

10.3.3   Service Based – SaaS, PaaS and IaaS

Table 10.2 outlines the features of cloud networks for different commercial service models.

10.4   Architecture of Cloud Networks

10.4.1   Overview

Cloud networks are built using network elements such as switches and routers. Network elements vary in capacity and features. Network elements are physically connected by copper or optical cables, which carry the data traffic. The arrangement of the network elements and the connections among them are designed by following standard architectural practices. The standard architectural practices are required to meet the objectives of cloud networks. Table 10.3 provides a brief overview of the objectives.

Table 10.2 *Cloud networks for different cloud computing service models*

Type of service	Features of cloud networks
Software as a service (SaaS)	Connects Internet to the virtual servers hosting SaaS applications Takes care of network security, load balancing, and quality of service (QoS) of the offered SaaS
Platform as a service (PaaS)	Connects the resources of the platform such as the virtual machines, storage, and network appliances like firewall. Allows flexible network connectivity so that the user can design that as per the application requirement
Infrastructure as a service (IaaS)	Connects the storage and virtual servers of cloud infrastructure and makes sure that the connectivity is configured among the chosen utility blocks in the desired way Makes sure the network isolation and security are maintained as required by the user Provides secure connectivity with the enterprise's on-premises network

Table 10.3 *Cloud networks' architecture overview*

Objective	Overview
Connecting the cloud infrastructure with the Internet	The **hierarchical architecture** with multiple layers helps to connect resources to the Internet. This architecture caters to the north-south data traffic.
Interconnecting the infrastructure resources	The **interconnection** architecture helps in fast connection of resources and in convergence when needed. This architecture caters to the east-west traffic.
Being elastic and dynamically configurable so that it can be leveraged in similar fashion to cloud computing infrastructure	**Virtualization** helps in achieving this objective by creating a logical network that is flexible and easily configurable
Keeping the infrastructure secured and keeping data securely isolated across users	**Security** aspects in cloud networks architecture take care of this objective
Managing the cloud networks and creating service APIs	**Management and orchestration** features of cloud networks architecture help in meeting the set of goals laid out by this objective

10.4.2 Hierarchical Architecture

Cloud networks architecture is hierarchical. There are multiple tiers in the hierarchy. Traditional three-tier architecture has core, aggregation and access layers as in Figure 10.1. Cloud networks are becoming flatter due to convergence happening in the cloud infrastructure. Convergence simplifies the cloud networks and optimizes the performance. The traditional three-tier architecture is made flatter by collapsing three tiers into two tiers. This is achieved by aggregating the access layer into high density aggregation switches. Two-tier architecture has only core layer and aggregation layer as in Figure 10.2. As a result of the convergence, design and management have become simpler. Network latency is also reduced because of reducing switching hops.

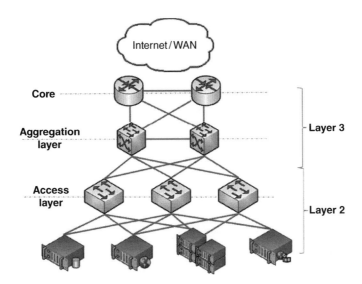

Figure 10.1   *Three-tier architecture of cloud networks. (Separate access and aggregation layers.)*

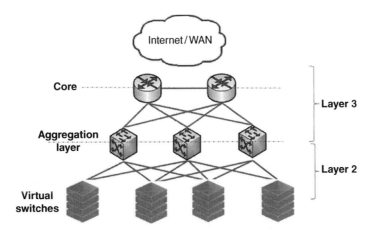

Figure 10.2   *Two-tier architecture of cloud networks. (Access layer is collapsed within high-density aggregation layer and virtual servers.)*

10.4.2.1   *Layer 2 and Layer 3 Networks*

The cloud networks' hierarchical architecture is based on layer 2 and layer 3 networks. The "layer 2" and "layer 3" refer to the data link and network layers respectively of the standard model of communication system called the Open Systems Interconnection (OSI) model as in Table 10.4.

A switch is a layer 2 network element. A network created by switches is referred as a layer 2 network. In a layer 2 network, network packets are sent to specific network interfaces based on the destination media access control (MAC) address. This network is typically used for high-speed connectivity among the infrastructure resources attached to it. A network with many infrastructure resources attached typically needs to be divided into multiple broadcast domains.

Table 10.4 *OSI model of layers of communication system*

Presentation (layer 6) – Formats data to be presented to application
Session (layer 5) – Handles session establishment between processes
Transport (layer 4) – Ensures that messages are delivered in sequence without any error, losses and duplication.
Network (layer 3) – Controls routing of packets which contains IP address
Data link (layer 2) – Provides transfer of data frames over the physical layer
Physical (layer 1) – Responsible for reception and transmission of bit stream over physical medium

Router and "layer 3 switch" are layer 3 network elements. A layer 3 network deals with the IP addresses configured on a network interface. The IP address for a network is designed to create multiple IP address subnets to reduce the broadcast domain. Routers help in forwarding traffic from one IP subnet to another IP subnet.

10.4.2.2 Core Network Layer of Cloud Networks

The hierarchical architecture of cloud networks is like a tiered pyramid. In the topmost tier of the pyramid there are routers operating at layer 3 and connecting the local area network (LAN) to the Internet. This tier is called the core network. The core network is known for high processing power, high capacity, high resiliency, and high availability. The core network layer's primary job is speedy routing of high bandwidth of traffic. This layer uses many routing protocols. The purpose of a routing protocol is to learn available routes in a network, build routing tables, and make routing decisions. When a route becomes unavailable, the protocol helps in finding the best alternative route and updating the routing tables. This process is called routing convergence.

10.4.2.3 Aggregation Layer of Cloud Networks

The middle layer of three-tier architecture is called the aggregation layer. It is also known as the distribution layer. It aggregates multiple access switches in such a way that multiple connections are combined to have better throughput and redundancy. Primarily, switches with layer 2 and layer 3 capabilities are used in the aggregation layer. This layer is the aggregation point for many other network functions such as firewalls and load balancers. The primary objective of the aggregation layer is fast routing convergence and high availability. Routing convergence is achieved by the routing protocols. High availability is designed by keeping a standby switch where the fail-over can happen.

10.4.2.4 Access Layer of Cloud Networks

The access layer is the layer 2 network, which interconnects all the physical and virtual resources of cloud infrastructure such as compute servers, storage servers, and security servers. Each resource in a cloud infrastructure has a network interface where the interconnection happens. Different network interfaces support different data transfer rates and cable types. Transfer rate means the maximum rate at which data can be transferred across the interface. The cable types are copper cable and optical cable. Primarily layer 2 switches with high port density are used in this layer.

10.4.3   Interconnection

10.4.3.1   Storage Interconnect Structure

Storage is one of the most important resources of cloud infrastructure. This infrastructure is made of virtualized pool of storage servers interconnected with other resources of infrastructure. Data is organized, stored and accessed from disks by different storage mechanisms. Cloud networks are configured for latency and quality of service according to the typical needs of cloud storage such as backup and restore. They also provide a gateway service to enterprise networks to connect and use the cloud storage services.

10.4.3.2   Server Interconnect Architecture

Compute servers have network interface cards (NICs) supporting various speeds from 3 mbps to 100 gbps. Rack-mounted servers use the standard network interface. Blade servers use integrated network cards. Servers can have multiple physical network interface cards grouped into one virtual NIC. It helps in balancing the traffic load across the NICs. The aggregated bandwidth available from the team of NICs is used effectively in this manner. It increases the performance. Network interface card teaming also makes the server fault tolerant.

10.4.3.3   Interconnection Using Network Fabric Architecture

Network fabric is a very high bandwidth interconnect system that connects loosely coupled compute, storage and network resources. It is also known as unified fabric. There are two components in a fabric. They are nodes and links, as in Figure 10.3. Nodes are the servers and the links are fiber-optic connections with high-speed switches. Due to the meshlike structure of the fabric, multiple links are available between two points. The network fabric takes advantage of the aggregated bandwidth of the multiple links. Links are aggregated and link aggregation protocols are used to carry the packets without any duplication and without being out of sequence. It increases the performance of the network.

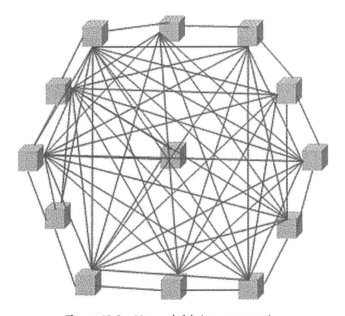

Figure 10.3   *Network fabric representation*

The switches in a fabric select the shortest path available out of the many links of a fabric to reach a destination and also forward the traffic to multiple paths. Network latency is reduced by shortest path and the heavy load of east-west traffic can be efficiently managed by multipath forwarding.

10.4.3.4   Interconnection in Modular Pod

Modular architecture allows a predefined set of datacenter resources to form a container, which can be added repeatedly. A pod is one such contained, discrete and modular unit of a datacenter. Pods contain a network layer along with compute, storage, and application components. Typically access layer switches are kept within a pod. With this modular architecture, the network interconnection happens typically in either of two models – top-of-the-rack (ToR) or end-of-the row (EoR).

In the top-of-the-rack model, one or two Ethernet switches are installed within the rack (not-necessarily at the top) and the servers in a rack connect to them, as in Figure 10.4.

In end-of-the-row model, server racks are arranged in a row and a chassis-based Ethernet switch is placed at one end of the row, as in Figure 10.5.

10.4.3.5   Software-Defined Networking

Software-defined networking (SDN) is a concept that decouples the control plane and data plane of a network (see Figure 10.6). It envisages a simplified networking where a centralized controller is able to make decisions about the traffic forwarding in a network. It introduces a communication mechanism between the control plane and data plane when they are decoupled. OpenFlow is an open-source implementation of this communication protocol.

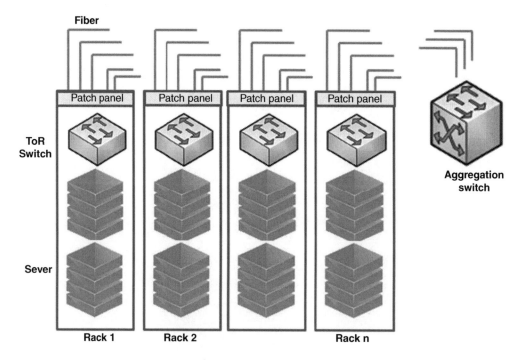

Figure 10.4   *ToR model*

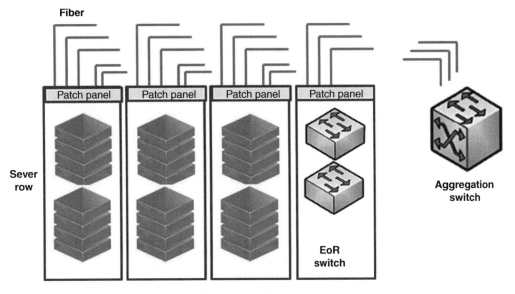

Figure 10.5 *EoR model*

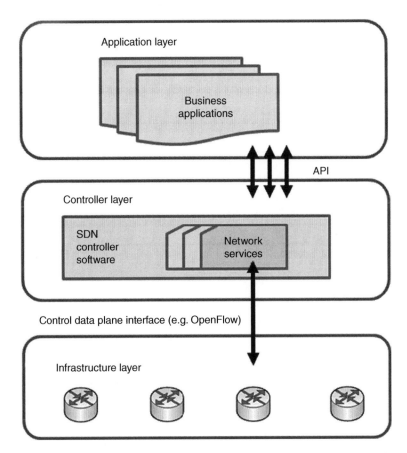

Figure 10.6 *Software-defined networking. Source: Open Networking Foundation*

An OpenFlow-capable switch maintains a table called a flow table. The flow table contains entries called flows. Packet forwarding happens according to the flow matching entries in the flow table. A software-based controller manipulates the flow table entries using the OpenFlow protocol to decide on how to forward the packets.

The controller opens up services on the northbound side to integrate with the rest of the system. As a result, the applications get the privilege of controlling the traffic.

The following points describe how software-defined networking helps cloud networks:

- it enables cloud networks to steer and route the application traffic dynamically across the cloud infrastructure using controller APIs;
- it simplifies the network operation and reduces the complexity of traditional network architecture;
- it helps network management systems in network automation;
- it helps network orchestration components to compose a service by consolidating with other resource management;
- it allows innovation in the field of cloud networks by exposing APIs for programmability;
- it enables network virtualization;
- it makes cloud networks flexible to scale out dynamically.

10.4.4   Virtualization

10.4.4.1   Network Virtualization

Network is virtualized to create a logical network on top of the physical network. Virtualization helps in many aspects. It helps in creating smaller logical networks for network isolation. It helps in making the network elastic. Elastic network can be scaled out or in dynamically. Virtualization helps in providing bandwidth on an on-demand basis. Virtualization also helps in moving virtual machines from one location to another by adjusting the logical configurations without making any changes in the physical network. The abstraction of the network is not possible without abstracting the network end points – the network interfaces. Hence NICs have been virtualized. The physical switches have also been virtualized. Hypervisors run a virtual switch and provide a virtual network interface card to the virtual machines created on hypervisor.

10.4.4.2   Overlay

An overlay network is a concept of creating an independent network on top of a different network. In cloud networks, overlay is important because it has been introduced with a promise to tunnel information from one datacenter to another datacenter in a direct and simplified way. Overlay connects the switches over the physical network as shown in Figure 10.7. The Ethernet fabric is the physical network infrastructure. The tunnel fabric is an overlay created on top of the physical layer. With help of overlay the cloud networks can scale up to millions of logical networks providing layer 2 adjacency across all of them.

10.4.4.3   Network Function Virtualization

Network function virtualization (NFV) is carried out by the software layers within the devices. It reduces the capital expenditure of cloud networks as the virtualized network functions can run on commodity servers.

10.4.5   Cloud Networks Security

Cloud computing infrastructure requires many security measures to minimize the vulnerability and risk exposures. Table 10.5 highlights some of the well known network security threats and countermeasures.

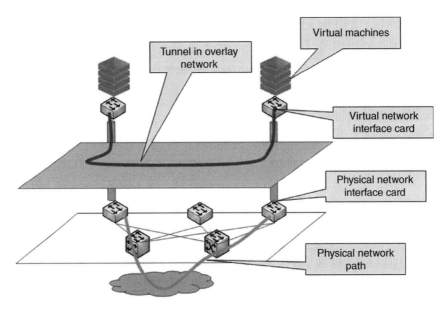

Figure 10.7 *Overlay network*

Table 10.5 *Threats to cloud networks*

Network threat	Countermeasure
Denial of Service (DoS) attack: in this attack an attempt is made to prevent a server to serve to its clients. The service is either crashed or flooded. There are various forms of DoS attack.	Access control list Filters used to prevent a DoS attack Firewalls Deep packet inspection system Intrusion prevention systems
Man-in-the-middle attack. The attacker sniffs a network, eavesdrops in a network communication and steals confidential information. This attack exploits the vulnerability of unencrypted or weakly encrypted packets.	Strong encryption and strong mutual authentication Authorization Authentication
Port scanning. The attacker sends requests to identify an active port of a server offering service and later uses that port to exploit a vulnerability of the service. This may not by itself be an attack but often leads to a network attack.	Packet-filtering firewalls and proxies are used to limit the request to go to only defined ports. This prevents the attempt to reach to a wide range of ports.
Cross-site scripting (XSS): The attacker injects malicious scripts into the client's computers through web applications. The script runs at the client computers and gain access to confidential information.	Security controls at the client side such as secured cookies, and script disablement.

10.4.6 Cloud Networks Management and Orchestration

The cloud network management system configures the cloud networks and monitors the network for performance and faults. It takes control action when a fault occurs or performance degrades. The management system provides a dashboard where the network operator can visualize the complete network. It generates alerts when a fault occurs. Network elements such as switches and routers provide information to the management system periodically. The management system also probes the network periodically to collect information. Management communication protocols are used to communicate the information. The primary challenge in managing cloud networks is its complex configuration. An operator needs to know how to configure a network configuration request on the physical or virtual network elements. The management system provides an easy and faster way to do the same without much effort. It provides graphical user interface and templates or wizards to do the configuration. Network management systems also provide scripting facilities where a customized configuration can be built and used on the network.

A cloud network management system is also used for configuration of authorization and authentication of network access. It involves communicating with the authentication-authorization-accounting (AAA) server for access policies.

The network management system has interfaces by which it can be integrated with other systems. Integration with trouble-ticketing systems is an example of such integration. A trouble-ticketing system creates a ticket when a fault occurs in the network. Support personnel attend to the ticket to resolve the fault. Integration with the order management system is another example. Service orders from users of a cloud infrastructure are sent using this integration mechanism to configure the network. Integration with the billing system helps in measuring the network service usage by user and charging the user for the same. Network elements maintain counters that measure usage, such as bandwidth. The usage information is collected by network management systems and then provided to the billing system. The billing system generates an invoice for users.

Advanced cloud network management systems self-manage the network. They collect information from all possible sources using the monitoring functionalities, and do this very frequently. A large volume of unstructured data is analyzed using big data techniques. "Big data" refers to the collection and analysis of really big and complex data sets. Patterns identified from the analysis are used to predict faults that may occur in the network. The network management system reconfigures the network to avoid such future faults.

There are associated systems with cloud network management systems that allow dynamic tuning of the network as per the application requirements. These associated systems are policy management systems, deep-packet inspection (DPI) systems, layer 4-7 switches, and load balancers. A DPI system inspects the packets and identifies which applications are running in the network. Layer 4–7 switches refer to the layers of the OSI model, as in Table 10.1, where these layers denote the application-specific headers of the packet. A DPI is also used to detect a security threat from the traffic pattern. A layer 4–7 switch is also used for identifying the applications. A policy management system allows policies to be set up for applications. So when an application is identified, the policies are triggered from the policy-management system. Policies reconfigure the network according to the needs of the application. The configuration parameters include quality of service requirements, dedicated bandwidth requirements, security requirements, and so forth. Load balancers are used to balance the traffic load across the host servers of the application. The layer 4-7 switch helps in load balancing as well.

Cloud orchestration means arranging multiple resources in a coordinated way so that the user gets an end-to-end environment with all resources needed. It helps the user to get all resources without much knowledge of the internal architecture of cloud infrastructure. It helps the provider to optimize the resource usage by controlling the provision requests from users. Cloud infrastructure hosts the orchestration layer as a software middleware. This layer can directly control and monitor all compute, storage and networking resources. It exposes REST APIs. Application programming interfaces are used to trigger orchestration across the infrastructure resources. The orchestration layer creates a service composition by coordinating all resources together. It takes automated network configuration into the composition. OpenStack is widely used as cloud infrastructure management software. It has a networking component called "Neutron" which manages the cloud networks, IP addresses, dynamic traffic routing, and so forth. It can automate network configuration. Orchestration uses this automated network configuration along with other automations and creates a workflow to provide an end-to-end coordination.

10.5 Conclusion

Cloud networks are in continuous flux in terms of introduction of novel concepts, emerging technologies, new solutions and new products. There is plethora of technology options available today and many of them are under ratification. However, it can be safely assumed that cloud networks have evolved to a position where, on the one hand, they are providing the network services to the cloud computing infrastructure and, on the other hand, they are emerging as a service model for business use. The evolution is made possible by technological evolution in networks such as virtualization, software-defined networking, next-generation network management capabilities, and networking protocols solving the challenges posed by the cloud computing infrastructure. One general trend being observed in cloud networks is that the network, which was primarily driven by hardware and ASIC designs so far, is now being inundated with software applications. This is promoting innovation as software provides the capability to program and creates value-added layer in networks. Researchers, standards bodies, universities, and business houses are investing in bringing in further advancement in cloud networks. The next-generation cloud-ready datacenter is going to see major traffic growth and agility in coming years. Cloud networks are maturing slowly to face the challenge of traffic growth and flexibility required to become nimble.

Additional Resources

Al-Fares, M., Loukissas, A., and Vahdat, A. (2008). A scalable, commodity data center network architecture. *ACM SIGCOMM Computer Communication Review* **38**(4), 63–74.

Armbrust, M., Fox, A., and Griffith, R., *et al.* (2010). A view of cloud computing. *Communications of the ACM* **53**(4), 50–58.

Arregoces, M. and Portolani, M. (2004) *Data Center Fundamentals*, Cisco Press, Indianapolis, IN.

Cisco (2013) *Cisco Global Cloud Index: Forecast and Methodology, 2012–2017*. Cisco Press, Indianapolis, IN.

Greenberg, A., Hamilton, J. R., Jain, N., *et al.* (2009) VL2: a scalable and flexible data center network. *ACM SIGCOMM Computer Communication Review* **39**(4), 51–62.

Greenberg, A., Lahiri, P., Maltz, D. A., Patel, P., and Sengupta, S. (2008) *Towards a Next Generation Datacenter Architecture: Scalability and Commoditization*. Proceedings of the ACM Workshop on Programmable Routers for Extensible Services of Tomorrow. ACM, pp. 57–62.

Gupta, M. (2002) *Storage Area Networks Fundamentals*, Cisco Press, Indianapolis, IN.

Josyula, V., Orr, M., and Page, G. (2011) Cloud Computing: Automating the Virtualized Data Center. Cisco Press, Indianapolis, IN.

Matias, J., Jacob, E., Sanchez, D., and Demchenko, Y. (2011). *An OpenFlow Based Network Virtualization Framework for the Cloud*. IEEE Third International Conference on Cloud Computing Technology and Science (CloudCom). IEEE, pp. 672–678.

McKeown, N., Anderson, T., and Balakrishnan, H., *et al.* (2008). OpenFlow: Enabling innovation in campus networks. *ACM SIGCOMM Computer Communication Review* **38**(2), 69–74.

Mell, P. M. and Grance, T. (2011) *The NIST Definition of Cloud Computing*. Special Publication 800-145. NIST, Gaithersburg, MD, http://www.nist.gov/customcf/get_pdf.cfm?pub_id=909616 (accessed November 25, 2015).

Niranjan Mysore, R., Pamboris, A., and Farrington, N., *et al.* (2009, August). PortLand: A scalable fault-tolerant layer 2 data center network fabric. *ACM SIGCOMM Computer Communication Review* **39**(4), 39–50.

Open Networking Foundation, https://www.opennetworking.org/ (accessed November 30, 2015).

Waldvogel, M. and Rinaldi, R. (2003) Efficient topology-aware overlay network. *ACM SIGCOMM Computer Communication Review* **33**(1), 101–106.

11

Wireless Datacenter Networks

Yong Cui[1] and Ivan Stojmenovic[1,2]

[1] Tsinghua University, China
[2] University of Ottawa, Canada

11.1   Introduction

As cloud-computing infrastructure, datacenters are built to provide various distributed applications such as Web search, e-mail, and distributed file systems. However, traditional datacenters, which evolved from local area networks (LAN), are limited by oversubscription and high wiring costs. Servers connected to the same switch may be able to communicate at full bandwidth (e.g., 1 gbps) but the communication bandwidth in traditional datacenters may become oversubscribed dramatically when the communicated servers are moving between switches, potentially across multiple levels in a hierarchy. Addressing this oversubscription issue requires solutions such as large 10 gbps switches and routers. The performance of current Ethernet-based DCNs solutions is also severely constrained by the congestion caused by unbalanced traffic distributions. The fixed topology of traditional datacenter networks (DCNs) connected with a large number of fabric wires is the root of the problem and influences the performance of the whole network.

Current trends in cloud computing and high-performance datacenter applications indicate that these issues are likely to become worse in the future. A flexible architecture can mitigate the issues. As a complement technology to Ethernet, wireless networking has the flexibility and capability to provide feasible approaches to handle the problems. Recently, a newly emerging 60 GHz RF communication technology characterized by high bandwidth (4–15 gbps) and short range (possibly 10 m) has been developed, to satisfy the transmission demands in DCNs efficiently.

Researchers have designed a hybrid architecture for DCNs that integrates the existent Ethernet-based DCNs and wireless networks. These architectures typically place radio transceivers on top of each rack and thus enable wireless communications among servers and routers. A multihop relay can be avoided by using a

Encyclopedia of Cloud Computing, First Edition. Edited by San Murugesan and Irena Bojanova.

recently emerging technology, called 3D beamforming, which uses the ceiling of the datacenter to reflect the wireless signals from each sender to its receiver. Completely wireless DCNs have also been investigated.

To deal with the scheduling problem of wireless transmissions in the hybrid DCN, approximation and genetic algorithms to tackle the problems of channel allocation (Cui *et al.*, 2011a) and wireless link scheduling (Cui *et al.*, 2011b, 2012b) have been designed. The wireless scheduling targets the unbalanced traffic distribution and maximizes the total network utility, under the constraints of limited wireless resources and co-channel interference.

One-to-many group communications (multicasts) are common in modern datacenters. However, current research is based on wired multicast transmissions, which are known to be susceptible to congestion problems in datacenters. Further, the number of group members in DCN multicast is often very large, which makes the construction of multicast trees difficult. The limited number of multicast addresses is an additional challenge of supporting IP multicasts in the datacenter. Based on the hybrid DCN architecture, Liao *et al.* (2013) have proposed wireless multicast mechanisms in DCNs, which address congestion due to heavy traffic.

Traffic redundancy originates from applications, software frameworks, and underlying protocols. For example, in Web-service applications, some contents are more popular than others and thousands of end users (such as PC or smartphone users) will want to download them very frequently, thus resulting in a large proportion of duplicate traffic content carried over the Internet. Although, for the sake of robustness, security, and so forth, some of the duplicate traffic content is essential, most is redundant and can be eliminated. To solve the traffic-redundancy problem in DCNs, cooperative redundancy elimination schemes in wireless DCNs have been proposed (Cui *et al.*, 2012a, 2013).

11.2 Challenges for Traditional DCNs

Cloud computing, enabled by datacenter platforms, provides various distributed applications such as search, e-mail, and distributed file systems. Datacenter networks are constructed to provide a scalable architecture and an adequate network capacity to support the services. Potentially thousands of servers communicate to each other across hundreds of racks. The applications, operated on DCN servers, can generate enormous traffic. The transmission optimization of DCNs is one of the most crucial challenges in cloud computing research.

Current DCNs are typically organized by scalable multiroot tree topology, as in Figure 11.1. The leaf nodes in this figure represent the racks in the DCNs and each rack contains 10 to 20 servers, which are connected to the top-of-rack switch (ToR Switch). The leaf nodes are connected to the routers through their parent nodes' switches hierarchically and finally linked to the root nodes, which are located in the core layer via the devices in the aggregation layer. In this hierarchical treelike organization, the DCNs can add new servers on the existing infrastructure.

However, this tree-like architecture is very vulnerable to one-point failure, as the servers connected to the leaf nodes will not work once the switches of the parent nodes fail. By adding multiple root nodes and backing up links at different levels, current DCNs can partly avoid one-point failure. Unfortunately, even under normal conditions, some parent nodes can become bottlenecks affecting the whole network because leaf nodes may generate a huge amount of traffic. This localized congestion can reduce the throughput of the whole network. Consequently, the throughput of the DCN may be far lower than its upper bound, even when the servers are equipped with a Gb-level Ethernet interface. This oversubscription is a major challenge to improving the performance of DCNs.

To solve this problem, some researchers focus on optimizing the parameters of the network's topology, like the bisection width and diameter, to increase the available bandwidth to the servers located in the root nodes, and reduce data-forwarding latency (Guo *et al.*, 2008, 2009). Some studies deal with routing mechanisms to maximize the utility of the bandwidth provided by the infrastructure of DCNs; other try to optimize the

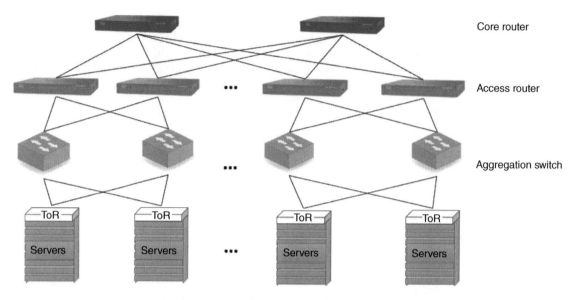

Figure 11.1   *The structure of current DCNs*

route-selection algorithm according to the requirements of applications and distribution of the traffic. Some studies attempt to eliminate the negative impact of the local congestion at the transport layer (Alizadeh *et al.*, 2010).

These solutions can alleviate the network performance bottleneck to some extent. However, it is impossible to build a wired (cable) DCN that can link all the servers and racks directly in an acceptable wiring cost range. Different leaf nodes can share common links during their communication, which can become a bottleneck. Moreover, the structure of current DCNs is a fixed symmetry and balanced topology for all the servers and the available bandwidth is equal for all servers. This symmetrical structure is not efficient in dealing with asymmetric flow.

The DCNs should meet the following design goals: scalability, cost effectiveness, and high throughput. However, having evolved from large-scale enterprise networks, traditional datacenters are typically constructed based on a hierarchical topology and this fixed and symmetric topology could not meet the demands of high dynamics of traffic in DCNs. Datacenters may contain tens of thousands of computers with significant aggregate bandwidth requirements. Unfortunately, the resulting topologies may only support 50% of the aggregate bandwidth available at the edge of the network, while still incurring the high wiring cost of switching fabric. The performance of current DCNs is far from satisfying the requirements of the modern application running in the cloud platform and it is very urgent to develop a new efficient underlying framework for DCNs.

11.3   The architectures of Wireless DCNs

11.3.1   Efficient wireless technology for DCNs

Datacenter networks can be enhanced by augmenting wireless links. The wireless technology should enable the Gb level throughput of DCNs. An important such technology is the utilization of the frequency domain with extremely high frequency (EHF). This usually refers to the frequency domain within the range of 30 GHz to 600 GHz and the frequency domain around 60 GHz (57 GHz ~ 64 GHz). The 60 GHz frequency domain has several unique characteristics. It can support high-bandwidth channels for Gb-level transmission, and has inherent directional characteristic because its signal has a very short wavelength and a high-frequency reuse. A 60 GHz signal attenuates dramatically during transmission in the air because of oxygen absorption and its valid transmission range is no more than 10 m. However, wireless transmission technology with 10 m cover range is already adequate for short-distance indoor wireless communications as a complementary technology. Therefore, new DCN architectures are built upon newly emerging directional, *beamformed* 60 GHz RF communication channels characterized by high bandwidth (4–15 gbps) and short range.

New 60 GHz transceivers based on standard 90 nm CMOS technology implement channels with low cost and high-power efficiency (<1 W). Directional short-range beams employed by these radios enable a large number of transmitters to communicate simultaneously with multiple receivers in tightly confined spaces.

11.3.2   Architectures for Wireless DCNs

Many problems need to be solved to build wireless DCNs, such as the tradeoff between wired connections and wireless connections, layout of the antenna, the security of the high-quality line-of-sight (LOS) transmission that 60 GHz RF requires, and so forth. When designing the network architecture, the basic requirements of a DCN, including scalability, high capacity, and fault tolerance, should be addressed. The limited transmission range of EHF and the interference in wireless communications pose challenges. In the next section, we will give brief outlines of several typical examples of wireless DCNs.

11.3.2.1   *Hybrid Ethernet/Wireless Architecture*

It is difficult for a wireless network alone to meet all the demands. For example, the capacity of wireless links is usually limited due to the interference and high transmission overhead. Thus wireless networks could not be employed to substitute for the Ethernet entirely. A hybrid architecture that uses wireless communication as an assistant technology has been proposed in Cui *et al.* (2011c).

One prerequisite for utilizing wireless communications in DCNs is to equip servers with radios. An intuitive approach is to assign radios to each server. However, this leads to a large number of radios, with high cost and a waste of wireless devices because wireless channel limitations allow only some of the radios to transmit simultaneously. Therefore, it is more reasonable to assign radios to groups of servers. In the following, "wireless transmission unit" (WTU) is used to refer to a group of servers supported by the same set of radios for communicating to the servers out of the group.

In practice, datacenters are mainly constructed by connecting racks of servers via Ethernet. Therefore it is reasonable to consider each rack as a WTU, as illustrated in Figure 11.2. Note that the racks do not block the LOS transmissions as the radios are located on top of them. This addition can be applied to various DCN topologies without rearranging the servers.

11.3.2.2   *3D Beamforming Architecture*

A 60 GHz wireless links is limited by line of sight, and can be blocked by even small obstacles. Even beamforming links leak power, and potential interference will severely limit concurrent transmissions in dense datacenters. To address these issues, a new wireless primitive for datacenters, 3D beamforming, where 60 GHz signals bounce off datacenter ceilings, to connect racks wirelessly, has been proposed in Xia *et al.* (2012). In 2D beamforming, sender and receiver only communicate with line-of-sight direct transmissions, in 3D beamforming (see Figure 11.3) a transmitter can bounce its signal off of the ceiling and communicate in this way with the receiver. This creates an indirect line-of-sight path between the sender and receiver by passing obstacles and reducing interference footprint. To align its antenna for a transmission, the sender only needs to know the physical location of the receiver rack, and point to a position on the ceiling directly between the two racks. This is because all racks (and their 60 GHz radio antennas) are of the same height.

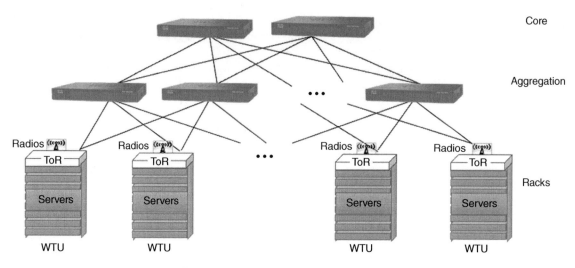

Figure 11.2   *A hybrid architecture of DCNs*

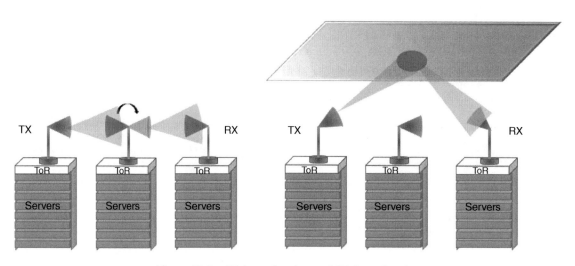

Figure 11.3   *2D beamforming and 3D beamforming*

This novel design can establish indirect line of sight between any two racks in a datacenter. In 3D beamforming, a top-of-rack directional antenna forms a wireless link by reflecting a focused beam off the ceiling towards the receiver. There are several advantages of 3D beamforming over prior "2D" approaches. First, bouncing the beam off of the ceiling allows links to extend the reach of radio signals by avoiding blocking obstacles. Second, the 3D direction of the beam significantly reduces its interference range, allowing more nearby flows to transmit concurrently. Third, the reduced interference extends the effective range of each link, allowing DCNs to connect any two racks using a single hop, and mitigating the need for multihop links. Xia *et al.* (2012) address limitations of 60 GHz beamforming that arise from signal blockage and interference caused by signal leakage. The reach and capacity of 60 GHz links are greatly expanded by 3D beamforming, making them feasible as flexible and reconfigurable alternatives to wired cabling.

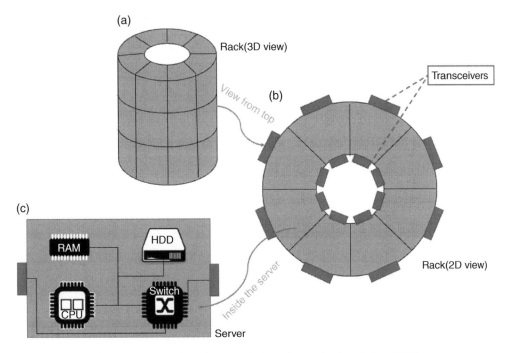

Figure 11.4 *Rack and server design of completely wireless DCNs*

11.3.2.3 Completely Wireless Architecture

Based on emerging 60 GHz RF technology, authors in Shin *et al.* (2012) investigated a radically new methodology for building wire-free DCNs. They proposed a novel rack design and a resulting network topology inspired by Cayley graphs that provided a dense interconnect. Because its network connectivity subgraphs belong to a class of Cayley graphs, they call their design a Cayley datacenter.

To separate the wireless signals for communications within a rack and among different racks, they propose cylindrical racks (Figure 11.4a) that store servers in prism-shaped containers. This choice is appealing, because it partitions the datacenter volume into two regions: intrarack and interrack free space. A single server can be positioned so that one of its transceivers connects to its rack's inner space and another to the interrack space, as illustrated in Figure 11.4b. The prism containers can hold commodity half-height blade servers. A custom-built Y-switch connects the transceivers located on opposite sides of the server (Figure 11.4c). The Y-switch multiplexes incoming packets to one of the outputs.

Although it is difficult for today's wireless technology alone to meet all the demands of the communications in DCNs, Cayley datacenter is an architecture with deep insight, as it provides a new direction to the design of DCNs.

11.4 Performance of Wireless DCNs

11.4.1 Channel Allocation in Wireless DCNs

There are some challenges in the design of hybrid DCNs. To begin with, wireless links should be arranged appropriately to improve the performance. A lot of factors are involved in the wireless scheduling. For example, wireless links should be set up to solve the congestion of hot servers. Channels should be allocated to avoid interference between wireless links. The scheduling of wireless communications should coordinate with the Ethernet transmissions. In other words, the performance of wireless networks (typically measured by throughput) and the global job completion time should be considered together.

To deal with these challenges, a centralized scheduling mechanism for wireless transmissions has been proposed (Cui *et al.*, 2011c, 2012b). They adopt a popular management model of open flow for their central controller to gather flow statistics from all the servers in DCNs when acquiring information. Open flow can help to achieve effective centralized scheduling for large datacenters with low overhead. By implementing the open flow protocol, switches in DCN can both obtain statistics of flows and open a secure channel to the central controller. The injection of a new flow or completion of a transmission can therefore be detected and signaled to the central controller, and up-to-date traffic distributions can also be estimated and sent to the central controller. Then a periodical or trigger-based schedule mechanism can be used to tackle the congestion problem. Their scheme can trigger scheduling when a new hot node is observed as it mainly focuses on the distribution of hot nodes, (i.e. the aggregate flow rate of a node grows beyond a specified threshold). Periodical scheduling can help balance the workload in DCN at a regular interval while hot-node-based triggering can adapt to the dynamism of the workload more effectively. Their scheduling mainly consists of two parts: the first step is to construct a wireless transmission graph based on the traffic information; and the second is to perform channel allocation in the wireless transmission graph. Their work has improved global performance by wireless transmissions considerably in terms of both throughput and job completion time.

11.4.2   Wireless Link Scheduling for DCNs

Besides the channel allocation problems, there are a number of other issues to be handled in order to provide a feasible wireless DCN. First, the requirements of scalability and network capacity should be considered in designing the network architecture. Second, the Ethernet DCN infrastructure and the overlay wireless network need to be carefully coordinated. Third, wireless scheduling is the key issue to determine when and where to establish wireless links.

To handle these problems, wireless link scheduling for DCNs (WLSDCN) has been proposed in Cui *et al.* (2011b). This considers various factors such as the traffic distribution, network topology, and interference. First, they have designed a hybrid architecture (as shown in Figure 11.5) that integrates the existent Ethernet-based DCNs and wireless networks to take advantage of the high capacity of Ethernet and the high flexibility of wireless networking. Second, they have presented a distributed wireless scheduling mechanism that is able to adapt wireless links to the dynamic traffic demands of the servers. Furthermore, they have introduced a novel method to organize the servers to effectively exchange traffic information through the network. Third, they formulate two wireless-scheduling problems based on different optimization objectives. Both the traffic distributions and the contention of wireless resources are considered. Additionally, they also have analyzed the complexity and designed solutions for each optimization problem. Their architecture approach leverages wireless connections to schedule links in DCNs. Both the network architecture and the scheduling mechanism are designed to provide an effective wireless DCN.

11.4.3   Multicast Optimization in Wireless DCNs

One-to-many group communications are common in modern datacenters. For example, distributed file systems such as Hadoop can generate a huge volume of one-to-many traffic. Multicast is known to be most efficient in supporting one-to many transmissions.

In BCube (Guo *et al.*, 2009), a server-based multicast tree-construction algorithm was proposed. Vigfusson *et al.* (2010) proposed a datacenter multicast mechanism called Dr. Multicast. Based on the hardware capacity,

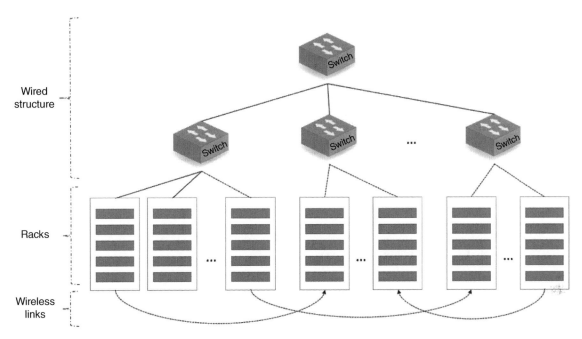

Figure 11.5   *A wireless DCN architecture*

some groups are supported by multicast while the remaining groups use unicast communications. Dan and co-workers (Li *et al.*, 2012) proposed a multicast scheme by exploiting the feature of DCNs and in-packet bloom filters to achieve scalability. However, all these mechanisms are based on wired multicast transmissions, which are known to be susceptible to congestion problems in the datacenter. Further, the number of group members in DCN multicast is often very large, which makes the construction of multicast trees difficult. The limited number of multicast addresses adds in the additional challenge of supporting IP multicast in the datacenter.

Based on the hybrid Ethernet/wireless architecture of DCNs, Liao *et al.* (2013) proposed multicast for wireless DCNs (MWDCN), a multicast mechanism that makes use of wireless links in DCNs to facilitate one-to-many transmissions more efficiently, especially when the wired transmissions in DCN experience congestion due to heavy traffic. They treat ToRs as the basic wireless broadcast units (WBUs), and data routed to different servers connected to the same ToR are treated as being transmitted to the same destination. Since tens of servers are often connected to each ToR, the number of ToRs is much smaller than that of servers. By selecting ToRs as WBUs, the number of multicast members can be greatly reduced, which would, in turn, reduce the possible wireless interference. Since DCNs are relatively well managed networks, they also have designed a centralized module (CM) (see Figure 11.6), which can facilitate the proposed scheme. Each WBU will periodically update the CM, through a wired connection, with the link conditions between itself and its neighbors, based on past transmission statistics. As all WBUs have fixed locations, the channel conditions are relatively stable in DCNs. The CM decides which WBUs broadcast data and the transmission rates used by WBUs (i.e., determining the forwarding set and corresponding rates) based on the packet delivery probabilities, transmission ranges, and interference ranges. To take into account the possible difference in loss probability between estimation and actual transmissions, some additional number of packets can be scheduled to transmit. The CM will update the schedule based on the transmission condition. Based the architecture with the CM, they also proposed a direction-based scheme to allocate the broadcast duration of these relay nodes based on the location of group members.

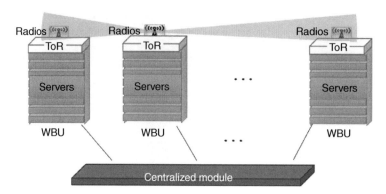

Figure 11.6   *Wireless DCNs' framework with centralized module*

11.4.4   Redundancy Elimination in Wireless DCNs

A proportion of the traffic in DCNs is redundant. It originates from applications, software frameworks, and underlying protocols. In Web-service application, some content is more popular than others, resulting in a proportion of duplicate traffic content carried over the Internet. According to research, the duplicate content of the traffic based on http (or http-opt) is up to 50%. MapReduce can have over 30% (Kim *et al.*, 2011) of duplicate content. Although, for the sake of robustness, security, and so forth, some of the duplicate traffic content is essential, most is redundant and can be eliminated. Hence, reducing the traffic in DCNs is desirable, to improve their performance.

A lot of work has been done by researchers to eliminate traffic redundancy. A well known approach to eliminate traffic redundancy is to use proxy servers to cache popular objects at locations close to clients. However, these existing mechanisms are not applicable to DCNs. They are highly centralized, and cause large communication overheads to gather the traffic matrix and some other information needed. They ignore the difference in utilities of caching the same data by different intermediate nodes, especially when the cache capacity is relatively small compared to the huge traffic volume in DCNs. Further, they do not take the redundancy of the data sent by different servers (cross-source redundancy) into account. Due to the similarity of the applications provided by the servers (e.g., file sharing), there is much redundancy between data sent by different servers. In existing work, servers register the cache state of the data sent by themselves but are not aware of the data sent by other servers. Moreover, they do not address the change in popularity of some content (e.g., Web searches) with time.

To solve the hot-point problems in DCNs, Kandula and Bahl (2009) proposed wireless links between servers to make use of the "hotspot" feature of DCN traffic and reduce the transmission time of DCN jobs. In their work, only servers are equipped with wireless network interface cards. In another architecture proposed in Cui *et al.* (2012a, 2013), there are wireless network interface cards and a cache in each router and each server. Routers cache the payload of the flow through them, and servers monitor what is cached by each router. Hence a server forwards encoded data, previously cached by a router downstream, and that router decodes the data and forwards it toward the destination.

Based on the designed architecture of wireless DCNs in Figure 11.7, authors in Cui *et al.* (2012a, 2013) have proposed a mechanism of redundancy elimination in the datacenter to schedule the channels for optimizing the performance of wireless DCNs. They have also designed solutions to choose which node caches the specific data in order to minimize the network-wide flow. Their solution caches data units with the largest caching utility among routers with remaining capacity, as illustrated in Figure 11.8. Servers and routers can update their status in a single wireless transmission, using an efficient prioritized schedule.

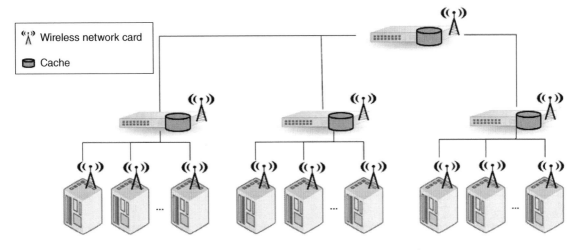

Figure 11.7 *Datacenter with wireless cards at servers and routers*

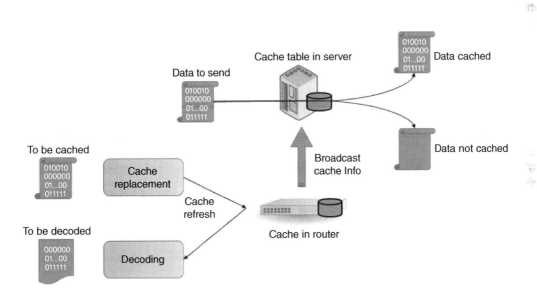

Figure 11.8 *Procedures of traffic redundancy elimination in wireless DCNs*

11.5 Conclusion

The bottlenecks in current DCNs typically originate from their fixed architecture. Fortunately, wireless networking, as a complementary technology to Ethernet, has the flexibility and capability to provide feasible approaches to handle the problem. Based on the wireless technology, researchers have done a lot of work designing a new, efficient architecture for wireless DCNs. Typically, their architectures are based on the newly emerging 60 GHz transmission technology. There is also plenty of research on improving the performance of wireless DCNs. This work has been a remarkable achievement and has made a great contribution to the development of DCNs.

References

Alizadeh, M., Greenberg, A. and Maltz, D. (2010) Data Center TCP (DCTCP). Proceedings of ACM SIGCOMM Computer Communication Review, Volume **40**, pp. 63–74.

Cui, Y., Liao, C., Stojmenovic, I., Li, D., and Wang, H. (2012a) *Cooperative Redundancy Elimination in Data Center Networks with Wireless Cards at Routers*. Proceedings of 32nd IEEE International Conference on Distributed Computing Systems Workshops, ICDCSW 2012, Macau, China. IEEE, pp. 35–42.

Cui, Y., Wang, H., and Cheng, X. (2011a) *Channel Allocation in Wireless Data Center Networks*. IEEE INFOCOM 2011, Shanghai, China. IEEE, pp. 1395–1403.

Cui, Y., Wang, H. & Cheng, X. (2011b) Wireless link scheduling for data center networks. Proceedings of the Fifth International Conference on Ubiquitous Information Management and Communication, ICUIMC, Seoul, Korea.

Cui, Y., Wang, H., Cheng, X. and Chen, B. (2011c) Wireless data center networking. *IEEE Wireless Communications* **18**(6), 46–53.

Cui, Y., Wang, H., Cheng, X., *et al.* (2012b) Dynamic scheduling for wireless data center networks. *IEEE Transaction on Parallel and Distributed Systems* **24**(12), 2365–2374.

Cui, Y., Xiao, S., Liao, C., *et al.* (2013) Data centers as software defined networks: Traffic redundancy elimination with wireless cards at routers. *IEEE Journal on Selected Areas in Communications (JSAC)* **31**(12), http://docplayer.net/6533108-Ieee-journal-on-selected-areas-in-communications-vol-31-no-12-december-2013-1.html (accessed December 1, 2015).

Guo, C., Lu, G. & Li, D. (2009) BCube: A high performance, server-centric network architecture for modular data centers. ACM SIGCOMM Computer Communication Review, Barcelona, Spain, pp. 63–74.

Guo, C., Wu, H. & Tan, K. (2008) Dcell: a scalable and fault-tolerant network structure for data centers. Proceedings of the ACM SIGCOMM (2008) Conference on Data Communication, New York, NY, pp. 75–86.

Kandula, S. & Bahl, P. (2009) Flyways to De-congest Data Center Networks. Proceedings of HotNets 09: the Eighth ACM Workshop on Hot Topics in Networks.

Kim, S., Han, H., Jung, H., *et al.* (2011) Improving mapreduce performance by exploiting input redundancy. *Journal of Information Science and Engineering* **27**(2), 789–804.

Li, D., Li, Y., Wu, J., *et al.* (2012) ESM: efficient and scalable data center multicast routing. *IEEE-ACM Transactions on Networking* **20**(3), pp. 944–955.

Liao, C., Liu., Y., Cui, Y., *et al.* (2013) Location-based multicast in wireless data center network. IEEE International Conference on Computer Communications and Networks (ICCCN). Nassau, Bahamas.

Shin, J., Sirer, E., Weatherspoon, H., and Kirovski, D. (2012). *On the Feasibility of Completely Wireless Datacenters*. Proceedings of the 8th ACM/IEEE Symposium on Architectures for Networking and Communications Systems, Austin, TX, pp. 3–14.

Vigfusson, Y., Abu-Libdeh, H., Balakrishnan, M., and Birman, K. (2010) Dr. Multicast: Rx for data center communication scalability. Proceedings of the Eurosys 2010 Conference, Paris, France, pp. 349–362.

Xia, Z., Zhang, Z., and Zhu, Y. (2012) Mirror mirror on the ceiling: flexible wireless links for data centers. *Computer Communication Review* **42**(4), 443–454.

12

Open-Source Cloud Software Solutions

G. R. Gangadharan, Deepnarayan Tiwari, Lalit Sanagavarapu, Shakti Mishra, Abraham Williams, and Srimanyu Timmaraju

Institute for Development and Research in Banking Technology, India

12.1 Introduction

The cloud computing paradigm is delivering information technology capabilities to users in the form of "services/utility" with elasticity and scalability, making it an attractive concept for enterprises requiring agility and cost-effectiveness. Various cloud software providers (commercial or open source) offer cloud software solutions that enable cloud service providers to offer one of the three services – infrastructure as a service (IaaS), platform as a service (PaaS), and software as a service (SaaS).

Open-source cloud software solutions are distributed along with their source code and are licensed so that the copyright holder has the right to study, change, and redistribute the software to anyone and for any purpose (Andrew, 2008). The primary motives for an organization to adopt open-source cloud software solutions include reduction in cost of investment, avoidance of vendor lock in, and security. Open-source cloud software solutions with lower initial investments are viable alternatives to commercial cloud software solutions (Endo *et al.*, 2010; Voras *et al.*, 2011a, b; Salih and Zang, 2012). The strengths, weaknesses, opportunities, and threats (SWOT) analysis of open source cloud software adoption is outlined in Figure 12.1.

For each cloud service delivery model, we select a set of criteria relevant to that service, generally covering the SWOT aspects. Then we briefly outline the features of few leading open-source cloud software solutions and compare those solutions based on the set of criteria chosen for the cloud service delivery model.

12.2 Criteria for Evaluating Open-Source IaaS Cloud Solutions

Infrastructure as a service (IaaS) refers to the capability of provisioning raw computing infrastructure, in the form of computing units, storage, and so forth, by a provider to a client. By using IaaS open-source solutions, organizations can take advantage of cloud interoperability and portability. We have

Encyclopedia of Cloud Computing, First Edition. Edited by San Murugesan and Irena Bojanova.
© 2016 John Wiley & Sons, Ltd. Published 2016 by John Wiley & Sons, Ltd.

Figure 12.1   *SWOT analysis of open-source cloud software adoption*

chosen the following criteria for evaluating open-source IaaS cloud solutions – the criteria are not listed in any specific order of preference:

- *Ease of deployment and management of cloud software*. Customers who have already invested in in-house infrastructure capacity want the investments to coexist with the cloud. They should have an option to migrate a portion of their workload or the entire workload to the cloud. Cloud service providers should address these needs by providing tools and services to ease migration. These tools and services should support virtual machine portability for the cloud platform. The cloud software stack should also support appropriate and commonly used infrastructure services such as identity and access-management systems.
- *Resource provisioning and service orchestration*. Resource provisioning is an important issue in cloud computing. It refers to how resources may be allocated to an application mix so that the service-level agreements (SLAs) of all applications are met in a cost-effective manner (Hu *et al.*, 2009). Service orchestration refers to the coordinating and provisioning of virtualized resources, as well as the running and coordination of resource pools and virtual instances. Service orchestration also includes static and dynamic mapping of virtualized resources to physical resources.
- *Monitoring*. Monitoring of IaaS is a task of paramount importance for both providers and consumers. Monitoring of IaaS cloud solutions involves capacity and resource planning, datacenter management, SLA management, billing, and security management.
- *Interoperability*. In IaaS, interoperability typically refers to the ability to move the workload and data from one cloud provider to another or between private and public clouds. A common criterion for analyzing interoperability is the use of open standards.

- *Network.* Network plays a key role in the delivery and performance of cloud services. In IaaS, there is a guarantee of the minimum amount of network bandwidth that can be expected for each virtual machine (VM) irrespective of the network utilization of other tenants. This type of guarantee is common for resources like CPU and memory, and having the same for the network is key to achieving lower bounds for the worst-case performance of an application. These criteria are referred to as min-guarantee (Popa *et al.*, 2012).
- *Storage.* Storage is a critical criterion when developing high-performance IaaS solutions with high availability, reliability, and scalability. Cloud-storage infrastructure needs to provide a distributed, shared solution, which eliminates single points of failure and also avoids high costs.
- *Security.* IaaS supports a multitenant architecture to share the computing resources between users. The security analysis for IaaS should include network security, VM repository security, data security, and security of cloud APIs.
- *Support.* Customer service is the key to success in any organization. Support for IaaS should include community forum, knowledge base, Web-based issue tracking, and documentation.

12.2.1   Comparison of Open Source IaaS Cloud Solutions

We consider the following leading open source IaaS cloud software solutions for comparison:

- Eucalyptus (Elastic Utility Computing Architecture for Linking Your Programs To Useful Systems): Eucalyptus (http://www8.hp.com/us/en/cloud/helion-eucalyptus-overview.html, accessed December 1, 2015; Nurmi *et al.*, 2008, 2009; Truksha, 2012) provides a scalable IaaS framework for implementation of clouds. Eucalyptus implements the Amazon Web Services (AWS) API allowing interoperability with existing services, creating the possibility of combining resources from internal private clouds and from external public clouds to create hybrid clouds. This capability presents seamless integration with Amazon EC2 and S3 public cloud services. Eucalyptus currently supports Xen and KVM virtualization. Eucalyptus can be deployed on all major Linux OS and Windows OS distributions.
- OpenStack: NASA and Rackspace jointly developed an open source project called OpenStack (www.openstack.org, accessed December 11, 2015) (Sefraoui *et al.*, 2012; Truksha, 2012) supported by renowned companies including AMD, IBM, HP, Intel, and Cloud.com. OpenStack has a modular architecture with various components including compute, object storage, block storage, networking, dashboard, identity service, image service, telemetry, orchestration, and database.
- CloudStack: Jointly developed by Cloud.com and Rackspace, CloudStack (Truksha, 2012; Sabharwal and Shankar, 2013) supports the Xen, KVM, and VMware vSphere virtualization approaches. The CloudStack architecture comprises two components: the management server and the compute nodes. The management server features a Web user interface for administrators and users. Other management server tasks are to control and manage the resources when distributing the instances to the compute nodes. Although this software is available in a community edition, an enterprise edition, and a service provider edition, only the community edition can be used under an open-source license.

A detailed comparison of the three open-source IaaS software solutions listed above is given in Table 12.1:

12.3   Criteria for Evaluating Open-Source PaaS Cloud Software Solutions

Platform as a Service (PaaS) refers to the capability in which development platforms and middleware systems hosted by a vendor are offered to application developers, allowing developers to simply code, generate executables and deploy their application, without directly interacting with the underlying infrastructure. By using open-source PaaS, organizations could move an application very easily from one platform to another.

Table 12.1 *Comparison between open-source IaaS cloud software solutions*

Criteria	Eucalyptus	OpenStack	CloudStack
Ease of deployment and management of cloud software	Eucalyptus offers a rich API compatible with Amazon web services API. It also provides support for management of cloud software by clustering and zoning, flexible network management, security groups, and traffic isolation.	OpenStack supports a set of deployment tools including OpenStack Native, DevStack, StackOps, RDO, Puppet, Chef, and Fuel for ease of deployment and management of software.	CloudStack deployment architecture consists of two parts: the management server and the cloud infrastructure. The management server helps to provision resources such as hosts, storage devices, and IP addresses, and also manages those resources. The cloud infrastructure is organized as follows: zone, pod, cluster, and host.
Provisioning and orchestration	The node controller (NC) and cloud controller (CLC) components of Eucalyptus, manage service provision and orchestration.	OpenStack orchestration is a service that orchestrates cloud applications onto cloud resources using the OpenStack heat orchestration syntax (HOT) template format, through an OpenStack ReST API. OpenStack orchestration also provides compatibility with the AWS CloudFormation template format.	The orchestration engine is used for configuring, provisioning and also in scheduling any operation.
Monitoring and alerts	Eucalyptus monitors running components, instantiated virtual machines and storage services with the integration of monitoring tools like Ganglia and Nagios.	OpenStack supports external tools like Nagios, Logmonitor, CopperEgg/ RevealCloud for monitoring and generating alerts.	CloudStack supports external tools like Zenoss or Nagios for monitoring and generating alerts.
Interoperability	Eucalyptus supports limited interoperability with Amazon Web Services (AWS) by compatible standardized Web services API.	OpenStack supports CloudBridge, which enables conversion between Amazon API and CloudStack API.	CloudStack supports a rich set of API, the platform supports CloudBridge, which enables converting an Amazon API into a CloudStack API.
Network	Eucalyptus supports three networking modes: system, static, and managed.	OpenStack networking is a pluggable, scalable and API-driven system for managing networks and IP addresses. OpenStack networking has an extension framework allowing intrusion detection systems, load balancers, firewalls, and virtual private networks (VPN) to be deployed and managed.	CloudStack supports two types of networking zones, namely basic zone and advanced zone.

Table 12.1   *(continued)*

Criteria	Eucalyptus	OpenStack	CloudStack
Storage	The storage controller (SC) communicates with the cluster controller and node controller, to allocate/ deallocate the memory on demand.	OpenStack support two types of storage namely as object storage and block storage. Object Storage is ideal for cost effectiveness, scaleout storage. It provides a fully distributed, API-accessible storage platform that can be integrated directly into applications or used for backup, archiving and data retention. Block storage allows devices to be connected to compute instances for expanded storage, better performance, and integration with enterprise storage platforms.	CloudStack has two major categories of storage – primary and secondary storage. Primary storage is used for providing disk volumes for guest VMs and is associated with a cluster of the CloudStack deployment. Secondary storage is associated with a zone, common to all the pods in the zone, and is used for storing templates, ISOs and the disk volume's snapshots.
Security	Eucalyptus provides two primary mechanisms for instance security: availability zones, and security groups. An availability zone receives a fixed amount of resources, and those resources can be controlled via quotas and access control lists. Security groups are sets of networking rules applied to all virtual machine instances associated with a group.	OpenStack composes security components for identity provisioning, authentication, password management, and authentication tokens.	CloudStack uses secured Web sessions to protect the access information and DES-encrypted access tokens.
Support	Eucalyptus provides support through community forum, knowledge base, Web-based issue tracking, and documentation.	OpenStack provides support through community forum, documentation, and Internet Relay Chat (IRC).	CloudStack provides support through Instant Guru, access by phone, e-mail or Web portal, etc.

The following are the criteria for evaluating open source PaaS cloud software solutions – the criteria are not listed in any specific order of preference:

- *Ease of Deployment*: The key factors to consider include ease of deploying an application and the steepness of the learning curve associated with its management for adoption of open source PaaS solution. Enterprises already have large investments in IT infrastructure. Leveraging existing investments like load balancers and management tools for PaaS will save money and accelerate the deployment.

- *Security.* As PaaS offers an integrated environment to design, develop, test, deploy, and support custom applications developed in the language that the platform supports, PaaS security includes security of the PaaS platform and security of customer applications deployed on a PaaS platform. The security of the PaaS platform should be considered from multiple perspectives including access control, privacy, and service continuity while protecting both the service provider and the consumer. Security of customer applications includes securing the software development lifecycle, Web services security, and secure coding practices.
- *Features.* An organization should look for the following (partial) list of features when selecting a PaaS solution:
  - multitenant architecture;
  - customizable / programmable user interfaces;
  - unlimited database customization;
  - robust workflow engine/capabilities;
  - granular control over security / sharing (permission / access control models);
  - flexible "services-enabled" integration model;
- *Service metering.* Platform as a service (PaaS) billing and metering are determined by actual usage, as platforms differ in aggregate and instance-level usage measures. Actual usage billing enables PaaS providers to run application code from multiple tenants across the same set of hardware, depending on the granularity of usage monitoring. For example, the network bandwidth, CPU utilization, and disk usage per transaction or application can determine the cost. The primary concept and criteria for service metering and billing include incoming and outgoing network bandwidth, CPU time per hour, stored data, high availability, and monthly service charges.
- *Auto sizing for resilience.* Resilient computing is a form of failover that distributes redundant implementations of IT resources across physical locations. Information technology resources can be preconfigured so that if one becomes deficient, processing is automatically handed over to another redundant implementation. Within cloud computing, resiliency can refer to redundant IT resources within the same cloud (but in different physical locations) or across multiple clouds. Cloud consumers can increase both the reliability and availability of their applications by leveraging the resiliency of cloud-based IT resources.

12.3.1   Comparison of Open-Source PaaS Cloud Software Solutions

The following are some of the leading open-source PaaS cloud software solutions:

- *Cloud Foundry.* Cloud Foundry (www.cloudfoundry.org, accessed December 11, 2015) (Iwasaki *et al.*, 2012) is an open-source PaaS platform developed by VMware Corporation. Cloud Foundry has an open architecture that consists of a self-service application execution engine, an automation engine for application deployment, lifecycle management, and a scriptable command-line interface (CLI). Cloud Foundry is integrated with various development tools to ease deployment processes.
- *Tsuru.* Tsuru (www.tsuru.io, accessed December 11, 2015) is an open-source cloud PaaS developed by Globo.com. Developers use Web-service architecture to develop applications in programming languages supported by Tsuru, which include Go, Java, Python, and Ruby. Tsuru uses JuJu for service orchestration to enable configuration, management, maintaining, deploying, and for scaling efficiently required services on the cloud.

A detailed comparison of the two open-source PaaS software solutions listed above is given in Table 12.2:

Table 12.2   *Comparison between open-source PaaS cloud-software solutions*

Criteria	Cloud Foundry	Tsuru
Ease of deployment	Cloud Foundry provides an easy way for developers to deploy, run and scale Web applications.	Tsuru supports deploy hooks to deploy an application. Deploy hooks gives a notification whenever a new version of an application is pushed to Tsuru. It is useful to integrate different systems together.
Security	The User Account and Authentication (UAA) service is the identity management service for Cloud Foundry.	Tusru uses cryptographic SSH-based requests to communicate with services.
Features	Cloud Foundry supports multiple languages and frameworks with flexible configurations.	Tsuru is built to be extensible and scalable.
Service metering	Cloud Foundry provides general-purpose metered usage API and specific add-on instances for add-on offerings with metered usage billing, such as overages, right scale, etc.	Tusru offers a general-purpose metered-usage API.
Auto sizing for resilience	The deployed applications can be scaled to a configured number of instances and can use more than their allocated memory.	Tusru does not support dynamic resource allocation at the time of initialization of a node. Required resources must be predefined at the time of the request to the node.

12.4   Criteria for Evaluating Open-Source SaaS Cloud Software Solutions

Software as a service (SaaS) refers to the capability provided to the user to use applications hosted on a cloud infrastructure of the provider. Buyers are freed from developing, possession, and maintenance issues of software and hardware. Open-source SaaS cloud solutions enable organizations to modify a cloud application according to their needs in a rapid manner. This capability can be accessed by users from various client devices.

The following criteria may be used for evaluating open source SaaS cloud solutions – the criteria are not listed in any specific order of preference:

- *Features and applicability*. Various open-source cloud vendors offer SaaS services to the consumers to analyze the performance of required services, including user-experience monitoring, scalability analysis, and projection, tracking Web transactions, and application analytics.
- *Security*. As SaaS may follow multitenant architecture to provide services to the consumers, security is the major criteria for SaaS applications. A SaaS provider provides extensive controls to manage and define the security of the application. The following key security criteria are considered for the security assessment of SaaS application development and deployment process:
  - data security;
  - network security;
  - data access;
  - authentication and authorization;
  - Web application security;
  - identity management;
  - sign-on process.

- *Scalability*. Scalability can refer to the capability of a system to increase total throughput under an increased load due to dynamic requests to the services. Scalability criteria can be categorized into two aspects. Load scalability deals with a system's ability to increase throughput effectively while increasing levels of demand for computing resources. Administrative scalability deals with the SaaS vendor's ability to manage multiple customers in a single environment.
- *Customization*. SaaS provides a rich set of functionality to achieve the functional needs of a customer. Customization is sometimes thought of as defining certain preferences that affect how an application behaves. The traditional concept for billing and metering SaaS applications is a monthly fixed cost. In some cases, the number of users determines the cost of the service and the organization dynamically allows the number of users to access the SaaS applications, which increases the monthly fee. The following criteria are used to evaluate the cost of SaaS application:
  ○ monthly subscription fees;
  ○ per-user monthly fees.
- *Support*. The open-source cloud vendors must have techniques to solve software conflicts and usability problems, and to supply updates and patches for bugs and security holes in the SaaS application.

12.4.1   Comparison of Open-Source SaaS Cloud Solutions

The following are some of the leading open-source SaaS cloud software solutions.

- *Acquia*. Acquia (www.acquia.com, accessed December 11, 2015) offers multitenant hosting of the Drupal open-source content management system and uses Amazon's EC2 cloud service.
- *SugarCRM*. Using SugarCRM's (www.sugarcrm.com, accessed December 11, 2015) SaaS application, organizations can integrate multiple systems within the cloud.

A detailed comparison of the two open-source SaaS software solutions listed above is given in Table 12.3.

12.5   Open-Source Cloud Infrastructure Automation Tools

Cloud software solutions are starting to have good ecosystem to automate repetitive tasks, quickly deploy, monitor, and manage virtual machines in the cloud. Some of the popular open-source automation and management tools are described in this section.

12.5.1   Hyperic

Hyperic HQ (www.hyperic.com/, accessed December 11, 2015) manages, monitors, and controls large IT environments, ranging from hundreds to thousands of machines. This translates to tens of thousands of managed resources, from CPUs and network interfaces to application servers and databases. With HQ autodiscovery, cloud monitoring and heterogeneous environments can be managed within minutes of installation in both physical and virtual environments.

CloudStatus is built on Hyperic HQ, Hyperic's flagship product designed to monitor and manage large-scale Web infrastructure. The Hyperic HQ Server aggregates multiple metrics from sources inside and outside the cloud to provide cloud availability and health status. Hyperic HQ then calculates the aggregate data to determine overall availability and normalized metrics across the cloud. The multiple metrics origination scheme assures users a relevant overall perspective on cloud performance. For each service, Hyperic collects specific metrics using production instances across the cloud and tailored exercises typical of each service.

Table 12.3   *Comparison between open-source SaaS solutions*

Criteria	Acquia	SugarCRM
Features and applicability	Acquia includes a content-management system with the facility of adding modules and custom code to meet the needs of a specific organization.	SugarCRM includes sales management, marketing management, news services, support automation, e-mail integration, reporting, team selling, advanced security and workflow automation.
Security	Acquia includes AWS control environment, physical security, customer segregation, system access controls, OS and LAMP stack security-patch management, antivirus upload scanning, file-system encryption, SSL and HTTPS, data and physical media destruction, and logging to support security.	SugarCRM uses multiple levels of protection and security.
Scalability	Acquia manages sudden spikes in traffic.	SugarCRM manages sudden spikes in traffic.
Customization and its implication on costing	Customizations can be done based on the requirements of the client.	Basic customizations can be done through its admin interface without touching its code base. Advanced customization and extensions to meet business requirements are offered by the SugarCRM development team.
Support	Acquia provides support through a community forum, documentation, and IRC.	SugarCRM provides support through a support portal, e-mail, and forums.

CloudStatus results reflect general service levels, and serve as an indicator of whether further investigation of application behavior or cloud performance is warranted.

12.5.2   Chef

Chef (www.getchef.com/chef/, accessed December 11, 2015) is a systems and cloud infrastructure-automation framework that makes it easy to deploy servers and applications to any physical, virtual, or cloud location, no matter the size of the infrastructure. The Chef client relies on abstract definitions (known as cookbooks and recipes) that are written in Ruby and are managed like source code. Each definition describes how a specific part of the infrastructure should be built and managed. The Chef client then applies those definitions to servers and applications, as specified, resulting in a fully automated infrastructure. When a new node is brought online, the only thing the chef-client needs to know is which cookbooks and recipes to apply.

Chef comprises three main elements: a server, one (or more) nodes, and at least one workstation. The Chef server acts as a hub that is available to every node in the organization. This ensures that the right cookbooks (and recipes) are available, that the right policies are being applied, that the node object used during the previous chef-client run is available to the current chef-client run, and that all of the nodes that will be maintained by the chef-client are registered and known to the Chef server. The workstation is the location from which cookbooks (and recipes) are authored, policy data (such as roles, environments, and data bags) are defined, data is synchronized with the Chef repo, and data is uploaded to the Chef server. Each node contains a Chef client that performs the various infrastructure automation tasks that each node requires.

Cookbooks are also a very important element and can be treated as separate components (alongside the server, nodes, and the workstation) across the documentation. In general, the cookbooks are authored and managed from the workstation, moved to the Chef server, and then are pulled down to nodes by the Chef client during each Chef client run.

12.5.3   Puppet

Puppet (http://puppetlabs.com, accessed December 10, 2015) is an IT automation tool to manage infrastructure from provisioning and configuration to orchestration and reporting. Puppet can automate repetitive tasks, deploy critical applications, and proactively manage change, scaling from tens of servers to thousands, on-premises or in the cloud.

The four-step process in using Puppet is as follows:

- define the desired state of the infrastructure's configuration using Puppet's declarative configuration language;
- simulate configuration changes before enforcing them;
- enforce desired state automatically, correcting any configuration drift;.
- report differences between actual and desired states and any changes made enforcing the desired state.

Puppet works for Unix-based systems as well as Microsoft Windows, and is built using Ruby.

12.5.4   Zenoss

Zenoss (http://www.zenoss.org/, accessed December 11, 2015) software is used to manage datacenter and cloud infrastructures for monitoring application, server, and network-management platforms. Zenoss provides an interface to system administrators for monitoring availability, inventory / configuration, performance, and events. Zenoss leverages SMIS and native APIs to extract performance, availability, and utilization KPIs on storage from leading vendors. It leverages standard SNMP interfaces to extract performance, availability, and utilization KPIs on a wide range of networking providers.

12.6   Concluding Remarks

There is an increasing awareness and adoption of open-source software across the industries. Even the governments of various countries are adopting and encouraging the adoption of open source. Open-source cloud software stack deployments in organizations will reduce concerns about data security and privacy and will also aid more efficient use of some of the existing resources. Open-source cloud software solutions can be customized for an organization's needs and can be integrated with existing monitoring and altering systems. As the source code is available with all open-source cloud software solutions, the interoperability with other hypervisors becomes relatively easy. The biggest concerns – vendor lock in and dependency on the vendor for every small change – are drastically reduced and even the total cost of ownership will come down.

References

Andrew, M. L. (2008) *Understanding Open Source and Free Software Licensing*, O'Reilly Media, Sebastopol, CA.
Endo, P. T., Gonçalves, G.E., Kelner, J., and Sadok, D. (2010) A survey on open-source cloud computing solutions. Proceedings of the Brazilian Symposium on Computer Networks and Distributed Systems.

Hu, Y., Wong, J., Iszlai, G., and Litoiu, M. (2009) Resource provisioning for cloud computing. Proceedings of the Conference of the Center for Advanced Studies on Collaborative Research.

Iwasaki, Y., Kurumatani, S., Nomoto, T., *et al.* (2012) *PaaS Software based on Cloud Foundry*. NTT Technical Report.

Nurmi, D., Wolski, R., and Grzegorczyk, C. (2008) *Eucalyptus: A Technical Report on an Elastic Utility Computing Architecture Linking Your Programs to Useful Systems*. UCSB Computer Science Technical Report Number 2008-10.

Nurmi, D., Wolski, R., Grzegorczyk, C. *et al.* (2009) The Eucalyptus open-source cloud-computing system. Proceedings of the 9th IEEE/ACM International Symposium on Cluster Computing and the Grid.

Popa, L., Kumar, G., Chowdhur, M., *et al.* (2012) FairCloud: Sharing the network in the cloud. Proceedings of the Tenth ACM Workshop on Hot Topics in Networks.

Sabharwal, N. and Shankar, R. (2013) *Apache CloudStack Cloud Computing*. Packt Publishing, Birmingham.

Salih, N. and Zang, T. (2012) Survey and comparison for open and closed sources in cloud computing. *International Journal of Computer Science Issues (IJCSI)* **9**(3), http://ijcsi.org/papers/IJCSI-9-3-1-118-123.pdf (accessed December 11, 2015).

Sefraoui, O., Aissaoui, M., and Eleuldj, M. (2012) OpenStack: Toward an open-source solution for cloud computing. *International Journal of Computer Applications* **55**(3), 38–42.

Truksha, V. (2012) Cloud Platform Comparison: CloudStack, Eucalyptus, vCloud Director and OpenStack. *Network World* (July), http://www.networkworld.com/article/2189981/tech-primers/cloud-platform-comparison--cloudstack--eucalyptus--vcloud-director-and-openstack.html (accessed December 11, 2015).

Voras, I., Mihaljevic, B., and Orlic, M. (2011a) Criteria for evaluation of open-source cloud computing solutions. Proceedings of the 33rd International Conference on Information Technology Interfaces.

Voras, I., Mihaljevic, B., Orlic, M., *et al.* (2011b) Evaluating open-source cloud computing solutions. Proceedings of the 34th International Convention of MIPRO.

13

Developing Software for Cloud: Opportunities and Challenges for Developers

K. Chandrasekaran and C. Marimuthu

National Institute of Technology Karnataka, India

13.1 Introduction

In recent years, cloud computing and its services have led the IT market, and it is one of the most promising technologies of the decade. "Everything as a service" is the concept that makes small and large organizations migrate from on-premises applications to off-premises cloud services. Smaller enterprises, in particular, are moving towards cloud for their computing, storage, networking, and application needs. According to the National Institute of Standards and Technology (NIST), cloud computing is an on-demand, rapid service provisioning model over the Internet in the form of compute, network, storage, and application with minimal management effort (Mell and Grance, 2011). The characteristics of cloud computing include on-demand self-service, broad network access, resource pooling, rapid elasticity, and measured services.

13.1.1 The Cloud Computing Ecosystem

Cloud computing is composed of three service-delivery models and four deployment models. Software as a service (SaaS), platform as a service (PaaS), and infrastructure as a service (IaaS) are the service-delivery models (Mell and Grance, 2011). The SaaS model changes the way in which the software is delivered to the customer, as the software products are delivered as a service on a pay-per use basis. Here, the SaaS consumers need not manage any development platform and on-premises infrastructure; a simple Web browser is enough to use the services. The PaaS model is for the developers who can access the development and testing platforms that are hosted by service providers. Here, the developers have no need to install any heavyweight development tools or integrated development environments (IDEs) on their infrastructure. In the IaaS model, consumers can access the required compute, storage, and networking resources from the infrastructure managed by service provider. Here, the IaaS consumers are free from managing huge datacenters. These three service-delivery models can be delivered to the customers in any of the deployment models (Mell and Grance, 2011), namely public cloud, private cloud, community cloud, and hybrid cloud. The public cloud is for public use, where all the customers share the services provided by the service provider. Customers who cannot invest more in infrastructure and are less concerned about security can go for a public cloud deployment model. The private cloud is for those who are more concerned about security and are ready to invest a huge amount in maintaining their own infrastructure. The community cloud will be suitable for consumers who want security with minimum investment. Here, two or more organizations with the same goals can maintain and share common resources where the expenses of maintaining the resources can be shared among themselves. The combinations of public, private and community deployment model are generally known as hybrid cloud.

13.1.2 Evolution of SaaS Applications

Software as a service applications have evolved through several stages as shown in Figure 13.1. After mainframe applications, stand-alone desktop applications were used in many enterprises for their internal operations. After the networking era there was a need for networked applications that connect all machines in the enterprises so that they could communicate with each other. These types of applications follow the client server architecture, which forces the enterprise to maintain huge servers. This resulted in high capital expenditure and maintenance costs for the enterprises. Thus application service providers (ASPs) came into the picture and provide on-demand applications to enterprises, which reduces the spending on maintaining on-premises infrastructure. Even though ASPs are helpful to the enterprises, they do not use the infrastructure fully, as separate servers are maintained for each enterprise. The enterprises started using shared hosting, which allows multiple applications to share the same infrastructure by server consolidation. In this model, the scalability and management of the application is a big issue. After the advent of cloud computing, multitenant SaaS applications are delivered as a service to the consumer on a pay-per use basis, which can be accessed from Web UI or Thin Clients. SaaS applications are maintained by service providers; thus, consumers need not worry about the infrastructure, development platform, and updates.

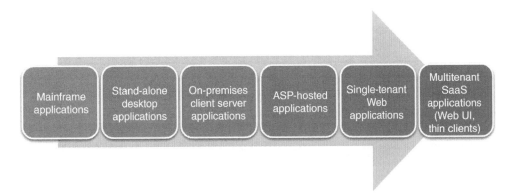

Figure 13.1 *Evolution of SaaS applications*

13.1.3   Software as a Service Benefits

Software as a service is an on-demand, easy, and affordable way to use an application without the need to buy it. The SaaS solutions are easy to adopt and integrate with other software or service. Some of the benefits of SaaS are mentioned below:

- *Pay-per use.* The traditional software forces the customers to buy full packages of software, even if consumers only use it occasionally. Unlike traditional software, SaaS is delivered based on subscription or pay-per-use, which includes upgrades, maintenance, and customer support.
- *Zero infrastructure.* With SaaS, the customers up-front cost is zero as the underlying infrastructure is maintained by the SaaS provider. Software as a service customers need not maintain infrastructure, operating system, development platforms, and software updates.
- *Ease of access.* A Web browser and Internet access is enough to use SaaS solutions. The browser-based user interface will allow business owners to use their business service without any training. Most modern SaaS uses a responsive user interface, which can be accessed from any device (laptops, mobiles, tablets, and so forth) with ease.
- *Automated updates.* As the SaaS solutions are managed by the service provider, there is no need for the customer to perform the software updates. With SaaS, customers are assured that, at any point in time, they will always use the most recent version of the SaaS solution.
- *API integration.* Most of the recent SaaS solutions allow the customers to use the APIs to easily integrate their applications with other software or Web service.
- *Scalability.* It is always possible to scale up and scale down rapidly. The SaaS model increases the resources of the application.
- *Green IT solutions.* Software as a service solutions are becoming smarter and have energy-aware features, which do not consume many resources for their operation. The SaaS applications also share the same infrastructure. This results in greener IT solutions with less energy consumption and electronic waste.

13.1.4   When and Where to use SaaS?

Cloud computing and SaaS applications are growing in the market and are used by many enterprises and individuals, mainly because of their cost-effective services. Even though SaaS is used in many places, there are some applications where SaaS may not be the best option (Kepes, 2013):

- some real-time applications where fast processing of data is needed;
- applications where the organization's data is confidential and where it does not want to host its data externally;
- applications where existing on-premises applications fulfill requirements.

SaaS services are the best option (Kepes, 2013) for applications:

- where the end user is looking for on-demand software rather than the full-term/licensing-based software;
- where a startup company cannot invest more money in buying licensed software such as tax or billing software;
- where there is a need for accessibility from handheld devices or thin clients;
- where there is an unpredictable dynamic load.

13.2   Challenges

Software as a service (SaaS) is one of the cloud service delivery models that changes the way in which software is developed and delivered to customers. Most small and medium-scale enterprises migrated to cloud from traditional Web applications. Transition from traditional Web applications to a cloud-based SaaS application imposes lots of challenges for consumers, developers, and providers. Successful Web developers are struggling to be successful in developing and delivering SaaS applications. To be successful in SaaS application development, traditional software developers should be aware of opportunities and challenges in developing cloud-based applications.

13.2.1   Multitenancy

Multitenancy is a one-to-many model that allows multiple customers to share a single instance of code and database of SaaS application (Fishteyn, 2008). The important benefit of multitenancy is effective utilization of resources, which results in greener IT solutions. Enabling multitenancy for an application with traditional development platforms is a difficult job. The developers need to acquire more knowledge about developing multitenant software. There are several multitenancy levels (Chong, 2012) available for databases, namely separate databases, shared databases with separate schema, and shared databases with shared schema. The developer has to choose the correct multitenancy level based on the customer's requirements before developing the application. There are many "platform as a service" (PaaS) solutions available, which will reduce the overheads in enabling the multitenancy features. Some of the PaaS solutions are discussed later in this chapter. Instead of sticking to traditional development platforms, the developer can start using the PaaS solutions that are already available for developing multitenant applications, with lower overheads.

13.2.2   Data Security

Most enterprises are afraid to migrate to cloud because of data security. There might be a possibility of data leakage as a single instance of an application is shared between multiple tenants. Isolating the tenants' data in a cloud environment is a biggest challenge for any SaaS developer. The developers can address this challenge by choosing an appropriate multitenant model based on user requirements. For example, if the customer data is more confidential, then the developer can choose a separate database for each customer. But, this kind of model is costly and more-or-less the same as traditional Web applications. A shared database and separate schema model will be the best option for the applications that need logical isolation. Such a model is also best suited for applications that do not have confidential data. Even though the data leakage between tenants can be prevented through different multitenant models, the security threats of traditional Web applications are still applicable to SaaS applications. These external security attacks on traditional applications can be mitigated by intrusion detection, strong encryption, auditing, RBAC, authentication, and authorization (Fishteyn, 2008). The developer can use these techniques to secure SaaS applications also, but the only problem is that most of the SaaS applications are accessed from resource-limited devices such as mobiles. If the SaaS application uses a strong encryption mechanism to protect user data, then it will definitely consume more computing power, energy, and time to process. Hence the response time to the user will increase because of limited bandwidth in mobile phones, which leads to SLA violation. When developers choose a security mechanism, they have to make sure that it will not lead to SLA violation and more resource consumption.

13.2.3   Scalability

Since social networking, online market places, and other online communities are growing exponentially, the number of users of any application is dynamic and unpredictable. Traffic and load on the application cannot be predicted correctly. This creates lots of challenge for the SaaS application developers to develop highly scalable applications. Before developing any SaaS applications, the application developer should ask the questions "what kind of application am I developing now?" and "what architecture should I follow to make my application highly scalable?" The developer has to sit with the system architect and should decide about the different tiers of the application architecture. For, example, the system architect can design architecture with a load-balancing tier, an application tier, a caching tier, and a database tier for a scalable SaaS application. Normally, scalability refers to how well the application is serving requests whenever there is an additional load or traffic on the system. Generally, the scalability of the application can be increased by adding additional resources to it. There are two types of scaling – vertical scaling and horizontal scaling. In general vertical scaling refers to increasing capacity and horizontal scaling refers to adding more resources. The developer should decide which scaling mechanism will best suit the application. The scaling mechanism should be applied in all tiers of the application for better scalability. Normally, horizontal scaling is used widely in many applications as it provides higher availability and better fault tolerance than the vertical scaling.

13.2.4   Backup and Recovery

Since SaaS users are storing their data in the service provider's datacenter, the SaaS provider should ensure high availability of the data. It is very difficult to predict the failure of a data store in advance, and one cannot take preventive measures. Ensuring that data is readily available is the biggest challenge for developers. A better way to address this challenge is to maintain a proper backup or replica mechanism, which will ensure that the data is highly available. Most of the traditional application developers rely on third-party tools to ensure backup and disaster recovery. A developer can use distributed NoSQL databases or modern relational databases that support automatic replication and disaster recovery (see https://www.mongodb.com/nosql-explained, accessed December 11, 2015).

13.2.5   Vendor Lock In

Vendor lock in is the major problem that the most of the cloud customers face, especially in public clouds. If the SaaS application is hosted with a public cloud provider it can be very difficult to migrate to other cloud providers. The worst part is that the SaaS application also cannot be migrated to an on-premises datacenter (da Silva and Lucredio, 2012). For example, many public PaaS providers like Google App Engine use their own way of developing and deploying applications online, which is different from other service providers. If an application is developed using Google App Engine, it is very difficult to port that application to other service providers. The SaaS developers can address this problem by using some open-source programming languages such as PHP, Perl and Python, which are supported by most of the PaaS providers. The vendor lock in can be eliminated by using standard APIs and following standard application development approaches, such as model-driven approaches (da Silva and Lucredio, 2012).

13.2.6   Usability

The SaaS application allows the end user to access applications from any device such as a mobile, tablet or laptop. This feature imposes the challenges to the developer that the SaaS application should be compatible with multiple devices. After Web 3.0, most of the Web applications are responsive. Software as a service developers can start using responsive application development platforms in order to provide high-quality and adaptable SaaS. Most importantly, the SaaS applications are used by a wide range of customers. The developer should provide some user-level customization (Chate, 2010).

13.2.7   Billing and Provisioning

Automated billing or metering (Chate, 2010) of the application based on customer use is an important component of SaaS applications. Most of the traditional applications do not have this feature (da Silva and Lucredio, 2012), and it is unavoidable in SaaS applications. The developer should develop the SaaS application in such a way that automated invoices or bills will be sent to the customer (Chate, 2010). Developers also have to provide the facility to customers to pay the bill through credit cards and electronic fund transfer. The other challenge for the SaaS developer is to enable quick service provisioning to customers as soon as they agreed to use the services. The application should also be able to allow customers to upgrade or downgrade their subscription plans. Automated billing and quick provisioning are still a problem with most SaaS applications.

13.2.8   Automatic Updates

In traditional applications the customers can become tired of updating the software and applying patches through the Internet. This is one of the reasons for customers to look for SaaS applications. As SaaS applications are managed by service providers, the customers have no need to worry about the updating and patching. The automated updating feature of the SaaS application benefits the customer – but there are a lot of challenges for the developer to develop software with automated updating features. Whenever the developer thinking of bulk updates, the update should not make the software unavailable, nor it should affect the normal behavior of the customer. Developers can solve this issue by keeping the update size small and scheduling updates for when there is less user activity.

13.3   SaaS Development Platforms

Platform as a service platforms can be used to develop SaaS application. The PaaS platforms change the way that the software is developed and deployed when compared to traditional application developments. Generally, PaaS offers a virtual application development platform (Lawton, 2008) to the developers and hides the complexity of maintaining the underlying infrastructure. The SaaS developers can use the benefits given by PaaS platforms and can concentrate on application development and addressing the challenges of SaaS application development. This subsection discusses some of the public and private PaaS platforms that can be used to develop SaaS applications.

13.3.1   Public PaaS Platforms

The applications that are developed using public PaaS platforms will be hosted in the service providers' infrastructure. The PaaS provider will maintain the development platform and the underlying infrastructure, which can be shared by different SaaS developers. Those SaaS developers who have limited infrastructure and are looking for off-premises applications can use public PaaS. Many public PaaS providers offers Web Command Line Interface (Web CLI), Web UI, IDE integration and APIs to access the development platform. One of the disadvantages of the most public PaaS platform is that they require Internet connectivity for access. However, some of the PaaS providers like Google App Engine and OpenShift Online allow the developers to develop applications offline and can push the application online whenever there is Internet connectivity. Most of the PaaS providers are charging customers based on their usage or monthly subscription plans. Some of the providers, like Microsoft Windows Azure, provide a free trial to familiarize potential customers with the platform. Table 13.1 gives a summary and comparison of some popular public PaaS providers. Public PaaS enables faster development and deployment of SaaS applications online, which is useful to many

Table 13.1 *Comparison of popular public PaaS providers*

Provider	Pricing / licensing	Supported languages	Supported frameworks	Supported databases	Client tools
Cloud Foundry	Open source and proprietary	Python, PHP, Java, Groovy, Scala, and Ruby	Spring, Grails, Play, Node.js, Lift, Rails, Sinatra, and Rack	MySQL, PostgreSQL, mongoDB, and Redis	cf. CLI, IDEs, and Build Tools
Google App Engine	Free quota and paid options available	Python, Java, Groovy, JRuby, Scala, Clojure, Go, and PHP	Django, CherryPy, Pyramid, Flask, web2py and webapp2	Google Cloud SQL, Data store, Big Table and Blob store	APIs
Heroku	Pay per use	Ruby, Java, Scala, Clojure, and Python, PHP, and Perl	Rails, Play, Django, and node.js	ClearDB, PostgreSQL, Cloudant, Membase, MongoDB, and Redis	CLI, and RESTful API
Microsoft Windows Azure	Free trial, pay per use, and subscription basis	.NET, PHP, Python, Ruby, and Java	Django, Rails, Drupal, Joomla, WordPress, DotNetNuke, and Node.js	SQL Azure, MySQL, mongoDB, and CouchDB	RESTful API, and IDEs
Red Hat OpenShift Online	Commercial	Java, Rupy, Python, PHP, and Perl	Node.js, Rails, Drupal, Joomla, WordPress, Django, EE6, Spring, Play, Sinatra, Rack, and Zend	MySQL, PostgreSQL, and MongoDB	Web UI, APIs, CLI, and IDEs

developers; but public PaaS fails to fulfill the needs of many private enterprises and independent software vendors (ISVs). Most public PaaS providers are not offering the variety of development platforms or programming languages that private enterprises and ISVs are looking for. The major disadvantage of public PaaS offerings is vendor lock in. They do not allow the developers to migrate the application from one PaaS platform to another PaaS platform. Because of vendor lock in, most well established enterprises started using private PaaS to develop their SaaS applications.

13.3.2 Private PaaS Platforms

It is very difficult to port or migrate the application from one public PaaS provider to other public PaaS providers. Most importantly, there is no security for data that is stored on a public shared platform. This forces the enterprises to go for a private PaaS instead of a public one. There are two ways to use a private PaaS: one is a private PaaS hosted and maintained by a third party and the other one is a self-managed private PaaS. If the enterprise does not have the required IT infrastructure and skilled IT staff, then it can go for a third-party-hosted private PaaS. If the enterprise has enough resources and skilled IT staff, then on-premises private PaaS will be the best option. Table 13.2 gives a summary and comparison of some of the popular tools available to enable a private PaaS platform in an on-premises datacenter. Once a private PaaS is built, and if it works

Table 13.2   *Comparison of popular private PaaS providers*

Provider	Pricing / licensing	Supported languages	Supported frameworks	Supported databases	Client tools
Activestate Stackato	Free for Micro Cloud, and Paid for enterprise	Java, Perl, PHP, Python, Ruby, Scala, Clojure, and Go	Spring, Node.js, Drupal, Joomla, WordPress, Django, Rails, and Sinatra	MySQL, PosgreSQL, MongoDB, and Redis	CLI and IDE
Apprenda	Free trial and Annual Subscription	.NET and Java	Most of the frameworks from .NET	SQL server	REST APIs
CloudBees	Free and monthly payment	Java, Groovy, and Scala	Spring, Jrails, Jruby, and Grails	MySQL, PostgreSQL, mongoDB, and CouchDB	API, SDK, and IDEs
Cumulogic	Free trial and monthly payment	Java, PHP, and Python	Spring, Grails	MySQL, mongoDB, and Couchbase	RESTful API
Gigaspaces Cloudify	Open Source	Any programming language specified by recipe	Rails, Play, and others	MySQL, mongoDB, Couchbase, Cassandra, and others	CLI, Web UI and REST API
Redhat OpenShift Enterprise	Free trial and paid for enterprise	Java, Ruby, Python, PHP, and Perl	Node.js, Rails, Drupal, Joomla, WordPress, Django, EE6, Spring, Play, Sinatra, Rack, and Zend	MySQL, PostgreSQL, and MongoDB	Web UI, CLI, and IDEs

perfectly, the enterprises experience secure deployment of applications and data. The private PaaS also can benefit the ISVs as most of the private PaaS tools support a variety of platforms and also allow the developers to build custom development environments. A private PaaS also allows sharing the platform with other enterprises, which will result in a community PaaS. The community PaaS will give a cost-effective and secured development platform for organizations with same goals. Ultimately, the decision about the PaaS deployment model will be based on the requirements and available infrastructure of the enterprise.

13.4   Multitenancy at Database Level

Multitenancy allows the service provider to deliver a single instance of service to multiple customers. It ensures effective resource utilization and reduces consumption of resources and energy, leading to green IT solutions. Multitenancy can be provided at infrastructure, platform and application levels of cloud computing. This subsection discusses the different multitenancy architectures at database level that the developer needs to consider when developing SaaS applications. However, the multitenant environment reduces the cost by sharing most of the resources; the customers are at risk of data leakage. Most of the SaaS vendors are struggling to ensure the security of customers' private information. Now the challenge for the SaaS developers is that they have to share most of their resources and at the same time they also have to protect the customer's data. When the developer shares more resources, it adds more complexity to the application. The developers have to use proper multitenancy and data isolation models based on the security of the user's data. The data isolation can be either logical or physical. Usually, in most of the SaaS applications, logical isolation will be preferred for efficient resource utilization. Multitenancy at database level in an SaaS application can be achieved in three different ways (Chong, 2012):

* separate database;
* shared database and separate schema;
* shared database and shared schema.

Table 13.3 shows the cost and data security of the different multitenancy models. The developer should choose any one of the multitenancy models based on the cost and security of the user data that will be stored in the application, as shown in Figure 13.2.

Table 13.3   *Cost and data security of different multitenancy models*

Multitenancy model	Cost	Data security
Separate database	High	High
Shared database and Separate schema	Moderate	Moderate
Shared database and Shared schema	Low	Low

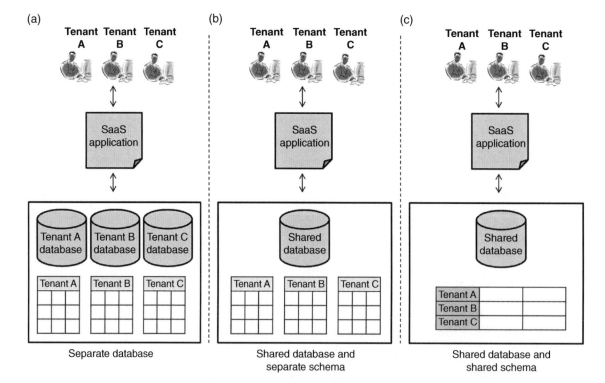

Figure 13.2   *Multitenancy at database level*

Multi-Tenancy Model 1: Separate Database

In this model, data isolation is good as a separate database is used for different customers. This is the preferred model when the data stored is highly confidential. It is the costliest of all the multitenancy models. The disadvantage of this model is that it does not encourage resource utilization. It requires more storage space because the same set of tables will be created in each customer's database. The developer has to execute separate database queries for the same set of operations because the databases and table structures are different. The complexity of the application is increased in terms of space and maintenance.

Multi-Tenancy Model 2: Shared Database and Separate Schema

Unlike Model 1, this model uses the same database for storing customer data and hence increases resource utilization. The data isolation is provided in the schema level as different schema will be used for different customers. The complexity of customizing the data is reduced as the same queries can be executed on the database with a different schema name. It reduces the cost and the difficulties in maintaining the data. The disadvantage of this model is that still it needs more space to store the data because the same tables will be repeatedly created in different customer schemas.

Multi-Tenancy Model 3: Shared Database and Shared Schema

This is the preferred model for a cost-effective SaaS application that does not contain confidential data. It increases resource utilization as the database and schema are shared by all customer data. The maintenance of data becomes easy because there is no need to write separate queries for different customer. The advantage of this model is it reduces cost and storage as both database and schema are shared. The main disadvantage is that it will affect all customer data if any one table is corrupted. The probability of data leakage is also high compared to other two multitenancy models.

13.5   Best Practice

There are many developers who have developed traditional Web applications and want to reuse their existing Web applications to develop SaaS applications. They cannot provide the normal Web applications as SaaS applications directly. There are some important constraints that the Web applications should satisfy to become SaaS applications. This subsection discusses some of the best practices (Petersson, 2011; Tier 3, Inc., n.d.) that developers should follow to make their traditional Web application a SaaS application.

Decide Which Application is to be Provided as a SaaS

The developers cannot provide all the applications as the SaaS applications. They should analyze the following before provide an application as SaaS:

- What is the business objective and expected outcome?
- What will the return on investment be (ROI)?
- What will be the risks and challenges?

- Who are all the targeted customers and what are their requirements?
- What will be the sensitivity of the user data?

   If the answer to all of these questions is positive with respect to a transformation to a SaaS application, then the service provider can start restructuring its existing applications as SaaS applications.

Architect the Application to Support Multitenancy

After deciding about the application, they need to sit with application architect and design application architecture by keeping all risks and challenges in their mind. For SaaS applications data security is the important constraint. Based on the sensitivity of the data the developer needs to choose the appropriate multi-tenancy model. They should architect the application for better scalability and high availability.

Use NoSQL Databases instead of Relational Databases

Normally, the developers may use relational databases as their back-end store for traditional applications. Since the SaaS application handles big data and big users, it is recommended to change the back end to NoSQL databases to ensure better scalability and high availability. With support for horizontal scalability, NoSQL databases ensure high scalability and automated replica mechanisms. The developers can use cache mechanisms like Memcached to increase the efficiency of the application.

Restructure the Application to Support Multiple Devices and User-Level Customization

The SaaS application may be accessed from any device such as a laptop, mobile, tablet, or others. Normally, traditional applications are compatible with laptops and desktop PCs. The developer should change the user interface to support multiple devices by following responsive Web design methodologies. Users also want to change or customize the application for their own comfort. The SaaS developers should also provide some mechanism to allow the users to customize the application according to their needs.

Select an Infrastructure Provider to Host the Application

Providers of SaaS can deploy the application in their own datacenters or they can obtain the service from any IaaS provider. Before choosing any third-party infrastructure provider, their network connectivity, traffic on their servers, and physical security measures should be reviewed. The service provider should be selected in such a way that the place of the infrastructure provider is nearer to the end user's community to increase the efficiency of the application. Before choosing the infrastructure provider, it is necessary to ensure interoperability with other service providers and to avoid vendor lock in.

Enable On-Demand Self-Service Signup

The SaaS application should allow the customers to sign up for the services without the manual intervention of the service provider. Normally, the service provisioning system should be automated, allowing customers to obtain their services as soon as they register. The service provisioning system should be on demand and should allow the customers to activate or deactivate their services whenever they want. Each SaaS application should have Web form-based registration mechanism for rapid provisioning of services to the customers.

Employ an Automated Billing Mechanism

The SaaS application should have the ability to automate the billing and payment mechanism. The application should calculate the bill based on use by the particular user and should notify the user. It also should be able to provide the mechanism for the administrator to have the summary of per tenant usage, per service usage. To increase customer satisfaction, the developer can incorporate a secured online payment mechanism in the application itself.

Incorporate the Functions to Monitor, Configure, and Manage the Application and Resources

As many customers are sharing the same instance of a single application, any misbehavior from one tenant will affect the other tenants and the underlying resources. The application should have a monitoring mechanism to detect and prevent the misbehaviors, security attacks, and disasters. One of the tedious jobs in managing the SaaS application is scheduling the update. The update should not affect the normal behavior of the application. The developers need to schedule the update when the application is being less used – and this can be seen by analyzing use history of the application.

Define and Manage the Service Level Agreements (SLAs)

Providers of SaaS should define the SLA clearly to the end users before providing any services. Normally, the SLA should include details of the availability, response time and level of support provided. Simulations and modeling could be used to define a reasonable SLA. The SLA also should clearly mention the behaviors of the users in the multitenant application. The service provided to the customers should be disconnected if any misbehavior or SLA violation is encountered with any of the users. The service providers also should ensure that they are also not violating the conditions in the SLA.

Make the Support Team Work for 24/7

Finally, end-user support is an important component of SaaS. The SaaS provider should have the customer support team working for 24 hours every day to respond to customer queries. The customer support team should be trained to solve any application-related issues. It also has to keep a summary of the type of issues that arise, time taken to fix the issues, and the degree of customer satisfaction, to improve the performance of the customer support team. Providers of SaaS also need to incorporate a robust ticketing mechanism as part of their customer support service.

13.6   Conclusion

This chapter has highlighted some of the challenges in developing SaaS, and ways of addressing those challenges. It has discussed some popular public PaaS technologies, which will help the developers to provide highly scalable SaaS applications. The major problems with public PaaS are vendor lock in and security, which make ISVs and private enterprises move towards private PaaS. Private PaaS gives the developers freedom to work with any programming languages of their choice. It also ensures the security of the data by deploying the application on-premises or in third-party-managed private datacenters. Additionally, most of the private PaaS are polyglot platforms that support multiple languages and frameworks, and allow the open-source community to extend them further. The underlying database of the SaaS applications plays an important role in the performance of those applications. Some of the popular relational and NoSQL databases that can be used to develop SaaS applications were discussed. There is a need to transform traditional Web applications to SaaS applications. Best practice for doing this was also discussed. Nowadays, green cloud computing is receiving attention from researchers, which motivates developers to develop SaaS applications that support green computing. In future there will be a need for developers to adopt energy-aware application development methodologies that result in greener IT solutions.

References

Chate, S. (2010) *Convert your Web Application to a Multi-Tenant SaaS Solution*. Technical Article. IBM developerWorks, December.

Chong, R. F. (2012) *Designing a Database for Multi-Tenancy on the Cloud: Considerations for SaaS Vendors*. Technical Article. IBM developerWorks, June.

da Silva, E. A. N. and Lucredio, D. (2012) Software engineering for the cloud: A research roadmap. Twenty-sixth Brazilian Symposium on Software Engineering (SBES), 23–28 September, pp.71, 80.

Fishteyn, D. (2008) *Deploying Software as a Service (SaaS)*. White paper. WebApps, Inc., http://www.saas.com/files/SaaS.com_Whitepaper_Part1.pdf (accessed December 11, 2015).

Kepes, B. (2013) *Understanding the Cloud Computing Stack: SaaS, PaaS, IaaS*, http://www.rackspace.com/knowledge_center/whitepaper/understanding-the-cloud-computing-stack-saas-paas-iaas (accessed December 11, 2015).

Lawton, G. (2008) Developing software online with platform-as-a-service technology. *Computer* **41**(6), 13, 15.

Mell, P. M., and Grance, T. (2011) *The NIST Definition of Cloud Computing*. Special Publication 800-145. NIST, Gaithersburg, MD, http://www.nist.gov/customcf/get_pdf.cfm?pub_id=909616 (accessed November 25, 2015).

Petersson, J. (2011) *Best Practices for Cloud Computing Multi-Tenancy*. White paper, IBM Corp., http://www.ibm.com/developerworks/cloud/library/cl-multitenantcloud/ (accessed December 11, 2015).

Tier 3, Inc. (n.d.) *Six Best Practices to Cloud Enable Your Apps*. White paper, https://www.ctl.io/lp/resources/Tier%203_cloud_enable_your_app.pdf (accessed December 11, 2015).

Part IV

Cloud Integration and Standards

14

Cloud Portability and Interoperability

Beniamino Di Martino, Giuseppina Cretella, and Antonio Esposito

Second University of Naples, Italy

14.1 Introduction

Cloud computing is emerging as the most promising technology for software development, changing the way in which customers interact with their data and applications. There are a lot of reasons behind the choice to move to cloud computing:

- Companies no longer need to buy, store, and maintain expensive infrastructures, reducing time and money involved in maintaining, updating, and repairing their own equipment.
- Hardware dimensioning does not have to be related to peak workload any more but the infrastructure can be scaled dynamically according to current needs. This results in a better use of resources.
- Customers pay only for the resources they actually use, following a "pay-as-you-go" paradigm, thus saving money.
- Using distributed resources, including datacenters and computing nodes, can enhance systems' resilience and improve recovery from disasters.
- The ability to choose among a broad range of available resources and services should lead to strong competition among cloud providers, resulting in better quality and lower prices for customers.

These are just a few of the possible benefits that we could derive from the adoption of a cloud-computing paradigm. But, despite the diffusion of cloud computing, differences in offerings from cloud providers still exist, which prevent changing or exploiting multiple cloud providers.

Two main issues that are currently preventing free exploitation of different or multiple cloud providers are portability and interoperability among cloud platforms at different service levels. A cloud provider is often interested in offering a technological solution that differentiates it from the others: these

Encyclopedia of Cloud Computing, First Edition. Edited by San Murugesan and Irena Bojanova.
© 2016 John Wiley & Sons, Ltd. Published 2016 by John Wiley & Sons, Ltd.

differences have the drawback of locking the customers in as no alternatives are provided. Even open-source technologies for setting up private clouds are often not compatible with each other.

The problems of interoperability and portability affect the cloud-computing panorama in different ways. The brokering, negotiation, management, monitoring, and reconfiguration of cloud resources are challenging tasks for the developer or user of cloud applications due to different business models associated with resource consumption and the variety of features that the cloud providers are offering. These points become very important when the landscape is a multicloud environment and the main concern is represented by the vendor lock-in problem.

This chapter addresses the problem of cloud interoperability and portability, offering an overview of different methodologies, solutions, and initiatives.

14.2   Interoperability Issues in Cloud Computing

In general terms, interoperability can be defined "as a measure of the ability of performing interoperation between two or more different entities (that can be pieces of software, processes, systems, business units, etc.)" (Vernadat, 2010). In the case of cloud computing platforms, the meaning of interoperability is not always clear.

To some, interoperability might represent applications that are running in different cloud environments being able to share information through a common set of interfaces. Achieving interoperability means to enable the use and exchange of information between systems so that they appear to be one, so the APIs for service implementation need to be vendor neutral (Panhelainen, 2012). For example, in the cloud, there is no common interface to access different relational database management systems (RDBMS) at the same time but cloud vendors have their own solutions and APIs for handling databases. In this case, the focus is on applications – thus we are at "platform as a service" (Paas) level.

Let's consider the simplified scenario shown in Figure 14.1.

In this example a generic application, running on a server hosted by Cloud Provider A, needs access to a database residing in the same Cloud. What if such an application needed to access data stored on a different

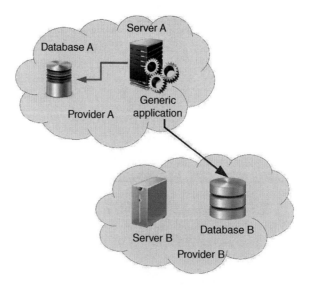

Figure 14.1 *Simple interoperability scenario*

database, hosted by Cloud Provider B? There can be several reasons to split data among storages hosted by different platforms:

- enhancing elasticity and failure recovery;
- exploiting better security policies (at higher prices) for sensitive data;
- exploiting certain capabilities not supported by all the providers;
- managing data in a hybrid cloud scenario (for example, A could be public and B private).

In a fully interoperable world, the application could access data from both databases through a common interface. In the real world, differences in APIs, data and message formats or communication protocols would represent a real obstacle for developers. The situation is worse if we consider management of access to both databases.

From another perspective, interoperability refers to the possibility for customers to use the same management tools, server images, and other similar software in a variety of cloud computing platforms. Here, the focus is on the infrastructure's management and monitoring, so we are at the "infrastructure as a service" (IaaS) level. If we consider the scenario drawn in Figure 14.2, we can have a better idea of this issue. Like in Figure 14.1, a generic application tries to access two databases residing on different cloud platforms; but now monitoring software, operating from cloud platform A, tries to access information regarding all the components involved in the application's work, including Database B. Again, there can be several reasons to deploy a monitoring component in a cloud environment:

- network traffic analysis;
- tracking of accesses to databases and or servers;
- workload balancing;
- failure detection.

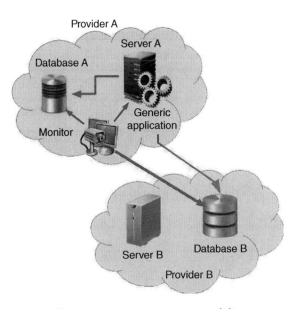

Figure 14.2 *Monitor interoperability*

If the cloud platforms involved were interoperable, the monitoring object would be able to access information regarding Database B freely: but rarely, if ever, would this happen in the real world because of the lack of common interfaces, for security or commercial reasons.

14.3   Portability Issues in Cloud Computing

Cloud portability is the ability of data and application components to be moved and reused easily regardless of the provider, operating system, storage, format or API (Cohen, 2009).

As for interoperability, portability issues differ slightly according to the target service level. At IaaS level we speak mostly of component and data portability. In particular, there are two kinds of component portability:

* *Runtime source portability.* Reuse of platform components across cloud IaaS services. An example (Figure 14.3) would be represented by the migration of an operating system (OS), with all the applications working on it, from one cloud environment to another. This would be possible if the OS supported the target hardware but this is not always the case.
* *Machine image portability.* Reuse of virtual images containing applications and data with their supporting platforms. In this situation it is possible to exploit virtualization techniques and standard image formats to avoid incompatibilities (Figure 14.4).

Porting data is much more difficult because storage models and formats often vary among platforms. We should also consider the effort needed to actually move data among cloud platforms, which is not free of charge. However, Cloud introduces no new technological problems: commercial agreements among vendors can make the existing technical problems more serious.

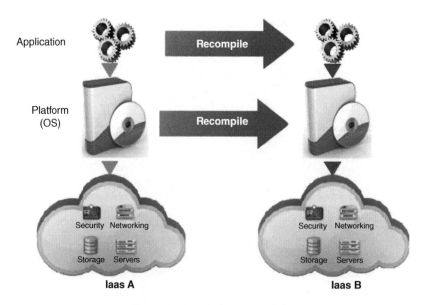

Figure 14.3   *Platform portability*

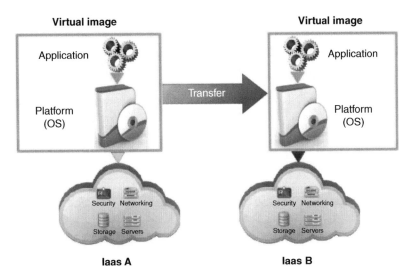

Figure 14.4   *Machine images portability*

At PaaS level we focus on application portability, which requires a standard interface, exposed by the supporting platform, enabling the application to use services and protocols provided by the platform, as well as providing access to the capabilities that support the application. This means that, even when standard languages are used, portability at PaaS level is not guaranteed, because implementations of platform services may vary between providers.

Generally, when portability is possible at IaaS level, application portability at PaaS level is automatically enabled.

PaaS portability issues are caused by:

- *Lack of a shared platform definition among PaaS providers.* Each provider chooses the operating system and middleware elements it will support and if competitors make different selections then applications using those features cannot be ported.
- *Lack of alternative providers for a platform.* A provider may make a cloud version of a server platform available, discouraging competition thanks to his established dominance. Providers of IaaS have been adding services steadily to their IaaS platform and, as there are no standards for these added services, using them could lock applications to a cloud provider.

Another important issue concerns portability between the development and operational environments. Platform as a service is particularly attractive because it avoids the need for investment in systems that will be dismissed once the development is complete. If a different environment will be used at run time, however, it is essential that the applications can be moved unchanged between such environments. Yet another aspect to be considered regards software modernization, which is still a significant challenge in general and even more ambitious when a change of the software delivery paradigm needs to be addressed, as in the case of cloud computing.

In general, the lack of portability can be seen as a barrier to the adoption of cloud computing because organizations fear "vendor lock in": once customers have selected a cloud provider, either they cannot move to another provider or they can do it but only at great cost. Risks of vendor lock in include reduced negotiation power in reaction to price increases and service discontinuation because the provider goes out of business.

14.4 Achieving Portability and Interoperability

As interoperability and portability issues can be seen from different perspectives, several approaches have been proposed to solve them. Such approaches can be divided into three main categories:

- *Framework and model-based approaches* try to exploit, extend or define frameworks and models, which, once applied to the cloud computing paradigm, can ease interoperability and or portability issues.
- *Adapting methodologies* try to solve interoperability and portability issues among different cloud platforms by interposing a level of indirection between incompatible interfaces, APIs or protocols, by means of "adapters" or "plug ins" working as mediators.
- *Standardization efforts* aim at defining homogeneous and shareable standards, which, once adopted by every cloud vendor, would completely eradicate incompatibility issues. This category includes *standard proposals* (resulting from research activities conducted by Standardization Committees) and *de facto* standards (imposed by leading cloud vendors).

14.4.1 Framework and Model-Based Approaches

14.4.1.1 Agent-Based Frameworks

Multi-agent systems (MASs) seem to offer one of the most effective approaches for reducing interoperability issues. An MAS can be described as a computerized system composed of interacting intelligent agents, collaborating within the same environment. The outcome of the mOSAIC research project, in particular, demonstrated, in the Cloud Agency (Venticinque *et al.,* 2012), the benefits of adopting this kind of cloud multiagent architecture. The Cloud Agency is a service for the deployment and execution of mOSAIC applications on multiple clouds.

14.4.1.2 Model-Driven Engineering Approaches

The *OMG model-driven architecture (MDA)* is a model-based approach for the development of software systems. The main benefits of MDA from the cloud perspective are the facilitation of portability, interoperability, and reusability of parts of the system that can be easily moved from one platform to another, as well as the maintenance of the system through human-readable and reusable specifications at various levels of abstraction. In the context of cloud computing, *model-driven development* can be helpful in allowing developers to design a software system in a cloud and to be supported by model transformation techniques in the process of instantiating the system into specific and multiple clouds.

For this reason, combining model-driven application engineering and the cloud computing domain is currently the focus of several research groups and projects: Model-Driven Approach for the Design and Execution of Applications on Multiple Clouds (MODACLOUDS) (Ardagna *et al.*, 2012), Advanced Software-Based Service Provisioning and Migration of Legacy Software (ARTIST) (Menychtas *et al.*, 2013), Model-based Cloud Platform Upperware (PaaSage) (Bubak *et al.*, 2013), Intercloud Architecture (ICA) (Demchenko *et al.*, 2012).

14.4.1.3 Semantic Modeling

One of the factors contributing to interoperability and portability problems is the difference in the semantics of the resources offered because no uniform representation exists. As Sheth and Ranabahu (2010) have stated, semantic models are helpful in three aspects of cloud computing:

- *Functional and nonfunctional definitions,* that is the ability to define application functionalities and quality-of-service details in a platform-agnostic manner;
- *Data modeling,* including metadata added through annotations pointing to generic operational models, which play a key role in consolidating API's descriptions;
- *Service description* enhancement, in particular as regards service interfaces that differ between vendors even if the operations' semantics are similar.

Existing technologies inherited from the semantic Web discipline can be useful to address these aspects:

- Web ontology language (OWL) can define a common, machine-readable dictionary able to express resources, services, APIs and related parameters, service-level agreements, requirements, offers and related key performance indicators (KPIs).
- OWL for Services (OWL-S) adds semantics to cloud services in order to enable users and software agents to automatically discover, invoke and compose them.
- SPARQL performs queries to retrieve resources according to particular constraints.
- Semantic Web rule language (SWRL) expresses additional rules and heuristics.

All these aspects are addressed by the mOSAIC project – in particular in two components of the mOSAIC framework: the *semantic engine* (Cretella & Di Martino 2012) and the *dynamic discovery and mapping system* (Cretella and Di Martino, 2013).

14.4.1.4 *"Write Once Run Anywhere" (WORA) approaches*

"Write once, run anywhere" is a slogan created by Sun Microsystems to illustrate the cross-platform benefits of the Java language. The approach followed by Sun was to provide a Java virtual machine (JVM), which, once installed on a target device, could interpret and execute any standard Java byte code, regardless of where such a byte code had been created. This approach is very similar to the one followed by some PaaS providers, such as CloudFoundry and OpenShift, which provide elastic application containers promising ubiquitous availability and portability of software. Other providers, instead, offer full API compatibility with the most common PaaS platforms, thus enabling developers to leverage a common programming interface. This is the case for AppScale, which is fully compatible with Google App Engine and virtually deployable anywhere that a virtual machine can exist (Bunch, *et al.*, 2011).

14.4.2 Adapting Methodologies

14.4.2.1 *OpenNebula*

OpenNebula is a cloud computing toolkit that enables consumers to manage heterogeneous and distributed datacenter infrastructures, specifically addressing IaaS platforms and focusing on virtualized infrastructures in datacenters or clusters (private clouds). Support for hybrid solutions, which combine local and public cloud infrastructures, is also provided, together with interfaces to expose the public cloud's functionalities for virtual machines, storage, and network management.

Apart from a set of native APIs, offered via XML-RPC and Java or Ruby bindings, OpenNebula also implements Amazon EC2, OGF OCCI, and vCloud APIs. Interoperability is supported by leveraging existing standards, which leads to the development of adapters and transformers for APIs provided by different vendors. There is a remarkable possibility to exploit adapters for DeltaCloud, introduced below, and Libcloud, a standard client library supporting popular cloud providers, written in Python.

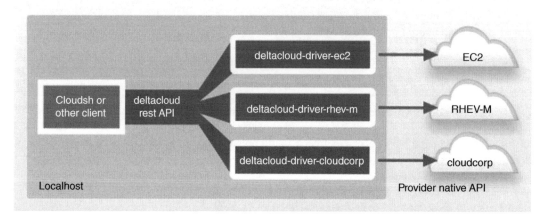

Figure 14.5   *Use of DeltaCloud API*

14.4.2.2   DeltaCloud

DeltaCloud is an open source project that aims to define a REST-based API to access any kind of cloud platform exposing its services at IaaS level. Written entirely in Ruby, it offers the user the opportunity to leverage native Deltacloud, DMTF, CIMI, and Amazon-EC2 APIs, representing the Deltacloud abstraction API. Such an API works as a wrapper around these platforms, abstracting and hiding their differences (see Figure 14.5. For each cloud provider there is a driver interpreting that provider's native API, so users do not need to deal with it directly.

14.4.2.3   Open-Source API and Platform for Multiple Clouds (mOSAIC)

The mOSAIC project offers an API to develop components that run on the top of its platform. This API was designed to be event driven and communications among mOSAIC components take place through message queues. The mOSAIC's basic component is the Cloudlet, an event-driven and stateless element, able to access cloud services through connectors, introduced to ensure the independence from the cloud service interfaces.

14.4.3   Standardization Efforts

Currently no standard has been accepted worldwide yet to solve interoperability and portability issues definitively. Instead, different efforts have been made towards the definition of such a standard, addressing the problem from different points of view. Studies have been carried out to collect existing standards and proposals, in order to determine the specific cloud issues they can solve or to define how such standards can be used to build a cloud infrastructure. Of course, cloud-specific standards are in development, both for IaaS and PaaS offers. In this scenario, leading cloud vendors are heavily influencing the development of new standards, sometimes even imposing *their* own standards in the market.

14.4.3.1   Cloud Standards Coordination Initiative

Following a direct request from the European Community, the European Telecommunications Standards Institute (ETSI) launched the cloud standards coordination (CSC) initiative. The main objective of the initiative was to identify a set of standards to apply in different cloud environments through a series of use cases, in order to support both service customers and administrators in choosing the correct technologies according to their requirements. The final report produced by the group, released in November 2013, describes in detail a set of cloud-related use cases. Each use case focuses on a specific cloud issue: in particular, the "Cloud Bursting" case deals with interoperability. Some of the standards described in the report are presented in the following sections.

14.4.3.2   Standard for Intercloud Interoperability and Federation

The Standard for Intercloud Interoperability and Federation (SIIF) project, aims at defining a topology, a set of functionalities, and a governance model for cloud interoperability and federation. The current standard, still in development, focuses on the description of the *inter-cloud topology*, which makes reference to the NIST definition of cloud computing, defining in detail its components and the relationships among them.

A set of protocols and standards, not exclusively related to the cloud, is reported in order to respond to a series of requirements:

- extensible messaging and presence protocol (XMPP), together with its extensions, XEP and XWS4J, is suggested for communications among InterCloud gateways;
- transport layer security (TLS), simple authentication and security layer (SASL) and security assertion markup language (SAML) are used to provide secure authentication in federated environments;
- XMPP-based RDF and SPARQL are investigated for service discovery.

14.4.3.3   Cloud Application Management for Platforms

OASIS Cloud Application Management for Platforms (CAMP) TC aims at defining models, mechanisms, and protocols for the management of applications in a PaaS environment, in order to develop an interoperable protocol for PaaS management interfaces that users can exploit to build, deploy, and administer their applications. To be more precise, CAMP's goal is to define a simple standard RESTful API, along with a JSON-based protocol, with an extensibility framework that enables interoperability across multiple vendors' offerings.

Currently the standard defines a set of basic APIs that a cloud vendor should provide as part of its PaaS offer, in order to manage the building, running, administering, monitoring, and patching of applications. This would enable interoperability among self-service interfaces to PaaS clouds through the definition of artifacts and formats shared among conforming platforms. This would also allow independent vendors to create tools and services that communicate with any CAMP-compatible platform, using the defined interfaces and vice versa.

14.4.3.4   Topology and Orchestration Specification for Cloud Applications (TOSCA)

The purpose of OASIS Topology and Orchestration Specification for Cloud Applications (TOSCA) is to enhance portability of cloud applications and related IT services by defining an interoperable description of cloud services, of the relationships existing among components of these services, and of their operational behavior, in a way that is independent of the cloud provider offering the services and of the technologies involved. In particular the TOSCA technical committee will leverage the existing Topology and Orchestration Specification for Cloud Applications document submitted by a number of cloud vendors (among which IBM and Red Hat are remarkable contributors), through a process of revision and extension of the proposed XML Schema. Scope of the work is the definition of a language for the specification of a service template that can describe the topology of a service, also by exploiting existing workflow languages (especially BPMN 2.0) to define orchestration. TOSCA mainly focuses on the description of services and of their relationships, not on the IaaS infrastructure. To manage infrastructure, other standards, like CIMI, are more suitable: a provider could easily manage the cloud infrastructure required by a service described using TOSCA with CIMI.

14.4.3.5   Cloud Infrastructure Management Interface

Cloud Infrastructure Management Interface (CIMI) is a standard proposed by the Distributed Management Task Force (DMTF), which specifies an interface to manage IaaS cloud platforms. The specification documents focus on the description of a RESTful interface but the standard separates the API design from the particular communication protocols to use.

The CIMI model defines a set of resources, associated templates, and configurations that can be accessed, operated and managed through basic HTTP methods in a RESTful fashion. These include cloud entry points to access lists of all available assets, virtual machines, storage, network, and monitoring resources. Security issues are also addressed by the interface, with a focus on the client's identification.

Using the CIMI interface, a consumer of an IaaS service can request and control the infrastructure offered by the provider. The consumer can create, wire up, and control the infrastructure for an entire system on an IaaS cloud but the interface is not intended to go beyond its management. Software may be preinstalled on the images CIMI instantiates but the interface does not reach beyond the initial instantiation. The consumer must use other means to install and manage complex services implemented on an IaaS cloud: in this case other standards like TOSCA can be useful.

14.4.3.6   Cloud Data Management Interface

The Cloud Data Management Interface (CDMI) is a standard for managing data on cloud platforms, proposed by the Storage Networking Industry Association (SNIA).

It defines a functional interface that users and applications can use to create, retrieve, update, and delete data elements from the cloud. Using the interface, clients can also discover the capabilities offered by the cloud platform and manage the containers and the data that is placed in them, together with metadata associated with both containers and data. Administrative and management applications can leverage the interface to manage containers, accounts, security access, and monitoring/billing information.

The CDMI interface is RESTful and no other implementations will be included in the standard in the near future. Unlike other standard interfaces using both XML and JSON to code messages, CDMI only supports JSON in order to reduce the dimension of the payloads.

14.4.3.7   Open Cloud Computing Interface

The Open Cloud Computing Interface (OCCI) is a community-specified standard published by the Open Grid Forum. The standard aims at describing an interface to manage IaaS cloud infrastructures, with goals similar to CIMI, but the available documentation points out its applicability to PaaS and SaaS. Much like other standards, OCCI defines a set of core elements that fully describe a cloud infrastructure but these elements are defined using a hierarchical structure. A user can define a new kind of resource or relationship just by extending one of the existing elements, as if working with an object-oriented language. Unlike CIMI, which offers predefined resources that are intended to meet most consumers' needs, the OCCI model doesn't provide templates for resources or links. While the model is able to define any kind of feature associated to a resource and is never limited by existing definitions, the developer will have to describe every implementation in detail in order to make it usable and understandable to all of its possible users. Without some kind of external agreement it could also be difficult for a provider to support extensions provided by another.

14.4.3.8   *De Facto Standards*

A de facto standard is a standard that has been accepted and adopted throughout the industry but has not been defined or endorsed by industry groups (such as W3 Consortium) or standards organizations (such as ISO). They are also known as market-driven standards. These standards arise when a critical mass simply likes them well enough to use them collectively. Market-driven standards can become *de jure* standards if they are approved through a formal standards organization.

As regards the IaaS offer, Amazon's AWS has been the market leader and comprehensively dominates its competitors, cementing its status as one of the world's leading options for cloud-based data storage and data warehousing. This is why many see AWS as the *de facto* standard in the public cloud. Their API is proven and widely used, their cloud is highly scalable, and they have by far the biggest traction of any cloud. The open-source counterbalance to Amazon's dominance is OpenStack. Managed by the OpenStack foundation, it is released under the Apache license and received a lot of support from large IT companies including Oracle, IBM, Red Hat, and RackSpace. These companies now include OpenStack-compliant solutions into their cloud offerings or they are starting to build their products completely around this open platform (IBM efforts are surely remarkable), thus suggesting OpenStack as a *de facto* standard for IaaS platforms.

Regarding PaaS, initially, the major players were Azure and Google and for this reason they are now widely adopted. However, recently other PaaS platforms that could be candidates to become a *de facto* standard emerged. Among those worthy of attention are OpenShift and Cloud Foundry.

Amazon AWS

This is a comprehensive cloud services platform that offers compute power, storage, network, content delivery and other functionalities.

The most popular service offered by Amazon is Elastic Compute Cloud (Amazon EC2) that provides resizable compute capacity in the cloud. Associated to EC2 Amazon also offers Auto Scaling and Elastic Load Balancing services. To store data on the cloud Amazon offers different ways: Simple Storage Service (S3 – a file system-like storage), Amazon Elastic Block Store (EBS – block level storage volumes), SimpleDB and DynamoDB (as a semistructured data store), Relational Database Service (RDS – a relational database SQL).

Amazon offers two types of integration services for system-to-system decoupling and messaging: Simple Notification Service (SNS) and Simple Queue Service (SQS).

OpenStack

OpenStack is a cloud operating system that controls computing, storage, and networking resources, managed by users through the available Web-based dashboard, command line tools, or RESTful APIs. The platform acts as an infrastructure as a service, providing services such as: *Compute,* enabling the provisioning and managing of large networks of virtual machines; *Object Storage,* a distributed storage system in which data are modeled as objects; *Block Storage,* providing access to block devices, which can be directly exposed and connected; and *Networking,* a scalable and API-driven system for managing networks and IP addresses,.

OpenStack Shared Services include a series of services exploited by the core components, providing a powerful means of integration.

The OpenStack community has worked to reduce incompatibilities with Amazon EC2 and S3, in an attempt to further enhance platform interoperability. However, interoperability is still a strong issue for the platform, since the OpenStack Foundation didn't have a policy or a set of standards that companies could adhere to in order to legitimately say they were OpenStack compliant. IBM's commitment to the project may lead to interesting developments, given its support for the OASIS Topology and Orchestration Specification for Cloud Applications (TOSCA) TC, together with the OpenStack Heat project, aiming at developing standards for workloads orchestration through templates.

Openshift

Openshift is a PaaS cloud computing platform provided by Red Hat whose objective is to provide a cloud environment in which developers can easily and quickly design, develop, build, host, and scale applications, by using one or more of the programming languages and frameworks supported. The Openshift platform is composed of one or more Brokers and a series of Nodes, which are grouped into sets managed by a specific Broker. The user interacts with the Broker only, using the available Openshift APIs, allowing developers to manage every aspect of the developed applications, from hosting to scaling.

An interesting feature of the platform is the option of choosing a specific programming language, database, or other services to use via standard or custom cartridges: developers can build applications in any programming language (given a standard or custom cartridge exists for that language), and then host them on a node through the broker's API.

One major advantage of this architecture is the possibility of running third-party applications on an Openshift node just by selecting the correct cartridge, thus strongly supporting portability.

Cloud Foundry

Cloud Foundry is an open-source PaaS, originally developed by VMware/SpringSource. Among supported languages and frameworks are Spring, Rails and Sinatra for Ruby, Node.js and Scala. To promote interoperability and portability among cloud providers, a set of common core services has been defined. Cloud vendors can support extra runtimes on top of the core set in order to compete, thus some companies have created public instances of cloud foundry implementations, differing in some features. While such implementations share a common set of services, they are not exactly the same: differences in programming platforms, and in some cases different versions of the same platform, lead to compatibility issues.

To cover offerings from different cloud providers, Cloud Foundry Core has been recently introduced. Core is a Web application that verifies public instances (Cloud Foundry Endpoints) against a common set of features, thus allowing users to know in advance if they can port their Cloud Foundry application to another cloud provider, by detecting the supported services and runtimes.

14.5   Conclusion

Due to the huge number of vendors, offers, and technologies involved, interoperability and portability issues among cloud environments are still far from a definitive solution. In this chapter, we presented the most relevant methodologies, research projects, standards and initiatives aiming to resolve interoperability and portability problems by addressing them from different points of view. Among methodologies, model-based approaches like the model-driven architecture (MDA) seem to be the most promising. European research projects produced very good results in defining new frameworks and standards: among these, mOSAIC clearly showed how semantic and agent-based technologies can ease interoperability and portability issues and lead to an effective and efficient solution. Commercial proposals distributed as open-source software and sustained by cloud vendors, like OpenStack, are steadily growing in importance and are being adopted by a growing number of consumers, also thanks to the support given by large communities of developers. Cloud vendors also support the creation and adoption of new standards, by proposing them to standardization groups: examples are CAMP for PaaS and TOSCA for IaaS, submitted to OASIS.

As no standard or framework to solve portability and interoperability issues stands above the others, research in this field is still needed, together with further investment. Of course it is also possible that, given all the challenges related to interoperability and portability that such a standard should address, together with all the different possible approaches, there will never be a unique solution: instead, vendors could adopt a set of interoperable standards, giving consumers the opportunity to choose among them according to their specific requirements.

References

Ardagna, D., Di Nitto, E., Mohagheghi, P., *et al.* (2012) *Modaclouds: A Model-Driven Approach for the Design and Execution of Applications on Multiple Clouds*. ICSE Workshop on Modeling in Software Engineering (MISE), June 2012. IEEE, pp. 50–56.

Bubak, M., Baliś, B., Kitowski, J., *et al.* (2011) *PaaSage: Model-Based Cloud Platform Upperware,* Department of Computer Science AGH, Krakow, Poland

Bunch, C., Chohan, N., and Krintz, C. (2011) *Appscale: Open-Source Platform-as-a-Service*. UCSB Technical Report 2011-01.

Cohen, R. (2009) *Examining Cloud Compatibility, Portability and Interoperability*, http://www.elasticvapor.com/2009/02/examining-cloud-compatibility.html (accessed December 2, 2015).

Cretella, G., and Di Martino, B. (2012) *Towards a Semantic Engine for Cloud Applications Development*. Sixth International Conference on Complex, Intelligent and Software Intensive Systems (CISIS), July 2012. IEEE, pp. 198–203.

Cretella, G. and Di Martino, B. (2013) Semantic and matchmaking technologies for discovering, mapping and aligning cloud provider's services. Proceedings of the 15th International Conference on Information Integration and Web-based Applications and Services (iiWAS2013), 2013.

Demchenko, Y., Makkes, M. X., Strijkers, R., and de Laat, C. (2012) *Intercloud Architecture for Interoperability and Integration*. Fourth International Conference on Cloud Computing Technology and Science (CloudCom), December 2012. IEEE, pp. 666–674.

Menychtas, A., Santzaridou, C., Kousiouris, G., *et al.* (2013) ARTIST Methodology and Framework: A novel approach for the migration of legacy software on the Cloud. Second Workshop on Management of Resources and Services in Cloud and Sky Computing (MICAS), September.

Panhelainen A. (2012) Interoperable and Portable Cloud Services. Seminar on Internetworking. T-110.5191. Aalto University, Spring 2012.

Sheth, A., and Ranabahu, A. (2010) Semantic modeling for cloud computing, part 2. *IEEE Internet Computing* **14**(4), 81–84.

Venticinque, S., Tasquier, L., and Di Martino, B. (2012) *Agents Based Cloud Computing Interface for Resource Provisioning and Management*. Sixth International Conference on Complex, Intelligent and Software Intensive Systems (CISIS), July 2012. IEEE, pp. 249–256.

Vernadat, F. B. (2010) Technical, semantic and organizational issues of enterprise interoperability and networking. *Annual Reviews in Control* **34**(1), 139–144.

15

Cloud Federation and Geo-Distribution

William Culhane,[1] Patrick Eugster,[1,2] Chamikara Jayalath,[1] Kirill Kogan,[3] and Julian Stephen[1]

[1] *Department of Computer Science, Purdue University, USA*
[2] *Department of Computer Science, TU Darmstadt, Germany*
[3] *IMDEA Networks, Spain*

15.1   Introduction: The Case for Federation

The term "cloud" nebulously defines a variety of services related to distributed storage and computation. There are many cloud providers, and the exact services provided vary based on the provider, even at the abstract level. For instance, a provider may provide any – or a combination of – infrastructure as a service (IaaS), platform as a service (Paas), or software as a service (Saas) (Lenk *et al.,* 2009). Consumers can choose one or several of these based on their own needs and development capabilities. Sometimes multiple offerings have to work together, be it because applications are built from different existing services or components hosted in different clouds, or by design. In fact, there are instances where one cloud service uses another. This is referred to as *vertical* federation, as one cloud offering takes precedence over another in a stack. It is different from *horizontal* federation, in which services do not strictly rely on accessing each other in the same order. Horizontal and vertical integration may happen in tandem, as when a processing service accesses multiple storage services of different clouds. Technologies including Eucalyptus (http://ieeexplore.ieee.org/stamp/ stamp.jsp?tp=&arnumber=5071863, accessed December 5, 2015) and OpenNebula (http://opennebula.org/, accessed December 5, 2015) can similarly be viewed as supporting vertical federation, building a service/ infrastructure on top of horizontally federated cloud services.

Some services are interchangeable, or nearly so. The most straightforward example of this is data storage. A consumer may choose to spread data between providers for security so that the full data is not known by any third party, or any individual breach does not gain access to the full data. Alternatively, a consumer may switch providers when a contract expires or a new product becomes available. In both these cases, data has to be accessible across clouds.

Sometimes the choice of which cloud offering to use is motivated by legal concerns, especially with regard to the location or accessibility of data (Winkler, 2011). For instance, the European Union issued a directive controlling where private data may traverse or reside. Even storing data within the EU may make that data subject to the relevant laws regardless of the origin of the data. In the United States, the Health Insurance Portability and Accountability Act and the Gramm-Leach-Bliley Act require strong protection for certain types of data, which limits the hosting options that comply with the Acts. It may also make sense to isolate such data, as it cannot be used for purposes beyond those stated when it was originally collected. Thus if there is any reason why the data may be considered sensitive, it is imperative to understand which data can and does fall under various jurisdictions to discern which laws govern its storage and use. If there are any disputes with the cloud provider, it is also necessary to know which laws apply.

In general, to support federation and to avoid a case where the consumer is locked in to a single provider ("vendor lock in") and a single set of tools it must be possible to (i) distribute data, (ii) compute on such data, and (iii) communicate it across clouds and thus most commonly across datacenters (Kurze *et al.*, 2011). Given the many different abstraction levels and even abstractions within the same level offered by different cloud vendors, we focus in the following on how to fundamentally solve these three issues efficiently; these are independent of mediation between specific technologies, APIs, and abstractions.

15.2   Distributed Data

There are several reasons for distributing data across datacenters and ways in which this can be done. For example, when global businesses offer cloud-based Web services on a global scale, they naturally want to serve customers from the closest datacenter. This is illustrated in Figure 15.1, where client C1 communicates with datacenter DC1, whereas C2 prefers DC3. This can lead to customer databases building up in different datacenters. However, many operations – for example, audits – have to involve logical datasets that are distributed across datacenters. Customers may not want to pay to move the data to one place, or such an operation may be infeasible, for example due to legal issues when different laws govern the data hosted at the different sites.

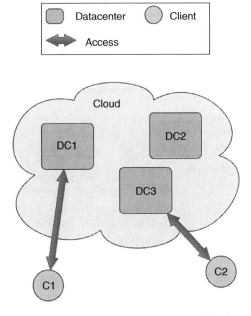

Figure 15.1   *Datacenters and the cloud*

15.2.1 Addressing

Distributed datasets used for a single application can be stored in individual files in distributed file systems such as the Hadoop Distributed File System (HDFS) deployed in these datacenters, or in alternative storage systems. (For simplicity we may refer in the following to such datasets as *files*, although they may be stored in other formats.) Such files need not share common formats.

The networks within cloud datacenters usually offer significantly different guarantees from the networks connecting them. Indeed, cloud providers typically control the network within their offerings and, as such, ensure some practical level of quality of service (QoS) for communication. Communication between such datacenters may, however, go through the Internet, making it harder to provide guarantees on latency or bandwidth. Consequently, distributed file systems such as HDFS typically only support deployments within single datacenters and perform poorly, if at all, across datacenter boundaries. Each datacenter might host its own deployment of such a file system, and accesses would happen only from within a respective datacenter. However, several geodistributed file systems and storage systems have been proposed recently, with a global perspective on data management. Two main possible approaches exist:

- Distribution-*aware* addressing consists of managing a different / addressing name space for every datacenter involved. Thus file operations, including creation and access, are explicitly parameterized by their location. This approach is similar to deploying a separate instance of a storage system in each datacenter and facilitating accesses from remote datacenters.
- Distribution-*agnostic* addressing. Here the level of abstraction is raised by addressing several datacenters as a whole. That is, users do not have to know a priori exactly where their files are located.

The resulting explicit and implicit distribution can also be combined. For instance, with a consolidated name space, file-creation operations can support optional arguments denoting the desired location(s) for files, or locations can be queried and explicitly modified by users independently of file names per se.

Access features and guarantees are a priori decoupled from the addressing model. For instance, transactional guarantees for multiple combined file accesses are possible with both approaches. However, such advanced features are usually more often found with single address spaces, in addition to other conveniences like automatic load distribution across datacenters, automatic (re)location of files based on access patterns, or automatic replication for fault-tolerance. We describe these services in more detail shortly.

GridFarm (Tatebe *et al.*, 2003) is an early example of a file system that supports geodistributed access with explicit denomination of the servers that are managing files. Files are addressed and accessed individually without guarantees across accesses.

15.2.2 Partitioning

Sometimes individual logical files are partitioned across datacenters. This can happen for different reasons besides the security concern mentioned in section 15.1:

- Human initiative. With large datasets, especially those that keep growing at a high rate, one management mechanism is a manual coarse-grained division of one dataset into several datasets or files which can be stored in a distributed manner. The resulting partitions typically follow a single partition criterion (based on the values for some primary key in the data such as "person with family names starting by A–D", or "logs from machines x–y"), or can apply several such criteria in a nested fashion with directories.
- Technical limitations. Despite the tremendous scalability of existing distributed file systems and storage systems, there are always limitations in how much data can be addressed or indexed with a given number

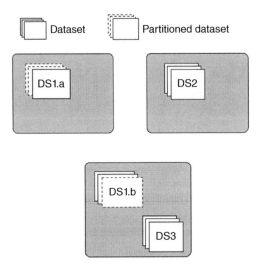

Figure 15.2 *Partitioned data*

of bits. The same goes for the datasets themselves, which can be limited in size, for example by the size of individual hard disks. Even without considering such limits, or pushing them very far in practice, file systems providing only restricted features for data access can yield bottlenecks in the applications that use them. For instance, append-only semantics for files in HDFS become increasingly cumbersome when attempting to access arbitrary parts of larger files.

Figure 15.2 shows an example where a single dataset DS1 is partitioned into two parts DS1.a and DS2.b, which are stored in two different datacenters respectively.

15.2.3 Migration

There are many reasons a file might be moved from one location to another, including a change in the rules governing the distribution of data or the need for low latency local accesses. Files may even be moved back and forth or across multiple sites. This is of particular interest in the case of a unique name space for files stored across datacenters. Such a global name space abstracts geodistribution, yet access times between datacenters can hardly be abstracted away. Thus it is of increased importance to keep such access times minimal, which can be achieved by taking into account access patterns of individual files, and performing automatic migration of files to optimize for some objective function such as minimizing the average access latency. Optimization may further take latency and bandwidth or cost of remote communication from a given datacenter to another into account, as well as legal constraints to disable certain migrations. Correlations between accesses to different files can yield additional cues. An example of a system for automatic relocation of files is Volley (Agarwal *et al.*, 2010). Volley analyzes logs of access requests from datacenters, and uses an iterative algorithm to optimize access latencies to individual files as well as costs by taking inter-datacenter communication bandwidths into account to achieve a user-defined desired trade-off between the cost of proactive migration and its benefits.

For files which are only read, or for which slight deviation from up-to-date versions can be tolerated, migration can be augmented with *caching* for faster access.

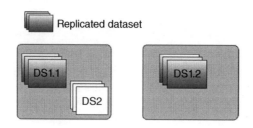

Figure 15.3   *Replicated data*

15.2.4   Replication and redundancy

Another common technique to orchestrate distributed accesses is to replicate files. Figure 15.3 shows an example where two replicas of a dataset DS1 – DS1.1 and DS1.2– are stored in two datacenters. The resulting redundancy can also be used to provide tolerance for failures of nodes hosting files, or possibly larger outages in datacenters. Inversely, such fault tolerance can motivate multidatacenter setups.

In the case of files that can be modified, replication immediately leads to the issue of synchronizing concurrent access, which is solved straightforwardly when there is a single copy for a file. Several storage systems support geodistributed replication and provide fault tolerance, differing in the guarantees that they provide.

15.2.5   Consistency

Consistency guarantees become of the utmost importance in the context of access to distributed shared data, especially with replication and failures. The CAP theorem states the impossibility of providing simultaneous consistency, availability, and partition tolerance (Brewer, 2000); only two of them can be implemented simultaneously.

Several systems thus focus on crash failures and/or provide weaker consistency guarantees. For example, Dynamo (DeCandia *et al.,* 2007) is a key-value storage, which provides only some consistency. Vivace (Cho and Aguilera, 2012) provides strong consistency in the presence of failures and congestion.

15.2.6   Security

Geodistributed file access exacerbates security problems; for example:

- Access control solutions need to be deployed across datacenters, which usually comes down to deploying such solutions in datacenters individually, and coordinating them manually.
- Data transfer: if the cloud provider is trusted (cf. private clouds), users may be willing to store sensitive data in datacenters. With inter-datacenter communication taking place across the Internet, most users will want to encrypt data that is thus transferred though, which further increases the performance overhead of cross-datacenter communication and motivates the consideration of distribution constraints in communication and computation.

15.3   Distributed Computation

With datasets being distributed across datacenters, one also needs to consider how best to perform computations on such datasets. This is of particular relevance in the context of datasets that are partitioned across multiple datacenters. While the concepts introduced and discussed in the following are generic in nature, we highlight them mostly in the light of a core application scenario for (geodistributed) cloud-based computation, namely that of (big) data analytics.

15.3.1   Centralized computation

With data being geodistributed, even with replication, there might not always be a copy of every relevant data item available at every site where computation occurs.

The simplest approach to deal with this situation is to copy – if possible – all data to a single location before performing any computation on it. In the example scenario given in Figure 15.4, dataset DS1 is copied to datacenter DC2 prior to performing a computation that involves multiple datasets. This approach is simple implementation-wise, yet there are a number of associated disadvantages:

- Performance: before executing a program, all data has to be transferred locally, whilst at least parts of the computation could possibly be performed locally to the data. Without specific support, this becomes especially inefficient when programs are run repeatedly (possibly after minor modification) on the same data, leading to repeated copying.
- Cost: with many cloud providers charging for inter-datacenter communication (while intra-datacenter communication is typically free of charge), transfers of large datasets can become financially expensive.
- User effort: in order to avoid the same data being transferred several times, a straightforward approach is for the user to deal manually with fetching data and copying it close by. This is, however, tedious, as it requires users to deal with updates to datasets manually.

While several systems have been proposed to store data in a geodistributed manner, computation has been given only little attention thus far.

15.3.2   Decentralized computation

Performing as much computation as possible close to the data usually improves performance as well as minimizing costs. This is particularly valid for big data analysis, where the amount of data tends to decrease as the computation proceeds – typically large datasets are used as inputs to the computations, which attempt

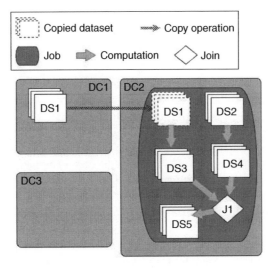

Figure 15.4 *Centralized computation*

to extract specific, more concise, knowledge from those datasets. Additionally, legal constraints or security concerns can prevent the copying of data to the site of the issuer of a query. Airavat (Roy *et al.,* 2010) thus supports remote statistical queries such as averages over large datasets via MapReduce, combined with access control and query analysis to allow precisely restricted third-party queries on protected datasets. However, Airavat still assumes that all data is located in a single datacenter.

Implementing generic decentralized computation is far from trivial, when the goal is to optimize performance and/or cost, as there are different *execution paths* according to which computational steps can be interleaved with consolidation of data from several datacenters. Consider an operation involving two datasets implemented through a single MapReduce task which, as its name suggests, consists of two phases: a first map phase, which outputs *intermediate* <key, value> pairs for its input datasets, followed by a reduce phase which atomically processes <key, list<value>> pairs where an instance of list <value> represents all values generated a same key. Considering that at some point in the computation all data has to be consolidated in a single datacenter, there are three straightforward execution paths, differing by when the consolidation happens:

- Premap: copying one of the datasets to another datacenter – for example, the datacenter hosting the smaller dataset – prior to computation yields an execution path similar to performing computation in a centralized fashion, and in some cases might be the only option due to the nature of the job.
- Postmap/prereduce: here the first phase can occur on datasets individually, with copying occurring before the reduction phase, by aggregating <key, list<value>> pairs for a same key across datacenters.
- Postreduce: consolidating data at the end of the reduce phase assumes that typically no two datacenters will generate intermediate values for a same key, or that some simple way exists to aggregate results after the reduction.

This is only a simplification: the abovementioned approaches could be combined in many different ways especially when more than two datasets are involved, for example consolidating some datasets following path (a), with further consolidation following (b). The situation becomes yet more complicated when – as is common in real-life MapReduce computations – several such tasks are performed in sequence. Relevant parameters for optimization beside latency, bandwidth, and cost of inter-datacenter communication as in data management include but are not limited to:

- amount of data present at individual points in the computation;
- complexity and costs of individual computational steps;
- (momentary) resource availabilities at datacenters.

As emphasized by the possibly transient nature of resource availability, determining an optimal execution path can require runtime information and conflict with other jobs executed simultaneously on a shared cluster of nodes. Rout (Jayalath and Eugster, 2010) is an example of a seminal system for geodistributed big-data processing based on MapReduce. Rout employs a lazy copying heuristic based on momentary resource availabilities in involved datacenters (or more precisely, clusters of nodes allocated for MapReduce tasks in respective datacenters). Computation proceeds individually at the sites of the datasets involved until consolidation becomes necessary, at which point cluster resource availabilities are balanced with amounts of data to be transferred and communication bandwidths in order to choose the datacenter in which computation is to proceed. Another example of a system supporting geodistributed computation is G-MR (Geo-distributed MapReduce) (Jayalath *et al.,* 2013b): G-MR achieves more fine-grained optimization by using sampling to determine the amounts of data at respective points in the computation, yet it uses an offline optimization technique that does not take resource availability into account at runtime, yet can support computation models other than (sequences of) MapReduce jobs. Legal policies can similarly be considered. Figure 15.5 shows how the computation of

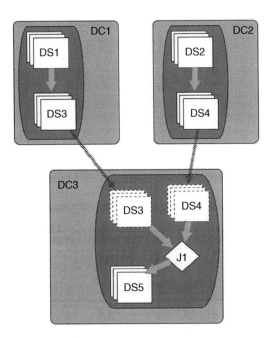

Figure 15.5   *Decentralized computation*

Figure 15.4 can be executed in a decentralized manner, involving three different datacenters. Note that computation may also be replicated, for instance to ensure high availability.

15.3.3   Associativity and distributed associativity

As hinted above, a reasonable consolidation heuristic during computation on geodistributed and possibly partitioned datasets involves consolidating data as late as possible, given that the amount of data typically decreases as computation proceeds. However, it is not always possible to perform computation on subdatasets in isolation. We can distinguish several scenarios:

- Associative computations: following the mathematical definition of associativity, some functions f can be performed on arbitrary subsets of data in isolation, before being applied to the outcome of some of these subcomputations, and so forth. Consider summing the values of some particular attribute across many records: we can create partial sums and sum these further in whatever order, as long as every partial sum is considered exactly once.
- "Distributed associativity" (Yu *et al.*, 2009): certain functions f can be "made associative," by describing them alternately in two steps $f = h \circ g$ where g is performed on subdatasets whilst the second h consolidates these results in an arbitrary number of stages in an associative manner. As an example, an average can be expressed as follows: a first function which computes for a subdataset (1) the respective average as well as (2) *the number of elements considered*; the second function outputs the sum of all the element counts (2) for any number of inputs in addition to the respective average computed by prorating the averages of its inputs by the respective number of elements (and dividing the result by the total number of elements considered thus far).
- We can consider the scenario where the first function is the "original" one, that is, the function that we want to apply to an entire dataset ($f = h \circ f$). In this case we can refer to the function h as a "merge" or "aggregation" function for the function f. An example consists of counting the number of occurrences of

some value across several datasets or a partitioned dataset: we can count the number of occurrences in each subdataset, before summing these counts in whatever order we like.

- Several authors have considered *inferring* merge functions (*h*) or decompositions (*g, h*) automatically or with little programmer input. Most notably, the third homomorphism theorem (Morihata *et al.,* 2009) states that if two sequential programs iterate a list from opposite sides and compute the same value, there exists a parallel program that can divide-and-conquer the program with the same results.

15.3.4   Programming

The impossibility of automatically inferring merge functions in general naturally leads to the question of how to aid the programmer in expressing such functions.

In the case of big-data analysis, several so-called "data-flow" languages have been proposed in the recent past, which represent data at individual stages by some form of *collection* or *data-structure*, and computational steps as operations performed on such data structures. Such languages are usually compiled to MapReduce jobs. Languages differ in the data structures that they propose, though most include some form(s) of sets or bags along with some kind(s) of associative maps. For instance, Pig Latin (Olston *et al.,* 2008) introduces bags and maps. While early languages tried to be mostly distribution transparent, more recent approaches for the sake of performance and advanced functionality (e.g., supporting iteration and increment computation) abandon transparency and provide richer APIs to give the programmer more explicit choices of how to perform operations. For instance, resilient distributed datasets (RDDs) (Zaharia *et al.,* 2012) go as far as exposing map and reduce operations that are parameterized by functions; the signature and semantics of the application of these functions are given by the types of their respective formal arguments and the operation through which they are applied. In a similar vein, Rout thus extends Pig Latin with support for merge functions for many of the language's built-in operators, while for others (e.g., COUNT), the runtime has built-in knowledge of how to perform these in an associative or distributed associative manner.

15.4   Distributed Communication

Executing across multiple datacenters requires generic support for communicating between nodes spread across these datacenters. This leads to various challenges beyond efficiency, due to the different mechanisms and features supported by cloud providers. The space for solutions to overcome these challenges is still very limited.

15.4.1   Mechanisms

Different cloud providers support different basic communication mechanisms, and to different extents. Two particular limitations are:

- Multicast. Few cloud providers support IP Multicast or UDP Broadcast, for simple reasons. First off, different users / applications could use the same IP Multicast address. Second, such clashes could be exploited to generate denial-of-service attacks. Since cloud host resources are often exploited in bunches, however, multicast addressing is a common and thus relevant scenario.
- Host addresses. Most cloud providers do not expose hardware addresses, or propose proprietary addressing mechanisms to avoid enabling arbitrary direct access to hosts from outside of clouds. Such access can lead to security vulnerabilities, inefficiencies, and to bypassing cost accounting.

To make up for these limitations, cloud providers typically offer specialized middleware services, which are, however, not portable or interoperable across cloud providers and often do not operate efficiently across datacenters of the same cloud provider. Examples of specialized services are Amazon EC2's CloudFront content delivery network or its Simple Notification Service.

15.4.2 Communication models

Several core models of cloud communication can be distinguished. Roughly:

- One-to-one (unicast): This communication scenario between two endpoints remains the most common one. One can further distinguish here between different semantics, and unidirectional versus bidirectional communication.
- One-to-many (multicast). We can further subdivide this scenario:
  - "Explicit multicast": here, a component producing data wants to share that data with an explicitly specified set of destinations.
  - "Implicit multicast": here the set of actual destination components is given by these components themselves, for example, based on matching of message attributes to desired ranges of values specified by them.
- One-of-many (message queuing). This is similar to implicit multicast, yet the system selects exactly one destination for every message based on characteristics of potential destination components (e.g., current load), specific policies (e.g., for "fair" load balancing), or others.
- One-to-all (broadcast): here, a message is sent to all "participating" components, typically of a given application instance.
- Many-to-many: Here we consider systems that allow for different messages to be aggregated within the middleware, thus allowing composite messages to be delivered directly to applications.

Figure 15.6 illustrates a system with communicating components distributed across multiple datacenters.

15.4.3 Abstractions

There are several traditional abstractions for programming distributed applications in general, which also play a role in cloud datacenter communication:

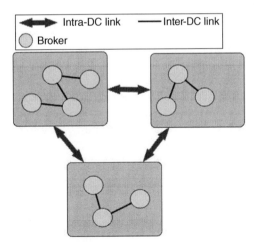

Figure 15.6 *Components communicating across datacenters*

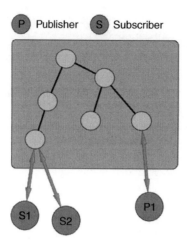

Figure 15.7   *Publish subscribe*

- Sockets: these are typical for programming unicast interaction (one to one) as well as simple one-to-many interaction
- Publish/subscribe: this is a common high-level abstraction for multicasting. Producers are termed *publishers* and consumers *subscribe* to specific messages. Figure 15.7 illustrates entities of a publish/subscribe system. Several variants for outlining messages of interest can be distinguished:
  - Topic-based publish/subscribe (TPS): here, publishers explicitly post messages under specific *topics*, whilst subscribers delineate topics of interest. Topics can exhibit a "flat" name-space or be in different relationships to each other such as hierarchical.
  - Content-based publish/subscribe (CPS): here subscriptions are expressed based on predicates on message content ("filters"). Most systems thus support messages with some form of property-value pairs, which are used for filtering.
- Message queuing: these systems are the most common implementation of one-of-many communication. Typically consumers pull messages from queues, but this can also be replaced by push-style communication.

Sockets are typically bound to IP addresses, which, given the problems with such addresses outside of clouds, makes them unsuited for intercloud communication.

15.4.4   Elasticity

Elasticity is a core tenet of third-party cloud computing, whose economic benefits depend precisely on the ability to tap into resources at need, pay as one goes, and release unneeded resources instantly to avoid any amortization or constant costs. This necessitates communication mechanisms to efficiently scale up and down: a component may have to communicate with hundreds of others to inform them of changes relevant to their individual subcomputations, and the next moment only ten such components may be necessary to proceed. Based on this requirement as well as other cloud-specific needs already mentioned, the publish/subscribe abstraction is a suitable candidate for elastic communication in and across clouds:

- Addressing: the logical addressing introduced by publish/subscribe abstracts cloud-specific addressing, which in turn enables inter-datacenter and inter-cloud communication.
- Scale: the publish/subscribe paradigm is able to capture both unicast (one-to-one) and multicast (one-to-many) communication scenarios.

- Integration: the property-name abstraction offered by content-based publish/subscribe integrates especially well with other systems such as many storage systems, which focus on key-value pairs.

However, typical publish subscribe systems focus on scaling *up*, and less on small scales. TPS systems typically scale up to many topics, and many consumers per topic, although many existing systems work well with small consumer numbers for given topics. Content-based publish/subscribe systems integrate better abstractionwise with other cloud services and are more expressive than their TPS counterparts, yet do not deal well with scenarios where few consumers are listening to a given producer. Yet, many scenarios exist where a producer's messages are relevant to a handful of consumers: a common pattern is threefold replication of components for fault-tolerance; CPS systems *performancewise* deal poorly with such scenarios.

Atmosphere (Jayalath *et al.*, 2013a) is a CPS system designed specifically for cloud communication in a way that supports multi-datacenter setups as well as scaling down and up. The system uses original protocols for tracking the number and distribution of subscribers for specific publishers, and splitting the traditional overlay networks used in CPS systems in order to facilitate efficient small-scale communication: the overlay is used primarily to keep components connected, while "shortcuts" are created over this overlay for more efficient propagation of messages.

15.5   Conclusions

No single cloud solution will be sufficient for all uses while still being wieldy enough to use, and some consumers will want to use different services or separate their data. Thus cloud federation is necessary. Several solutions exist to distribute and migrate data, as well as to provide services across clouds. Clouds may rely on each other, in vertical federation, or coexist to be accessed by a third service in a horizontal federation. New services may be introduced to remap proprietary services to a new abstraction, or services may be tied together to create seamless interaction.

Acknowledgements

This work is financially supported by Cisco Systems grant "A Fog Architecture," Google grant "Geo-Distributed Big Data Processing," DARPA grant "Large-Scale Cloud-based Data Analysis," and ERC grant number 617805.

References

Agarwal, S., Dunagan, J., Jain, N., *et al.* (2010) *Volley: Automated Data Placement for Geo-Distributed Cloud Services,* USENIX, San Jose, CA, pp. 17–32.

Brewer, E. A. (2000) *Towards Robust Distributed Systems,* ACM, Portland, OR.

Cho, B. and Aguilera, M. K., 2012. *Surviving Congestion in Geo-Distributed Storage Systems,* USENIX, Boston, MA, pp. 439–451.

DeCandia, G., Hastorun, D., Jampani, M., *et al.* (2007) *Dynamo: Amazon's Highly Available Key-Value Store.* Twenty-first ACM SIGOPS Symposium on Operating Systems Principles, Stevenson, Washington, DC, pp. 205–220.

Jayalath, C. and Eugster, P. (2013) *Efficient Geo-Distributed Data Processing with Rout.* Thirty-third International Conference on Distributed Computing Systems, Philadelphia, PA. IEEE, pp. 470–480.

Jayalath, C., Stephen, J. and Eugster, P. (2013a) *Atmosphere: A Universal Cross-Cloud Communication Infrastructure.* Proceedings of the 14th ACM/IFIP/USENIX Middleware Conference, Beijing, China. Springer, pp. 163–182.

Jayalath, C., Stephen, J. and Eugster, P. (2013b) From the cloud to the atmosphere: Running MapReduce across datacenters. *IEEE Transactions on Computers* (May 27), pp. 74–87.

Kurze, T., Klems, M., Bermbach, D., *et al.* (2011) Cloud federation. The Second International Conference on Cloud Computing, GRIDs, and Virtualization. Rome, Italy.

Lenk, A., Klems, M., Nimis, J., *et al.* (2009) What's inside the cloud? An architectural map of the cloud landscape. ICSE Workshop on Software Engineering Challenges of Cloud Computing. Washington, DC.

Morihata, A., Matsuzaki, K., Hu, Z., and Takeichi, M. (2009) *The Third Homomorphism Theorem on Trees: Downward and Upward Lead to Divide-and-Conquer.* Thirty-Sixth ACM SIGPLAN-SIGACT Symposium on Principles of Programming Languages, Savannah, GA. ACM, pp. 177–185.

Olston, C., Reed, B., Srivastava, U., *et al.* (2008) *Pig Latin: A Not-so-foreign Language for Data Processing.* ACM SIGMOD International Conference on Management of Data. New York, NY. ACM, pp. 1099–1110.

Roy, I., Setty, S. T., Kilzer, A., *et al.* (2010) *Airavat: Security and Privacy for MapReduce.* Seventh USENIX Symposium on Networked Systems Design and Implementation, San Jose, CA. USENIX, pp. 297–312.

Tatebe, O., Sekiguchi, S., Morita, Y., *et al.* (2003) Worldwide fast file replication on grid datafarm. Talk given at Thirteenth International Conference for Computing in High Energy and Nuclear Physics, La Jolla, CA.

Winkler, V. (2011) *Securing the Cloud: Cloud Computer Security Techniques and Tactics,* Elsevier, Waltham, MA.

Yu, Y., Gunda, P. K., and Isard, M. (2009) *Distributed Aggregation for Data-Parallel Computing: Interfaces and Implementations.* ACM SIGOPS Twenty-Second Symposium on Operating Systems Principles, Big Sky, MT. ACM, pp. 247–260.

Zaharia, M., Chowdhury, M., Das, T., *et al.* (2012) *Resilient Distributed Datasets: a Fault-Tolerant Abstraction for In-memory Cluster Computing.* Ninth USENIX Conference on Networked Systems Design and Implementation, San Jose, CA. USENIX, pp. 15–28.

16

Cloud Standards

Andy Edmonds,[1] Thijs Metsch,[2] Alexis Richardson,[3] Piyush Harsh,[1] Wolfgang Ziegler,[4]
Philip Kershaw,[5] Alan Sill,[6] Mark A. Carlson,[7] Alex Heneveld,[8] Alexandru-Florian Antonescu,[9]
and Thomas Michael Bohnert[1]

[1] *Zurich University of Applied Sciences, Switzerland*
[2] *Intel, Germany*
[3] *VMWare, UK*
[4] *Fraunhofer-Institute SCAI, Germany*
[5] *STFC, UK*
[6] *Texas Tech University, USA*
[7] *Independent*
[8] *CloudSoft, UK*
[9] *SAP/University of Bern, Switzerland*

16.1 Introduction

This chapter will focus on the standards that are produced and published by organized bodies known as standards defining organizations (SDOs). Such organizations have a wide variety of internal processes and membership rules, which range from completely open access to closed formal representation that can in some cases require the approval of national governments and international coordinating bodies.

The standards that are dealt with in this chapter are primarily those that are most mature within the infrastructure as a service (IaaS) and platform as a service (PaaS) layers. While software as a service (SaaS) standards also exist, they tend to be specialized to their areas of application, and so are not amenable to more than a brief overall summary. The chapter starts by motivating the practical use of standards, what types of communities issue them, who they are, their primary cloud-related outputs, and their organizing rules for participation.

Encyclopedia of Cloud Computing, First Edition. Edited by San Murugesan and Irena Bojanova.
© 2016 John Wiley & Sons, Ltd. Published 2016 by John Wiley & Sons, Ltd.

16.2 Why Standards?

Among the goals for any robust software, the following are key basic aspects (taken from Edmonds *et al.*, 2011):

- Interoperability. Describes how two cloud services can interoperate on the fly. This demands a standardized API and protocol (e.g. live migrating a virtual machine from one host to another, when they are in different management domains).
- Integration. Describes how a service provider can bring together different technologies and interconnect them within his domain (e.g. integrate a virtual machine management tool with an identity-management system).
- Portability. This is mostly about porting between cloud service providers. In comparison with interoperability, there is no direct connection between the service providers. Portability demands that there are standardized data formats that providers can understand (e.g. porting a virtual machine from one hypervisor to another).

   As services are the key component in today's cloud-service offerings, the development and deployment of a service should be made as simple as possible. Hence service developers know that incorporation of appropriate standards can provide powerful tools to reduce the risk of vendor lock in and allow for incorporation of up-to-date new methods without the complete rewriting of the entire package.

   This use of standards facilitates interoperation with other cloud-based software services, components, and infrastructures. Standardized APIs and interfaces intrinsically promote interoperability and portability. This built-in interoperability allows service developers, vendors, and users to focus more on the higher level capabilities of their services and therefore less on reinventing common aspects and features of their APIs and interface modules.

16.3 What Sort of Standards?

There is not just one type of standard. There are a number of types and those considering or using standards should know the difference. The following types of standards are of relevance to most users of the cloud:

- *De jure* standards are those driven by government-related entities. Community driven *de facto* standards can become *de jure* standards if an official governing body (e.g. the government) officially embraces them.
- Community standards are those driven by a mostly open community of people, who by a certain process reach agreement on a set of features that a standard should have. Most community standards are open and driven by individual people for the common good.
- Industry standards are defined and driven either by all or a subset of vendors active in an industry. Most such standards are developed in closed groups and require a voting mechanism to be adopted.
- *De facto* standards are those defined by a (single) entity and are adopted by the community so quickly that they gain huge popularity. This popularity means that they essentially become the accepted standards, as the community expects these standards to apply to all products of the same kind.

16.3.1 Open Standards

Some of the standard types mentioned above can be deemed to be open standards. However, the only way to know this is by studying the exact specification and which service-defining organization it comes from. To date there have been three separate efforts to define what an open standard is: the Open Cloud Initiative (http://www.opencloudinitiative.org/, accessed December 5, 2015), OpenStand (http://open-stand.org/, accessed December 5, 2015), and the work in Krechmer (2008). When looking at the three initiatives it makes sense to see how each can help with a wider definition. To consider if a particular standard is an open standard, four dimensions of the standard and its SDO should be considered. These are the governing organization, the specification in question, how the specification might have been implemented, and adoption of the specification. The final two aspects are key to all the standards.

16.4   What Sort of Organizations?

There are many organizations defining standards. The organizations that oversee the standards described in this chapter are discussed below.

- The Distributed Management Task Force (DMTF) (http://www.dmtf.org/, accessed December 5, 2015) creates standards that enable interoperable IT management. Its management standards are critical in enabling management interoperability among multivendor systems, tools, and solutions within the enterprise. The DMTF is the steering body behind both Cloud Infrastructure Management Interface (CIMI) and Open Virtualization Format (OVF), which are described later

- The Organization for the Advancement of Structured Information Standards (OASIS) (https://www.oasis-open.org/, accessed December 5, 2015) is a nonprofit consortium that drives the development, convergence and adoption of open standards. It promotes industry consensus and produces worldwide standards for security, cloud computing, service-oriented architecture (SOA), Web services and other areas. It is the steering body behind both the Cloud Application Management for Platforms (CAMP) and the Topology and Orchestration Specification for Cloud Applications (TOSCA), which are described later.

- The Open Grid Forum (OGF) (http://www.ogf.org/, accessed December 5, 2015) is a leading standards development organization operating in the areas of grid, cloud and related forms of advanced-distributed computing. The OGF community pursues these topics through an open process for development, creation and promotion of relevant specifications and use cases. The OGF is the steering body behind the Open Cloud Computing Interface (OCCI) and WS-Agreement, which are described later.

- The Storage Networking Industry Association (SNIA) (http://www.snia.org/, accessed December 5, 2015) is a not-for-profit global organization, whose mission is to lead the storage industry worldwide in developing and promoting standards, technologies, and educational services to empower organizations in the management of information. The SNIA is the steering body behind the Cloud Data Management Interface (CDMI), which is described later.

- The Internet Engineering Task Force (IETF) (http://www.ietf.org/, accessed December 5, 2015) is a body that supports the development of new standards related to the Internet of which OAuth and NFS – mentioned in this section – are two of many. The IETF is organized under the Internet Society (ISOC), itself a nonprofit organization.

- The OpenID Foundation (http://openid.net/foundation/, accessed December 5, 2015) is a nonprofit organization of individuals and companies. It was founded back in 2007 as an open community of vendors and developers. The role for the foundation is to drive the OpenID technologies that are described later.

- The Cloud Security Alliance (CSA) (https://cloudsecurityalliance.org/, accessed December 5, 2015) is again a nonprofit organization trying to promote the security assurance in the cloud. It gained significant impact in later years when, for example, the White House used the CSA summit as the venue for announcing the federal government's cloud computing strategy. The Cloud Security Alliance recommends usinf SAML, and OAuth, which are described later in this chapter.

16.5   Cloud, Standards and Management

The standards that are discussed in this chapter are primarily those that are related to some aspect of service management. This set of resources is created and / or orchestrated by the service provider, which then gives access to the requesting client. What the client sees is not just a collection of resources or anything to do with the internal workings of those resources but a limited set of features organized through a management inter-face. The organization of this interface is managed by the service provider, either directly or through orches-tration. From this perspective, the client works and manages interaction with a service instance by controls exposed by the management interface, either directly through a user interface (UI) and / or by coding or interaction of other services through an application programming interface (API). As the UI itself is often just an easily coded wrapper to simplify user interaction with the API or with services that talk to the target API, it is the API layer that is the focus of most of the cloud standardization efforts described in this chapter. One of the core principles of such services is that clients have broad network access through which service provid-ers make their offers. These cloud services are "capabilities … available over the network and accessed through standard mechanisms" (Mell and Grance, 2011). From a technical perspective, clients generally access their service instance and service provider management interfaces over a TCP-based protocol. For the service and service-instance management interfaces, the TCP-based protocol, HTTP, is the preferred approach with many service providers choosing to implement a REST-based (Fielding, 2000) API (note that RESTful approaches do not mandate TCP – it is just the most common). Indeed, cloud-standard-defining organizations take this approach.

16.6   Individual Standards

There are many cloud-service offerings available today, from IaaS through PaaS to SaaS, which show their adherence to the NIST definition of cloud computing (Mell and Grance, 2011). In the following sections, individual standards will be detailed and related to the layer they best fit.

16.6.1   OAuth

OAuth (Hammer-Lahav, 2010) is an open standard for authorization. It enables users to delegate access to resources they own to a trusted third party to access in a constrained manner. Version 1.0 of the protocol is published as IETF RFC 5849 (Hammer-Lahav, 2010). It should be noted that the recent version 2.0 of OAuth has made significant changes to its predecessor.

16.6.2   WS-Agreement and WS-Agreement Negotiation

This sub-section describes the open standards WS-Agreement and WS-Agreement Negotiation, which define languages and protocols for negotiating and creating service level agreements. The development of WS-Agreement (Andrieux *et al.*, 2011) initially started 2005 in the environment of computational grids. However, the specification is both generic and domain agnostic such that WS-Agreement can generally be used (and is used) in environments where an agreement between a provider and a customer needs to be created. WS-Agreement is not tied to a specific application domain. Constraints used to express particular terms of services can be expressed using domain-specific languages as long as their definition is realized using XML. WS-Agreement itself is fully symmetric and allows either party to enter the roles of agreement initiator and agreement responder.

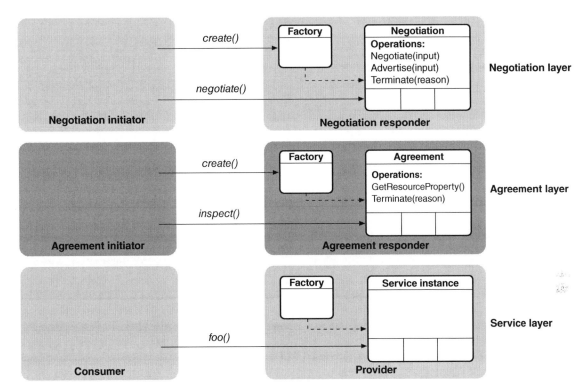

Figure 16.1   *Layered model of WS-Agreement (lower two layers) and WS-Agreement Negotiation. Source: Fraunhofer Institut SCAI*

WS-Agreement Negotiation (Wäldrich *et al.*, 2011) is an extension of WS-Agreement adding multiround negotiation capabilities to WS-Agreement. While for many use cases the discrete-offer protocol of WS-Agreement is sufficient, for some use-cases, a multiround negotiation protocol is more efficient. In order not to break compatibility with WS- Agreement, an independent additional layer for negotiations on top is defined. An existing agreement can be renegotiated by either of the parties if the circumstances demand.

WS-Agreement and WS-Agreement Negotiation are based on a three-layer (negotiation layer, the agreement layer and the service layer) model depicted in Figure 16.1.

There is a clear separation between these layers. The negotiation layer is decoupled from the agreement layer and the service layer. In that way, the negotiation layer may change independently of the agreement layer and can be replaced by another negotiation layer, which might be better suited for a specific negotiation scenario.

WS-Agreement provides a template-based approach, where usually the service provider makes predefined agreement templates describing properties of the service available to potential service consumers. The service customer (the agreement initiator) may download a template suitable for its desired service, probably make some modifications if allowed by the constraints defined in the template, and send the template back to the provider (the agreement responder) as an agreement offer.

Once the provider receives the offer it can decide, to either accept the offer or to reject it. In the first case the agreement is created and binding with obligations for both parties. In the second case no agreement is created and the customer is no longer bound to its offer.

WS-Agreement Negotiation is also based on templates. Similar to the Agreement templates these Negotiation templates define negotiation constraints that allow both parties to restrict the scope of a negotiation.

Implementations of WS-Agreement consist of a server and a client where the server is realized as a Web service running in a tomcat container. The implementation is using the Web Service Resource Framework (WSRF) to provide stateful resources.

There are a large number of implementations, many of them provided by European co-funded projects, such as PHOSPHORUS (http://www.ist-phosphorus.eu/, accessed December 5, 2015), SLA@SOI (Theilmann, J., C., A., K., and J., 2010), OPTIMIS (Ziegler, 2012). Moreover, there are first implementations in commercial and pre-commercial environments, for example as part of a product of the UK company Cybula (http://www.cybula.com, accessed December 5, 2015), or in the cloud broker of CompatibleOne (http://www.compatibleone.org, accessed December 5, 2015).

16.6.3 Open Cloud Computing Interface (OCCI)

OCCI (http://www.occi-wg.org, accessed December 5, 2015) comprises a set of open community-led specifications delivered through the Open Grid Forum. The OCCI is a protocol and API for the management of cloud-service resources, primarily for IaaS model-based services; however, it has since evolved into a flexible API with a strong focus on integration, portability, interoperability, and innovation while still offering a high degree of extensibility (Edmonds *et al.*, 2012). The main design tenets of OCCI are simplicity, extensibility, discoverability and modularity. A key aim of OCCI is to leverage existing SDO specifications and use them to leverage the existing work, so where an OCCI feature is needed, a more capable one can be brought into play. An example of this is the integration of both CDMI and OVF. In particular those two standards, when combined with OCCI, provide a profile for open and interoperable infrastructure cloud services (Edmonds *et al.*, 2011). The specification itself currently comprises of three modular parts:

- Core. This specifies the basic types and presents them through a metamodel. It is this that dictates the common functionality and behavior that all specializations of it must respect (Nyren, Edmonds, Papaspyrou, and Metsch, 2011). It specifies how extensions may be defined.
- Infrastructure. This specification is an extension of Core (it provides a good example of how other parties can create extensions) (Edmonds and Metsch, 2011) and defines the types necessary to provide a basic infrastructure as a service offering.
- HTTP Rendering. This document (Metsch and Edmonds, 2011) specifies how the OCCI model is communicated both semantically and syntactically using the RESTful architectural-style.

From an architectural point of view OCCI sits on the boundary of a service provider. It does not seek to replace the proprietary protocols/APIs that a service provider may have as legacy. The communication between the service consumer and the OCCI implementation can be carried out over HTTP and both exchange serialized renderings of the OCCI model.

The OCCI Core Model (Nyren *et al.*, 2011) is the basis of understanding anything about OCCI and its extensions. The core model is an abstract model. It defines a "type" system for cloud resources. In this system, a cloud resource has a fixed kind with a set of capabilities. Over the lifecycle of the resource, mix ins can be associated giving it additional capabilities. Capabilities are expressed through a set of attributes and actions. The core model also defines the means to link resources to each other.

The Infrastructure model defines the capabilities of compute, storage, and types of network. Next to this the links are defined, which, for example, express network interfaces, storage attachments, and so forth.

The rendering documents are free of information that is particular to an extension such as the infrastructure model. While being modular and extensible, OCCI offers a unique query interface through which the capabilities of a service provider can be inspected.

Clinets interact with an OCCI implementation over HTTP following a RESTful approach and uses the appropriate request to the implementation as detailed in the HTTP Rendering specification (Metsch and Edmonds, 2011). The OCCI supports two rendering types – the first is a text-based rendering and the second is a JSON-based rendering.

There are numerous implementations of the OCCI specifications (http://occi-wg.org/community/implementations/, accessed December 5, 2015). The available implementations include those that cover the main open-source infrastructure service frameworks, including OpenStack (http://www.openstack.org, accessed December 5, 2015), CloudStack (http://www.cloudstack.org, accessed December 5, 2015), and OpenNebula (http://www.opennebula.org, accessed December 5, 2015). It is also the core technology for the cloud broker of CompatibleOne. More recently, implementations of OpenShift, and other solutions, such as an OCCI interface for the Internet-of-Things (IoT), have appeared.

16.6.4 Open Virtualization Format (OVF)

A software package is to a computer system what a virtual appliance is to a cloud. More formally, a virtual appliance is a "pre-built software solution, comprised of one or more VMs that are packaged, maintained, updated and managed as a unit" (DMTF, 2010). In this era of increasing use of virtualization and clouds, it is important to find a standard way to package any virtual appliance to ensure that it can be deployed on various cloud platforms, and DMTF's OVF is a packaging standard that addresses the "portability and deployment of virtual appliances."

Open Virtualization Format is a vendor-neutral, open metadata standard that includes all necessary virtual machine specification and configuration parameters needed to deploy the application successfully over any virtualization technology of the customer choice. It is also extensible and allows inclusion of vendor specific data if required. It allows for automatic verification of the virtual appliance using industry-standard public-key infrastructure, thereby enabling a safe distribution of virtual solutions from vendors to their customers. The OVF standard specifies the packaging format along with the OVF descriptor format, which contains the hardware specifications, disks, and other characteristics of a set of VMs.

The detailed OVF specification is available online (DMTF, OVF, 2010). The descriptor contains the details of each virtual machine along with disks, network interfaces, and contextualization parameters of that machine. It also contains optional startup order information and the virtual appliance scale-out rules to govern elasticity.

There are several other types of metadata sections described in the specification; together they are sufficient to specify completely all the characteristics of a virtual appliance, thus enabling seamless portability and distribution over heterogeneous virtualization tools. The OVF standard enjoys wide industry support. Popular open-source virtualization software Oracle's VirtualBox (http://www.virtualbox.org/, accessed December 5,2015) has full support for OVF packages. It allows import/export of virtual appliances from/to OVF packages. Commercial virtualization products from VMware (http://www.vmware.com/, accessed December 5, 2015), IBM SmartCloud (http://www.ibm.com/cloud-computing/, accessed December 5, 2015), and several other major cloud players provide full OVF standard support.

16.6.5 Cloud Data Management Interface (CDMI)

CDMI (SNIA, CDMI, 2012) from the Storage Networking Industry Association (SNIA) was first released as a standard in April 2010, saw increasing adoption across the storage industry, and is now an ISO standard (ISO/IEC 17826:2012, http://www.iso.org/iso/catalogue_detail.htm?csnumber=60617, accessed December 5, 2015).

It standardizes both a data path and control path for cloud-storage products and offerings. The data path is a superset of the existing cloud storage APIs currently being deployed across the industry. The data path supports single-level or unlimited nesting of containers for storage objects. The standard also specifies a globally unique object identifier for each storage object. In addition to objects and containers, CDMI also supports a first-in first-out queue type of resource that allows for cross-cloud coordination and is used by CDMI itself to provide logs and audit trails. It also standardizes fine-grained access control for storage containers and objects leveraging Access Control Lists from the NFS standard. Full control over the creation of cloud users and their access rights is standardized through CDMI Domains, which also standardize summary information for usage of the cloud by each user. CDMI leverages the web technologies by using RESTful semantics and the HTTP protocol for all operations.

The CDMI control path uses metadata tagging on containers and objects to control the type and configuration of data services such as backup, archive, encryption, retention, and others at granularity down to individual storage objects. It permits implementations to advertise their unique capabilities, allowing a CDMI client to discover programmatically whether any given cloud can meet the requirements of the data they want to store. The client then marks the data with those requirements using the metadata. During operation, CDMI also provides how the cloud can tell the client how well those requirements are being met in real time. Combined, this amounts to a dynamic service discovery and negotiation of service levels that are only now being recognized as important in the cloud industry. The CDMI provides a query syntax for searching indexed object content as well as metadata that is stored and accessible to the client. Results are available using the queue resource, allowing powerful and distributed processing.

Finally, CDMI also standardizes a format for transporting the data and metadata between clouds. The transport can be done not only through the network but also via removable media such as flash, disk, or tape. The metadata that represents the data requirements is preserved during this operation so that, upon import into the new cloud, no new setup is required for the data to receive the same service levels of the previous cloud.

16.6.6 Cloud Infrastructure Management Interface (CIMI)

DMTF's CIMI is an standard (DMTF, CIMI, 2012) that allows management of infrastructure as a service (IaaS). It allows for management of cloud resources including virtual machines, network, and storage but is not intended for management of applications and services that the cloud users may start. It uses RESTful messages over HTTP protocol with messages sent and received in either JSON or XML formats. The standard supports cloud application portability by allowing importing an OVF package, and creation of CIMI resource elements based on the OVF descriptor elements.

The CIMI standard contains representation for all necessary elements required for an effective management of a virtual environment provisioned over IaaS cloud. The resources supported in CIMI are broadly organized into cloud entry point, machine resources, volume resources, network resources, system resources, and monitoring resources. Some of the supported CIMI resources are:

- Cloud entry point – this represents the entry point for the IaaS cloud defined by the CIMI model. It has catalog of resources including systems, system templates, machines, and machine templates, and can be queried and browsed by the end users.
- Machine resources – these represent the set of resources along with the representation needed to specify all necessary parameters involved in defining a virtual machine or a set of virtual machines. They also define all the states a virtual machine can be in, along with a set of operations allowed in a given VM state.

- Volume resources – these resources represent storage, at block- or file-system level, which can be connected to a virtual machine. The volume resource specification also includes the state information, type information, capacity, list of volume snapshots, meters associated, and so forth.
- Network resources – these define how network entities, ports, and their relationship can be represented in CIMI. The element's state, service class, type, and so forth, can be represented easily. A valid list of permitted operations can also be represented, along with several other aspects in order to represent the virtual networking capabilities and resources of a cloud provider properly.

The Apache project ∂-cloud (http://deltacloud.apache.org/cimi-rest.html, accessed December 5, 2015) is a cloud API that supports several IaaS clouds, including OpenStack, OpenNebula, IBM SmartCloud, and many more. It exposes three interfaces to application developers and end users including DMTF's CIMI. Leading cloud providers such as Broadcom, CA Technologies, Citrix, Fujitsu, Hewlett-Packard, IBM, Oracle, Red Hat, VMware, and SunGard have announced support for the CIMI v1.0 standard in their products.

16.6.7   Cloud Application Management for Platforms (CAMP)

CAMP standard from OASIS (OASIS, CAMP, 2012) focuses on making it possible to package and operate applications across different platforms and cloud services. The principal components of CAMP are the REST API, YAML (http://www.yaml.org/, accessed December 5, 2015).

Its REST API specifies how to create, inspect, manage, and terminate applications hosted in a CAMP-compliant platform. The Platform resource is the principal entry point. From here, a client can follow links to individual Assembly resources, representing applications deployed and managed by a CAMP platform. New Assembly resources can be created by POSTing a Plan (format described in the next section). At an individual Assembly resource, a client can traverse Component resources, which comprise the Assembly, and those Components in turn can comprise further Components. Each Assembly and Component can also expose Sensor resources, which report status and metrics through GET requests, as well as Operation resources, which allow management of resources through POST requests. DELETE requests can be used to terminate an Assembly. CAMP specifies a schema for application "plans," in YAML, together with an archive format. Both are designed to be simple to create whilst permitting complex descriptions where required.

Multiple archive formats are supported, with the sole stipulation that a file "camp.yaml" be included. In simple cases, this YAML file may describe the application completely with no reference to external items. For more sophisticated scenarios, the YAML may refer to other files contained in the archive or hosted externally, ranging from WAR files and SQL scripts to Chef recipes, RPM packages, and OASIS TOSCA plans.

The CAMP YAML schema for Plans centers on the concepts of Artifacts and Services. In simple cases, a Plan may refer to one or more Artifacts, such as a WAR file. Artifact entries in the schema can include requirements to inform the platform how it should be handled. Requirements may be monolithic and specific, such as indicating that a WAR file requires a precise version of a Tomcat container, or they may follow a more general mix-in approach, such as indicating capabilities expected in the deployed environment but without specifying a precise type.

The CAMP specification defines the REST API and the Plan YAML for management but it deliberately does not define concrete subtypes for individual artifacts or services, or for characteristics, requirements, and components. Instead it provides the schema by which such types can be used in an interoperable way, on the basis that the ecosystem of types is best developed by expert communities in the domains being provided by CAMP platforms.

Nonetheless, clear extension points are defined for types, endpoints, and the plan schema. Concrete subtypes can be defined by a provider and discovered through the REST API. Namespaces are defined in such a manner as to avoid conflict, such as using inverted domain names (e.g. com.java:WAR).

While early providers and adopters will have to work with custom types, the capacity for interoperability is present where platform implementers choose to offer this or where clients or brokers perform plan resolution. Due to the aspect/mix-in approach adopted in the Plan schema, it is possible, in many cases, for different vocabularies to be supported simultaneously.

The only open-source, freely available CAMP Server is the Apache licensed Brooklyn CAMP Server. This has been integrated with the codebase for the Apache-licensed project Brooklyn (http://brooklyn.io/, accessed December 5, 2015) to make available a wide range of application servers, data stores, messaging solutions, and PaaS systems. A proprietary implementation, called n-CAMP, has been developed by Oracle and integrated with a selection of application servers and SQL databases.

16.6.8   Topology and Orchestration Specification for Cloud Applications (TOSCA)

TOSCA, defined by OASIS (OASIS, TOSCA, 2013), is used for defining both the service components of distributed applications (topology), as well as the service management interfaces (plans). Service orchestration is realized by describing the interactions between the services using workflows, called plans. TOSCA's objective is to ensure the semiautomatic creation and management of application layer services, while guaranteeing applications' portability across various cloud implementations.

The purpose of TOSCA is to create and declaratively describe the components of distributed applications, in terms of services, infrastructure requirements, and interactions between services. It effectively combines both declarative and imperative approaches. A declarative model can be defined as describing the desired end-state, while providing the means for adjusting the state until the desired end-state is achieved. TOSCA's model is composed of the key entities shown in see Figure 16.2.

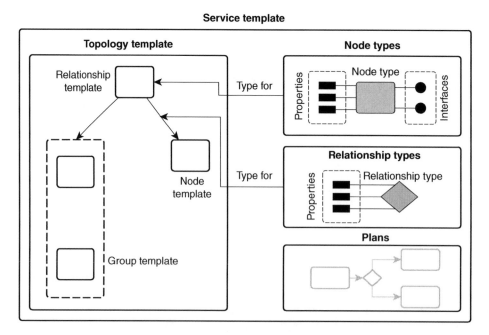

Figure 16.2   *TOSCA model entities. Source: OASIS 2013. Reproduced with permission of OASIS*

Nodes represent components of an application or service and its properties. Example nodes include infrastructure, and platform. TOSCA nodes include operations as management functions for the nodes, such as deploy, start, stop, or connect. The nodes export their dependencies on other nodes as requirements and capabilities.

Relationships represent the logical relations between nodes, such as "hosted on", and "connects to". They describe the valid source and target nodes that they are designed to couple, and have their own properties and constraints.

Service templates group the nodes and relationships that make up a service's topology. This enables modeling of subtopologies, by means of composition of applications from one or more service templates, and substitution of abstract node types with service templates of the same type.

TOSCA relies on service management platforms for interpreting the service templates and associated management plan and for converting them into infrastructure operations. Such a cloud management platform is IBM SmartCloud Orchestrator (http://www-03.ibm.com/software/products/en/smartcloud-orchestrator/, accessed December 15, 2015). TOSCA plans and service topologies can be created with a variety of tools, such as Vnomic's Application Landscape Designer (http://www.vnomic.com, accessed December 15, 2015).

16.7   Outlook and Conclusion

Each standard presented in this chapter does its job rather well. However, this is from the individual standard perspective. If we look from a wider perspective and consider how all of these standards can be used in unison, the picture becomes a little more challenging. Given that a cloud-native application will use many different service types, it would need to interface with many standards as described in this chapter. Hence it would be important to also understand wider interoperability issues between standards. An initial foray into this area was made in (Edmonds *et al.*, 2011); however, it did not include one very important aspect of not just interoperability but federation and that authentication and access (AA). From a standardization perspective there are a set of AA standards yet these are not common between the IaaS and PaaS oriented standards. For a true suite of complementing standards such agreement on AA should be considered.

When considering the use of standards, decision makers should understand what they don't do and where they are best applied. Standards are an excellent means to arrive at interoperability and compliance. Given the time taken to create standards they are often well tested and thought out. This makes them excellent tools for procurement and compliance within large IT organizations, corporations, and governments. A good example of this is report from the German Federal Ministry of Economics and Technology (German Federal Ministry of Economics and Technology, 2012) on how they see the market in respect to cloud standards. From the perspective of interoperability and how standards can facilitate this, an excellent example is the European Grid Initiative's Federated Cloud task force (http://www.egi.eu/infrastructure/cloud/, accessed December 5, 2015). In this work, the goal was to federate a set of cloud providers (sites) and their related institutions. There were 23 providers and between all of these there were 15 different IaaS technologies. The goal that the cloud task force had was to federate all of these and essentially have them all speak the same "language." The resulting solution was given through the selection of a set of cloud standards (including CDMI and OCCI) through which interoperability could be achieved. Without standards this large task of interoperating different providers would be much more difficult.

Considering activities listed above such as prototyping or development within startups, and indeed within in the general ICT industry, SDOs need to adopt a faster cycle in order to stay up with the industry and community, otherwise they face becoming left behind. This is especially true when considering the pace of development within industry, and community-led initiatives such as OpenStack (http://www.openstack.org, accessed December 5, 2015) or OpenDaylight (http://www.opendaylight.org/, accessed December 5, 2015).

An interesting development within the world of standardization is the decision that the European Telecommunications Standards Institute's (ETSI) Network Function Virtualization (NFV) (http://www.etsi.org/technologies-clusters/technologies/nfv, accessed December 5, 2015) has chosen to take. Rather than enter into a possibly long process of multiple technical interface specifications, the group has chosen to provide best practices coupled with open-source software.

One must keep in mind that not considering standard interfaces leaves such activities open to the risk of lock in. Where a cloud service must be moved from one service provider there can then be large costs and time associated with the move. It is for such a reason that standards exist, and considering them for adoption in a service's or product's development lifecycle is a means to minimize risk of lock in. Standards afford risk mitigation and, as such, provide a type of interoperability insurance.

Acknowledgements

This work was supported in part by the European Community Seventh Framework Programme (FP7/20012013) project "MobileCloud Networking" under grant agreement no.318109.

References

Andrieux, A., Czajkowski, K., Dan, A., *et al.* (2011) *Web Services Agreement Specification*. Technical Report, http://www.ogf.org/documents/GFD.192.pdf (accessed December 6, 2015).

DMTF (2010) *OVF Overview Document*, http://dmtf.org/sites/default/files/OVF_Overview_Document_2010.pdf (accessed December 5, 2015).

DMTF, CIMI (2012) *Cloud Infrastructure Management Interface (CIMI) Model and RESTful HTTP-based Protocol an Interface for Managing Cloud Infrastructure*, http://dmtf.org/sites/default/files/standards/documents/DSP0263_1.0.1.pdf (accessed December 5, 2015).

DMTF, OVF (2010) *Open Virtualization Format Specification*, http://www.dmtf.org/sites/default/files/standards/documents/DSP0243_1.1.0.pdf (accessed December 5, 2015).

Edmonds, A., and Metsch, T. (2011) *Open Cloud Computing Interface – Infrastructure*, http://www.ogf.org/documents/GFD.184.pdf (accessed December 5, 2015).

Edmonds, A., Metsch, T. and Luster, E. (2011) *An Open, Interoperable Cloud*, http://www.infoq.com/articles/open-interoperable-cloud (accessed December 5, 2015).

Edmonds, A., Metsch, T., and Papaspyrou, A. (2011) Open cloud computing interface in data management-related setups, in *Grid and Cloud Database Management* (eds. S. Fiore, and G. Aloisio), Springer, Berlin, pp. 23–48.

Edmonds, A., Metsch, T., Papaspyrou, A., and Richardson, A. (2012) Toward an open cloud standard. *IEEE Internet Computing* (Jul./Aug.), 15–25.

Fielding, R. (2000) Architectural Styles and the Design of Network-based Software Architectures. Dissertation, University of Cailfornia, Irvine, Computer Science.

German Federal Ministry of Economics and Technology (2012) *The Standardisation Environment for Cloud Computing*, from http://www.bmwi.de/English/Redaktion/Pdf/normungs-und-standardisierungsumfeld-von-cloud-computing,property=pdf,bereich=bmwi,sprache=en,rwb=true.pdf (accessed December 5, 2015).

Hammer-Lahav, E. (2010) *The OAuth 1.0 Protocol*, retrieved December 8, 2013, from http://tools.ietf.org/html/rfc5849 (accessed December 5, 2015).

Krechmer, K. (2008) *Open Standards: A Call for Action*. First ITU-T Kaleidoscope Academic Conference, Innovations in NGN: Future Network and Services. IEEE, Geneva, pp. 15–22.

Mell, P. M., and Grance, T. (2011) *The NIST Definition of Cloud Computing*. Special Publication 800-145. NIST, Gaithersburg, MD, http://www.nist.gov/customcf/get_pdf.cfm?pub_id=909616 (accessed November 25, 2015).

Metsch, T., and Edmonds, A. (2011) *Open Cloud Computing Interface – RESTful HTTP Rendering*, http://www.ogf.org/documents/GFD.185.pdf (accessed December 5, 2015).

Nyren, N., Edmonds, A., Papaspyrou, A., and Metsch, T. (2011) *Open Cloud Computing Interface – Core*, http://www.ogf. org/documents/GFD.183.pdf (accessed December 5, 2015).

OASIS, CAMP. (2012) *Cloud Application Management for Platforms*, https://www.oasis-open.org/committees/download. php/47278/CAMP-v1.0.pdf (accessed December 5, 2015).

OASIS, TOSCA (2013) *Topology and Orchestration Specification for Cloud Applications Version 1.0*, http://docs. oasis-open.org/tosca/TOSCA/v1.0/cos01/TOSCA-v1.0-cos01.pdf (accessed December 5, 2015).

SNIA, CDMI (2012) *Cloud Data Management Interface*, from http://snia.org/sites/default/files/CDMI%20v1.0.2.pdf (accessed December 5, 2015).

Theilmann, W., Kotsokalis, C., Edmonds, A. *et al.* (2010) A reference architecture for multi-level SLA management. *Journal of Internet Engineering* **4**(1), 289–298.

Wäldrich, O., Battré, D., Brazier, F., *et al.* (2011) *WS-Agreement Negotiation Version 1.0. OGF*, http://www.ogf.org/ documents/GFD.193.pdf (accessed December 6, 2015).

Ziegler, W. (2012) SLAs for Energy-Efficient Data Centres: The Standards-Based Approach of the OPTIMIS Project, in *Energy Efficient Data Centers* (eds. S. Klingert, X. Hesselbach-Serra, M. Perez Ortega, and G. Giuliani). Springer, Berlin, pp. 37–46.

Part V

Cloud Security, Privacy, and Compliance

17

Cloud Security: Issues and Concerns

Pierangela Samarati and Sabrina De Capitani di Vimercati

Università degli Studi di Milano, Italy

17.1 Introduction

Rapid advances in information and communication technology (ICT) have enabled the emergence of the cloud as a successful paradigm for conveniently storing, accessing, processing, and sharing information. With its significant benefits of scalability and elasticity, the cloud paradigm has appealed to companies as well as individuals, which are increasingly resorting to the clouds to store and process data. Unfortunately, this convenience comes at the price of loss of control by the owners of the data, and consequent security threats, which can limit the potential widespread adoption and acceptance of the cloud-computing paradigm. On one hand, cloud providers can be assumed to employ security mechanisms for protecting data in storage, processing, and communication, devoting resources to ensure security that many individuals and companies may not be able to afford. On the other hand, data owners, and users of the cloud, lose control over their data and their processing. The European Network and Information Security Agency (ENISA) has listed loss of control and governance as a top risk of cloud computing (European Network and Information Security Agency, 2009). The Cloud Security Alliance (CSA) lists data breaches and data loss as two of the top nine threats in cloud computing (Cloud Security Alliance, 2013). Security threats can arise because of the complexity of the cloud scenario (e.g., dynamic distribution, virtualization, and multitenancy), because data or computations might be sensitive, and should be protected even from the provider's eyes, or because providers might be not fully trustworthy and their – possibly lazy or malicious – behavior should be controlled.

The cloud encompasses a variety of distributed computing environments, varying in the architectural or trust assumptions and the services offered. In particular, the US National Institute of Standards and Technology (NIST) distinguished four deployment models and three service models (National Institute of Standards and Technology, 2011). The deployment models range from a *private* cloud, where the infrastructure and services are operated for a single organization and are maintained on a private network, to a *public* cloud, where the

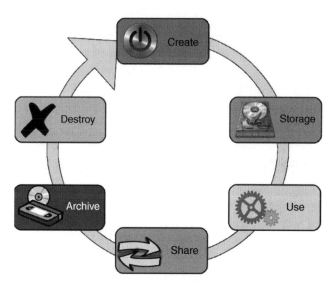

Figure 17.1   *Data security lifecycle. Source: Adapted from Cloud Security Alliance (2011)*

infrastructure is made available to the public and is owned by an organization offering cloud services. Ownership and operation models between these two extremes are also possible, such as in a *community* cloud, where different companies with common objectives (e.g., business goals and security requirements) share the cloud infrastructure, and a *hybrid* cloud, composed of multiple clouds, which can be private, public, or community, under the control of one or more cloud providers, and with more stringent security requirements than a public cloud. Similarly, different service models, namely infrastructure as a service (IaaS), platform as a service (PaaS), and software as a service (SaaS), entail different responsibilities in enforcing security. The security and privacy issues to be addressed and the challenges involved can vary in different deployment and service models.

In this chapter, we highlight security issues that need to be considered when using the cloud to offer or enjoy services in the different models above. We discuss security aspects that are more affected by the cloud paradigm, in particular in relationships to the *data security lifecycle*, reported in Figure 17.1 (Cloud Security Alliance, 2011). Of course, complete protection also requires the use of other, perhaps more traditional, security techniques, on which we do not elaborate.

The chapter is organized in two main sections. Section 17.2 discusses how the classical confidentiality, integrity, and availability properties translate in the cloud. Section 17.3 presents an overview of the security issues and concerns to be addressed to ensure confidentiality, integrity, and availability. For each issue, we provide a description of the problem and challenges to be addressed together with possible existing solutions or directions.

17.2   Confidentiality, Integrity, and Availability in the Cloud

Security problems can be classified with the classical CIA (*confidentiality*, *integrity*, and *availability*) paradigm, which in the cloud can be interpreted as follows. Confidentiality requires guaranteeing proper protection of confidential or sensitive information stored or processed in the cloud. Depending on the requirements of the scenario in question, this can relate to any or all of: the data externally stored, the identity/properties of the users accessing the data, or the actions that users perform with the data. Integrity requires guaranteeing the authenticity of the parties (users and providers) interacting in the cloud, the data stored at external providers, and the responses returned from queries and computations. Availability requires providing the ability to define and verify that providers satisfy requirements expressed in service-level agreements (SLAs) established between data owners/users and providers. The issues to be tackled, the challenges to be addressed, and the specific guarantees to be provided ensuring satisfaction of the security properties above depend on the characteristics of the different scenarios. For instance, in a simple scenario, where an individual or a company uses the cloud simply for *archival/storage* purposes, problems to be addressed concern protecting confidentiality or integrity of data in storage and assessing satisfaction of SLAs, also ensuring correct enforcement of *create* and *destroy* operations. In a more complex scenario requiring execution of queries over data (*use*), the problem arises of executing queries as well as guaranteeing confidentiality and integrity of the dynamically computed results. Where people other than the owner (or a restricted set of trusted users) access the data (*share*), this entails further complications such as the need to enforce access-control restrictions over the data, ensure data integrity in the presence of concurrent independent operations, and even ensure confidentiality of a user's actions with respect to other users. A further aspect that affects the issues to be addressed and possible applicable techniques are the trust assumptions – and consequent potential threats – regarding the providers involved in the storage and processing of the data, which could be *fully trusted, curious, lazy,* or *malicious.* "Fully trusted providers" can be assumed in cases of private clouds (or parts of them) under complete and full control of the data owner. "Curious providers" refers to scenarios where the storage or processing involves sensitive information (data or actions on them) that should be maintained confidential to the providers themselves. "Lazy providers" refers to scenarios where the storing or processing providers might not be considered fully trustworthy for ensuring data or computation integrity or for providing the availability promised in the service level agreements. Finally, "malicious (or byzantine) providers" refers to cases where providers may intentionally behave improperly in the management, storage, and processing of the data, possibly compromising their confidentiality, integrity, or availability (this case accounts also for insider threats at the provider's side).

17.3   Issues and Challenges

The discussion in the previous section makes it clear that there is not a one-size-fits-all solution (or even a one-size-fits-all problem definition). There are instead different aspects, with related issues, challenges, and security controls that need to be considered and that can find application in different scenarios. In this section, we illustrate these issues and challenges, summarized in Table 17.1.

17.3.1   Protection of Data at Rest

A first basic problem that needs to be addressed when relying on the cloud for storing data is to guarantee protection (i.e., confidentiality, integrity, and availability) to the stored data. With current solutions, users typically need to completely trust the cloud providers. In fact, although cloud providers apply security measures to the services they offer, such measures allow them to have full access to the data. For instance, Google Docs or Salesforce support encryption of the data both in transit and in storage but they also manage the encryption keys, and therefore users do not have direct control on who can access their data. Whenever data confidentiality needs to be guaranteed, even to the extent of hiding it from the provider's eyes, other solutions have to be considered. Solutions for protecting confidentiality in this *honest-but-curious,* scenario typically require encrypting data before releasing it to the cloud providers (Figure 17.2 (a)). For instance, services like

Table 17.1 *Summary of cloud security issues*

Issue	Description
Protection of data at rest	Guarantee confidentiality, integrity, and availability of data
Fine-grained access	Enable fine-grained retrieval and query execution on protected data
Selective access	Enable owner-regulated access control and authorization enforcement
User privacy	Support privacy of users accessing data and performing computations
Query privacy	Support privacy of users' actions in the cloud
Query and computation integrity	Enable assessment of correctness, completeness, and freshness of queries and computations
Collaborative query execution with multiple providers	Enable controlled data sharing for collaborative queries and computations involving multiple providers
SLA and Auditing	Specification and assessment of security requirements to be satisfied by providers
Multi-tenancy and virtualization	Provide confinement of different users' data and activities in the shared cloud environment

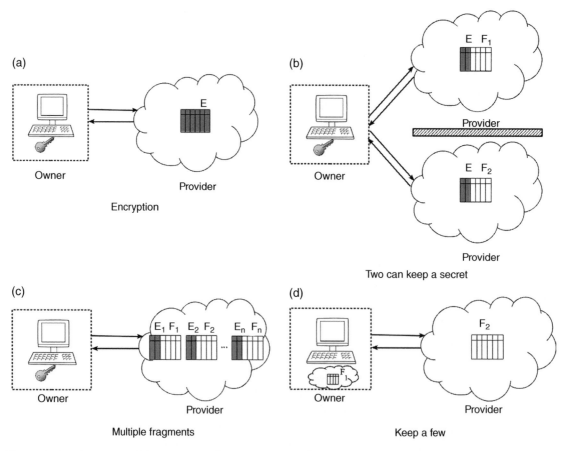

Figure 17.2 *Protection of data at rest with: encryption (a), fragmentation over two independent providers (b), fragmentation with un-linkable fragments (c), and fragmentation with the owner storing some of the data (d)*

Boxcryptor allow users to encrypt their files locally before releasing them to a cloud provider such as Dropbox, Google Drive, and Microsoft SkyDrive. Encryption guarantees both confidentiality and integrity (as data tampering can be easily detected). For performance reasons, symmetric encryption is usually adopted. While encryption can be effective in many environments, it brings in several complications in scenarios where fine-grained retrieval of data needs to be supported (see section 17.3.2). For this reason, recent approaches have put forward the idea of using *fragmentation*, instead of encryption, when what need to be kept confidential are the associations among data values, in contrast to the values themselves (Ciriani *et al.*, 2010). Fragmentation protects sensitive associations by splitting the concerned pieces of information and storing them in separate unlinkable fragments. Fragmentation can be applied in conjunction with encryption or by itself, resulting in different approaches (Figure 17.2 (b–d)). In the *"two can keep a secret"* approach (Figure 17.2 (b)), the data owner relies on two independent noncommunicating providers, each of whom stores a portion of the data, as much as possible in plaintext form, with encryption applied only to data values that either are sensitive by themselves or cannot be stored in the clear at any of the two providers without disclosing some sensitive associations. In the *"multiple fragments"* approach (Figure 17.2 (c)), only attributes with sensitive values are encrypted, while all other attributes are stored in the clear in as many fragments as needed, trying to avoid excessive fragmentation. In the *"keep a few"* approach (Figure 17.2 (d)), nothing is encrypted and there is instead the involvement of a trusted party (typically the data owner) for storing and processing a limited amount of data that are sensitive by themselves or whose visibility would disclose some sensitive associations.

Ensuring integrity and availability of data in storage requires providing the data owners/users with the ability to verify that data have not been improperly modified or tampered with, and that their management at the provider side complies with the service-level agreements. Integrity of data can be verified by employing signature schemes, where data is digitally signed to make improper modifications on them detectable. Signatures provide a deterministic guarantee of data integrity. Probabilistic guarantees can be provided by the use of checks, such as *sentinels* used in "proof of retrievability" (POR) solutions, which apply to encrypted data, or *homomorphic verifiable tags* used in "provable data possession" (PDP) solutions, which apply to generic datasets. Availability of data despite failures or noncompliance of providers can be guaranteed by employing classical replication techniques, distributing data at different providers.

Protection of data also entails ensuring correct destruction of the data at the owner's demand. The use of encryption under the control of the owner can provide such a guarantee because possible remaining data copies would be unintelligible without the proper key (Cachin *et al.*, 2013).

17.3.2 Fine-Grained Access to Data in the Cloud

Maintaining confidentiality of the data even with respect to the providers storing or processing it implies, when data is protected with encryption, that the providers cannot decrypt the data for query execution. In applications where fine-grained access, typically query execution, needs to be supported, queries should then be evaluated on the encrypted data. There are two lines of approach for providing this ability. The first approach consists in performing *queries directly on the encrypted data,* where such a capability is made possible by specific cryptographic techniques (e.g., homomorphic encryption). The main drawbacks of these approaches, applicable typically for keyword searches or very basic operations, remain the limited kinds of accesses supported and the computational complexity of the execution, which make them inapplicable in many real-life scenarios. Other solutions enabling execution of SQL queries directly on encrypted data while guaranteeing more support for operations and efficiency rely on different layers of encryption, each supporting specific operations. An example is CryptDB (Popa *et al.*, 2011), where each relation is encrypted at the column level with different onion layers of encryption, each supporting the execution of a specific SQL operation. Whenever the CryptDB proxy server receives an SQL query, it determines the onion layer needed for its execution. If the encrypted data is not already at the required onion layer, the proxy sends the provider the key of the onion layer, enabling the provider to strip off the other layers and execute the query. The second approach involves attaching to the encrypted data some metadata (*indexes*), which is then used for fine-grained information retrieval and query execution. For instance, in a relational table where tuples are encrypted, different indexes can be specified for the different attributes on which conditions might need to be evaluated. Indexes should be well related to the data behind them, so to be precise and effective for query execution, but at the same time should not leak information on such data. Such a protection should be guaranteed from *static observations* (observation of the encrypted and indexed data in storage) as well as *dynamic observations* (observation of the queries in execution on such data). Different kinds of indexes have been investigated, including *direct indexing* (providing a one-to-one correspondence between plaintext and index values), *bucket-* or *hash-based indexing* (providing a many-to-one correspondence between plaintext and index values), and *flat indexing* (providing a one-to-many correspondence between plaintext and index values). Other types of indexes have been investigated in relation to tree-based data structures, and order-preserving or homomorphic encryption solutions, for providing support of range queries or aggregate functions. Different approaches to indexes provide different protection guarantees as well as different support for, and performance in, query execution. For instance, the many-to-one correspondence in bucket and hash-based approaches, where multiple plaintext values collide to the same index, and the flat indexing, where all different index values have the same number of occurrences, provide better protection of the confidentiality of the indexing with respect to direct indexing, at the price of a more complex query process. Indexing approaches based on order-preserving encryption also provide support for range queries, but are exposed to some information leakage.

Query execution over encrypted and indexed data typically involves a trusted client application translating the plaintext query Q in a query Q_p to be sent to the provider and query Q_c performing some postprocessing for decrypting data and removing possible tuples originated by collisions in the index function and not belonging to the result (Figure 17.3).

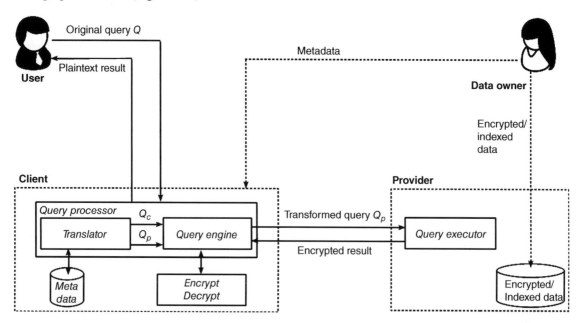

Figure 17.3 *Query evaluation over outsourced (encrypted/indexed) data: the user query Q is translated by a trusted client in a query Q_p to be executed by the provider and a query Q_c to be executed at the client side over Q_p's result once decrypted*

17.3.3   Selective Access to Data in the Cloud

In many scenarios, access to data is selective, meaning different users (or groups thereof) should enjoy different views of and access to the data. When data is stored in the cloud, the problem arises of how to enforce such access control restrictions. For instance, some cloud storage services (e.g., Amazon S3 and Google Cloud Storage) support the definition of access-control lists for regulating access to data. The enforcement of such access-control policy is, however, delegated to the cloud provider. In many scenarios this solution is not possible because the access-control policy, just like the data, might be confidential and therefore should not be disclosed to the provider (note also that even authorization to access data could leak information about the data, therefore potentially compromising the protection enforced by encryption). Outsourcing access control to the cloud requires complete trust in the enforcing providers, as data protection would be completely in their hands (and providers could collude with users to acquire – and improperly grant – unauthorized access to data). On the other hand, having the data owner mediate every access request, to ensure only authorized accesses are granted, is clearly impractical and inapplicable. A promising approach to delegate access control to the cloud while not requiring complete trust in the providers relies on combining access control and encryption, that is, encrypt data with different keys, depending on the user's authorization. Enforcing access control policies via encryption entails some challenges: users should not be required to hold many keys for the different resources they can access; at the same time every resource should be maintained only once (different replicas encrypted with different keys should be avoided as their management would clearly be impractical). This problem can be solved by employing *key derivation methods*, by which users can derive keys from a single key assigned to them, and public tokens. Access control can then be enforced by properly organizing the keys in a hierarchy reflecting authorization, or better the access control lists (ACLs) of resources, where the key corresponding to an ACL allows access – via one or more tokens – to the keys associated with all ACLs that are superset of it. This way users are able to derive, from their keys and public tokens, all (and only) the keys that are needed to access resources that they are authorized to access (see Figure 17.4).

Updates to the access control policy can require changing the key with which resources have been encrypted, and therefore the need to download the resources from the cloud and release a newly encrypted version of them. Such a burden can be avoided by assuming some collaboration from the external providers in enforcing policy changes, having the providers apply a further level of encryption, called *overencryption* (De Capitani di Vimercati *et al.*, 2010) in addition to – and on top of – the one applied by the owner. To access a resource *r* (see Figure 17.5), a user needs to pass both the encryption imposed by the provider (SEL, surface encryption layer) and the encryption imposed by the owner (BEL, base encryption layer).

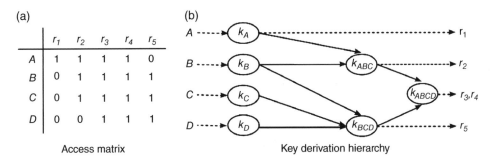

Figure 17.4   *An example of access control policy (1 represents authorized accesses) with four users and five resources (a) and of key derivation hierarchy enforcing it: solid lines represent public tokens, dotted lines represent the keys associated with users and resources (b)*

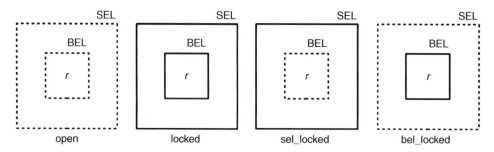

Figure 17.5   *Protection of resources with overencryption. Every resource is encrypted first by the owner (BEL, base encryption layer) then by the provider (SEL, surface encryption layer). A resource is accessible (open) to users only if they can pass both levels of encryption*

Alternative solutions to enforce access control in the cloud use *attribute-based encryption* (ABE) techniques, possibly combined with other cryptographic techniques such as proxy and lazy re-encryption (Yu, Lou, and Ren, 2012). Attribute-based encryption is a public-key encryption that regulates access to data according to descriptive attributes associated with the data itself and/or users, and to policies defined over these attributes. It can be implemented either as Ciphertext-Policy ABE (CP-ABE) or as Key-Policy ABE (KP-ABE), depending on how attributes and policies are associated with data and/or users.

17.3.4   User Privacy

In a cloud scenario there might be need to grant access to data to users not registered in the system without requiring such users to declare their identity. In these scenarios, access-control authorizations and enforcement should be based on properties of users (in contrast to their identity), typically provided by means of attributes within digitally signed certificates. Access control solutions supporting this new paradigm are referred to as *attribute-based, credential-based,* or *certificate-based access control*, to stress the departure from identity to consider instead certified properties in the access decisions, or *privacy-enhanced access control*, to stress the privacy offered by departing from user authentication. Several proposals have investigated different issues to be addressed in this context, including: the language for expressing authorizations, the access control engine for evaluating users' requests, the possible dialog and negotiation to be supported between providers and users, the support for users' preferences with respect to properties to be released for acquiring services, and possible secondary use restrictions. As for languages, early proposals typically investigated the use of logic-based approaches, whereas later approaches aimed at balancing the tradeoff between expressiveness of the language and simplicity of (and hence ability to maintain control on) the specifications. Different strategies for the dialog between users and providers have been investigated, including multistep negotiations. Even in this case, later proposals aimed at balancing the need to exchange information to establish trust between users and providers, and the simplicity of the dialog to make it suitable for practical applications. As for user preferences, whereas earlier approaches assumed users to regulate release of their credentials and properties, with an access-control approach similar to one adopted by the providers, more recent proposals have been investigating solutions specifically targeted at users and their natural way of thinking about preferences (Foresti and Samarati, 2012). Standards, such as XACML, have also been developed in these contexts supporting interoperation of access control policies.

17.3.5 Query Privacy

In some scenarios what is confidential is not (or not only) data, or users' identities/properties but also the access that users make to such data. In particular, confidentiality should be guaranteed, even from the provider's eyes, with respect to the fact that an access aims at a specific data (*access confidentiality*) or the fact that two accesses aim at the same data (*pattern confidentiality*). Traditional approaches for protecting access and pattern confidentiality are based on *private information retrieval* (PIR) techniques that, assuming a database modeled as an N-bit string, provide protocols for users to retrieve the i-th bit in the string without disclosing to the provider the specific bit accessed. In addition to the limitations of such modeling and of the fact that they do not consider data confidentiality, PIR solutions suffer from high computational complexity and communication costs. Recent efforts, trying to make PIR more practical, have investigated the application of the Oblivious RAM, in particular with recent *practical ORAM* and *Path ORAM* solutions (Stefanov *et al.,* 2013), and of a key-based hierarchical and dynamic data structure, called *Shuffle index* (De Capitani di Vimercati *et al.,* 2011b). These proposals protect data confidentiality with encryption and protect access and pattern confidentiality by dynamically changing (*shuffling*), at every access, the physical location of the data, thus destroying the otherwise static correspondence between data and the physical blocks where it is stored. These approaches also employ a *cache* to maintain some data at the client side. Besides caching and dynamic allocation, Path ORAM assumes a tree-shaped data structure where nodes can contain, in addition to actual blocks, *dummy blocks* to guarantee that nodes have always the same size. The Shuffle index assumes that, at every access, additional fake searches, called *cover searches*, are executed together with the actual target search. Cover searches provide confusion to the provider with respect to the targeted block. At every access, the content of the blocks accessed (target/cover) and in cache is shuffled and rewritten. This dynamically changes the allocation of nodes, and the provider can only observe that some blocks have been read and written (Figure 17.6). By assuming a hierarchical value-based organization of the data (B+-tree with encrypted node content and with no pointer between leaves), the Shuffle index is also able to support range-based queries.

17.3.6 Query and Computation Integrity

In scenarios where queries/computations are performed by providers that are not fully trustworthy, the problem arises of providing data owners and/or users with the ability to assess that the result returned from a query/computation is *correct*, *complete*, and *fresh*. Correctness means that the result has been computed over the original data and the query/computation performed correctly. Completeness means that no data is missing from the result. Freshness means that the query/computation has been performed on the most recent version of the data. Most of the current approaches focus on providing guarantees of completeness and correctness, with some proposals complementing them with timestamps or periodical refreshing to provide freshness guarantees. Current solutions can be roughly classified in two categories: *deterministic* and *probabilistic*. Deterministic approaches are provided by authenticated data structures that, like signature schemes for static data, permit integrity violations to be detected with certainty. Examples of deterministic approaches for correctness/completeness are *signature-chaining schemas* and *Merkle hash trees*. Signature-chaining schemas allow the verification of the ordering among tuples and can then be used to verify the integrity of range queries where the selection condition is based on the attribute on which the signature schema has been applied. Merkle trees and their variations organize data within a tree-based structure over a given attribute (e.g., a search key). The result of a query with selection conditions on the attribute includes, in addition to the tuples belonging to the result, a verification object that allows the assessment of the integrity of the query (Figure 17.7). These authenticated data structures provide deterministic integrity guarantees but only for queries about the specific attribute/s on which the data structure has been

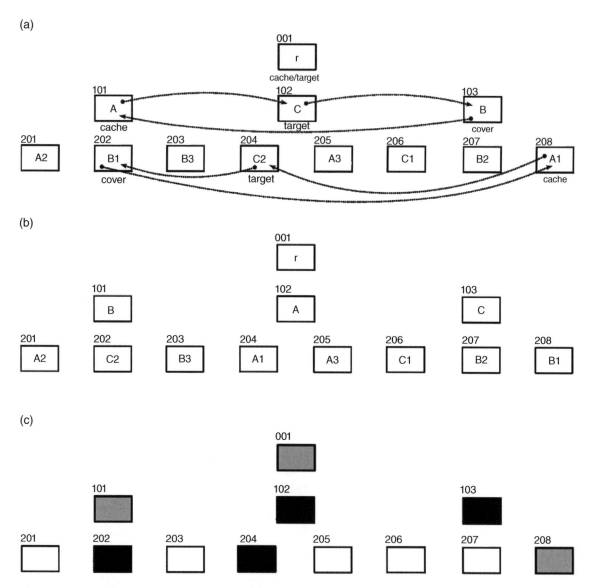

Figure 17.6   *Shuffle index: original structure with cache/target/cover and shuffling operations due to an access (a); resulting structure at the end of an access (b); provider's view: blocks written (gray) or read and written (black) in the access execution (c)*

organized. Techniques that have been applied, individually or in combination, for providing probabilistic guarantees include: insertion of *fake tuples* in the data, which, if not retrieved in the query result, signals an integrity compromise; *replication* of a portion of the data with replicas not recognizable as such, so that the presence of duplicated data where the replica is missing signals an integrity compromise; and *precomputation of tokens* associated with chosen query results, which allows the verification of the integrity of such queries. Probabilistic approaches, as their name says, provide only probabilistic guarantees: while the absence of an expected fake tuple or replica signals an integrity problem, its presence does not imply the

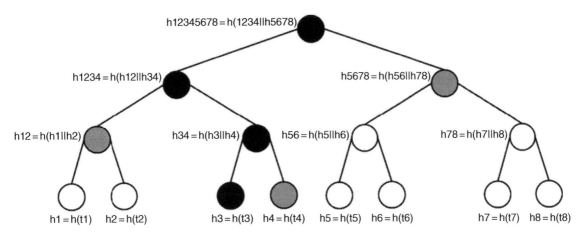

Figure 17.7 *A Merkle tree: every leave node is a hash of a tuple, internal nodes are hashed over the concatenation of their children. Colored node represent the integrity check assuming a query with result tuple t3: in gray the verification objects returned by the provider together with tuple t3, in black the hash computed for verification*

integrity of the result because the providers might have just been lucky in not missing any of the checks inserted by the data owner. The probability of detecting an integrity compromise typically depends on the amount of controls enforced (e.g., fake tuples, replicas inserted, or precomputed tokens), where the more the control the higher the guarantees, but also the higher the performance overhead imposed for the verification. The involvement of multiple providers in the storage or computation complicates the scenario and requires devising additional controls. A possible solution to assess integrity of joins computed by an untrusted provider over data stored at two trusted storage providers assumes the cooperation of the storage providers to insert control information consisting of fake control tuples (*markers*) and duplicate tuples (*twins*) that if not present in the join result signals its incompleteness (De Capitani di Vimercati *et al.*, 2014).

17.3.7 Collaborative Query Execution with Multiple Providers

Data stored and managed by different cloud providers may need to be selectively shared and accessed in a cooperative way. This scenario may see the presence of different providers as well as of different data owners. Exchange of data and collaborative computations should be controlled to ensure that information is not improperly accessed, released, or leaked. For instance, data stored at one provider might be released selectively only to specific providers and within specific contexts. Some solutions have addressed the specific problem of private and *secure multiparty computation*, which provide the ability of different parties to perform a collaborative computation learning only the query results and nothing on the inputs. Along the same line are solutions for computing *sovereign joins* over data, retrieving the result of a join operation over different tables, while guaranteeing confidentiality of the information not belonging to the join result. Recent approaches have also addressed a more general scenario where different parties (data owners or cloud providers) need to collaborate and share information for performing a *distributed query computation* with selective disclosure of data. The problem has also been investigated of determining an efficient and safe execution plan for the query computation in which different providers collaborate releasing to others the information authorized and needed to compute the query result (De Capitani di Vimercati *et al.*, 2011a).

17.3.8   Service Level Agreement and Auditing

An SLA is a contractual agreement that specifies the performance and availability guarantees that a cloud provider promises to deliver, as well as penalties in the case of violations of the SLA. Due to the shared and dynamic nature of the cloud, cloud providers have to address several issues related to offering and managing SLAs, with different requirements coming from different users. Also, while in the past SLAs mainly focused on aspects related to the quality of the services offered (e.g., availability, response time, and fault resolution time), today they may also include the specification of the security guarantees, such as evidence of: the integrity of the stored data, its possession, its handling, or the application of specific security mechanisms (e.g., encryption or perimeter protection). In this contest, the auditability of cloud providers refers to the ability of users to verify full respect of the security guarantees declared in a SLA. Some proposals have presented solutions for verifying; for example, whether cloud providers are correctly storing data or correctly executing computation-intensive tasks on behalf of the users. In fact, lazy providers could delete some rarely accessed data or omit some computations to save resources. Some approaches apply proof-of-retrieval solutions as building blocks to allow users to verify that their data is properly secured via encryption, intact, and retrievable. The correctness of the result of outsourced computations can be verified by applying the techniques for assessing integrity that we have discussed previously.

17.3.9   Multitenancy and Virtualization

Multitenancy refers to the ability to provide computing services to different users by using a common cloud infrastructure. Each user or company (i.e., a *tenant* of the cloud infrastructure) shares computation, memory, network, and storage resources, thus reducing the costs and improving the utilization of resources as well as the scalability and reliability. A basic mechanism enabling multitenancy in the cloud is *virtualization*, which creates a virtual version of, for example, an operating system, a storage device, or network resources, within a single physical system. Although virtualization brings great flexibility, it also introduces several security concerns that may have the *hypervisor* and/or the resident *virtual machines* as the main target. The hypervisor is a software component whose goal is to create and run the virtual machines. A compromised hypervisor can put at risk the confidentiality and integrity of the data managed by the virtual machines. Other security concerns can be related to the allocation and deallocation of resources associated with virtual machines. In fact, improper leakages can result if the memory allocated to a virtual machine is not properly wiped before being reallocated to another virtual machine. The communication, monitoring, modification, and migration of virtual machines can also be a source of security concerns. In fact, due to the multitenant nature of cloud environments, there is the risk of improperly leaking information if the virtual resources allocated to different users are not perfectly isolated. Other aspects can be related to placement of virtual machine instances in the cloud, also supporting security constraints imposed by users, such as the request not to allocate given virtual machine instances to the same provider (Jhawar *et al.*, 2012).

17.4   Conclusion

With the rapid growth of cloud computing platforms and services, cloud security is becoming a key priority for all players (individuals, companies, and cloud providers). In this chapter, we presented an overview of cloud security issues and concerns, illustrating their impact on confidentiality, integrity, and availability properties and describing solutions to address them. We also discussed possible challenges and future directions.

Acknowledgements

This work was supported in part by: the EC within the 7FP under grant agreement 312797 (ABC4EU) and within the H2020 program under grant agreement 644597 (ESCUDO-CLOUD); and the Italian Ministry of Research within PRIN project "GenData 2020" (2010RTFWBH).

References

Cachin, C., Haralambiev, K., Hsiao, H. C., and Sorniotti, A. (2013) *Policy-Based Secure Deletion*. Proceedings of the ACM Conference on Computer and Communications Security (CCS 2013), Berlin, Germany. ACM.

Ciriani, V., De Capitani di Vimercati, S., Foresti, S, *et al.* (2010) Combining fragmentation and encryption to protect privacy in data storage. *ACM Transactions on Information and System Security (TISSEC)* **13**(3), 22:1–22:33.

Cloud Security Alliance (2011) *Security Guidance for Critical Areas of Focus in Cloud Computing V3.0*, http://www.cloudsecurityalliance.org/guidance/ (accessed December 7, 2015).

Cloud Security Alliance (2013) *The Notorious Nine – Cloud Computing Top Threats in 2013*, http://www.cloudsecurityalliance.org/topthreats (accessed December 7, 2015).

De Capitani di Vimercati, S., Foresti, S., Jajodia, S., *et al.* (2010) Encryption policies for regulating access to outsourced data. *ACM Transactions on Database Systems (TODS)* **35**(2), 12:1–12:46.

De Capitani di Vimercati, S., Foresti, S., Jajodia, S., *et al.* (2011a) Authorization enforcement in distributed query evaluation. *Journal of Computer Security (JCS)* **19**(4), 751–794.

De Capitani di Vimercati, S., Foresti, S., Jajodia, S., *et al.* (2014) Integrity for join queries in the cloud. *IEEE Transactions on Cloud Computing (TCC)* **1**(2), 187–200.

De Capitani di Vimercati, S., Foresti, S., Paraboschi, S., *et al.* (2011b) Efficient and private access to outsourced data. Proceedings of the 31st International Conference on Distributed Computing Systems (ICDCS), Minneapolis, MN.

European Network and Information Security Agency (2009) *Cloud Computing: Benefits, Risks and Recommendations for Information Security*, http://www.enisa.europa.eu/activities/risk-management/files/deliverables/cloud-computing-risk-assessment/at_download/fullReport (accessed December 7, 2015).

Foresti, S., and Samarati, P. (2012) Supporting user privacy preferences in digital interactions, in *Computer and Information Security Handbook* (ed. J. R. Vacca). 2nd edn. Morgan Kaufmann, Burlington, MA.

Jhawar, R., Piuri, V., and Samarati, P. (2012) Supporting security requirements for resource management in cloud computing. Proceedings of the Fifteenth IEEE International Conference on Computational Science and Engineering (CSE 2012), Paphos, Cyprus.

National Institute of Standards and Technology (2011) The NIST Definition of Cloud Computing. Special publication 800-145, http://csrc.nist.gov/publications/nistpubs/800-145/SP800-145.pdf (accessed December 7, 2015).

Popa, R. A., Redfield, C. M. S., Zeldovich, N., and Balakrishnan, H. (2011) CryptDB: Protecting confidentiality with encrypted query processing. Proceedings of the Twenty-Third ACM Symposium on Operating Systems Principles (SOSP 2011), Cascais, Portugal.

Stefanov, E., van Dijk, M., Shi, E., *et al.* (2013) Path ORAM: An extremely simple oblivious RAM protocol. Proceedings of the Twentieth ACM Conference on Computer and Communications Security (CCS 2013), Berlin, Germany.

Yu, S., Lou, W. and Ren, K. (2012) Data security in cloud computing, in *Handbook on Securing Cyber-Physical Critical Infrastructure* (eds. S. K. Das, K. Kant, and N. Zhang). Morgan Kaufmann, Burlington, MA.

18

Securing the Clouds: Methodologies and Practices

Simon Liu

National Agricultural Library, USA

18.1 Cloud Security Management

The fundamental challenge of cloud security is the loss of hands-on control of systems, applications, data security, and other resources. Many of the existing best practice security controls that security teams have come to rely on are not available in cloud environments, are stripped down in many ways, or cannot be controlled by security teams. To manage cloud security effectively, organizations should focus on major areas such as cloud governance, security policies, service contracts, and service level agreements (SLAs).

18.1.1 Cloud Governance

Cloud governance is a set of processes, policies, technologies, and services affecting the way an enterprise's cloud solutions are directed, administered, or controlled. Cloud governance is essential for enterprises to maintain control over increasingly complex and integrated systems, services, and human resources environments. There are many models of cloud governance; however, all follow the same basic principle – applying sound policies and following mature processes when using cloud technologies and services. An effective cloud-governance framework must include the following components.

- Workforce education – many security breaches and attacks stem from negligence or ignorance of internal workforce in an organization (Petruch *et al.,* 2011). Many breaches are a result of something that internal users have done or failed to do. To prevent such things from happening again, or at all, internal users must be made aware of the dangers of some actions and must be educated about security measures.

Encyclopedia of Cloud Computing, First Edition. Edited by San Murugesan and Irena Bojanova.
© 2016 John Wiley & Sons, Ltd. Published 2016 by John Wiley & Sons, Ltd.

- Identity and access management – this is one of the most effective ways to keep track of people who have access to sensitive systems or data. It prevents, or at least mitigates, breaches and attacks from internal sources. Access management must be paired with a data-logging solution that allows administrators to know who does what, when, and where, and that all changes are logged and audited properly.
- Risk and event management – all businesses face uncertainty and natural or human make events. Businesses need to follow established processes to determine how they can detect, manage, and mitigate that uncertainty and quickly respond to expected or unexpected events to minimize the negative impact to business.
- Compliance audit – proper cloud governance naturally includes a compliance audit. The audit must be independently conducted and should be robustly designed to reflect best practice, appropriate resources, and tested protocols and standards. Use audit tools to assess and view the organization's vulnerabilities across the board.

A sound governance framework is essential for cloud computing. There is no one-size-fits-all solution as no two organizations are alike. Despite the differences, all organizations need a security governance framework for any cloud solution that they may be using.

18.1.2   Security Policies

A cloud security policy is a document that states in writing how a company plans to protect the company's cloud solutions and information assets. A security policy is often considered to be a "living document," meaning that the document is never finalized but is continuously updated as technology and business requirements change. Building cloud security policies is a crucial step to take before diving into the cloud, to ensure maximum benefits are achieved and data is secure. Developing a good cloud security policy involves thinking of the right questions and answering those questions about why organizations are moving information assets to the cloud, finding a good fit in a provider, and understanding organizations' cultures and needs. Below are major questions every organization moving to the cloud needs to ask beforehand:

- Existing policies – what existing policy does the organization have that can be applied to the cloud solutions?
- Best practices – are there proven policies from standards bodies such as ISO, SANS, NIST or CSA? There is no need to reinvent the wheel.
- What to put in the cloud? Based on this, an organization can identify criteria to determine the best cloud provider and service required. Does the organization allow sensitive corporate data, protected data, and day-to-day operational data in the cloud? Does the organization have a good data-classification policy and mechanisms?
- What is the exit strategy? Having a clear exit strategy before starting prevents an organization from potentially large operational costs or downtime later.
- Who is allowed to enter into agreements with cloud providers? Who has authority to negotiate SLAs? Who can set up an application in the cloud or move data to it, and with whom should it be approved beforehand? Who is allowed to modify settings on the cloud that affect performance? Define who, when, and under what circumstances changes can be made.
- Where are data and applications physically located? The location of the organization's data and where it could be moved could have legal and privacy implications.

18.1.3   Contract Requirements and Languages

A cloud-computing contract is a formal written agreement between cloud provider and customer. Before entering into a cloud-computing arrangement, a company should evaluate its own practices, needs, and restrictions, in order to identify the legal barriers and compliance requirements associated with a proposed cloud solution. In addition, the company should conduct due diligence of the proposed cloud service provider, in order to determine whether the offering will allow the company to fulfill its continued obligation to protect its assets.

Depending on the nature of the services, the contract may commonly be in the form of a click-wrap agreement, which is not negotiated; or the parties may negotiate a more suitable written document that is tailored to the specific situation (National Institute of Standards and Technology, 2011). If a click-wrap agreement is the only agreement available, the cloud customer should balance the risks from forgoing negotiations against the actual benefits, financial savings, and ease of use promised by the cloud service provider. If the parties can negotiate a contract, they should ensure that the provisions of this contract address the needs and obligations of the parties both during the term of the contract and upon termination. Detailed, comprehensive provisions, addressing the unique needs and risks of operating in a cloud environment, should be negotiated. While there are multiple variations of cloud computing delivery models, the contract issues associated with each are similar and the contract language should cover at least the following areas:

- Pricing – the contract should have a specific price model, including caps to eliminate ballooning costs after the initial investment.
- Data – the contract should state clearly that the customer owns all data residing within the cloud environment; mandate that the customer be able to access and retrieve its data stored in the cloud at its sole discretion; provide a mechanism for the customer to require the cloud provider to destroy specified records as requested; provide clear instructions on how customer data will be returned or retrieved in the event of contract termination, and specify the cloud provider's obligations in the event of data breach or unauthorized access.
- Audit – the customer has the right to request independent audits or certification related to infrastructure and security. The customer has a right to perform an onsite inspection of the cloud provider's infrastructure and security practices on a specified basis; review the infrastructure and security specifications in written format if it so chooses; audit the performance records of the cloud provider, and have access to daily and weekly service quality statistics.
- Service level agreements – the contract should specify service-level parameters, minimum levels, and specific remedies and penalties for noncompliance with SLAs. The SLA should clearly define uptime, performance and response time, and error-correction time. Remedies for violation of the SLA should include corrections and / or penalties.
- Business continuity / disaster recovery (BC/DR) – the contract should specify minimum disaster recovery and business continuity requirements, and penalties for failures in complying with the minimum requirements.
- Termination – the contract should state that the customer can terminate the contract "at any time without having to show cause and without additional fees or penalties."

Customers should always aim to avoid a situation where the provider's liability is severely limited, especially in relation to business-critical services because the customer may be left without an effective remedy in the case of a serious service breakdown.

18.1.4   Service-Level Agreements

Service levels are an important way of ensuring that a provider meets the level of service expected by the customer. The contract should specify service-level parameters, minimum levels, and specific remedies and penalties for noncompliance with SLAs. Service-level agreements are important to set clear expectations for service between the cloud consumer and the cloud provider. Consideration must also be given to the different models of service delivery: infrastructure as a service (IaaS), platform as a service (PaaS), and software as a service (SaaS), as each model brings different requirements. The SLAs must be enforceable and state specific remedies that apply when they are not met.

Before evaluating any cloud SLA, customers must first develop a strong business case and strategy for their cloud computing solutions. This includes identifying specific services that will be deployed in the cloud along with a clear understanding of the importance of these services to the business. A check on the exit clauses of current hosted services contracts is also important. Only after this strategic analysis has been completed can the consumer effectively evaluate and compare SLAs from different providers. The following steps should be taken by cloud consumers to evaluate cloud SLAs with the goal of comparing cloud service providers or negotiating terms with a provider:

- understand roles and responsibilities;
- evaluate business-level policies;
- understand service and deployment model differences;
- identify critical performance objectives;
- evaluate security and privacy requirements;
- prepare for service-failure management;
- understand the disaster recovery plan;
- define an effective management process;
- understand the exit process.

18.2   Cloud Security Mechanisms and Techniques

It is critical to recognize that security is a cross-cutting aspect of the architecture that spans all layers of cloud-computing services. Therefore, security concerns in cloud-computing architecture is not solely under the purview of the cloud providers, but also cloud consumers. The three service models (i.e. IaaS, PaaS, and SaaS) present consumers with different types of service management operations and expose different entry points into cloud systems, which in turn also create different attacking surfaces for adversaries (National Institute of Standards and Technology, 2011). It is therefore important to consider the impact of cloud service models and their different issues in security design and implementation. For example, SaaS provides users with accessibility of cloud offerings using a network connection, normally over the Internet and through a Web browser. The variations of cloud deployment models (public, private, community, and hybrid) have important security implication as well. One way to look at the security implications from the deployment model perspective is the differing level of exclusivity of tenants in a deployment model. Regardless of service and deployment models, three major cloud security mechanisms are essential across models. They are network, data, and application security.

18.2.1   Network Security

Network security addresses risks relating to the use of, and access to, businesses networks. Network security encompasses protecting data as it traverses networks, including public networks such as the Internet, protecting systems and data from network-based attacks, and protecting the networking components themselves (Cloud Security Alliance, 2012). In general, all ingress and egress points to the cloud environment need to inspect traffic and log network activity at specified periods of time. Major cloud network security mechanisms are discussed below.

- Network access controls – network access is the fundamental security control point that ensures basic attack vectors are mitigated by effective controls. Controls can be implemented in physical, converged, or virtual appliances.
- Perimeter firewall controls – these are the first layer of defense which provide real-time protocol inspection and detection of known attacks. Place systems or services within a perimeter of security provided by the firewall to ensure that known attacks and anomalies are detected and blocked.
- Subtier firewall controls – these provide a separate security boundary within the virtualization layer of the cloud, to secure the virtual machines and tiers of network created within the cloud network.
- Access control lists – These provide a basic security control layer to support securing virtual machines from standard layer-2 security threats like flooding and scanning.
- Content inspection and control – various technologies exist to protect the network, business systems, and business data from both external attacks and internal data theft. These include intrusion detection, intrusion prevention, data-loss prevention, and proxy servers.
- DDOS protection/mitigation – the distributed denial of service (DDoS) attack can best be mitigated at the cloud service-provider backbone or other network that has significantly more bandwidth than the sum of the bandwidth of all attack traffic (Lohman, 2011). Once an attack condition is identified, monitoring entities trigger a reroute of suspicious traffic through a cleansing instance that attempts to filter out the attack traffic while allowing legitimate packets to pass through (Cloud Security Alliance, 2012).

18.2.2   Data Security and Privacy

Data is at the heart and center of the cloud. Three major issues need to be addressed with regard to data security and privacy in a cloud environment: data availability, data encryption, and data protection. When choosing the best way of protecting data privacy keep in mind how valuable that data is to an organization and to what extent it is reasonable to protect it. Therefore, the first thing an organization should do is to define the level of privacy it needs and thus a level of protection for it (Securosis, 2011). Here are five data-privacy protection tips to help an organization tackle the issue of cloud data privacy:

- Avoid storing sensitive information in the cloud – if it has a choice, an organization should opt for keeping its sensitive information away from the cloud environment or use appropriate alternative solutions.
- Review the contract agreement – this is to find out how provider's cloud service storage works. If organizations are not sure what cloud storage to choose or if they have any questions as for how different cloud services work, they need to review carefully the contract agreement of the service they are planning to sign up for.
- Use strong passwords – weak passwords can be cracked within seconds. A great part of all the sad stories about someone's account getting broken is caused by an easy-to-create-and-remember password. Strong passwords with a combination of alpha, numeric, and upper and lower cases are strongly recommended.
- Use an encrypted cloud service – encryption is, so far, the best way to protect corporate data. There are some cloud services that provide local encryption and decryption of customer files in addition to storage and backup. It means that the service takes care of both encrypting organization files on its own computer and storing them safely on the cloud.

18.2.3   Application Security

Cloud computing is a particular challenge for applications across the layers of SaaS, PaaS, and IaaS. Cloud-based application security must be provided by the application without any assumptions being made about the cloud environment. Developing applications for a cloud environment is different than the traditional hosting environments in the following areas (see http://resources.infosecinstitute.com/intro-secure-software-development-life-cycle/, accessed December 7, 2015):

- Control over physical security is substantially reduced in public cloud scenarios.
- Potential incompatibility between providers when services are migrated from one provider to another.
- Protection of data through the lifecycle must be considered. This includes transit, processing and storage.
- The combinations of Web services in the cloud environment can potentially cause security vulnerabilities to be present.
- Fail-over for data and data security in the cloud has to be more detailed and layered than traditional environments.
- Assuring compliance with relevant industry and government regulations is typically more difficult within a cloud environment.

This creates the need for rigorous practices that must be followed when developing or migrating applications to the cloud. Essential cloud application security best practices involve a secure software development life cycle, effective application security assurance programs, and reliable application monitoring mechanisms.

18.2.3.1   Secured Software Development Life Cycle

A Secure Software Development Life Cycle (SSDLC) has assumed increased importance when migrating and deploying applications in the cloud and organizations should ensure that the best practices of application security, identity management, data management, and privacy are integral to their development programs and throughout the life cycle of the application (http://resources.infosecinstitute.com/intro-secure-software-development-life-cycle/, accessed December 7, 2015). In implementing a SSDLC, organizations must adopt best practice for development, either by having a good blend of processes, tools, and technologies of their own or adopting one of the maturity models such as the Building Security in Maturity Model (http://bsimm.com/, accessed December 7, 2015), the Software Assurance Maturity Model (http://www.opensamm.org/, accessed December 7, 2015), and the Systems Security Engineering Capability Maturity Model (http://www.iso.org/iso/catalogue_detail.htm?csnumber=44716, accessed December 7, 2015).

18.2.3.2   Application Security Architecture

Traditional noncloud enterprise applications could be protected with traditional edge security controls like firewalls, intrusion detection, and proxies. In a cloud environment, all these traditional controls are no longer effective enough to protect as these applications are running on untrusted networks. Applications could be residing with other co-tenants of same service provider and could be accessed from anywhere through any type of device. This changes the very nature of security requirements for cloud applications. Effective application security in the cloud should include the following main components:

- Authentication – authentication refers to establishing/asserting the identity to the application. This is usually done in two phases. The first phase is disambiguating the identity and the second phase is validating the credentials already provided to the user. Enterprises should plan to use risk-based authentication for their cloud applications. This type of authentication is based on device identifier, Internet service provider, heuristic information, and so forth. A cloud application should not only perform authentication during the initial connection but should also perform risk-based authentication based on the transactions being performed within the application.

- Authorization – authorization refers to enforcing the rules by which access is granted to the resources. There are different types of authorization models namely, role-based, rule-based, attribute-based access, claims-based, and authorization-based access control. The enterprise should plan for how the users are authenticated seamlessly across all these cloud applications and how the users' profiles, such as group association, entitlements, and roles are shared across these cloud applications for granular access controls. Enterprises are recommended to use open standards such as SAML, OAuth, or XACML in this case (http://resources.infosecinstitute.com/intro-secure-software-development-life-cycle/, accessed December 7, 2015; National Institute of Standards and Technology, 2011).
- Administration – administration refers to managing users and managing access policies for enterprise applications. Effective administration provides not only timely access to the users but also timely revocation of access when the user leaves or timely management of access when the user moves to a different role. In addition to users, administration also manages cloud application/services identities, access control policies for these cloud applications/services, and privileged identities for the applications/services (National Institute of Standards and Technology, 2011).

18.2.3.3   Cloud Application Monitoring

As with other aspects of cloud security, how an organization monitors a cloud-based application varies with the type of cloud under consideration. What it means to monitor applications in the cloud and how to monitor different types of cloud applications are explained in detail below.

- Log monitoring – archiving logs is only the first step for compliance. Understand the potential entries that could be sent to these logs, and monitor for actionable events. An application logging entry is of zero use unless a process exists to detect and respond to those entry events.
- Performance monitoring – a significant change in the performance of one application could be a symptom of another customer using more than his fair share of a limited resource (e.g., CPU, memory, storage), or it could be the symptom of malicious activity, either with the application being monitored or with another applications in the shared infrastructure (National Institute of Standards and Technology, 2011).
- Monitoring for malicious use – an organization must understand what happens when a malicious user attempts to gain access, or use permissions that they do not have. Audit logs must log login attempts. If an application experiences a significant increase in traffic load, verify whether it is an alert created from other applications in the cloud environment.
- Monitoring for policy violations – it is also important to monitor and audit how a policy decision point came to a decision. This is in line with a general policy-driven monitoring approach that avoids the typical monitoring problems of false-positives and incident overload.

With an IaaS-based application, monitoring the application is almost "normal," compared to noncloud applications. The customer needs to monitor issues with the shared infrastructure or involving attempted unauthorized access to an application by a malicious co-tenant. Monitoring PaaS-based applications requires additional work. Unless the cloud platform provider also provides a monitoring solution capable of monitoring the application deployed, the customer should either write additional application logic to perform the monitoring tasks within the platform or send logs to a remote monitoring system. As SaaS-based applications provide the least flexibility, it should not come as a surprise that monitoring the security of these types of applications is the most difficult (National Institute of Standards and Technology, 2011).

18.3   Cloud Security Audit and Assessment

Cloud solutions has to be auditable and assessable in order to enable continuous evaluation of whether the security level of a cloud provider's specific solution is sufficient to be used for a given system or solution. At the same time, a cloud provider must be able to provide adequate information about audit and assessment to meet the customers' risk assessment and be compliant with laws or regulations.

18.3.1   Auditing

Auditing ensures that organizations cloud solutions work according to expectations. The auditing could be done either internally by internal IT or business teams, or externally by a third-party service. The type of information provided by the provider is dependent on the type of cloud (IaaS, PaaS, SaaS). Access to audit and assurance information is primarily the cloud provider's responsibility in SaaS solutions but it is a shared responsibility between provider and customer in IaaS and PaaS solutions. In general, the following information should be provided by the provider for security audit (National IT and Telecom Agency, Denmark, 2015).

- Written documentation for procedures, standards, policies, etc.
- Information about standard configurations and documentation for the current configuration of the customer's systems.
- Continuous logging and monitoring information.

In IaaS there are more opportunities in terms of setup, configuration and installation of programs in the virtual servers such as installing their own surveillance programs (National Institute of Standards and Technology, 2011). In PaaS, the customers can set requirements for the logging facilities of their own programs, whereas in SaaS solutions the customers depend on the logs and surveillance functionalities provided by cloud providers. In SaaS solutions, access to information about audit and assurance is, therefore, primarily the responsibility of the provider. In IaaS and PaaS solutions, responsibility is shared between provider and customer.

Regardless of the solution considered, the cloud provider must always be able to document physical security, the underlying policies, procedures and configurations to meet the customer's risk assessment, and be compliant with legislative requirements. The available information about configuration and logging depends on the type of solution (IaaS, PaaS or SaaS) and the functionality offered by the individual provider. As a minimum, the cloud provider is always held responsible for the host (the underlying physical server), the hardware, network units, and the physical buildings (National Institute of Standards and Technology, 2011). The same goes for procedures concerning operation of the hardware. In terms of administration and providing functionality to the customers, the cloud provider's responsibility increases when moving from IaaS to PaaS to SaaS.

18.3.2   Assessments

Cloud-based security assessments provide the information necessary for an intelligent, risk-based decision-making process, while relieving IT staff of the operational burdens of managing the assessment tool infrastructure. Organizations must establish policies, processes, and procedures, and implement controls to ensure the confidentiality, integrity, and availability of the information and information technology upon which their critical business processes depend.

Cloud assessment involves primarily vulnerability and compliance scanning. Network and system vulnerability assessments attempt to identify vulnerabilities through the use of IP scanning techniques, combined with a detailed understanding of vulnerabilities. The assessments include the identification of systems and associated vulnerabilities, as well as information about the potential impact on the network or enterprise. Compliance scanning attempts to verify the compliance status of the devices in the cloud solution. Server and workstation compliance assessments allow for the discovery of server/workstation configurations, as well as the comparison of the results against industry best practices and any customized configuration standards.

Key business processes and the assets that run them should be identified prior to the start of an assessment. All parties should agree to, and document, rules of engagement to be used during the assessment, such as asset exclusions, time windows, level of attacks, social engineering techniques, and others. Assessments are usually safe to run even on production systems, but a service provider should have mechanisms in place to be able to stop an assessment at any time if something happens that could be destructive to a server or workstation, or disruptive to key business processes (National Institute of Standards and Technology, 2011). As with all assessment activities, understanding the tools that the assess team will use is important. Coordinate with the cloud provider to conduct testing in accordance with best practices. Determine if the cloud provider offers such a service. If not, understand the limitations of performing such actions in hosted cloud environments. Scope and limitation will vary based on which stack of the SPI model the cloud-hosted application seems to fit (National Institute of Standards and Technology, 2011).

The data that results from cloud-based security assessment services is very sensitive in nature and must be properly safeguarded. Cloud-based security assessment services must ensure the confidentiality of this sensitive data and ensure that only authorized parties can access the data. Both cloud provider and customer should sign an appropriate nondisclosure agreement (NDA). The NDA should limit disclosure of all information obtained, either in preparation for the assessment or through the assessment results, to only those with a verified need to know. Data must be secured at all stages: at creation, in storage, in transit in processing, or at deletion.

18.3.3 Penetration Testing

Penetration testing is a security testing methodology that gives the tester an insight into the strength of the target's system security by simulating an attack from a malicious source. The process involves an active analysis of the cloud system for any potential vulnerability that might result from poor or improper system configuration, known and/or unknown hardware or software flaws, or operational weaknesses in process or technical countermeasures (National Institute of Standards and Technology, 2011). This analysis is carried out from the position of a potential attacker, and can involve active exploitation of security vulnerabilities. In general, penetration testing involves the following three major phases: preparation, execution, and reporting.

The type of cloud model has a huge impact on penetration testing and deciding if penetration test is possible. Generally, IaaS and PaaS clouds are likely to permit penetration testing. However, SaaS providers are not likely to allow customers to penetration test their applications and infrastructure, with the exception of third parties performing the cloud providers' own penetration tests for compliance or security best practices.

When conducting penetration testing assessments of a cloud environment, special consideration must be given to the multitenant nature of the environment and the potential disruption to other organization's systems. Organizations may not be able to conduct a full end-to-end internal penetration test of a cloud hosting provider, and may have to rely on attestation by the provider that testing is conducted. Organizations should understand what type of testing they will be allowed to conduct before selecting a cloud provider's security initiatives.

18.4 Cloud Intrusion Detection and Incident Response

The purpose of intrusion detection is to monitor the enterprise environment at key vantage points to uncover malicious activity aimed at degradation, disruption, infection, or exfiltration of data, applications, and the systems that host or transmit them. The main purpose of incident response is taking actions to avoid, block, contain, disrupt, or continue to operate in the face of attack. A comprehensive intrusion service combines detection and response, management infrastructure for control and reporting, and interfaces with the rest of the security architecture in order to have a more holistic view of events and better uncover anomalous activity.

18.4.1 Intrusion Detection Techniques and Strategies

Intrusion detection is defined as the process of identifying and responding to malicious activity targeted at computing and networking resources. Intrusion-detection mechanisms are employed at opportune cross-connects or shared points, where protected and foreign traffic cross paths: a network administrative boundary, end system network interfaces, within hosts at virtualized container boundaries, or directly inside a guest environment (National Institute of Standards and Technology, 2011). There are two flavors of intrusion detection techniques:

- Network-based detection – this looks for binary or behavioral patterns and anomalous activity in network traffic. Network-based techniques employ strategies for signature detection, behavior heuristics, and traffic pattern correlation. Each of these relies on some form of deep packet inspection, which is the ability of the detection device or software to understand various headers and components of network datagrams (see https://www.sans.org/security-resources/idfaq/, accessed December 7, 2015). Network-based intrusion detection systems are usually placed at ingress and egress points of the network in order to detect and prevent anomalous traffic, usually based on a combination of signatures, heuristic behavioral analysis, and statistic protocol anomaly detection. They can be deployed using existing network equipment, specialized appliances and interfaces, or software that runs on the host.
- Event-based detection – this looks for activity or events on the host at the system, virtual, and application layers. Event-based techniques use access to, or reporting of, events and configurations to determine potential activity leading to or resulting from malicious attack, compromise, or resultant degradation, corruption, or exfiltration (https://www.sans.org/security-resources/idfaq/, (accessed December 7, 2015). Events are detected through analysis of centrally reported logs or by software running directly on the system, in the virtual layer, or in the guest OS monitor for particular behaviors that indicate potential intrusion: policy violations, changes in configuration, workload changes, foreign processes or system calls, changes to the integrity of the OS and file systems, etc. (National Institute of Standards and Technology, 2011).

Intrusion detection in a cloud environment can be much more arduous depending on the available resources in the cloud and the level of management or control of the devices, services, or configurations required (National Institute of Standards and Technology, 2011). Service level agreements should define the locations to be monitored, specify service and performance levels, and how rules are added and managed. Secure management, transport location, segregation and analysis of collected data must be considered and defined in any contracts with the cloud service provider. Any intrusion detection device must be capable of handling the volume of traffic that is expected to pass through it in order to be effective.

18.4.2 Intrusion Detection

Intrusion detection systems (IDSs) are an essential component of defensive measures protecting computer systems and networks against harm and abuse. They become crucial in the cloud computing environment. One key feature of intrusion detection systems is their ability to provide a view of unusual activity and to issue alerts notifying administrators and / or blocking a suspected connection. Differences in the target environment have tremendous impact on the available functions and features of the service delivered.

In a cloud environment, event-based detection requires more visibility at more layers in the system. When that environment moves more to a multitenant cloud, additional complexities such as cloud APIs, guest-process interactions, and the management plane are introduced. In these cases, some specific events or checks might include: virtualization layer events, VM image repository monitoring, integrity monitoring, interaction among interdependent workloads, and other API activity. To manage the protection of hosts, applications, and data, an intrusion service must have control of, or at least visibility into, these points of interest; interface with information and event correlation capabilities, and provide the infrastructure for communications in and among the components of the service within the target enterprise, as well as back to the intrusion service-provider environment. Delivering this capability from the cloud often requires administrative relationships, elevated user rights, and end-to-end transactional access between hosted elements and central control and reporting.

18.4.3   Incident Response

The nature of incident response will be affected when services are moved to the cloud. According to the Cloud Security Alliance, the customer must consider what must be done to enable efficient and effective handling of security incidents in the Cloud (National Institute of Standards and Technology, 2011). Given the possibility that log information may not be directly accessible to the customer, the incident response will need to take into consideration the type of service being utilized (i.e. IaaS, PaaS, SaaS) and craft a security SLA to address responsibilities of the cloud service provider.

If IaaS is being used, then the provider is responsible for the infrastructure-related logs such as storage, networks, and hypervisors. Customers will have access to their VM logs and IDS logs during an incident. If PaaS is being utilized, then the incident response team will have access to application logs but the provider will still maintain server logs. The customer has more opportunity to retrieve log information from a PaaS provider by communicating to the provider what triggers an event (National Institute of Standards and Technology, 2011). Examples of triggers can be failed authentication attempts or application errors. If SaaS is being used, then the provider incident response team will internally respond to triggers from their security incident and event-management capability, intrusion-detection tools or other log management tools. In this scenario, the customer has no responsibility. In general, services such as PaaS and SaaS may make incident response easier because the burden rests upon the cloud service provider. Incidents that require obtaining an image or snapshot of the virtual machine for forensics are also easier because a virtualized environment is designed to copy or clone images, including memory states. Special software is no longer needed when the inherent capability of your virtual platform provides these functions.

18.5   Cloud Business Continuity and Disaster Recovery Planning

The majority of business continuity and disaster recovery (BC/DR) considerations and best practices for cloud-based services are very similar to those for traditional noncloud solutions in terms of what is required, and how the business must prepare prior to implementation. Key benefits of using a reliable, highly available cloud-based service for BC/DR of systems include the following (www.hostway.com, accessed December 7, 2015):

- Secure backup – The cloud service provider will host all replicated systems in the cloud, making them available should disaster recovery be invoked.
- Scalable infrastructure – A key component of any cloud-based service is its elasticity and effective unlimited scalability. This ensures that the cloud service consumer's BC/DR systems will scale up as required.

- Pay per use – Consumers pay only for the actual use of the service. This translates to lower service costs during times of normal operation, when minimal capacity and performance are required for storage and replication of systems and data to the BC/DR solution. When events invoke a BC/DR response, requisite supporting services ramp up to meet the need.
- Reduced BC/DR expertise – There is a significant reduction in the expertise and effort required on the part of the consumer versus traditional BC/DR solutions. The provider can manage scaling up capacity, public records, and provide guidance for BC/DR best practices. This is especially true when looking at DR in the cloud for cloud-based systems, as the provider usually will have automated fail-over facilities protecting their systems and infrastructure.

There are numerous business drivers when planning, executing and testing BC/DR solutions either for traditionally hosted systems or those already in the cloud. The chosen combination of these benefits discussed above will influence a business's BC/DR strategy. The considerations and concerns presented below should be part of any discussion regarding cloud-based BC/DR (www.hostway.com, accessed December 7, 2015):

- What is the value of an enterprise's data to the business and how much will it cost to replace it, if it is lost or stolen? What data should go into the cloud?
- How effective is the cloud provider's own BC/DR planning? Elasticity of the cloud provider – can they provide all the resources if BC/DR is invoked?
- How is DR testing achieved? Does the cloud provider support DR testing?
- How are the DR services accessed if invoked?

Beside the logical components of BC/DR, there also is a need to consider physical locations, such as alternative sites of operation, geographically distributed datacenters / infrastructure and relevant jurisdictions, network survivability, and the incorporation of third-party ecosystems in planning and testing. Data-protection requirements may restrict data that constitutes personally identifiable information (PII). A careful choice of vendor and a clear understanding of where the vendor's datacenters are, along with their data-movement policies, can mitigate this type of risk.

18.6   Conclusion

Cloud computing is profoundly changing the IT landscape. Unfortunately, some of these changes have created new security challenges. While enterprises have been focusing on securing their internal computing systems for decades, working with external cloud computing providers has exposed new vulnerabilities that must be addressed both internally and externally by the cloud computing providers and customers.

Security controls in cloud computing are, for the most part, not greatly different from security controls in any noncloud IT environment. However, because of the cloud types adopted, the delivery models employed, and the technologies used to enable cloud services, cloud computing presents different challenges and risks to an organization than traditional noncloud solutions. This chapter presents fundamental methodologies, best practices, and practical techniques to address security issues among cloud models (IaaS, PaaS, and SaaS), which vary substantially.

The security threats associated with cloud computing are continuing to evolve, while the perceived and actual level of certain security threats has also changed over the last several years. Understanding what major threats exist in cloud computing, adopting effective multilayered defense mechanisms, and following matured processes and best practices will help organizations to make the successful shift towards the cloud.

References

Cloud Security Alliance (2012) *SecaaS Category 10 // Network Security Implementation Guidance*, https://cloudsecurityalliance. org/download/secaas-category-10-network-security-implementation-guidance/ (accessed December 7, 2015).

Lohman, T. (2011) *DDoS is Cloud's Security Achilles' Heel*, http://www.computerworld.com.au/article/401127/ddos_ cloud_security_achilles_heel/ (accessed December 7, 2015).

National Institute of Standards and Technology (2011) *NIST Cloud Computing Reference Architecture*, http://www.nist. gov/customcf/get_pdf.cfm?pub_id=909505 (accessed December 7, 2015).

National IT and Telecom Agency, Denmark (2011) *Cloud Audit and Assurance Initiatives*, http://www.digst.dk/ Servicemenu/English/IT-Architecture-and-Standards/Cloud-Computing/~/media/Files/English/Cloud%20Audit%20 and%20Assurance%20EN_cagr.pdf (accessed December 7, 2015).

Petruch, K., Stantchev, V. and Tamm, G. (2011) *Cloud Computing Governance Aspects*, http://www.srh-hochschule-berlin. de/fileadmin/srh/berlin/pdfs/Publikationen/Cloud-Computing-Governance_2011.pdf (accessed December 7, 2015).

Securosis (2011) *The Data Security Lifecycle*, http://www.securosis.com/blog/data-security-lifecycle-2.0 (accessed December 7, 2015).

19

Cloud Forensics

Shams Zawoad and Ragib Hasan

University of Alabama at Birmingham, USA

19.1 Introduction

Cloud computing opens a new horizon of computing for business and IT organizations. Since 2007, we have seen an explosion of applications of cloud-computing technology, both for enterprises and individuals seeking additional computing power and more storage at a low cost. Small- and medium-scale industries find cloud computing highly cost effective as cloud infrastructures replace the need for costly physical and administrative infrastructures, and offer a flexible, pay-as-you-go payment structure. Opportunities provided by the cloud have led to estimates that the growth of global cloud computing market could be 30% compound annual growth rate (CAGR), reaching $270 billion in 2020 (http://www.marketresearchmedia. com/?p=839, accessed December 7, 2015).

Although cloud computing offers numerous facilities and contributes extensively to the advancement of information technology, cloud security is not transparent. Clouds use the multitenant usage model and virtualization to ensure better utilization of resources. However, these fundamental characteristics of cloud computing are a double-edged sword – the same properties also make it difficult to prevent and investigate cloud-based crimes and attacks on clouds and their users. Besides attacking cloud infrastructures, adversaries can use clouds to launch attacks on other systems. For example, an adversary can rent hundreds of virtual machines (VMs) to launch a distributed denial of service (DDoS) attack. After a successful attack, the adversary can erase evidence that is important to trace the attack by turning off the VMs. Criminals can also keep their secret files (e.g., child pornography, terrorist documents) in cloud storage and remove the files from local storage to remain clean. To investigate such crimes involving clouds, investigators have to carry out a digital forensic investigation in the cloud environment. This particular branch of forensic has become known as *cloud forensics*.

While computer forensics itself is not well matured yet, cloud forensics imposes greater challenges on digital forensics. As many of the assumptions of traditional digital forensics are not valid in the cloud computing model, the traditional digital forensics tools and procedures are not suitable for investigating crimes involving clouds.

Encyclopedia of Cloud Computing, First Edition. Edited by San Murugesan and Irena Bojanova.
© 2016 John Wiley & Sons, Ltd. Published 2016 by John Wiley & Sons, Ltd.

As a result, cloud forensics brings new challenges from both technical and legal points of view and has opened new research areas for security and forensics experts. In this chapter, we examine the cloud forensics problem and explore the challenges and issues in cloud forensics, provide a comprehensive analysis of proposed solutions for cloud forensics, discuss the advantages and usability of cloud computing to expedite the digital forensic procedures, and, finally, we highlight outstanding problems and future directions in cloud-forensics research.

19.2   Background

19.2.1   Digital Forensics

The National Institute of Standards and Technology (NIST) defines digital forensics as "The application of science to the identification, collection, examination, and analysis, of data while preserving the integrity of the information and maintaining a strict chain of custody for the data" (Kent *et al.*, 2006). From the definition, we can say that digital forensics is comprised of four main processes:

- *Identification.* The identification process has two main steps: identification of an incident, and identification of the evidence. Evidence identification depends on the nature of an incident.
- *Collection.* In the collection process, an investigator extracts the digital evidence from different types of media, for example, a hard disk, e-mail, or a network router. He also needs to preserve the integrity of the evidence.
- *Organization.* There are two main steps in the organization process: examination, and analysis of the digital evidence. In the examination phase, an investigator extracts and inspects the data and its characteristics. In the analysis phase, investigators interpret and correlate the available data to come to a conclusion that can prove or disprove a criminal allegation.
- *Presentation.* In this process, an investigator makes an organized report to state his/her findings about the case. This report should be admissible and comprehensible to the jury.

Figure 19.1 illustrates the flow of these digital forensic processes.

19.2.2   Cloud Forensics

Ruan *et al.* (2011) defined cloud forensics as a subset of network forensics, because cloud computing is based on extensive network access and network forensics handles forensic investigation in private and public networks. However, besides network forensics, investigators may need to execute disk forensics, memory forensics, and so forth, for a proper forensics investigation. Hence, we define cloud forensics as *the application of digital forensics principles and procedures in a cloud-computing environment to establish facts about civil, administrative, or criminal allegations.* Digital forensics procedures vary according to the service and deployment model of cloud

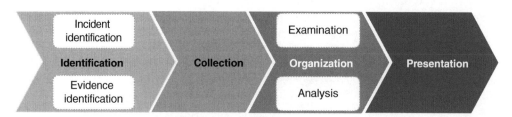

Figure 19.1   *Digital forensics process flow*

computing. For software as a service (SaaS) and platform as a Service (PaaS), forensics investigators have very limited control over process or network monitoring. A forensics investigator may be able to gain enough control in infrastructure as a service (IaaS) to deploy a forensic-friendly logging mechanism. For SaaS, investigators depend solely on cloud service providers (CSP) to get the application log, while for IaaS, investigators may be able to acquire a VM image from customers and enter into the examination and analysis phase.

Cloud forensics is applicable in a number of possible scenarios that are mentioned below:

- Cloud computing can be used as a tool for an attack; for example, to launch a DDoS attack using VMs running inside clouds.
- Virtual machines running inside a cloud or the cloud host machines are under attack – for example, they may be hacked or compromised – and the method of attack, extent of damage, and so forth, need to be determined.
- While investigating a crime, where cloud is not directly related, such as murder or identity fraud, the existing evidence may lead to a cloud that needs to be examined. The cloud can store further crucial evidence, such as images or documents, to settle the case.
- The computing resources of the cloud can be used to expedite a forensic investigation.

19.3 Challenges of cloud forensics

Most of the factors that make digital forensics challenging in clouds are related to the fundamental natures of clouds. We present our analysis by looking into the challenges faced by investigators in each of the stages of computer forensics.

19.3.1 Cloud Data Storage

19.3.1.1 *Volatile Data*

Volatile data cannot be sustained without power. Logs, registry entries, or temporary Internet files that reside within the virtual environment will be unavailable when users power off the VM instances and no snapshots of the instances have been saved or data have not been synchronized continuously in a persistent storage, such as Amazon S3 (http://aws.amazon.com/s3/, accessed December 8, 2015) or EBS (http://aws.amazon.com/ebs/, accessed December 8, 2015). With extra payment, customers can have persistent storage. However, this is not common for small- or medium-scale business organizations. Moreover, a malicious user can exploit this vulnerability. After carrying out a malicious activity (e.g., launching a DoS attack, or sending spam e-mail), an adversary can power off her VM instance, which could lead to a complete loss of the valuable evidence inside the VM and make the forensic investigation almost impossible. Moreover, because of the volatile data, a malicious owner of a cloud instance could fraudulently claim that her instance was compromised by someone else and had launched a malicious activity. Later, it will be difficult to prove her claim as false by a forensic investigation (Birk and Wegener, 2011).

19.3.1.2 *Multitenancy*

In cloud computing, multiple VMs can share the same physical infrastructure – data for multiple customers are co-located. This aspect of clouds is different from the traditional single-owner computer system. While acquiring evidence for any adversarial case, two issues can arise because of the multitenancy nature. First, we need to prove that data of a suspect user were not co-mingled with other users' data. Secondly, we need to preserve the privacy of other tenants while performing an investigation. The multitenancy characteristic also raises the issue of side-channel attacks, which are difficult to investigate.

19.3.2   Forensics Data Acquisition

19.3.2.1   *Physical Inaccessibility*

The physical inaccessibility of digital evidence makes the evidence collection procedure harder in forensics of remote or public cloud environments ((Birk and Wegener, 2011; Dykstra and Sherman, 2011). The established digital forensic procedures and tools assume that we have physical access to the computing resources, for example the hard disk, or network router. However, in cloud forensics, the situation is different. Sometimes, we do not even know where the data is located as it is distributed among many hosts of multiple geodispersed datacenters. This issue also poses problems in preparing a search warrant as the search warrant must specify a location (Dykstra and Sherman, 2011).

19.3.2.2   *Less Control in Clouds*

In traditional digital forensics, investigators often have full control over the evidence (e.g., a confiscated physical hard drive). In a cloud, unfortunately, the control over data varies in different service models. Figure 19.2 shows the limited amount of control that customers have in different layers for the three service models – IaaS, PaaS, and SaaS. In IaaS, users have more control than SaaS or PaaS. The lower level of control has made data collection in SaaS and PaaS more challenging than in IaaS; sometimes it is even impossible. For IaaS, the availability of VM images can make the investigation process smooth. Conversely, investigators need to depend on cloud providers to collect digital evidence from SaaS and PaaS.

19.3.2.3   *Issues with Log Acquisition*

Analyzing logs from different sources, such as networks, operating systems, and applications, plays a vital role in digital forensic investigation. However, gathering this crucial information from the cloud environment is not as simple as it is in a privately owned computer system. In a cloud infrastructure, log information is not located at any single centralized log server; rather logs are decentralized among several servers. Multiple users' log information may be co-located or spread across multiple servers. There are several layers and tiers

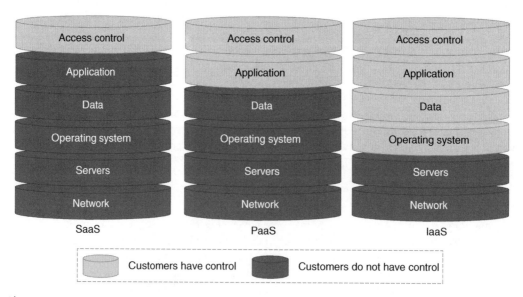

Figure 19.2   *Customers' control over different layers in different service models (Zawoad, et al., 2013)*

in cloud architecture. Logs are generated in each tier. For example, application, network, operating system, and database – all of these layers produce valuable logs for forensic investigation. Collecting logs from these multiple layers is challenging for the investigators. Logs generated in different layers need to be accessible to different stakeholders of the system. System administrators need relevant logs to troubleshoot the system. Developers need the required logs to fix bugs in an application. Forensic investigators need logs that can help in their investigation. Hence, there should be some access-control mechanism, so that everybody can get only what they need and, obviously, in a secure way.

19.3.3 Trustworthiness of Evidence

Dependence on CSPs to acquire evidence inevitably affects trust and evidence integrity. After issuing a search warrant, the examiner needs a technician from the cloud provider to collect data. However, the employee of the cloud provider, who collects data, is most likely not a licensed forensics investigator and it is not possible to guarantee his integrity in a court of law. The date and timestamps of the data are also questionable as a malicious VM owner or a dishonest cloud employee can tamper with the system clock. Researchers identified that it is not possible to verify the integrity of the forensic disk image in Amazon's EC2 cloud, because Amazon does not expose the functionality of collecting checksums of volumes, as they exist in EC2 (Dykstra and Sherman, 2011). Moreover, CSPs can collude with adversaries and tamper with the evidence to save a criminal, or a dishonest forensics investigator can collude with a CSP to frame an honest user (Zawoad and Hasan, 2012; Zawoad *et al.*, 2013).

19.3.4 Forensics Data Organization

19.3.4.1 Absence of Reliable Forensics Timeline Analysis Support

In the investigation of a criminal case involving computers, the timeline of events can provide critical information relating to the prosecution of a suspect. Reliable timelines can help to pinpoint the location of certain individuals, can assist with the determination of alibis, can uncover conversations and correspondences, and can possibly help ultimately to determine the guilt or innocence of suspects. Conversely, alteration of timeline can lead the investigation in a wrong direction (Kent *et al.*, 2006). The first prerequisite for generating reliable timeline is trustworthy logs, which is itself a challenging task in clouds (Zawoad *et al.*, 2013). Moreover, the system clock of the VM or cloud host can also be altered leading to an untrustworthy timeline. To the authors' knowledge, none of the existing CSPs provide reliable forensics timelines, which hinders the forensics analysis.

19.3.4.2 Absence of Critical Information in Logs

Organizing logs collected from clouds is challenging, as there are no standard formats for different types of logs. Logs are available in heterogeneous formats – from different layers to different service providers. Moreover, not all the logs provide crucial information for forensic purpose – for example, who, when, where, and why some incident occurred (Marty, 2011).

19.3.5 Legal Issues

19.3.5.1 Chain of Custody

Chain of custody (CoC) should clearly depict how the evidence was collected, analyzed, and preserved in order to be presented as admissible evidence in court. Chain of custody is defined as a verifiable provenance or log of the location and possession history of evidence from the point of collection at the crime scene to the point of presentation in a court of law. However, in clouds, maintaining CoC is not possible as investigators can acquire the available data from any workstation connected with the Internet and the actual location of data cannot be determined properly. As multiple people may have access to the evidence and investigators need to depend on CSPs to acquire the evidence, preserving CoC throughout the investigation process is problematic (Dykstra and Sherman, 2011). Moreover, multi-jurisdictional laws, procedures, and proprietary technology in clouds also make CoC preservation challenging (Zawoad and Hasan, 2012).

19.3.5.2 Crime Scene Reconstruction

Reconstructing the crime scene is important when investigating a malicious activity. It helps investigators to understand how adversaries launch attacks. However, shutting down a virtual instance after a malicious activity can make the reconstruction task impossible. Moreover, many CSPs are closed and proprietary and the investigator does not know what the cloud environment looks like. Without internal knowledge of clouds, it will be very difficult for an investigator to recreate a crime scene.

19.3.5.3 Cross Border Law

Data centers of the service providers can be distributed worldwide. It may happen that an attacker is accessing the cloud computing service from one jurisdiction, whereas the data she is accessing resides in a different jurisdiction. Differences in laws between the two locations can affect the whole investigation procedure, starting from evidence collection to capture the attacker. Because of the multitenancy nature of clouds, we need to preserve the privacy of the tenants while collecting data of a suspect tenant. However, the privacy and privilege rights may vary among different countries or states, which makes cloud forensics a challenging task.

19.3.5.4 Presentation

The final step of digital forensic investigation is presentation, where an investigator gathers his/her findings and presents them to the court as the evidence of a case. Proving the evidence in front of the jury for traditional computer forensics is relatively easy compared to the complex structure of cloud computing. Presenting admissible and comprehensible evidence from clouds is challenging due to the technicalities of a cloud datacenter, running thousands of VMs accessed simultaneously by hundreds of users.

19.3.6 Regulatory and Compliance Issues

Large business and healthcare organizations are not moving towards clouds because today's cloud infrastructures do not comply with regulatory requirements. Trustworthy data retention is one of the mandatory compliance issues that have a direct impact on digital forensics. Trustworthy data retention should provide long-term retention and disposal of organizational records to prevent unwanted deletion, editing, or modification of data during the retention period. It should also prevent recreation of a record once it has been removed. While there are still some problems in ensuring secure data retention at storage level, the cloud computing model imposes some new challenges. For example, who enforces the retention policy in a cloud, and how are exceptions, such as litigation holds managed? Moreover, how can CSPs assure us that they do not retain data after destruction (Popovic and Hocenski, 2010)? There are several laws in different countries that mandate trustworthy data retention. For example, the Sarbanes–Oxley Act (https://www.sec.gov/about/laws/soa2002.pdf, accessed December 8, 2015), the Health Insurance Portability and Accountability Act (HIPAA) (http://www.hhs.gov/ocr/privacy, accessed December 8, 2015), The Gramm–Leach–Bliley Act (http://www.business.ftc.gov/privacy-and-security/gramm-leach-bliley-act, accessed December 8, 2015), and European Commission data protection legislation (http://ec.europa.eu/justice/newsroom/data-protection/news/120125_en.htm, accessed December 8, 2015). Building clouds compliant with all of these laws is challenging, which consequently affects the investigation process in clouds.

19.4   Towards Reliable Forensics in Clouds

In the public interest, law enforcement first contacts the cloud provider with a temporary restraining order to suspend the offending service and account, and a preservation letter to preserve evidence pending a warrant. There are some proven digital forensics tools used by forensic investigators, for example, Encase, Accessdata FTK, and others that can be used for forensics data acquisition from clouds. Dykstra *et al.* were able to collect data remotely from the guest OS layer of Amazon EC2 cloud using Encase servlets and FTK agents as the remote programs (Dykstra and Sherman, 2012). They also prepared a Eucalyptus cloud platform and collected data from the virtualization layer. They tested evidence acquisition from the host operating system layer by Amazon's export feature and found that, although it is possible to export data from S3, it is not possible from EBS. However, collecting data using these tools tells us nothing about who put the data in the clouds. Moreover, a VM or host machine can be compromised or adversaries can terminate VMs. To overcome such challenges, researchers have proposed the following solutions to support reliable forensic investigation in clouds.

19.4.1   Cloud Data Storage

19.4.1.1   *Continuous Synchronization*

Forensics data acquisition is challenging from cloud storage due to the volatile nature of data stored in VMs. To overcome this problem, CSPs can provide a continuous synchronization API to customers. Using this API, customers can preserve the synchronized data to any cloud storage, for example, Amazon S3, or to their local storage. Unfortunately, the attacker will not be interested in synchronizing a malicious VM. Hence, CSPs by themselves can integrate the synchronization mechanism with every VM and preserve the data securely within their infrastructures.

As a solution to continuous synchronization of cloud data, Zawoad *et al.* proposed the Proof of Past Data Possession (PPDP) scheme, which stores cryptographic proof of cloud data periodically but without preserving the data itself (Zawoad and Hasan, 2012). The proofs of data possession are accessible by forensics investigators and cloud users. PPDP is a tamper-evident scheme that can detect any modification of data by dishonest CSPs and investigators. Preserving only proofs saves significant amount of storage cost but can still serve the purpose of continuous synchronization. In this solution, the CSP is responsible for the synchronization task. Hence, even if an adversary terminates a VM after some malicious activity, investigators can still prove the data possession using the PPDP scheme.

19.4.1.2   *Isolating a Cloud Instance*

Because of the multitenant architecture, a cloud instance must be isolated if any incident takes place on that instance. Delport *et al.* (2011) presented some possible techniques for cloud isolation. One way of isolating a suspect instance is to move other honest instances residing in the same node. The second technique is server farming, which can be used to reroute a request between user and node. The third technique is failover, where there is at least one server that is replicating another. We can also isolate an instance by placing it in a sandbox.

19.4.2 Forensics Data Acquisition

19.4.2.1 Centralized Repository of Logs

Marty (2011) proposed a centralized log management solution for the SaaS cloud model. He implemented an application logging library that can be used in Django. This library can export logging calls for each severity level, such as debug, info, error, and others. By tuning the Apache configuration, he was able to collect logs from the load balancer in the desired format. For PaaS, Birk and Wegener (2011) proposed a central log server, where customers can store the log information. In order to protect log data from possible eavesdropping and altering actions, customers can encrypt and sign the log data before sending it to the central server.

19.4.2.2 Application Programming Interface

To get the necessary logs from IaaS cloud model and to preserve the integrity and confidentiality of the logs, Zawoad *et al.* (2013) proposed Secure Logging-as-a-Service (SecLaaS). By using SecLaaS API services, investigators can collect various important logs, such as network, process, registry, and application logs. SecLaaS can also detect any alteration of logs by a malicious CSP, or a malicious forensic investigator.

19.4.2.3 Cloud Management Plane

Cloud service providers can play a vital role in data acquisition steps by providing a Web-based management console like AWS management console. Dykstra and Sherman (2013) implemented FROST, a forensic data-collection tool for OpenStack. Using FROST, cloud users / investigators can acquire an image of the virtual disks associated with any of the users' virtual machines, and validate the integrity of those images with cryptographic checksums. It is also possible to collect logs of all API requests made to CSP and OpenStack firewall logs for users' VMs.

19.4.3 Trustworthiness of Evidence

19.4.3.1 Trust Model

Dependence on CSPs poses trust issues in investigation procedures. Dykstra *et al.* proposed a trust model with six layers: guest application and data, guest OS, virtualization, host OS, physical hardware, and network (Dykstra and Sherman, 2012). The further down the stack is the less cumulative trust is required. For example, in the guest OS layer, we require trust in the guest OS, the hypervisor, the host OS, the hardware, and the network layer. While in network layer, we require trust in only the network. Examiners can examine evidence from different layers to ensure the consistency of the digital evidence. While executing a forensics investigation in clouds, investigators need to choose the appropriate layer, which depends on the data available in the layer and trust in the available data.

19.4.3.2 Integrity Preservation

Generating a digital signature on the collected evidence and then checking the signature later is one way to validate the integrity. Hegarty *et al.* (2009) proposed a distributed signature detection framework that will facilitate the forensic investigation in cloud environment. The proposed system uses peer-to-peer analysis nodes to validate the signature of data distributed in clouds. Zawoad and Hasan (2012) and Zawoad *et al.* (2013) proposed bloom filter-based integrity preservation techniques for cloud evidence, such as logs and files.

19.4.3.3   Trusted Platform Module (TPM)

To preserve the integrity and confidentiality of cloud evidence, researchers proposed TPM as the solution (Birk and Wegener, 2011; Dykstra and Sherman, 2012). Krautheim *et al.* (2010) proposed the Trusted Virtual Environment Module (TVEM) to establish trust in clouds. The root of trust in the solution is the TPM. By using the TPM, we can get machine authentication, hardware encryption, signing, secure key storage, and attestation. The TPM can provide the integrity of a running virtual instance, trusted log files, and trusted deletion of data to customers.

19.4.4   Forensics Data Examination and Analysis

19.4.4.1   Virtual Machine Introspection

Virtual machine introspection (VMI) is the process of externally monitoring the runtime state of a VM from either the virtual machine monitor (VMM), or from a virtual machine other than the one being examined. Runtime state refers to processor registers, memory, disk, network, and other hardware-level events. Through this process, we can execute a live forensic analysis of the system, while keeping the target system unchanged (Hay and Nance, 2008).

19.4.4.2   Guideline for Log Structure

There is no standard format for logs, so it is difficult to examine and analyze the log evidence. Marty (2011) proposed a guideline to overcome this problem. The proposed guideline tells us to focus on three things: when to log, what to log, and how to log. At minimum, he suggested logging the timestamps record, application, user, session ID, severity, reason, and categorization, so that we can get the answer of what, when, who, and why.

19.4.5   Legal Issues

19.4.5.1   Service Level Agreement

At present, there is a massive gap in the existing service level agreement (SLA), which neither defines the responsibilities of CSPs at the time of a malicious incident, nor their role in forensic investigations. A robust SLA should state how the providers deal with the cybercrimes – how and to what extent they help in forensic investigation procedures. To ensure the quality of SLA, we can take help from a trusted third party (Birk and Wegener, 2011).

19.4.5.2   Secure Provenance for Chain-of-Custody Preservation

Provenance provides the history of an object. Provenance can ensure the chain of custody in cloud forensics as it can provide the chronological access history of evidence, how it was analyzed, and preserved. Hence, we can maintain the CoC by preserving provenance records. Proper provenance in clouds should have four properties (Muniswamy-Reddy *et al.*, 2010): (i) data coupling to ensure that an object is described properly from its provenance; (ii) multiobject causal ordering to maintain causal relationships among objects; (iii) data-independent persistence – even if data is removed; we need to store its provenance record, and (iv) efficient query support on multiple objects' provenance. However, as the entire provenance records are under the control of CSPs, they can always tamper with the provenance records and, from the provenance data in clouds, an attacker can learn confidential data about cloud users. To prevent provenance from these types of attacks, Lu *et al.* (2010) introduced the concept of secure provenance in clouds. They proposed a trusted third-party (TTP) based scheme for secure cloud provenance. This scheme ensures some important properties, such as the confidentiality of the data, unforgeability and full anonymity of the signature, and full traceability from a signature.

19.5   Advantage of Clouds for Digital Forensics

19.5.1   Advantage

Cloud forensics is a complicated process and imposes new challenges in digital forensic procedures but it offers some advantages over traditional computer forensics. We can use the VM image as a source of digital evidence as we can acquire the computing environment of a VM instance for forensics investigation through the image. The computation and storage power of cloud computing can also speed up the investigation process by reducing the time for data acquisition, data copying, transferring, and data cryptanalysis. Forensic image verification time will be reduced if a cloud application generates a cryptographic checksum or hash. Integrating forensic facilities in a cloud environment or offering forensics as a service to customers by using the immense computing power will make cloud forensics cost effective for small- and medium-scale enterprise. Currently, Amazon replicates data in multiple zones to overcome the single point failure. In the case of data deletion, this data abundance can be helpful in collecting evidence. Amazon S3 automatically generates an MD5 hash of an object when we store the object in S3, which removes the need for external tools and reduces the time for generating the hash. Amazon S3 also provides versioning support. From the version log, we can get some crucial information for investigation, such as who accessed the data, and when, what the requestor's IP was, and what the changes in a specific version were.

19.5.2   Cloud Computing Use in Digital Forensics

The use of cloud-computing technology can also facilitate the traditional digital forensic investigation. Lin *et al.* (2012) proposed an RSA signature-based scheme to transfer data safely from mobile devices to cloud storage. It ensures the authenticity of data and thus helps in maintaining a trustworthy chain of custody in forensic investigations. By using the RSA signature protocol, a verifier can verify the evidence in the court. In this process, the cloud computing center computes the RSA signature and sends the signature to cloud storage center, which preserves the final output. The final output can later be downloaded to check the integrity of the data.

19.6   Open Problems

Cloud forensics is a new research area and has recently gained popularity among researchers. There are number of open problems that have not been addressed yet or some of the solutions proposed so far are not feasible enough to incorporate with real-world cloud infrastructures. Only a few of the solutions discussed above have been tested in real-world scenarios. To the best of the authors' knowledge, CSPs have not adopted any of the proposed solution yet.

Maintaining a CoC is one of the basic requirements for reliable and trustworthy forensics investigation. Since CoC depends on the location of data, we need to know the precise location of data in clouds at a given time. However, there has been no solution that can identify the location of data precisely. To maintain the trustworthiness of the CoC and protect eavesdropping, we need secure cloud provenance schemes. A secure provenance for cloud infrastructures is also necessary to make clouds more accountable and compliant with the regulatory rules, such as HIPAA and SOX (Zawoad and Hasan, 2012). However, the existing TTP-based secure cloud provenance solution (Lu *et al.*, 2010) is not robust enough because TTP increases the attack surface and introduces a single point of failure. Besides data, we also need the provenance of applications running inside clouds and states of cloud instances, which have not been addressed yet in existing cloud-provenance works. Existing works on accountable-clouds attack some specific problems, such as proof of data possession, proof of retrievability, or proof of data redundancy. A cloud computing framework that complies with all the major regulatory rules has not been proposed yet. Cloud computing will not be embraced universally unless we integrate regulatory compliance with clouds. For example, as we are yet to offer SOX-compliant cloud storage, business organizations invest significant amount of money to host their own private SOX-compliant storage rather than go to cheap cloud storage.

Trusted platform modules have been considered as a solution to establish trust in cloud infrastructures. However, the TPM is not totally secure and it is possible to modify a running process without being detected by the TPM (Dykstra and Sherman, 2012). Moreover, at present CSPs have heterogeneous hardware and few of them have TPMs. Hence, CSPs cannot ensure a homogeneous hardware environment with TPMs in the near future.

Crime-scene reconstruction is crucial for forensic investigation. However, reconstructing a criminal incident is impossible in clouds when a user turns off a VM after some illegal activity. There has been no work that can tackle this situation. For crime-scene reconstruction, forensics timeline analysis is also an important step. However, the system clock of a cloud host or a VM can be altered by adversaries and dishonest employees of a CSP. Such alteration of a system clock will produce a fake timeline of events and impede the investigation process.

Challenges about legal issues and regulatory compliance have not been resolved yet. To mitigate the cross-border issue, researchers have proposed global unity but there is no guideline about how this will turn into reality. Similarly, there is no solution about preparing a robust SLA that can cover all the legal bindings of CSP in the case of an adversarial situation. Modifying the existing forensic tools or creating new tools to cope with the cloud environment is another big issue that has not been resolved yet. If the cloud storage is too high, then limited bandwidth is a big challenge for time-sensitive investigations. We need an efficient solution to acquire evidence from clouds that can give us high throughput using low bandwidth. About the logging issue in cloud forensics, Marty (2011) proposed some open research topics in application-level logging: security visualization, log review, log correlation, and policy monitoring.

19.7   Conclusion

With the wide adoption of cloud computing there is an increasing emphasis on providing trustworthy cloud forensics. In this chapter, we have summarized current challenges, solutions, and open research problems in cloud forensics. By analyzing the challenges and existing solutions, we argue that CSPs need to come forward to resolve most of the issues. There is very little to do from the customers' point of view other than application logging. All other solutions are dependent on CSPs and the policy makers. For forensics data acquisition, CSPs can shift their responsibility by providing a robust API or management panel to acquire evidence. Legal issues also hinder the smooth execution of forensic investigation. We need a collaborative attempt from public and private organizations as well as research and academia to overcome these issues. Solving all the challenges of cloud forensics will clear the way for making a forensics-enabled cloud and allow more customers to take advantage of cloud computing.

References

Birk, D. and Wegener, C. (2011) *Technical Issues of Forensic Investigations in Cloud Computing Environments*. Sixth International Workshop on Systematic Approaches to Digital Forensic Engineering (SADFE). IEEE.

Delport, W., Kohn, M., and Olivier, M. S. (2011) Isolating a cloud instance for a digital forensic investigation. In Proceedings of the Information and Computer Security Architecture (ICSA).

Dykstra, J. and Sherman, A. (2011) Understanding issues in cloud forensics: Two hypothetical case studies. Journal of Network Forensics b(3), 19–31.

Dykstra, J. and Sherman, A. (2012) Acquiring forensic evidence from infrastructure-as-a-service cloud computing: Exploring and evaluating tools, trust, and techniques. DoD Cyber Crime Conference, January, http://www.cisa.umbc.edu/papers/DFRWS2012_Dykstra.pdf (accessed December 8, 2015).

Dykstra, J. and Sherman, A. (2013) Design and implementation of FROST: Digital forensic tools for the openstack cloud computing platform. *Digital Investigation* **10**, S87–S95.

Hay, B. and Nance, K. (2008) Forensics examination of volatile system data using virtual introspection. *ACM SIGOPS Operating Systems Review* **42**(3), 74–82.

Hegarty, R., Merabti, M., Shi, Q., and Askwith, B. (2009) *Forensic Analysis of Distributed Data in a Service Oriented Computing Platform.* In Proceedings of the Tenth Annual Postgraduate Symposium on the Convergence of Telecommunications, Networking & Broadcasting. PG Net.

Kent, K., Chevalier, S., Grance, T., and Dang, H. (2006) *Guide to Integrating Forensic Techniques into Incident Response.* NIST Special Publication 800-86. NIST, Gaithersburg, MD, http://csrc.nist.gov/publications/nistpubs/800-86/ SP800-86.pdf (accessed December 8, 2015).

Krautheim, F., Phatak, D., and Sherman, A. (2010) Introducing the trusted virtual environment module: A new mechanism for rooting trust in cloud computing, in *Trust and Trustworthy Computing* (eds. Acquisti, A., Smith, S. W., and Sadeghi, A.). Springer, Heidelberg, pp. 211–227.

Lin, C., Lee, C. and Wu, T. (2012) A cloud-aided RSA signature scheme for sealing and storing the digital evidences in computer forensics. *International Journal of Security and its Applications* **6**(2), 241–244.

Lu, R., X. Lin, X. Liang, and X. Shen (2010) *Secure Provenance: The Essential of Bread and Butter of Data Forensics in Cloud Computing.* Proceedings of the Fifth ACM Symposium on Information, Computer and Communications Security. ACM, pp. 282–292.

Marty, R. (2011) *Cloud Application Logging for Forensics.* In Proceedings of the 2011 ACM Symposium on Applied Computing. ACM, pp. 178–184.

Muniswamy-Reddy, K., Macko, P., and Seltzer, M. (2010) *Provenance for the Cloud.* Proceedings of the Eighth USENIX Conference on File and Storage Technologies. USENIX Association, pp. 15–14.

Popovic, K. and Hocenski, Z. (2010) *Cloud Computing Security Issues and Challenges.* In Proceedings of the 33rd International Convention MIPRO, 2010. IEEE, pp. 344–349.

Ruan, K., Carthy, J., Kechadi, T., and Crosbie, M. (2011). Cloud forensics: An overview. In Proceedings of the Seventh IFIP International Conference on Digital Forensics.

Zawoad, S., Dutta, A. K., and Hasan, R. (2013) *SecLaas: Secure Logging-as-a-Service for Cloud Forensics.* In Proceedings of the Eighth ACM Symposium on Information, Computer and Communications Security (ASIA CCS). ACM.

Zawoad, S. and Hasan, R. (2012) Towards building proofs of past data possession in cloud forensics. *Academy of Science and Engineering Journal* **1**(4), 195–207.

20

Privacy, Law, and Cloud Services

Carol M. Hayes and Jay P. Kesan

University of Illinois College of Law, USA

20.1   Introduction

Consumers often worry about how to protect the privacy of their information when they are using cloud services. They may be worried about identity theft and government surveillance, or the possible privacy invasions of targeted marketing. What steps do cloud services take to secure customer information and prevent data security breaches that could lead to identity theft? When might the service share the customer's information with third parties? What should consumers do if they are worried about their online activities being the subject of government surveillance?

Currently, being informed is the best way to preserve privacy online. Consumers can refrain from posting personal information, they can customize their browser settings to disallow third-party cookies, and they can route all of their Internet traffic through an anonymizing system like Tor, among other precautions. But if consumers are not sufficiently informed about the practices of the companies with which they deal, they may still lose control of their information.

Consumers can familiarize themselves with the privacy policies and terms of service (TOS) of the services that they use, and these documents will give them information about things like what information is collected, with whom the information will be shared, and whether and to what extent the service provider obtains a license to use the consumer's intellectual property. However, consumers often click "I agree" out of habit, without actually reading these documents. This chapter examines a variety of issues relating to privacy, with a primary focus on theories and law, with the goal of applying this understanding to cloud services. As the following sections discuss, there are a number of possible legal ramifications from the often disregarded legal contracts that consumers accept in order to use a service.

Encyclopedia of Cloud Computing, First Edition. Edited by San Murugesan and Irena Bojanova.
© 2016 John Wiley & Sons, Ltd. Published 2016 by John Wiley & Sons, Ltd.

20.2   The Evolution of Privacy Theory

Samuel Warren and Louis Brandeis published *The Right to Privacy* in 1890, which substantially influenced privacy law in the United States in the twentieth century (Richards and Solove, 2010). The publication of this piece was spurred by the authors' concerns about intrusions into personal privacy by the press, especially considering the technological improvements that had enabled the production of small, affordable cameras (Solove, 2006a). Warren and Brandeis framed their concept of privacy as a "right to be let alone" that relates to human dignity. By the time the Warren and Brandeis article was 50 years old, privacy was still a very minor doctrine in tort law. Only 12 states recognized the right of privacy by common law, and only two recognized it by statute (Richards and Solove, 2010).

The next major development of privacy theory in the United States was the 1960 publication of an article by William Prosser on privacy. In this article, Prosser argued that there were four categories within privacy tort law: appropriation privacy, intrusion privacy, unauthorized public disclosure of private facts, and false light. Today, modern discussions of privacy violations often assume that the privacy violation will be addressed through civil suits grounded in one of these four torts.

There are several different types of privacy. Informational privacy is the category of privacy that is the most relevant to the cloud context. The nature of the torts proposed by Prosser makes it difficult to apply these torts to modern informational privacy issues. The privacy tort of invasion typically requires the invasion to be of an offensive nature but a lot of information collection appears largely innocuous (Richards and Solove, 2010). In one Ohio case, for example, the sale of information about a magazine's subscribers was found not to meet the injury requirement for the sale to be an invasion of privacy. It is likely that legal redress for violations of privacy online will require either the revision of privacy torts or the introduction of new alternatives. Solove, a leading informational privacy scholar, divides information privacy problems into four categories: information collection, information processing, information dissemination, and invasion (Solove, 2006b). Adoption of these new categories could potentially fill in some of the gaps that Prosser's theories do not address.

20.3   Privacy Law

20.3.1   Stored Communications Act

The Electronic Communications Privacy Act (ECPA) is a federal statute that covers several topics relevant to electronic privacy. The ECPA was passed in 1986, partly in response to the findings of the Office of Technology Assessment that the protections of e-mails were "weak, ambiguous, or nonexistent" (Kattan, 2011).

The ECPA consists of three federal statutes (see Figure 20.1): the Stored Communications Act (SCA), the Pen Register statute, and the Wiretap Act. Its protections supplement those of the Fourth Amendment (Bagley, 2011). The application of each depends on what type of information is sought and where it is in the transmission process (Kerr, 2004). The Wiretap Act covers interception of wire, oral, and electronic communications. Under the Wiretap Act, obtaining e-mail content in real time requires a Title III order to be issued with Department of Justice (DOJ) approval and a grant by a federal judge, and the order must be renewed every 30 days. Under the Pen Register statute, obtaining real-time subscriber data requires an *ex parte* pen register order. Stored electronic information and the requirements for obtaining each type are addressed under the SCA.

Explaining the SCA is difficult because much of the language is very unclear or outdated, and interpretations of the statute by courts have varied significantly (Kerr, 2004). The two most important sections for understanding the SCA's relevance to online privacy are: (i) § 2702, which addresses the circumstances under which a provider can voluntarily disclose customer information to others, and (ii) § 2703, which addresses

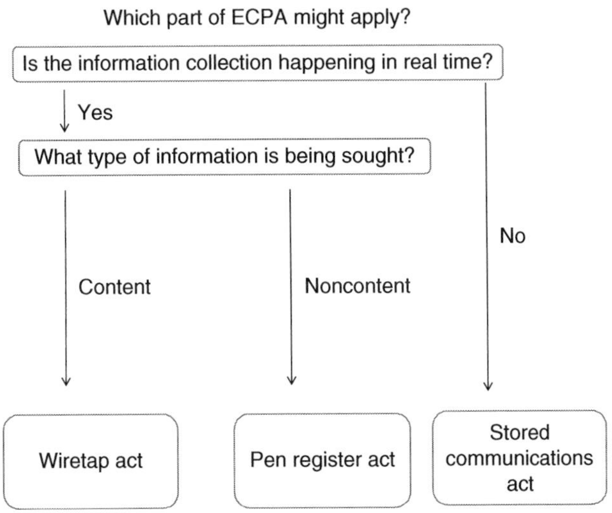

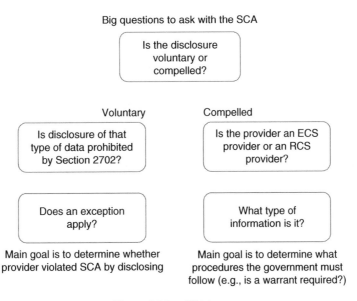

Which part of ECPA might apply?

Is the information collection happening in real time?

Yes

What type of information is being sought?

No

Content

Noncontent

Wiretap act

Pen register act

Stored communications act

Figure 20.1 *Different aspects of the ECPA*

Big questions to ask with the SCA

Is the disclosure voluntary or compelled?

Voluntary

Compelled

Is disclosure of that type of data prohibited by Section 2702?

Is the provider an ECS provider or an RCS provider?

Does an exception apply?

What type of information is it?

Main goal is to determine whether provider violated SCA by disclosing

Main goal is to determine what procedures the government must follow (e.g., is a warrant required?)

Figure 20.2 *SCA issues*

how the government can compel a provider to produce stored information (see Figure 20.2). However, it is not always clear whether a disclosure was voluntary or compelled, and therefore it is sometimes difficult to determine which set of exceptions or requirements will apply.

Section 2702 lists situations when a provider may disclose stored information without violating the SCA. For example, service providers may disclose information when the subscriber or customer (depending on the type of service) consents to disclosure. But what actions can amount to consent? Is it consent under the SCA to accept the terms of a very broad privacy policy without reading these terms?

To determine the propriety of a disclosure under the SCA, the government must first determine whether the sought information is stored as part of an electronic communications service (ECS) or a remote computing service (RCS). Section 2702 prohibits ECS providers from knowingly disclosing communication contents that the provider holds in "electronic storage," and prohibits RCS providers from knowingly disclosing communication contents that the provider maintains for the sole purpose of providing the subscriber or customer with "storage or computer processing services." Even though "storage" and "electronic storage" sound very similar, distinguishing between the two can make a difference in the outcome of a case.

The process required to compel information can also be affected by whether the provider is ECS or RCS, and varies with the type of information sought. Notice is required prior to the disclosure of some information types. Some types of information require a warrant, while others require a special court order under § 2703(d), and still others require only a subpoena. These three methods of compelling information are listed in descending order of the strength of the showing required to obtain them. To obtain a § 2703(d) court order, the governmental entity must establish "reasonable grounds" to believe that the information sought is "relevant and material to an ongoing criminal investigation." This standard is less than the "probable cause" standard for obtaining a warrant, but greater than the "reasonable relevance" standard for obtaining a subpoena. If information is wrongfully disclosed, the injured party has a right under the SCA to sue the provider for the wrongful disclosure.

The SCA is a complex statute that Congress wrote based on how early computer networks operated, so its application to cloud computing is sometimes unclear. The category of RCS provider was intended to address the business model in which companies outsourced a lot of storage and processing functions due to the high cost of doing this in house (Robison, 2010). It is thus likely that the RCS category would easily apply to commercial cloud services that provide options for outsourcing IT, but in most other contexts, there is substantial overlap between RCS and ECS.

The overlap between RCS and ECS is most easily seen in e-mails. Currently, the degree of privacy that the SCA assures for an e-mail likely depends on whether the e-mail is stored on a hard drive or in the cloud. If e-mails downloaded from a service provider are unopened and are fewer than 180 days old, the government must obtain a warrant, but it is unclear what requirements apply to webmail (Bagley, 2011).

Companies with services that are supported by advertising often access communication contents for targeted advertising purposes (Kattan, 2011). This may prevent these services from being considered RCS providers because the provider is authorized to access communication contents for purposes other than rendering storage and computer processing services (Robison, 2010); TOS agreements and privacy policies thus have potentially significant effects on the extent to which the SCA protects the customer's privacy because these terms may give the provider explicit authority to take actions that would disqualify the provider from being considered an RCS provider.

20.3.2 Fourth Amendment

The Fourth Amendment declares that people have a right "to be secure in their persons, houses, papers, and effects, against unreasonable searches and seizures."

Fourth Amendment cases often focus on the need for a warrant, the presence of probable cause, and the existence of a reasonable expectation of privacy (REOP). Courts recognize a REOP in papers and effects sent in the mail, and also recognize a REOP in locked containers. Courts have analogized e-mail to postal mail to conclude that there is a REOP in e-mail. Similarly, courts have analogized password protection to locked containers to conclude that there is a REOP when passwords are required to access specific data. However, if a person is careless with password security, password protection may not lead to a REOP. The Fourth Amendment likely protects a lot of digital content, but noncontent information is often not protected. Building on the e-mail analogy, because courts do not recognize a REOP in noncontent address information on the outside of envelopes, they generally will not recognize a REOP in similar noncontent address information in the "To" field of an e-mail.

If a warrantless search is conducted where a REOP exists, a court will examine if one of the exceptions to the warrant requirement applies. If no exception applies, the search violated the Fourth Amendment, and the evidence derived from the violation may be suppressed at trial. Similar to the SCA, where consent to disclosure makes the disclosure legal, consent to a search can validate an otherwise illegal search.

The "third-party doctrine" of the Fourth Amendment prevents a REOP from being found in papers and effects turned over to a third party. However, the third-party doctrine might not apply when items are entrusted to a third party who does not have the authority to view the items or to consent to search of the items by others. Thus, much will turn on the authority that the third party has with respect to the entrusted items. If a private carrier explicitly retains the right to inspect a package for any reason, customers may not have a REOP in packages sent using this private carrier.

Online, users must reveal their information to third parties in order for it to be transmitted or processed, so the third-party doctrine may limit the application of the Fourth Amendment on the Internet (Bagley, 2011). On the other hand, one recent federal appellate court case distinguished e-mail interception from other third-party doctrine cases by holding that the e-mail provider was an intermediary in the communication, not a recipient. However, the court also noted that if an agreement with a service provider gave the service provider the authority to "audit, inspect, and monitor" the e-mails of its subscribers, that might cause the subscriber to lose a REOP in those e-mails.

20.3.3   Consumers, the SCA, and the Fourth Amendment

In the past few years, controversies have arisen concerning government monitoring of online activity. Consumers who are concerned about their privacy in the cloud should keep these issues in mind when considering different cloud service providers. Many people do not read TOS agreements and privacy policies, but these agreements can affect whether there is a reasonable expectation of privacy in information stored using the service. These agreements also may affect the application of the SCA. Thus, consumers who wish to preserve their rights against unreasonable searches of digital information should become familiar with the policies of the companies with whom they deal.

20.3.4   Other Privacy Laws

Federal privacy law focusing on consumer protection is typically narrow, often focusing on the type of records in issue or a particular industry (Solove, 2006a). Early congressional action on privacy includes the Fair Credit Reporting Act (FCRA) in 1970 and the Family Educational Rights and Privacy Act (FERPA) in 1974. Other federal statutes addressing specific privacy issues include the Children's Online Privacy Protection Act (COPPA), the Health Information Portability and Accessibility Act (HIPAA), the Electronic Communications Privacy Act (ECPA), and the Gramm–Leach–Bliley Act (GLBA). Several federal statutes focus on the presence of personally identifiable information (PII), while others focus on transparency and access to information, on protecting consumers from inappropriate use of their personal data, or on imposing duties of confidentiality (Solove, 2006b). If directed by statute, federal agencies will enact rules to regulate the privacy practices of specific industries.

States also adopt their own privacy laws to protect consumers. For example, Massachusetts requires detailed data security procedures, and 45 states require customer notification in the event of a security breach (Rhodes and Kunis, 2011). Minnesota has a merchant liability statute, under which a merchant can be held liable if there was a security breach and customer credit-card information was insufficiently protected. California's Song–Beverly Act protects PII by prohibiting merchants from requiring customers to give personal information like their address and phone number "as a condition to accepting the [customer's] credit card."

20.3.5   European Privacy Law

The United States and the European Union take very different approaches to privacy. In the United States, most privacy laws are sector-specific, and privacy concerns may be trumped by other important interests like free speech and national security. In the European Union, privacy is considered to be a fundamental right (Lanois, 2010).

Information privacy laws in the EU are much stricter than similar laws in the United States. Because of the significant differences between data privacy law in the EU and the United States, American companies must often be more careful with EU customers' data than US law requires. Some cloud providers have segregated EU clouds, and many companies promise to follow the Safe Harbor framework.

The first European data-protection laws were enacted in the 1970s, followed by the adoption of the Convention for the Protection of Individuals with regard to Automatic Processing of Personal Data in 1981, and the enactment of the EU Data Protection Directive 95/46 (DPD 95/46) in 1995. DPD 95/46 focuses on the protection of "personal data," which it defines as "information relating to an identified or identifiable natural person." By its terms, it applies EU law to data controllers that use "equipment" within the EU for the processing of personal data, and the term "equipment" has been read broadly to apply to things like cookies and JavaScript (Kuner, 2010).

20.3.5.1   The Safe Harbor Framework

DPD 95/46 permits data transfers only to other countries with adequately protective privacy laws (Stylianou, 2010). The United States does not have sufficient privacy laws but the data of European users can nonetheless be transferred to the United States if the company handling the transfer complies with the Safe Harbor agreement between the United States and the European Union (Lanois, 2010).

The Safe Harbor framework allows companies to certify compliance with European privacy standards without necessarily using segregated clouds. Under the Safe Harbor Privacy Principles, organizations must: (i) provide notice about data collection; (ii) give individuals a choice to opt out (or to opt in, if the personal information is considered sensitive); (iii) extend these standards to onward transfers – that is, ensure that third parties to whom personal information is transferred also adhere to the Safe Harbor Privacy Principles or have comparable controls in place; (iv) provide individuals with access to their personal information held by the organization; (v) take reasonable security precautions to protect personal information; (vi) take reasonable steps to protect data integrity; and (vii) provide adequate measures for enforcement of the principles.

Adhering to the Safe Harbor Privacy Principles provides a mechanism for companies in the United States to preserve the status quo of a self-regulatory approach to privacy while still being eligible to serve customers in the European Union.

20.4   TOS Agreements and Privacy Policies

20.4.1   As Binding Contracts

As noted above, excessively permissive TOS agreements can affect users' protections under US privacy law. This makes determining the validity of these contracts extremely important.

Contract law is typically state law, so standards will vary across cases. Under the common law of contracts, forming a contract requires mutual assent. When a contract is not subject to negotiation and is offered by the more powerful party on a "take it or leave it" basis, the contract is often referred to as a contract of adhesion. Privacy policies and TOS agreements typically meet this definition for an adhesion contract (Bagley, 2011). Such contracts are not automatically invalid but they may be subject to greater scrutiny.

If a court finds that the terms of a TOS agreement are excessively unfair, the court may invalidate the contract. In addition to the possibility of courts invalidating TOS agreements, the FTC may intervene in some situations to prevent unfair business practices. If a company violates its own privacy policy, the FTC may require the company to follow its own stated policy.

Excessively unfair TOS terms may be invalidated if the court concludes that the terms are unconscionable. Courts might be more willing to find unconscionability when there are no market alternatives but the diverse reality of the cloud market makes it unlikely that a lack of market alternatives will be a persuasive argument (Bagley, 2011).

20.4.2 Common Terms

TOS agreements set forth terms governing the relationship between a service provider and its customers. Generally, cloud-based services targeted at individual users are accompanied by non-negotiable TOS agreements that favor the service provider over the end user (Wittow and Buller, 2010). The agreements will generally address things like metering, monitoring, and data backup, and often include clauses in which the provider disclaims liability for harm and forbids customers from using the company's intellectual property without authorization (Bagley, 2011). Some also include terms concerning the retention, control, and ownership of a user's information (Kesan *et al.*, 2013). TOS agreements take a variety of approaches to customer information. Some include terms that allow providers to access customer information for advertising and other purposes relating to the business, whereas others are less transparent about what the company may do with customer information, and still others make explicit promises in their TOS agreements that the companies will not access customers' data (Robison, 2010).

One of the aspects of TOS agreements that many customers are interested in is whether companies claim ownership of uploaded material. In their analysis of twelve TOS agreements, Kesan *et al.* (2013) found a very wide variety of approaches to intellectual property. Table 20.1 summarizes these findings.

It is very important that consumers be aware of the terms of cloud services' TOS agreements because of the large amounts of sometimes sensitive information stored with these services (Soma *et al.*, 2010). Consumers should pay special attention to how the TOS agreements address customer data, including the information that the company claims rights in, and how the consumer can terminate his relationship with the cloud provider (Wittow and Buller, 2010). Consumers might be storing information solely in the cloud, making it very important for the TOS agreements to include provisions protecting customers' ability to retrieve their content if, for example, a service is shut down.

Table 20.1 *Findings from study by Kesan et al. (2013)*

Intellectual property provisions	Out of 12	Percentage of total
The company retains full rights in its intellectual property that it is licensing to the customer	12	100
The company owns all rights in any e-mails, suggestions, or ideas that the customer sends to the company	11	91.67
The customer retains full rights in his intellectual property that is maintained on the company's servers	6	50
The company obtains a license to use the customer's content to provide the service to others	5	41.67
The company obtains a license to use the customer's content for marketing purposes	1	8.33

Table 20.2   *Findings from study by Kesan et al. (2013)*

Security measures mentioned in privacy policies	Out of 19	Percentage of total
Industry standard or commercially reasonable security measures	13	68.42
SSL is used for data transmission	12	63.16
Says that the customer should be managing security from their end as well (change passwords often, etc.)	12	63.16
Lists organizational security measures	11	57.89
Lists other technical security measures	9	37.37
Language warning that nothing is 100% secure	6	31.58
Describes how the company will respond if customers' information is compromised	1	5.26
Disclaims responsibility for securing or backing up users' data	1	5.26
Policy does not address	1	5.26

Privacy policies and TOS agreements often overlap, but privacy policies tend to be more focused on making the customer aware of the company's policies regarding their data instead of the customer's obligations concerning the service. Terms in a provider's privacy policy might address things like the quantity and nature of collected data and the company's policies on data retention and customer control over data (Stylianou, 2010). Privacy policies often also address data security issues, like the use of SSL encryption during data transmission (Kesan *et al.*, 2013). However, many of these providers insert provisions in their privacy policies or TOS agreements that repudiate any liability for data loss, and reserve to the provider the right to discontinue the service at the provider's sole discretion.

Table 20.2 summarizes the findings of Kesan *et al.* concerning security measures mentioned in 19 privacy policies. Companies generally listed more than one of these provisions.

Privacy policies typically include information about how the provider may gather, use, disclose, and manage the personal information of its customers. Privacy policies, like TOS agreements, often reserve significant advantages for the service provider, such as the right to amend its privacy policy unilaterally with little notice to its customers (Kesan *et al.*, 2013). Privacy policies may include broad permissions to allow the provider to access information for its own marketing purposes and to disclose customer information to its business partners for business-related purposes (Soma *et al.*, 2010; Kesan *et al.*, 2013).

However, few consumers actually read a company's privacy policy, and even fewer understand it (Schwartz and Solove, 2011). Solove criticizes many privacy policies as being "written in obtuse prose," containing large amounts of extraneous information.

Cloud services collect a lot of data, both through the customer's voluntary disclosure of data and through the provider's automatic collection of information through its operations or advertising policy (Stylianou, 2010). Many privacy policies assure limited use of customer information. Some, however, are vague, leaving ambiguities and loopholes. Transparency in privacy policies is very important, and consumers should be informed about how their data will be collected and used. If consumers had more control over their data, this could raise consumer awareness of privacy issues, and have a positive effect on the market for cloud services (Kesan *et al.*, 2013).

The possible legal effects of TOS agreements and privacy policies are troubling but it is unlikely that most consumers have either the time or the expertise to analyze and understand the variety of terms contained in these documents. However, as interest in this topic grows there will be an increase in the demand for services that track and summarize key points of these agreements.

20.5   Data Control

Data control, which includes the ideas of data mobility and data withdrawal, is important for informational privacy (Kesan *et al.*, 2013). Data control proposals can focus on PII and on "course-of-business" data that is stored as part of the customer's use of the service. Secondary use of this information is also an important consideration. Adequate data mobility and data withdrawal provisions would attract consumers who are more risk averse and who would not use these services in the absence of these protections, thus leading to a net benefit to the industry.

Data mobility is an important aspect of data control. Ideally, customers would not lose everything if a service provider became inoperable or if the data had to be moved to a new service provider. However, some analysis of TOS agreements and privacy policies indicates that companies often do not address the handling of such data after the contract has terminated (Kesan *et al.*, 2013). Data should be exportable so that the customer is not locked in to a particular service provider, and could easily move their data from one provider to another. When deciding between cloud services, customers should note whether the company provides a method for former customers to withdraw and transfer their data.

Current data mobility issues are similar to issues surrounding mobile number portability (Kesan *et al.*, 2013). In November 2003, an FCC regulation became effective that required cell phone carriers to allow numbers to be ported from one carrier to another. Service providers resisted, claiming that this rule would be too costly to carriers and would not benefit consumers. The FCC, however, concluded in 2006 that the number portability requirement did not significantly increase "wireless churn," and did in fact have a positive impact on service quality due to the need that it created for carriers to devote extra effort to customer retention. Based on this history, Kesan *et al.* (2013) conclude that data mobility in the cloud would similarly facilitate consumer participation and reduce transaction costs for consumers when moving from one provider to another.

20.6   Conclusion

Privacy online is a very important topic that causes concern to many people. Massive data breaches could lead to identity theft, so consumers typically demand strong security measures. Recent controversies over government surveillance have also increased interest in the issue of protecting the rights of consumers. Lax intellectual property policies could give a company permission to use a customer's creations for the company's marketing purposes. Consumers should educate themselves on topics like these, which could enable them to make informed decisions with regard to services that they use in the cloud.

Consumers can take a number of steps to keep their information private, including using traffic anonymizers and simply not uploading personal information online. This chapter stresses one of the most important tools for protecting privacy online: information. Consumers may currently be underinformed about possible privacy issues online, and may not be aware of their legal rights and how to protect them. One way that consumers can become better informed is by becoming familiar with the policies of companies they choose to deal with online. Grassroots public education campaigns designed to increase awareness of privacy issues could assist with this goal.

Additional Resources

Birnhack, M. and Elkin-Koren, N. (2011) Does law matter online? Empirical evidence on privacy law compliance. *Michigan Telecommunications and Technology Law Review* **17,** 337.

Soghoian, C. (2010) Caught in the cloud: Privacy, encryption, and government back doors in the Web 2.0 era. *Journal on Telecommunications and High Technology Law* **8,** 359.

Acknowledgements

The subject matter of this chapter was addressed more fully by the authors in an article published in Volume 70 of the Washington and Lee Law Review (cited in References).

References

Bagley, A. W. (2011) Don't be evil: The Fourth Amendment in the age of Google, national security, and digital papers and effects. *Albany Law Journal of Science and Technology* **21**, 167.

Kattan, I. R. (2011) Cloudy privacy protections: Why the Stored Communications Act fails to protect the privacy of communications stored in the cloud. *Vanderbilt Journal of Entertainment and Technology Law* **13**, 617.

Kerr, O. S. (2004) A user's guide to the Stored Communications Act, and a legislator's guide to amending it. *George Washington Law Review* **72**, 1208.

Kesan, J. P., Hayes, C. M., and Bashir, M. N. (2013) Information privacy and data control in cloud computing: Consumers, privacy preferences, and market efficiency. *Washington and Lee Law Review* **70**, 341.

Kuner, C. (2010) Data protection law and international jurisdiction on the Internet (Part 2). *International Journal of Law and Information Technology* **18**, 227.

Lanois, P. (2010) Caught in the clouds: The Web 2.0, cloud computing, and privacy? *Northwestern Journal of Technology and Intellectual Property* **9**, 29

Rhodes, K. L. and Kunis, B. (2011) Walking the wire in the wireless world: Legal and policy implications of mobile computing. *Journal of Technology Law and Policy* **16**, 25

Richards, N. M. and Solove, D. J. (2010) Prosser's privacy law: A mixed legacy. *California Law Review* **98**, 1887.

Robison, W. J. (2010) Free at what cost? Cloud computing privacy under the Stored Communications Act. *Georgetown Law Journal* **98**, 1195.

Schwartz, P. M. and Solove, D. J. (2011) The PII problem: Privacy and a new concept of personally identifiable information. *New York University Law Review* **86**, 1814.

Solove, D. J. (2006a) *The Digital Person: Technology and Privacy in the Information Age*, New York University Press, New York.

Solove, D. J. (2006b) A taxonomy of privacy. *University of Pennsylvania Law Review* **154**, 477.

Soma, J., Gates, M. M., and Smith, M. (2010) Bit-wise but privacy foolish: Smarter e-messaging technologies call for a return to core privacy principles. *Albany Law Journal of Science and Technology* **20**, 487.

Stylianou, K. K. (2010) An evolutionary study of cloud computing services privacy terms. *John Marshall Journal of Computer and Information Law* **27**, 593.

Wittow, M. H. and Buller, D. J. (2010) Cloud computing: Emerging legal issues for access to data, anywhere, anytime. *Journal of Internet Law* **14**(1), 1.

21

Ensuring Privacy in Clouds

Travis Breaux[1] and Siani Pearson[2]

[1] *Carnegie Mellon University, USA*
[2] *Hewlett Packard Enterprise, UK*

21.1   Introduction

In this chapter, we examine how privacy relates to the fundamentals of a cloud-based computing architecture. We first introduce important terminology. Later, we will define privacy referring to prevailing legal definitions and principles that influence architecture-dependent assumptions in cloud services. With an understanding of how cloud systems affect privacy, we will identify common risks to privacy and discuss tools to ensure privacy in the cloud.

Cloud-based computing architecture consists of one or more service offerings that enable data processing over a telecommunications network, and the responsibility for the data primarily resides with, or is at least shared with, the cloud consumer. There are a few cloud service models (software, platform, and infrastructure as a service, abbreviated SaaS, PaaS and IaaS, respectively) and deployment models (private, public, community, hybrid) (Mell and Grance, 2011) that affect privacy risks differently. In Figure 21.1, we present an example cloud-based architecture for an online video rental web site that offers entertainment and instructional videos for children, adults, and classroom instructors and their students. In this view, the *cloud consumer* is the video rental web site, which purchases streaming media services from a *cloud broker*; the broker packages other cloud-based services from *cloud providers*, such as advertising, payment processing, and media content services, into a single composite service. The video rental web site works directly with an advertising network that tracks what users watch and their browsing behavior at other web sites, in addition, to providing banner advertisements that appear in the user's web browser. The streaming media service contracts with a separate advertising service that supplies in-media advertising. In Figure 21.1, personally identifiable information (PII) is collected from the users and shared with different services to deliver the end-user experience.

Encyclopedia of Cloud Computing, First Edition. Edited by San Murugesan and Irena Bojanova.
© 2016 John Wiley & Sons, Ltd. Published 2016 by John Wiley & Sons, Ltd.

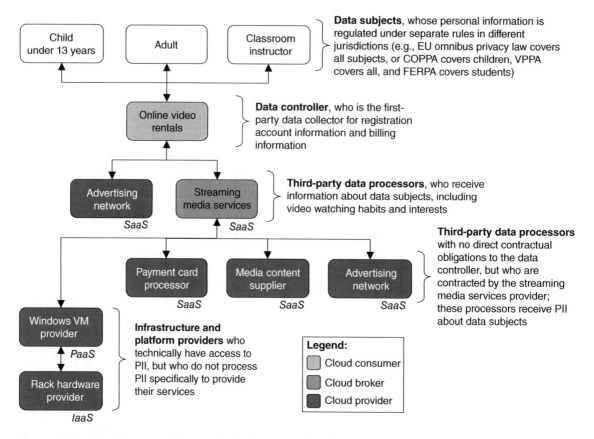

Figure 21.1   *Cloud-based architectures include relationships that may overlap with data privacy relationships: this example architecture shows a data subject whose personal data is collected by the cloud consumer and later transferred to one or more third-party cloud providers who go on to store or process the data*

With respect to data privacy, three important distinctions have been widely accepted that we discuss throughout this chapter: *personal data*, also called *personally identifiable information* (PII), is any data relating to an identified natural person, called the *data subject*; the *data controller*, which is the entity who (perhaps with others) determines the purposes and means of processing personal data; and the *data processor*, which is any entity who processes personal data on behalf of the controller. In Figure 21.1, the video rental company, acting as the data controller, collects personal and other data from the data subjects. The company further outsources its data processing to third parties, who may then use any number of cloud providers to deliver these third-party services. Generally, outsourcing is assumed to reduce costs while affording companies the ability to scale processing to meet changing demand. For example, the streaming media services in Figure 21.1 rely on third-party platform and infrastructure services to provide basic operating system and hardware to meet video watching demand dynamically. A major privacy challenge, however, is in determining who is responsible for protecting privacy. Moving across service models from SaaS to PaaS to IaaS, the cloud provider has increased control over the data processing and thus takes on increased responsibility; however, cloud providers who offer only infrastructure still have access to the data, and thus must implement privacy and security controls. The IaaS providers may not know who the end users are or what services are being offered, while the design and infrastructure of the cloud are not transparent to the data controllers, yet data controllers are primarily responsible for the purposes for which data is used.

The economies of scale attributed to the cloud that arise from streamlining and co-locating offerings of the same kind of service generally lead to reduced transparency or control afforded to cloud consumers and their data once it is stored in the cloud (European Commission, Directorate General of Justice, 2012). Thus, to mitigate privacy risks, the cloud provider and consumer must ensure certain privacy practices and principles are guaranteed by written contract and by design. In Figure 21.1, the data controller would generally use service level agreements (SLAs) to acquire assurances that any direct and indirect data processors are protecting privacy in a manner consistent with the online video rental's privacy promises to its customers.

Cloud consumers primarily ensure privacy through their contracts with the cloud provider. These contracts can include privacy policies, which are typically written for end users, and various service agreements, which are broadly written to cover all customers or may be tailored to service offerings personalized for specific customers. For small customers, and data subjects in particular, there is often little to no leverage to negotiate these contracts, which can expose individuals to increased privacy risk. Thus, data subjects largely depend on data controllers to ensure that cloud providers have adequate privacy safeguards and controls. In addition to contracts, cloud providers can design their services to provide cloud consumers (and data subjects) with increased control over their data. These service contracts and cloud designs can be developed to address specific privacy risks that arise in the cloud, which include:

- Where is the data geographically stored and how does this location affect the rights of the cloud consumer and the data subjects?
- What is the scope of third-party access to the data (e.g., to what extent can the data be repurposed or data mined by the cloud provider or its affiliates)?
- What security practices are used by the cloud provider to minimize access and ensure availability, confidentiality and integrity of the data?
- How long is the data retained for and how are backups managed? This includes competing rights and duties to minimize and prolong data retention.
- How is individual consent and access managed by the service? This includes allowing data subjects to receive copies of the data under certain conditions, or allowing them to correct erroneous data.

In the second section of this chapter, we will define privacy in three important ways: what are the prevailing legal definitions of privacy, what is the difference between privacy and security, and what are the common, high-level privacy principles that drive regulation and the emerging paradigm of privacy by design? In the third section of this chapter, we will examine how answers to the above questions affect privacy risk to individuals and provide developers with available controls to address some of those risks in part four. With knowledge of the privacy risks and controls in mind, there are several tools available to cloud providers, data controllers and data processors that can be used to help them ensure privacy in their clouds. These tools include mapping Information technology (IT) privacy and security controls to their data practices and conducting privacy threat analyses and impact assessments. We will describe these tools in more detail in the fourth section of this chapter. In the fifth section, we review emerging research results that hold promising, new opportunities for addressing particular aspects of privacy in the cloud, such as anonymity, secure storage, and privacy-preserving information flow.

21.2 What is Data Privacy?

The concept of privacy is multifaceted, pluralistic, and has roots that are well over a century old. A common misconception among IT professionals and computer scientists is that privacy is equivalent to data protection or confidentiality, which is only one part of what constitutes personal privacy. In the United States, privacy scholars frequently begin with the Warren and Brandeis (1890) definition of privacy as "the right to be let alone," which is complemented by Alan Westin's (1967) four states of individual privacy: *solitude*, which is experienced by individuals who are separated from the group, free from the observation of peers; *intimacy*, which is the shared experience of a small unit of two or more people; *anonymity*, which is experienced by the individual in public, yet still free from identification and surveillance; and *reserve*, which is the creation of a psychological barrier against unwanted intrusion, including holding back communication. Protecting privacy includes ensuring that individuals can obtain solitude, intimacy, and anonymity, while diminishing the need for frequent uses of reserve. In practice, individuals engage in reserve when they decide not to use a service, or when they filter their communications due to fear of unwanted surveillance. In our example in Figure 21.1, customers of the online video rental service may not watch certain videos, or may not share their viewing history with friends, if they feel the list could expose them to unwanted attention. Users may engage in reserve by fabricating or obfuscating their data to conceal their preferences and avoid exposure.

Brandeis, Warren, and Westin emphasize individual control, which minimally affords a person the ability to restrict access to information about him or herself, or to restrict interference with freedom of movement and speech. Individual control allows people to explore their identity, make mistakes and build strong personal relationships without the threat of social or political pressure to conform (Cohen, 2000). With respect to the cloud, personal freedom and autonomy can be limited in many ways. Inaccurate data can lead to decisions that restrict a person's freedom, such as denying a person credit or denying a person employment based on false financial data. Poorly designed privacy settings can lead to unwanted information exposure, in which data from an intimate online communication leaks into public spaces. The public reaction could be forced isolation or cyberbullying depending on the perception of the data. In cloud architectures composed of third-party services, responsibility for reducing threats to privacy may be shared across multiple business relationships but ultimately accountability resides with the data controller. For a taxonomy of privacy harms that can arise in information systems in general, see Solove (2010).

Privacy harms have been further characterized by Calo (2010) as either *objective harms*, which result when a person experiences emotional or financial distress due to actual third-party observations, and *subjective harms*, which occur when the person fears that a revelation can occur without specific knowledge that it has or will occur. Subjective harms, which are perceived but unconfirmed, can create the kind of reserve that Westin envisions and that Cohen warns could restrict experimentation, innovation, and creativity. Cloud providers can reduce subjective harms by increasing transparency in their data practices and offering assurances to data subjects that their personal data is protected and in their control. In third-party architectures, these assurances must be propagated through to the data subject, as the cloud provider may not have a first-party relationship with the data subject. In cloud-based services, the lack of transparency, the lack of affirmative statements about data protection and the lack of privacy controls can all increase subjective harms when sensitive personal information is collected. In addition to the difficulty in measuring appropriate compensation for privacy harms, there are different assumptions about whether privacy is a human right (e.g., in the European Union) or whether privacy is only a statutory requirement (e.g., in the United States).

Privacy has been tacitly framed as a "subset of security" by some computer security experts. Computer security concerns the data properties of confidentiality, integrity and availability, which are ensured through functionalities such as authentication, access control, data retention, storage, backup, incident response and recovery, among others. Confidentiality is sometimes confused with privacy and assumed to be guaranteed through access control and encryption; however, as we discussed above, privacy is only partly about protecting the confidentiality of personal data and can include increasing data subjects' awareness and control over their personal data and avoiding misrepresentation by protecting the integrity of their data. While security functionalities can be used to limit unwanted surveillance, security controls alone cannot address the broader privacy design challenges: how to determine who should have access (e.g., family, friends, employers, the government) and when access can lead to privacy harms (objective or subjective), and how much participation is appropriate for individuals in the use and disclosure of their personal data.

As an alternative to relying solely on security controls as a means to implement privacy, system architects and developers can rely on standard privacy principles as high-level guidelines to focus their design decisions on privacy. We highlight three prominent standards that contain commonly accepted privacy principles:

- the OECD Guidelines on the Protection of Privacy and Transborder Flows of Personal Data enacted in 1980;
- the Fair Information Practice Principles (FIPPs) first articulated in the US Department of Health, Education and Welfare report *Records, Computers and the Rights of Citizens* (1973).
- the Generally Accepted Privacy Principles (GAPPs) promoted by the American and Canadian professional associations for certified public accountants.

Each of these standards is similar, with significant overlap in the kind and number of topics. The OECD guidelines underpin most of the world's data protection laws and serve as an international standard, whereas the FIPPs are primarily used in US government and industry self-regulatory practices. The common privacy principles described in these standard are:

- notice, or openness: prior to collection, notify individuals about how their personal information will be used;
- consent: whenever possible, avoid using personal information without first obtaining an individual's consent;
- purpose specification: specify the valid purposes for which personal data will be used;
- data minimization, or collection limitation: limit the collection and storage of personal data to minimal, necessary purposes;
- use and transfer limitation: limit the use and transfer of personal data to those purposes for which it was originally collected;
- individual participation, or access: whenever possible, allow individuals to access and correct data about themselves;
- security: protect personal data against unauthorized access and loss;
- data quality: ensure data is accurate and up to date;
- accountability: implement privacy practices and mechanisms, monitor privacy and take corrective action when principles are violated.

There are several challenges that arise when implementing privacy principles in the cloud. In a simple cloud architecture, the cloud provider has a first-party relationship with the data subject, so the provider can more easily manage the consumer-facing privacy principles of notice, consent, and access. This is because the cloud provider has "design control" over the cloud service implementation. In more complex architectures, however, the cloud provider is a third party and the cloud consumer contracts with the provider to ensure that certain privacy principles are properly implemented. While the cloud consumer and data controller are responsible for specifying the purposes for which personal data will be used (the purpose specification principle), the cloud provider must still implement several principles, such as use and transfer limitation, security, data quality and accountability, as appropriate. Moreover, if significant data processing is conducted by the cloud provider, such as combining data from other sources, then the cloud consumer may require additional control and transparency into the cloud provider's practices to ensure individual notice, consent, and access are consistent with the service implementation. In summary, there can be significant tension between a service provider's desire to offer a set of generic, streamlined services to the largest possible market wherein privacy is treated homogenously, and cloud consumers' desire to customize privacy for their business practices and first-party consumer relationships. That said, it is reasonable for cloud providers to tailor offerings with custom IT controls for privacy and security that are appropriate for the kind of personal data being processed, whether it be financial, health, retail, and so on.

21.3   Risks to Privacy in the Cloud

Privacy threats differ according to the particular cloud scenario and sensitivity of the data involved. In particular, the Article 29 Working Party (European Commission, Directorate General of Justice, 2012) highlights the loss of control and transparency that results from insufficient information, such as the risks of unauthorized access to personal data, and of vendor lock in and vendor demise. This makes the task more difficult of selecting a suitable service with an appropriate risk tolerance from the vast choice of cloud offerings.

As discussed above, reasonable security must be used to protect personal data and many security issues for cloud computing may be viewed as privacy issues. These security issues include the difficulties of enforcing data protection within cloud service provision systems, lack of security awareness and expertise, and unauthorized usage. Indeed, there are a number of security issues for cloud computing, some of which are new, some of which are exacerbated by cloud models, and others that are the same as in traditional service provision models – as considered in detail for example within (Catteddu and Hogben, 2009; Cloud Security Alliance, 2011). The security risks depend greatly upon the cloud service and deployment model being used. At the network, host and application layers, security challenges associated with traditional IT computing are generally exacerbated by cloud computing; they are usually not specifically caused by cloud computing, although there are cases where this can arise, such as when security breaches are caused by reducing cloud service costs through multitenancy, which is the operation of multiple cloud consumers on the same cloud infrastructure. Other new security vulnerabilities within the cloud that can affect privacy include compromise of the management interface and hackers using public cloud systems to launch massive attacks.

It could be argued that the fundamental challenges tend to be more about the business model rather than the technical issues. In particular, an organization moving to a public cloud is similar to an organization deciding to move to an outsourcing arrangement, and the associated complexities, including service-level agreements, vendor lock in, e-discovery, auditability, and so on, are often more difficult to address than the technical challenges, which are generally similar to datacenter security problems (apart from issues associated with isolation failure, which occur when resource demands of one customer affect another). Many current problems are related to immature technology but this is expected to change over time.

Major privacy risks include the following concerns:

- *Lack of transparency, assurance and accountability*, especially within complex cloud service ecosystems where it is difficult to assess the adequacy of the cloud provider's security controls. Although many find the management system in the ISO 27001 standard to be appropriate for cloud suppliers, some cloud consumers find inadequacies in this standard with respect to security in cloud services as it is not designed for a cloud context. This immaturity has been a barrier to the wide adoption of cloud services, but it is starting to be overcome, for example with the CSA Open Certification Framework (OCF).
- *Lack of clear responsibility* arises in the division of responsibility within cloud supply chains. There is uncertainty about who must ensure that data has been properly destroyed, who controls retention of data, how to know that privacy breaches have occurred and how to determine fault and liability in such cases. In the case of data deletion, care must be taken to delete data and virtual storage devices, especially with regards to device reuse. This division of responsibility is hampered by the fact that cloud application programmer interfaces (APIs) are not yet standardized.
- *Lack of trust*, in part, is triggered by lack of user control and the fear of unauthorized secondary usage of or third-party access to personal data. For example, government surveillance is a major cloud customer

concern, especially after the exposure of two previously undisclosed mass surveillance programs (i.e., the PRISM and Tempora programs) that have a connection to cloud computing in that information collected by certain US-based companies about European Union (EU) citizens was made available to the US and UK security services.

- *Regulatory challenges*. Cloud computing involves environments with data proliferation and global, dynamic data flows that create compliance challenges when meeting complex regulatory requirements and upholding privacy rights. Cloud computing faces the same privacy issues as other service delivery models but it can magnify existing issues, especially transborder data flow restrictions, liability and the difficulty in knowing the geographic location of data processing and which specific servers or storage devices will be used. Individual data subjects' rights to access data and consent management are often not adequately addressed within cloud service provision, and the resulting threats to privacy are especially an issue when combined with big data analytics. Indeed, big data predictive analysis techniques can produce personal data by making inferences based on observational behavior that might be very sensitive and even damaging – for example, leading to discrimination. At present, these analytics are not protected by data protection measures in their current form and the data subjects may not even be aware that this is happening.

In addition to the above risks, see Ko *et al.* (2012) for a survey of cloud failures that can impact privacy, from insecure APIs to data loss and hardware failure. Within a cloud ecosystem, issues from one cloud provider may have ramifications further upstream in a supply chain, for example, in terms of loss of governance. Loss of governance may arise as the cloud consumer cedes control to the cloud provider and service level agreements do not offer the level of commitment necessary to provide critical security services on the part of the provider, thus leaving a gap in security. There are many ways in which there can be data loss or leakage involving IaaS, PaaS and SaaS providers due to security gaps.

21.4 IT Best Practices for Addressing Privacy

Information technology (IT) best practices have emerged to support privacy, including privacy impact assessments (PIA), privacy risk analysis, IT control sets and mappings, and model contract language to comply with the EU Data Protection Directive. These practices aim to address privacy across the data lifecycle, from data collection, use, retention and transfer to destruction. Standard security techniques should be leveraged to achieve privacy: not only mechanisms for confidentiality of information (e.g. using encryption in transit) and integrity provision (e.g. using digital signatures) but also access control, identity management, incident management and data breach notification, among others (Cloud Security Alliance, 2011).

Accountability can be further supported through the use of auditing mechanisms, to create a high integrity record of who had access to the data, at what times, and for what purposes. Among technical measures, cloud consumers may utilize private or virtual private clouds, anonymization, encrypted storage and just-in-time decryption, among others (Mowbray and Pearson, 2012). With encrypted storage, current best practices recommend that the cloud consumer retain the encryption keys as opposed to letting these keys reside with the cloud provider. In addition, PaaS and SaaS providers can partition data to reduce the risk that one service can gain access to the data of another service, which is a multitenancy issue. Overall, these practices and measures target different stages of the software development lifecycle and can be used to ensure privacy in the cloud to varying degrees.

The PIA can be used as an audit tool from the early design stage of requirements and architecture to well after deployment. The typical PIA consists of a list of questions that a privacy analyst completes through interviews with key stakeholders, including IT managers, IT administrators, developers and users, or anyone with knowledge of the low-level technical design and operations of the system. Typical questions are designed to map the privacy principles, such as the OECD guidelines and Fair Information Practice Principles, onto the system as well as to identify all the relevant information flows within the system and to connecting systems. Example questions include:

- *What is the purpose of the system?* This includes for what business practices and other activities the system will be used (e.g., processing employee timecards, payroll and benefits, or processing retail sales transactions).
- *From whom and to whom does the system collect, use, retain and disseminate information?* This includes internal and external actors or systems that communicate with the system under assessment, such as third parties.
- *What kinds of information about individuals can be collected, created, and retained?* This includes information stored in databases as well as log files and appropriate references to those locations.

The PIA is a starting point for conducting the privacy risk analysis that consists of four basic elements (different names may be used in other risk models): the *privacy threat*, which is an agent or activity that can exploit a *privacy vulnerability*; exploited vulnerabilities can lead to *privacy harms* of individuals, who are typically data subjects; and the *response* taken by the data controller or processor to avoid, transfer, mitigate or accept the risk. For example, a company's act to share their customer purchase histories with a third party is a vulnerability, if the third party should repurpose the information to conduct analysis that reveals a medical condition of an individual (the threat). If the third party uses this revelation to market to the individual in ways that further disseminate knowledge of the medical condition to others, then the individual would suffer an exposure, which is a kind of privacy harm. Solove (2010) introduces several privacy harms that can be used to identify this and other privacy risks.

Responding to risks may include a combination of administrative and technical controls. Administrative controls include policy decisions, such as appointing a chief privacy officer or privacy manager, conducting routine privacy training and audits (e.g., PIAs), or ensuring policies enforce the least privilege principle. Technical controls include steps to implement software and hardware-based measures, such as using access control and encryption to reduce unwanted disclosure, or reducing the retention of personally identifiable information by sanitizing log files. The US National Institute of Standards and Technology publishes a privacy control catalogue in Appendix J of Special Publication 800-53, Revision 4. This catalogue includes several administrative and technical controls that are common best practices in privacy. In addition, the NIST guidelines provide a broad overview of best practices for addressing privacy in public cloud computing (Jansen and Grance, 2011).

As part of their larger compliance program, many companies maintain control mappings between their system requirements, business practices, privacy laws and international standards, such as the ISO 27002, which include security techniques for security management. Privacy laws vary widely: the EU has established an omnibus privacy law, which covers any personal data, whereas the United States employs a patchwork of laws that separately cover specific information types or industries (Gellman, 2009). Security standards and control mappings may be reused to implement a subset of privacy (i.e., use limitation through access control, or accountability through logging access to personal data); however, as we discussed above, privacy is much broader than security and requires a specific focus on those privacy harms that are relevant to a particular cloud-based service to identify the most appropriate response. Consider the following requirements REQ-47 and REQ-48, which restrict access to data in a fictional cloud-based service:

REQ-47: The system shall restrict uses of personal data to only those purposes for which the information was collected

REQ-48: The system shall not retain customer transaction data longer than 60 days, unless required by subpoena or court order

Table 21.1 *Example control matrix that maps privacy laws and standards to system requirements*

Privacy laws and standards	REQ-47	REQ-48	...
Data minimization			
NIST SP 800-53, App. J, DM-2 Data Retention and Disposal		X	
OECD Use Limitation Principle	X	X	
GAPP 5.2.1 Use of personal information	X	X	X
Government access			
Stored Communications Act (SCA) U.S.C. §2704 (a)(1)		X	
Gramm-Leach-Bliley Act (GLBA) §313.15(a)(4)			

Note: Itemized controls appear along the left-hand side, while requirement identifiers appear along the top; where a control maps to a requirement, an "X" appears in the matrix.

Table 21.2 *Example trace link from a source requirement to a target control*

Source	Target	Rationale description
REQ-48	GAPP 5.2.1	The GAPP 5.2.1 provides an exception for any disclosure that is required by law or regulation
REQ-48	SCA §2704(a)(1)	An administrative subpoena can be used to require a company to preserve backups for an extended period of time

Note: Each link includes the appropriate references and a rationale to explain why the requirement supports the target control.

The cloud designer can map these requirements to relevant privacy laws and standards and record these mappings in matrices similar to the example in Table 21.1. These matrices complement the traditional requirements of trace matrices that are normally used in software development to trace requirements to software and hardware specifications.

In addition to control mappings, privacy analysts can preserve the rationale that they used to justify the mapping. Table 21.2 presents an example rationale to justify the mapping between REQ-48 and the GAPP 5.2.1 control and the SCA legal requirement from Table 21.1.

In addition to privacy controls and trace matrices, companies may use model contract language to govern information flows to third parties, such as between a data controller and a data processor (European Commission, 2015). Model contracts are used as one mechanism to meet the adequacy standards under the EU Data Protection Directive for transfers of personal data outside the EU and are an alternative to US Safe Harbor membership. An alternative approach to model contracts is to use binding corporate rules (BCRs), which were adopted by the Article 29 Working Party – the body that defines data protection rules in Europe – by which a corporate group can ensure adequate protection for transfers of personal data between EU and non-EU group members in compliance with the EU Data Protection Directive. The BCRs comprise a binding internal agreement or contract that obligates all legal entities within a corporate group who have access to EU personal data to adhere to all obligations of the EU Data Protection Directive, including evidence of supporting measures to ensure compliance. Both data controllers and data processors that participate in a corporate group may use BCRs to comply with the directive. Without BCRs and due to transborder data flow restrictions, it is often necessary to use many different model contracts (up to several hundred contracts in a single cloud configuration), which can be cumbersome: each contract can require 1–6 months to establish, and they must be maintained during the entire lifetime of data processing. Hence, while model contracts afford certain benefits, they are not well suited to global and dynamic business environments. The Asia-Pacific Economic

Cooperation (APEC) adopted a similar approach to BCRs in what are called Cross-Border Privacy Rules (CBPRs). More broadly, there is ongoing development of privacy and security certification for cloud service providers to help cloud consumers with the selection process.

21.5 Recent Research in Privacy

Research in data anonymity, secure storage and information flow modeling may hold promise for protecting personal privacy in future cloud services. Sweeny introduced *k-anonymity*, which is an algorithm for determining whether a person described in a dataset can be distinguished by at least *k-1* other individuals also described in the dataset (Sweeney, 2002). The algorithm makes several assumptions that have been challenged and further improvements have been proposed, such as *l-diversity* (Machanavajjhala *et al.*, 2007) and *t-closeness* (Li *et al.*, 2007). These early approaches all seek to address a common threat to privacy in databases, which is the use of auxiliary information known by an attacker to reidentify previously anonymized records. In *k*-anonymity, for example, Sweeny shows that a public voter registration database can be used to reidentify patient diagnoses, procedures, and medications from an anonymized health database. More recently, Dwork proposed *differential privacy*, which is based on statistical databases and aims to provide statistically accurate answers to database queries without releasing information that is specific to any one individual (Dwork, 2006). In other words, the answer is statistically the same, regardless whether any one individual's personal information were removed from the database. Each of these approaches is computationally intensive; however, practical implementations are possible for datasets with specific characteristics.

Protecting data in the cloud often begins with encryption, such as full disk encryption and encrypted network connections using the secure socket layer (SSL). Gentry (2009) proposed an implementation of *homomorphic encryption* as an alternative to full disk encryption, which is the ability to perform queries on encrypted data without decrypting the data. At present, homomorphic encryption is computationally impractical for many applications. Alternatively, Song *et al.* (2012) argue for a comprehensive approach that integrates access control, key management, and audit logs into a single, reusable, and scalable cloud service. If encryption is still desirable, however, Stefanov and Shi (2013) discovered an oblivious RAM protocol that allows a cloud consumer to store sensitive data across multiple, noncolluding clouds in a way that is confidential, asynchronous, distributed, and scalable. The approach relies on shuffling storage blocks in a seemingly random pattern so that the cloud provider cannot infer which blocks correspond to what type of data or how the data is being used. This approach has been evaluated using Amazon E2 cloud services to demonstrate its practical feasibility.

As discussed above, accountability in practice often begins with privacy impact assessments and mapping privacy laws and standards to system requirements. Researchers have sought to formalize related dimensions of accountability to prove privacy guarantees for a system. Contextual integrity was first described by Helen Nissenbaum as a means to align the flow of personal information with acceptable norms (Barth *et al.*, 2006). For example, patients are likely comfortable sharing their medical information with their doctor and her assistants for the purpose of diagnosing an illness but they would be less comfortable sharing this information with an employer. Barth *et al.* (2006) formalize contextual integrity in temporal logic in the form of allow and deny rules, similar to role-based access control, with the aim of proving that data is not shared in violation of stated norms. In cloud services, a significant challenge arises because individual service providers have little access to third-party practices. Breaux *et al.* (2013) introduce a formal language based on description logic for specifying privacy requirements governing data collection, use and transfer that can be used to trace data flow across multiple services and check whether third parties are repurposing data solely through the use of these specifications (Breaux *et al.*, 2013). This approach protects the confidentiality of third-party practices but allows developers to identify potential conflicts needed to take corrective action.

21.6   Conclusion

In this chapter, we introduced the topic of ensuring privacy in the cloud. Emerging privacy challenges largely depend on the type of cloud architecture (service and deployment model) used and the relationship between the cloud provider and consumer, and the roles of data controller, data processor, and data subject. Evaluating how well a particular cloud configuration and implementation addresses privacy depends on prevailing definitions of privacy, which go beyond traditional notions of confidentiality, which is commonly found in security. In addition to controlling who has access to our personal data, privacy includes the ability to isolate oneself and control one's image, to freely express one's opinions and develop a sense of identity without unwanted intrusions or fear of surveillance or other forms of exposure. To simplify matters, cloud providers may rely on established privacy principles, such as the OECD guidelines, which provide general guidance in developing privacy-preserving data practices. While these principles have been relatively timeless, the cloud introduces specific risks that must also be considered, such as lack of transparency, accountability, responsibility, and trust, and the complex regulatory environment that arises from geographically distributing data processing across multiple jurisdictions. Lack of customer trust and regulatory complexity in global business environments are difficult issues to tackle because of the underlying complexity across multiple dimensions of cloud computing and the interdisciplinary nature of the problem (i.e., legal, IT systems, security). While location matters from a legal point of view, data processing and data flows are dynamic, global, and fragmented: there are restrictions about how information can be sent and accessed across boundaries, but in cloud computing data will flow along chains of service providers both horizontally between SaaS providers and vertically, down to infrastructure providers, where the information can be fragmented and duplicated across databases, files and servers in different jurisdictions. Finally, we briefly described several IT best practices that can be used to align one's system requirements with privacy laws, standards, and guidelines, and we briefly surveyed emerging research in privacy.

References

Barth, A., Datta, A., Mitchell, J. C., and Nissenbaum, H. (2006) *Privacy and Contextual Integrity: Framework and Applications*. Proceedings of the IEEE Symposium on Security and Privacy. IEEE, pp. 184–198.

Breaux, T. D., Hibshi, H., and Rao, A. (2013) Eddy, a formal language for specifying and analyzing data flow specifications for conflicting privacy requirements. *Requirements Engineering Journal*. doi: 10.1007/s00766-013-0190-7

Calo, M. R. (2010) The boundaries of privacy harm. *Indiana Law Journal* **86**(3), 1131–1162.

Catteddu, D., and Hogben, G. (eds.) (2009) *Cloud Computing: Benefits, Risks and Recommendations for Information Security*. ENISA, Heraklion.

Cohen, J. E. (2000) Examined lives: Informational privacy and the subject as object. *Stanford Law Review* **52**, 1373–1438.

Cloud Security Alliance (2011) *Security Guidance for Critical Areas of Focus in Cloud Computing V3.0*, http://www.cloudsecurityalliance.org/guidance/ (accessed December 7, 2015).

Cloud Security Alliance (2013) *The Notorious Nine – Cloud Computing Top Threats in 2013*, http://www.cloudsecurityalliance.org/topthreats (accessed December 7, 2015).

Dwork, C. (2006) *Differential Privacy*. Proceedings of the 33rd International Colloquium on Automata, Languages and Programming (ICALP '06). LNCS, pp. 1–12.

European Commission (2015) Model Contracts for the transfer of personal data to third countries. http://ec.europa.eu/justice/data-protection/international-transfers/transfer/index_en.htm (accessed December 21, 2015).

European Commission, Directorate General of Justice (2012) *Opinion 05/12 on Cloud Computing*, http://ec.europa.eu/justice/data-protection/article-29/documentation/opinion-recommendation/files/2012/wp196_en.pdf (accessed December 21, 2015).

Gellman, R. (2009) *Privacy in the Clouds: Risks to Privacy and Confidentiality from Cloud Computing*. World Privacy Forum, http://www.worldprivacyforum.org/www/wprivacyforum/pdf/WPF_Cloud_Privacy_Report.pdf (accessed December 15, 2015).

Gentry, C. (2009) Fully homomorphic encryption using ideal lattices. Proceedings of the 41st Annual ACM Symposium on the Theory of Computing (STOC 09), pp. 169–178.

Jansen, W. and Grance, T. (2011) *Guidelines on Security and Privacy in Public Cloud Computing.* Special Publication 800-144, NIST, http://csrc.nist.gov/publications/nistpubs/800-144/SP800-144.pdf (accessed December 21, 2015).

Ko, R. K. L., Lee, S. S. G., and Rajan, V. (2012) Understanding Cloud Failures. *IEEE Spectrum* **49**(12), 84.

Li, N., Li, T., and Venkatasubramanian, S. (2007) t-closeness: Privacy beyond k-anonymity and l-diversity. Proceedings of the International Conference on Data Engineering (ICDE '07), pp. 106–115.

Machanavajjhala, A., Kifer, D., Gehrke, J., and Venkitasubramaniam, M. (2007) l-diversity: Privacy beyond k-anonymity. *ACM Transactions on Knowledge Discovery from Data* **1**(1), Article 3.

Mell, P. M., and Grance, T. (2011) *The NIST Definition of Cloud Computing.* Special Publication 800-145. NIST, Gaithersburg, MD, http://www.nist.gov/customcf/get_pdf.cfm?pub_id=909616 (accessed November 25, 2015).

Mowbray, M., and Pearson, S. (2012) Protecting personal information in cloud computing, in *On the Move to Meaningful Internet Systems: OTM 2012* (eds. R. Meersman, H. Panetto, T. Dillon *et al.*). Springer, Berlin.

Stefanov, E., and Shi, E. (2013) ObliviStore: High performance oblivious cloud storage. Proceedings of the IEEE Symposium on Security and Privacy, pp. 253–267.

Solove, D. J. (2010) *Understanding Privacy*, Harvard University Press, Cambridge MA.

Song, D., Shi, E., and Fischer, I. (2012) Cloud data protection for the masses. *IEEE Computer* **45**(1), 39-45.

Sweeney, L. (2002) k-anonymity: a model for protecting privacy. *International Journal of Uncertainty, Fuzziness and Knowledge-Based Systems* **10**(5), 557–570.

Warren, S. D. and Brandeis, L. D. (1890) The right to privacy. 4 *Harvard LR* 193.

Westin, A. (1967) *Privacy and Freedom*, Atheneum, New York, NY.

22

Compliance in Clouds

Thorsten Humberg and Jan Jürjens

Fraunhofer Institute for Software and Systems Engineering, Germany

22.1 Introduction

The requirement to ensure security and compliance of systems and processes has become an important consideration when leveraging information technology for business operations. Violations of compliance requirements can have serious consequences, including legal and contractual penalties, loss of business growth, and loss of reputation among customers.

Ensuring and managing compliance is a complex multidisciplinary challenge (Rath and Sponholz, 2009). It incorporates the management of, for example, general legal regulations, IT-specific standards and planning, and management of business processes within a company. We will provide an overview of general as well as IT-related compliance and security regulations in section 22.2.

Cloud-based services have seen a rapid growth in popularity in recent years. The cloud has significant potential in terms of increased efficiency of IT administration while reducing costs. Nevertheless, many companies are reluctant to outsource existing services to cloud providers, largely due to the concerns regarding security and compliance issues. In this chapter, we address these concerns and introduces the concept of certificates as a means to evaluate service providers with respect to their compliance.

22.2 IT-Related Compliance

The term "compliance" is not specific to the information technology (IT) context, and can be defined differently depending on the way it is approached. In its broadest sense, it means adherence to regulatory specifications of any kind.

Encyclopedia of Cloud Computing, First Edition. Edited by San Murugesan and Irena Bojanova.

From a commercial perspective, the frequently encountered financial and accounting issues related to compliance requirements are predominantly from the Sarbanes-Oxley Act (SOX) and the Gramm–Leach–Bliley Act (GLB). Such regulations do not address information technology aspects directly but IT plays an important role in providing the necessary means to implement the requirements (Herrmann, 2007).

In addition to these general requirements, businesses in some domains are subject to specific regulations, like the Health Insurance Portability and Accountability Act (HIPAA) for the healthcare domain.

As per the general definition, "IT compliance" means the implementation and operating of IT systems according to the relevant regulations. Those might be the general requirements mentioned above if information technology is used in their implementation, or IT-specific regulations, for example regarding security aspects.

The IT compliance regulations that have recently seen a significant rise in public awareness are those concerning data privacy. These regulations are a complex field in themselves, as they strongly depend on the jurisdiction where IT infrastructure is located, which can lead to inconsistent or even conflicting requirements (Mather *et al.*, 2009). The dynamic nature of the physical distribution of data in cloud computing environments further complicates this problem.

IT-relevant compliance regulations can be divided into four groups, depending on their scope and issuer (Strasser and Wittek, 2012):

- Legal regulations and laws, issued by a public authority, which have an effect on IT. Typically, the regulations in this category concern IT aspects only indirectly, like the Sarbanes–Oxley Act mentioned above.
- Contracts with suppliers or customers, either concerning IT aspects explicitly, or merely influencing them.
- Internal regulations, issued by the company itself. This includes procedures and regulations established in order to meet external requirements. For example, creating and deploying a policy for the use of passwords is a compliance requirement in implementing the information security standard ISO 27002. Many companies define their own ethical codes as a high-level orientation for their personnel.
- External regulations, such as standards and best practices, which were not issued by an official legislative authority. Examples are standards by the ISO and equivalent organizations. The significant compliance regulations related to security standards and certificates are discussed in section 22.4.

22.3   IT-Compliance in Cloud Environments

Cloud services can be considered as an additional layer on top of a classic IT infrastructure. This means that compliance and security for clouds can only be achieved if the underlying systems are compliant and secure in themselves. Compliance, especially when it comes to personally identifiable information (PII), is in many cases concerned with the restriction of access to certain types of data. In cases where an attacker can gain control over the systems processing the data, the attacker can access any cloud software running on them directly and the restrictions cannot hold. This implies that information security is an important prerequisite for IT-compliance.

The extra layer introduced by cloud software can also raise new issues regarding compliance, some of which will be presented in this section.

In order to achieve compliance, it is necessary to define clearly the kind of data that is processed by the service, who is allowed to access them, and for what purposes. With cloud service providers, and possibly even more subcontractors, more stakeholders are involved in storing and processing data, leading to an increased need to define and enforce rules regarding ownership and accountability of data.

In general, it is the cloud customer's responsibility to make sure that data is handled in an appropriate way. A cloud service provider (CSP) often does not even have enough knowledge about the kind of data stored and processed by its customers to determine which compliance regulations are applicable (Golden, 2010). It is therefore not realistic to expect the CSP to offer a guarantee for compliance, or take responsibility for it.

To ensure that a cloud service provider is sufficiently reliable to be chosen, it is generally necessary for the customer to thoroughly evaluate the level of compliance and security the CSP offers. But in most cases, the actual systems of the provider are inaccessible for the end user, and thus cannot be assessed by him directly. The dynamic nature of resource provisioning in cloud concepts makes assessments by individual customers even more impractical. Furthermore, the CSP that the customer interacts with might work with even more subcontractors. For example, a common business model is for providers of SaaS solutions to host their services on systems that are themselves rented from an IaaS provider.

From the customer's perspective, one of the benefits of the cloud concept is flexibility in choosing the provider, which contrasts with the need to invest much effort into assessing individual cloud providers for their respective compliance levels.

Both problems can be mitigated by the use of certificates for approved compliance standards, as discussed in the next section.

In cloud environments, compliance issues arise from the physical locations of servers, which could be opaque to the end user. Especially for data privacy regulations, physical locations of servers play an important role. Inappropriate locations can violate certain requirements in themselves, but can also lead to more regulations becoming applicable. Again, this problem is intensified when subcontractors are involved, whose data-centers' locations might not be known or restricted appropriately. A common way to handle this concern is to define acceptable locations or regions in the SLA of the particular service explicitly.

So far, we have considered a number of reasons that complicate security and compliance management in cloud environments. The implication would be that staying with the traditional model of internal IT infrastructure (or at most a cloud following the private cloud deployment model) is the superior approach. This is not necessarily the case.

Maintaining security and compliance for any IT environment is a challenging, time-consuming, and thus expensive task. Neglecting this task, for example by not being able or willing to provide the resources necessary, generally leads to insecure systems.

On the other hand, the cost scales well with larger systems: the bigger the IT infrastructure, the more efficiently security can be handled. This is the point where cloud concepts can lead to a gain in security and compliance. By offering similar setups to a number of customers, a cloud service provider can develop and apply a sophisticated security concept to a (possibly large) number of systems in a standardized way.

Whether using a cloud service will lead to an increase or decrease in security strongly depends on the actual case at hand. In particular, the question of which requirements have to be met at all strongly depends on the way a customer would use the services.

22.4   Security Standards and Certificates

Within the context of information technology, and especially regarding information security and compliance, certificates are an established mechanism to create trust between a provider and its (potential) customers.

Note that in the field of information technology, the term "certificate" is ambiguous. In this article, we use it to mean a way to assure compliance with a given regulatory standard. We use it neither in its cryptographic context, nor as a proof of personal qualification.

For cloud computing, the concept of certificates can be used to mitigate some of the problems that arise in terms of IT compliance. In most cases, a customer has little to no opportunity to examine the cloud service provider's facilities or its business processes thoroughly. Nevertheless, he is responsible for choosing a provider that offers a sufficient level of security and compliance. An appropriate certificate can be a way to offer the trust needed.

Obtaining a certificate can also be beneficial for the CSPs themselves, despite the substantial effort involved on their part. By obtaining a certificate, they can avoid discussing their security measures with each individual customer and also eliminate the need to disclose confidential internal details, or data of other customers, in order to provide assurance of security levels (Mather *et al.*, 2009).

Certificates can even be a prerequisite for using a service for specific business activities at all. A prominent example is the Payment Card Industry Data Security Standard (PCI DSS) that every company has to implement in order to be allowed to process credit card payments (PCI Security Standards Council, 2013).

Certificates can roughly be distinguished with regard to their domains:

- general information security standards and certificates;
- cloud-specific certificates;
- domain-specific certificates.

The distinction cannot always be clearly drawn, one reason being that cloud services are based on traditional IT infrastructure, as described above, and must thus adhere to general security requirements. Which standards are relevant for a CSP or its customers depends on the intended use. Some examples for the different categories will be given below.

22.4.1   General Information Security Standards and Certificates

Most regulations and standards for information technology systems are not specific to cloud environments, but are relevant to the IT systems they are based on. Probably the most commonly acknowledged standards are those of the ISO 27000 series, which cover different aspects of information security. ISO 27001 describes how an information security management system (ISMS) can be established (ISO/IEC, 2013). It also includes a set of controls that can be implemented in order to mitigate risks for an IT system; those controls are further detailed in standard ISO 27002.

Certification according to ISO 27001 is a common requirement for a provider to guarantee an adequate level of security. It is issued by accredited third-party certification bodies (ISO/IEC, 2014). How audits are to be performed is specified in a separate standard, ISO 19011 (ISO/IEC, 2011).

22.4.2   Cloud-Specific Standards and Certificates

As discussed above, using the cloud computing model for outsourcing of IT resources imposes additional concerns regarding security and compliance. In order to open up its possibilities, nonetheless, several organizations have proposed a number of cloud-specific standards. For some of those standards, corresponding certificates can be issued.

22.4.2.1   FedRAMP (Federal Risk and Authorization Management Program)

In line with the Federal Cloud Computing Strategy, US federal government agencies use cloud computing resources to reduce costs for IT services. The FedRAMP standard is intended to define a consistent level of information security for cloud service providers that is sufficient for federal agencies to use these services in particular applications (Figliola and Fischer, 2014). Compliance with FedRAMP has to be evaluated by an accredited third-party assessment organization (3PAO) (US General Services Administration, 2014).

22.4.2.2   *Security, Trust and Assurance Registry (STAR)*

The Cloud Security Alliance (CSA) has developed STAR to establish a registry of cloud service providers that offer specified levels of security. STAR defines three stages of certification, based on the Open Certification Framework, which has also been developed by the CSA (Cloud Security Alliance, 2014):

- STAR Entry – Self Assessment: providers answer a questionnaire, which is then published by the CSA.
- STAR Certification: the certification is granted by a third-party assessment organization (3PAO); the necessary criteria are based on those needed for an ISO 27001 certificate.
- STAR Continuous: the third stage requires a continuous monitoring of the CSP's security, based on the CSA Cloud Trust Protocol (CTP).

22.4.2.3   *Cloud-Specific ISO Standards*

In addition to the general security standard ISO 27001, several cloud-specific standards are currently under development:

- ISO 27017: *Information Technology – Security Techniques – Code of Practice for Information Security Controls for Cloud Computing Services based on ISO/IEC 27002.*
- ISO 27018: *Information Technology – Security Techniques – Code of Practice for PII Protection in Public Cloud Acting as PII Processors.*
- ISO 27036-4: *Information Technology – Information Security for Supplier Relationships – Part 4: Guidelines for Security of Cloud Services.*

These documents are not redundant to other standards of the ISO 27000 series but extend those by cloud-specific controls.

22.4.3   Domain-specific certificates

For some domains, there exist specific standards that any company has to implement and verify in order to act in that field. If a cloud customer intends to use a cloud service provider to perform such tasks, or the CSP plans to offer services directed at such a domain, those certificates have to be established.

A common example is the Payment Card Industry Data Security Standard (PCI DSS), which is required for systems that handle credit-card transaction data.

Similarly, US companies involved in the healthcare domain have to adhere to the Health Insurance Portability and Accountability Act (HIPAA).

22.4.4   Examples of Certificates offered by Cloud Service Providers

This section gives an overview of the current situation of popular cloud service providers regarding the information security and compliance certificates they offer. We focus on those certificates that have been presented in the previous sections.

Note that even though a CSP might meet the requirements of a specific standard, this does not imply that every usage of those services is secure and compliant as well. As mentioned in section 3, and stressed by the CSPs themselves, certification only relates to the underlying infrastructure and services. Determining whether a given service is compliant to use for a specific purpose is still the responsibility of the cloud user.

When choosing a CSP based on required certification, a customer also has to pay attention to the scope of the certificate. Cloud service providers commonly offer a number of different services. Issued certificates do not necessarily include every one of these services; their scope can instead be restricted to a subset.

In the following paragraphs, we will take three popular providers of IaaS products as examples: Amazon AWS, Microsoft Windows Azure and Google Compute Engine.

Security management, according to ISO 27001, can be regarded as a basic requirement that is fulfilled for the underlying infrastructure of any of the major providers.

All three services also claim to be compliant for healthcare applications subject to the HIPAA standard, although an additional contract in the form of a "business associate agreement" (BAA) has to be signed.

For FedRAMP, CSA STAR Audit and PCI DSS, only the services provided by Amazon AWS and Microsoft Azure have been certified. For STAR, both providers have been certified for the first-stage "STAR Entry – Self Assessment." For PCI DSS, both have been certified for the highest level 1, sufficient for merchants processing more than 6 million payment transactions per year.

22.5   Conclusion

Identifying the relevant regulations for a given business case and fulfilling the requirements is a complicated task for classic IT systems. The nature of cloud deployment models adds further difficulties:

- More stakeholders are involved, whose responsibilities have to be evaluated and defined. This not only concerns providers directly interacting with the customer but also includes potential subcontractors.
- Actual physical locations of servers and data matters a lot but become harder to determine, verify or restrict to allowed regions. This can lead to the violation of compliance requirements, especially concerning data privacy.
- The cloud provider's datacenters are generally inaccessible for customers, so they cannot examine its level of security.

On the other hand, using the experience and resources a cloud service provider can offer for establishing security and compliance can lead to an increase in security with little effort from the customer.

By establishing adequate approved certificates, it is also possible at least to mitigate the problem of not being able to access the provider's facilities. Instead of every customer having to assess a CSP separately, a single certification body asserts that the necessary measures are implemented to ensure an appropriate level of compliance and security.

Most of the existing standards make no assumptions about the computing model of the IT infrastructure used. Several cloud-specific standards have been developed, and more are currently under development. Nevertheless it is still the task of cloud customers to determine which requirements have to be met for their particular applications, and choose appropriate cloud service providers.

Additional Resources

Resources by Cloud Service Providers

Amazon and Microsoft give broad overviews of a number of compliance requirements for their AWS and Azure products, respectively, some of which have been described in this article:

http://aws.amazon.com/compliance/ (accessed December 22, 2015).
http://azure.microsoft.com/en-us/support/trust-center/compliance/ (accessed December 22, 2015).

Information for the Google Compute Engine is part of the listed features of the product:

https://cloud.google.com/products/compute-engine/ (accessed December 22, 2015).

Security and Compliance Standards

ISO 27001 series:

http://www.iso.org/iso/home/standards/management-standards/iso27001.htm (accessed December 22, 2015).

FedRAMP, and the NIST standard 800-35 it is based on:

http://cloud.cio.gov/fedramp (accessed December 22, 2015).
http://csrc.nist.gov/publications/PubsSPs.html (accessed December 22, 2015).

CSA STAR certification and the registry of participating cloud service providers:

https://cloudsecurityalliance.org/star/certification/ (accessed December 22, 2015).
https://cloudsecurityalliance.org/star/#_registry (accessed December 22, 2015).

Payment Card Industry Data Security Standard (PCI DSS):

https://www.pcisecuritystandards.org/security_standards/ (accessed December 22, 2015).

References

Cloud Security Alliance (2014) *CSA Security, Trust and Assurance Registry*, https://cloudsecurityalliance.org/star/ (accessed December 22, 2015).

Figliola, P. M. and Fischer, E. A. (2014) *Overview and Issues for Implementation of the Federal Cloud Computing Initiative: Implications for Federal Information Technology Reform Management*, s.l. Congressional Research Service, Washington DC.

Golden, B. (2010) *Data Compliance and Cloud Computing Collide: Key Questions*, http://www.cio.com/article/612063/Data_Compliance_and_Cloud_Computing_Collide_Key_Questions (accessed December 22, 2015).

Herrmann, D. (2007) *Complete Guide to Security and Privacy Metrics: Measuring Regulatory Compliance, Operational Resilience, and ROI*, Auerbach Publications, Boca Raton, FL.

ISO/IEC, 2011. ISO 19011:2011 (2011) *Guidelines for Auditing Management Systems*, ISO, Geneva.

ISO/IEC, 2013. ISO/IEC 27001:2013 (2013) *Information Technology – Security Techniques – Information Security Management Systems – Requirements*, ISO, Geneva.

ISO/IEC, 2014. (2014) *Certification – ISO*, http://www.iso.org/iso/home/standards/certification.htm (accessed December 22, 2015).

Mather, T., Kumaraswamy, S. and Latif, S. (2009) *Cloud Security and Privacy: An Enterprise Perspective on Risks and Compliance*, O'Reilly Media, Inc., Sebastopol, CA.

PCI Security Standards Council (2013) *Payment Card Industry Data Security Standard*, Version 3.0, PCI, Wakefield, MA.

Rath, M. and Sponholz, R. (2009) *IT-Compliance: Erfolgreiches Management regulatorischer Anforderungen*. Erich Schmidt Verlag, Berlin.

Strasser, A. and Wittek, M. (2012) *IT-Compliance – GI – Gesellschaft für Informatik e.V.*, https://www.gi.de/service/informatiklexikon/detailansicht/article/it-compliance.html (accessed December 22, 2015).

US General Services Administration (2014) *About FedRAMP*, http://www.gsa.gov/portal/category/102375 (accessed December 22, 2015).

Part VI

Cloud Performance, Reliability, and Availability

23

Cloud Capacity Planning and Management

Yousri Kouki,[1] Frederico Alvares,[2] and Thomas Ledoux[2]

[1] *Linagora, France*
[2] *Mines Nantes, France*

23.1 Introduction

Capacity management is a process used to manage the capacity of IT services and the IT infrastructure. Its primary goal is to ensure that IT resources (services and infrastructure) are right-sized to meet current and future requirements in a cost-effective and timely manner.

Capacity planning is the subactivity within capacity management that determines the optimal capacity of resources while capacity management is the complete management process including monitoring the performance, estimating the capacity and triggering the reallocation of resources, amongst other things. In the literature, capacity management and capacity planning are frequently used interchangeably.

In traditional computing environments, resources have to be purchased in advance, which may lead to over/underprovisioning. With the revolutionary promise of computing as a utility, cloud computing has the potential to transform the way IT services are delivered and managed. Indeed, cloud computing is a model for providing ubiquitous, convenient, on-demand network access to a shared pool of configurable computing resources that can be rapidly provisioned and released with minimal management effort. As a consequence, resources can be seamlessly and dynamically requested/released to face demand spikes/lulls while maintaining the required level of quality of service (QoS), usually formalized by service level agreements (SLAs).

Capacity management makes sure that adequate resources are available for the business at the right time and at optimal cost. With respect to the time resources should be added/released, capacity management can be conceived in three ways: (i) reactive, (ii) proactive, or (iii) hybrid. According to the last values obtained using a monitoring service, reactive solutions manage on-the-fly the capacity whereas proactive solutions predict future demand to plan the capacity. Hybrid solutions combine reactive and proactive approaches.

Encyclopedia of Cloud Computing, First Edition. Edited by San Murugesan and Irena Bojanova.

Thanks to the on-demand service-provisioning model of cloud computing, capacity management can be achieved using a simple capacity-planning method by allocating/releasing resources based on a set of rules and conditions previously defined, (e.g., Amazon Auto Scaling). However, doing it right, in a way that respects SLAs while minimizing service cost is not a trivial task because several parameters should be taken into account such as multiple resource types (e.g., physical/virtual machines), nonignorable resource initiation time and billing model granularity (hourly, daily, monthly, etc.). For that purpose, effective capacity planning may require the combination of several solutions out of different domains. For example, it may be necessary to rely on more theoretical models (such as queuing theory or reinforcement learning) to profile services' performance. Based on those models, operational research techniques can be applied to find optimal solutions so that the costs are minimized.

This chapter presents a comprehensive overview of capacity planning and management for cloud computing. First, we state the problem of capacity management in the context of cloud computing from the point of view of several service providers. Next, we provide a brief discussion about when capacity planning should take place. Finally, we survey a number of methods for capacity planning and management proposed by both practitioners and researchers.

23.2   Problem Statement

The cloud architecture is usually composed of several XaaS (Anything as a Service) layers and SLAs are characterized at various levels in this stack to ensure the expected QoS for different stakeholders. In Figure 23.1, any cloud layer, except the end user, plays a provider-consumer role: it is a provider for the upper layers and a consumer for the lower layers. Its main challenge is to maintain its consumers' satisfaction while minimizing the service costs due to resource costs and SLA penalties (if there are violations of the SLA).

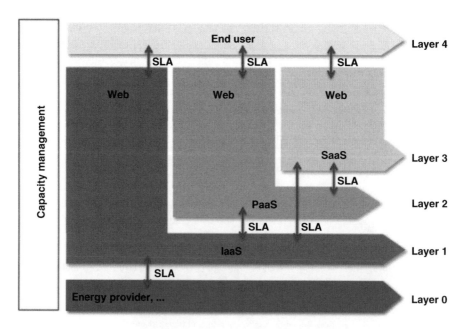

Figure 23.1   *Cloud layered architecture*

Capacity management can be seen as a generic problem and thus can be applied at each layer of the cloud stack. For instance, the energy provider aims to provide as much energy as required by its clients while minimizing the costs due to distribution and storage. The objective of IaaS providers is to allocate physical resources requested by their clients in the form of virtual resources in a way to maximize the availability and throughput and minimize costs due to energy consumption. The aim of SaaS providers is to adjust the amount of virtual resources needed to cope with varying workloads (from end users) to maintain the required level of response time and availability.

In this chapter, to better understand capacity planning and management in the context of cloud computing, we only focus on the SaaS layer, although the same ideas may be applied to services at other layers. We consider a Web application as an example, whose architecture consists of three tiers: one for load balancing, one for computing, and one for storage. In order to offer services with different QoS levels (e.g., for gold, silver, or bronze clients), the SaaS provider proposes several SLAs to its clients based on response time, availability, and financial cost. The objective of the SaaS provider is to determine the optimal capacity (number of instances) that meets SLAs at optimal cost.

For the SaaS provider, this raises the following questions:

- When should capacity planning and management take place? Should the process of allocating/releasing resources react to the recent past rather than anticipate the future? There are three main approaches: reactive, proactive, and a hybrid one.
- How should resources be adjusted? There are two main categories for scaling resources: (i) vertical scaling, also described as scale up, typically refers to resizing existing resource instances such as CPU, RAM, Disk, and (ii) horizontal scale, or scale out, usually refers to adjusting the number of instances.
- How many resources are needed to meet demand (i.e., workload requests)?
- What is the target resource? Storage instances, compute instances, balancer instances, and so forth? For the sake of simplicity, we focus only on the capacity of compute instances.

Later in this chapter, we illustrate how you can answer these questions.

23.3   When Should Capacity Planning and Management Take Place?

Capacity management is the process tasked with defining the capacity of a system, to meet changing demands, at the right time and at optimal cost. In order to calculate the capacity plan, we distinguish two main approaches: we either react when changing demands arise, or anticipate these demands. In this regard, capacity management can be implemented in three ways: (i) reactive, (ii) proactive, or (iii) hybrid (a mix of the previous approaches).

23.3.1   Reactive

This kind of capacity management is performed in reaction to events (e.g., if workload exceeds a threshold value), and involves taking decisions on the amount of resources based on the system state and monitoring data from the environment. The cloud computing provisioning model suits this approach very well because it enables on-demand and rapid resource allocation so as to adjust the capacity while respecting SLAs. This approach can be designed and implemented easily when decisions are taken based on monitored performance metrics and predefined policies. In that case, monitoring systems, event-condition-action rules, vertical/horizontal scaling make up the standard toolbox (such as Amazon Auto-Scaling) to implement capacity management. However, it may be difficult to recognize oscillating and/or unstable phases and hence to avoid the "ping-pong effect" while requesting/releasing resources that are not desirable due to a non-negligible instance initiation time (ranging from a few seconds to several minutes). In addition, as early resource reservations generally cost less than on-demand ones, proactive capacity management could plan capacity better in terms of financial cost.

23.3.2   Proactive

Proactive approaches can be used to predict future demands in order to overcome issues that may be raised by the reactive approach. We distinguish three categories: (i) *cyclic*: periodic demand that occurs at fixed interval (daily, weekly, monthly, quarterly, etc.), (ii) *event based*: demand due to a scheduled business event (new product releases, marketing campaigns, and so forth), and (iii) *prediction based*: use past history to predict future value of demand.

The later one needs a prediction method. Time-series analysis (Box and Jenkins, 1994) offers a number of methods to predict values at certain specific future times (prediction interval) based on a set of previously observed values (prediction window). A time series is a sequence of data points, typically measured at successive points in time, spaced at uniform time intervals. Time-series methods analyze time series data in order to extract meaningful statistics. Examples of time-series analysis methods are: averaging methods (e.g., moving average, weighted moving average and exponential smoothing), autoregression, autoregressive moving average (ARMA) and machine learning-based methods (e.g., neural networks).

A drawback of proactive approaches is the fact that they may be misleading due to bad prediction results. In fact, the accuracy of prediction methods depends on the input window size and the prediction interval.

23.3.3   Hybrid

In order to benefit from the advantages of both strategies and thus be capable of finely adjusting capacity, a hybrid approach (i.e., a combination of both reactive and proactive solutions) can be employed. For instance, proactive capacity management can be applied for long-term demand prediction. If there is a prediction error, the capacity can be reactively adjusted for the short term. An alternative solution is to use the full-instance period before terminating any instance (Kouki and Ledoux, 2013). These strategies not only absorb workload differences due to prediction error but also fit a full-hour billing model. In this way, the SaaS provider uses what he/she actually pays for, avoiding waste due to partial usage.

23.4   Capacity Planning and Management: Industrial Solutions

In this section, we provide a brief discussion on the industrial solutions considered in this chapter. They are essentially grounded on threshold-based rules methods, as their management must be as simple as possible so that customers can set them up and parametrize them by themselves easily (see discussion in section 23.4.4).

23.4.1   An Ad Hoc Method

The threshold-based rules method is the most popular method for capacity planning. The main idea behind this method is that the capacity might vary according to a set of rules. Each rule is based on one or more metrics such as response time, availability or CPU usage. A rule is like a pattern in the sense that it is defined by set of parameters. For example, a rule may be composed of an upper threshold *UPPER*, a lower threshold *LOWER*, and two time values (time$_{upper}$, time$_{lower}$), during which the metric is greater/lower than the corresponding threshold, and two calm durations: $calm_{add}$ and $calm_{remove}$ during which no scaling decisions can be committed in order to prevent system oscillations. Based on those parameters, the rule can be defined as follows:

If $m>UPPER$ for $time_{upper}$ then capacity=capacity+k_{add}
Do no thing for $calm_{add}$

If $m<LOWER$ for $time_{lower}$ then capacity=capacity-k_{remove}
Do no thing for $calm_{remove}$

where m is the metric, *capacity* is the current capacity, and k_{add} and k_{remove} are the capacity to add and the capacity to remove respectively.

A rule can be static or dynamic, which means that the parameters (metrics, time intervals and actions) involved in it may be modified at runtime in order to better fit to a given context. However, the static threshold-based rules method is the most popular for capacity planning and it has been used by cloud providers like Amazon EC2 and Microsoft Azure. Although most solutions use only two thresholds per metric, recent work (Hasan *et al.*, 2012) has considered using a set of four thresholds to prevent further spurious auto-scaling decisions: *UPPER*, the upper threshold; *UPPER⁻*, which is slightly below the upper threshold; *LOWER*, the lower threshold; *LOWER⁺*, which is slightly above the lower threshold; and two durations used for checking persistence of metric value above/below *UPPER/LOWER* and *UPPER⁻/LOWER⁺*. It is obvious that this kind of solution better follows changing demands.

23.4.2   Examples

This section considers the two most popular mainstream ad hoc solutions for capacity planning, namely Amazon Auto Scaling, from Amazon EC2 and Autoscaling Application Block, from Microsoft Azure. Other industrial solutions are Scalr and RightScale.

23.4.2.1   *Amazon Auto Scaling*

Amazon Auto Scaling (http://aws.amazon.com/autoscaling/, accessed December 12, 2015) is a Web service that enables a consumer to launch or terminate Amazon Elastic Compute Cloud (EC2) instances automatically based on user-defined policies, health status checks, and schedules. It is particularly well suited for applications that experience hourly, daily, or weekly variability in usage. It is enabled by Amazon CloudWatch and available at no additional charge beyond Amazon CloudWatch fees.

The basic concepts of Amazon Auto Scaling are: (i) launch configuration that specifies the type of Amazon EC2 instance that Auto Scaling creates for a consumer, (ii) Auto Scaling Group, which is a collection of Amazon EC2 instances to which a consumer wants to apply certain scaling conditions, (iii) policies that describe scaling actions, and (iv) alarms that define conditions under which a specific policy is triggered. The following commands exemplify those concepts:

Create load balancer
```
$> elb-create-lb LB --headers --listener "lb-port=80,instance-
port=9763,protocol=http" --availability-zones us-east-1a
```

Create launch configuration
```
$> as-create-launch-config LC --image-id ami-49db1a20 –instance-type
m1.small
```

Create Auto-scaling group

```
$> as-create-auto-scaling-group SG --availability-zones us-east-1a
--launch-configuration LC --min-size 2 --max-size 6 --load-balancers LB
--region us-east-1
```

Create scaling policies

```
$> as-put-scaling-policy scaleOutPolicy --auto-scaling-group SG
--adjustment=1 --type ChangeInCapacity --cooldown 300 --region us-east-1

$> as-put-scaling-policy scaleInPolicy --auto-scaling-group SG
--adjustment=-1 --type ChangeInCapacity --cooldown 300 --region
us-east-1
```

Create alarms for each condition

```
$> mon-put-metric-alarm scaleOutAlarm --comparison-operator
GreaterThanThreshold --evaluation-periods 1 --metric-name CPUUtilization
--namespace "AWS/ELB" --period 60 --statistic Average --threshold 70
--alarm-actions scaleOutPolicy --dimensions "LoadBalancerName=LB"
--region us-east-1

$> mon-put-metric-alarm scaleInAlarm --comparison-operator
LessThanThreshold --evaluation-periods 1 --metric-name CPUUtilization
--namespace "AWS/ELB" --period 60 --statistic Average --threshold 30
--alarm-actions scaleInPolicy --dimensions "LoadBalancerName=LB"
--region us-east-1
```

In this example there are two policies: one for scale out and one for scale in. These policies operate on a load balancer *LB*. The policy (*scaleOutPolicy*) increments the instance by one if the average CPU utilization over the last minute has been greater than or equal to 70%. The policy (*scaleInPolicy*) reduces the number of instances by one, if the average CPU utilization has been lower than 30% over the last minute.

23.4.2.2   *Autoscaling Application Block*

The Autoscaling Application Block (WASABi) is a part of the Enterprise Library Integration Pack for Windows Azure (https://msdn.microsoft.com/en-us/library/hh680892(v=pandp.50).aspx, accessed December 12, 2015). It can automatically scale Windows Azure applications based on rules that the consumer defines specifically for his application. The Autoscaling Application Block uses rules and actions to determine how the application should respond to changes in demand. There are two types of rules – constraint rules and reactive rules – each with its own actions:

- constraint rules enable consumer to set minimum and maximum values for the number of instances of a role or set of roles based on a timetable;
- reactive rules allow consumers to adjust the number of instances of a target based on aggregate values derived from data points collected from Windows Azure environment or application.

Rules are stored in XML documents. The following code sample shows an example rule:

```xml
<?xml version="1.0" encoding="utf-8" ?>
<rules xmlns="http://schemas.microsoft.com/practices/2011/entlib/
autoscaling/rules">
  <constraintRules>
    <rule name="default" enabled="true" rank="1" description="The
default constraint rule">
      <actions>
        <range min="2" max="6" target="ApplicationRole"/>
      </actions>
    </rule>
  </constraintRules>
  <reactiveRules>
    <rule name="ScaleOut" rank="10" description="Scale out the web role"
enabled="true" >
      <when>
        <any>
          <greaterOrEqual operand="CPU_Avg_5m" than="70"/>
        </any>
      </when>
      <actions>
        <scale target="ApplicationRole" by="1"/>
      </actions>
    </rule>
    <rule name="ScaleIn" rank="10" description="Scale In the web role"
enabled="true" >

      <when>
        <all>
          <less operand="CPU_Avg_5m" than="30"/>
        </all>
      </when>
      <actions>
        <scale target="ApplicationRole" by="-1"/>
      </actions>
    </rule>
  </reactiveRules>
  <operands>
    <performanceCounter alias="CPU_Avg_5m" performanceCounterName=
"\Processor(_Total)\% Processor Time"   source = "ApplicationRole"
timespan="00:05:00" aggregate="Average"/> </operands>
</rules>
```

In this example there are three autoscaling rules: one constraint rule and two reactive rules. These rules operate on a target named *ApplicationRole*. This role is defined in the service model. The constraint rule is always active and sets the minimum number of role instances to 2 and the maximum number of role instances to 6.

Both reactive rules use an operand named *CPU_Avg_5m*, which calculates the average CPU usage over the last 5 minutes for an *ApplicationRole*. The reactive rule (*ScaleOut*) increases the instance counter of the target role by one if the average CPU utilization over the last 5 minutes has been greater than or equal to 70%. The reactive rule (ScaleIn) decrements the instance counter of the target role by one if the average CPU utilization over the last 5 minutes has been less than 30%.

23.4.3   Limits

Threshold-based rules method is the most popular method for capacity planning. In fact, the simplicity of this method makes it more attractive. However, the definition of thresholds is a per-application task, and requires a deep understanding of the workload. This understanding is reflected in the choice of metrics and associated thresholds. In addition, this method can lead to system instability: capacity oscillations. These oscillations can be absorbed by methods using more than two static metric thresholds or dynamic thresholds (Hasan *et al.*, 2012).

Another limitation is that most solutions focus on low-level (infrastructure) performance metrics such as CPU utilization. Using low-level performance metrics in the scaling rules is a good indicator for system utilization information, but it cannot clearly reflect the QoS provided by a cloud service or show whether performance meets users' requirements. In addition, the mapping of high-level metrics (expected by the end user) to low-level metrics (provided by IaaS provider) is difficult and requires some infrastructure expertise.

23.4.4   Discussion

It is important to state that here we are focusing on capacity management at the SaaS level, and in particular capacity management offered as a service such as Auto Scaling service from Amazon. In this case, it needs to be as simple as possible so that end users can parametrize by themselves, which is the case of the threshold-based techniques. Moreover, the variety of applications makes it hard to create and/or maintain more complex models such as the ones based on Queueing Networks or Reinforcement Learning. For instance, not all types of applications can be modeled as a closed network.

That said, we believe that major cloud providers do rely on complex models at a certain level, for instance, to determine the number of physical machines required for running the incoming jobs (e.g., virtual machines). However, this kind of information is generally not provided so no assumptions can be made in this regard.

23.5   Capacity Planning and Management: Research Solutions

The ad hoc method enables capacity planning and management to be carried out simply but it may be insufficient to model performance goals effectively and thus it may be unable to optimize the allocation of resources with respect to the demand and service levels previously agreed. For that purpose, it is necessary to use more sophisticated methods. Capacity planning can be modeled in different ways, including queuing networks, stochastic processes, control theoretical methods, game theory, artificial intelligence and optimization methods. For reasons of space and clarity, we will focus here only on approaches that have been widely adopted: queuing networks, reinforcement learning, and linear and constraint programming.

23.5.1   Queuing Theory

23.5.1.1   Definition

Queuing theory (Daniel *et al.*, 2004) makes reference to the mathematical study of waiting lines, or queues. Kendall's notation is the standard system used to describe and classify queuing models, in which a queue is described in the form *A/S/c/K/N/D* where *A*: inter-arrival time distribution, *S*: service time distribution *c*:

number of servers, *K:* system capacity, *D*: queuing discipline and *N:* calling population. The elements K, N and D are optional. When these parameters are not specified, it is assumed $K = \infty$, $N = \infty$ and $D = $ FIFO. The most typical values for both *A* and *S* are Markovian (M), deterministic (D) and general distribution (G).

A queuing network (QN) is a network consisting of several interconnected queues. Efficient algorithms to compute performance measures of queuing networks are proposed such as mean value analysis MVA (Reiser and Lavenberg, 1980). This is based on the arrival theorem and allows the efficient computation of mean response times, availability, and throughput, while avoiding numerical issues such as the explicit computation of the normalizing constant.

23.5.1.2   *Capacity Planning Modeling*

The capacity planning problem can be formulated using a simple queuing model or a queuing network model (QNM). A multitier system (implementing the SaaS application) can be modeled as an M/M/c/K queue (see Figure 23.2) where their services are considered as closed loops to reflect the synchronous communication model that underlies these services. That is, a client waits for a request response before sending another request. Then, we rely on MVA for evaluating the performance of the queuing network.

The MVA algorithm predicts the response time and the availability based on the service workload and the current capacity. A utility function is defined to combine performance, availability and financial cost objectives:

$$\theta(t) = \frac{\varsigma(t)}{\omega(t)} \qquad (23.1)$$

where $\omega(t)$ corresponds to the number of instances allocated to the service at time t and $\varsigma(t) \in [0,1]$ corresponds to the SLA function that checks SLA objectives. The demanded capacity is the one that provides the highest utility (Eq. 23.1). To compute it, we can use any search algorithm.

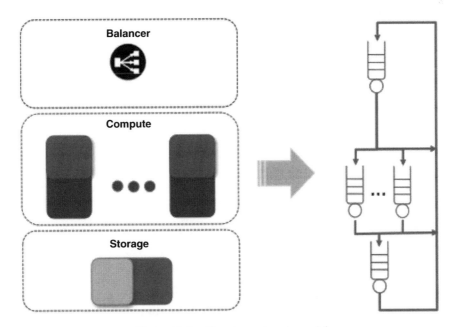

Figure 23.2   *Queuing network model*

23.5.1.3   Limits

The queuing network model (QNM) is a popular method for the modeling and analysis of systems. A well known technique for the analysis of single as well as multiclass separable closed QNM is the MVA algorithm. In addition to being efficient, this algorithm is also numerically stable. However, the computation times and memory requirements of MVA grow exponentially with the number of classes. To deal with networks with many classes, approximative algorithms were proposed (e.g., to compute the solution without iterating over the full set of N customers).

23.5.2   Reinforcement Learning

23.5.2.1   Definition

Reinforcement learning (RL) (Richard and Andrew, 1998) is an area of machine learning in computer science where the environment is typically formulated as a Markov decision process (MDP). The main difference between the classical techniques and reinforcement learning algorithms is that the latter do not need any a priori knowledge.

An MDP provides a mathematical framework for modeling decision making in situations where outcomes are partly random and partly under the control of a decision maker. It is defined by a 4-tuple (S, A, P, R), where S: state space, A: action space, P: transition rate function and R: reward function.

23.5.2.2   Capacity Planning Modeling

The capacity planning problem can be modeled via basic elements of a MDP. We can model capacity planning in the following way:

- $S = (N, Cap, Perf)$ is the state where N is the workload in number of requests, *Cap* corresponds to the current capacity, and *Perf* is the performance (e.g. response time).
- $A = (a)$ is the action set, which consists of increasing, reducing or maintaining capacity.
- $P = S \times A \times S \to [0, 1]$ is the probability distribution $P(s'|s, a)$ of transition from state s to new state s' using action a.
- $R: S \times A \times \mathfrak{R}$ is reward function when the system is in state s and action a is taken. This function takes into account the SLA and the financial cost.

At each step t, the decision maker receives the current state s and the available actions set A. The value function of taking action a in state s can be defined as:

$$Q(s, a) = \sum_{k=0}^{\infty} \gamma^k r_{t+k+1} \Big| s_t = s, a_t = a \qquad (23.2)$$

where r_{t+k+1} denotes the reward function and γ is a discount factor helping $Q(s, a)$'s convergence $(0 \leq \gamma < 1)$.

The objective is to find the best action that maximizes expected return. To this end, any search algorithm can be used.

23.5.2.3   Limits

Reinforcement learning seems a promising method for capacity planning, even in complex systems. In addition, this method is able to learn a scaling policy from experience, without any a priori knowledge. However, this method has two main limitations: (i) bad initial performance and (ii) large state spaces. The first leads to a long training time whereas the latter produces bad performance and very slow adaptation to change. Using parallel learning, agents can reduce the training time. A nonlinear function such as neural networks can cope with larger state spaces.

23.5.3   Linear Programming and Constraint Programming

23.5.3.1   *Definition*

Linear programming (LP) (Dantzig, 1963) and constraint programming (CP) (Rossi *et al.*, 2006) are techniques for solving constraint satisfaction problems and optimization problems (CSP/CSOP). A CSP model is defined by a set of variables, a set of constraints, and an objective function to be maximized or minimized. The basic idea is that the end-users define a set of variables, impose a set of constraints and an objective function, and a general-purpose solver tries to find a solution that meets these constraints and optimizes the objective function. Examples of solvers include JOptimizer and lpsolve, for LP; and Choco and IBM CP Optimizer, for CP.

Linear programming problems are about maximizing or minimizing a linear function subject to linear constraints, which can be linear equalities or inequalities, or in the standard form:

maximize $c^T x$

subject to $Ax \leq b$

and $x \geq 0$

where x corresponds to the vector of unknown variables, c and b correspond to known coefficients, and A to a matrix of known coefficients. $Ax \leq b$ and $x \geq 0$ are the linear constraints and $c^T x$ the objective function.

Similarly, a CP problem is defined as a triplet *(X, D, C)*, where $X = \{X_1, X_2, ..., X_n\}$ is the set of decision variables of the problem. *D* refers to a function that maps each variable $X_i \in X$ to the respective domain $D(X_i)$. A variable X_i can be assigned to integer values (i.e. $D(X_i) \subseteq Z$), real values (i.e. $D(X_i) \subset \Re$) or a set of discrete values (i.e., $D(X_i) \subseteq P(Z)$). Finally, C corresponds to a set of constraints *{C_1, C_2, ..., C_m}* that restrain the possible values assigned to variables. So, let $(v_1, v_2, ..., v_{nj})$ be a tuple of possible values for subset $X^j = \{X^j_1, X^j_2, ..., X^j_n\} \subseteq X$. A constraint C_j is defined as a relation on set X_j such that $(v_1, v_2, ..., v^j_n) \in C_j \cup (D(X^j_1) \times D(X^j_2) \times ... \times D(X^j_{nj}))$. Solving a CP problem *(X, D, C)* is about finding a tuple of possible values $(v_1, v_2, ..., v_n)$ for each variable $X_i \in X$ such that all the constraints $C_j \in C$ are met. In the case of a CSOP, that is, when an optimization criterion should be maximized or minimized, a solution is the one that maximizes or minimizes a given objective function $f : D(X) \rightarrow \Re$.

It may be noticed that the ways problems are solved in both LP and CP are very similar. They are distinguished, firstly, by their purposes: CP's main goal is to find feasible solutions (i.e. to meet constraints), whereas LP focuses on finding optimal solutions (i.e. solutions that optimize the objective). Regarding the modeling, CP may rely on predefined global constraints (see the *Global Constraint Catalogue* at http://sofdem. github.io/gccat/, accessed December 22, 2015). Of course global constraints can be translated as linear inequalities and equations such as the LP constraints. Another advantage is that global constraints can be composed so as to generate new constraints.

23.5.3.2   *Capacity Planning Modeling*

Capacity planning can be modeled as a set of variables $X = \{K_1, K_2, ..., K_n\}$, where $K_i \in [0, N] \; \forall i \in [1, n]$ corresponds to number of compute/storage units that should be allocated to each client class $c_i \in C$ of the SaaS application. A first constraint states that the sum of compute/storage allocated should not exceed a certain

K_{max} (cf. Eq. 23.3). There might be several client classes $C = (c_1, c_2, ..., c_n)$, each one with a specific arrival rate $\lambda_i \in \Lambda$ and a SLA $\varsigma : \Lambda \rightarrow [1, K_{max}] \rightarrow \{true, false\}$ to be met. Hence, the constraint expressed by Eq. 23.4 states that the all the SLAs (ς_i) must hold for a given K. It should be remembered that function ς_i is provided by the predefined contracts along with any performance model such as the one obtained in section 23.5.1.2.

$$\sum_{i=1}^{n} K_i \leq K_{max} \tag{23.3}$$

$$\forall i \in [1, n] \varsigma_i (\lambda_i, K_i) \neq 0 \tag{23.4}$$

The utility is application specific – that is, it depends on the application domain. For example, for a given application and a given workload, the utility function may take into account the requirements on the response time, whereas for another workload it may be based on other criteria such as availability. Because of this, we do not provide any further details about how the functions that verify the SLA are calculated.

The last part of the LP/CP model is the objective function. It should be said that this component of the model, although optional in the case of CP, is very important. In fact, it restrains the number of possibilities that eventually meet the constraints into the ones that optimize a given criteria. As a consequence, the solving time may be drastically increased. In the capacity planning scenario, it is always needed to minimize the costs related to the resource allocation. So, the sum of all variables K_i should be minimized, as formalized in Eq. 23.5.

$$\theta (\lambda, K) = \sum_{i=1}^{n} \frac{\varsigma_i (\lambda_i, K_i)}{K_i} \tag{23.5}$$

23.5.3.3   Solving Algorithm

Simplex (Murty, 1983) is the most popular algorithm for solving LP problems. It involves constructing a feasible region (a convex polyhedron) formed by the intersection of the constraints (inequalities and equalities), as it is exemplified in Figure 23.3(a). A maximum of the objective function corresponds to the set of variables satisfying the constraints, which resides in a vertex of the convex polyhedron.

Constraint satisfaction problems are usually solved with search algorithmic approaches, which can be either complete (systematic) or incomplete (nonsystematic). The former performs systematic search (e.g. backtracking or branch-and-bound) that guarantees that a solution is found, if at least one exists, whereas the latter performs a sort of local search and therefore should not be used to show that such a solution does not exist or to find an optimal solution. Nonetheless, nonsystematic approaches might take less time to find a solution (if there is any) and an approximation of the optimal solution. Constraint programming also makes use of inference to propagate the impact of variable assignments on the possible values the other variables can in the following subtrees search space. Hence, the search tree can be constantly pruned as the search goes on. Figure 23.3 (b) shows an example of such a mechanism.

23.5.3.4   Limits

It is straightforward that the main advantage of LP/CP is its declarative characteristics – users state variables and constraints and a general-purpose engine/algorithm solves it. Hence, users are free from the burden of thinking about and building solvers for each CSP. However, it may be a very hard task to model a CSP correctly in a way that it can be solved in an efficient fashion, as CSOPs are generally NP-complete (Rossi *et al.*, 2007).

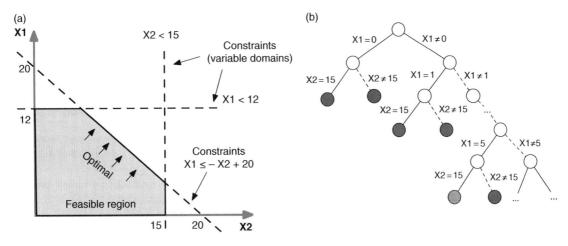

Figure 23.3 Solutions

23.6 Conclusion

In this chapter, we presented the essentials of cloud capacity planning and management. Capacity management is the process tasked with defining the capacity of a system, to meet changing demands, at the right time and at optimal cost. Since in cloud computing, upon request, resources are almost instantly provisioned, it is possible to perform capacity management in reaction to events that have already happened without taking them into consideration a long time ahead (reactive approaches). On the other hand, these approaches (i) may misbehave in unstable or oscillating environments; (ii) on-demand resources generally cost way more than previously reserved ones. Therefore, there may be times where adjusting the amount of resources in a proactive manner could also pay off. Capacity management approaches may rely on hybrid adaptation strategies to improve the cost-effectiveness in the long term (a proactive approach) and be capable of reacting to unanticipated demand variations in the short term (a reactive approach).

Capacity management and planning may be achieved in a completely ad hoc manner by dynamically adjusting the allocation of resources (e.g., by adding or releasing resources) according to a set of predefined rules. Examples of such an approach are implemented, for example, by Amazon Auto Scaling and Auto Scaling Application Block (Microsoft Azure). However, it becomes harder to profile systems' performance as their level of complexity increases, which, consequently, compromises the accuracy of estimations of resource requirements. For that reason, capacity planning may combine techniques from different fields in order to be effective. For example, in order to model systems' performance effectively, queuing theory can be applied. Alternatively, reinforcement learning and game theory might also be used to give a more effective performance profiles or to deal with economic equilibrium and thus improve estimations of resource requirements. Those techniques can be combined along with operational research techniques like constraint and linear programming to find solutions that respect the constraints (e.g., imposed by SLAs) or even a solution that optimizes certain aspects of quality of service or cost. Table 23.1 summarizes the advantages and limitations for each method.

To conclude, we can imagine that it is not easy for a cloud service provider to compare these methods and choose the most appropriate for its needs. We devised a tool – or a platform as a service – to help any cloud provider to select the best capacity planning and management method from a catalog according to the service's context (service topology, workload characteristics, and SLA). This service is based on capacity planning and

Table 23.1 *Capacity planning methods: A summary*

Method	Advantages	Limitations
Threshold-based rules	Simple to design and develop	Definition of thresholds/rules per application
Queuing theory	Performance modeling and analysis	Scalability
Reinforcement learning	Does not need any a priori knowledge	Bad initial performance, large state spaces
Constraint programming	Declarative method, catalog of constraints	Scalability
Linear programming	Optimal solution	Definition of objective function
Game theory	Economy equilibrium	Definition of strategies
Control theory	Support for dynamic systems	May lead to system instability

management methods evaluation criteria (e.g., system stability, method scalability, solution optimality, and implementation simplicity) applied to the cloud-provider context, and it proposes the correct method to choose.

References

Box, G. E. P. and Jenkins, G. M. (1994) *Time Series Analysis: Forecasting and Control*, Prentice Hall, Upper Saddle River, NJ.

Daniel, A. M., Lawrence, W. D. and Virgilio A. F. A. (2004) *Performance by Design: Computer Capacity Planning by Example*, Prentice Hall, Upper Saddle River, NJ.

Dantzig, G. B. (1963) *Linear Programming and Extensions*, Princeton University Press, Princeton, NJ.

Hasan, M. Z., Magana, E., Clemm, A. *et al.* (2012) *Integrated and Autonomic Cloud Resource Scaling*. Network Operations and Management Symposium (NOMS), 2012. IEEE, pp. 1327–1334.

Kouki, Y. and Ledoux, T. (2013) RightCapacity: SLA-driven cross-layer cloud elasticity management. *International Journal of Next-Generation Computing* **4**(3), 250–262.

Murty, K. G. (1983) *Linear Programming*, John Wiley & Sons, Inc., New York.

Reiser, M. and Lavenberg, S. S. (1980) Mean-value analysis of closed multichain queuing networks. *Journal of the ACM* **27**(2), 313–322.

Richard, S. S. and Andrew G. B. (1998) *Introduction to Reinforcement Learning*, MIT Press, Cambridge, MA.

Rossi, F., Beek, P. V and Walsh, T. (2006) *Handbook of Constraint Programming (Foundations of Artificial Intelligence)*, Elsevier Science Inc., New York, NY.

Rossi, F., Beek, P. V. and Walsh, T. (2007) *Constraint Programming*, Elsevier Science, San Diego, CA, pp. 189–211.

24

Fault Tolerance in the Cloud

Kashif Bilal,[1] Osman Khalid,[1] Saif Ur Rehman Malik,[1] Muhammad Usman Shahid Khan,[1] Samee U. Khan,[1] and Albert Y. Zomaya[2]

[1] *North Dakota State University, USA*
[2] *Sydney University, Australia*

24.1 Introduction

Cloud computing is an emerging and innovative platform, which makes *computing* and *storage* available to the end users as *services*. The cloud is a "blob" of unstructured resources that are classified into three domains: (i) applications (or software), (ii) platforms, and (iii) infrastructure. The cloud is a merger of business and computing models, which makes it a very important scientific and business medium for the end users. Cloud computing has been widely adopted in various domains, such as research, business, health, e-commerce, agriculture, and social life. It is also being increasingly employed for a wide range of research activities in domains such as agriculture, smart grids, e-commerce, scientific applications, healthcare, and nuclear science.

As cloud computing systems continue to grow in scale and complexity, it is of critical importance to ensure its stability, availability, and reliability. But varying execution environments, addition and removal of system components, frequent updates and upgrades, online repairs, and intensive workload on servers, to name a few can induce failures and faults in the large-scale, complex, and dynamic environments of cloud computing. The reliability of the cloud can be compromised easily if proactive measures are not taken to tackle the possible failures emerging in cloud subsystems. For instance, Google reported a 20% revenue loss when an experiment caused an additional delay of 500 ms in response times (Greenberg *et al.*, 2009). Amazon reported a 1% sales decrease for an additional delay of 100 ms in search results (Greenberg *et al.*, 2009). A minor failure in the O2 network (a leading cellular service provider in the United Kingdom) affected around 7 million customers for 3 days. Similarly, a core switch failure in BlackBerry's network left millions of customers without Internet access for 3 days.

Encyclopedia of Cloud Computing, First Edition. Edited by San Murugesan and Irena Bojanova.
© 2016 John Wiley & Sons, Ltd. Published 2016 by John Wiley & Sons, Ltd.

To achieve reliability, and as a countermeasure for faults and failures, cloud service providers adopt various mechanisms to implement *fault tolerance* at the system level. Fault tolerance is a vital issue in cloud computing platforms and applications. It enables a system to continue operation, possibly at a reduced level, rather than failing completely, when some subcomponent of the system malfunctions unexpectedly. The significance of interconnection networks is obvious from the discussion above.

The diversity in the needs of various services deployed in a cloud environment means that fault tolerance also has a pivotal role in maintaining service level agreements (SLAs), as well as, the desired levels of quality of service (QoS). The SLAs define various rules to regulate the availability of the cloud service to the end users. Virtualization in clouds assigns various services at different levels of access to numerous subscribers. Virtualization and multiple subscriptions with diversifying SLAs and QoS requirements significantly raise the complexity and unpredictability of cloud environments.

24.1.1   Types of Faults

Faults can be of various types including: (i) transient, intermittent, or permanent hardware faults, (ii) software bugs and design errors, (iii) operator errors, and (iv) externally induced faults and errors. In a typical cloud environment, faults appear as a failure of resources, such as applications/hardware storage, which are being used by the end users. The two most commonly occurring faults in cloud environment are byzantine failures and crash failures.

24.1.1.1   Byzantine Failures

In these faults, the system components fail in arbitrary ways, causing the system to behave incorrectly in an unpredictable manner. The system may process requests incorrectly and produce inconsistent outputs.

24.1.1.2   Crash Failures

When crash failures occur, they cause system components to stop functioning completely or remain inactive during failures – for instance, failures due to power outages or hard-disk crashes.

As discussed earlier, cloud computing is divided into several operational layers, such as PaaS, SaaS, and IaaS. If a failure occurs in one of these layers, then this layer affects the services offered by the layers above it. For instance, failure in PaaS may produce errors in the software services offered by SaaS. However, if a failure occurs in a physical hardware layer (IaaS) then this may negatively affect both the PaaS and SaaS layers. This implies that the impact of failures occurring at hardware level is significantly high, and it is of critical importance to devise fault-tolerant strategies at hardware level. Supporting research in fault-tolerance computing involves system architecture, design techniques, coding theory, testing, validation, proof of correctness, modeling, software reliability, operating systems, parallel processing, and realtime processing.

24.1.2   Redundancy

A practical approach of implementing fault tolerance is through redundancy that involves duplication of hardware and software components, such that if a component or a process fails, the backup process or component is available to take the place of the primary one. Some vendors have been involved in developing computing solutions with built-in capability of fault tolerance. For instance, Stratus Computers produces duplex-self checking computers, where each computer belonging to a duplex pair is internally duplicated and runs synchronously. If one of the machines fails, the duplication allows the other machine of the pair to continue the computations without delay. Similarly, Tandem Computers uses a number of independent identical processors and redundant storage devices and controllers to provide automatic recovery in the case of a hardware or software failure.

24.1.3   Fault-Tolerance Validation

Cloud service providers need to perform fault tolerance / availability analysis of the services they provide to end users. The fault-tolerance validation of a service is of critical importance to ensure proper adherence to SLAs. However, due to the numerous stochastic factors involved, it is quite difficult to verify that a fault-tolerant machine will meet reliability requirements. To aid in the assessment of system reliability, a great deal of research has been conducted recently in experimental testing by making use of a methodology known as *fault injection*. Fault injection is an important method to mimic the occurrence of errors in a controlled environment to make the necessary measurements. A number of stochastic models based on probability computations have been developed that use Markov and semi-Markov processes to perform the availability analysis of a fault-tolerant machine. These models have been implemented in several computer-aided design tools. Some well known tools are: (i) the hybrid automated reliability predictor (HARP) (developed at Duke University), (ii) the system availability estimator (SAVE) (IBM), (iii) the Symbolic Hierarchical Automated Reliability and Performance Evaluator (SHARPE) (Duke University), and (iv) UltraSAN– (University of Illinois, UIUC), and **(e)** DEPEND – (UIUC). These tools are used to perform various measures, such as latency, coverage, and fault rates. We define some of the fault-tolerance measurements in the next subsection.

24.1.4   Fault-Tolerance Measures

Fault-tolerance measurement is a crucial aspect of cloud paradigm. Fault tolerance measures can be used to quantify the dependability of the cloud system. Two of the major legacy measures for fault tolerance of the system are: (i) availability and (ii) reliability (Koren and Krishna, 2010). Availability is the ratio between the uptime and sum of the uptime and downtime of the system. Availability can be quantified as:

$$availability = \frac{uptime}{uptime + downtime} \tag{24.1}$$

The uptime and downtime values are either predicted using the mathematical modeling techniques, such as the Markov availability model, or can be calculated from actual field measurements (Bauer and Adams, 2012). Availability can also be measured as a percentage of agreed service time and downtime of the system. The agreed service time is the expected operational time of the system per month. The planned downtime of the system is explicitly excluded from the agreed service time. The availability of the system can be calculated as:

$$availability(\%) = \frac{agreed\,service\,time - downtime}{agreed\,service\,time} \times 100 \tag{24.2}$$

Availability of the system can also be quantified using metrics, such as mean time to failure (MTTF), mean time between failures (MTBF), and mean time to repair (MTTR). The MTTF is the average time of the system operating accurately until a failure occurs. The MTBF is the average time between two consecutive failures of the system. The MTTR measure predicts the average time required to replace the faulty component of the system to bring the system back to operational mode. Availability of the system in terms of MTBF and MTTR can be computed as:

$$availability = \frac{MTBF}{MTBF + MTTR} \tag{24.3}$$

Reliability, denoted as R(t), is the probability of the system to work accurately as a function of time "t" (Koren and Krishna, 2010). The TL9000 measurement handbook defines reliability as "the ability of an item to perform a required function under stated conditions for a stated time period" (Bauer and Adams, 2012). Service reliability can be formulated as:

$$service\ reliability = \frac{successful\ responses}{total\ requests} \times 100 \tag{24.4}$$

As most of the services are highly reliable, services are quantified by defective/unsuccessful transactions per million attempts. Defects per million (DPM) can be formulated as:

$$DPM = \frac{(total\ requests - successful\ transactions)}{total\ request} \times 1\,000\,000 \tag{24.5}$$

The availability and reliability of the system hold a pivotal role in the cloud paradigm. A small fraction of downtime has severe financial impacts. It has been reported that a single hour of downtime costs around $50\,000 in a datacenter (Bilal *et al.*, 2014). Therefore, round the clock availability of cloud services are vital.

In this chapter, we discuss fault tolerance in the cloud and illustrate various fault-tolerance strategies existing in the literature. The rest of the chapter is as follows. In section 24.2, we discuss the different fault tolerance strategies for cloud and provide taxonomy of fault-tolerance approaches. Section 24.3 concludes the chapter.

24.2   Fault-Tolerance Strategies in Cloud

Crashes and failures, such as disk failures or link failures, are very common in the cloud. Moreover, the number of nodes (servers) involved in the cloud has an order of tens of thousands or more servers that increases the probability and cost of the failures. Jobs executing over the cloud may have a time span that may evolve for few days. The effect of failure on medium or long-running jobs can jeopardize the fulfillment of SLA contracts and waste computation time. For instance, if a task requires 24 hours to complete and if, after 23 hours, the node that was executing the task crashes, then almost one day of execution will be wasted; it will also lead to the violation of the SLA. One way to tackle such a problem is to execute a backup process on a different system for each primary process. The primary process, in this case, is responsible for checkpointing the current state to redundant disks. Should the primary process fail, the backup process can restart from the last checkpoint. Generally, fault-tolerant systems characterize recovery from errors as either roll forward or roll back. Roll-forward mechanisms take the system state at the time when the error was detected in order to correct the error, and from there the system moves forward. Alternatively, the roll-back mechanism uses checkpointing to revert the system state to some earlier correct state, and the system moves forward from that state. The operations between the checkpoint and the erroneous state can be made idempotent, as a requirement of roll-back recovery. Some systems make use of both roll-forward and roll-back recovery for different errors or different parts of one error. We discuss checkpointing in detail in the next subsection.

24.2.1   Checkpoint-Based Fault Tolerance

To avoid the problem mentioned above, checkpoint mechanisms have been proposed and implemented on the cloud. This mechanism records the system state periodically after a certain time limit so that, if a failure occurs, the last checkpoint state of the system is restored and the task execution is resumed from that point.

However, significant overheads are associated with the application of checkpoint strategy as it can be expensive in terms of performance. In case of virtualized environment, such as the cloud, the checkpoint strategy becomes more challenging, where huge virtual machine (VM) images needs to saved and restored (Goiri *et al.*, 2010). In the said perspective, several researchers have proposed different approaches to make the use of checkpoint efficient in the cloud. In the following subsections, we will discuss and highlight some of the recent checkpoint-based fault-tolerant approaches that are deployed in the cloud.

24.2.1.1   *Disk- and Diskless-Based Checkpointing Schemes*

Message-passing interface (MPI) is generally used to achieve parallelism in high-performance computing. This is a language-independent communications protocol that supports both point-to-point and collective communications. The goal of MPI is to attain high performance, scalability, and portability. To achieve fault tolerance in MPI-based applications, several strategies have been proposed, such as BlobCR (Nicolae and Cappello, 2011) that are specifically optimized for tightly coupled scientific applications written using MPI and that need to be ported to IaaS cloud. Two techniques have been widely adopted to achieve fault tolerance in cloud: replication (redundancy) and checkpoint. For tightly coupled applications, redundancy implies that all components of the process (which is itself part of a distributed application) must also be replicated. The reason for such replication is that a failure of one process results in a global failure of all processes and eventually leads to process termination. Therefore, in tightly coupled applications, checkpoint approaches provide a more feasible solution than replication. The BlobCR use a dedicated repository of checkpoints that periodically takes the snapshots of the disk attached to the virtual machine. The aforesaid approach allows use of any checkpointing protocol to save the state of processes into files, including application-level mechanisms (where the process state is managed by the application itself) and process-level mechanisms (where the process state is managed transparently at the level of the message-passing library). Moreover, the BlobCR also introduces a support that allows I/O operations performed by the application to rollback. The BlobCR brings a checkpointing time speedup of up to 8x as compared to the full VM snapshotting based on qcow2 over a parallel virtual file system (PVFS) (Gagne, 2007). The checkpoints performed on the whole state of the VM are expensive. However, using only a snapshot of the portion of disk is smaller and faster, even if a reboot is needed.

Disk-based checkpointing strategies are widely used. However, the applications that require checkpointing frequently lead to several disk accesses that result in a performance bottleneck. Several diskless-based checkpoint strategies have been proposed, such as a multilevel diskless checkpointing (Hakkarinen and Chen, 2013). In multilevel checkpointing, there are *N*-level of diskless checkpoints. Multiple levels of checkpoints reduce the overhead for tolerating a simultaneous failure of *N* processors by layering the diskless checkpointing schemes for a simultaneous failure of *i* processors. For example, an *N*-failure checkpoint can recover any number of failures from *1, 2,..., N*. Multilevel diskless checkpointing scheme can significantly reduce the fault tolerance time as compared to a one-level scheme. To perform multilevel diskless checkpointing, a schedule for diskless checkpoints must be developed for each level of recovery, such as one, two, or *N* simultaneous failures. When the checkpoints are scheduled, the processor takes specific steps to perform the recovery. For instance, if a one-level checkpoint is scheduled, then processor will take only one step back for the recovery. The coordination of the checkpoints among the processors is important for the consistency of the checkpoints. If a failure is detected, then an *N*-level diskless checkpointing mechanism will attempt to use the most recent checkpoint to recover the state. However, if another failure occurs during the recovery, then the mechanism will use most recent two-level checkpoint to restore the state. If the number of failures exceeds the number that is supported, then the system needs to restart the computation. For any diskless checkpoint there is an overhead of both communication and calculation. The difference between the disk and diskless checkpointing is significant when a failure occurs during a recovery. The system simply restarts from the same checkpoint in disk-based checkpointing. However, in diskless checkpointing, an earlier checkpoint is used to restore the state of the system.

24.2.1.2   *Checkpoint Placement Scheme*

The number of checkpoints inserted can significantly decrease the performance of the system during recovery. Moreover, the number of checkpoints must be optimal to minimize the storage. Several optimized and efficient checkpoint placement strategies have been proposed. One such optimal checkpoint strategy, which maximizes the probability of completing all tasks within the deadline, is presented in Kwak and Yang (2012). In real-time systems, the occurrence of faults is normal rather than exceptions. Every fault is detected at the checkpoint that comes first after the fault occurrence. In such strategies, the checkpoints are inserted constantly after certain interval at the execution of the tasks. However, the time limit for the interval may vary from task to task. The slack time of each task is identified first and then the maximum number of re-executable checkpoints are determined that can meet the deadline of the task. In a multitasking environment, the slack time is calculated not only by the execution time of the task but also of execution of the other tasks. Based on the information about the slack time, formulas are derived that compute the number of checkpoints that need to be re-executed to meet the deadline of the task. The significance of such checkpoint schemes are that they provide an integrated solution to multiple realtime tasks by solving a single numerical optimization problem.

Some smart infrastructures are also proposed, for example in Goiri *et al.* 2010, which uses Another Union File System (AUFS), which differentiates between the read-only and read-write parts of the virtual machine image file. The goal of such infrastructures is to reduce the time needed to make a checkpoint that will, in return, reduce the interference time on task execution. The read-only parts can be checkpointed only once and the read-write checkpoints are incrementally checkpointed, which means the modifications from the last checkpoints are restored. This checkpoint mechanism can also be implemented in a virtualized environment using Xen hypervisor, where the tasks can be executed on VMs created on demand. Making the checkpoint for tasks running within a VM may involve moving tons of GBs of data as it may include the information to resume the task on another node, such as task and memory content, and disks information (Malik *et al.,* 2013). Some checkpoint mechanisms mount the base system as a read-only file system, because only a small portion of the data changes with respect to the VM startup. Moreover, the user modifications are stored in an extra disk space called delta disk. Besides the delta disk, which contains user modification data, there is another read-only disk, which contains the base system. These disks are merged to form a root-file system to start the VM. Once the VM is booted, then the user can work with the file system without worrying about the underlying structure. The checkpoints are compressed and are stored in the Hadoop file system (HDFS) so that the checkpoints are distributed and replicated in all the nodes. Moreover, by doing this the possibility of a single point of failure can also been eliminated.

24.2.1.3   *Failures/Faults that are Hard to Recover*

Among the different kind of failures or faults, such as node failures and link failures, the faults that are really costly and take a longer time to mitigate are hardware failures. In the case of cloud computing, several VMs are running on a single physical machine. In such an environment, if a hardware failure occurs, then all the VMs have to be migrated. This requires longer downtime than a software or application failure. Moreover, the hardware device may need to be replaced, resulting in a longer repair time. Hardware failures can have a significant because they may involve (i) device replacement, and (ii) migrations, including VM migrations, which cause the recovery time to increase. Some faults and errors are hard to detect, such as routing and network misconfigurations. It is difficult to recover from such faults as it is hard to detect them. However, once detected they are easy to address.

24.2.2 Adaptive Fault-Tolerance Techniques

Adaptive fault-tolerance techniques help the system to maintain and improve its fault tolerance by adapting to environmental changes. Adaptive fault-tolerance techniques for cloud computing monitors the state of the system and reconfigures the cloud computing system for the stability of the system if errors are detected. In this subsection, we will overview some of the recently proposed adaptive fault-tolerance techniques for the cloud computing paradigm.

24.2.2.1 *Fault Tolerance for Realtime Applications (FTRT)*

The realtime applications that execute on cloud environments range from small mobile phones to large industrial controls. The highly intensive computing capabilities and scalable virtualized environment of the clouds help the systems to execute the tasks in realtime. Most of the realtime systems require high safety and reliability. A fault tolerance model for realtime cloud computing is proposed in (Malik *et al.*, 2011). The realtime cloud fault tolerance model revolves around the reliability of the virtual machines. The reliability of the virtual machines is adaptive and changes after every computing cycle. The proposed technique depends on the adaptive behavior of the reliability weights assigned to each processing node. The increase and decrease in reliability depends on the virtual machines to produce the results within the given time frame. The technique uses a metric to evaluate reliability. The metric assesses the reliability level of the node against a given minimum reliability threshold. The nodes are removed if the processing nodes fail to achieve the minimum required reliability level. The primary focus of the system is on the forward recovery mechanism.

24.2.2.2 *Dynamic Adaptive Fault-Tolerant Strategy (DAFT)*

The dynamic adaptive fault-tolerant strategy (Sun *et al.*, 2013) observes a mathematical relationship between the failure rates and the two most common fault-tolerance techniques, checkpoints and replications. Historical data about failure rates helps the cloud computing system to configure itself for the checkpoints or the replicas. A dynamic adaptive checkpoint and replication model is made by combining checkpoints and replications to achieve the best level of service availability and to attain the service-level objectives (SLOs). The dynamic adaptive fault-tolerant strategy was evaluated in a large-scale cloud datacenter with regard to level of fault tolerance, fault-tolerance overheads, response time, and system centric parameters. The theoretical and experimental results presented in Sun *et al.*, 2013, demonstrate that DAFT provides highly efficient fault tolerance and excellent SLO satisfaction.

24.2.2.3 *Fault- and Intrusion-Tolerant Cloud Computing Hardpan (FITCH)*

The novel fault-tolerant architecture for cloud computing, FITCH, supports the dynamic adaptation of replicated services (Cogo *et al.*, 2013). It provides a basic interface for adding, removing, and replacing replicas. The FITCH interface also provides all the low-level actions to provide end-to-end service adaptability. The technique was originally designed for two replication services: a crash fault-tolerant Web service and a Byzantine fault-tolerant (BFT) key-value store based on state machine replication. Both the services, when deployed with FITCH, are easily extendable and adaptable to various workloads through horizontal and vertical scalability. The number of computing instances that are responsible for providing the service are increased or decreased through horizontal scalability. When there is a requirement to handle peak users' requests or to handle as many faults as possible, the number of computing resources is increased. The number of resources is reduced when there is a requirement to save resources and money. Vertical scalability is achieved by increasing or decreasing the size and capacity of allocated resources. The FITCH adapts horizontal and vertical scalability depending on requirements.

24.2.2.4   *Byzantine Fault-Tolerance Cloud (BFTCloud)*

The BFTCloud is a fault tolerant architecture for voluntary-resource cloud computing (Zhang *et al.*, 2011). In voluntary-resource cloud computing, infrastructure consists of numerous user-contributed resources, unlike the well managed architecture provided by a large cloud provider. The architecture of BFTCloud is based on a Byzantine fault-tolerance approach. The architectures operate on five basic operations: primary node selection, replica selection, request execution, primary node updating, and replica updating. The primary node is selected based on QoS requirements. The request for the service is handled by the primary node. The primary node also selects the *3f+1* replicas from the pool based on QoS requirements. All the replicas and primary node perform the operation on the request and send back the result to the primary node. Based on the result, the primary node decides to update the other primary node or update the replicas. In primary updating, one of the replicas is updated to primary node. In replica updating, the faulty replica is replaced with a new one. The BFTCloud provides high reliability and fault tolerance along with better performance.

24.2.2.5   *Low-Latency Fault Tolerance (LLFT) Middleware*

Low-latency fault tolerance (LLFT) middleware uses the leader/follower replication approach (Zhao *et al.*, 2010). The middleware consists of a low-latency messaging protocol, a leader-determiner membership protocol, and a virtual determiner framework. The low-latency message protocol provides a reliable and ordered multicast service by communicating a message ordering information. The ordering is determined by the primary replica in the group. The technique involves fewer messages than earlier fault-tolerance systems. A fast reconfiguration and recovery service is provided by the membership protocol. The reconfiguration service is required whenever the fault occurs at the replica or some replica joins or leaves the group. The membership protocol is faster as it finds the primary node deterministically based on the rank and the degree of the backups. The virtual determiner framework takes the ordering information from the primary replica and ensures that all the backups receive the same ordering information. The LLFT middleware provides a high degree of fault tolerance and achieves low end-to-end latency.

24.2.2.6   *Intermediate data fault tolerance (IFT)*

Intermediate data is the data that is generated during the parallel dataflow program (Ko *et al.*, 2010). The technique considers the intermediate data as of high priority (first-class citizen). The other techniques either use the store-local approach or distributed file system (DFS) approach. In store-local approach the data is not replicated and is used in Map outputs in Hadoop. Although the approach is efficient but is not fault tolerant. If there is a failure of a server that stores the intermediate data, this results in the re-execution of the tasks. In the DFS approach the data is replicated, but causes too much network overhead. The network overhead results in the delay of jobs completion time. The DFS approach is used for reduce outputs in Hadoop. There are three techniques for intermediate data fault tolerance: (i) asynchronous replication of intermediate data, (ii) replication of selective intermediate data, and (iii) exploiting the inherent bandwidth of the cloud datacenter topology. A new storage system, the intermediate storage system (ISS), implements the techniques for Hadoop mentioned above. Hadoop with ISS outperforms the base Hadoop under failure scenarios.

24.2.2.7   *MapReduce Fault Tolerance with Low Latency (MFTLL)*

MapReduce fault tolerance with low latency is a passive replication technique on the top of re-execution of the MapReduce jobs to improve overall execution time (Zheng, 2010). The technique uses the extra copies for the cloud tasks to improve MapReduce fault tolerance while keeping the latency low. The technique is referred to as a passive replication technique because, in passive replication, not all copies need to be in a running state as compared to the active replication technique. The proposed technique allocates a few (k) backup copies of the tasks. The backup assignment for each task is based on data locality and on rack locality. The placement of the backup in the locality avoids heavy network traffic. The backup copy is only executed if the primary task fails. The resources that take a longer time to execute (stragglers) are also identified and, for these stragglers, backups are executed in parallel. The MapReduce users or cloud providers decide the value of k based on failure statistics. The technique also uses a heuristic to schedule backups, move backup instances, and select backups upon failure for fast recovery.

24.2.2.8   *Adaptive Anomaly Detection System for Cloud Computing Infrastructures (AAD)*

An adaptive anomaly detection (AAD) system for cloud computing infrastructure ensures the availability of the cloud (Pannu *et al.*, 2012). The framework uses cloud performance data to discover future failures. Predicted possible failures are verified by the cloud operators. The failures are marked as true or false failures on verification. The algorithm recursively learns and improves future failure prediction based on the verified data. The framework also takes into account the actual failures that were not previously detected.

In Table 24.1, we provide a summary of various fault tolerance techniques discussed above. The schemes are categorized on the basis of methodology, programming framework, environment, and the type of faults detected.

Table 24.1   *Summary of fault tolerance strategies*

Strategy	Faulty tolerance technique	Programming framework	Environment	Faults detected
(Nicolae and Cappello, 2011)	Disk based Checkpoint	MPI	IaaS cloud	Node/network failure
(Hakkarinen and Chen, 2013)	Diskless Checkpoint	NA	HPC	Process/application failure
(Kwak and Yang, 2012)	Checkpoint	Probability analytic framework	Real-time systems	Process failures
(Goiri *et al.*, 2010)	Checkpoint	Java	Virtual Machine	Node failure
(Malik *et al.*, 2011)	FTRT (adaptive)	–	Real-time	–
(Sun *et al.*, 2013)	DAFT (adaptive)	Java	Large scale Cloud	Works on historical failure rate
(Cogo *et al.*, 2013)	FITCH (adaptive)	Java	Large scale Cloud	–
(Zhang *et al.*, 2011)	BFTCloud (adaptive)	Java	Voluntary-resource cloud	Byzantine problems
(Zhao *et al.*, 2010)	LLFT (adaptive)	C++	Middleware	Replication faults
(Ko *et al.*, 2010)	IFT(adaptive)	Hadoop	Hadoop	Intermediate data faults
(Zheng, 2010)	MFTLL (adaptive)	MapReduce	MapReduce	Replication faults, stragglers detection
(Pannu *et al.*, 2012)	AAD (adaptive)	–	Local cloud	Discovers future failures

24.3 Conclusion

In this chapter we studied fault tolerance and illustrated various state-of-the-art fault-tolerant strategies for cloud environments. Fault tolerance is a major concern in a cloud environment because of the need to guarantee availability of critical services, application execution, and hardware. As cloud-computing systems continue to grow in scale and complexity, it is of critical importance to ensure the stability, availability, and reliability of such systems. Cloud environments are susceptible to failure because of varying execution environments, addition and removal of system components, frequent updates and upgrades, online repairs, and intensive workload on the servers. The reliability of such systems can be compromised easily if proactive measures are not taken to tackle possible failures emerging in cloud subsystems. We discussed various types of faults, and the different methods that are in use to tackle such faults. The chapter also mentioned some methods for validating the fault tolerance of a system and the various metrics that quantify fault tolerance, and discussed state-of-the-art techniques for fault tolerance in cloud computing. A taxonomy of fault-tolerant schemes was also presented.

References

Bauer, E., and Adams, R. (2012) *Reliability and Availability of Cloud Computing*, John Wiley & Sons, Inc., Hoboken, NJ.

Bilal, K, Malik, S., Khan, S. U., and Zomaya, A. (2014) Trends and challenges in cloud data centers. *IEEE Cloud Computing Magazine* **1**(1), 1–20.

Cogo, V. V., Nogueira, A., Sousa, J., *et al.* (2013) FITCH: Supporting adaptive replicated services in the cloud, in *Distributed Applications and Interoperable Systems* (eds. R. Meier and S. Terzis). Springer, Berlin, pp. 15–28.

Gagne, M. (2007) Cooking with Linux – still searching for the ultimate Linux distro? *Linux Journal* **9**(161).

Goiri, I., Julia, F., Guitart, J., and Torres, J. (2010) Checkpoint-based fault-tolerant infrastructure for virtualized service providers. *IEEE Network Operations and Management Symposium (NOMS)*, pp. 455–462.

Greenberg, A., Hamilton, J., Maltz, D., and Patel, P. (2009) The cost of a cloud: Research problems in data center networks. *ACM SIGCOMM Computer Communication Review* **39**(1), 68–79.

Hakkarinen, D. and Chen, Z. (2013) Multi-level diskless checkpointing. *IEEE Transactions on Computers* **62**(4), 772–783.

Ko, S. Y., Hoque, I., Cho, B., and Gupta, I. (2010) *Making Cloud Intermediate Data Fault-Tolerant*. Proceedings of the 1st ACM symposium on Cloud Computing. ACM, pp. 181–192.

Koren, I., and Krishna, C. M. (2010) *Fault-Tolerant Systems*, Morgan Kaufmann, Burlington, MA.

Kwak, S. W., and Yang, J. M. (2012) Optimal checkpoint placement on real-time tasks with harmonic periods. *Journal of Computer Science and Technology* **27**(1), 105–112.

Malik, S. and Huet, F. (2011) *Adaptive Fault Tolerance in Real Time Cloud Computing*. Proceedings of the IEEE World Congress on Services (SERVICES), pp. 280–287.

Malik, S. R., Khan, S. U., and Srinivasan, S. K. (2013) Modeling and analysis of state-of-the-art VM-based cloud management platforms. *IEEE Transactions on Cloud Computing* **1**(1), 50–63.

Nicolae, B., and Cappello, F. (2011) BlobCR: efficient checkpoint-restart for HPC applications on IaaS clouds using virtual disk image snapshots. Paper presented at the ACM International Conference for High Performance Computing, Networking, Storage and Analysis.

Pannu, H. S., Liu, J., and Fu, S. (2012) *AAD: Adaptive Anomaly Detection System for Cloud Computing Infrastructures*. Proceedings of the IEEE 31st Symposium on Reliable Distributed Systems (SRDS). IEEE, pp. 396–397.

Sun, D., Chang, G., Miao, C., and Wang, X. (2013) Analyzing, modeling and evaluating dynamic adaptive fault tolerance strategies in cloud computing environments. *Journal of Supercomputing* **66,** 193–228.

Zhang, Y., Zheng, Z., and Lyu, M. R. (2011) *BFTCloud: A Byzantine Fault Tolerance Framework for Voluntary-Resource Cloud Computing*. Proceedings of the IEEE International Conference on Cloud Computing (CLOUD). IEEE, pp. 444–451.

Zhao, W., Melliar-Smith, P. M., and Moser, L. E. (2010) *Fault Tolerance Middleware for Cloud Computing*. Proceedings of the IEEE 3rd International Conference on Cloud Computing (CLOUD), pp. 67–74.

Zheng, Q. (2010) *Improving MapReduce Fault Tolerance in the Cloud*. Proceedings of the IEEE International Symposium on Parallel and Distributed Processing, Workshops and PhD Forum (IPDPSW). IEEE, pp. 1–6.

25

Cloud Energy Consumption

Dan C. Marinescu

University of Central Florida, USA

25.1 Introduction and Motivation

Cloud computing has revolutionized our thinking about information processing and data storage. Its impact on activities of many organizations, large and small, and on individual application developers, is a proof of the economic benefits of the new paradigm. In addition to low cost and the advantages of the utility model (users pay only for the resources they consume, "elasticity" – the ability of the system to supply to an application the precise amount of resources needed, reliability, security); cloud computing is promoted as a realization of the "green computing" ideal, an IT infrastructure with a considerably smaller carbon footprint than the traditional ones.

Indeed, cloud computing has the potential to reduce the energy consumption for computing and data storage, thus shrinking the carbon footprint for IT-related activities. This potential is far from being realized by existing cloud infrastructures. To understand the reasons for this state of affairs, we discuss cloud energy consumption and its relationship with other aspects of cloud resource management.

The rapid expansion of the cloud computing community helps us realize the full impact of cloud computing on energy consumption in the United States and the world. The number of cloud service providers (CSPs), the spectrum of services offered by the CSPs, and the number of cloud users have increased dramatically during the last few years. For example, in 2007, Elastic Cloud Computing (the EC2) was the first service provided by Amazon Web Services (AWS); 5 years later, in 2012, *AWS* was used by businesses in 200 countries. Amazon's Simple Storage Service (*S3*) has surpassed two trillion objects and routinely runs more than 1.1 million peak requests per second. The Elastic MapReduce has launched 5.5 million clusters since the start of the service in May 2010 (ZDNet, 2013).

The infrastructure for supporting cloud services is continually growing. A recent posting on ZDNet reveals that, in January 2012, *EC2* was made up of 454 600 servers. When one adds the number of servers supporting other *AWS* services, then the total number of Amazon systems dedicated to cloud computing is much larger.

Encyclopedia of Cloud Computing, First Edition. Edited by San Murugesan and Irena Bojanova.
© 2016 John Wiley & Sons, Ltd. Published 2016 by John Wiley & Sons, Ltd.

Table 25.1   *The costs for a medium utilization instance for quadruple extra-large reserved memory in two different regions for EC2*

Costs/ region	US East (N. Virginia)	South America (Sao Paolo)
Upfront payment for a year	$2604	$5632
Per hour cost	$0.412	$0.724

Amazon, Google, and Microsoft were the first to offer infrastructure as a service (IaaS), software as a service (SaaS), and platform as a service (PaaS), respectively. In recent years, a fair number of IT organizations began offering SaaS and PaaS services. There are differences in energy consumption among the three cloud delivery models IaaS, SaaS, and PaaS; there are also differences among public, private, and hybrid clouds.

A public cloud provides services to a large user community with diverse applications; the hardware is shared by multiple customers and applications. Users access the public cloud from anywhere on the Internet and only pay for the services they use. A private cloud serves a single organization, typically maintained behind a firewall and accessed through an intranet. A set of diverse applications may run on a private cloud. An organization may purchase its own private, isolated group of servers from public cloud service providers such as Amazon or Rackspace.

The costs of maintaining the cloud-computing infrastructure are significant. A large fraction of these costs is for datacenter housing and cooling and for powering the computational and storage servers and the interconnection networks.

The energy costs are different in different counties and in different regions of the same country. Most of the power for large datacenters, including cloud computing datacenters, comes from power stations burning fossil fuels such as coal and gas but in recent years the contribution of solar, wind, geothermal and other renewable energy sources has steadily increased.

The energy costs are passed down to the users of cloud services and differ from one region to another as we can see from Table 25.1 (Amazon, 2013), which shows the cost in two regions: United States East and South America. The energy costs for the two regions differ by about 40%. The higher energy and networking costs are responsible for the significant price difference in this example.

We have witnessed improvements in the *energy efficiency of computing* – the number of computations completed per kWh of electricity. We have also seen an improvement in datacenters' energy efficiency. A measure of the energy efficiency of a datacenter, the so-called power usage effectiveness (PUE), is the ratio of the total energy used to power a datacenter to the energy used to power computational servers, storage servers, routers, and other IT equipment. The PUE has improved from around 1.93 in 2003 to 1.63 in 2005. Recently, it was reported that Google's ratio is as low as 1.15 (Baliga *et al.*, 2011).

The increased energy efficiency of the cloud infrastructure, though significant, had a relatively small impact on the total energy used for cloud-related activities. Indeed, in recent years we have witnessed a dramatic increase in the number of mobile devices that access data stored on the cloud; this leads to a significant increase in the energy used to transfer data to and from the clouds.

The energy used for IT activities represents a significant and continually increasing component of the total energy used by a nation; its ecological impact is of serious concern. Thus, it is no surprise that energy optimization is an important dimension of the cloud resource management policies, discussed in section 25.2.

Economy of scale, analyzed in section 25.5, affects the energy efficiency of data processing. For example, Google reports that the annual energy consumption for an e-mail service varies significantly depending on the business size and can be 15 times larger for a small business than for a large one (Google, 2013). Cloud computing can be more energy efficient than on-premises computing for many organizations (Baliga *et al.*, 2011;

Table 25.2   *Estimated average power use of volume, mid-range, and high-end servers (in Watts)*

Type/year	2000	2001	2002	2003	2004	2005	2006
Volume	186	193	200	207	213	219	225
Medium	424	457	491	524	574	625	675
High end	5534	5832	6130	6428	6973	7651	8163

NRDC and WSP, 2012). The energy used to transport the data is a significant component of the total energy cost and according to (Baliga *et al.*, 2011) "a public cloud could consume three to four times more power than the private cloud due to increased energy consumption in transport."

One of the main appeals of utility computing is cloud elasticity; additional resources are allocated when an application needs them and released when they are no longer needed. The user ends up paying only for the resources it has actually used. Elasticity currently comes at a stiff price as cloud resource management is based on *overprovisioning*. This means that a cloud service provider has to invest in a larger infrastructure than the "typical" cloud load warrants. It follows that the average cloud server utilization is low (Abts *et al.*, 2010; Ardagna *et al.*, 2012; Google, 2013; Marinescu, 2013); the low server utilization affects negatively the common measure of energy efficiency, the performance per Watt of power and the ecological impact of cloud computing, the topic of sections 25.4 and 25.5.

The power consumption of servers has increased over time (Koomey, 2007). Table 25.2 shows the evolution of the average power consumption for volume servers (servers with a price less than $25 000), mid-range servers (servers with a price between $25 000 and $499 000), and high-end servers (servers with a price tag greater than $500 000).

Reduction in energy consumption and thus of the carbon footprint of cloud-related activities is increasingly more important for society. Indeed, more and more applications run on clouds, and cloud computing uses more energy than many other human-related activities. Reduction of the carbon footprint can only be achieved through a comprehensive set of technical efforts. The hardware of the cloud infrastructure has to be refreshed periodically and new and more energy-efficient technologies have to be adopted; the resource management software has to pay more attention to energy optimization; the housing and cooling cost have to be reduced, and last, but not least, more energy for powering the cloud infrastructure should come from renewable energy sources.

Finally, a word of caution: the predictions formulated by organizations such as the Department of Energy (DOE), the National Laboratories, the National Resource Defense Council, or partnerships between industry and universities such as CEET, a joint venture of the Bell Labs and the University of Melbourne, regarding many aspects of cloud computing differ, as they are based on different models. Typically such models assume several scenarios for each one of the model parameters – worst case, average, and best case. The data on cloud resource utilization and energy consumption reported by different sources are often slightly different.

25.2   Cloud Resource Management Policies and Energy Optimization

Resource management is a core function of any system; it critically affects the performance, the functionality, and the cost, the three basic criteria for the evaluation of a system. An inefficient resource management has a direct negative effect on the performance and the cost and an indirect effect on the functionality of the system; indeed, some functions provided by the system may become too expensive or may be avoided due to the poor performance (NRDC and WSP, 2012).

A cloud is a complex system with a very large number of shared resources subject to unpredictable requests and affected by external events it cannot control. Cloud resource management requires complex policies and decisions for multiobjective optimization. Cloud resource management is extremely challenging because of the complexity of the system, which makes it impossible to have accurate global state information and because of unpredictable interactions with the environment (Marinescu, 2013).

The strategies for resource management associated with the three cloud delivery models, IaaS, PaaS, and SaaS, are different. In all cases the cloud service providers are faced with large fluctuating loads, which challenge the claim of cloud elasticity. The resources can be provisioned in advance when a spike can be predicted, for example, for Web services subject to seasonal spikes.

Energy optimization is one of the five cloud resource management policies; the others are:

- admission control;
- capacity allocation;
- load balancing; and
- quality of service (QoS).

The explicit goal of an admission-control policy is to prevent the system from accepting workload in violation of high-level system policies; for example, a system may not accept additional workload that would prevent it from completing work already in progress or contracted. Limiting the workload requires some knowledge of the global state of the system. This becomes more challenging as the scale of the system increases. Moreover, in a dynamic system such knowledge, when available, is at best obsolete due to the fast pace of state changes.

Capacity allocation means allocating resources for individual instances; an instance is an activation of a service. Locating resources subject to multiple global optimization constraints requires a search of a very large search space when the state of individual systems changes rapidly.

Load balancing and energy optimization can be done locally but global load-balancing and energy-optimization policies encounter the same difficulties as the ones we have already discussed. Load balancing and energy optimization are correlated and affect the cost of providing the services.

There are strong interdependencies among these policies and the mechanisms for implementing them. Many mechanisms are concentrated on system performance in terms of throughput and time-in-system but rarely include either energy optimization or QoS guarantees (Ardagna *et al.,* 2012, NRDSC and WSP, 2012; Paya and Marinescu, 2013). Self-management and self-organization could overcome the limitations of these mechanisms (Marinescu *et al.,* 2013).

25.3 Energy Proportional Systems and Server Utilization

There is a mismatch between server workload profile and server energy efficiency (Barosso and Hölzle, 2010). In an ideal world, the energy consumed by an idle system should be near zero and should grow linearly with the system load. In real life, even systems whose power requirements scale linearly, when idle, use more than half the power they use at full load (Abts *et al.,* 2010). Indeed, a 2.5 GHz Intel E5200 dual-core desktop processor with 2 GB of RAM consumes 70 W when idle and 110 W when fully loaded; a 2.4 GHz Intel Q6600 processor with 4 GB of RAM consumes 110 W when idle and 175 W when fully loaded (Baliga *et al.,* 2011).

An *energy-proportional system* is one when the amount of energy used is proportional to the load of the system. Different subsystems of a computing system behave differently in terms of energy efficiency; while many processors have reasonably good energy-proportional profiles, significant improvements in memory and disk subsystems are necessary. The processors used in servers consume less than one-third of their peak power at very-low load and have a dynamic range of more than 70% of peak power; the processors used in mobile and / or embedded applications are better in this respect.

The dynamic power range of other components of a system is much narrower (Barosso and Hölzle, 2010):

- less than 50% for DRAM;
- 25% for disk drives; and
- 15% for networking switches.

The power consumption of such devices is: 4.9 kW for a 604.8 TB, HP 8100 EVA storage server, 3.8 KW for the 320 gbps Cisco 6509 switch, 5.1 kW for the 660 gbps Juniper MX-960 gateway router (Baliga *et al.,* 2011).

A number of proposals have emerged for energy-proportional networks; the energy consumed by such networks is proportional to the communication load. For example, a datacenter network based on a flattened butterfly topology is more energy and cost efficient (Abts *et al.,* 2010).

Servers running a single application may be 5% to 15% utilized. Virtualization increases the energy efficiency of a system because it allows multiple virtual machines to share one physical system. The effects of virtualization are captured by a recent model developed by the NRDC (National Resource Defense Council) and WSP Environment and Energy, based on a wealth of data from multiple sources. The model gives the server utilization in several environments and for each it considers the worst case, the average case, and the best case scenarios (NRDC and WSP, 2012), see Table 25.3.

Virtualization leads to better utilization for on-premises operation but utilization is only slightly better as we move from on-premises with virtualization, to private, and then to public clouds.

The utilization of cloud servers can be low in spite of the economy of scale and the advantages of virtualization. Typical server utilization is in the 10% to 50% range, as we can see in Figure 25.1 which illustrates the relationship between the power used by a system and the utilization of the system (Barosso and Hölzle, 2010). A 2010 report (Abts *et al.,* 2010) shows that a typical Google cluster spends most of its time within the 10–30% CPU utilization range.

Similar behavior is also seen in the datacenter networks; these networks operate in a very narrow dynamic range, and the power consumed when the network is idle is significant compared to the power consumed when the network is fully utilized. High-speed channels typically consist of multiple serial lanes with the same data rate; a physical unit is stripped across all the active lanes. Channels commonly operate plesiochronously and are always on, regardless of the load, because they must still send idle packets to maintain byte and lane alignment across the multiple lanes. An example of an energy-proportional network is the Infiniband, a switched fabric interconnection network used by many supercomputers. The main reasons why Infiniband is also used as the interconnection network of datacenters are: scalability, high throughput, low latency, QoS support, and failover support. A network-wide power manager called Elastic Tree, which dynamically adjusts the network links and switches to satisfy changing datacenter traffic loads is described in Heller *et al.,* 2010).

Table 25.3 *Server utilization in several environments*

Environment/case	Worst case (%)	Average case (%)	Best case (%)
On-premises nonvirtualized	5	10	25
On-premises with virtualization	6	30	60
Private cloud	7	40	60
Public cloud	7	30	70

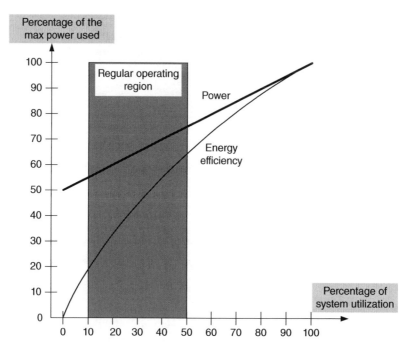

Figure 25.1 *The relationship between the power used and the system utilization; typical utilization of a server is in the 10% to 50% range*

Energy saving in large-scale storage systems is also of concern. One strategy to reduce energy consumption is to concentrate the workload on a small number of disks and allow the others to operate in low-power modes. Public and private clouds are more energy efficient than local storage of data provided that data is accessed infrequently (Baliga *et al.*, 2011).

In cloud computing, a critical goal is minimizing the cost of providing the service and, in particular, minimizing the energy consumption. This leads to a different meaning of the term "load balancing"; instead of having the load evenly distributed amongst all servers, we wish to concentrate it and use the smallest number of servers while switching the others to a standby or sleep mode, a state where a server uses very little energy. This strategy is discussed in Preist and Shabajee (2010), Qian and Medhi (2011) and Paya and Marinescu (2013).

Dynamic voltage and frequency scaling (DVFS) is a power-management technique to increase or decrease the operating voltage or frequency of a processor to increase the instruction execution rate, to reduce the amount of heat generated, and to conserve power. Techniques such as Intel's SpeedStep and AMD's PowerNow lower the voltage and the frequency to decrease the power consumption. Motivated by the need to save power for mobile devices, these techniques have migrated to processors used for high-performance servers.

25.4   Energy-Aware Load Balancing and Server Consolidation

Recently, Gartner Research reported that the average server utilization in large datacenters is 18% (Snyder, 2010) while the utilization of x86 servers is even lower, 12%. These results confirm earlier estimations that the average server utilization was in the 10–30% range. A 2010 survey (Blackburn and Hawkins, 2010) reported that idle servers contributed 11 million tons of unnecessary CO_2 emissions each year and that the total yearly costs for idle servers was 19 billion.

A strategy for resource management in a computing cloud is to concentrate the load on a subset of servers and switch the rest of them to a sleep state whenever possible. This strategy aims to reduce the power consumption and, implicitly, the cost of providing computing and storage services.

A comprehensive document produced by Hewlett-Packard, Intel, Microsoft, Phoenix Technologies, and Toshiba (Hewlett-Packard *et al.*, 2011) describes the advanced configuration and power interface (ACPI) specifications, which allow an operating system (OS) to manage the power consumption of the hardware effectively. Several types of *sleep states* are defined:

- C-states (C1-C6) for the CPU;
- D-states (D0-D3) for modems, hard drives, and CD-ROM; and
- S-states (S1-S4) for the basic input-output system (BIOS).

The C-states allow a computer to save energy when the CPU is idle. In a sleep state, the idle units of a CPU have their clock signal and the power cut. The higher the state number, the deeper the CPU sleep mode, the larger the amount of energy saved, and the longer the time for the CPU to return to the state C0, which corresponds to the fully operational CPU.

Some of the questions posed by energy-aware load balancing discussed in Paya and Marinescu (2014) are:

- Under what conditions a server should be switched to a sleep state?
- What sleep state should the server be switched to?
- How much energy is necessary to switch a server to a sleep state and then switch it back to an active state?
- How much time does it take to switch a server in a sleep state to a running state?
- How much energy is necessary to migrate a VM running on a server to another one?
- How much energy is necessary to start a VM on the target server?
- How to choose the target for the migration of a VM?
- How much time does it take to migrate a VM?

Two basic metrics ultimately determine the quality of an energy-aware load-balancing policy:

- The amount of energy saved.
- The number of violations it causes.

In practice, the metrics depend on the system load and other resource management policies such as the admission-control policy and the QoS guarantees offered. The load can be slow- or fast-varying, can have spikes or be smooth, can be predictable or totally unpredictable; the admission control can restrict the acceptance of additional load when the available capacity of the servers is low. What we can measure in practice is the *average energy* used and the *average server setup time.* The setup time varies depending on the hardware and the operating system and can be as long as 260 s (Gandhi *et al.*, 2012b); the energy consumption during the setup phase is close to the maximal one for the server.

The time to switch the servers to a running state is critical when the load is fast varying, the load variations are very steep, and the spikes are unpredictable. The decisions when to switch servers to a sleep state and back to a running state are less critical when a strict admission control policy is in place; then new service requests for large amounts of resources can be delayed until the system is able to turn on a number of sleeping servers to satisfy the additional demand.

Energy-optimization algorithms decide when a server should enter a sleep state, the type of sleep state, and when to wake up. Such algorithms have been proposed by Gandhi and Harchol-Balter (2011) and Gandhi *et al.* (2012a, b, c).

25.5   Economy of Scale and Energy Consumption

Large enterprises often have significant cost advantages due to their scale, as fixed costs are spread out over more units of output. Often, operational efficiency is greater with increasing scale leading to lower variable cost as well. It is thus reasonable to expect that economy of scale will manifest itself in cloud computing where large datacenters are able to operate more efficiently than smaller or medium-size ones. The actual cost for cloud services indicates that this expectation is justified.

We argued earlier that energy costs are a significant component of the total costs for providing cloud services so we should not be surprised that a cloud infrastructure is more energy efficient than a small or medium-size datacenter. A 2013 study led by the Lawrence Berkeley National Laboratory with funding from Google (LBL, 2013), considers three common business applications – e-mail, customer relationship management software, and productivity software such as spreadsheets, file sharing, and word processing – and concludes that: "moving these applications from local computer systems to centralized cloud services could cut information technology energy consumption by up to 87% – about 23 billion kilowatt-hours. This is roughly the amount of electricity used each year by all the homes, businesses, and industry in Los Angeles."

The September 12, 2012 issue of the *New York Times* quotes a senior Google executive: "Google's servers refresh 20 billion pages a day, process over 100 billion search queries a month, provide e-mail for 425 million Gmail users and process 72 hours of video uploaded per minute to YouTube. And yet we're able to do all that work with relatively little energy, compared to other industries."

A Google report (Google, 2013) discusses costs for an e-mail service for organizations of different sizes; this data is summarized in Table 25.4.

The energy consumption per user per year for powering the servers is almost 15 times lower for a large center compared to a small center, while the energy used for housing and cooling is 23 times lower. The carbon emissions are more than 20 times smaller for a large center compared with a small one.

A recent report (NRDSC and WSP 2012) compares the carbon emissions for on-premises versus clouds. One of the main conclusions of this report is that "running a business application on the cloud is generally more energy and carbon efficient than running it on-premises because cloud computing offers greater diversity and can serve more customers at the same time, achieving better economy of scale than small and medium organization." The same report concludes that a private cloud offers essentially the same benefits as a public cloud in terms of energy consumption.

The carbon emissions per GB of storage for three modes of operation: (i) on-premises with virtualization; (ii) private cloud; and (iii) public cloud and three scenarios: (a) worst case; (b) average, and (c) best case are summarized in Table 25.5.

The carbon emissions for data storing on-premises with virtualization and private clouds are similar and are significantly lower in case of the public clouds.

Table 25.4   *The costs for an e-mail service for small organizations, with up to 50 users, medium ones with up to 500 users, and large ones with up to 10 000 users*

Annual energy and CO_2 emissions / e-mail service	Small	Medium	Large
Annual energy/user for powering the servers (kWh)	70	16	4.7
Annual energy/user for heating and cooling (kWh)	175	28.4	7.6
Annual CO_2 emissions/users (Kg)	103	16.7	1.23

Table 25.5   *Carbon emissions per GB of storage for three modes of operation and three scenarios*

Operation mode /scenario	Worst case (kg)	Average case (kg)	Best case (kg)
On-premises with virtualization	15.9	7.9	1.7
Private cloud	12.7	5.0	1.4
Public cloud	11.1	1.4	0.6

25.6   Energy Use and Ecological Impact of Large Datacenters

A conservative estimation of the electric energy used now by the information and communication technology (ICT) ecosystem is 1500 TWh of energy, 10% of the electric energy generated in the entire world. Energy consumption for ICT equals the total electric energy used for illumination in 1985 and represents the total electric energy generated in Japan and Germany.

The ICT ecosystem consist of:

- datacenters;
- broadband wired and wireless communication networks;
- end-user devices such as PCs, tablets, smartphones, and digital TVs;
- manufacturing facilities producing ICT hardware.

The number of end-user devices is increasing. In 2013 there were 1.5 billion PCs and an equal number of smart mobile devices. It was predicted that the number of PCs would level off, whereas the number of smart mobile devices would increase rapidly (Mills, 2013).

The US demand for electricity grew around 30% per year since the mid-1990s (EIA 2013). The energy consumption of large-scale datacenters and their costs for energy used for computing and networking and for cooling are significant now and are expected to increase substantially in the future. In 2006, the 6000 data centers in the United States reportedly consumed 61×10^9 kWh of energy, 1.5% of all electricity consumption in the country, at a cost of $4.5 billion (Vrbsky *et al.*, 2010).

The predictions are dire. The energy consumption of data centers and the network infrastructure is predicted to reach 10 300 TWh/year in 2030, based on 2010 levels of efficiency (Vrbsky *et al.*, 2010).

Cloud computing is only possible because the Internet allows user access to the large datacenters; but the Internet itself consumes a fair amount of energy as we shall see shortly. This amount is likely to increase dramatically. The hourly Internet traffic will shortly exceed the annual traffic in 2000 (Mills, 2013)!

A study produced by CEET (2013), a joint venture of the Bell Labs and the University of Melbourne, warns that [ex]access networks, not data centers, are the biggest threat to the sustainability of cloud services. This is because more people are accessing cloud services via wireless networks. These networks are inherently energy inefficient and a disproportionate contributor to cloud energy consumption …

The support for network centric content consumes a very large fraction of the network bandwidth; according to the CISCO VNI forecast, in 2009 consumer traffic was responsible for around 80% of the bandwidth used and is expected to grow at a faster rate than business traffic. Data intensity for different activities varies widely (Preist and Shabajee, 2010) as shown in Table 25.6.

The same study reports that if the energy demand for bandwidth is 4 W/h per MB and if the demand for network bandwidth is 3.2 GB/day/person or 2570 EB/year for the entire world population, then the energy required for this activity will be 1175 GW. These estimates do not count very high bandwidth applications that may emerge in the future, such as 3D-TV, personalized immersive entertainment, such as Second Life,

Table 25.6   *Data intensity for different types of activities*

Activity	Data intensity (MB/minute)
HDTV streaming	20
Standard TV streaming	10
Music streaming	1.3
Internet radio	0.96
Internet browsing	0.35
E-book reading	0.0024

or massive multiplayer online games. As we have seen in section 25.3, the power consumption of the networking infrastructure is significant; as the volumes of data transferred to and from a cloud increase, the power consumption for networking will also increase.

We now take a closer look at the *energy efficiency of computing* measured as the number of computations completed per kWh of electricity used. In 1985, Richard Feynman, the 1965 Nobel laureate in Physics for his contributions to quantum electrodynamics, estimated that computing technology which uses electrons for switching could theoretically improve by a factor of 10^{11} relative to the technology used at that time. In reality, the actual performance per kWh improved only by a factor of 4×10^4 since 1985, far from Feynman's theoretical limit (Koomey *et al.*, 2011).

The popular summary of Moore's law (not really a "physical law" but an "empirical observation" stating that the number of transistors on a chip, and thus the complexity of the circuit, doubles every 18 months) is that computing performance doubles approximately every 1.5 years. Koomey *et al.* (2011) reported that electrical efficiency of computing doubles also about every 1.5 years. Thus, performance growth rate and improvements in electrical efficiency almost cancel each out. It follows that the energy use for computing scales linearly with the number of computing devices.

The power consumption required by different types of human activities is partially responsible for greenhouse gas emissions. According to Vrbsky *et al.* (2010), the greenhouse gas emission due to datacenters was estimated to increase from $116\,Mt$ of CO_2 in 2007 to $257\,Mt$ in 2020, due primarily to increased consumer demand.

A comprehensive report released in 2012 (NRDSC and WSP, 2012) identifies four major factors that affect the carbon emissions of different types of IT activities:

- PUE – the ratio of the total energy consumption of a center to the fraction of energy used to power the servers and the routers of the center.
- The average server utilization (U) – the lower U is, the lower the energy efficiency is.
- The refresh period of the equipment (P) – the time between successive replacements of equipment with newer equipment.
- The ratio (R) of virtualized servers versus nonvirtualized ones.

For example, for a public cloud, assuming an average scenario, the *grid emission factor measured in kg of CO_2 equivalent per kilowatt-hour* is 0.554. This factor improves to 0.268 for a best case scenario but worsens to 0.819 for a worst case scenario as shown in Table 25.7.

The CO_2 emissions are identical for a private cloud but under more relaxed conditions – see Table 25.8.

It is reported that only $3\,W$ out of every $100\,W$ of power consumed in a typical datacenter end up contributing to useful computation. One can only conclude that datacenters should increase their PUE, adopt newer technologies, and improve their resource-management policies.

Table 25.7 The grid emission factor measured in kg of CO_2 equivalent per kWh for a public cloud

Grid emission factor ($Kg\ CO_2$ / kWh)	PUE	U (%)	P (years)	R
0.819	2.0	7	3	5/1
0.544	1.5	40	2	8/1
0.268	1.1	70	1	12/1

Table 25.8 The grid emission factor measured in kilograms of CO_2 equivalent per kWh for a private cloud

Grid emission factor ($Kg\ CO_2$ / kWh)	PUE	U (%)	P (years)	R
0.819	2.5	7	5	3/1
0.544	1.8	30	3	5/1
0.268	1.3	60	2	10/1

The study (NRDSC and WSP, 2012) shows that the source of electric power has a dramatic effect on the carbon footprint of a datacenter. The carbon footprint of a center using power generated from renewable energy sources can be four times smaller than the one of a center using high-carbon sources. The *key variables* that determine the carbon emissions of a datacenter are:

- the PUE;
- the server utilization; and
- the carbon emission factor of the electricity used to power the center, in kg of CO_2.

For example, in the United State the carbon emission is:

- 70–80 kg of CO_2 per year – when PUE = 3.0, the average server utilization is 5%, and the electricity used by the cloud is generated by a high carbon-intensity source.
- 30–35 kg of CO_2 per year – when PUE = 1.5, server utilization is 40%, and the electricity used by the cloud is generated by an average carbon-intensity source.

25.7 Summary and Conclusions

Optimization of energy consumption, along with improved security and privacy, is critical for the future of cloud computing. There are natural tensions between basic tenets of cloud computing. A public cloud is attractive for the users due to its elasticity and the ability to provide additional resources when needed, but the price to pay is overprovisioning. In turn, overprovisioning leads to lower server utilization, and thus to higher energy costs. The physical components of the cloud infrastructure are not energy proportional; even under a low load or when idle, their energy consumption represents a large fraction of the energy consumed at full load.

Virtualization by multiplexing allows multiple virtual machines to share the same physical system. In principle, this can lead to higher server utilization but virtualization adds a considerable overhead. Recall that the hypervisor and the management operating system (OS), running on one of the virtual machines, control the allocation of local resources. Communication and the I/O requests of an application have to be channeled through the management OS; system calls issued by an application result in hypercalls to the hypervisor.

The overhead due to virtualization cannot be easily quantified; it differs from one hypervisor to another, and it depends on the hardware and the operating system. For example, Xen pays a high penalty for context switching and for handling a page fault. The CPU utilization of a VMware Workstation system running a version of Linux was five to six times higher than that of a native system in saturating a network (Marinescu, 2013).

At the same time, virtualization allows migration of virtual machines for workload consolidation but there is a considerable overhead and, thus, additional energy consumption associated with virtual machine migration. Multitenancy, sharing a physical platform among a number of virtual machines, adds to the system overhead. Data replication is another source of overhead and increased energy consumption.

It is expected that hardware will continue to improve, power management techniques such as DVFS will improve the efficiency of computing, and new technologies, such as solid-state disks, will use much less energy than mechanical disks. The scheduling algorithms and the system software running on the clouds will continue to improve and make their own contribution to energy saving.

Computer clouds contribute indirectly to energy consumption and the number of smart mobile devices used to access data stored on the cloud is increasing rapidly. Research on energy-aware communication protocols and energy efficient switches should lead to new communication software and hardware capable of sustaining the dramatic increase in the volume of the Internet traffic to/from a cloud.

It is likely that the SaaS cloud-delivery model will continue to attract a larger user community than the other two models due to the wide range of SaaS applications and the ease of access to these services. Even though the cloud service providers are faced with large fluctuating loads, the resource management seems less challenging in the SaaS case. Energy can be saved by requiring all running systems to operate in an optimal region. The load from lightly loaded servers should be migrated to other servers of the cloud and these systems should then be switched to a sleep state to save energy, and when the load increases, they should be switched back to a running state (Paya and Marinescu, 2013). Some of these services involve data streaming thus, increased network traffic and energy consumption.

The spectrum of PaaS services will also continue to grow as more and more e-commerce and business applications migrate to the cloud and demand sophisticated analytics. The number of CSPs supporting this cloud delivery model, the number of applications, and the user population, will most likely increase in the future. Performance optimization of applications running under this model and the corresponding energy optimization will continue to be challenging as performance isolation cannot be ensured due to multitenancy. Energy optimization in the context of the IaaS delivery model is probably the most challenging but, as this model appeals only to a smaller population of sophisticated users, its impact on the overall cloud energy consumption will be modest.

While public clouds will continue to attract the largest number of applications and users, private clouds and hybrid clouds will develop their own audience. Organizations with very strict security and privacy concerns will opt for these types of clouds. A private cloud that groups together the computing resources of an organization can offer the same benefits in terms of energy optimization as a public cloud (Marinescu, 2013).

Hybrid clouds combine the advantages of public cloud – elasticity and low cost – with the security and privacy characteristic of private clouds. Applications in many areas, including healthcare and engineering, could benefit from operating hybrid clouds. For example, it is conceivable that a smart power grid will be based on a hybrid cloud where each electric utility company will maintain proprietary data and code on its private cloud, while "big data" will be stored on a public cloud along with information needed by all utilities to operate in concert.

We expect the CSPs to place their datacenters in regions with low energy consumption to reduce costs. At the same time, it seems likely that energy from renewable sources such as solar, wind, and geothermal will represent a significant fraction of the overall energy consumption of big datacenters. Already solar energy is used to power some of Google's cloud-computing infrastructure.

Better policies and mechanisms addressing all aspects of resource management should be coupled with server upgrades to reduce the overall energy consumption. Newer technologies such as solid-state storage devices, or communication systems with a larger dynamic range, coupled with short refresh periods for the equipment, will lower the energy consumption and reduce the carbon footprint attributed to computing and communication activities.

References

Abts, D., Marty, M., Wells, P.M., *et al.* (2010) Energy proportional datacenter networks. Proceedings of the International Symposium on Computer Architecture (ISCA '10), pp. 238–247.

Amazon (2013) *AWS EC2 Pricing*, http://aws.amazon.com/ec2/pricing/ (accessed December 14, 2015).

Ardagna, D., Panicucci, B., Trubian, M., and Zhang, L. (2012) Energy-aware autonomic resource allocation in multi-tier virtualized environments. *IEEE Transactions on Services Computing* **5**(1), 2–19.

Baliga, J., Ayre, R. W. A., Hinton, K., and Tucker, R. S. (2011) Green cloud computing: Balancing energy in processing, storage, and transport. *Proceedings of the IEEE* **99**(1), 149–167.

Barosso, L. A. and Hölzle, U. (2010) The case for energy-proportional computing. *IEEE Computer* **40**(12), 33–37.

Blackburn, M. and Hawkins, A. (2010) *Unused Server Survey Results Analysis*, http://www.thegreengrid.org/~/media/WhitePapers/Unused%20Server%20Study_WP_101910_v1.ashx?lang=en (accessed December 14, 2015).

CEET (2013) *The Power of Wireless Cloud*, http://www.ceet.unimelb.edu.au/publications/downloads/ceet-white-paper-wireless-cloud.pdf (accessed December 15, 2015).

EIA (2013) *Annual Energy Outlook 2013*, US Department of Energy, Washington DC.

Gandhi, A. and Harchol-Balter, M. (2011) How data center size impacts the effectiveness of dynamic power management. Proceedings of the 49th Annual Allerton Conference on Communication, Control, and Computing, Urbana Champaign, IL, pp. 1864–1869.

Gandhi, A., Harchol-Balter, M., Raghunathan, R., and Kozuch, M. (2012a) Are sleep states effective in data centers? Proceedings of the International Conference on Green Computing, pp. 1–10.

Gandhi, A., Harchol-Balter, M., Raghunathan, R., and Kozuch, M. (2012b) AutoScale: dynamic, robust capacity management for multi-tier data centers. *ACM Transactions on Computer Systems* **30**(4), 1–26.

Gandhi, A., Zhu, T., Harchol-Balter, M., and Kozuch, M. (2012c) *SOFTScale: Stealing Opportunistically for Transient Scaling*. Proceedings of Midlleware 2012. Lecture Notes in Computer Science. Springer Verlag, Berlin, pp. 142–163.

Google (2013) *Google's Green Computing: Efficiency at Scale*, http://static.googleusercontent.com/external_content/untrusted_dlcp/www.google.com/en/us/green/pdfs/google-green-computing.pdf (accessed December 14, 2015).

Heller, B., Seetharaman, S., Mahadevan, P., *et al.* (2010) ElasticTree: Saving energy in data center networks. Proceedings of the Seventh USENIX Conference on Networked Systems Design and Implementation, pp. 17–17.

Hewlett-Packard, Intel, Microsoft *et al.* (2011) *Advanced Configuration and Power Interface Specification*. Revision 5.0, http://www.acpi.info/DOWNLOADS/ACPIspec50.pdf (accessed December 14, 2015).

Koomey, J. G. (2007) *Estimating Total Power Consumption by Servers in the US and the World*, http://www-sop.inria.fr/mascotte/Contrats/DIMAGREEN/wiki/uploads/Main/svrpwrusecompletefinal.pdf (accessed December 14, 2015).

Koomey, J. G., Berard, S., Sanchez, M. and Wong, H. (2011) Implications of historical trends in the energy efficiency of computing. *IEEE Annals of Computing* **33**(3), 46–54.

LBL (2013) *Moving Computer Services to Cloud Promises Big Energy Savings*, http://crd.lbl.gov/news-and-publications/news/2013/study-moving-computer-services-to-the-cloud-promises-significant-energy-savings (accessed December 14, 2015).

Marinescu, D. C. (2013) *Cloud Computing: Theory and Practice*, Morgan Kaufmann, New York, NY.

Marinescu, D. C., Paya, A., Morrison, J. P., and Healy, P. (2013) *An Auction-Driven, Self-Organizing Cloud Delivery Model*, http://arxiv.org/pdf/1312.2998v1.pdf (accessed December 14, 2015).

Mills, M. P. (2013) *The Cloud Begins with Coal*, http://www.tech-pundit.com/wp-content/uploads/2013/07/Cloud_Begins_With_Coal.pdf (accessed December 14, 2015).

NRDC and WSP (2012) *The Carbon Emissions of Server Computing for Small- to Medium-Sized Organizations – A Performance Study of On-Premise vs. The Cloud*, https://www.wspgroup.com/Globaln/USA/Environmental/Sustainability/Documents/NRDC-WSP_Cloud_Computing.pdf (accessed December 14, 2015).

Paya, A. and Marinescu, D. C. (2013) *Energy-Aware Application Scaling on a Cloud*, http://arxiv.org/pdf/1307.3306v1.pdf (accessed December 14, 2015).

Paya, A. and Marinescu, D. C. (2014) *Energy-Aware Load Balancing Policies for the Cloud Ecosystem*, http://arxiv.org/abs/1401.2198 (accessed December 14, 2015).

Preist, C. and Shabajee, P. (2010) Energy use in the media cloud. Proceedings of the IEEE International Conference on Cloud Computing Technology and Science, pp. 581–586.

Qian, H., and Medhi, D. (2011) Server operational cost optimization for cloud computing service providers over a time horizon. Proceedings of USENIX Workshop on Hot Topics in Management of Internet, Cloud, and Enterprise Networks and Services (HotICE 2011), Boston, MA, http://citeseerx.ist.psu.edu/viewdoc/download?doi=10.1.1.190.6851&rep=rep1&type=pdf (accessed December 14, 2015).

Snyder, B. (2010) *Server Virtualization has Stalled, Despite the Hype*, http://www.infoworld.com/article/2624771/server-virtualization/server-virtualization-has-stalled--despite-the-hype.html (accessed December 14, 2015).

Vrbsky, S. V., Lei, M., Smith K., and Byrd, J. (2010) *Data Replication and Power Consumption in Data Grids*. Proceedings of the IEEE International Conference on Cloud Computing Technology and Science. IEEE, pp. 288–295.

ZDNet (2013) *Amazon CTO Werner Vogels: "Infrastructucture is Not a Differentiator,"* http://www.zdnet.com/amazon-cto-werner-vogels-infrastructure-is-not-a-differentiator-7000014213 (accessed December 14, 2015).

26

Cloud Modeling and Simulation

Peter Altevogt,[1] Wolfgang Denzel,[2] and Tibor Kiss[3]

[1] *IBM Germany Research & Development GmbH, Germany*
[2] *IBM Research GmbH, Zurich Research Laboratory, Switzerland*
[3] *IBM Hungary Ltd., Budapest, Hungary*

26.1   Introduction

Cloud computing is perceived as a game-changing technology for the provisioning and consumption of data-center resources (Kavis, 2014). Various nonfunctional attributes like performance, scalability, cost, energy efficiency, resiliency, high availability, and security are considered vital to enable the further growth of cloud computing (Armbrust *et al.,* 2009).

An engineering approach, limited only to measurements, tuning, and fixing defects, is not sufficient to ensure a cloud design would address these attributes. In general, these activities happen late in the development cycle. Therefore, they cannot adequately address cloud-design issues like scalability of algorithms to assign virtual machines (VMs) to certain compute nodes or the performance impact of various network topologies. Furthermore, only a very limited set of scenarios can be handled this way due to time and resource limitations. Although most development projects face these limitations, their impact on cloud computing is especially severe due to the huge resources and costs required to create and manage large-scale clouds.

Modeling and simulation technologies can help address these issues. They can be applied to model various cloud attributes like energy dissipation or to evaluate different pricing models. This chapter will focus on performance and scalability. The discrete-event simulation approach described here could be augmented to address nonfunctional cloud features like energy efficiency or costs too, but this is beyond the scope of this chapter.

Performance modeling and simulation techniques enable performance analysis of cloud designs and capacity planning early in the development cycle on a large scale, with moderate costs, and respecting the strict time constraints of an industrial development project. Despite being widely used and well established in

Encyclopedia of Cloud Computing, First Edition. Edited by San Murugesan and Irena Bojanova.
© 2016 John Wiley & Sons, Ltd. Published 2016 by John Wiley & Sons, Ltd.

various branches of the information and telecommunication industries, the application of these techniques to cloud computing provides some new challenges due to complexity, diversity, and scale.

In the following sections, we will first provide a short introduction to performance modeling and simulation technologies and provide a rationale to choose simulation technologies in the industrial context. We will also use this section to introduce some important terminology. Then we will discuss specific requirements regarding the modeling and simulation of clouds, followed by a survey of publications and tools currently available here.

Next, we will identify the various objects of a cloud that need to be modeled. This includes hardware and software components as well as workloads consisting of various types of requests. We will also propose appropriate levels of modeling abstractions for various cloud objects. We will discuss the modeling of some key performance-related concepts like contention and segmentation in the cloud. Some challenges associated with a cloud-simulation project are then described, especially the specification of objectives, parameterization, calibration, execution, and result analysis. Finally, a real-life cloud simulation project will be described, including all project phases, execution characteristics, and some typical results.

26.2   Modeling and Simulation Overview

A model of a system consists of general artefacts like resources and algorithms used to capture the behavior of the system. Such a model can then be implemented and solved in terms of mathematical equations (e.g. using analytic queueing modeling) or in the form of a simulation that imitates the operation of the system using a computer. A request is a general unit of work posted against the system to be handled. Typical examples of such requests in clouds are the provisioning of images, transferring data packets over networks or compute jobs. Workloads then consist of patterns of such requests characterized, for example, by request types, the distribution of interarrival times between requests, or the concurrent number of requests in the system.

We provide a short survey of the two major approaches used to model performance and scalability of hardware and software components, namely solving analytic queueing models and executing discrete-event simulations.

Analytic queueing modeling uses mathematical equations to describe the performance of systems consisting of hardware and software components as well as workload requests interacting with them (Bloch *et al.,* 2006). In fact, analytic modeling can be considered as a branch of applied mathematics, namely probability and stochastic processes. It has a long tradition and was first applied by Erlang to model the performance of telephone systems at the beginning of the last century (Erlang, 1917). As a simple example here we may think of requests (SQL queries) posted by clients and arriving at a database server. Both the arrival of requests and their handling at the server can be modeled as stochastic processes using probability distributions for their interarrival times and the service times. In some especially simple cases, these stochastic processes can then be solved analytically (using a paper-and-pencil approach and not relying on approximations or numerical methods) to obtain explicit expressions for observables like the average response time as a function of service time and device utilization – see Eq. 26.1, which shows the typical results of analytic queueing modeling: the average response time R and the average number of requests Q in the system as a function of the service time S and the utilization ρ.

$$R = \frac{S}{1-\rho}$$

$$Q = \frac{\rho}{1-\rho}$$

(26.1)

Unfortunately, in the majority of real-life cases, such solutions are not available, so significant idealizations are required for the system components and workloads under consideration. In fact there are quite a few well known and highly important performance-relevant features in the context of clouds where analytic methods are difficult to apply (Bloch *et al.*, 2006), for example:

- Blocking. The amount of hardware resources like processor cycles and network bandwidth as well as software resources like database connection pool sizes or number of operating system threads in a cloud is finite. When a request has to wait for such resources due to contention, its execution blocks. This may also impact the execution of other resources. For example, a request waiting for a database connection currently in use by other requests cannot proceed using available network bandwidth to access the database server. (Therefore sizes of connection pools are always good candidates for performance tuning in clouds.)
- Bulk arrivals. A single request, for example a "launch instances" infrastructure as a service (IaaS) request, may contain several subrequests (various instances to be provisioned) arriving at the same moment at the cloud.
- Segmenting/reassembling. Basically all data transfers in clouds take place in the form of packets requiring a segmentation of large data blocks into smaller packets at the sender; these packets have to be reassembled at the receiver.
- Load balancing. Adaptive load balancing is heavily used within clouds to ensure optimal utilization of resources.
- Mutual exclusion. Frequently data structures at cloud components require atomic access of requests for updating data to ensure their consistency (e.g. network configuration data at hypervisors). This is frequently implemented by so-called "critical sections" of code allowing the execution of a single request only. In general, "critical sections" are protected by a lock that can only be owned by one request at a time.
- Nonexponential service times. The usage of exponential probability distributions to model service times is key in analytic modeling but in clouds service times will in general violate this assumption.
- Simultaneous resource possession. One of the key methods to enhance performance in clouds is the use of asynchronous I/O to overlap computation with storage or network access. This results in requests owning several resources at the same instant, namely processor cycles and storage or network resources.
- Transient states. Significant workload fluctuations (e.g. depending on the time of the day) result in cloud states far away from a steady state.

On the other hand, various approximation schemes are available for analytical models, which are highly efficient in terms of resource consumption and execution time (Bloch *et al.*, 2006) but their impact on the accuracy of results is often unclear.

A simulation is an imitation of the operation of a real-world process or system over time implemented on a computer system (Banks *et al.*, 2010; Steinhauser, 2013). In science and engineering simulation technologies have a long tradition and are extensively used in processes or systems that change their state continuously over time and can be described by differential equations (Steinhauser, 2013) – see the left side of Figure 26.1. However, this approach is of limited value when applied to cloud computing, because of the inherently discrete, noncontinuous behavior of cloud components and workloads (for example, at a processor level, request execution is driven by the discrete processor clock and at the cloud level by events like the arrival and departure of launch instance requests at compute nodes). This motivates the application of another simulation technology, namely discrete-event simulations (also known as event-driven simulations), which only model state changes at discrete times, see the right side of Figure 26.1.

Although a detailed discussion of discrete-event simulations (Banks *et al.*, 2010) is beyond the scope of this chapter, we provide a brief description here. The key idea of discrete-event simulations is to model the

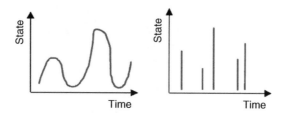

Figure 26.1   *The left figure shows a system with state changing continuously as a function of time. Such systems frequently occur in physics and engineering and are modeled by (systems of) differential equations. The right figure shows a system with state changing only at discrete times. Basically all systems used in information technology, including clouds, fall into this category. Differential equations are not well suited to model such systems because of their discrete nature*

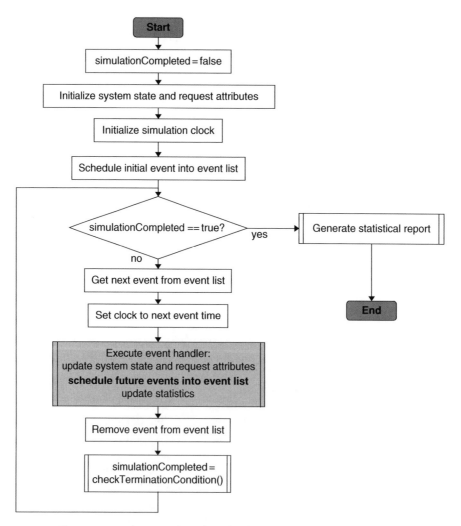

Figure 26.2   *The core algorithm of a discrete-event simulation engine*

system operation only by the sequence of events that causes changes in system state at discrete points in time. Typical events are, for example, arrivals or departures of requests occurring at system components. Future events that are already scheduled at a particular point in time are kept in a future-event list. As illustrated in the core discrete-event algorithm in Figure 26.2, the simulator repeatedly picks from this event list the earliest next event, hands it over to the particular event handler that is associated with the event, and sets the simulation clock to the event's occurrence time. A particular event handler, for example, might be responsible for performing the necessary state update for a specific system component at arrival of a specific request. Typically, a handler also schedules new upcoming events into the future event list, which may be events modeling the next arrival of the same kind or a departing request targeted to another system component. This core process shown in the grey box at the center of Figure 26.2 is repeated until the simulation ending condition is reached. This way, the dynamic system behavior is correctly modeled by immediate time jumps across the nonrelevant time periods between subsequent events. These time jumps make discrete-event simulation superior in efficiency to the activity-based simulation method (also known as time-driven simulations) where time is incremented in sufficiently small equidistant steps thereby wasting computing time during nonrelevant time periods.

In our cloud simulation scenarios, the time between two events in the real system is typically several seconds, while it is simulated in approximately a millisecond on contemporary computer systems. Leveraging parallel simulation technologies (Wu *et al.*, 2002) allows us to handle many events concurrently enabling the accurate simulation of large clouds for long time intervals.

Furthermore, the remarkable simplicity of the core discrete-event algorithm results in high flexibility and enables a straightforward simulation beyond the capabilities of analytic queueing modeling including the features mentioned in the bulleted list of section 26.2. Still, significant efforts are required to implement meaningful cloud-simulation models and their execution may require significant resources and execution time.

To sum up, for cloud modeling, discrete-event simulations seem to be superior to analytic modeling technologies in terms of accuracy and flexibility. This is essential in an industrial context. The greater efficiency of analytic methods can be offset (at least partially) by leveraging parallel simulation technologies on appropriate contemporary computer systems. Therefore we have selected discrete-event simulations as our primary tool to model clouds with the option to use simulation models as a basis for analytic queueing modeling efforts.

26.3   Cloud Modeling and Simulation Requirements

Although performance modeling and simulation technologies as introduced in the previous section are widely used and well established in various branches of information and telecommunication industries (Bertsekas and Gallager, 1992; Pasricha and Dutt, 2008), their application to clouds provides some new challenges due to complexity, diversity, and scale (Altevogt, *et al.*, 2011):

- The hardware infrastructure of clouds consists of servers, networking devices, and storage subsystems. All of these components need to be taken into account on an equal footing. This is in contrast to most of the performance simulation work focusing on (parts of) just one of these components.
- A cloud is a complex system with intricate interactions between hardware and software modules. Therefore, both software workflows and hardware infrastructure must be treated with equal emphasis when simulating end-to-end performance.
- In general, the software workflows for managing and using a cloud change at a much higher rate than the available cloud hardware infrastructures; therefore it is important to introduce separate modules for simulating software heuristics and the hardware infrastructure to support a rapid implementation of new cloud software heuristics for unchanged hardware infrastructure and vice versa.

- The market for cloud solutions being highly dynamic, simulations of new clouds must be provided in a timely manner – we need to support a rapid prototyping.
- We need to allow for selectively and rapidly adding details to the simulation of specific hardware or software components to increase the credibility of the simulation effort if required by the stakeholders of a simulation effort.
- Last but not least, the enormous size of contemporary cloud datacenters requires highly scalable simulation technologies.

Besides addressing these challenges, a key point is to find a useful degree of abstraction for modeling the system components and workloads under investigation, for example to identify the features of them that need to be included in the model and how they should be simulated. This is largely determined by the goals of the modeling effort, but must ensure modeling of typical cloud level objects like compute nodes, switches, routers, hypervisors, VMs, load balancers, firewalls, and storage with an appropriate level of detail. All these objects must be treated as first-class citizens in a cloud simulation in contrast to, for example, a network simulation that will focus on simulating network devices and use compute nodes only as workload generators posting packets against the network.

In fact there are quite a few simulation frameworks for clouds available – see Wei *et al.* (2012) for a review. We will focus on our cloud simulation approach (Altevogt *et al.*, 2011, 2013) to address the challenges mentioned above in more detail.

26.4   Modeling and Simulation of Clouds

Although clouds are complex systems, they consist of a rather small set of fundamentally different high-level hardware and software building blocks – see Figure 26.3. The key modeling challenges here are to:

- identify the appropriate building blocks;
- model these building blocks on an appropriate abstraction level;
- enable their modular and scalable combination to build complex clouds;
- support the implementation of request workflows on various levels of detail.

Figure 26.3 shows various hardware and software building blocks of clouds that need to be modeled. Besides the hardware components like compute nodes and disk arrays, we use VMs as the building blocks to implement software workflows. We distinguish between hardware resources like processor cores, disks, and switches where requests spend time for execution. Software resources represent artefacts like database connections or thread pools. By definition, requests do not spend time for execution at these software resources, but frequently need to own one or several software resources to access a hardware resource.

As a typical example for a model of a hardware resource, we take a closer look at the model of a network device – see Figure 26.4. It consists of basic building blocks ("Lego bricks") providing bandwidth to model the ports and a crossbar switch, another "Lego brick" providing processor cycles and a module implementing the software workflow details including the creation of routing tables at initialization of the simulation. Using different parameterizations of the components (e.g., processing times, link bandwidths or number of ports), this simulation module can be used to model a wide variety of network devices like switches, routers and firewalls. A packet enters the network device at a port, traverses the crossbar, spends some time at a processor core, the switch or router looks up the appropriate port connected to the next target in the routing table of the workflow module and finally the packet leaves the device via this port. (This low-level request workflow

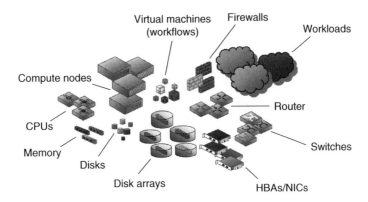

Figure 26.3 *Samples of modeled cloud hardware and software components. The software components are implemented as workflows within VMs*

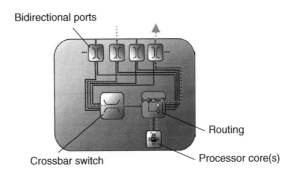

Figure 26.4 *A general network device. The dotted line shows the flow of a packet traversing the network device (e.g. a switch or a router)*

within a network device is in general initiated by a high-level workflow request implementing, for example, a cloud-level software application.)

A typical scenario for the application of software resources is the simulation of so called critical sections – see Figure 26.5. Such sections represent processing phases that allow only one concurrent request to proceed and are used, for example, to ensure consistency of data updates. In this case a request has to queue for a token (phase 1) provided by a token pool to access the critical section and spend some time for execution within it (phase 2). After leaving the critical section, the request releases the token again and proceeds (phase 3). Other requests waiting may then try to obtain the token.

Models of large-scale clouds are created by combining more basic modules, such as processor cores, switches disks, and VMs to create a compute node, compute nodes and switches to create a rack, and several interconnected racks and storage units to create a datacenter. This is shown in the left part of Figure 26.6. Modules may be replaced by more fine-grain or coarse-grain models at any level of this building-block approach – for example, we may add a simulation of various RAID policies at the disk arrays if required. Finally, these modules can be combined to build a cloud model consisting of a number of world-wide distributed datacenters. The key design concepts of a simulation framework supporting this approach are modularity and the ability to replicate ("copy-and-paste") and connect objects at any level of complexity. This capability significantly supersedes that of ordinary Lego bricks, so one might think of it as a "Lego++ approach" – see Figure 26.6.

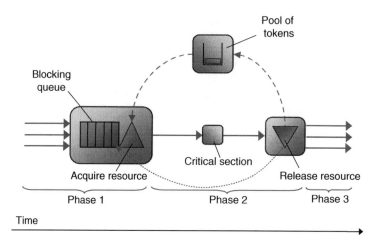

Figure 26.5   *Modeling a critical section allowing only one request (thread) concurrently in flight*

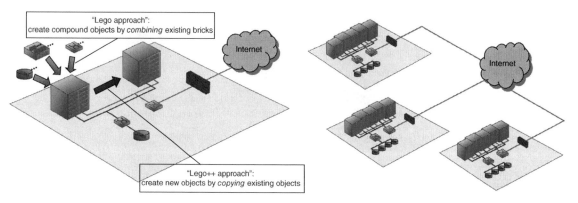

Figure 26.6   *Complex compound modules can be created by combining more basic modules, for example a server rack, by combining compute nodes with various VMs, disk arrays and network components. Using these racks, a datacenter can then be created by a copy-and-paste approach. Finally, a cloud consisting of a number of worldwide distributed datacenters can also be created by applying the copy-and-paste approach, this time applied at datacenter level*

For a simple taxonomy of cloud requests, see Figure 26.7. At the root of the taxonomy graph, we have a general request. Its child requests represent cloud management respectively application (user) requests. The children of these requests represent specific cloud management operations respectively application requests. This taxonomy supports the implementation of request-type specific workflows well.

Furthermore, we can associate various levels with requests. Requests at a higher level (e.g. requests at cloud level to provision images) are then initiating request workflows at a lower level (e.g. requests associated with sending packets over the network traversing various NICs and switches). The highest level request workflows are in general associated with applications at cloud level and are implemented in the context of VMs, the lowest level request workflows with accessing hardware components like disk media. This allows the implementation of new cloud-level application workflows without the need to take details of low-level device-related workflows into account and vice-versa.

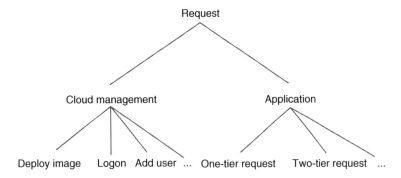

Figure 26.7 *Request taxonomy with the request of type "Request" at its root*

In the workload-generator module (the "Internet" in Figure 26.6), all functionality related to generating, initializing, and posting requests of various types against the cloud is implemented. Furthermore, it may also be used to collect all request-related statistics. (Device-related statistics like utilization and queue lengths are in general collected at the device-simulation modules.)

We will now focus on some key cloud modeling aspects that are quite independent of the concrete resources and requests under consideration. All cloud resources are limited and demand may well be bigger than resources available in the cloud, so arbitration (scheduling) of resources to requests is essential. We frequently use a simple FCFS (first-come, first-served) arbitration heuristic to model this in our simulations (Altevogt, *et al.*, 2013), but it is straightforward to implement more complex schemes if required. Another key feature when modeling clouds is segmentation and reassembly of requests, especially of requests modeling the transfer of data over the network. This refers to the process of fragmenting requests into smaller ones (e.g. splitting a request to transfer a long message into a bunch of small packets) and reassembling these after the completion of processing. Such segmentation allows a request to be in flight concurrently at several devices, resulting in an increased throughput. Therefore it is essential to model segmentation even if some compromises concerning granularity have to be made given the finite time and resources provided to execute simulations (for example, it is hardly feasible to model a MTU size of 1500 bytes in large cloud networks due to the excessive number of events associated with processing such fine granular data segmentation).

26.5 Cloud Modeling and Simulation Challenges

Modeling and simulation of cloud performance is associated with some specific challenges caused by their highly volatile nature and scale:

- Architectural and workload details of clouds are changing rapidly. This is caused, for example, by new processes of delivering software like continuous delivery (Swartout, 2012). Besides requiring fast updates of the simulation models themselves, this makes their precise parameterization difficult. The problem can be alleviated by focusing on the prediction of *relative* values of performance metrics against a known baseline, for example modeling the *increase* of request throughput when updating the cloud.
- For calibrating models, high-quality measurement data with a detailed specification of workloads and cloud infrastructure components are of great value. Unfortunately, such data is rare (this is especially the case for end-to-end measurements on large scale clouds due to constraints in time and resources), and this is significantly impacting model accuracy.

- Due to constraints in time and available resources, measurement data is in general available only for small or medium-sized clouds. Using this data for parameterization and calibration, there is a considerable risk to overlook resources of importance for clouds at large scale. Not taking these resources into account will most likely result in significant modeling errors.
- To enable nonexpert users to execute cloud simulations, a limited set of key simulation scenarios should be made available as an easy-to-use Web service. The challenge here is to identify the most useful scenarios and a small set of associated key parameters to be exposed to a non-expert user.

26.6   Simulation Project Case Study: Openstack Image Deployment

We will use the simulation of image deployment in OpenStack-managed clouds as an example of a concrete cloud simulation project and describe its phases below. For a detailed description see Mirantis Inc. (2012).

26.6.1   Specification of Objectives

Objectives of a simulation project should be specified and agreed on with the stakeholders at the beginning of a simulation project. In our case, the objective is to study the impact of increasing the number of concurrent image deployment requests on throughput and response time for various cloud architectures. Furthermore, we want to learn about various device utilizations and identify potential bottlenecks.

26.6.2   Design and Specification

In this phase, the appropriate cloud architectures to be simulated are specified as well as key workload characteristics. These are in our case, for example, network topologies, the number of concurrent image deployment requests, the type of images to be deployed, and overall cloud sizes.

26.6.3   Implementation

The OpenStack image-deployment workflow (Mirantis Inc., 2012) is implemented for various cloud architectures, e.g. for an OpenStack managed cloud consisting of a hierarchical 10 Gbps network fabric, a managed-from and a managed-to system. The managed-from system supports most of the OpenStack software components, while the VMs are deployed on the nodes of the managed-to system. Storage may be attached to the network fabric. Our implementation uses an OMNEST based cloud simulation framework (Simulcraft Inc., 2014).

26.6.4   Parameterization

Key workload-related parameters like service times and resource consumption (e.g. CPU utilization) of an image deployment request at the managed-from and the managed-to system are to be extracted from available measurements. Parameters of infrastructure components are based on vendor specifications or benchmark results. When no appropriate data is available, reasonable values based on past experience are used. Furthermore, various tradeoffs have to be made (e.g. concerning size of data packets) to balance between execution time, resource consumption and simulation accuracy.

26.6.5   Calibration

Measurements for a single image deployment are used for parameterization (see the second bullet of section 26.5), while measurement results for concurrent image deployments are used for verification and calibration of the simulation. These latter measurements already demonstrate the impact of resource contention and queueing on response times and throughput, so they provide invaluable feedback on the modeling approach, for example whether contention at various resources has been modeled appropriately. To factor out various unknown details of the benchmark setups and executions, we frequently normalize the times for concurrent image deployments with the time for a single image deployment and compare these normalized response times with the simulation results.

26.6.6 Execution

The key performance metric characterizing the execution of discrete-event simulations is the event throughput. We observe a throughput of approximately 1.2 million events per second on one contemporary compute node using the OMNEST simulation software. This number, as well as the size of the cloud that can be simulated, is largely determined by the type and amount of available memory. This is because discrete-event simulations need to update data structures representing system state very frequently resulting in a high rate of I/O operations. At the current level of detail, a single compute node with 96 GB main memory is sufficient to simulate various image-deployment scenarios on cloud datacenters with approximately 1000 compute nodes and associated storage and network devices. Larger clouds would require either a higher level of abstraction or parallel simulations on a cluster. The execution time depends critically on the required accuracy: to ensure a deviation of less than a few percent for most simulation results, the number of executed requests should be at least two orders of magnitude greater than the number of concurrent requests in the cloud.

26.6.7 Result Analysis and Visualization

Visualization is mandatory to convey simulation results – for example, plotting image deployment throughput response times as a function of the number of concurrent image deployment requests (Altevogt *et al.*, 2013). The simulation results demonstrate the impact of an increased request concurrency on response time and throughput as well as their dependency on the number of available compute nodes at the managed-to system. Throughput first increases linearly because the cloud can easily handle all the concurrent image deployments. This is true until the first bottleneck shows up and associated queueing of requests sets in. Beyond this saturation point, a further increase in the number of concurrent image deployment requests does not result in an associated growth of throughput but instead the response times start to increase linearly. Studying the impact of increasing the number of concurrent image deployment requests on queue lengths and device utilizations allows us to identify bottlenecks (Altevogt *et al.*, 2013). We may then simulate the impact of various architectural changes (like increasing the number of nodes of the managed-to system) to analyze how the bottlenecks can be eliminated.

26.7 Conclusion

Cloud modeling and simulation are of great value to enhance the design of workload-optimized clouds, especially because of the prohibitive costs and time associated with the creation of large-scale test clouds. The challenges here lie in addressing cloud diversity and scalability, treating all hardware and software components of a cloud as first-class citizens, and making cloud modeling and simulation technologies accessible easily and timeously for non-experts. These challenges require quite a different approach than in traditional simulation domains like microprocessor design or networking. Progress has been made by leveraging modular and parallel simulation technologies and a "cloud simulations as a service" approach. Further research and innovations are required to enable a more widespread use of these highly valuable technologies by combining simulation accuracy and scalability with ease of use and flexibility to deliver relevant results more quickly.

References

Altevogt, P., Denzel, W., and Kiss, T. (2011) Modular performance simulations of clouds. Proceedings of the Winter Simulation Conference, pp. 3300–3311.

Altevogt, P., Denzel, W., and Kiss, T. (2013) *The IBM Performance Simulation Framework for Cloud*, IBM Research Technical Paper, http://domino.research.ibm.com/library/cyberdig.nsf/papers/041264DDC6C63D8485257BC800504EA2 (accessed December 14, 2015).

Armbrust, M., Fox, A., Griffith, R., *et al.* (2009) *Above the Clouds: A Berkeley View of Cloud Computing*, http://www.eecs.berkeley.edu/Pubs/TechRpts/2009/EECS-2009-28.pdf (accessed December 14, 2015).

Banks, J., Carson II, J. S., Nelson, B. L., and Nicol, D. M. (2010) *Discrete-Event System Simulation*, 5th edn. Pearson, Harlow.

Bertsekas, D. and Gallager, R. (1992) *Data Networks*, 2nd edn. Pearson Education, London.

Bloch, G., Greiner, S., de Meer, H., and Trivedi, K. S. (2006) *Queueing Networks and Markov Chains*, 2nd edn. Wiley Interscience, New York, NY.

Erlang, A. (1917) Solution of some problems in the theory of probabilities and telephone conversations. *Nyt Tidsskrift for Matematik B* **20**, 33–40.

Kavis, M. J. (2014) *Architecting the Cloud*, John Wiley & Sons, Inc., Hoboken, NJ.

Pasricha, S. and Dutt, N. (2008) *On-Chip Communication Architectures*, Elsevier Inc., Amsterdam.

Simulcraft Inc. (2014) *High-Performance Simulation for all Kinds of Networks*, http://www.omnest.com/ (accessed December 15, 2015).

Steinhauser, M. O. (2013) *Computer Simulation in Physics and Engineering*, Walter de Gruyter, Berlin.

Swartout, P. (2012) *Continous Delivery and DevOps: A Quickstart Guide*, Packt Publishing, Birmingham.

Mirantis Inc. (2012) *OpenStack Request Flow*, http://www.slideshare.net/mirantis/openstack-cloud-request-flow (accessed December 15, 2015).

Wei, Z., Yong, P., Feng, X., and Zhonghua, D. (2012) *Modeling and Simulation of Cloud Computing: A Review*. IEEE Asia Pacific Cloud Computing Congress (APCloudCC). IEEE, pp. 20–24.

Wu, D., Wu, E., Lai, J., *et al.* (2002) *Implementing MPI Based Portable Parallel Discrete Event Simulation Support in the OMNeT++ Framework*. Proceedings of the 14th European Simulation Symposium. SCS Europe.

27

Cloud Testing: An Overview

Ganesh Neelakanta Iyer

Progress Software Development, BVRIT, and IIIT-H, India

27.1 Introduction

The technology explosion – with cloud, mobile and big data – demands modern software to be mobile, agile, and accessible. More and more software is developed and deployed as a "software as a service" (SaaS) application using cloud platforms ("platform as a service" or PaaS). Several industry sectors such as healthcare and procurement increasingly use customized SaaS applications provided by cloud vendors. Most information technology companies are either developing and offering new cloud solutions or adopting cloud-based solutions. These solutions have several advantages such as resource availability on demand, faster time to market, and reduced capital and operational expenses.

Other key characteristics of cloud computing include interoperability, browser compatibility, multitenancy and autoelasticity. This paradigm shift in software development poses several challenges when assessing and quantifying the quality of the products developed before they are offered for customers. These challenges include security, availability, elasticity, and multitenancy.

Even for a legacy software, testing could be done on a cloud-based solution. Software testing in the cloud changes the traditional testing scenario by leveraging the resources provided by the cloud computing infrastructure to reduce test time, increase the execution cycles available, and thereby increase the efficacy of testing to improve the quality of the application (Tilley and Parveen, 2012). Testing in the cloud relies on underlying technologies such as a distributed execution environment, service-oriented architecture (SOA), and hardware virtualization.

James Whittaker, engineering director at Google says:

> In the cloud, all the machines automatically work together; there's monitoring software available, and one test case will run anywhere. There's not even a test lab. There's just a section of the datacenter that works for you from a

---

Encyclopedia of Cloud Computing, First Edition. Edited by San Murugesan and Irena Bojanova.
© 2016 John Wiley & Sons, Ltd. Published 2016 by John Wiley & Sons, Ltd.

testing point of view. You put a test case there and it runs. And all of the different scheduling software that any datacenter uses to schedule tasks can be used to schedule tests. So, a lot of the stuff that we used to have to write and customize for our test labs, we just don't need anymore. (Shull, 2012)

This chapter describes various research findings on major test dimensions to be assessed by any quality-assessment team for any cloud-based platforms/applications that are developed. It also details the challenges introduced by cloud for software testing, various cloud-test dimensions, and benefits of using cloud-based solutions for software testing. The chapter also describes the challenges associated with automating cloud testing and some ways to mitigate these challenges.

Often, cloud-based systems will be integrated with each other to deliver a cloud-based offering. For example, for an SaaS application, the single sign-on (SSO) mechanism may be handled by a third-party system and payment mechanisms might be handled by another third-party system such as PayPal. In such systems, the testing offers unique challenges. This chapter discusses testing in such systems. It concludes by considering future directions.

27.2   Challenges Introduced by Cloud for Software Testing

As described in the first section, various characteristics of cloud computing introduce several unique challenges for the software development process – specifically for software quality assurance. Some important challenges are given in this section.

27.2.1   Paradigm Shift

With cloud, Web-driven SaaS applications are becoming more popular. Traditionally, applications are designed completely upfront, then developed, tested, and distributed for installation and use. With cloud, users can just create an account for a particular software product, provide online payment details, and start using the product through the browser in a seamless manner. Everything happens in a few mouse clicks rather than the user purchasing the software online or offline, downloading it, or getting it in some electronic form such as CD-ROM, installing it, and setting it up. Thus, the emergence of cloud computing resulted in a revolution for the software industry and hence for software testing.

27.2.2   Seamless Upgrades

Traditionally, a software upgrade is a lengthy process in which one needs to get the new version of the software in the same way the initial version is obtained, take the system down for a particular period of time (which depends on the complexity of the installation and setup process), and perform the upgrade. With cloud, an upgrade should happen live, with minimal or no down time. Often customers might see that the software has been upgraded next time they log in – without any down time. Google products are a common example. So the quality-assurance team needs to find ways to test the upgrade process of software products seamlessly.

27.2.3   Sharing of Resources: Multitenancy

With cloud, the resources used for software development and deployment might be in publicly shared resources as against the traditional way of having dedicated resources for individual users. These resources are often shared among multiple customers. For example, a company might be developing, deploying, and running its software in Amazon Web Services (AWS) using AWS CPU machines and the storage services. These resources are often virtualized and are shared among multiple software developers across the globe.

This introduces several challenges for the quality-assurance team including the need to have specific testing for systems such as multitenant penetration testing.

27.3   Key Benefits of Testing in the Cloud

All levels of testing can be carried out in cloud environments and, indeed, some types of testing benefit greatly from a test environment in the cloud. The primary benefit of cloud-based testing is reduced costs for putting up, maintaining, and licensing internal testing environments (Riungu et al., 2010). For example, Amazon's cloud tested a network management system for a voice over IP telephony system for less than US$130 (Ganon and Zilberstein, 2009). Other benefits include the flexibility to acquire a testing environment as needed and global market access for both vendors and customers (Riungu et al., 2010).

Cloud computing can carry out performance, scalability, and stress testing in a timely and economical manner (Riungu-Kalliosaari, 2012). With cloud computing, an engineer can acquire the necessary number of servers as well as different variations of the operating system and testing environments; in other words, engineers can test faster and can decommission the servers when they are not using them. Key benefits of testing in the cloud are given in this section.

27.3.1   Pay-as-You-Go (Cost Savings)

Test systems need to be configured and used only when testing needs to be performed. At all other times, use and maintenance of such systems can be avoided. This results in no or minimal capital expenses and reduced operational expenses. It also leads to green testing due to the improved efficiency in using existing resources and the reduced use of new resources.

27.3.2   Faster Test Execution

One can configure more test environments and run the tests in parallel with ease, resulting in faster test execution. The advantages are quick resource provisioning and concurrent test execution, resulting in fast execution and hence quick feedback.

27.3.3   Test as you Develop

With cloud-based software development, testing can be done along with the development life cycle to uncover potential bugs in the system rapidly. Moreover, the test environment can be identical to (or close to) the production environment to perform real-time testing along with the development.

27.3.4   Better Collaboration

A test running in cloud environments can be accessed from anywhere on the globe. Hence geographically distributed business units can collaborate better and work better by running the tests in the cloud.

27.4   Cloud Test Dimensions

From this section onwards the chapter focuses on testing software products that are completely or partially developed using some of the cloud resources. These applications could be pure SaaS applications,

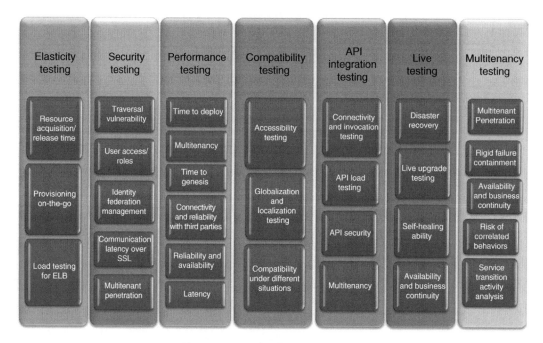

Figure 27.1   *Cloud test dimensions*

developed, deployed, and running in the cloud, or applications that are built using cloud platforms, or they could be in-house applications integrated with the cloud.

This section describes important characteristics of cloud computing and its impact on the software testing. Some of the important cloud dimensions include elastic scalability, multitenancy, security and performance (Iyer *et al.*, 2013). The cloud test dimensions and their components are summarized in Figure 27.1.

27.4.1   Elasticity and Scalability

Autoelasticity is the key for cloud-based business solutions – software solutions can be developed and deployed in cloud where resource provisioning should happen on demand within a span of few minutes with ease. So the test team should identify various possible resource requirement patterns and perform scalability testing for all such patterns. One has to test for cases where users can elastically increase and reduce their resource requirements quickly and easily.

Some elastic load-generation patterns are illustrated in Figure 27.2. The top left graph shows the constant load increase pattern that is typical of applications such as Gmail and Facebook where new users are coming at a steady rate. The top right pattern illustrates typical step increases in load with time, with sudden dips in resource use at times. The bottom left pattern represents typical use of resources that follows exponential resource use. Finally, the bottom right pattern represents applications where the resource use is constant in general with spikes in between.

Scenarios involving both vertical and horizontal scalability need to be tested. With vertical scalability, users should obtain better performance by replacing current resources with a more powerful resource in order to satisfy the increasing demand. In the case of horizontal scalability, a customer should experience increased performance by adding more resources of same type in the platform.

One also needs to test the functionality of fully automated scripts for automatic resource provisioning "on the go." Further, testing is required to discover the worst-case elasticity that can be achieved (in terms of

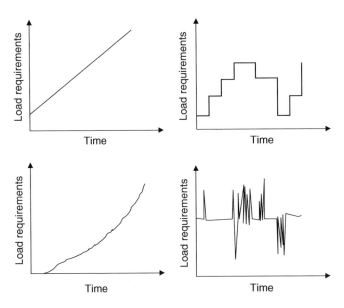

Figure 27.2 *Sample load patterns for testing*

scale up). Often, IaaS providers offer load balancers to balance the load among multiple resources based on the incoming traffic flow (Elastic Load Balancer in the case of Amazon Web Services, for example). Such providers should test the load balancing capability with different load patterns.

27.4.2 Security Testing

With cloud, security is considered as one of the great challenges and hence it is a challenge for testing as well. In addition to regular security issues, cloud imposes a set of unique security constraints such as multitenant penetration testing, application development user interface (UI) testing, and identity federation management testing. A summary of various security attacks that need special focus by the software testing team is illustrated in Figure 27.3.

User access and various user roles need to be tested to make sure that those roles work properly as expected. Identity federation management is another important aspect to be tested. This is called a single sign-on (SSO) mechanism, with which a user, once logged in successfully, can access other components of the systems without being prompted for credentials again. For example, cloud software may have default application areas as well as community support for help and support. In such an environment, once logged into the application area, users should be able to access the forums without being prompted for their credentials again.

The SSO concept can be understood using this example. Consider that you have provided your credentials to log into Google's Gmail service. Then, if you want to open your Google drive in another tab, it will do this without asking you to log in again. Then assume you have opened Blogger (another Google application) in the next tab. Blogger will also be opened automatically, detecting the credentials provided for Gmail. Now assume that you have logged out from your Google drive. You will observe that entire session is terminated, including the sessions for Gmail and Blogger. This is a typical SSO implementation mechanism.

Other important security vulnerabilities to be tested are the top issues identified by the Open Web Application Security Project (OWASP) (OWASP, 2013) such as SQL injection, URL manipulation, and cross-site scripting (XSS). A penetration test is a method of evaluating security by simulating an attack by a malicious user. The intent of a penetration test is to determine the feasibility of an attack and the business

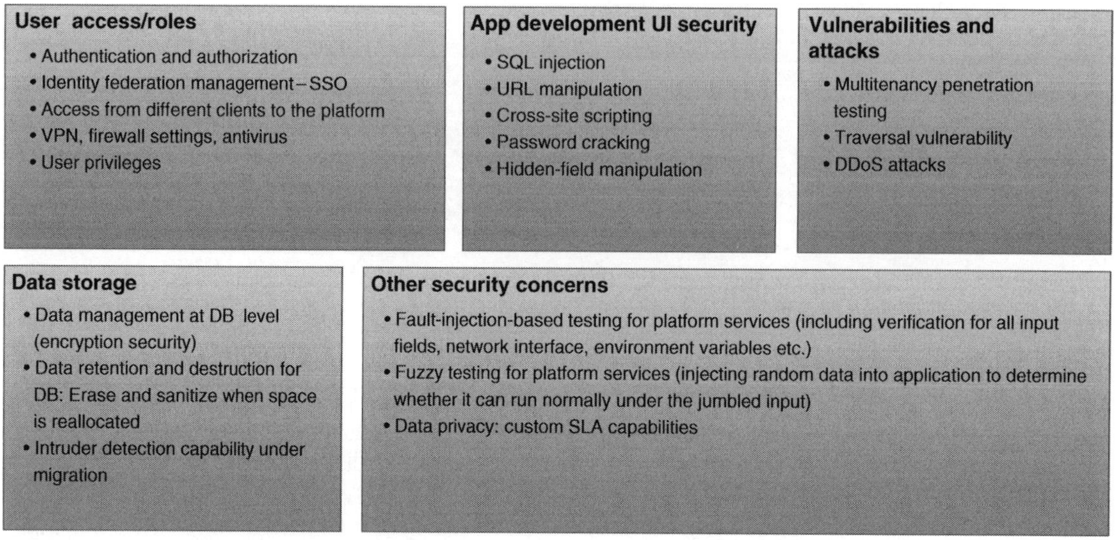

User access/roles

- Authentication and authorization
- Identity federation management – SSO
- Access from different clients to the platform
- VPN, firewall settings, antivirus
- User privileges

App development UI security

- SQL injection
- URL manipulation
- Cross-site scripting
- Password cracking
- Hidden-field manipulation

Vulnerabilities and attacks

- Multitenancy penetration testing
- Traversal vulnerability
- DDoS attacks

Data storage

- Data management at DB level (encryption security)
- Data retention and destruction for DB: Erase and sanitize when space is reallocated
- Intruder detection capability under migration

Other security concerns

- Fault-injection-based testing for platform services (including verification for all input fields, network interface, environment variables etc.)
- Fuzzy testing for platform services (injecting random data into application to determine whether it can run normally under the jumbled input)
- Data privacy: custom SLA capabilities

Figure 27.3 *Various security vulnerabilities to be considered*

impact of a successful exploit. Attacks like SQL injection and cross-site scripting are possible to a greater extent in a multitenant application. Test cases should be written to attack the system from the point of view of a rogue user who possesses the ID of a tenant of the application, with that user's ability to penetrate the system and view the information of other tenants.

Some important security test considerations are multitenant database security testing and validating the storage and retrieval of confidential information such as credit-card information and passwords.

27.4.3 Cloud Performance Testing

With the unique characteristics of cloud environments, a quality team needs to test the accuracy of various data present in the cloud – latency and throughput. Elastic load testing and multitenant performance testing are other key items to be considered. High availability and failover testing are also required, to test the behavior of the platform and applications for resilience.

Resilience is very important in the cloud context. It is the ability for the cloud to work normally under extreme and adverse conditions. Testing possible security threats such as DoS attacks is one way to measure resilience; measuring the system's behavior when hardware fails is another. A cloud should be able to adapt quickly to such failures in various ways, such as spawning new instances, load replication, and so forth. While an activity is going on in a particular cloud instance, the test team should kill the VM and measure the time taken to recover from such failures and the impact on the ongoing activity.

Another possible resilience issue could result from the presence of multiple tenants sharing the same resources. In order to test such scenarios, as one tenant, create and run a process that can continuously allocate some memory or disk space. Then measure the performance of other tenants in the system and discover any issues that they experience caused by such activities. Refer to Cartlidge and Sriram (2011) for a simulation model to evaluate resilience for cloud-scale datacenters

Performance test metrics differ for different types of clouds. For example, the main performance metrics for an SaaS application could be the time it takes to customize the application based on the business requirements, the time it takes to load the application, and so forth. For a PaaS solution, it could be the time it takes

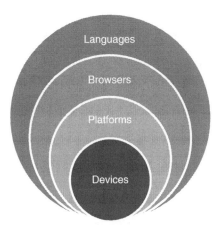

Figure 27.4 *Compatibility circles*

to develop a solution from scratch or using given templates, the response time for executing database-related queries, and so forth. Finally for an IaaS solution, such as CPU or memory, it would include throughput, provisioning time, memory, and CPU utilization.

With an integrated cloud-based product, quality teams have to test other aspects such as latency, reliability, and connectivity with other third-party components. Quality teams must also carry out special performance testing such as high availability testing, longevity testing, and scalability testing.

27.4.4 Compatibility Testing

Cloud-based products are primarily driven by Web browsers. They are designed to work across any platform and device. So compatibility testing needs to be performed for browser compatibility, platform compatibility, and device compatibility. In addition many cloud-based solutions support multiple languages and hence internationalization (commonly referred to as i18n) should be performed for all supporting languages. These compatibility circles are illustrated in Figure 27.4. An interesting observation is that, with a cloud operating system, compatibility is out of scope. This is because the applications are designed to run from the browser itself and OS details are abstracted out.

27.4.5 Application Program Interface Integration Testing

Application program interfaces (APIs) are direct links between the client code and the infrastructure. They run only when requested by client code. In order to test such systems, one might need to develop a client shell, write sample programs to invoke such APIs and then verify the output generated by such APIs. Further, APIs need to be tested for different types of user loads. Application program interfaces are exposed to customers and hence are vulnerable to security attacks. So all necessary security testing should be performed for all the APIs. Under a multitenant environment, APIs could be tested to make sure that tenants are isolated from each other through the APIs.

27.4.6 Live Upgrade and Disaster Recovery Testing

This is closely related to understanding the performance of the system when an upgrade of the software / platform happens, and understanding its capability to continue its business services for users even when the upgrade is going on. Companies need to ensure business continuity even when software / hardware maintenance / upgrades are performed.

It is often necessary to perform live testing for disaster recovery and related issues (Naganathan and Sankarayya, 2012). This includes testing the robustness of the platform to disasters, measuring recovery time in case of disaster, and self-healing ability in case of disaster. It also includes live upgrade testing.

It is necessary to test cases where an application developed in one PaaS is migrated to another PaaS (Cunha *et al.,* 2012). For example, Progress Pacific (http://www.progress.com/products/pacific, accessed December 15, 2015) is a rapid application development platform that allows users to develop and deploy high-end business solutions quickly. It also allows users to migrate applications built using the Force.com platform in few mouse clicks. Test engineers need to identify such scenarios and to test all such cases. Progress Pacific also offers both cloud-based hosting services and on-premises self-hosting services. It then allows users to develop applications in one of the two environments and import them in another environment. Quality analysis teams must identify such cases and test all of them.

27.4.7 Multitenancy Testing

It is necessary to test for rigid failure containment between tenants (Bauer and Adams, 2011). This means that failure of one tenant instance should not cascade to other tenant instances, and that service transition activities should apply to individual application instances without inadvertently impacting multiple tenant application instances. A quality team also needs to test for availability and business continuity in a multitenant environment. Further, multitenancy introduces the risk of correlated or synchronized behaviors (Bauer and Adams, 2011), which can stress the underlying virtualized platform, as when multiple application instances execute the same recovery action or periodic maintenance actions simultaneously.

Moreover, if the application supports multitenancy, then the service transition activity analysis should also verify that no service transition activity impacts active application instances that are not the explicit target of the activity. In addition to traditional service transition activities, the multitenancy analysis should also verify that there is no service impact on other tenant instances when each and every tenant-specific configuration parameter is changed.

A unique issue to be tested in the virtual environment is the "traversal vulnerability" (Owens, 2010). One is able to traverse from one VM (virtual machine) client environment to other client environments being managed by the same hypervisor. Imagine that many customers are being managed by a single hypervisor within a cloud provider. The traversal vulnerability might allow a customer to access the virtual instances of other customers' applications. Consider the impact if your bank or particularly sensitive federal government or national defense information happen to be managed in this sort of environment, and the cloud provider does not immediately deal with, or even know about, a vulnerability of this nature. It is clear that providing adequate administrative separation between virtual customer environments will be a significant security challenge with elasticity.

Another security challenge (Owens, 2010) that develops out of this scenario is that of enforcing proper configuration and change management in this elastic, dynamic, and scalable model. Even where a portal is capable of granular access controls that control which actions a given user is able to perform, it also needs to enforce when and under what circumstances a user is allowed to perform certain actions. Without this ability, untested code or system changes could result in business-impacting (or even devastating) results. Even something as "slight" as rolling a new system into production without ensuring that proper server and application patches have been applied could result in significant damage to an organization. Therefore, a mechanism within self-service portals for enforcing an organization's change policies becomes necessary. One cloud test framework that takes into account many aspects of cloud testing can be found in Iyer *et al.* (2013).

27.5   Test Challenges and Approaches for Cloud Integration Testing

Often, cloud software will not be a standalone system. It will have the core product components integrated with many back-office systems and third-party components. For example, accepting credit card details and processing invoices need special compliance certification based on security guidelines. In such cases, a software product may integrate itself with third-party payment gateways such as Bill Desk or PayPal. Hence, the test team needs to get sandbox environments (which are identical to the production environment) for such payment gateways in order to obtain the same behavior in the test and live environments.

There are several issues that need to be resolved when automating an integrated cloud-based product testing. They are:

- In an integrated cloud-based product, multiple systems behave differently. Further, verification processes for different systems differ each other.
- Some of the systems do not allow automated deletion of data created for testing. So it imposes a unique requirement to have unique users created every time an engineer performs such test automation
- Unpredictable delays in updating various systems.
- Different types of testing environments. For example, a requirement to perform Web UI testing and runtime testing in one test scenario poses its own unique challenges.

27.5.1   Integrated Test Automation with Cloud

Software test automation in cloud imposes its own unique set of challenges. Such software systems consist of both run time components as well as UI components. In order to automate such products, one might need to either develop a test framework that can efficiently test both run time and UI components or use a combination of two frameworks one for run time component testing and the other one for UI component testing. In the latter case, some mechanisms are required to initiate the test written in one framework from the other one as well as approaches to generate a combined test result so that the test engineer does not need to analyze two set of results.

Typical test automation framework architecture for such cases is illustrated in Figure 27.5. In this architecture the outer harness is the run time automation framework and the inner harness is the UI automation framework. In such cases, the runtime component starts executing the test cases with runtime calls such as API calls and wherever required it will trigger the test cases in UI framework for the UI testing. After UI testing has been completed, the runtime framework will continue its execution. Then the runtime component will generate the combined test results and will be stored in the result repository.

For example, consider a cloud-based SaaS application, CRM. The user is required to authenticate using a third-party system and then lands in the application. Similarly, payment is made through PayPal system. One automated test case could first call an API of the authentication system using runtime framework with appropriate credentials as parameters and receive a token. Then it could use the UI test framework to test the CRM application dashboard for all user interface testing. Finally when it comes to payment, the runtime test framework will make the appropriate API calls or runtime actions to check if payment is correct and then complete the test execution by compiling the test results from both frameworks.

In addition, one might have to use the exposed APIs provided by other third party systems integrated as part of the product. Sometimes, it is also required to develop stubs to simulate certain functionalities of other systems in order to achieve an end to end flow in the automation.

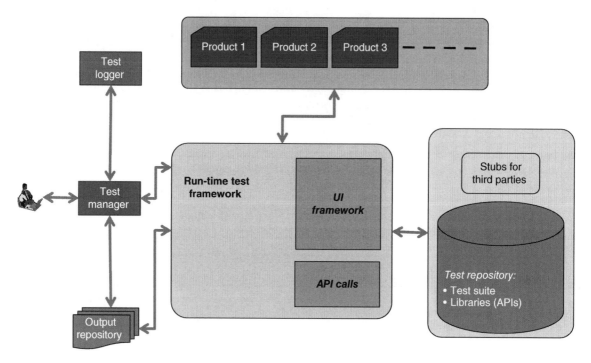

Figure 27.5   *Test Automation Architecture*

27.6   Conclusions and Remarks about the Future

This chapter first described several key challenges introduced by cloud-computing paradigms for software testing. These include the paradigm shift in software development from legacy on-premises software to Web-driven, dynamic, cloud-based applications. The chapter then described several benefits of software testing in the cloud. The key benefits achieved by testing in the cloud are reduced cost for software testing, faster test execution, ability to test in environments that are close to production environments, and better collaboration.

Various cloud test dimensions were then described, such as elasticity, scalability, security testing, performance testing, compatibility testing, API integration testing, live upgrade testing, disaster recovery testing, and multi-tenancy testing. Finally, challenges and automation approaches for cloud integration testing were discussed.

There are no known frameworks that can test all the issues described in this chapter. However, combinations of two or more frameworks can test all the issues described. Future research could attempt to identify such frameworks and develop strategies to combine them so that efficient, reliable, and easy software testing can be carried out for modern applications.

References

Bauer, E. and Adams, R. (2011) *Reliability and Availability of Cloud Computing*, Wiley-IEEE Press, Hoboken, NJ.
Cartlidge, J. and Sriram, I. (2011) Modelling resilience in cloud-scale data centres. Proceedings of the 23rd European Modeling and Simulation Symposium (Simulation in Industry).
Cunha, D., Neves, P., and Sousa, P. (2012) Interoperability and portability of cloud service enablers in a PaaS environment. Proceedings of the Second International Conference on Cloud Computing and Services Science (CLOSER 2012), Porto, Portugal, pp. 432–437.

Ganon, Z. and Zilberstein, I. E. (2009) *Cloud-Based Performance Testing of Network Management Systems*. Proceedings of the IEEE 14th International Workshop on Computer Aided Modeling and Design of Communications Links and Networks (CAMAD 09). IEEE CS, 2009, pp. 1–6.

Iyer, G. N., Pasimuthu, J., and Loganathan, R. (2013) PCTF: An integrated, extensible cloud test framework for testing cloud platforms and applications. Thirteenth International Conference on Quality Software (QSIC), Nanjing, China, pp. 135–138.

Naganathan, V. and Sankarayya, S. (2012) *Overcoming Challenges associated with SaaS Testing*. Infosys, https://www.infosys.com/IT-services/independent-validation-testing-services/white-papers/Documents/overcoming-challenges-saas-testing.pdf (accessed December 15, 2015).

OWASP (2013) *OWASP Top Ten Project*, https://www.owasp.org/index.php/Category:OWASP_Top_Ten_Project (accessed December 15, 2015).

Owens, D. (2010) Securing elasticity in the cloud. *ACM Queue* **8**(5), https://queue.acm.org/detail.cfm?id=1794516 (accessed December 15, 2015).

Riungu, L. M., Taipale, O., and Smolander, K. (2010) Software testing as an online service: Observations from practice. Proceedings of the Third International Software Testing, Verification, and Validation Workshop (ICSTW 10), pp. 418–423.

Riungu-Kalliosaari, L. M., Taipale, O, and Smolander, K. (2012) Testing in the cloud: Exploring the practice. *IEEE Software* **29**(2), 46–51.

Shull, F. (2012) A brave new world of testing? An interview with Google's James Whittaker. *IEEE Software* **29**(2), 4–7.

Tilley, S. and Parveen, T. (2012) *Software Testing in the Cloud: Migration and Execution*, Springer, Berlin.

28

Testing the Cloud and Testing as a Service

Nitin Dangwal, Neha Mehrotra Dewan, and Sonal Sachdeva

Infosys Limited, India

28.1 Introduction

Testing is an important phase of the software development life cycle. It can be regarded as a process of verification and validation of the application being tested, to ensure that it is built as per the design and specifications, and that the application meets its intended requirements.

Testing is considered as a necessary evil as it consumes a fair amount of time and resources to configure test labs and test application but it is unarguably the phase that uncovers functional problems, bugs, and gaps in nonfunctional requirements, and helps to build a more predictable application. No amount of testing is considered sufficient to find all the defects in a system/application; it is always desirable to have an organized and systematic testing framework/approach in order to minimize the number of test cycles and to create an application whose behavior is more predictable.

Cloud computing has impacted the IT industry and all the business sectors in general, and testing, or the quality assurance domain, has also been affected. With the growing demand for cloud deployments, quality-assurance strategies also need to be formalized to cater for the areas associated with testing. The two areas that are referred here are testing of applications deployed on cloud ("testing a cloud") and leveraging cloud for testing for applications deployed on-premises or cloud ("cloud-based testing"). This chapter discusses the impact of cloud computing in the area of testing and significance of testing for cloud-based services as well.

Encyclopedia of Cloud Computing, First Edition. Edited by San Murugesan and Irena Bojanova.
© 2016 John Wiley & Sons, Ltd. Published 2016 by John Wiley & Sons, Ltd.

28.2   Testing of Cloud-Based Services

Testing of cloud-based services refers to the verification and validation of applications, environment, and infrastructure that are available on demand and deployed on the cloud infrastructure (Dewan, 2010), which ensures that applications, environment, and infrastructure conform to the expectations of the cloud computing business model. Like any other software application, the "testing phase" holds equal importance for cloud applications as well. This section discusses why cloud testing is important, how testing cloud applications differs from testing traditional applications, and what quality-assurance strategies can be employed for testing cloud-based applications.

28.2.1   Testing Traditional Deployments versus Testing Cloud Deployments

Organizations need to be aware of risks associated with their cloud deployments and accordingly chalk out the testing strategy to mitigate the risks through comprehensive testing methodologies. The areas to be considered for testing a cloud offering encompass both functional and the nonfunctional areas. Both these areas are applicable for traditional deployments as well; however, there is a need to test certain aspects that are specific to cloud deployments. Apart from these two areas, cloud deployments also need to be tested for operational aspects of the cloud. We will cover all of these areas in detail as follows.

28.2.2   Testing for Functional Aspects

This section includes testing to assure the quality of applications deployed in the cloud around functional services and business processes of the software. The functional testing aspects of cloud applications are more or less similar to traditional deployments. Just like in testing of traditional deployments, cloud-based deployments will test the application based on its functional specifications, which include types of testing like black box testing, integration testing, system testing, user acceptance testing, and regression testing.

28.2.3   Testing for Nonfunctional Aspects

This section includes testing to assure the quality of applications deployed in the cloud around nonfunctional aspects of the software. The common testing aspects (Katherine and Alagarsam, 2012) that are applicable to both cloud and traditional deployment are performance testing, high-availability testing, and security testing; however, distinctive ones include compatibility or interoperability testing, and elasticity testing.

28.2.3.1   *Performance and Load Testing*

This includes testing of performance parameters like response time and throughput under defined user load and progressively evaluating it with increasing levels of load to ensure that the performance does not degrade with an increased volume of data and transactions.

28.2.3.2   *High Availability Testing*

This ensures application uptime according to the service-level agreement (SLA) and having a defined disaster-recovery mechanism in place. Cloud application providers need to design their architecture for high availability in the cloud by creating redundancy across servers, zones, and clouds. They should test for high availability by disabling individual servers to simulate a failure scenario.

28.2.3.3   Security Testing

This is important for both traditional and cloud applications. However, for the latter it holds more significance as in a cloud application the data may be stored on cloud and that may be outside of the organization's boundaries. So cloud applications need to follow globally accepted security standards for maintaining data confidentiality and integrity, and authorized access and testing methodologies need to align to test these aspects of the application (Dangwal *et al.*, 2011).

28.2.3.4   Elasticity Testing

This ensures that cloud deployments adhere to economics of scale as and when desired – i.e. when the demand for the application increases or decreases, the cloud readjusts the resources to meet the demand without impacting the performance of the application. This is same as scalability in the traditional application, although with the difference that the resources used by the application conform to the "elastic" nature of the cloud.

28.2.3.5   Compatibility or Interoperability Testing

This ensures the interoperational capability of the cloud application across cloud platforms and applications on premises and for legacy systems and applications. It should also ensure that the cloud application can be easily migrated from one provider to another, thus avoiding vendor lock in.

28.2.4   Testing for Operational Aspects

There are certain operational aspects of the cloud deployments that do not apply to traditional deployments. They are:

- multitenancy, customizability/configurability, provisioning;
- metering and billing.

With regard to multitenancy and customizability/configurability, provisioning testing is carried out to ensure that the same application deployment is able to cater for multiple tenants. This is specifically applicable to deployment by ISVs (independent software vendors) that have multiple customers (tenants) for their application. Multitenancy testing should ensure that, even in a shared deployment model, the application is customizable and configurable for each tenant. The provisioning and deprovisioning of a tenant and the ability to meter and bill a tenant based on amount of its usage of shared applications are aspects that are covered by the operational testing of the cloud.

With regard to metering and billing, a cloud application may have inbuilt billing, licensing, or user subscription management or it may use or rely on cloud APIs to provide the necessary data. For example, the application may rely on cloud to provide different licensing models like pay per use, or to track bandwidth usage data. Comprehensive testing of all interface interaction must ensure that all the interaction with the environment (cloud) is happening as per expectations.

The weight and applicability of each of these areas may vary from application to application. Different cloud-based applications may give different weights to these parameters; depending on the application, some parameters may not be applicable. For example, for an online banking application security may be more important, whereas response time will have more weight for a gaming application. Therefore, the testing strategy for individual cloud applications needs to be tailored.

28.3   Why Test a Cloud Deployment?

Consider the following three examples that use cloud deployment for hosting applications. Each of the following scenarios focuses on some of the important testing areas that were discussed in the previous section.

28.3.1   Scenario One

Alice needs to transfer funds to a vendor immediately in order to make sure that her shipment arrives on time, just before the holiday season. Alice opens her computer to transfer the funds using her online banking application, only to find out that her online application is running very slow because it is experiencing a heavy load. She patiently waits for the application, fills out all the required details to transfer funds, and clicks on the "transfer funds" button. Unfortunately, the application goes offline because of the increased load but before it goes down the money is deducted from her account and is not transferred to the vendor's account.

The relevant testing areas are availability, response time, and function testing. Thorough function testing would have ensured that transactions are autonomous. High availability and response-time testing would have ensured that the application is constantly available under high load, and that it responds within acceptable limits. Thorough testing of these areas would ensure a more predictable behavior.

28.3.2   Scenario Two

"ABC Gaming" is a small startup company that develops online games. Its existing cloud-based servers are running at 20% load capacity. "ABC Gaming" launched a new game, which became an instant hit all across the world. Suddenly its subscription rises from half a million to 10 million with users from all across the world. This new game gave "ABC Gaming" the exposure that it needed. However, its application couldn't sustain such a high load, leading to frequent crashes and a slowdown in performance. This resulted in a bad press and a lot of users being turned away.

The relevant testing areas are dynamic scalability (elasticity), response time, and availability across geographical regions. Thorough dynamic scalability testing would have ensured that additional servers are added automatically to the deployment. Availability and response testing would have allowed the company to know the application's behavior when accessed across geographical regions and thus providing information for better planning.

28.3.3   Scenario Three

"XYZ Docket" is a company that provides a cloud-based online document repository, with unlimited data backup. Multiple users have registered with the application to use the service for backup and restore. One day John, a registered user of "XYZ Docket," loses his personal external hard drive and decides to restore data from the backup using "XYZ docket." He logs onto the online application and clicks to download with a full backup. After the download completes, he checks his hard drive only to find that, along with his data, the backup has also downloaded the financial statements and personal files of Mr. Franc.

The relevant testing areas are multitenancy and data security. Thorough multitenancy and data security testing would have ensured that multiple clients can use the same application and that every client's data is private and protected.

In all these scenarios, if the cloud application had been tested more thoroughly in all aspects (functional and nonfunctional), as discussed in sections 28.2.3 and 28.2.4, it would have yielded more predictable results and thus it would have allowed the organizations to know, ahead of time, what to expect and to plan accordingly.

Now that the importance of elaborate testing in cloud deployments is established, it is important to understand that the testing strategy also requires special considerations. Based on the stack (software, platform, or infrastructure) at which the cloud deployment is to be incorporated, the quality assurance strategy needs to be framed out accordingly. Listed below are the testing strategies that need to be followed at each stack level. along with some hybrid scenarios:

- Adoption of software as a service (SaaS) – replacing on-premises applications with their equivalent on-demand versions. Quality assurance strategies, in this case, should involve black-box functional testing, concentrating on networking, user interface, business critical requirements, and multitenancy testing.
- Adoption of infrastructure as a service (IaaS) to deploy business applications that were deployed on premises earlier. Apart from all the traditional assurance strategies and testing business workflows, availability, disaster recovery as well as security testing should be the emphasis for this type of scenario.
- Adoption of a platform as a service (PaaS) to build and host an application. Quality assurance for applications built using platform as a service mainly depends on the features of the service. Security APIs, storage services, and uptime and virtualization efficiency are governed by their underlying service. It is therefore important for consumers to ascertain that the service meets their requirements before choosing the PaaS vendor for the development of applications.
- Adoption of a hybrid model in which a part of the business application is moved to cloud and the rest stays on premises. This is a special type of scenario in which comprehensive interoperability and integration testing are required. Security and data-integrity testing are also critical because of constant transmission of critical data over the network.

There is no standard strategy or framework to perform quality assurance for cloud deployments. Vendors and consumers should take into account cloud computing features and relevant scenarios during the strategic planning phase. The following section discusses a framework that might be adopted to instill more confidence in cloud offerings.

28.4 Generating Confidence in the Outcomes of Cloud Testing

As we discussed in previous sections, different testing areas can be assigned weights to indicate their priority. For example, in Table 28.1 a reference weight is a number assigned to each area.

Such data should provide an insight into the level of testing and the quality of the product being tested. To be able to achieve an accurate insight it requires individual test stages to have properly weighted test cases. For example, most of the applications generally have priorities assigned to the test cases. A priority 1 test case will have a higher weighting than a priority 2 test case, and similarly, priority 2 test cases will hold a higher weighting than priority 3 test cases, and so forth.

The number of passed and failed test cases for a stage will form a cumulative weighted score for an individual stage; a weighted average of cumulative scores for all stages can then be calculated.

Every testing stage must have a clearly defined input and output criteria. For example it could be as simple as successful execution of all P1 and P2 test cases.

Multiple stages of deployment can also be tested before making a cloud application available on a public cloud. For example, a cloud application can be tested on a private or a hybrid cloud first before making the application available on a public cloud. Once the application has passed on a private cloud/hybrid cloud, it can then be deployed on a public cloud for final round of testing.

28.5 Cloud-Based Testing

Cloud-based testing or "testing as a service" (TaaS) is defined as a cloud-based model for offering testing services. Information technology organizations that deal with testing products and services are making use of

Table 28.1   *Weighted scores of test areas*

Functionality	Test case	Weight	Weighted % passed test cases
Module 1		M1	...
	Test case 1	w1	...
	Test case 2	w2	...
	Test case n	wN	...
Module 2		M2	...
	Test case 1	w1	...
	Test case 2	w2	...
Total			Average of all stages.

a cloud-based licensing model to offer their testing products and services from the cloud for their end clients. The offerings cover functional as well as nonfunctional testing of various applications/products.

In this section we will discuss how cloud testing differs from traditional testing, what benefits cloud testing brings, and possible challenges associated with it. This is followed by a case study explaining the benefits of cloud testing and giving information about some prominent cloud-testing providers.

28.5.1   Traditional Testing and Cloud-Based Testing

As soon as the test planning begins in traditional testing, the organization starts planning for:

- Procurement of hardware for testing. This includes budgeting and approvals for acquiring new hardware for test servers and could range from low-end test servers to high-end load/stress/performance test servers.
- Procurement of software. This includes budgeting and approvals for various testing software, and could include penetration testing, code coverage, load testing, etc.
- Procurement of hardware and software, which is a time-consuming process and may require approvals from multiple stakeholders, and thus have a direct or indirect impact on initial investment and on time to market the application.
- Installing and configuring testing software and configuring test beds.

28.5.2   What Cloud Brings to a Table

All the testing stages that can be run using a traditional testing approach can be executed through cloud testing as well. From an organization's perspective, cloud brings simplicity and flexibility in the testing process, resulting in lower investment costs by providing:

- Ease of deployment. Complete hardware and software can be hosted on a cloud environment. It reduces the time required to set up the servers and helps in reducing the time to market the application.
- Scalable model. Cloud allows organizations to add or remove hardware or software dynamically, according to their needs. For example, a high-end server may not be needed until the load test or stress test or performance test starts. The server can be decommissioned as soon as the testing is complete. Similarly, in cloud testing, a license for a load-test application may be required only for the duration when load test is being run.
- Reduced initial capital expenditure. The capability to scale test labs dynamically and the provision to deploy or remove software dynamically allows organizations to reduce the initial investment cost. The availability of different licensing models, like pay per use, can further help in reducing cost.

- Real-world simulation. Cloud testing can also help organizations to simulate real-world production servers more effectively. In addition, the load on servers can simulate a real world scenario more closely both from a scale of load perspective and from a geographical distribution perspective. The simulation of different browsers and platforms is also much easier when using the cloud infrastructure than when using the traditional one.
- Effective use of resources. In general, cloud testing will require less administration effort compared to traditional testing. The organization can thus use resources more effectively.
- More accuracy. The ready-to-use standard infrastructure and preconfigured tools and utilities in the cloud lead to more accuracy in testing and save a lot of effort that could be wasted due to the wrong test bed or tool configuration.

28.5.3   Testing as a Service

As explained above, testing using cloud presents inherent advantages for enterprises, and various test-tool vendors have added cloud-based testing to their portfolio to benefit from the evolution of this model. Vendors have tailored their conventional test-tool software to adapt to the SaaS model. This, coupled with utilization of an on-demand infrastructure (using IaaS) for hosting their test-tool software, has led cloud-based testing services to be transformed into a TaaS model.

A typical TaaS architecture (Neotys, 2013) is shown in Figure 28.1. It is built using machines with high computing power acquired from the cloud, set up for testing tools and executing scripts, all of this automatically. There is also a test controller that controls the TaaS setup remotely over the Web using the TaaS interfaces

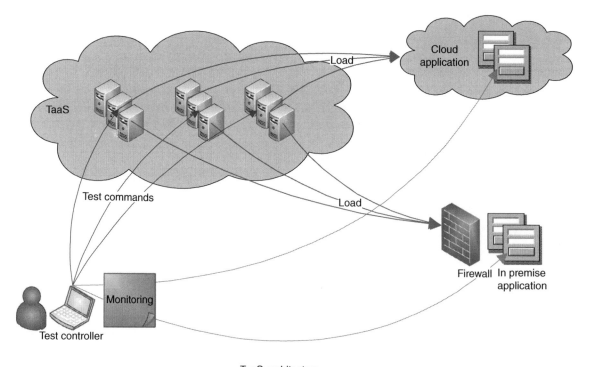

TaaS architecture

Figure 28.1   *TaaS architecture. Source: SOASTA (2012). Reproduced with permission of SOASTA Inc.*

that issues commands using these interfaces. The TaaS setup generates the load on any cloud application or on the on-premises application. Monitoring of the test results is also done remotely through the test controller.

Some of the cloud-based testing providers and their tools are listed in the following subsections.

28.5.3.1 SOASTA CloudTest

SOASTA CloudTest (http://www.soasta.com/products/cloudtest/, accessed December 16, 2015) is used for Web and mobile performance testing. CloudTest allows users to build, execute, and analyze load tests on a single powerful platform to and run the tests on SOASTA's Global Test Cloud to any traffic level. It is built on leading cloud provider platforms and enables the simulation of traffic for load and performance testing of Web and mobile applications by leveraging the elasticity and power of Cloud Computing.

28.5.3.2 Blazemeter

Blazemeter (http://blazemeter.com/, accessed December 16, 2015) is a commercial, self-service load-testing platform, which is fully compatible with open-source Apache JMeter, the performance-testing framework by the Apache Software Foundation. Blazemeter allows massive load tests to be run in the cloud and can simulate any user scenario for Web apps, web sites, mobile apps or Web services.

28.5.3.3 LoadStorm

LoadStorm (http://loadstorm.com/, accessed December 16, 2015) is a Web-based load-testing tool for simulating user load for Web sites or Web applications. It is an on-demand load and performance testing tool that uses cloud infrastructure for massive scalability. LoadStorm leverages the power of Amazon Web Services to scale on demand with processing power and bandwidth as needed to test the largest Web projects. It automatically adds machines from Amazon's server farm to handle the processing and releases the machines when they are not needed.

28.5.3.4 Neotys NeoLoad

NeoLoad (http://www.neotys.com/, accessed December 16, 2015) is a load-testing software solution designed for Web and mobile applications, to simulate user activity realistically and monitor infrastructure behavior. Neoload supports load generation from local infrastructure as well as from the cloud through the Neotys cloud platform. It allows use of the cloud load generators in a few minutes; they are paid for with a pay-as-you-go model.

28.5.4 Case Study

The following case study highlights how cloud testing can help companies to prepare to validate their product offerings for real-life load situations, to handle the anticipated demand for their product, and potentially to build a more stable and reliable product offering, thus leading to a superior user experience.

The case study (SOASTA, 2012) is for a company named Myspace that provides social networking service and drives social interaction by providing a highly personalized experience around entertainment and by connecting people to the music, celebrities, TV, movies, and games that they love.

Myspace launched a new music video series in New Zealand, which included streaming video, favorite lists, and search. Anticipating a huge increase in server load, they wanted to supplement existing live traffic with performance-test traffic to get an idea of the impact of the new video offering. Myspace decided that adding the load of one million virtual users to the existing, prelaunch load would be an adequate performance test for its infrastructure. So they decided to supplement their live traffic in New Zealand by a million simultaneous users.

They were now confronted with a challenge to get the required infrastructure for simulating traffic for such a huge user population to test the performance of their application. Realizing that the only source for such a huge supplement lies in the cloud, the company turned to the SOASTA CloudTest platform.

The Myspace operations team specified the load that they wanted to impose during testing:

- a million concurrent virtual users;
- test cases split between searching for and watching music videos, rating videos, adding videos to favorites, and viewing artists' channel pages;
- transfer rate of 16 Gb/s;
- 6 TB of data transferred per hour;
- Over 77 000 hits per second, in addition to live production traffic.

Using the CloudTest platform, the calls were made to Amazon EC2 API, requesting 800 large EC2 instances in groups of 25 servers at a time. It also invoked two extra-large EC2 instances to act as the test controller and the results database. All this took about just 20 minutes to spin up. Once CloudTest had instantiated the EC2 servers, it performed stability checks on them, then discarded and replaced any dead units until it had a healthy population. So within minutes, the CloudTest platform could provide a test setup that generated more than 77 000 hits per second. While the tests ran, the load generators sent data back to a single analytics service connected to a PostgreSQL database, which aggregated performance test metrics.

28.5.5   Testing as a Service: Challenges and Considerations

Although there are several benefits of utilizing the cloud for testing, there are a few challenges associated with TaaS:

- Pure cloud-based test beds. Providers of TaaS provide online automated interfaces to create the required test infrastructure. The chosen TaaS vendor should provide a cloud-based way of meeting all the requirements and configuration settings for the test environments. In order to overcome this challenge, TaaS vendors can partner with IaaS providers (or provide cloud-enabled processes) to create on-demand test beds. Infrastructure as a service can be used to lower significantly the costs and the time required to create test beds and to reduce complexity.
- Testing data integrity and security. Data security is a major concern in case of public clouds. Critical data might have to be stored at remote locations, which might not comply with the organization's security requirements. Providers of TaaS must ensure that data-security requirements are met and should provide a way to validate that privacy requirements are met, and applications being tested are secured.
- Choosing a cost-effective model. With so many license models being available, it is important to choose the correct model, which suits the organization's requirements. In addition to ensuring security for the organization, test managers may have to plan for any additional costs that might incur for securing and encrypting data to be sent to the TaaS provider. Similarly, creating additional disaster recovery environments, or requesting additional test beds for an updated configuration, may need to be planned ahead.
- Test tools and understanding and interpreting test reports. Testing teams may need to acquire knowledge about running different TaaS tools and may need to understand which tool suits their application and requirements best. Different tools have different reporting mechanisms for publishing the test report. The testing team should be able to understand and interpret the test reports accurately to be able to take

any corrective action. It may require thorough analysis to determine a reported problem (for example a performance bottleneck) at the TaaS provider end, in a network interface, or in the actual product. It is becomes essential to understand how much coverage TaaS provides. The better the coverage, the better are the chances to identify problems.

28.6 Cloud Testing Benchmarks

Cloud testing benchmarks are not yet standardized. Some of the popular benchmarking standards are discussed below:

- The *Yahoo Cloud Serving Benchmark* (YCSB) system uses a group of different loads to benchmark the performance of various IaaS (specifically "storage as a service") providers like Google BigTable, Microsoft Azure, and Amazon. It provides a framework that can be used to find the best possible offering that meets the organization's requirements, especially around service operation levels. It involves a customizable load generator and proposes a set of loads that can be applied to various systems under similar configurations and compares performance on different parameters.
- The *Cloud Spectator* benchmarking system uses its custom framework to measure and compare server-level performance for various public IaaS providers. Although it does not provide a customizable framework like YCSB, it publishes comprehensive comparative reports detailing performance for each provider on various parameters like storage efficiency, networking, and server. It also publishes the price versus performance metrics, which can be very helpful for calculating return on investment while choosing a cloud offering.
- *CloudStone* benchmarking is an open-source framework to measure performance for various cloud offerings based on different parameters. It uses an automated workload generator to apply loads on various offerings using open-source social media tools.

28.7 Conclusion and Future Scope

In order to understand the difference between cloud testing and testing cloud, recollect the terminology discussed in previous sections. "Testing Cloud" implies testing cloud-based offerings emphasizing key nonfunctional parameters to ascertain that the offering is meeting the contractual requirements and its required intents. On the other hand, TaaS is sometimes referred to as "cloud testing," which implies cloud-based testing where various genres of testing – test beds as well as test management suites – are provided on demand.

Cloud computing is playing a key role in making "anywhere anytime" access possible. Private clouds, public IaaS and even some SaaS service vendors have a significant role here, especially in lowering prices.

Thorough periodic testing for cloud services is essential to establish an operation's continuity and to minimize the risks. Testing the cloud offering at agreed intervals will also help in determining whether contractual obligations are being met. A cloud offering vendor whose offering is benchmarked for its ability to meet requirements, its availability, performance, and security, is more likely to be trusted and partnered.

Testing as a service is also becoming essential with the advent of a slew of devices, everyday launches of smart device-based applications, and worldwide high-speed access. The leading vendors providing integrated software quality suites are successful precisely because they provide testing services on demand, especially in the areas of performance testing and mobile testing. On-demand test beds, automated test execution, and online test reports are some of the key features provided by these vendors.

With the growing popularity of PaaS, there is a plethora of applications being developed every day that are anticipated to be executed on various devices globally. Consequently, there will be an enormous number of test configurations, including platforms, browsers, devices, networks, and ability to mirror real-time user access. These are becoming challenging for traditional testing suites and processes. Automated test execution and management suites and on-demand testing infrastructure need to be considered when making a choice of test vendors.

Over a period of time, testing in cloud computing is expected to reach a more mature stage. As cloud testing standards start evolving, testing strategies will be modified and tailored to test quality of service parameters, based on defined standards. Standardization will also allow existing concerns, like security and interoperability, to be addressed with more objective data. It is evident that cloud computing is here to stay and will grow over a period of time. The testing domain has thus made, and will continue to make, a significant contribution to the growth of cloud computing, and vice versa.

References

Dangwal, N., Dewan, N. M., and Vemulapati, J. (2011) *SaaS security testing: Guidelines and Evaluation Framework*, http://www.infosys.com/engineering-services/white-papers/Documents/SaaS-security-testing-cloud.pdf   (accessed December 16, 2015).

Dewan, N. M. (2010) Cloud-Testing vs. Testing Cloud. Published paper, Tenth Annual Software Testing Conference, http://www.infosys.com/engineering-services/white-papers/Documents/cloud-testing-vs-testing-cloud.pdf (accessed December 16, 2015).

Katherine, A. V. and Alagarsam, K. (2012) Conventional software testing vs. cloud testing. *International Journal of Scientific and Engineering Research* 3(9), http://www.ijser.org/researchpaper%5CConventional-Software-Testing-Vs-Cloud-Testing.pdf (accessed December 15, 2015).

Neotys (2013) *NeoLoad Cloud Platform Architecture*, http://www.neotys.com/product/neoload-cloud-testing.html (accessed December 16, 2015).

SOASTA (2012) *Case Study on MySpace*, http://cdn.soasta.com/wp/wp-content/uploads/2012/09/CS-Myspace-061312.pdf (accessed December 16, 2015).

29

Cloud Service Evaluation

Zheng Li,[1] Liam O'Brien,[2] and Rajiv Ranjan[3]

[1] *Australian National University and NICTA, Australia*
[2] *Geoscience Australia, Australia*
[3] *CSIRO Computational Informatics, Australia*

29.1 Introduction

Along with the boom in cloud computing, more and more cloud services have become available on the market, supplied by an increasing number of providers. Different cloud providers have their own idiosyncratic characteristics when developing services (Li *et al.*, 2010), so cloud services may be offered using different terminology, and with different qualities and cost models. Furthermore, even the same provider can supply different kinds of services with similar functionalities for different purposes. For example, Amazon has provided several options for a storage service, such as S3, EBS, and the local disk on EC2 (Chiu and Agrawal, 2010).

Given the diversity of cloud services, cloud usage should require a deep understanding of how the different service candidates may (or may not) match particular demands. Unfortunately, the indicators such as compute units, memory size, and service price often do not provide comprehensive information about services and their suitability for a specific task (Lenk *et al.*, 2011). As a result, on the one hand, customers have little knowledge and control over the precise nature of cloud services even in a "locked-down" environment (Sobel *et al.*, 2008); on the other hand, there could be uncertainty in the runtime regarding cloud services due to the various quality of service (QoS) consumption requirements of different applications (Zhang *et al.*, 2012). Consequently, service evaluation should be one of the prerequisites of employing cloud computing.

In general, the evaluation of cloud services would be beneficial and important for both service customers and providers. For example, proper performance evaluation of a candidate cloud service can help customers carry out cost-benefit analysis and decision making for service selection, while it can also help providers improve their service quality against competitors.

This chapter presents a summary of the state of the practice of cloud services evaluation. The summary covers the current evaluation purposes (cf. section 29.2), the service features that have been evaluated (cf. section 29.3), the *de facto* benchmarks (cf. section 29.4), and more importantly, a practical methodology for implementing evaluations (cf. sections 29.5 and 29.6). Note that, here, we pay attention to infrastructure as a service (IaaS) and platform as a service (PaaS), without considering software as a service (SaaS). As SaaS is not used to build individual business applications (Binnig *et al.*, 2009), various SaaS implementations may comprise infinite and exclusive functionalities to be evaluated, which could make the summary complicated.

29.2   Current Purposes of Cloud Services Evaluation

Given different application requirements, there may be various motives for evaluating cloud services. However, we can roughly identify four main purposes of the current cloud services evaluation studies (Li *et al.*, 2013b), as shown in Figure 29.1 and specified below.

- *Cloud resource exploration* investigates the available resources (e.g., computation capability) supplied by cloud services.
- *Business computing in the cloud* investigates applying cloud computing to solve business issues.
- *Scientific computing in the cloud* investigates applying cloud computing to solve scientific issues.
- *Comparison between computing paradigms* compares cloud computing with other computing paradigms.

Interestingly, although public cloud services are provided mainly to meet the technological and economic requirements of business enterprises, commercial cloud computing is still regarded as an encouraging potential paradigm to deal with scientific issues. For example, existing evaluation studies have revealed that smaller scale scientific computations can particularly benefit from the moderate computing capability of the cloud; on-demand resource provisioning in the cloud can satisfy some high-priority or time-sensitive requirements of scientific work when in-house resource capacity is insufficient; through appropriate optimization, the current commercial cloud can be improved for scientific computing; furthermore, once commercial cloud vendors pay more attention to scientific computing, they can make the current cloud more academia-friendly by slightly changing their existing infrastructure.

29.3   Cloud Service Features that have been Evaluated

Although there could be various theoretical concerns about applying cloud computing, in current evaluation practices the most evaluated cloud-service features are performance, economics, and security (Li *et al.*, 2013b), as illustrated in Figure 29.2.

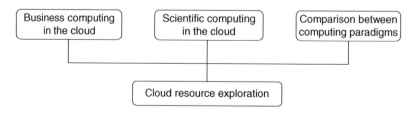

Figure 29.1   *Purposes of cloud services evaluation*

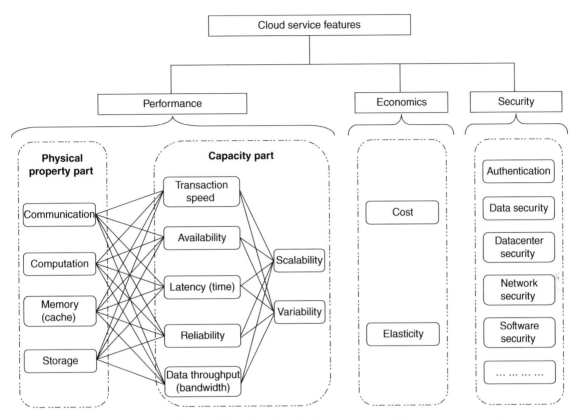

Figure 29.2   *Cloud service features evaluated*

29.3.1   Performance

In practice, one evaluated performance feature is usually represented by a combination of a physical property of cloud services and its capacity – for example communication latency, or storage reliability. We therefore split the performance feature dimension into two parts: a *physical property* part, and a *capacity* part.

In the *physical property* part, communication refers to the data/message transfer between the client and the cloud, different cloud services, or different service instances. Computation refers to computing-related data/job processing in the cloud. Memory and cache are treated as a unified physical property of cloud services, because they are both designed to temporarily save and provide fast access to data. Contrasted with memory (cache), storage of cloud services can be used to store users' data permanently, until the data is removed or the services are suspended intentionally. Note that one of the popular IaaS styles in the market, virtual machines, can be viewed as an integration of the above four types of physical properties.

In the *capacity* part, transaction speed refers to the frequency of independent jobs or predefined operations related to a physical property of cloud services. Availability is driven by the time lost, which describes the probability that a system works in a functioning condition during a specific period of time. Latency is mainly related to the measure of time delay for a particular job, which can be further distinguished between different time windows. Reliability is driven by the number of failures, which describes the probability that a system can properly perform its intended function during a specific period of time. Throughput, here, is treated as a general concept that describes an amount of data processed in a particular period of time (from input to output) by any physical property of cloud services.

In particular, although both scalability and variability have been viewed as aspects of cloud service independent of performance, in some literature, they are inevitably reflected by a change in performance features. Considering their close relationship with performance, and for the convenience of discussion, here we also regard scalability and variability as two elements in the *capacity* part but distinguished from the other capacities.

29.3.2   Economics

Economics has generally been considered a driving factor in the adoption of cloud computing. According to some views, the economics aspect of a commercial cloud service comprises two properties: cost and elasticity. Cost is an important and direct indicator to show how economic the application of cloud computing is. In theory, cost may cover a wide range of factors when moving computing to the cloud. However, in the primary studies that we reviewed, we found that the current evaluation work mainly concentrated on the real expense of using cloud services.

Elasticity describes the capability to add and remove cloud resources rapidly in a fine-grained manner. In other words, an elastic cloud service concerns both growth in and reduction of workload, and particularly emphasizes the speed of response to changed workload. Although evaluating the elasticity of a cloud service is not trivial, we considered a metric to be elasticity related as long as it measures the time of resource provisioning or releasing.

29.3.3   Security

The security of commercial cloud services has many dimensions and issues with which people should be concerned. For example, authentication has been identified as a weak point in hosted and virtual services and has been targeted frequently, therefore suitable mechanisms should be employed to validate the identity of services, service providers, and cloud consumers. Data security is one of the key criteria for using the cloud, which involves appropriately protecting customers' data from the outside world. Datacenter security may refer to internal security, which requires infrastructural protection from artificial damage or natural disasters. Since clouds are inherently associated with intranet and Internet – that is a shared environment – network security may face multitenancy challenges, and it mainly relies on various encryption techniques. Software security takes into account systemwide issues ranging from antivirus mechanisms to robust design.

Unfortunately, not many evaluations of cloud service security were reflected in the existing empirical studies. Even in the limited number of studies that exist, security evaluation was realized mainly by qualitative discussions. The main reason could be that security is hard to quantify, and there is a lack of suitable metrics for measuring cloud service security (Li *et al.*, 2013b). Consequently, a practical approach to security assessment can be using a preidentified risk list to discuss the security strategies supplied by cloud services.

29.4   De Facto Benchmarks for Cloud Service Evaluation

As traditional benchmarks were considered to be insufficient to meet the idiosyncratic characteristics of cloud computing, Binnig *et al.* (2009) described what an ideal cloud benchmark should be. Several new cloud benchmarks have been developed – for example Yahoo! Cloud Serving Benchmark (YCSB) (Cooper *et al.*, 2010) and CloudStone (Sobel *et al.*, 2008). In particular, six types of emerging scale-out workloads were collected to construct a benchmark suite, CloudSuite (Ferdman *et al.*, 2012), to represent today's dominant cloud-based applications, such as Data Serving, MapReduce, Media Streaming, SAT Solver, Web Frontend, and Web Search.

Table 29.1 *Popular traditional benchmarks for evaluating cloud service performance*

Cloud service performance property	Popular traditional benchmarks
Communication	iperf, ping, Operate/Transfer Data
Computation	HPCC: DGEMM, HPCC: HPL, LMBench
Memory/Cache	HPCC: STREAM
Storage	Bonnie/Bonnie++, IOR, NPB: BT/BT-IO, Operate/Transfer Data
Overall Performance	BLAST, HPCC: HPL, Montage, NPB suite, TPC-W

However, Li *et al.* (2013b) show that traditional benchmarks have been used extensively in cloud services evaluation. Given the many different aspects of cloud services, a normal strategy is to employ benchmark suites to perform holistic evaluations, as it would be impossible to use only one benchmark to fit all evaluation scenarios. Traditional benchmarks can therefore still satisfy at least some of the requirements of cloud service evaluation. Furthermore, past evaluation experiences can be used to indicate the applicability of those traditional benchmarks. For example, we may define a benchmark's "applicability" as the number of related studies. Through the applicability of different traditional benchmarks (Li *et al.*, 2013b), we can list the popular benchmarks as recommendations for cloud services evaluation, as shown in Table 29.1. For the sake of brevity, we do not elaborate these benchmarks in this chapter. Note that these traditional benchmarks are particularly related to the cloud service performance properties. On the one hand, existing cost evaluation studies were generally based on the corresponding performance evaluation. On the other hand, there is a lack of distinct benchmarks for evaluating elasticity and security.

29.5 Methodology for Cloud Service Evaluation

It has been recognized that the evaluation of cloud services belongs to the field of experimental computer science (Stantchev, 2008), which requires a suitable evaluation methodology to direct experimental studies (Blackburn *et al.*, 2008). Stantchev (2008) extended the ASTAR method and specifically suggested a five-step methodology (identify benchmark, identify configuration, run tests, analyze, and recommend) for evaluating cloud services. A more detailed evaluation methodology was specified in Lenk *et al.* (2011), which used business process modeling notation to describe the general steps of developing, executing, and evaluating a cloud benchmark suite.

This chapter introduces an evidence-based methodology for cloud services evaluation, namely cloud evaluation experiment methodology (CEEM) (Li *et al.*, 2013a). This evolved from three knowledge resources, as shown in Figure 29.3. Firstly, it borrowed from evaluations of traditional computing systems. As cloud computing emerged as a computing paradigm, individual cloud services can be viewed as concrete computing systems within such a paradigm. Thus, traditional approaches to evaluation would be also useful for evaluating cloud services. Secondly, it referred to the design of experiments (DOE) guidelines (Montgomery, 2009). Although DOE is normally applied to the agricultural, chemical, and process industries, considering the natural relationship between experiment and evaluation, the various DOE techniques of experimental design and statistical analysis can also benefit cloud services evaluation. Thirdly, it summarized existing experience of evaluating cloud services. Based on this experience, CEEM supplies specific, practical suggestions in steps, for example prelisting experimental factors and metrics.

The procedure of cloud services evaluation driven by CEEM can be roughly divided into two parts: pre-experimental activities (a linear process) and experimental activities (a spiral process). Experimental activities follow a

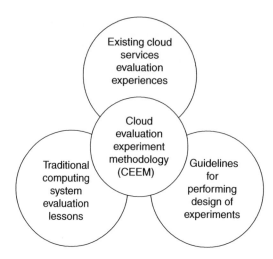

Figure 29.3   *Evolution of the cloud evaluation experiment methodology (CEEM)*

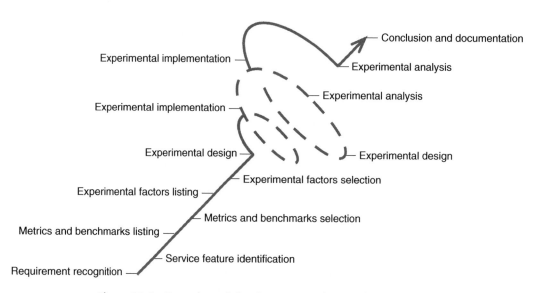

Figure 29.4   *Procedure of cloud services evaluation driven by CEEM*

spiral process because an evaluation could be composed of a set of experiments, while the latter experimental design could have to be determined by the former experimental results and analyses. For example, imagine the scalability evaluation of Amazon EC2; we may prefer the most cost-wise VM type for horizontal scalability experiments, while the selection of VM type should be based on the vertical scalability experiments among a set of candidate VM types. Overall, CEEM summarizes activities of cloud service evaluation into ten generic steps ranging from *Requirement Recognition* to *Conclusion and Documentation*, as specified below (also cf. Figure 29.4). Compared to existing studies of cloud services evaluation, the validation study shows that CEEM would be able to help evaluators achieve more rational experimental results and draw more convincing conclusions (Li *et al.*, 2013a).

1. *Requirement recognition.* Evaluation requirements are recognized not only to enable users to understand problems related to cloud service evaluation but also to achieve a clear statement of the evaluation's purpose, which is an obvious but nontrivial task (Montgomery, 2009). A clearly understood evaluation requirement can help to drive the remaining steps in an implementation of a cloud services evaluation properly. To help recognize a requirement, it has been suggested that a set of specific questions be prepared to be addressed by potential evaluation experiments (Montgomery, 2009). Moreover, it is normally helpful to replace one comprehensive question with a list of separate and easily answerable questions, so that evaluators can conveniently define specific evaluation objectives, and then employ the strategy of sequential experiments to satisfy the overall evaluation requirement.

2. *Service feature identification.* Given the recognized evaluation requirement, evaluators can identify the relevant cloud services and the features to be evaluated. Although it is hard to outline the scope of cloud computing, and various cloud services are increasingly available in the market (Li *et al.*, 2010), it is still possible to list a suite of general service features in advance. By exploring the existing practices of cloud services evaluation, Li *et al.* (2013b) show that mainly three service features have been of concern, namely performance, economics, and security, as specified in section 29.3. Thus, in most cases, evaluators may conveniently identify relevant service features in the general feature list.

3. *Listing of metrics and benchmark listing.* The choice of right metrics depends on the cloud service features to be evaluated. However, one service feature may be measured by different metrics with different benchmarks, and the selection of particular metrics and benchmarks may also have other constraints or tradeoffs. To facilitate the metric/benchmark selection, it is helpful, first, to list all the candidate metrics and benchmarks for a proposed cloud services evaluation. By using different cloud service features as the retrieval keys, evaluators may conveniently lookup suitable metrics and benchmarks.

4. *Selection of metrics and benchmarks.* According to research on the evaluation of traditional computer systems, the selection of metrics plays an essential role in evaluation implementation. A suitable metric would play a *response variable* role (Montgomery, 2009) in applying DOE to cloud services evaluation. Although traditional evaluation treats the selection of metrics as a prerequisite of benchmark selection, there were always tradeoffs between metrics and benchmark selection when evaluating cloud services. For example, only two metrics (*benchmark runtime* and *benchmark FLOP rate*) are available to measure computation latency and transaction speed when adopting NAS Parallel Benchmarks to evaluate cloud services. Therefore, it is suggested that metrics and benchmarks could be determined together in one step.

5. *Listing of experimental factors.* Before evaluating a cloud service feature, knowing all factors (also called parameters or variables) that affect the service feature is a tedious but necessary task. Although a complete listing of experimental factors may not be achieved easily, evaluators should keep the factor list as comprehensive as possible at all times, for further analysis and decision making about factor selection and data collection. Li *et al.* (2013b) have proposed a framework to capture the state-of-the-practice of experimental factors that people currently take into account when evaluating cloud services. This factor framework can in turn help facilitate the identification of suitable factors for designing evaluation experiments. Moreover, the factor framework offers a concrete and rational base for further discussion and listing of factors by expert judgments.

6. *Selection of experimental factors.* When applying DOE techniques, the determination of factors and their levels/ranges is a prerequisite for factor-based experimental design (Montgomery, 2009). For an evaluation experiment, it is better to start by distinguishing a limited number of design factors from those that are not of interest, and the factors that are expected to have high impacts should be preferably selected. As mentioned above, evaluators may refer to existing evaluation experiences (the proposed factor framework) to look up and identify design factors quickly. Note that the suggestion is to use the factor framework to supplement, but not replace, expert judgment in experimental factor selection. This would be particularly helpful for cloud services evaluation when there is a lack of experts.

7. *Experimental design.* Given the selected input-process variables (experimental factors) and output-process responses (metrics), evaluators can design cloud-service evaluation experiments using suitable DOE techniques. In particular, three basic principles, namely *randomization, replication,* and *blocking* (Montgomery, 2009) should be taken into account no matter what DOE technique is employed. Moreover, small-scale pilot experiments can often benefit the experimental design. For example, trial runs of an evaluation experiment may help evaluators become familiar with the experimental environment, optimize the experimental sequence, and even decide the sample size.

8. *Experimental implementation.* Implementing an experiment is to carry out a series of experimental actions ranging from preparing the environment to running benchmarks. As any error in the experimental procedure might spoil the validity of the experimental results, the implementation process should be monitored carefully to ensure every step of the experiments follows the design (Montgomery, 2009). Note that pilot experimental runs are regarded as the activities in *experimental design* rather than *experimental implementation.*

9. *Experimental analysis.* In DOE, statistical methods are strongly suggested for experimental analysis (Montgomery, 2009). Although such methods do not directly prove any factor's effect, statistical analysis adds objectivity to an evaluation's conclusions and the potential decision-making process. Moreover, visualizing experimental results using various graphical tools may significantly facilitate data analysis and interpretation. In addition to those statistical techniques, machine learning techniques like mining association rules are also useful for experimental analysis in some circumstances, for example evaluation results involving many experimental factors.

10. *Conclusion and documentation.* Drawing practical conclusions is important after analyzing the experimental results (Montgomery, 2009). In addition, it is worth paying attention to reporting the cloud services evaluation work. Complete evaluation reports, and not just conclusions, are vital for other people to learn from or replicate/confirm previous evaluation practices. However, the quality of the existing cloud services evaluation reports varies (Li *et al.*, 2013b), which implies a lack of evaluation reporting guidelines. Therefore, the first suggestion is to use these ten evaluation steps as basis for documentation structure. Considering the close relationship between cloud services evaluation and experimental computer science (Stantchev, 2008), moreover, it is also possible to adapt the well formulated structure for reporting generic experiments or case studies to reporting cloud-services evaluation studies.

29.6 A Sample Case of CEEM in Practice

Here a case study of evaluating Google Compute Engine (GCE) instances is used to demonstrate how CEEM can facilitate cloud services evaluation in practice. For conciseness and convenience, the evaluation activities are briefly described using six subsections instead of elaborating the individual steps one by one.

29.6.1 Requirement Recognition and Service Feature Identification

Given the recent availability of GCE in the public cloud market, Li *et al.* (2013c) performed an early stage evaluation of different types of GCE instances. A surprising observation in their work is that there seems to be no clear difference between the VM type *n1-standard-2-d* and *n1-highcpu-2-d* in terms of computation, as shown in Figure 29.5.

Considering the different specifications and prices (cf. Table 29.2), it is worth further investigating these two types of GCE instances. As a continuation of the original study, the evaluation requirement is still early observations on the performance of GCE, while the service feature to be evaluated is narrowed down to computation transaction speed only. Note that, although the linear-process evaluation activities are also mentioned here, this study essentially belongs to the spiral-process part of CEEM.

(a)

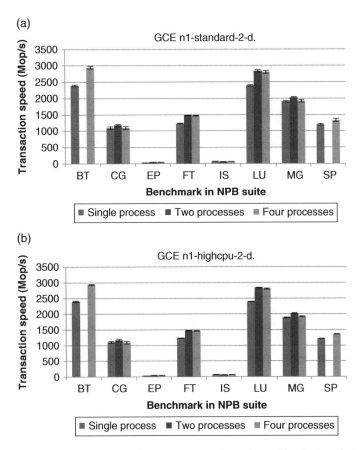

(b)

Figure 29.5 *Computation benchmarking results using NPB-MPI with workload Class A. Error bars indicate the standard deviations of the corresponding computation transaction speed*

Table 29.2 *Specifications and prices of two GCE instance types*

Specification	n1-standard-2-d	n1-highcpu-2-d
Virtual cores	2	2
Memory	7.5 GB	1.8 GB
GCEUs	5.5	5.5
Local disk	870 GB	870 GB
Price (USD/Hour)	$0.265	$0.163

29.6.2 Metrics and Benchmarks Listing and Selection

To make this study consistent with Li *et al.*'s (2013c) work, the benchmark suite and metric are directly determined as NPB-MPI 3.3 and Benchmark Operation Rate (Mop/s) respectively. In addition, since LU is the most typical pseudoapplication benchmark in the NPB-MPI suite, this chapter employs it to simplify the demonstration.

29.6.3 Experimental Factors Listing and Selection

Only one service feature (computation transaction speed) is to be investigated in this study, so the experimental input factors are correspondingly limited to *instance type* (n1-standard-2-d vs. n1-highcpu-2-d), *thread Number* (1 vs. 2), and *workload size* (Class A vs. Class B). Moreover, the LU Operation Rate (Mop/s) metric for measuring the service feature is the response factor.

29.6.4 Experimental Design

Given the two-level settings for the three input factors, it is natural to adopt the most straightforward design technique, namely full-factorial 2^3 design (Montgomery, 2009). This design technique adjusts one factor at a time, which results in an experimental matrix comprising eight trials, as shown in Table 29.3. Note that the sequence of experimental trials has been randomized to reduce the possible biases within the designing process.

29.6.5 Experimental Implementation and Analysis

Following the experimental design outlined above, the evaluation experiments should be implemented trial by trial. After filling the response column with experimental results (cf. Table 29.4), the significance levels of

Table 29.3 A full-factorial (2^3) design matrix for this case study (generated by Minitab)

Trial	Instance type	Thread number	Workload size	Response: LU operation rate
1	n1-highcpu-2-d	2	Class B	?
2	n1-standard-2-d	1	Class B	?
3	n1-standard-2-d	2	Class B	?
4	n1-highcpu-2-d	1	Class A	?
5	n1-highcpu-2-d	2	Class A	?
6	n1-standard-2-d	2	Class A	?
7	n1-standard-2-d	1	Class A	?
8	n1-highcpu-2-d	1	Class B	?

Table 29.4 Experimental results corresponding to the full-factorial (2^3) design

Trial	Instance type	Thread number	Workload size	Response: LU operation rate
1	n1-highcpu-2-d	2	Class B	2836.786 Mop/s
2	n1-standard-2-d	1	Class B	2349.052 Mop/s
3	n1-standard-2-d	2	Class B	2825.901 Mop/s
4	n1-highcpu-2-d	1	Class A	2394.74 Mop/s
5	n1-highcpu-2-d	2	Class A	2836.059 Mop/s
6	n1-standard-2-d	2	Class A	2833.513 Mop/s
7	n1-standard-2-d	1	Class A	2389.469 Mop/s
8	n1-highcpu-2-d	1	Class B	2384.797 Mop/s

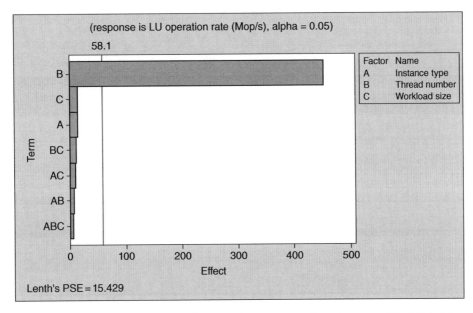

Figure 29.6 *The Pareto plot of factor and interaction effects (generated by Minitab)*

these factors can be further analyzed to reveal more comprehensive information. For example, by setting the significance level α as 0.05, a Pareto plot can be drawn to detect the factor and interaction effects that are important to the performance of running LU on GCE instances, as illustrated in Figure 29.6.

Given a particular significance level, a Pareto plot displays a red reference line beside the effect values. Any effect that extends past the reference line is potentially important (Montgomery, 2009). In Figure 29.6, since the effect of the *thread number (B)* factor is beyond the reference line, it is apparent that *thread number (B)* dominates the computation transaction speed when running LU. On the other hand, the factor *instance type (A)* and *workload size (C)* have little influence on the benchmark operation rate in this case.

29.6.6 Conclusion and Documentation

As explained previously, this work can be regarded as a supplement to the original study (Li *et al.*, 2013c). Interestingly, although GCE's hyperthread mechanism has been claimed to be unsuitable for multithread applications, increasing thread numbers is still an effective way to improve performance on multicore GCE instances. Furthermore, the experimental analysis statistically confirms that GCE n1-standard-2-d and n1-highcpu-2-d are replaceable with each other for running applications like LU. Considering the price difference, employing n1-highcpu-2-d would be a more economical choice.

Finally, driven by CEEM, evaluators may maintain a live and structured document to report the evaluation details. The demonstration in this chapter provides a sample of such documentation.

29.7 Conclusion

As cloud computing becomes one of the most promising computing paradigms, numerous vendors have started to supply public cloud infrastructures and services. In general, cloud services are different from traditional computing systems. For example, cloud service indicators are usually insufficient for service selection with regard to specific application scenarios, while customers have little knowledge or control over public cloud services except for those indicators. Thus, it would be necessary to carry out appropriate evaluations before employing particular cloud services. Moreover, by comparing competitors, cloud services evaluation can also help cloud vendors improve their services (Iosup *et al.*, 2011). To facilitate evaluation work, engineers and researchers may refer to this chapter to review quickly the state of the practice of IaaS and PaaS evaluation with regard to evaluation purposes, service features, and benchmarks.

Furthermore, there is always a need to evaluate a specific cloud service in different contexts over time. For example, cloud providers may gradually upgrade their service infrastructures at different times and in different datacenter locations, without necessarily notifying customers. To satisfy this need, a rigorous, systematic, and repeatable evaluation methodology would be required. In other words, unlike the evaluation of traditional computing systems with its emphasis on results, cloud services evaluation should be more methodology oriented. The ten-step CEEM suggested in this chapter would be able to play a generic and strategic role in guiding the implementation of cloud services evaluation.

References

Binnig, C., Kossmann, D., Kraska, T., and Loesing, S. (2009) *How is the Weather Tomorrow? Towards a Benchmark for the Cloud*. Proceedings of the 2nd International Workshop on Testing Database Systems. ACM, pp. 1–6.

Blackburn, S. M., McKinley, K. S., Garner, R., *et al.* (2008) Wake up and smell the coffee: Evaluation methodology for the twenty-first century. *Communications of the ACM* **51**(8), 83–89.

Chiu, D., and Agrawal, G. (2010) *Evaluating Caching and Storage Options on the Amazon Web Services Cloud*. Proceedings of the 11th ACM/IEEE International Conference on Grid Computing. IEEE Computer Society, pp. 17–24.

Cooper, B. F., Silberstein, A., Tam, E., *et al.* (2010) *Benchmarking Cloud Serving Systems with YCSB*. Proceedings of the First ACM Symposium on Cloud Computing. ACM Press, pp. 143–154.

Ferdman, M., Adileh, A., Kocberber, O., *et al.* (2012) *Clearing the Clouds: A Study of Emerging Scale-Out Workloads on Modern Hardware*. Proceedings of the 17th International Conference on Architectural Support for Programming Languages and Operating Systems. ACM Press, pp. 37–48.

Iosup, A., Ostermann, S., Yigitbasi, N., *et al.* (2011) Performance analysis of cloud computing services for many-tasks scientific computing. *IEEE Transactions on Parallel and Distributed Systems* **22**(6), 931–945.

Lenk, A., Menzel, M., Lipsky, J., *et al.* (2011) *What are You Paying For? Performance Benchmarking for Infrastructure-as-a-Service Offerings*. Proceedings of the Fourth International Conference on Cloud Computing. IEEE Computer Society, pp. 484–491.

Li, A., Yang, X., Kandula, S., and Zhang, M. (2010) *CloudCmp: Comparing Public Cloud Providers*. Proceedings of the 10th ACM SIGCOMM Conference on Internet Measurement. ACM Press, pp. 1–14.

Li, Z., O'Brien, L., and Zhang, H. (2013a) *CEEM: A Practical Methodology for Cloud Services Evaluation*. Proceedings of the 3rd IEEE International Workshop on the Future of Software Engineering for/in cloud in conjunction with IEEE 9th World Congress on Services. IEEE Computer Society, pp. 44–51.

Li, Z., Zhang, H., O'Brien, L., Cai, R., and Flint, S. (2013b) On evaluating commercial cloud services: A systematic review. *Journal of Systems and Software* **86**(9), 2371–2393.

Li, Z., O'Brien, L., Ranjan, R., and Zhang, M. (2013c) *Early Observations on Performance of Google Compute Engine for Scientific Computing*. Proceedings of the 5th International Conference on cloud Computing Technologies and Science. IEEE Computer Society, pp. 1–8.

Montgomery, D. C. (2009) *Design and Analysis of Experiments*, 7th edn. John Wiley & Sons, Inc., Hoboken, NJ.

Sobel, W., Subramanyam, S., Sucharitakul, A., *et al.* (2008) Cloudstone: Multi-platform, multi-language benchmark and measurement tools for Web 2.0. Paper presented at the First Workshop on Cloud Computing and its Applications, Chicago, IL.

Stantchev, V. (2008) *Performance Evaluation of Cloud Computing Offerings*. Proceedings of the Third International Conference on Advanced Engineering Computing and Applications in Sciences. IEEE Computer Society, pp. 187–192.

Zhang, M., Ranjan, R., Haller, A. *et al.* (2012) *Investigating Decision Support Techniques for Automating Cloud Service Selection*. Proceedings of the IEEE 4th International Conference on cloud Computing Technology and Science. IEEE Computer Society, pp. 759–764.

Part VII
Cloud Migration and Management

30

Enterprise Cloud Computing Strategy and Policy

Eric Carlson

TechTrend, Inc., USA

30.1 Introduction

A well developed cloud-computing strategy will help to accelerate and advance the adoption of cloud computing while simultaneously enabling tangible benefits to be captured and attained. The strategy should use a structured engineering approach that balances requirements, schedule, cost, and risk. The cloud strategy needs to take an approach that is realistic and achievable for the individual organization, to help position the enterprise towards a trajectory of success.

When developing a cloud strategy, the CTO should consider how the strategy will apply throughout the organization. The strategy should present a case to stakeholders, explaining why cloud computing is advantageous and approach for implementing the new technology in a way that helps to support the organization's core competencies and mission.

The cloud computing strategy will likely be the foundation block upon which the implementation of cloud computing is built. The strategy should be an overarching document that briefly touches on the major components of the organization that will be impacted by the introduction of cloud computing.

30.2 Foundations for success

In developing a strategy, it is suggested that a risk-based incremental approach be used that will gradually introduce cloud to the enterprise at large. A gradual process will allow for challenges that arise from the

Encyclopedia of Cloud Computing, First Edition. Edited by San Murugesan and Irena Bojanova.
© 2016 John Wiley & Sons, Ltd. Published 2016 by John Wiley & Sons, Ltd.

cloud's implementation to be resolved on a micro scale. Developing a strategy that plans for a few low-risk, low-hanging fruit systems to be the initial users of the cloud will lead to quick wins and consequently provide the momentum that will lead to a wider acceptance and growth of the user base. Candidates that are low-hanging fruit can be characterized by systems that have traits such as low security, low system dependency, and low reliability requirements. Although the low-hanging fruit systems are easier to migrate to the cloud, there are unknown risks that may result in complications; thus these system owners should be proponents of cloud computing and should be willing to address challenges that they might encounter. A general framework for cloud computing implementation phases could include:

- establishing an initial cloud computing capability;
- forming a coalition of cloud supporters that embrace cloud computing;
- maturing the practices and technologies by migrating low-risk systems first;
- moving additional systems to the optimized environment;
- having continuous improvement to optimize cloud usage.

30.3 Policy

It is critical that enterprise cloud computing policies be enacted to help preserve a well functioning system. Without proper policies there is the potential that ideal candidate systems for the cloud will not utilize the environment due to refusal by system owners who are averse to change. A policy will provide a mandate that can be referenced to system owners who do not wish to adopt cloud computing due solely to their resistance to change. Alternatively, there are other system owners who will be too anxious to use cloud technologies and will procure cloud services through improper channels that can lead to cloud sprawl and result in a Wild West of unmanaged cloud environments that are not secure and are inoperable with other systems in the enterprise. A policy will help to prevent unauthorized cloud-computing implementations.

Like the cloud strategy, cloud polices should be developed with inputs from all stakeholders to ensure they take into account the policies and requirements of the enterprise. The policies should be issued from the desk of the CIO, or from an equivalent level, to ensure that they carry the weight required by system owners.

Enforcement and monitoring of policies can be accomplished by several means, including:

- management signoff on IT procurements;
- contracting officer approval of noncloud versus cloud services, and
- enterprise checkpoints for integration of cloud services with back-end systems.

Policies for cloud computing are important, but ideally system owners should aspire to move their applications to the enterprise cloud without any type of strong arming. For this to happen, effective and frequent communication is needed to gain the trust and support of the user base.

30.4 Identify Goals

An effective cloud strategy should include the identification of realistic and achievable goals for introducing cloud computing into the enterprise. The success or failure of the adoption of cloud computing can be measured by whether these goals are achieved. Specific, measureable, attainable, realistic, timely (SMART) goals should be developed that will help determine if the investment is staying on a path towards success or if course corrections are needed. The goals should span the lifecycle of the anticipated operational timeframe to ensure that continual progress is being made. Doing so will help to demonstrate that cloud computing is not only properly integrated, but that the benefits that were used to sell the concept are achieved.

As an example, the United States Department of Defense (DoD) overarching cloud-computing goal is to "Implement cloud computing as the means to deliver the most innovative, efficient, and secure information and IT services in support of the Department's mission, anywhere, anytime, on any authorized device" (http://dodcio.defense.gov/Portals/0/Documents/Cloud/DoD%20Cloud%20Computing%20Strategy%20Final%20with%20Memo%20-%20July%205%202012.pdf, accessed December 18, 2015).

While an overarching goal should be identified, it is also important to define sufficiently goals that can be tracked and measured. For example, the FAA identified the a few key goals for monitoring as outlined in Table 30.1.

Table 30.1 *FAA cloud computing goals*

Cloud goal	Rationale
Increase cost efficiency	The multisharing and elastic characteristics of cloud computing present an opportunity to increase the cost efficiency of IT programs across the FAA. Under the cloud computing model, the costs of acquiring, maintaining, and refreshing software, tools, development platforms, hardware, storage, etc. are shifted to the cloud service providers (CSPs). These can serve a high number of customers and they are able to offer lower individual pay-per-use rates by leveraging economies of scale, commoditizing infrastructure, and automating datacenter processes. In FAA IT organizations, operational costs represent a significant percentage of the IT budget, and the opportunity to optimize and reduce operational costs by using cloud services represents a very attractive option to IT organizations and the FAA.
Increase provisioning speed	The self-serve automated characteristic of cloud computing presents an opportunity to increase the provisioning speed of IT services to FAA programs. Providing new applications, setting up technical environments, developing new software, etc., often take a long time and it is usually not aligned with changing business needs and customer expectations. A key characteristic of cloud computing is the automated provisioning of computing resources and it has the potential to increase the provisioning speed dramatically.
Increase flexibility and scalability	The multisharing characteristic of cloud computing provides an opportunity to provide a highly flexible and scalable IT environment. To take advantage of economies of scale, CSP datacenters have massive amounts of computing resources available automatically to cloud-service consumers. Individual program needs for computing resources may vary for several reasons. An unexpected event, new policies or regulations may trigger increased demand for computing resources that was not forecast in advance. Cloud computing offers the potential benefit of scaling computing resources up and down as needed.
Support and enable net-centricity and information sharing	CSP datacenters can be seen as massive repositories of data, applications and computing resources, accessed via a network with the potential for interconnectivity that creates a network of networks and accelerates net-centric operations and information sharing across the FAA.
Support innovation, research and development	A more affordable and agile access to massive cloud computing capabilities can be seen as an innovation lab in the desktop of every FAA personnel that may spur innovation across the FAA.
Support sustainability	Optimal utilization levels and the elimination of redundancies and in the FAA IT environment will reduce energy consumption and support a more sustainable environment.

Source: Adapted from FAA Cloud Computing Strategy (2012).

Each organization needs to establish goals that are not too broad or general and ensure they align with the core mission of that particular organization. If the goals are too broad or generic, they will result in objectives that do not bolster the case for cloud computing. The following are general examples of goals that could be built upon, but will need customization to be useful for the individual organization:

- adopt an enterprise approach to cloud computing;
- increase the efficiency of current and future IT investments;
- minimize the risks associated with introducing cloud computing to the enterprise.

30.5   Identifying the Approach

An approach needs to be established within the cloud strategy that identifies the major steps and phases that the organization will go through to make cloud computing a success. The approach should use a logical sequence of activities that enable the gradual introduction of the technology into the organization. Gartner, a highly respected information technology research and advisory firm, recommends that

> IT leaders follow five major phases when investigating cloud computing (the phases may vary, depending on your organization and the extent of your interest in cloud computing):
>
> - *Build the business case.* Link the key initiative to the overall drivers or objectives of the business. Gain support from senior business leaders and senior stakeholders. Set a baseline for assessing the impact of the investigation. Estimate costs and resource requirements.
> - *Develop the strategy.* Align the investigation with the business strategy, and show how it can deliver business value. Show how the investigation might lead to changes that will affect your business environment. Work with key stakeholders to identify business needs.
> - *Assess readiness.* Identify the budgetary, staffing, technology and other requirements necessary to prepare the business for the investigation. Develop a total cost of ownership analysis framework. Review established policies for assessing risk and managing governance.
> - *Pilot or prototype.* Identify a group to pilot, or develop a prototype for the investigation. Develop and communicate detailed requirements. Manage the pilot/prototype. Assess and communicate the results.
> - *Gain approval.* Analyze findings of the readiness assessment and pilot or prototype effort, and revise the strategy and business case accordingly. Present findings of the investigation to senior stakeholders and business leaders. (Gartner, 2014).

30.6   Enterprise Architecture

> Enterprise architecture (EA) is a discipline for proactively and holistically leading enterprise responses to disruptive forces by identifying and analyzing the execution of change toward desired business vision and outcomes. EA delivers value by presenting business and IT leaders with signature-ready recommendations for adjusting policies and projects to achieve target business outcomes that capitalize on relevant business disruptions. EA is used to steer decision making toward the evolution of the future state architecture. (Gartner, 2014)

Enterprise architecture provides a way for programs across an organization to gain an understanding of how the various elements of their business operate in relation to the overall ecosystem. With that in mind, adding cloud computing to the overarching enterprise architecture will help to convey a message that cloud computing is a key function that needs to be considered in various systems architectures.

Target dates for cloud computing operational capability should be added to an organization's roadmap to notify potential customers of when this technology will be available within the organization. Often program offices have to develop system requirements and plan their budget requirements years in advance of going operational. The roadmaps help to serve as a guide for early employment of cloud computing, which can be integrated into a program's systems requirements.

30.6.1   Assess the Current State of the Enterprise (As-Is Architecture)

Understanding the as-is architecture of the enterprise is important to get an understanding of current cloud computing offerings and how it may need to be altered in order to achieve the long-term vision. Ideally, the architecture of the enterprise will be well documented and understood within the organization. Key information to be gathered in the discovery process should include:

- *Inventory of current datacenters.* Knowing the organization's assets and where they are located will help to determine which systems are candidates for moving to the cloud environment.
- *Migration roadmap.* A roadmap will help to identify potential candidates and timelines for systems to move to the cloud. Strong candidates for moving to cloud computing will be systems that are currently in development or are approaching a tech-refresh cycle.
- *Define system dependencies.* The dependencies between systems are important to understand as the impact of moving a dependent system to the cloud could have major implications on other systems.
- *Network infrastructure.* Because cloud computing relies heavily on the network, it is important to gain an understanding of bandwidth, gateways and security policies of the organization moving to the cloud.

30.6.2   Develop a Long-Term Vision (To-Be Architecture)

The long-term vision, also referred to as the to-be architecture, is the concept of what the architecture will be at the end state, after cloud computing has been fully integrated into the enterprise. The architecture needs to present a model of how various systems within the organization will integrate into the cloud environment. It is challenging to identify all potential clients for the cloud environment, but the overarching end-state architecture needs to identify how cloud computing will impact the overall processes of the organization.

30.7   The Cloud Champion – Gaining Senior Executive Support

The success of cloud computing within the enterprise requires the support of senior executives. Too often, great ideas are tabled before they have a chance to flourish due to the lack of support by the right executives. Cloud computing has tremendous buzz around the IT world but that hype does not always translate outside of the IT domain.

There is always resistance to change; however, with the right executive support, the roadblocks to transformation can be made to crumble. The enterprise cloud strategy should lay out the direction to bring the executives on board. It is critical that an executive champion be found who fully understands the cloud computing concept and is willing to translate/sell this strategy to his peers and subordinates.

Executives are looking for tangible benefits in order to provide backing for a concept or project. Cloud computing has many benefits but the concept needs to be brought to a level that can show a direct impact to a given mission. For example, IT professionals and CIOs are often troubled by inefficient investment in IT infrastructure. Linthicum notes that while some estimate that only 10% of server capacity is used, this may be an overgenerous estimate and warns that many enterprises may actually have hundreds or thousands of servers running only at 3% to 4% of capacity. Presenting statistics that cite specific benefits can be used as a way to gain the support needed to establish cloud computing. Every new IT revolution has its hurdles; however, with the right support from the right people, the obstacles can be overcome.

30.8 Avoiding Cloud Silos

Cloud computing will not be successful if it is implemented in isolation. In order to avoid silos, the strategy should develop an approach to gain input and support from all anticipated stakeholders. While it is suggested that cloud computing be sponsored by the CIO's office, or an equivalent champion, there are many satellite organizations that will need to be involved, including finance, security, program offices, contracting offices, and enterprise engineering.

30.8.1 Keeping the Customer Front and Center

It is important to keep in mind that the success of a cloud computing integration relies on meeting the needs and desires of its potential customers. To achieve this, the cloud services must be implemented using a methodology that integrates them into the given ecosystem in such a way that they do not become a hindrance but are instead looked upon as an opportunity.

Effective communication is one of the keys to make cloud computing popular within an organization. To help develop an approach for communication, the strategy should include the development of a communications plan that will identify an approach for properly distributing the message of cloud computing to the appropriate channels.

Included in this plan should be activities such as holding symposiums, developing a web site, organizing brown-bag luncheons, and conducting meetings with key enterprise players.

- *Symposiums.* Holding a symposium is a great way to generate buzz around cloud computing. A symposium provides a place for the program offices implementing cloud computing to communicate with its stakeholders on the importance of this new technology and the impact it will have on the enterprise.
- *Web site.* In today's society, many look to the Internet to gather information on topics of interest. A well publicized and frequently updated web site on the organization's cloud-computing status will help keep parties informed and involved.
- *Brown-bag luncheons.* Brown-bag luncheons are short informational sessions that are held on specific areas of cloud computing. This type of forum provides a place for stakeholders of targeted groups and/or topics to gather information, such as "security in the cloud" or "estimating the cost of moving your systems to the cloud." The meetings should be kept short to accommodate busy schedules. Think in-terms of TED talks, which are popular and effective presentations where the "speakers are given a maximum of 18 minutes to present their ideas in the most innovative and engaging ways they can" (https://en.wikipedia.org/wiki/TED_%28conference%29, accessed December 18, 2015).
- *Meeting with key enterprise players.* Individual meetings should be held with key stakeholders on a regular basis to keep them informed and to provide a forum for addressing any new concerns.

30.9 Governance

The cloud computing strategy should outline a framework that identifies the governance processes that will need to be established within the organization. "IT governance is the processes that ensure the effective and efficient use of IT in enabling an organization to achieve its goals" (Gartner, 2014). "Cloud Computing

Governance and Integration activities ensure an effective cloud computing governance structure. They also ensure a smooth integration of the evolving Cloud Environments to the existing legacy IT environment" (FAA, 2012).

As an example, the United States DoD Cloud Strategy states that the "DoD CIO will establish a joint enterprise cloud computing governance structure to drive the policy and process changes necessary to transition to the DoD Enterprise Cloud Environment and oversee the implementation of the DoD Enterprise Cloud Strategy" (Department of Defense Cloud Computing Strategy, 2012).

30.10   Portfolio Analysis

It is important that an organization be able to survey the enterprise at large to identify potential opportunities for moving to a cloud-computing environment. Cloud computing provides the benefit of being able to scale up or down, but depending on the nature of the organization, there are upfront costs to an organization to start using cloud. This upfront cost will vary widely depending on the spectrum of cloud implemented, ranging from a private cloud to a public cloud. Some suggested startup costs to consider include items such as security authorizations, network modifications, and staffing.

There are two broad categories of system that need to be assessed for moving to a cloud environment: existing systems and future systems.

Existing systems will have already been deployed and are likely operating at some type of enterprise-owned datacenter. An important potential cost savings of cloud computing resides in the reduction of the human labor to maintain legacy datacenters. As a result, datacenters need to be assessed to identify which can be closed down as a result of moving tenant systems to the cloud environment and/or consolidated into another datacenter.

Systems that are in the planning or development lifecycle phase are prime candidates for moving to a cloud computing environment. The major reason for their prime candidacy is that the systems are still being architected and can take the cloud architecture into account while being designed.

A consistent assessment approach should be developed for assessing systems in either lifecycle stage. The assessment needs to take into account security, policies, and timelines to help to determine if a system is a good candidate for moving to the cloud. Table 30.2 presents examples of criteria that can be used in the cloud assessment.

An important element to keep in mind when assessing a system for cloud suitability is to look at the system in pieces. A system often has multiple aspects, where one portion of the system could be a bad candidate for a cloud environment and another portion of the same system might be an excellent candidate. For example, a system might be too critical to be hosted in a cloud in its operational state, but be a great candidate for the developmental phase.

Following the completion of the overall cloud assessment of the enterprise, the strategy should identify an approach to analyze the collected data. The data gathered should help to identify overall enterprise requirements and conclude which types of cloud deployment model are most beneficial to the organization.

30.10.1   Metrics

The organization's cloud computing strategy should include metrics that can be used to measure the benefits of cloud computing as it grows and matures within the enterprise. Table 30.3 outlines a few key cloud metrics.

Table 30.2   *Cloud assessment criteria*

Cloud assessment criteria	Criteria explanation
What is the availability requirement of the system?	The up-time requirements of the system.
What is the service restoral requirement of the system?	The requirement that determines how quickly the system must be restored after a failure.
Does the system process personally identifiable information (PII)?	Any information about an individual maintained by an agency, including, but not limited to, education, financial transactions, medical history and information which can be used to distinguish or trace an individual's identify, such as name, SSN, date and place of birth, mother's maiden name, biometric records, etc., including any other personal information which is linked or linkable to an individual. (Defined in OMB M-06-19)
What is the FISMA security level of the system?	FISMA-Low indicates that loss of confidentiality, integrity, or availability would have a limited adverse effect (e.g. minor financial loss) FISMA-Moderate indicates that loss of confidentiality, integrity, or availability would have a serious adverse effect (e.g. significant financial loss) FISMA-High indicates that a loss of confidentiality, integrity, or availability would have a severe or catastrophic adverse effect (e.g. harm to individuals including loss of life or life-threatening injuries or major financial loss) (Defined in NIST 800-60)

Table 30.3   *Cloud computing metrics*

Metric	Rationale
Number or percentage of systems assessed for cloud suitability	The number of cloud computing assessments indicates that the organization is proactively identifying systems with the potential to capture cloud computing benefits
Number or percentage of systems in the cloud	The number of systems moving to the cloud indicates that programs have identified cost efficiencies and other benefits during the investment analysis process, and selected cloud computing to support and enable the selected alternative. It also measures the multi sharing level of the environment which translates into economies of scale
Number of billing units	The number of cloud service units in use by systems or billed by the cloud service provider give an indication of the cloud adoption and use by systems taking advantage of cloud computing
Billing amount	The total billing amount per cloud service provides information on cost efficiency of cloud services. They can be compared and benchmarked, and unit costs over time can be calculated to measure efficiency over time
Cloud availability	Availability of the cloud indicates to what extent cloud risks have been reduced or mitigated

Source: Adapted from FAA (2012).

Table 30.4 *Potential risks*

Risk description	Risk response
If the cloud's availability levels are not being met by the vendor, then the system will be unable to use the cloud environment.	Develop a SLA with the cloud vendor that specifics the required availability levels. Monitor the cloud vendor to ensure that the service levels are being met.
If a malicious insider compromises the cloud environment, then the organization's data will be compromised.	Maintain strict security controls to reduce the chance of a malicious insider.
If the cloud vendor goes out of business, then another vendor will need to be selected.	Award the cloud contract to a vendor that is stable and demonstrates the ability to be a sustainable organization.
If there are not enough customers to support the cloud environment, then the initiative may be canceled.	Develop an outreach and communication plan to generate a customer base.
If there is a breach of sensitive intellectual property or a customer's personal information then there could be lost customers, fines and legal sanctions.	Develop a security approach that minimizes the risk of a security breach.
If a cloud account is hijacked, then undetected fraud and abuse could occur.	Develop a security approach that minimizes the risk of a security breach.

30.11 Anticipating the Challenges

There will be many challenges while implementing cloud computing. The organization's cloud computing strategy should address the risks and challenges that it could potentially encounter. Challenges to keep in mind include: "technical and management risks from cloud computing including security, performance and data latency, interoperability, portability, technology maturity, legacy application migration, and enterprise planning" (FAA, 2012).

Some examples of risks that an organization might encounter regarding cloud computing are identified in Table 30.4.

30.12 Conclusion

It is important to recognize that a cloud computing strategy is just that – a *strategy*. A well crafted cloud computing strategy will help develop an approach for the implementation of cloud computing. That said, it may be necessary to deviate from even the best of strategies as technologies evolve, priorities change, and lessons are learned. With that in mind, a sound strategy will lay the groundwork that will take an enterprise from a concept of utilizing cloud computing to the reality of embracing it and realizing its benefits.

References

FAA (2012) *Cloud Computing Strategy*, https://faaco.faa.gov/attachments/Attachement_3_FAA_Cloud_Computing_Strategy.pdf (accessed December 18, 2015).
Gartner (2014) *IT Glossary*, http://www.gartner.com/it-glossary/ (accessed December 18, 2015).

31

Cloud Brokers

Ganesh Neelakanta Iyer[1] and Bharadwaj Veeravalli[2]

[1] *Progress Software Development, India, and BVRIT, India*
[2] *National University of Singapore, Singapore*

31.1 Introduction

The emergence of cloud computing has provided access to infrastructure for developing, deploying, and running software services at reduced cost. Features such as on-demand availability and ability to share resources (known as multitenancy) along with the availability of a variety of services for various business ventures to enable them to improve their business offerings is very attractive for small and medium software organizations. With the vast number of independent cloud service providers (CSPs) currently existing, it is challenging for users to choose an appropriate CSP. Other challenges include addressing security, privacy, trustworthiness, and vendor lock-in. CSPs also face challenges such as understanding the market, adapting to market conditions, and monitoring user profiles. This suggests the need for a cloud broker that can act as an intermediary between cloud customers and CSPs, to connect them and to help them make their business-critical decisions.

There are five essential characteristics for cloud environments (Mell *et al.*, 2011). Firstly, resources can be provisioned rapidly on demand as and when needed. Secondly, the cloud allows broad network access so that users can access and use cloud resources through the network using various heterogeneous client devices such as mobile phones and laptops. Thirdly, CSPs use virtualization techniques to pool computing resources to serve multiple consumers based on their demand. Fourthly, the autoelasticity of the cloud allows users to configure resources in minutes and enables them to scale capacity elastically based on their immediate resource requirements. Finally, cloud computing has an attractive utility computing-based pay-as-you-go policy in which the user needs to pay only for the capacity that is actually being used.

Encyclopedia of Cloud Computing, First Edition. Edited by San Murugesan and Irena Bojanova.
© 2016 John Wiley & Sons, Ltd. Published 2016 by John Wiley & Sons, Ltd.

Despite these advantages, cloud environments pose considerable challenges to prospective users as well as to the CSPs, which offer different types of services. In this chapter, we describe various challenges for the users in moving their services to the cloud and various challenges faced by providers in offering their services and increasing their revenue. Then we portray cloud brokers as a unified solution to most of the problems and describe cloud brokers in detail.

31.2 Key Challenges in Cloud Computing and the Need for Cloud Brokers

There are several key challenges for both users and providers when entering and becoming established in the cloud paradigm.

31.2.1 Challenges from the Users' Perspective

Key challenges faced by the users in moving their data/services to cloud platforms include the following:

- Choosing the right provider. With the variety of services offered by several CSPs, users may find it difficult to choose the right provider to fulfill their requirements. At present there is no platform providing information about the capabilities of all the CSPs.
- Security and privacy issues. As several users may share the same physical infrastructure in a virtualized manner simultaneously, users are often concerned about the security and privacy of their data in the cloud platform.
- Trustworthiness of CSPs. Users are concerned about the trustworthiness of the CSPs. This aspect is different from security because trustworthiness conveys information pertaining to task execution such as adhering to service-level agreements (SLAs) and reliability of task execution (such as handling node failure, and meeting task deadlines).
- Dealing with lock-in. In economics, vendor lock-in makes a customer dependent on a vendor for specific products and/or services, rendering it difficult for users to choose a different CSP without substantial switching costs. The switching cost includes possible end-of-contract penalties, charges for data format conversion and application switching and possible additional fees for bandwidth usage.

31.2.2 Challenges from the Providers' Perspective

From the providers' perspective there are many challenges to be addressed for exploiting various features of cloud platforms. These include:

- Understanding the market. New CSPs may need to understand current market dynamics in terms of the competitors in the domain, user preferences in terms of the products/services in demand, and user preferences for various features such as security and trust requirements.
- Adapting to the market. Current CSPs charge a fixed price per resource for their products and services, with some minor exceptions like Amazon spot pricing. Dynamic pricing strategies are required to improve their performance and to attract more customers based on the market situation.
- Monitoring user profiles. With competition among different CSPs, they may need to monitor the reliability of users – one needs to check if the feedback given by the users is reliable or not to decide user acceptance criteria. It also helps to avoid any unhealthy competition among the CSPs and users.

31.2.3 Cloud Orchestration: Cloud Broker as a Unified Solution

Clouds are emerging as a competitive sourcing strategy. With this, there is clearly a demand for the integration of cloud environments to create an end-to-end managed landscape of cloud-based functions. A broker-based cloud orchestration mechanism can solve most of the issues faced by users and the CSPs. Cloud orchestration enables connectivity of IT and business-process levels between cloud environments. Major benefits of cloud orchestration are:

- It helps users to choose the best service;
- It helps providers to offer better services and adapt to market conditions;
- It has the ability to create a "best of breed" service-based environment in which tasks can be deployed dynamically among multiple CSPs to reduce task execution time and to meet budget requirements;
- It helps users and providers to make their business decisions based on several parameters, such as trust, reputation, security, and reliability, which are difficult to handle in the absence of a broker;
- It helps users to designate a broker to make some decisions on their behalf so that users can focus on their core businesses rather than focusing on task-deployment strategies and other system administration jobs.

31.3 Cloud Brokers: An Overview

Cloud brokers play an intermediary role, to help customers locate the best and the most cost-effective CSP for their needs. A cloud broker is by far the best solution for multiple cloud orchestration (including aggregating, integrating, customizing, and governing cloud services) for SMEs and large enterprises. Major advantages are cost savings, information availability, and market adaptation. As the number of CSPs continues to grow, a single interface (broker) for information, combined with service, could be compelling for companies that prefer to spend more time with their clouds than devising their own strategies for finding a suitable CSP to meet their needs.

We can broadly classify cloud brokers into three groups based on the services offered by them (see Figure 31.1). The first group is cloud service arbitrage, where the brokers supply flexible and opportunistic

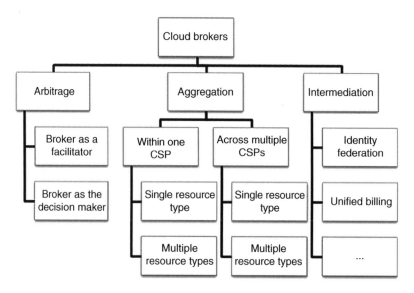

Figure 31.1 Classification of cloud brokers

choices for users and foster competition between clouds. If a broker acts as a source of information and provides a way for users to interact with CSPs, then we call this broker a facilitator. If the broker takes decisions in addition to giving information, and assisting in communication on behalf of users, then the broker is a decision maker.

The second group is broker aggregation, in which the broker deploys customer services over one or multiple CSPs. On many occasions users have a huge number of tasks that need to be deployed among multiple resources belonging to one or more CSPs. This may include the reduction of overall task execution time to meet deadlines or meet other complex objectives. In another case, aggregation might be required because one CSP alone is unable to meet the task requirements imposed by the user. Therefore, based on user requirements, the broker will perform aggregation among one or more CSPs.

Finally, there is cloud service intermediation, where brokers can build services on top of the services offered by the CSPs, such as additional security features and/or management capabilities. The most common services that a broker can offer include identity federation among multiple CSPs and unified billing services, abstracting the complexities of the CSPs underneath.

31.4   Architecture of a Cloud Broker

We describe a comprehensive cloud-broker architecture and strategies for multiple-cloud orchestration based on the broker architecture to solve several key issues faced by users and CSPs in cloud computing environments. Our model consists of a set of users and CSPs, which are connected through a broker. The broker maintains a few databases about the current system to aid users and CSPs to make their business decisions. A detailed architecture for the broker is illustrated in Figure 31.2 (Iyer, 2012). This broker architecture consists of three major components:

- job distribution manager (JDM);
- operations monitor (OM); and
- price manager (PM).

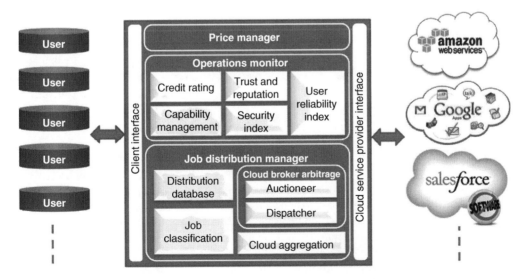

Figure 31.2   *Typical cloud broker architecture*

31.4.1   Job Distribution Manager

The job distribution manager (JDM) is responsible for receiving the user's job requirements, choosing an appropriate CSP selection strategy, informing appropriate CSPs about the jobs, and maintaining the job distribution statistics. When the user submits the job requirements, the job-classification module analyzes them and decides the preferred CSP selection strategy (either cloud service arbitrage or cloud aggregation).

This broker can handle several situations such as cloud broker arbitrage, aggregation, and intermediation. For example, if the user prefers an auction-based cloud arbitrage mechanism, then the auctioneer module will control the auction process and decide the winners. If the user wants to aggregate its application across multiple clouds, then the cloud aggregation unit will perform the necessary action. The dispatcher module dispatches the job to corresponding CSPs after the CSP selection process is completed.

The distribution database maintains a database about the job distribution statistics such as the winning CSP. This helps CSPs and users to analyze their performance in the past with respect to other competing players in the market.

31.4.2   Operations Monitor (OM)

The operations monitor (OM) monitors, manages, and maintains information pertaining to users and CSPs. The capability management module maintains databases about different resources and services offered by various CSPs. It also updates this information periodically when it notices any changes in the services offered by existing CSPs or when new CSPs enter the market. The JDM makes use of this information to shortlist the CSPs for participating in the competition based on the conditions provided in the new job requests.

The module's reputation index and security index maintain information about the reputation and security values of various CSPs periodically. These values are supplied to users when requested. Based on their preferences, users use these values to choose the appropriate CSP. A cumulative credit value is also based on these indices and user feedback, which is stored by the credit rating module.

A user reliability index is maintained by the user reliability index module. This is based on the trustworthiness of the feedback received from the users. It is used by CSPs in making their price offer and by other users in forming their utility functions.

31.4.3   Price Manager (PM)

The price manager maintains the price offers supplied by the CSPs. It is also responsible for calculating the current market price for different resources. This is used by the CSPs to adjust their price offers. It also maintains other financial matters such as maintenance of integrated billing information which can collectively calculate, display, and manage the billing information from all the CSPs for the users.

31.5   Cloud Broker Arbitrage Mechanisms

In contrast to traditional arbitrage mechanisms, which involve the simultaneous purchase and sale of an asset to make profit, a cloud service arbitrage aims to enhance the flexibility and choices available for users with different requirements and to foster competition between CSPs. For example, a user may want to choose the most secure e-mail provider (meeting certain security standards) whereas another user may want to choose the cheapest e-mail service provider.

In a typical broker arbitrage mechanism, the broker helps consumers to compare and contrast different CSPs according to their specific requirements. It also often helps them to test benchmark applications on different

CSPs and to estimate the cost and performance. The performance parameters considered might include the task execution time and quality of service (QoS) offered by the provider. In addition, the broker might track and provide the security mechanisms supported by the CSPs, the reputation of CSPs as perceived by its customers, and so forth. These help new customers to evaluate various providers in making appropriate decisions.

In many cases, the broker helps in discovering all available resource configurations (Clark *et al.*, 2012), choosing the desired configuration, negotiating an SLA, monitoring the SLA and assisting in the migration of services between CSPs. A broker can either facilitate users in making appropriate decisions or the broker could itself be the decision maker on behalf of users and CSPs.

In (Ganghishetti, 2011), the authors presented a scheme called as MCQoSMS, which collects QoS specifications from CSPs and QoS requirements from users, and finds a suitable match based on rough set theory. The model considers various QoS parameters such as availability and security. In (Calheiros *et al.*, 2012) the authors proposed architecture and a scheme for a cloud coordinate to improve the performance of various entities in a cloud ecosystem. They modeled a market that trades VMs. Cloud exchange is presented to help in negotiation services and a cloud coordinator is available for SOA implementation.

31.5.1 The Broker as a Decision Facilitator

When users are the decision makers (which is the typical case), the broker helps them to make their decisions. At any point of time a user can submit a job request with a demand to all CSPs through the broker. Each CSP quotes an offer price to the broker and the broker informs the user of the offers from all CSPs. Then the user selects the CSP that maximizes its utility, based on various parameters such as the offer price, reputation, and level of security. Based on individual requirements, different users may have different weightings for these factors. The job is submitted to the selected CSP and the corresponding databases are updated by the broker. The entire flow described above is illustrated in Figure 31.3. See (Iyer, 2012) for one such algorithm using the concept of incentives.

31.5.2 The Broker as a Decision Maker

A cloud broker who supports arbitrage mechanisms can be compared with a system such as eBay where game theoretic principles such as auctions are followed for the interactions between the seller and the buyer. Auctions are very powerful tools used in modeling such buyer-seller interactions. Specifically, continuous

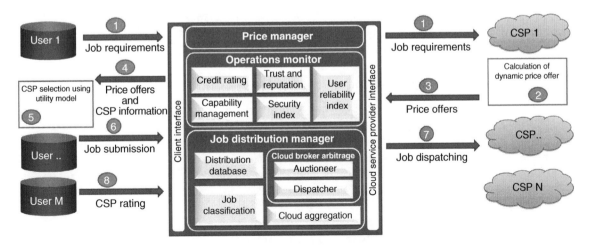

Figure 31.3 *Message flow with the broker as a facilitator*

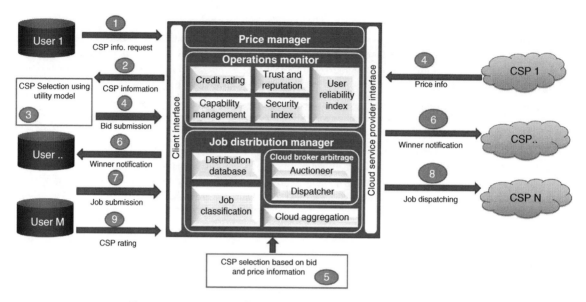

Figure 31.4 *Message flow when the broker is the decision maker*

double auctions are used (Iyer, 2012) in systems like this to enable user interactions. In such cases, sellers and buyers submit their bid information to the broker. Sellers submit their price offers and buyers submit their bid values and the mediator decides the winner(s) by matching these two values. The choice of bids reflects the user's strategic attempts to influence the selling price.

The user chooses a particular CSP based on its utility function and the user submits a bid price to the broker for the chosen CSP. The broker also maintains the minimum acceptable price offered by all the CSPs for the same resource. Using this information, the broker can select the user bid that is the smallest value above the minimum value as the winning bid for the auction. A flow in such a system is illustrated in Figure 31.4.

In such cases, the CSPs learn market demand using the auctioning process through the broker and alter their prices accordingly. The process of deciding prices is carried out at the beginning of every auction. Cloud service providers keep track of the number of auctions in which they have not been picked. If this value exceeds a certain number, the CSPs attempt to put an end to the dry spell by lowering their offer price. If the CSPs have been picked, they raise their prices in the next auction. In order to facilitate this process, CSPs keep track of the average winning price. Every time the winning price is updated by the broker, the CSPs modify their average winning price to keep it up to date.

31.5.3 Reputation of CSPs Maintained by the Broker

Every time a user uses one of the CSPs, it submits feedback to the broker about its experience with the CSP. The broker calculates and maintains a weighted average reputation value about the CSP, which can then be used by other users to evaluate a CSP. Mathematically, the reputation value is a combination of the performance reviews submitted by the users at the end of each auction and the minimum bid values submitted by the CSPs (Iyer, 2012). The reputation of a CSP j after k auctions can be calculated as follows:

$$R_{ij} = \frac{\sum_{\forall k} \frac{\sum_{i=1}^{N} Y_{ij}\varphi_i}{N}}{k} + \frac{1}{X_{ij}}$$

Here, the first term is the average value of the product of the performance review submitted by the users about a particular CSP under consideration and the reliability index of those users calculated for the last k auctions. Y_{ij} is the reputation for CSP j as perceived by user i and φ_i is the reliability index for that particular user. It means that reviews submitted by reliable users are given more consideration compared to the reviews submitted by unreliable users. The second parameter is the inverse of the minimum bid value (X_{ij}) submitted by the corresponding CSP. If a CSP believes that its performance is rendering its reputation value too low, it can increase its reputation by lowering the prices.

In order to enforce truthful feedback by the users after each auction, the broker maintains a reliability index, φ_i, for each user. Users are not aware of the reliability value assigned to them by the broker. In each auction, the average performance of a particular CSP is calculated. If a user reports a performance value that is different from the average value beyond a threshold value, that user's review is viewed as a special case. If this user consistently submits such reviews, that user's reliability is reduced.

31.6   Cloud Broker Aggregation Mechanisms

Customer applications can be deployed across multiple CSPs with the help of cloud aggregation brokers. Users may have certain requirements such as deadline and budget that need the application to be distributed across multiple CSPs, or users may require a service that can be obtained only by combining services from multiple CSPs. This second category is also known as cloud broker integration. A typical cloud broker aggregation process is illustrated in Figure 31.5.

There are two types of broker aggregation. One is aggregation among datacenters within one CSP and the other is aggregation among datacenters belonging to multiple CSPs. The former category is easier as the datacenters are within a CSP. In such cases the broker will most probably reside within the CSP firewalls and will be managed by the CSP itself. Based on the customer's requirements and resource availability, the broker will perform the aggregation.

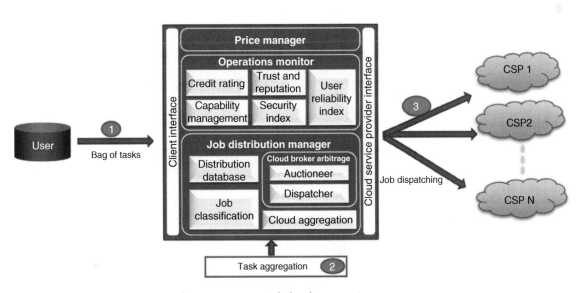

Figure 31.5   *Typical cloud aggregation process*

31.6.1   Task Aggregation in a Private Compute Cloud

An example of task aggregation in a private compute cloud running on a platform in which all computing resources have identical characteristics is described in this subsection. Based on requests received for resource instances, the broker needs to employ efficient strategies to allocate the optimal number of resources to achieve certain goals such as minimizing task-execution time or maximizing revenue, which is beneficial to CSPs and users. In our model, users submit parameters pertaining to the tasks such as task deadline or budget requirements and the broker calculates the optimal resource assignment.

Bargaining theory (Muthoo, 1999) can be used for task aggregation. In such cases the tasks to be executed in the compute cloud are known a priori. This is a reasonable assumption for a cloud because tasks to be executed in the cloud are generally submitted well in advance and they usually have a long processing time (of the order of hours or days). Further, tasks within certain applications such as compute-intensive workflow applications are known in advance. However, they may not know the exact amount of resources needed until the runtime.

We consider computation-intensive tasks that demand many virtual CPU instances (VCIs) and need to meet certain requirements such as deadline and/or budget. Further, without any loss of generality, we assume that the scheduling details such as the task distribution values for different CSPs (if the aggregation is performed on multiple CSPs), the individual billing information, and so forth, are abstracted from the user, that the broker manages this information, and the broker maintains information such as integrated billing.

Two types of bargaining approaches can be applied in such situation: the Nash bargaining solution (NBS) and the Raiffa bargaining solution (RBS) (Iyer, 2012). The NBS maximizes the use of resources and guarantees proportional fairness, whereas the RBS considers maximum requirements for resources for allocation, which is useful when we want to consider the cloud, where tasks, which are either independent or from workflow schemes, arrive in a dynamic fashion. Bargaining approaches consider bargaining power, which is a crucial parameter in determining the aggregation. Further, by changing the parameters used for bargaining power and the minimum and maximum resource requirements by users, brokers can come up with different aggregation vectors.

31.6.2   Task/Data Aggregation among Multiple CSPs

An example of broker aggregation among multiple clouds with heterogeneous compute capabilities is described in (Iyer, 2012). In such case CSPs can be modeled as independent M/M/1 queues (Bertsekas and Gallager, 1992), formulating an optimization problem to minimize the task execution time and derive optimal solutions for task distribution. A heuristic algorithm is also proposed, which considers not only the task-execution time but also the user's budget requirements in order to derive the task distribution.

We can observe that large-scale data-intensive divisible load applications such as image processing and biological computing applications can also use this model for execution to satisfy various constraints such as budget constraints and application execution time. In the case of divisible load applications we calculate the optimal amount of data to be distributed to each CSP instead of determining the number of tasks to be distributed. In addition to application execution time and budget constraints, users have many other considerations such as security, trust, and the reputation of the CSP. Using the operations monitor in our broker we can seamlessly integrate these features. Brokers can consider a subset of all the CSPs satisfying such user criteria to derive the optimal task distribution.

Other cases of broker aggregation are more complex and a complex process is needed to handle them. For example, when a set of tasks needs to be handled among multiple CSPs, some of the tasks might need computing resources whereas other tasks might need storage resources. Aggregation among multiple CSPs also imposes a unique set of challenges for the broker. These challenges include establishing appropriate

SLAs between the CSPs and/or the broker, unified billing for the users to abstract the presence of multiple CSPs in the system, data communication between CSPs through their network, and security restrictions and network latency for communication between the applications managed by different CSPs.

31.7 Cloud Broker Intermediation Mechanisms

Intermediation brokers customize and build add-on services on top of cloud services to incorporate additional features in current cloud services. They may include services to enhance the experience with clouds, such as add-on security features (e.g. single sign on), integrated billing and monitoring services, and financial services.

Another broker intermediation service is to offer an identity brokerage and federation model (Dimitrakos, 2010) to facilitate distributed access, license management, and SLA management for software as a service (SaaS), and infrastructure as a service (IaaS) clouds. In most cases, broker intermediation services are designed for specific services/application areas. For example, a service could be to find the optimal route between the user and the CSPs. Another example is to offer an identity federation mechanism (single-sign-on) on top of the services offered by the provider (He Yuan Huang, 2010). The intermediation services might be offered by the CSP itself or by external third-party services.

31.8 Conclusions and Remarks about the Future

In this chapter, we first described several key challenges faced by cloud users for moving their business into cloud platforms and the challenges faced by the CSPs to adapt to market conditions and to take critical business decisions. Then we proposed the need for a cloud broker to solve these challenges and we conducted a comprehensive classification of existing cloud broker mechanisms into three categories: cloud broker arbitrage, aggregation, and intermediation. We proposed a cloud broker architecture to solve several key issues existing in the current cloud computing scenario.

An ideal cloud broker should support more than one type of service. For example, the broker architecture that we discussed in this chapter supports auction-based broker arbitrage, bargaining-based broker aggregation and a few intermediation services such as unified billing, security index, and reputation index.

Brokers that can deal with vendor lock-in need to be explored to handle all issues associated with lock-in effectively. Mathematical models are required for the switching-cost model. It includes possible end-of-contract penalties, charges for format conversion and data/application switching and possible additional charges for bandwidth usage.

An interesting research project could be to use a cloud broker that can facilitate a spot market. Such a broker should aid in creating a market to offer its unused resources for a short period of time. Mathematical models are required to formulate the spot-pricing strategies based on supply of, and demand for, resources in a competitive environment.

References

Bertsekas, D. and Gallager, R. (1992) *Data Networks*, Prentice Hall, Englewood Cliffs, NJ.

Calheiros, R. N., Toosi, A. N., Vecchiola, C., and Buyya, R. (2012) A coordinator for scaling elastic applications across multiple clouds. *Future Generation Computer Systems* **28**(8), 1350–1362.

Clark, K. P., Warnier, M., and Brazier, F. M. T. (2012) *An Intelligent Cloud Resource Allocation Service – Agent-Based Automated Cloud Resource Allocation using Micro-Agreements*. Proceedings of the Second International Conference on Cloud Computing and Services Science (CLOSER 2012). SciTePress, 37–45.

Dimitrakos, T. (2010) *Common Capabilities for Service Oriented Infrastructures and Platforms: An Overview.* Proceedings of the 2010 Eighth IEEE European Conference on Web Services, Europe. IEEE, pp. 181–188.

Ganghishetti, P., Wankar, R., Almuttairi, R. M., and Rao, C. R. (2011) *Rough Set Based Quality of Service Design for Service Provisioning through Clouds.* Proceedings of the 6th International Conference on Rough Sets and Knowledge Technology (RSKT 2011), Berlin. Springer, pp. 268–273.

Huang, H. Y., Wang, B., Liu, X. X., and Xu, J. M. (2010) Identity federation broker for service cloud. Proceedings of the 2010 International Conference on Service Sciences, pp. 115-120.

Iyer, G. N. (2012) Broker-Mediated Multiple-Cloud Orchestration Mechanisms for Cloud Computing. PhD thesis. National University of Singapore.

Mell, P. M. and Grance, T. (2011) *The NIST Definition of Cloud Computing.* Special Publication 800-145. NIST, Gaithersburg, MD, http://www.nist.gov/customcf/get_pdf.cfm?pub_id=909616 (accessed November 25, 2015).

32

Migrating Applications to Clouds

Jyhjong Lin

Ming Chuan University, Taiwan

32.1   Introduction

Advantages of cloud applications include the fact that users can utilize them in a low-cost and risk-free way. Applications can be deployed quickly on clouds so that developers can focus on enhancing their quality of service (QoS) to improve core competitiveness. As such, cloud applications are recognized as a growing trend that will affect the next generation of business applications.

In terms of the architecture for on-premise applications (e.g., Web information systems), client-server models were most commonly used in previous decades; almost all kinds of existing on-premise applications were constructed using this style of architecture. How to migrate these applications to the clouds to take advantage of cloud applications is a subject of interest. Some discussion on the migration of on-premise applications into the clouds was presented in Banerjee (2010), Cisco Systems (2010), Huey and Wegner (2010), and Mallya (2010).

These works have clarified some important issues about migration and have proposed tips. There are thus many ideas about matters such as how on-premises applications can be migrated smoothly to any of the three service models – software as a service (SaaS), platform as a service (PaaS), and infrastructure as a service (IaaS) – in various cloud environments. Nonetheless, methods that take into consideration the architecture and characteristics of both on-premise applications and clouds to provide guidance on their migration are still missing. Moreover, little has been said about, for instance, how on-premise applications can be arranged to take advantage of cloud applications via proper virtual mechanisms so that any of the three service models – SaaS, PaaS, and IaaS – can be virtually supported in these mechanisms. A well guided process is critical for directing the migration of on-premises applications in a systematic and managed manner.

In this chapter, we present a method for directing the migration process. The method starts with the identification of the architecture and profile of the on-premise application to be migrated, and continues with discussion on

Encyclopedia of Cloud Computing, First Edition. Edited by San Murugesan and Irena Bojanova.
© 2016 John Wiley & Sons, Ltd. Published 2016 by John Wiley & Sons, Ltd.

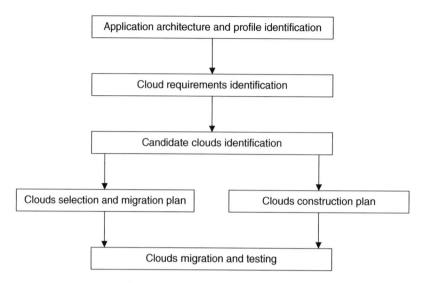

Figure 32.1   *The migration method*

the requirements for clouds, identification of the configurations of the available clouds, and the selection of the candidate clouds whose service models – SaaS or PaaS or IaaS – satisfy cloud requirements. It concludes with the deployment of the application into selected clouds where a deployment and test plan is specified. The situation in which no candidate clouds can be found for smooth migrations is specifically considered; some proposals are made to address how on-premises applications in such situations can be arranged via virtual mechanisms where either of the three service models – SaaS, PaaS, and IaaS – can be virtually supported in these mechanisms. As an illustration, the method is applied to the migration of a customer support system (CSS) application (Orman, 2007; Lin, 2009) to its cloud version, which particularly emphasizes both the collection of customer information (i.e., knowledge about / from customers) for enterprises and, conversely, delivering services information from enterprises to benefit customers.

This chapter is organized as follows. Section 32.2 presents the migration method, which encompasses three processes: an application-description process, a cloud-identification process, and an application-deployment process. The method is illustrated in section 32.3 by applying it to the migration of a CSS application to its cloud version. Finally, section 32.4 draws some conclusions and discusses future work.

32.2   Cloud Migration Roadmap

As shown in Figure 32.1, migrating an on-premises application to the clouds has the following six steps:

1. Identification of application architecture and profile. This determines (i) the architecture of the on-premises application, and (ii) the profile of the on-premises application including data about, for example, its CPU, memory, storage, I/O, and network usage data, as well as data about its users (e.g., the number of active users, request rates, transaction rates, and request/transaction latencies).
2. Identification of cloud requirements. This clarifies the cloud requirements for the on-premises application, based on its architecture and profile, including (i) the requirements for the deployment of its components on the prospective configuration elements in clouds such as virtual machines, data storage,

and a/synchronous message channels, and (ii) requirements for QoS in clouds – for example customized user interfaces and access modes, performance, reliability, security, and scalability.

3. Identification of candidate clouds. This identifies the candidate clouds whose configurations and services (i.e., service models – SaaS or PaaS or IaaS – provided in clouds) satisfy the cloud requirements identified above.

4. Cloud selection and migration plan. This determines which of the candidate clouds will be selected for the migration of the on-premises application. A plan for the migration into selected clouds will then be specified. In general, it includes (i) deploying the application components on the configuration elements in clouds; (ii) deploying the interaction mechanisms in clouds; and (iii) refactoring/restructuring components to satisfy user requirements such as customized user interfaces and access modes, performance, reliability, security, and scalability.

5. Cloud construction plan. This identifies and schedules alternatives for situations in which no candidate clouds can be found for smooth migrations. In such a situation, the on-premises application may achieve its cloud-based version via virtual mechanisms so that any of the three service models – SaaS, PaaS, and IaaS – can be virtually supported in these mechanisms.

6. Cloud migration and testing. This involves the migration of the on-premises application into selected or constructed clouds in accordance with the migration or construction plan identified above. As is usual, testing of the migration proceeds in accordance with the activities involved in the migration process.

The first step addresses an application description process, the middle two encompass a cloud identification process, and the last three cover an application deployment process.

32.2.1   Application Architecture and Profile Identification

This step addresses (i) the architecture of the on-premises application, and (ii) the profile of the on-premises application, which includes data about its users.

32.2.1.1   Application Architecture

The architecture of the on-premises application specifically addresses the architectural components desired by the user. As an example, Figure 32.2 shows the architecture of a customer support system (CSS) (Orman, 2007; Lin, 2009):

- It is a four-layer architecture of collaborative components, where *customers* interact with *enterprises* via three intermediaries: *community*, *customer knowledge agent*, and *task service provider*.
- It emphasizes *community* to help *customers* share information about their desired tasks (e.g., buying or renting services from *enterprises*).

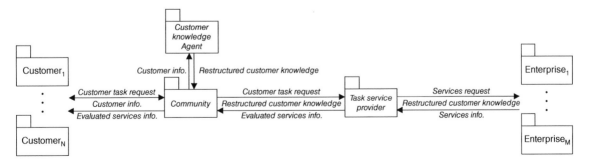

Figure 32.2   *The architecture for the customer support system*

- It emphasizes the collection of knowledge about the customer by the *customer knowledge agent* to help *enterprises* meet their needs (e.g., provide services satisfying their desired tasks).
- It focuses on the *task service provider* delivering information about services from *enterprises* to help *customers* make comparisons.

32.2.1.2   *Application Profile*

Given the architecture of the on-premise application, the profile of the on-premises application can then be captured. This may help to size the application before it is migrated to the clouds. In general, the application data should be collected for at least 10 to 14 days to allow any variances in daily or weekly usage patterns to be noted. There are two kinds of data about the application: (i) usage data (e.g., CPU, memory, storage, I/O, and network usage data), and (ii) data about its users (e.g., the number of active users, request rates, transaction rates, and request/transaction latencies). With such data, it is feasible to make an initial estimate of the cloud resources for the application to be migrated.

32.2.2   Identification of Cloud Requirements

The second step is to identify cloud requirements for the on-premise application based on its architecture and profile, including (i) requirements for the deployment of its components on the prospective configuration elements in clouds such as virtual machines, data storage, and a/synchronous message channels; and (ii) requirements for their QoS in clouds such as customized user interfaces and access modes, performance, reliability, security, and scalability. For the CSS, its five components may require deployment in various cloud environments. To collect customer knowledge for enterprises and deliver services information to benefit customers, it may require these deployed clouds to provide customized user interfaces and access modes, performance, reliability, security, and scalability.

32.2.3   Identification of Candidate Clouds

The third step is to identify the candidate clouds whose configurations and services (i.e., service models – SaaS or PaaS or IaaS – provided in clouds) satisfy the cloud requirements identified above. For this, it is good to consider all of the cloud environments available online whose service models may satisfy those cloud requirements. The following describes the possible service models in clouds:

- Software as a service (SaaS). In this model (Li *et al.*, 2012) the cloud provides application services that may replace those provided by the on-premises application. With such SaaS services, many QoS features need to be evaluated to determine their replacement with the on-premises application as below.
  - Their service-level-agreements (SLAs) for availability, scalability, security, and performance. Note that specific SLAs such as those for availability, scalability, and performance can be evaluated by assessing the profile of the on-premises application.
  - The compatibility of the application services with those offered by the SaaS.
  - The portability of the application data into the SaaS so that it can be accessed by the SaaS services.
  - The portability of access control by the application users into the SaaS for access control by the SaaS users.
  - The portability of the application interoperability with other services into the SaaS for interoperable operations by SaaS services.
- Platform as a service (PaaS). In this model (Cunha *et al.*, 2013), the cloud provides platform services on which the on-premises application may be deployed based on certain platforms such as JEE and MS.NET.

o   Service-level agreements for availability, scalability, security, performance, and configuration (e.g., platform versions, APIs). Note that, as in the case of SaaS, SLAs for availability, scalability, and performance can be evaluated by assessing the application's usage data and data about the application users.

o   The deployment of application components and their interaction mechanisms on the PaaS.

o   The portability of application services into the PaaS for access by the application users.

o   The portability of the application data into the PaaS for access by the application.

o   The portability of access control on platforms (e.g., virtual servers) by the application users into the PaaS for access control on clouds (e.g., virtual machines) by the application users.

o   The portability of application interoperability with other services into the PaaS for interoperable operations by the application.

o   The portability of the application management into the PaaS to monitor and manage the application.

o   Infrastructure as a service (IaaS). In this model (Baun and Kunze, 2011), the cloud provides infrastructure services such as servers, storage, and networks, which the on-premises application and its platforms may use with some QoS features.

o   Service-level agreements for availability, scalability, security, performance, and configuration. Note that, as with PaaS, SLAs for availability, scalability, and performance can be evaluated by assessing the profile of the on-premises application.

o   The portability of application services into the IaaS for access by application users.

o   The portability of application data into the IaaS to be stored in IaaS storage.

o   The portability of the access control on infrastructure services (e.g., physical servers) by application users into the IaaS for access control on clouds (e.g., physical machines) by the deployed application users.

o   The portability of application interoperability with other services into the IaaS for interoperable operations by the application.

After considering all possible cloud environments that provide either of the above three service models, it is expected that it would be possible to identify some of them whose service models satisfy the cloud requirements and then become the candidates to be selected for the migration. For the CSS example, Figure 32.3 shows the possible candidate clouds with service models that satisfy the cloud requirements for the CSS.

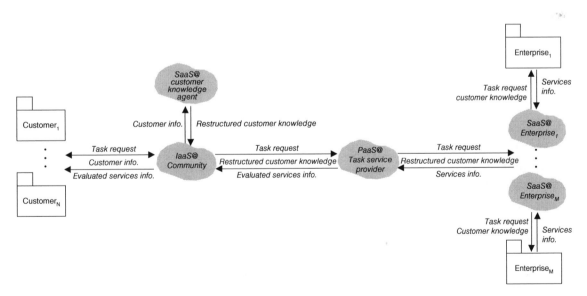

Figure 32.3   *The candidate clouds and service models for the CSS*

32.2.4   Clouds Selection and Migration Plan

The fourth step is to determine, from the candidates identified above, which clouds are selected for the migration of the on-premises application. In general, the determination can be clarified by the use of evaluation criteria that can rank these candidates in terms of how well they satisfy the cloud requirements. For example, based on the QoS features for the three service models, a candidate whose service models gain the best weighted assessments on them may be selected as the cloud environment for the on-premises application to be migrated. After determining the selection of targeted clouds, the plan about migration into these selected clouds will then be specified. In general, actions include (i) deploying the application components on the configuration elements in respective clouds; (ii) deploying the interaction mechanisms; and (iii) restructuring components to meet user requirements such as requirements for customized user interfaces and access modes, performance, reliability, security, and scalability.

32.2.5   Cloud Construction Plan

The fifth step is to identify and schedule alternatives for situations in which no candidate clouds can be found at step 3 for smooth migrations. In such a situation, alternatives may be considered; for instance, the on-premises application may be tailored to achieve a cloud-based version via virtual mechanisms (http://ventem.com.tw/DM11.aspx, accessed December 19, 2015). Any of the three service models – SaaS, PaaS, and IaaS – can be provided virtually with the support of these mechanisms.

32.2.6   Clouds Migration and Testing

The last step is the migration of the on-premises applications to selected or constructed clouds in accordance with the migration or construction plan identified above. As is usual, testing of the migration proceeds in accordance with the activities involved in the migration process.

32.3   Migration of the CSS into Cloud Environments

In this section, the method is illustrated by applying it to the migration of a CSS application to its cloud version.

32.3.1   Application Architecture and Profile of the CSS

In this step, two descriptions of the CSS are addressed: (i) the architecture of the CSS, and (ii) the profile of the CSS including data about its use and activity data about its users.

32.3.1.1   Architecture of the CSS

As shown in Figure 32.2, the CSS has a four-layer architecture where *customers* are interacting with *enterprises* via three intermediaries: *community, customer knowledge agent,* and *task service provider.* The constituents of each component need to be clarified. For example, *community* is organized for *customers* to share information about their desired tasks (e.g., to buy or rent services from *enterprises*). It is also responsible for forwarding the shared information to the *customer knowledge agent* for restructuring into specific styles of knowledge (e.g., knowledge of customers) (Mobasher *et al.*, 2002). It then sends the restructured knowledge to the *task service provider* for forwarding to *enterprises* to identify their needs (e.g., provide services to carry out their desired tasks). Finally, it also cooperates with the *task service provider* to receive services information relevant to customer requests.

In summary, these requirements for *community* can be described as follows:

1. Share customer information among *customers*$_{1...N}$.
2. Process shared information – forward shared information to *customer knowledge agent* for restructuring into knowledge, and then send restructured knowledge to the *task service provider*.
3. Process task request – receive task requests from and return evaluated information about task-relevant services to *customer*$_{1...N}$.
4. Cooperate with the *task service provider* to receive evaluated information about task-relevant services.
5. Present services information that provides *customer*$_{1...N}$ with rich user interface controls for visualizing the information from the *task service provider*.

Based on the above requirements for *community*, Figure 32.4 shows its five constituent parts that meet these requirements. In particular, an "interface manager" is imposed to customize and personalize user interfaces for

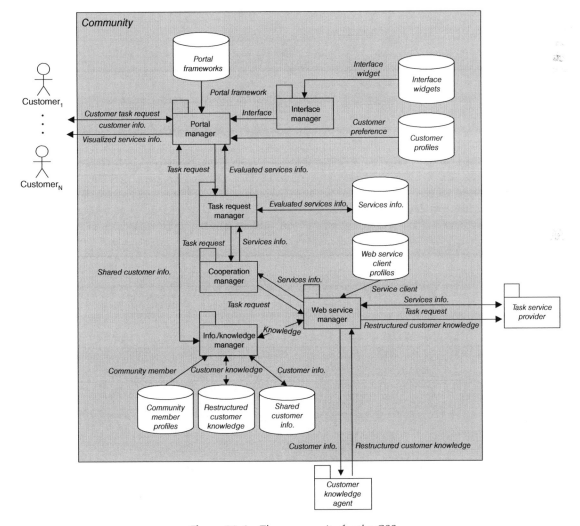

Figure 32.4 *The community for the CSS*

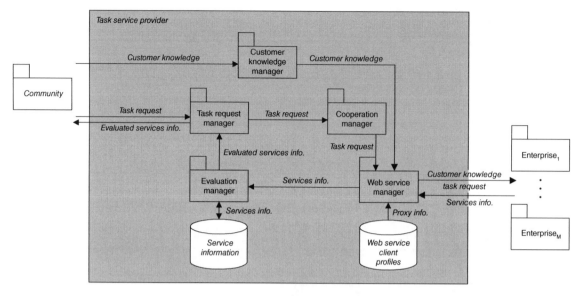

Figure 32.5   *Task service provider component of the CSS*

$customer_{1...N}$ where *customer profiles* are used to determine which interface components are preferred by them. The portals are then used to deliver to $customers_{1...N}$ the information that they require. The "Info./knowledge manager" accesses *community member profiles* and *shared customer info.* to help share information among interested customers; *shared customer info.* is also retrieved for restructuring into knowledge by *customer knowledge agent.* In addition, the "task request manager" forwards task requests from $customers_{1...N}$ to the "cooperation manager," which cooperates with the *task service provider* to receive evaluated information about these requests; the evaluated information is then visualized and returned to $customers_{1...N}$ through the "portal manager." Finally, the "Web service manager" is responsible for interoperating with the two external architectural components through *Web service client* APIs for accessing those remote services provided by the two components.

Another example is the *task service provider*. This is an important intermediary between *community* and *enterprises.* It receives customer knowledge from *community* and then forwards that knowledge to *enterprises* that use it to provide information about services that are useful for *customers.* In addition, based on the task requests received from *community*, it cooperates with *enterprises* to provide information about these tasks. It also helps to evaluate the task-relevant information in a comparative model for presenting to *customers* (via *community*) to aid their decision making. In summary, the requirements for the *task service provider* can be described as follows:

1. Process customer knowledge – receive customer knowledge from *community* and forwards it to $enterprise_{1...M}$.
2. Process task request – receive task requests from and return evaluated information about task-relevant services to *community.*
3. Cooperate with $enterprise_{1...M}$ to provide information about task-relevant services.
4. Evaluate services information so that task-relevant services from $enterprise_{1...M}$ can be compared.

Figure 32.5 shows the five constituents of the *task service provider* that meet these requirements. The "task request manager" forwards task requests to the "cooperation manager," which cooperates with $enterprise_{1...M}$ to provide information about task-relevant services; the services information from $enterprise_{1...M}$ is then evaluated by the "evaluation manager" for returning to *community*. Finally, the "Web service manager" is responsible for interoperating with various external architectural components through the *Web service client* APIs for those remote services provided in these components.

32.3.1.2   *Profile of the CSS*

With the CSS architecture, a CSS profile can then be identified that may help to size the CSS before it is migrated to the clouds. In general, the profile should be collected for at least 10 to 14 days to allow any variance in daily or weekly usage patterns to be noted. There are two kinds of profile data about the CSS: (i) data about its use (e.g., CPU, memory, storage, I/O, and network usage data), and (ii) data about its users (e.g., the number of active users, request rates, transaction rates, and request/transaction latencies). With such profile data, it is feasible to have an initial estimate of the cloud resources for the CSS to be migrated.

32.3.2   Cloud Requirements for the CSS

The second step is to identify the cloud requirements for the CSS based on its architecture and profile. Initially, considering its five distributed components, various cloud environments may be required.

32.3.3   Candidate Clouds for the CSS

The third step is to identify the candidate clouds whose configurations and services (i.e., service models – SaaS or PaaS or IaaS – provided by the clouds) satisfy the cloud requirements for the CSS. To do this it is common to consider all of the cloud environments available online whose service models may satisfy the requirements. There is usually more than one cloud that satisfies the requirements; such clouds hence become the candidates from which specific ones are then selected for the migration.

32.3.4   Cloud Selection and Migration Plan for the CSS

The fourth step is to determine, from the candidates identified above, which clouds are to be selected for the migration of the CSS. In general, this can be achieved by using specific evaluation criteria that may rank the candidates in the context of the satisfaction of the cloud requirements. For example, a candidate whose service models gain the best weighted assessments on the QoS features for the three service models may be selected as the cloud environment for the migration of the CSS. As an illustration for the CSS: (i) the SaaS clouds hosted by the collaborative agents/enterprises are undoubtedly selected for the migration of its *customer knowledge agent* and *enterprises* components; (ii) from the available PaaS clouds, which include Google GAE (https://appengine.google.com/, accessed December 19, 2015) and Microsoft Azure (http://www.windowsazure.com/zh-tw/, accessed December 19, 2015), the GAE, shown in Figure 32.6 is selected for the migration of its *task service provider* component due to its well known intercloud and analysis capabilities for the cooperation and evaluation requirements of the component; and (iii) from the available IaaS clouds, which include Google GCE (https://cloud.google.com/products/compute-engine, accessed December 19, 2015) and Amazon EC2 (http://aws.amazon.com/ec2/, accessed December 19, 2015), the EC2 as shown in Figure 32.7 is selected for the migration of its *community* component due to its well known storage and intercloud capabilities for information sharing among customers and cooperation with other components.

After selecting the cloud, the plan to carry out the migration can then be specified. In general, this includes (i) deploying the CSS components in the cloud; (ii) deploying the mechanisms for interactions among CSS components; and (iii) refactoring/restructuring components to meet requirements. As an illustration, for the *task service provider* to be migrated into the GAE PaaS cloud, denoted as *PaaS@Task Service Provider*,

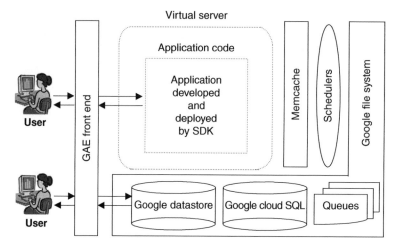

Figure 32.6 *The Google GAE PaaS cloud*

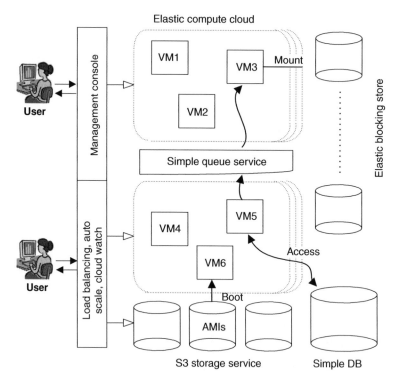

Figure 32.7 *The Amazon EC2 IaaS cloud*

Figure 32.8 shows the deployment of its five constituents on the three virtual servers (i.e., Mashup, MapReduce, and Web Service) in GAE and each one may use some storage services such as Datastore and Cloud SQL. As another illustration, for *community* to be migrated into the EC2 IaaS cloud, denoted as *IaaS@ Community*, Figure 32.9 shows the deployment of its five constituents on the four virtual machines (VMs) in EC2 and each one may use some storage services such as S3 storage, EBS storage, and simple DB.

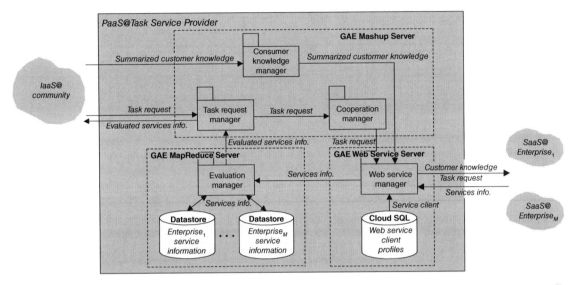

Figure 32.8   *The Google GAE-based deployment of PaaS@Task Service Provider*

32.3.5   Cloud Construction Plan for the CSS

The fifth step is to identify and schedule alternatives for a situation in which no candidate clouds can be found at step 3 for smooth migrations. In such a situation, some alternatives may be considered. For instance, with the virtual mechanism in Li *et al.* (2012), the migration of the *community* component, denoted as *IaaS@ Community*, can be virtually constructed by installing the Vas platform of the mechanism that supports the storage and manipulation of the information shared among *customers* via the dynamic allocation of shared information into data stores.

32.3.6   Cloud Migration and Testing for the CSS

The last step is to migrate the CSS into selected or constructed clouds in accordance with the migration or construction plan identified above. As usual, testing of the migration proceeds in accordance with the activities involved in the migration process.

32.4   Conclusions and Future Work

In this chapter, we presented a method for directing the migration of on-premise applications to selected clouds. The method takes into consideration the architecture and characteristics of both on-premises applications and clouds to provide guidance on the migration. It starts with the identification of the architecture and profile of the on-premises application, proceeds by discussing cloud requirements for the application and the identification of the configurations of selected clouds, and ends with the deployment of the application in selected clouds where a deployment and test plan is specified. To illustrate this, the method was applied to the migration of a CSS application to its cloud version, emphasizing both the collection of customer knowledge for enterprises and, conversely, delivering information about services from enterprises to benefit customers.

As cloud applications have been recognized in recent years as a trend for the next generation of business applications, how to migrate the many existing on-premises applications so that they can take advantage of

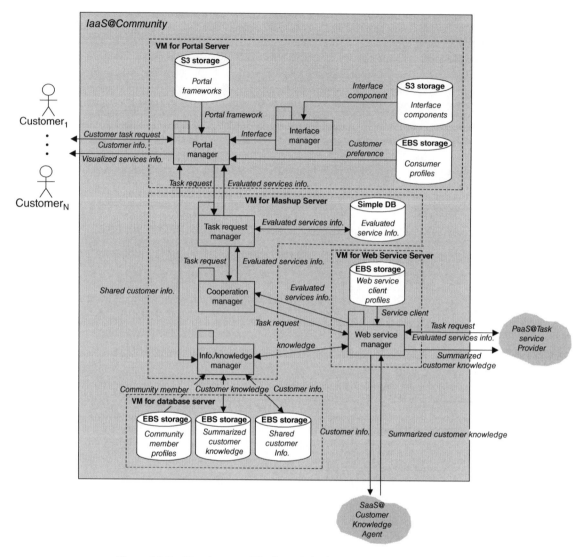

Figure 32.9   *The Amazon EC2-based deployment of IaaS@Community*

cloud applications has become a subject of interest. However, current discussions about this need mainly focus on important issues about the migration and then present tips for addressing such issues. Methods for the migration process that take into consideration the architecture and characteristics of both on-premises applications and clouds to provide guidance on their migration are still missing. Such methods, in my opinion, are important because a well guided process is critical in directing the migration of the many on-premises applications in a systematic and managed manner.

We will continue to explore the migration of existing on-premises applications to the clouds where specific PaaS and IaaS offerings such as Google GAE and Amazon EC2 are selected as the platforms used. While migrating these applications, experience can be gained about the usefulness and effectiveness of the method used. With my systematic and managed steps for gradually identifying application/cloud features and then conducting the deployment of the applications on the clouds, the quality of these migrated applications can be expected.

References

Banerjee, U. (2010) *Five Examples of Migration to Cloud*, https://udayanbanerjee.wordpress.com/2010/05/28/5-examples-of-migration-to-cloud/ (accessed December 19, 2015).

Baun, C. and Kunze, M. (2011) *A Modern Approach for Cloud Infrastructure Management*. Proceedings of the First International Workshop on Cloud Computing Platforms (CloudCP 2011). ACM Press.

Cisco Systems (2010) *Planning the Migration of Enterprise Applications to the Cloud*, https://www.cisco.com/en/US/services/ps2961/ps10364/ps10370/ps11104/Migration_of_Enterprise_Apps_to_Cloud_White_Paper.pdfUS/services/ps2961/ps10364/ps10370/ps11104/Migration_of_Enterprise_Apps_to_Cloud_White_Paper.pdf (accessed December 19, 2015).

Cunha, D., Neves, P. and Sousa, P. (2013) *A Platform-as-a-Service API Aggregator*. Proceedings of the 2013 World Conference on Information Systems and Technologies. Advances in Intelligent Systems and Computing Series. Springer, pp. 807–818.

Huey, G. and Wegner, W. (2010) *Tips for Migrating Your Applications to the Cloud*, http://msdn.microsoft.com/ en-us/magazine/ff872379.aspx.

Li, Q., Liu, S., and Pan, Y. (2012) *A Cooperative Construction Approach for SaaS Applications*. Proceedings of the 2012 IEEE 16th International Conference on Computer Supported Cooperative Work in Design. IEEE Computer Society, pp. 398–403.

Lin, J. (2009) An object-oriented development method for consumer support systems. International Journal of Software Engineering and Knowledge Engineering **19**(7), 933–960.

Mallya, S. (2010) Migrate Your Application to Cloud: Practical Top 10 Checklist, http://www.prudentcloud.com/cloud-computing-technology/migration-to-cloud-top-10-checklist-24042010/ (accessed December 20, 2015).

Mobasher, B., Cooley, R., and Srivastava, J. (2002) Five styles of customer knowledge management, and how smart companies use them to create value. *European Management Journal* **20**(5), 459–469.

Orman, L. (2007) Consumer support systems. *Communications of the ACM* **50**(4), 49–54.

33

Identity and Access Management

Edwin Sturrus and Olga Kulikova

KPMG Advisory N.V., the Netherlands

33.1 Introduction

In recent years, cloud technologies have introduced new methods to attack organizations and individuals, broadening their threat landscape. The digital identities that individuals and organizations use, in order to access cloud resources, are one of the main areas at risk. Past incidents with LastPass, Google and Evernote, where a number of user accounts became compromised, show how challenging it is to protect digital identities. Insecure management of identities and their access can cause a lot of trouble for organizations and individuals, resulting in data breaches, and noncompliance with important standards and regulations (such as HIPAA, PCI-DSS, EU GDPR), and inability to access resources, services, and critical data.

When dealing with identity and access management in cloud context, the main questions often posed by users are security related, such as "are my passwords stored securely?", "are there any privileged accounts that can be used to access my data?" and "how vulnerable am I or my organization to hacking attacks?"

The core of the challenge lies in the nature of cloud computing. When switching to the cloud, part of the data is no longer stored on devices managed by the owners of the data. This, combined with the growing number of users and roles in modern organizations and stricter regulations imposed by governments on privacy and data protection, further complicates the situation and raises the importance of data access controls. Robust identity and access management (IAM) is one of the approaches to minimize security risks of cloud computing.

Encyclopedia of Cloud Computing, First Edition. Edited by San Murugesan and Irena Bojanova.

33.2   IAM Explained

IAM refers to the processes, technologies, and policies that manage access of identities to digital resources and determine what authorization identities have over these resources.

For an individual user, IAM generally concerns several processes. The user can create, remove or adjust a user account within an application. Users also have a measure of authentication to prove their identity. Authentication measures can range from a combination of username and password to multifactor authentication where smartcards, generated tokens and/or biometric data can be combined to make the authentication stronger.

For organizations, IAM is generally used much more intensively as organizations represent multiple users (employees) using multiple digital resources. This requires extensive propagation of user accounts and better monitoring and audit capabilities. Even though the IAM scale differs based on user type, both organizations and individuals are affected by the same processes when accessing digital resources:

- Management of identities. Every identity requires a valid user account, with certain requirements assigned to it, in order to be able to access digital resources. These authentication requirements and authorizations may change during the lifecycle of an identity, up to the point where a user account has to be removed from the digital resource. The identity management process here is to reflect the changes in a timely manner.
- Management of access to resources. Every identity, when attempting to access a resource, needs to prove that he or she is who he or she claims to be. If the identity is proven correct and it has the right roles and authorizations assigned to access the resource, the requested resource is provisioned. If the validation of identity or authorization fails, the identity will be unable to access the requested resource.

For most organizations, managing identities and access means implementing a directory service (e.g. Microsoft Active Directory). This directory service allows users to verify their identity with the organization. Applications that cannot be tied to this directory service have to use a separate stand-alone authentication system, which generally means that users have to login to that system with separate credentials.

Figure 33.1 depicts the key IAM services that deliver IAM capabilities to manage identities and access to (cloud) IT services within an organization.

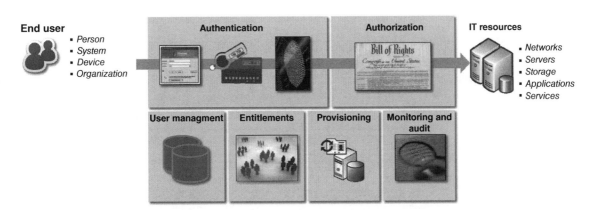

Figure 33.1   *Enterprise IAM functional architecture*

- Authentication – this service covers the processes and technology for determining that users are who or what they claim to be.
- Authorization – this service covers the processes and technology for determining that a user has the correct permissions to access IT resources.
- User management – this service covers the activities that effectively administer the lifecycle of identities (creating, changing, inactivating). Authoritative sources of identity information (e.g. HR information on the enterprise employees and customers) are required to govern and manage the identities lifecycle.
- Entitlements – this service covers the mapping of authorizations to identities and the related attributes.
- Provisioning – this service covers the propagation of identity and authorization data to IT resources via automated or manual processes.
- Monitoring and audit – this service covers the monitoring, auditing, and reporting of the compliance of users' access to IT resources, within the scope of IAM, based on the defined policies.

33.3   IAM and the Cloud

Customers of cloud services are usually looking for the same benefits that cloud can bring – increased innovation and productivity, reduced costs. The challenges presented by the adoption of cloud technologies can however be quite different and depend on the specific requirements of an enterprise, individual user, or cloud service provider (CSP). Identity management and access control are a part of the cloud challenge landscape, as the new ways of consumption and delivery of services introduce new scenarios in managing users and their access to the systems on and off premises. Especially when it comes to security, identity is a key focus area as secure provision and access to cloud-based services is the first step on the road to building a secure enterprise in the cloud.

Different types of stakeholders will face different IAM challenges:

- Organizations – extending as well as leveraging existing and new IAM solutions to tie together on premise and cloud-based systems.
- Individual users – ensuring access to information at any time, anywhere, from any device.
- Cloud service providers – securing customers' data and ensuring continued access to cloud services.

Still, despite these various challenges, IAM can play an enabling role in the adoption of cloud services. For example, existing on-premises IAM solutions can help an organization start moving to the cloud gradually with the step-by-step implementation of controls required for the shift. Later the company can choose whether to continue with an on-premises deployment, start using hybrid IAM solution, or completely shift IAM to the cloud.

33.3.1   IAM Architecture for the Cloud

In a traditional IT environment, users have to be added, changed, or removed from a system and also be assigned with certain authorizations in order to access digital resources. The general processes of managing identities or access remain the same in a cloud computing environment. Even if the management of identities is done at the CSP, the users must be recognized by the system in order to access its authorized resources. Depending on the requirements and type of cloud model used (IaaS, PaaS or SaaS) the access to different layers has to be managed on different layers:

- Network layer. This layer concerns the access to the network on which the cloud environment runs. Without access to the network, it will not be possible to connect to the cloud system at all.
- System layer. This layer concerns the access to the system. Access to this layer generally goes via protocols (e.g. TCP/IP) to access the server (e.g. Web server) on which the cloud environment runs.

- Application layer. This layer concerns the access to the application. Access to this layer determines whether a user can access the application.
- Process layer. This layer concerns the processes that run within an application. Access to this layer determines what processes (e.g. making a payment) a user can access within an application.
- Data layer. This layer concerns the data that is accessed via an application. Access to this layer determines which data a user can view or edit via the application.

33.3.2 IAM Interfacing Models

Roughly speaking, the three interfacing models can be distinguished to manage identities and access in cloud computing environment: local management of IAM services, use of an identity service provider (IdSP), or use of IAM-as-a-service (IAMaaS).

33.3.2.1 *Local IAM management*

In the local IAM management model, illustrated in Figure 33.2, business users can access the in-house resources of their organization, such as data and applications, using local authentication. To access resources, a user account is created at the CSP for each authorized user. There are software packages on the market to propagate user accounts to the CSPs automatically. This is done using standard or custom-made connectors based on the APIs of the CSPs.

Besides that, single sign-on (SSO) services can be used to reduce the number of identities a user should own to access on-premises resources and cloud services. Single sign-on is based on the implementation of federated

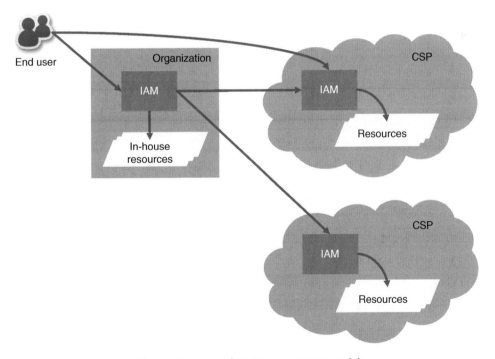

Figure 33.2 *Local IAM management model*

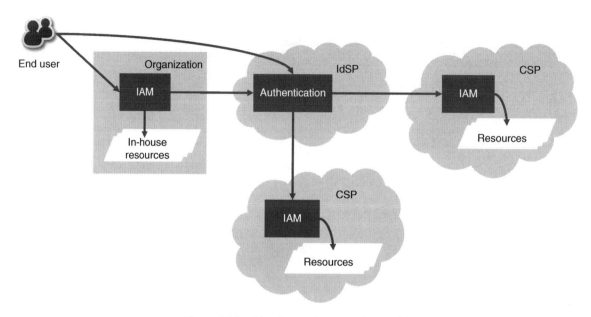

Figure 33.3   *Identity service provider model*

identity between multiple domains – end users can operate with a single identity to authenticate for multiple services in the cloud. Single sign-on will require the usage of secure federation standards such as AML and OAuth.

The benefit of this model is that it provides the most control over IAM services. However, this model requires a lot of costly local maintenance in order to connect to CSPs. For organizations, this limitation may make this model unsuitable for the long term, when adopting even more cloud services.

33.3.2.2   *Use of an Identity Service Provider*

The IdSP model, showed in Figure 33.3, allows use of a third-party identity provider to access digital resources. An IdSP will facilitate individual and business users with means to authenticate – prove their identity – in order to access local or cloud resources. This model can be beneficial for both types of users as they do not have to maintain multiple identities and a means of authentication to gain access to their authorized resources. Examples of organizations that provide identity services are DigiD, Facebook, or SURFnet. When a user attempts to access the cloud resource, the IdSP verifies the identity of the user and confirms the status to the CSP.

The main benefit of this approach is that the authentication for different cloud resources and the alignment of all different authentication mechanisms becomes the responsibility of the IdSP. Existing IAM solutions of organizations can also interact with the IdSP by using federation standards, such as SAML and OAuth. This reduces the overhead for IT administrators and employees and enables the opportunity to use SSO among various digital resources.

One of the main concerns with this model is that authentication for digital resources is controlled by a third party. This reduces the control of the organization or individual user on the strength and means of authentication.

33.3.2.3   *IAM as a Service (IAMaaS)*

The last option is to delegate IAM processes to a CSP and utilize it as a service - IAMaaS. The access to all other cloud service providers will be managed and controlled through this IAMaaS (see Figure 33.4). For organizations, the access to in-house IT resources will also be conducted via this IAM service. The

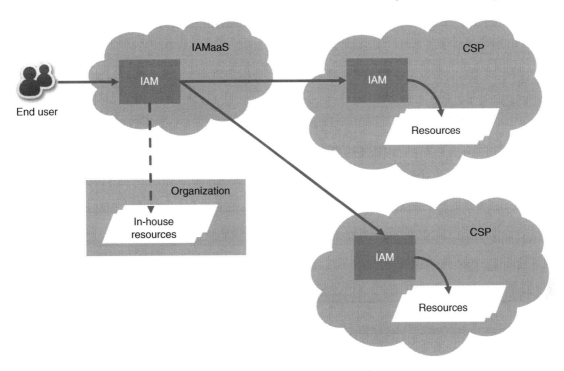

Figure 33.4 *IAM-as-a-service model*

CSP will store all user accounts, provide authentication options and authorize authenticated users to digital resources in the cloud. This gives users the opportunity to use SSO, as the CSP can be the single point to authenticate. In most cases, when using this model, organizations choose to outsource monitoring and audit to the same CSP.

This approach would result in the least responsibilities for organizations and individual users. However, it also means the biggest loss of control over IAM services, as visibility reduces significantly with outsourcing to the cloud.

Although this approach is not widely used by consumers yet, it is possible for them to use IAMaaS and have their IAM facilitated by a CSP. There are various providers on the market offering IAMaaS solutions, for example SailPoint, Okta, and Ping Idenity.

33.3.3 IAM-Related Standards

The choice of interfacing model determines how and where the identities and access to the CSP are managed. In order to share authentication and authorization data with the applicable CSPs, often, one or more of the following standards and protocols are used:

- OAuth – OAuth is an open standard that allows users to authorize software programs or web sites to access their data. It enables users to provide these authorizations without having to share their credentials (e.g. username and password).
- OpenID Connect - Open ID Connect is an open standard based on OAuth 2.0 protocol that allows users to validate their identity using an IdSP. This enables users to be authenticated at a different site than the site from which they are requesting access to a digital resource. This eliminates the need for sites to create and maintain their own authentication mechanisms. Users can maintain their digital identity with the preferred

IdSP and use this identity to access sites supporting this IdSP. Compared to the previous standard OpenID 2.0, OpenID Connect provides higher interoperability and is easier for developers to implement.

- SAML - Security Assertion Markup Language (SAML) is an XML-based message-exchange protocol that specifies the rules for exchanging authentication and authorization data between parties. Assertions contain statements that service providers use to make access-control decisions. Three types of assertions are provided by SAML:
  - authorization assertions – the rights assigned to a user or software program;
  - authentication assertions – the information required to verify the identity of the user or software program;
  - attribute assertions – information about the user or software program (e.g. "Name").
- SPML – Service Provisioning Markup Language is an XML-based framework that enables organizations to provision and manage user accounts.
- XAML – eXtensible Access Control Markup Language enables organizations to share and align authorization and entitlement data and methods across multiple CSPs. It consists of the following components:
  - Policy Enforcement Point (PEP) – enforces the entitlements based on the available policies in response to a request for access to a digital resource.
  - Policy Information Point (PIP) – provides information related to the policies in order for the policies to be evaluated.
  - Policy Decision Point (PDP) – makes the decision on whether the user is allowed to gain access to the digital resource that is requested.
  - Policy Administration Point (PAP) – enables the creation and administration of policies.

33.4   Challenges and Risks of IAM in the Cloud Context

When adopting a cloud service, a proportion of the IAM processes are managed by the CSP instead of an organization or individual user. The extend to which the IAM processes are managed by a CSP is determined by the cloud model and IAM interfacing model. This change to IAM introduces risks and challenges that should be taken into account when using cloud computing. For example, a CSP can decide to modify authentication mechanisms, such as password encryption, without informing its end users. This can affect the level of security and operability of the used cloud resource. The risks caused by shifting towards cloud computing are examined throughout this section, for each IAM service.

33.4.1   Authentication management

In most cases the CSP forces users to verify their identity by using the authentication mechanisms of the CSP. However, there are examples of CSPs that allow access through shared authentication mechanisms. Examples of providers of shared authentication mechanisms are Facebook, Google and IdSPs supporting OpenID Connect.

Having authentication managed by the CSP makes it difficult for end users to have any influence and control of the strength and security of the authentication mechanisms that are in place. If the authentication mechanism can be circumvented, the data that is stored in the cloud may be accessible to unauthorized users. Besides that, the authentication mechanisms of the CSP may not comply with laws, regulations, or policies of the organization. However, most of the time it is an enterprise that bears full responsibility in case of a security breach.

Operability can be affected by mismanagement or changes to the authentication mechanisms. For example, SSO may no longer work if the authentication mechanism is changed by the CSP, which can possibly result in end users not being able to access their cloud resources.

33.4.2   Authorization Management

The main difference for authorization management, when using cloud services, is the inability to modify and adjust available authorizations. Especially when using SaaS, users often cannot make changes to the authorization model and verify whether the appropriate authorizations are enforced. This way it is difficult to guarantee that users only access authorized resources and perform authorized actions on them. For example, a user can have access to a cloud-based service to view his monthly salary; however he may not see salaries of other employees. His manager at the same time is allowed to see the salaries of other employees. The CSP has to be able to manage these authorizations in the same way as the organization using the cloud service. Incorrect authorization management can lead to noncompliance with laws and regulations or data breaches.

33.4.3   User Management

In most cases it is the CSP that manages user accounts allowed in the system. This way it becomes difficult for end users to verify if changes to user accounts are correctly reflected within the system. This increases the unauthorized users accessing cloud resources, without the data owners being aware. Ineffective management of privileged users is another relevant risk, bringing compliance issues into play. Laws and regulations on protecting and managing user information have become stricter in recent years. Organization may face charges if the CSP does not enable compliance with the applicable laws and regulations. For example, organizations in Europe are not allowed to export personal data to country outside of Europe.

33.4.4   Entitlements

The main challenge related to entitlements when using cloud computing is manageability. When using more cloud resources managing entitlements tends to become time consuming and prone to errors. The task of translating applicable security policies into security implementation becomes more complex when the organization has to deal with multiple CSPs that are not hosted within the local environment, as most CSPs limit the amount of options to change an implementation. The inability to control the entitlements can lead to breaches of the segregation of duties or violations of security policies related to identity and access management.

33.4.5   Provisioning

The provisioning and deprovisioning of user accounts – the responsibility of the CSP – has to be performed quickly and accurately. End users generally have no control over these processes. Incorrect deprovisioning of user accounts can result in unauthorized access to the cloud resources by users that should no longer have access. At the same time, incorrect provisioning can make it impossible for authorized users to access their cloud services.

33.4.6   Monitoring and Audit

When using cloud computing, the data and assets of the organization or individual are stored at the servers of the CSP. In addition, IAM services could be (partly) managed by the CSP. As such, it becomes very important for cloud users to be able to monitor or audit access to their cloud resources. Without proper monitoring in place it is difficult to detect unauthorized access to data. Plus, in case of a technical problem with a cloud service, the cause of this problem cannot be easily found without using monitoring. Users are mostly dependent on the CSP to determine the extent to which they are able to monitor the cloud services that are used, and fix problems adequately.

For organizations these risks stretch further. Most organizations have to perform periodical audits on their processes, systems, and networks, as prescribed by laws and regulations. However, the organization cannot always perform an audit at the CSP, which runs a part of the IAM processes and stores organizations' data and assets. This could result in noncompliance with laws and regulations.

33.5 Considerations

From IAM perspective, what considerations should individual users and organizations keep in mind prior to shifting its data and IT assets to the cloud? When dealing with cloud adoption, it's key to find a CSP that incorporates robust IAM capabilities. This can be achieved by defining IAM requirements enforcing those requirements and by monitoring and auditing.

33.5.1 Define

An enterprise would predictably have much broader spectrum of IAM requirements than an individual user. For example, individuals might not go beyond deciding whether to find a CSP that provides two-factor authentication. At the same time organizations will need a larger pack of security measures, for example continuous security monitoring and incident response in order to ensure the enterprise business data is compliant with laws and regulations, accessible at all times and stored in a secure and controlled fashion.

In order to define the IAM requirements, data and asset classification should take place in combination with risk analysis. Data and asset classification are important to determine what an organization or individual wants to protect and to understand where their "crown jewels" are. Risk analysis helps to consider all the threats arising from off-premises data processing and storage, multitenant architecture of cloud environments, and dependence on public networks.

By assigning a risk rating to classified data and assets it will become easier for organizations and also individual users to determine what level of security and what IAM requirements are demanded from the future CSP.

The range of IAM requirements can vary significantly depending on the outcome of risk analysis. For instance, it might be required to verify whether or not a CSP complies with the applicable laws and regulations. If the CSP is unable to store organizational data within the region as required by law, the CSP might not be an option to consider. Another requirement that could be imposed on a CSP might be to use secured authentication mechanisms and authorization models, for example to provide authentication to various resources and data via a central authentication point. The technology used by the CSP also has to be reviewed for compatibility with the devices, for example the CSP should be able to support the preferred authentication mechanisms of users.

33.5.2 Enforce

The defined requirements will help organizations or individuals to ensure a sufficient level of control over IAM and eventually select the most suitable CSP(s). Requirements should be either covered in existing agreements or added to new agreements with the CSP. Individual users can refer to the general agreements and read these before accepting the conditions. Organizations can sign an agreement or contract that will include details about compliance with laws and regulations and the use of security requirements. For example, if the organization agrees to perform continuous monitoring, it has to be governed by an agreement or contract stating that the CSP should maintain this requirement. Another example can be to agree that the CSP does not make changes to the authentication mechanisms without informing the organization using the cloud services beforehand.

33.5.3 Monitor and audit

When the cloud service has been adopted, it is recommended that use of and access to the CSP should be monitored or audited on a periodical basis. For individual users it may be sufficient to review the CSP settings, read the news and take note of changes to the general agreements. Organizations should focus on monitoring and periodically auditing CSPs. Using a "right to audit" will allow an organization to verify that IAM controls are in place and effective and that the CSP is compliant with the applicable laws and regulations. In some cases a "right-to-audit" is not possible, in these situations third party certifications or assurance can help to gain more trust in the effectiveness of operation by the CSP. Periodical monitoring and auditing will help to find the causes of technological issues and verify changes to the mechanisms and errors made during IAM processes. This is an essential process to manage identified IAM risks caused by the nature of a cloud computing environment.

33.6 Conclusion

Cloud computing is susceptible to new methods of attack, making users vulnerable to potential cyber security incidents. One of the key risks lay with the digital identities that individuals and organizations use in order to access their cloud resources.

IAM refers to the processes, technologies and policies that manage access of identities to digital resources and determine what identities are authorized to do with these resources. Users require the ability to create, remove, or adjust user accounts within an application. Moreover, users need a measure of authentication (e.g. username/password, smartcard, biometrics) to prove their identity. The key processes, technologies and policies are gathered in the 'Enterprise IAM functional architecture'.

In this chapter, we described three interfacing models to use IAM in a cloud computing environment. The first model leaves the consumers and organizations in charge of their in-house IAM solution and connect this to the CSP. In the second model, an IdSP is used to allow consumers or organizations to authenticate – prove their identity – in order to access the cloud. The third model concerns consumers and organizations that fully delegate IAM processes to a CSP and make use of IAMaaS. Often, one or more standards or protocols such as OAuth, OpenID Connect, SAML, SPML or XAML are used for sharing authentication and authorization data with the applicable CSPs. In order to manage risks it is essential to define and enforce relevant IAM requirements prior to adoption a cloud service. After the cloud service has been adopted it is recommended that use of and access to the CSP should be monitored or audited periodically.

The demand for cloud applications, storage, and computing power will continue to grow, as will the cloud threat factor, and this will strengthen the demand for solid IAM and also create business opportunities for new IAM solutions.

Additional Resources

Blount, S. and Maxim, M. (2013) *CA Technologies Strategy and Vision for Cloud Identity and Access Management*. CA Technologies, Islandia, NY.

Carter, M. (2013) *Secure Identity in Cloud Computing*. Aerospace Corporation, http://gsaw.org/wp-content/uploads/2013/06/2013s11b_carter.pdf (accessed December 27, 2015).

Chung, W. S., and Hermans, J. (2010) *From Hype to Future: KPMG's 2010 Cloud Computing Survey*, KPMG Advisory, Amstelveen, Netherlands.

Cloud Security Alliance (2011) Security Guidance for Critical Areas of Focus in Cloud Computing v3.0, https://cloudsecurityalliance.org/guidance/csaguide.v3.0.pdf (accessed December 27, 2015).

Gopalakrishnan, A. (2009) Cloud computing identity management. *SETLabs Briefings* 7(7), 45–54.

Harauz, J., Kaufman, L. M., and Potter, B. (2009) Data security in the world of cloud computing. *IEEE Security and Privacy* 7(4), 61–64.

Jansen, W., and Grance, T. (2011) *Guidelines on Security and Privacy in Public Cloud Computing*. Special Publication 800-144. National Institute of Standards and Technology, Gaithersburg, MD, http://csrc.nist.gov/publications/nistpubs/800-144/SP800-144.pdf (accessed December 27, 2015).

Krelzman, G. (2011) *IAMaaS Adoption is Increasing*, Gartner, Stamford, CT.

Mather, T., Kumaraswamy, S., and Latif, S. (2009) *Cloud Security and Privacy: An Enterprise Perspective on Risk and Compliance*, O'Reilly Media, Inc., Sebastopol, CA.

Perkins, E., and Glazer I. (2013) *Identity and Access Management Key Initiative Overview*, Gartner, Stamford, CT.

Sturrus, E. (2011) Identity and Access Management in a Cloud Computing Environment. Master thesis. Erasmus University, Rotterdam.

Sturrus, E., Steevens, J. J. C., and Guensberg, W. A. (2012) Access to the cloud: identity and access management for cloud computing. *Compact*, http://www.compact.nl/artikelen/C-2012-0-Sturrus.htm (accessed December 27, 2015).

34

OAuth Standard for User Authorization of Cloud Services

Piotr Tysowski

University of Waterloo, Canada

34.1 Introduction: The Need for Secure Authorization

Many kinds of data are increasingly being outsourced to cloud computing systems because of the many benefits this brings. It is often more economical for data to be uploaded to and stored in the cloud by a client, as the client is typically charged only according to the actual storage space used. The cloud storage can instantly scale to match increased storage requirements whenever necessary. The outsourced data may also be shared readily with other clients and services if it is centrally located within a cloud. The originators of the data, especially mobile devices, may have limited local storage means and connectivity options available for retaining and sharing their data. Data that is stored within the cloud is easily accessible and secure due to automatic replication and backup mechanisms.

With the proliferation of cloud-based services, there is a need for applications running on the cloud or other servers to access user data stored in the cloud. However, security is a prime concern, as cloud systems share computing resources with multiple clients. It is important to safeguard adequately client data that is stored in the cloud from security threats by allowing only authorized parties to access it. Security policies may also dictate what data is accessible by whom, and for what amount of time, to mitigate risk. Nevertheless, such access may be frequent, as applications from multiple vendors may interoperate to provide useful services.

For instance, consider a user that wishes to generate, store, and process multimedia content in the cloud. A user may capture photos and videos on a mobile device or workstation and upload them to a multimedia storage service hosted in a public cloud to enjoy highly scalable capacity. Storing the media library in the cloud ensures safekeeping of the content and the ability to share it with others. A storage application in the cloud will

Encyclopedia of Cloud Computing, First Edition. Edited by San Murugesan and Irena Bojanova.

manage access to the library and provide indexing and search functions. The user, however, may wish to use a separate and complementary third-party cloud service that performs some postprocessing function on the content. This may include performing image enhancement or video editing, posting to a social site, and dispatching the content to a printing or disc-burning service. The user must grant permission for the post-processing site to access the content from the storage service. The two services may be completely independent and thus require user facilitation to collaborate.

It is unwise for the user simply to share secret login credentials with the postprocessing site because granting such unconditional access to the entire media library would be considered insecure. If the postprocessing site proves malicious, or an attacker hacks into its password store, then the media collection of the user would be placed at significant risk. If the user elected to choose a new password following the storage access, then it would need to be shared with all other sites that need access to the files of the user, which would be onerous. The alternative, requiring the user to transfer the content between cloud services manually, by downloading the files of interest from one then uploading to another as needed, is impractical in the long run.

Unfortunately, existing identity management systems prove insufficient in this scenario, where the goal is not for users to authenticate their identity with the photo sharing site; instead, the user wishes to authorize the postprocessing service to access content from the storage service directly, in a way that protects the account of the user and limits access to the data specified by the user. This authorization process requires its own dedicated host of solutions.

34.2   History of OAuth

The problem of secure authorization emerged after the adoption of OpenID, a federated identity management system. OpenID solved the problem of authentication (proving the user's identity to gain access to a service). It allows a user to log into a site or cloud service by being verified through another trusted service that acts as a trusted OpenID provider. This technique is a form of federated authentication, which allows any third party to authenticate a user based on an existing account with a single identity provider. The use of such authentication precludes the user from having to enter a separate user ID and password to log into every site that is visited. Most users tend to reuse the same password on multiple sites, which gives rise to an unnecessary security vulnerability.

The difficulty is that OpenID, by itself, does not address the problem of authorization, where one site gains access to the resources of another through a process that is facilitated and controlled by the user. Authorization alleviates the need for users to share passwords with third-party applications to allow access to their protected resources resident in the cloud. In fact, contemporary cloud services are often interoperable and must share client data; hence, they require both authentication and authorization functions.

Based on such considerations, the authentication protocol OAuth was born. Its basic purpose is resource API access delegation, which is not found in OpenID. The OAuth 1.0 protocol was initially formed by a working group, and the protocol was incorporated into the Internet Engineering Task Force (IETF) as a standard.

The OAuth 1.0 protocol evolved into its follow-up version: OAuth 2.0. The WRAP Web Resource Authorization Protocol (WRAP) profile of OAuth 1.0, which is now deprecated, became the basis for OAuth 2.0. The focus of OAuth 2.0 is primarily on simplicity of implementation for the client developer. New features introduced in 2.0 include the specification of authorization flows for various application contexts, including desktop, Web, and mobile devices. It defines a framework, rather than a strict protocol, with many options available with tradeoffs between convenience and security. Current publications include the specification of the core framework as well as the use of bearer tokens that are integral to the protocol. The OAuth 2.0 specification is being developed as an open Web standard for secure authorization by the IETF OAuth WG (Working Group).

34.3 Overview of OAuth 1.0

Third-party client applications use OAuth 1.0 to access protected resources on behalf of the user, who adopts the role of the resource owner. This is done without the need to divulge and exchange credentials such as a user name and password. In order to achieve authorization, OAuth provides an access token whose function resembles a valet car key. The token can be exchanged for any supported assertion that is compliant with a resource API. In practical terms, this includes querying for information and gaining access to protected files.

The high-level workflow of OAuth 1.0 is now presented, showing how a client application requests a limited access token to access resources on the behalf of the user. Refer to Table 34.1 for an explanation of the terminology used throughout. Note that all communication in the example occurs through Hypertext Transfer Protocol (HTTP).

Suppose that the end user logs into a third-party client, and wishes to authorize the client to access the resources of the user, which are resident on a server hosted on the cloud. For instance, the user, a photographer, might wish to allow an image postprocessing site to gain access to his photo collection stored in the cloud. The following steps are taken, as shown in the general workflow in Figure 34.1.

1. To carry out the authorization task, the client, which is the postprocessing site, first sends its client credentials, which were previously created and agreed upon, and requests temporary credentials from the server, which is the cloud on which the photos are stored. The server validates the request, and supplies the temporary credentials in the form of an identifier and shared secret. The purpose of the temporary credentials is to identify the access request from the client in further stages of the protocol.
2. Next, the client requires approval from the user, the photographer, to access the desired resources, which are the photos. To obtain this approval, the client automatically redirects the user to an authorization service on the server. The user signs into the server as the resource owner, and approves the granting of access, to the client, to the protected resources in the cloud, which are the images to be processed. The access request is identified by the temporary credentials obtained earlier.
3. Once approval occurs and the client is notified of it, the client then uses its temporary credentials to request a set of token credentials over a secure channel to the server. The server validates the request and replies with a set of token credentials, which will now allow the client access to the protected resources on the server. Thus, the image postprocessing site gains access to the photo collection in the cloud without knowing the password to the collection.

Table 34.1 *Definitions*

Client	A third-party HTTP client, or consumer, that initiates authorization requests in order to access the resources of an end user. The requests are authenticated under OAuth.
Server	An HTTP server, or service provider, that accepts authorization requests, typically located in the cloud.
Protected resource	A resource belonging to the end user for which authorization may be granted. A resource will typically consist of data that is permanently stored in the cloud.
End user	The end user, or resource owner, who normally accesses and controls protected resources in the cloud by using credentials, such as a user name and password, to authenticate with the server.
Credentials	A pair of a unique identifier and matching shared secret, which are used in various communication exchanges in the OAuth protocol. The credentials are used to identify and authenticate the client, the authorization request made by the client, and the granting of access in response.
Token	A unique identifier utilized by the client that associates authenticated resource requests with the resource owner who provides authorization.

4. The server continues to validate all data access requests from the client until the authorization window automatically expires, or until the user revokes access by the client. Thus, the postprocessing facility continues to retrieve images until its task is complete.

Once access has been granted by the server, the temporary credentials that were previously used by the client to obtain the token are revoked. Furthermore, a token can be issued with a limited lifetime, and may be independently revoked by the resource owner before it expires. Not allowing a token to be left permanently active mitigates the overall security risk for the user, and is a recommended practice.

Additionally, tokens may be issued to the client with restrictions on the scope of data that may be accessed. It would not be possible to specify limitations on the duration and the degree of permission if OAuth were not

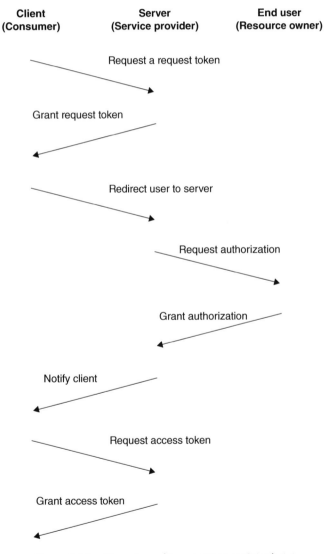

Figure 34.1   *Overview of transactions in OAuth 1.0*

used. The alternative of supplying owner credentials directly to the client would imply unrestricted access, which is unacceptable in most applications.

An OAuth-compliant protocol, much like an OpenID-compliant one, occupies a layer on top of HTTP. It typically relies upon Web redirection logic. For instance, the user is redirected by the client to an authorization endpoint on the server to grant access to resources, and then the client is notified of completion of the transaction via a callback mechanism.

34.4   Comparison of OAuth 1.0 to OpenID

In contrast to the authorization function provided by OpenID, the purpose and mechanism of the authentication function of the OpenID protocol are fundamentally different. It is instructive to compare the workflow of OpenID with that of OAuth 1.0 to understand how the former addresses the problem of user authentication, and the latter addresses the problem of resource authorization.

The high-level workflow of OpenID is now presented, where users provide their self-supplied identity to a third-party client application, called a relying party, through a certificate mechanism. To do so, the client first asks users for their identity, and the end user provides a reference to a known OpenID provider in response. The client then requests a referral for the user from the identity provider, which is trusted. Upon receiving the referral, the client completes the authentication of the user. The transaction sequence is shown in Figure 34.2.

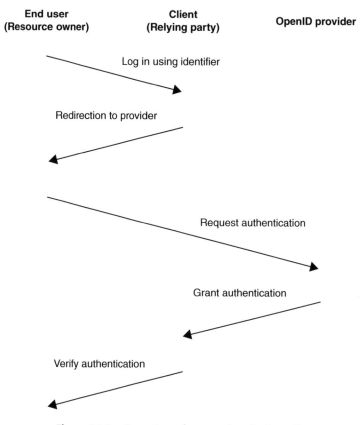

Figure 34.2   *Overview of transactions in OpenID*

OpenID authentication provides a way to prove that an end user owns an identifier without the user having to provide credentials such as a user name, e-mail address, or password. Its goal is to enable digital identity management in a portable and decentralized manner. However, proving the identity of the user to the client is not equivalent to allowing access to the user's resources, which reside on another server. Thus, OpenID provides an authentication function that is complementary, but not equivalent to, the authorization function of OAuth.

In many applications, the two protocols are implemented in tandem to provide both identity control and resource access. Recall the example of a multimedia library stored on a cloud server. Users could first use OpenID to authenticate their identity with the postprocessing client by using a known OpenID provider. Once authenticated, users could then allow access to the multimedia library by authorizing the postprocessing client using OAuth.

It is not necessary, however, to utilize OpenID to address the design issue of OAuth being intrinsically unable to authenticate the user. To reduce the complexity of implementation, it is possible to rely solely on an OAuth implementation and authorize access to additional resources such as the identity of the user, which is not otherwise provided by default in the basic protocol.

34.5   Overview of OAuth Version 2.0

The OAuth protocol underwent a significant transformation in its second iteration. One of the principal differences between versions 1.0 and 2.0 of the OAuth standard concerns the underlying security mechanism that is mandated. The security of the OAuth 1.0 protocol fundamentally relies upon signatures. In order for the client to access protected resources via an API, the client must generate a signature using a token secret that is shared with the server. The server will generate the same signature, and grant access if the signatures match. Although this scheme is considered secure, in practice it requires very careful implementation, which is prone to error.

Changes to the security model were deemed necessary in the next version of OAuth in order to reduce the complexity of development and to lessen the chance of vulnerabilities being unintentionally introduced that could be exploited. Thus, the OAuth standard evolved from a reliance on signatures in version 1.0 and other cryptographic features to the use of Secure Sockets Layer (SSL) or Transport Layer Security (TLS) in version 2.0. SSL/TLS is a cryptographic protocol that is required for all communications in OAuth, including during token generation and usage. Client-side cryptography through the use of certificates, as found in OAuth 1.0, is no longer required or recommended.

The workflow in OAuth 2.0 is now generalized as shown in the sequence of interactions in Figure 34.3:

1. A client requests authorization from the end user, acting in the role of the resource owner. Such a request can be made directly to the user or via an authorization server acting as an intermediary.
2. The client receives an authorization grant, in one of the various forms supported in the new specification. Authorization codes are transmitted over a secure channel.
3. The client authenticates with the authorization server and presents the authorization grant. If the grant is valid, the authorization server issues an access token to the client.
4. The client proceeds to access the protected resource from the resource server by presenting the access token, which is validated before the resource is served up.
5. The client continues to access the protected content on the resource server using the access token while it remains valid.

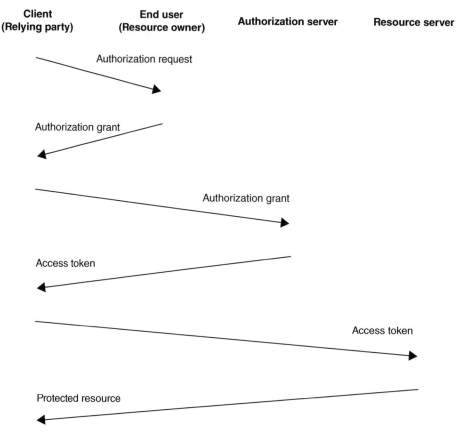

Figure 34.3   *Overview of transactions in OAuth 2.0*

There are four types of grants available in OAuth 2.0. An extensibility mechanism allows additional types of grants to be defined. The four basic ones are as follows:

- Authorization code: The client directs the resource owner to an authorization server. The authorization server authenticates the resource owner, obtains access authorization, and directs the resource owner back to the client with an authorization code. The authorization code is later used to obtain the access token. This technique is useful from the point of view of security because the client is properly authenticated, and the access token is transmitted directly to the client without being exposed.
- Implicit: The client is issued an access token without the use of an authentication code. Although this grant is useful in reducing the number of required transactions in the protocol, it is done at the cost of a lack of client authentication.
- Resource owner password credentials: The access token is provided on the basis of the client supplying the password credentials of the resource owner. The client must have direct access to the password, and therefore must be highly trusted by the user.
- Client credentials: The credentials of the client can be used for the authorization grant. This method is useful in allowing the client to gain access to its own protected resources, in the special case where the client is also the resource owner.

34.6   New Features in OAuth 2.0

Numerous new features are supported in OAuth 2.0, including new application contexts. In some contexts, client credentials may be stored confidentially in a location that is inaccessible to the resource owner. On the other hand, they may need to be stored on the device belonging to the resource owner, in which case confidentiality and secure client authentication may not be achieved. For instance, a client may run on a Web server where the credentials are stored, or the client may execute as an application running natively or within a Web browser on a desktop or mobile device belonging to the user. In the latter case, the client is executed locally, and so the credentials could theoretically be extracted by the resource owner. To achieve security in this case, client credentials may need to be issued dynamically as needed, rather than being fixed, in order to protect them.

The access token itself is a credential used to access protected resources. As in OAuth 1.0, the token may be limited by a particular scope and lifetime, which are enforced by the resource server. To permit continuation of access, OAuth 2.0 introduces the optional use of refresh tokens. Refresh tokens are credentials used to obtain access tokens once they expire or become invalid. Another reason to use refresh tokens is to obtain additional access tokens with narrower scope, without requiring additional authorization by the resource owner.

Refresh tokens are optionally issued alongside access tokens by the authorization server after authentication of the client and validation of the authorization grant. The client makes repeated requests against the resource server using the access token. Once the resource server signifies that the access token has become invalid, the client authenticates with the authorization server and presents the refresh token obtained earlier. The authorization server validates the refresh token and issues a new access token. The server may also supply a new refresh token so that the cycle of token requests can be repeated.

Additional protocol suites have been devised to address shortcomings in OAuth 2.0, including its inherent lack of identity control. For instance, OpenID Connect provides an identity layer on top of OAuth 2.0 to verify the identity of the user and obtain basic profile information. It allows clients of all types to query information about authenticated sessions and end users. OAuth 2.0 capabilities are actually integrated into the OpenID Connect protocol.

34.7   General Security Considerations

In OAuth 2.0, client-based credentials are no longer supplied by the client as part of the authorization process, as they were in version 1.0 of the protocol. Consequently, the client is no longer authenticated against the authorization server. Furthermore, the absence of mandatory client credentials in version 2.0 means that resource access tokens are no longer bound to clients as they were in version 1.0. The new protocol relies upon the concept of bearer tokens that do not require clients to prove that they should possess them in order to use them. Tokens may be misappropriated, and so the client must exercise care in choosing which resource owner to trust the tokens with. Otherwise, a token that falls into the wrong hands could result in resources being illegitimately accessed for the duration of the token's existence without means of revocation. As authorization codes and access tokens are sent in plain text, communication in OAuth 2.0 must be protected by SSL/TLS, which must be correctly configured in the first place.

34.8   Possible Threats and Countermeasures

Various threats may be directed against OAuth 2.0 clients. For instance, an attacker could steal and then replay a valid refresh token (that is, resend it to the authorization server pretending that it is the original). Or, the attacker could pose as a valid client and obtain a token on its behalf. By doing so, the attacker could illegitimately bypass client authentication and gain an access token permitting access to a resource without the knowledge and permission of the true client.

Many kinds of countermeasures may be employed to limit the efficacy of such attacks. For instance, the client identity associated with a refresh token may be validated with every refresh request, to be able to detect the presence of an attacker. Additionally, refresh tokens may require special protection to avoid being read and stolen by an attacker, such as being recorded to secure storage. Furthermore, the OAuth 2.0 framework allows a token to be revoked by the user if the user discovers that the token has been compromised at any point.

Access tokens may also be similarly revealed to an attacker if improperly secured. Similar safeguards apply in this case, including the use of secure or transient memory, as well as limitations on the permitted access scope and lifetime of tokens, to limit the potential damage that may occur with a compromised token. To prevent eavesdropping on access tokens transmitted over the network, from the authorization server to the client, transport-layer security that ensures end-to-end confidentiality is required.

Phishing attacks are a potential problem not only with regular web-site passwords, but also OAuth artefacts. A client could be tricked into revealing credentials by an attacker appearing to be a credible and trusted party in some form of attack that relies upon social engineering. However, such an attack appears to be less likely in principle than one that seeks to obtain the original user password; OAuth is deliberately designed to not reveal the password in the first place.

Attacks may also originate at the server end. For instance, an attacker may illegitimately pose as an authorization server, and OAuth provides no facility for verifying its authenticity. Another possibility is that a malicious client could pretend to be a valid client in order to obtain authorization; to prevent this scenario, it is necessary for the authorization server to authenticate the client if possible, after the user has authorized access. Many other attacks and countermeasures have been specified by the IETF, some of which are very unique to OAuth 2.0.

In general, it is recommended that transport-layer security such as TLS be used for client requests, as OAuth 2.0 provides no guarantee of request confidentiality. Furthermore, server authentication such as Hypertext Transfer Protocol Secure (HTTPS) can be used to authenticate the identity of servers. Credentials must be securely stored to prevent compromise, and the resource owner should remain in control of the authorization process at all times. The security of OAuth 2.0 continues to be actively researched by industry and the academic community. It is reasonable to expect that new attacks and countermeasures may be discovered, and new recommendations on best practices proposed, in coming years.

34.9   Application Support

Applications supported by OAuth are numerous and continue to grow. Major commercial vendors, including Facebook and Google, have implemented and endorsed the new version 2.0 specification of OAuth. Other popular service providers such as Microsoft and LinkedIn are also providing support for the protocol. Facebook's Open Graph social graph requires the use of OAuth's app access tokens. Like OpenID, OAuth is a decentralized protocol, in that no central authority manages information and authorization for all Internet users. Any server can elect to support OAuth, and likewise any entity can request to act as a resource consumer.

Possible applications of OAuth 2.0 in a cloud computing context are too numerous to list. Possible examples include: an advertising service posting a purchase transaction to a social network's feed, a consumer recommendation service performing big data analysis on a data warehouse, or an auditor accessing cloud server logs to test for compliance.

34.10   Criticism

Although OAuth 2.0 is gaining widespread use and backing in industry, its current state of specification and apparent trajectory has had its share of criticism. Areas of particular concern include the following:

- Interoperability. The standard resembles a highly extensible framework, resembling a blueprint, which leaves many implementation questions open by design. Many implementation details such as the capabilities of entities in the system, the format of tokens, and processes such as registration and discovery, are not fully defined. The lack of strict protocol definition is expected to lead to potential challenges in achieving interoperability between implementations as well as permitting backward compatibility. In fact, clients may need to be configured against specific authorization and resource servers in order for the authorization to work. For now, the interoperability issue has been left open for future work in the form of recommended extensions and example profiles that will be published eventually by the working group.
- Security. Another key challenge relates to the apparent strength of security present within OAuth 2.0. It has been argued that weaker cryptographic controls are found in the second version due to a lack of signatures and client-side encryption. The absence of mandatory client credentials means that resource access tokens are no longer bound to clients, leading to a tradeoff between flexibility and security. The new bearer tokens expire and must be refreshed, and no explicit revocation mechanism is specified. The new token handling logic results in its own complications in implementation. It remains to be seen whether security issues can be adequately enforced through compliance with the standard, without greater exactness in the protocol specification.
- Performance. Although SSL/TLS has gained maturity, its use in OAuth 2.0 requires configuration of a sufficient level of security, and may introduce performance overheads that did not exist previously.

It is expected that compatibility work, threat modeling, security analysis, performance analysis, and recommendations on security practices will continue to be a primary focus of work in defining the OAuth protocol.

34.11   Conclusions

OAuth is a highly useful standardized service that enables secure authorization for Web services to access online resources. OAuth uses a token scheme that makes it unnecessary for the resource owner to share its credentials with the client seeking access. The OAuth standard has evolved from one based on signatures to one based on the use of SSL/TLS and bearer tokens, which simplifies client development. It is anticipated that, through these changes, OAuth will gain more widespread use. At the same time, the new OAuth version 2.0 is presented as a framework with many available options, leading to concerns about interoperability and security, and suggesting that further guidelines on its implementation and usage are needed for practitioners.

Additional Resources

Hammer-Lahav, E. (2010) *The OAuth 1.0 Protocol*. Internet Engineering Task Force (IETF), http://tools.ietf.org/html/rfc5849 (accessed December 28, 2015).

Hardt, D. (2012a) *The OAuth 2.0 Authorization Framework, Request for Comments: 6749*. Internet Engineering Task Force (IETF), http://tools.ietf.org/html/rfc6749 (accessed December 28, 2015).

Hardt, D. (2012b) *The OAuth 2.0 Authorization Framework: Bearer Token Usage, Request for Comments: 6750.* Internet Engineering Task Force (IETF), http://tools.ietf.org/html/rfc6750 (accessed December 28, 2015).

Lodderstedt, T. (2013) *OAuth 2.0 Threat Model and Security Considerations, Request for Comments: 6819.* Internet Engineering Task Force (IETF), http://tools.ietf.org/html/rfc6819 (accessed December 28, 2015).

OpenID Foundation (2007) *OpenID Authentication 2.0 – Final*, http://openid.net/specs/openid-authentication-2_0.html (accessed December 28, 2015).

OpenID Foundation (2013) *OpenID Connect Core 1.0 – Draft 14*, http://openid.net/specs/openid-connect-core-1_0.html (accessed December 28, 2015).

35

Distributed Access Control in Cloud Computing Systems

K. Chandrasekaran and Manoj V. Thomas

National Institute of Technology Karnataka, India

35.1 Introduction

In cloud computing or services computing, users access various resources or services after verification of their identity by the service provider. Access control is concerned with determining which user has which access rights towards a service or a resource. The aim of an access-control system is to protect the system resources against unauthorized or illegal access by users. Access control in the domain of distributed applications, in collaborative, distributed, cooperative environments like cloud computing, where several users access the resources and services, with different access rights, is called distributed access control. Different users have different access rights towards the available resources in the system, which need to be concisely specified and correctly enforced.

35.1.1 Need for Distributed Access Control in Cloud

Unless there is a foolproof access-control mechanism enforced in the cloud, users would be reluctant to use the cloud platform to meet their resource requirements and hence the cloud service providers (CSPs) and the consumers may not be able to make use of the advantages this paradigm offers. In the cloud environment, which uses service-oriented architecture (SOA), the service providers and the service consumers generally do not have a pre-established trust relationship between them. Therefore, the authentication of strange users and the authorization of their access rights are extremely important in handling the access requests of cloud service consumers.

Encyclopedia of Cloud Computing, First Edition. Edited by San Murugesan and Irena Bojanova.
© 2016 John Wiley & Sons, Ltd. Published 2016 by John Wiley & Sons, Ltd.

35.2   Features and Functionalities of DAC

Achieving distributed access control to safeguard resources, information, or data is of extreme importance in cloud computing, where sharing of digital resources with different sensitivity levels is involved.

35.2.1   Major Functionalities of DAC

Enforcing DAC makes cloud computing safe, secure and scalable. A distributed access control system in cloud computing involves the following basic functionalities.

35.2.1.1   Identification

This process establishes or assigns an identity with a particular user or subject within the cloud by methods such as giving a username to a specific user.

35.2.1.2   Authentication

This is the process by which the identity of an entity requesting some resources is verified by the CSP. The entity could be a person, process, program or software agent requesting access to some resources for its computation or task.

35.2.1.3   Authorization

This is the process by which the access rights or privileges of a user or entity are verified by the cloud system against predefined security policies in order to avoid unauthorized access by malicious users.

35.2.1.4   Accountability

This involves tracing or recording the actions performed by a user in the cloud system by collecting the identification details of the user, time of access, resource accessed, and so forth. The collected details are entered in a log file which could be stored for future security auditing purposes.

35.2.2   Major Features of the DAC

The DAC system in the cloud should provide the following additional features to the CSPs and cloud service consumers (CSCs).

35.2.2.1   Confidentiality

Confidentiality ensures that there is no unauthorized access to the data, information, or resources. Physical isolation of virtual machines and cryptography techniques can be adopted to achieve confidentiality.

35.2.2.2   Integrity

Integrity ensures that there is no unauthorized modification of data or information in the cloud system. In order to provide integrity to the user's data stored in the cloud, message authentication code (MAC) and digital signature (DS) methods could be used.

35.2.2.3   *Availability*

Availability ensures that the system resources or information are accessible to the legitimate user on time.

35.3   Distributed Access Control in Cloud Environments

Cloud computing is a service-oriented distributed computing paradigm, where users access various services and shared resources hosted by the service providers, to carry out their tasks efficiently. The access control of distributed resources is most important in securing the cloud scenario.

35.3.1   Access-Control Models and Approaches in Cloud Computing

Many researchers have been working in this area of access control and some of the work carried out by them are highlighted here. The work carried out in Wei *et al.* (2010) shows an attribute- and role-based access control (ARBAC) model. In order to access the services, service consumers provide their attribute information to the service providers. When the service providers receive the access requests, they determine whether to permit or to deny these requests according to their access-control policies. How to enforce access control on the numerous users who are not defined in the system in the distributed computing environment is discussed in Lang *et al.* (2007). Feng *et al.* (2008) propose a trust- and context-based access control (TCAC) model, extending the RBAC model, for open and distributed systems. Turkmen *et al.* (2011) presented a framework for the verification of run-time constraints and security properties for role-based access control (RBAC) systems, considering the dynamic behavior of users during an active session. Pereira (2011) describes a role-based access-control mechanism for distributed high-performance computing (HPC) systems where both users and resources can be dynamic and can belong to multiple organizations, each with its own diverse security policies and mechanisms. Gunjan *et al.* (2012) discuss the issue of identity management in the cloud computing scenario. Loss of control, lack of trust, and multitenancy issues are identified as the major problems in the present cloud computing model.

35.3.2   Authorization Requirements in Distributed Access Control

An inadequate or unreliable authorization mechanism will result in the unauthorized use of cloud resources and services. Hence, there is a need for an effective and secure distributed access-control architecture for multitenant and virtualized environments of cloud computing. While designing secure distributed access-control architecture for cloud computing, the authentication and authorization requirements to be considered include those listed in the following subsections (Almutairi *et al.*, 2012).

35.3.2.1   *Multitenancy and Virtualization*

Virtualization in the cloud helps the service provider to achieve multitenancy, which makes resource utilization effective through resource sharing. In distributed clouds, an untrusted client might use side-channel attacks in order to access data or information pertaining to other tenants, exploiting any virtualization flaws in the cloud environment. Side-channel attacks arise due to lack of authorization mechanisms for accessing the shared cloud resources, and hence access-control architecture should be made secure against these attacks.

35.3.2.2   Decentralized Administration

Decentralized administration is characterized by the principle of local autonomy and this is an important feature in cloud computing. In this case, each cloud service provider or its individual service has full administrative control over all of its resources. The access-control policies should be designed effectively by identifying the required authorization rules in order to allow fine-grained access control over the resources in multicloud environments.

35.3.2.3   Secure Distributed Collaboration

In cloud computing, the clients obtain required services by integrating multiple services either from the same cloud service provider or from multiple cloud service providers. The security infrastructure should allow the cloud service providers to share the services (infrastructure, platform, or software services) effectively either within the same CSP or among multiple CSPs. In the decentralized environment, the individual access policies of the CSP should allow such horizontal (resource sharing among different cloud service providers at the same service level) and vertical (resource sharing among different service levels of the same or different cloud-service provider(s)) policy interoperation for efficient service delivery. These collaborative access policies must be correctly specified, verified, and enforced. In this collaborative environment, a service-level agreement (SLA) is made between the CSPs, which give an assurance that services are provided according to mutually agreed rules and regulations.

35.3.2.4   Credential Federation

Single sign-on should be supported in the cloud environment where a user invokes services from multiple clouds. The access-control policies of the CSPs must support a mechanism to transfer a customer's identity credentials across the service layers of CSPs to access the resources and services.

35.3.2.5   Constraint Specification

In the collaborative model of the cloud computing environment, semantic, and contextual constraints should be verified to protect services and resources from illegal access. Semantic constraints such as separation of duties, and contextual constraints such as temporal or environmental constraints should be evaluated while determining the access rights of cloud customers. Semantic and contextual constraints should be specified in the access-control policy.

35.3.3   Distributed Access-Control Architecture

The distributed access-control architecture that incorporates the authorization requirements mentioned above consists of the following three components: virtual resource manager (VRM), access control manager (ACM), and the SLA established between the CSPs (Almutairi *et al.*, 2012).

35.3.3.1   Virtual Resource Manager (VRM)

This module manages the resource requirements of the cloud customers. It is a part of the access-control architecture of each CSP and this module handles the resource requirements of each service layer of a cloud. The VRM is responsible for the provisioning and deployment of the virtual resources. Local resources are directly handled by this module, and if the requested resources are from a different CSP (remote resource), as per the agreed SLA between the CSPs, the VRM contacts the corresponding entity of the remote cloud in

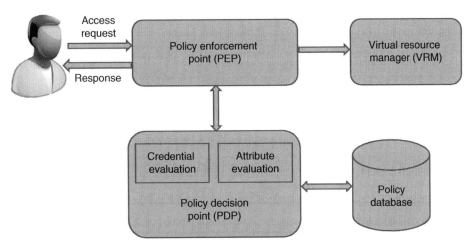

Figure 35.1   *ACM architecture*

order to provide access to the requested remote resources. The VRM monitors the deployed resources and allocates or releases them in order to ensure quality of service as per the agreed SLA between the CSPs.

35.3.3.2   *Access-Control Manager (ACM)*

This module of distributed access-control architecture is present at each service layer of the CSP to enforce the access-control policy at the corresponding layer. As shown in Figure 35.1 (Almutairi *et al.*, 2012), the architecture of ACM consists of the following components:

- policy decision point (PDP);
- policy enforcement point (PEP);
- policy database.

Access requests of the cloud customers, which include the requesting subject, the requested service or resource, and the type of permissions requested for that service or resource (such as read or write privileges) are given to the PEP. The access request might also include the identity credentials needed for authentication and authorization. The PEP extracts the authentication and the context information from the authorization request and forwards them to the credential evaluator and attribute evaluator of the PDP. The PDP takes the final decision regarding the access request by evaluating the policy rules from the policy database considering the attributes and identity credentials of the requesting user. The PEP receives the decision from the PDP regarding the access request, and subsequently it either grants or denies the request. If it is permitted, it is forwarded to the virtual resource manager (VRM) of the CSP for the deployment of requested resources.

35.3.3.3   *Role Mapping*

If the access request from the user contains an authorization credential, the credential evaluator, after verification of the identity, assesses if the role corresponds to a local role (if RBAC is the deployed access-control model) and assigns the local role to the user based on the user-to-role assignment rules stored in the RBAC policy base. The process of user-to-role assignment takes its input from the context evaluator. If the role does not correspond to a local role, it is assumed that this is a single sign-on request and hence requires role mapping by a relevant SLA. Role mapping is a function that maps a local role to a role in a remote cloud and grants access to all the mapped role's permissions. Subsequently, the user acquires the access rights of the locally assigned role or that of a mapped role in a remote cloud. The ACM module invokes the SLA if the requested resources are stored on a remote CSP.

35.3.4   Distributed Authorization Process

In the distributed access-control architecture, three types of interoperations related to authorization flow can occur at various layers of the CSPs as shown in Figure 35.2 (Almutairi *et al.*, 2012).

Type 1 depicts a horizontal (peer-to-peer) interoperation between the same service levels of different cloud providers; type 2 represents a vertical interoperation between service layers within the same cloud; and type 3 indicates a crosslayered interoperation between different clouds at different service layers. Both type 1 and 3 interoperations involve SLAs among the participating clouds for their effective operation. As shown in the figure, each service layer of the CSP has both the ACM module and VRM module to deal with the authorization and resource management at the corresponding layer. Effective authorization mechanisms should be implemented by the ACMs at all the service layers in order to make the interoperation secure. The VRM module that is present at each service layer deals with the resource provisioning and de-provisioning at the corresponding layer.

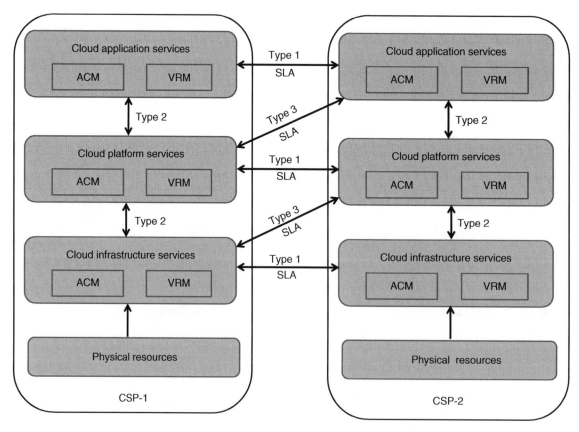

Figure 35.2   *Distributed security architecture of the multicloud*

Request rejected

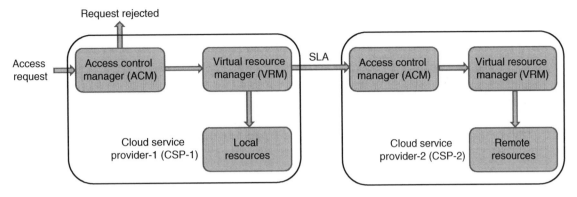

Figure 35.3   *Distributed authorization process in the multicloud*

35.3.4.1   *Authorization process in the cloud environment*

The overall distributed authorization process is shown in Figure 35.3 (Almutairi *et al.*, 2012).

When a service consumer requests a service or virtual resource, the request goes to the local ACM of the CSP. If the ACM grants this request, it forwards the request to the local VRM in order to deploy the required resources. If the required resources are in a remote cloud, the local VRM contacts the ACM of the remote cloud as the appropriate SLA is already established between them. After verifying its own access policies and constraints, if the access request is permitted, the ACM of the remote cloud contacts its local VRM to allocate the requested resources. Finally, the VRM identifies and configures the required resources for the cloud customer.

35.4   Access-Control Policies and Models

Any computing system enforcing access control over its resources should deal with access-control policies, access-control models, and access-control mechanisms (Hu *et al.*, 2006).

35.4.1   Access-Control Policies

In the cloud model, access-control policies are high-level statements that specify who is allowed to access the information or the resources in the system, with what access rights, and also the conditions under which the access is allowed. An example for a security policy is "separation of duty" (SoD), which prevents a person from assuming more than one role or associated operation in the cloud at a time as that could result in security breaches when abused.

35.4.2   Access-Control Mechanisms

An access-control mechanism is the means by which the access-control policies of a computing system are implemented or enforced. One simple example could be the access control lists (ACLs) maintained in the system.

35.4.3   Access-Control Models

Access-control models bridge the gap between an access-control policy and the corresponding mechanism. Security models formally present the security policy applicable to a computing system, and also discuss and analyze any theoretical limitations of a particular system in ensuring the proper access control.

35.4.4   Types of Access-Control Policies

Access-control policies implemented in any cloud model could be broadly classified into discretionary access-control (DAC) policies and nondiscretionary access-control (NDAC) policies.

35.4.4.1   Discretionary Access-Control (DAC) Policies

In the case of DAC policies, object access can be controlled as per the discretion of the owner of the object or resource. For example, the owner of a file decides who else can use that file and with what access rights (read, write, execute, or delete). Discretionary access-control policies are implemented using identity-based mechanisms such as ACL.

35.4.4.2   Nondiscretionary Access-Control (NDAC) Policies

This type of access-control policy depends on a rule-based access-control mechanism for the purpose of implementation. Generally, all the access-control policies other than the DAC policies are considered as NDAC policies. In this case, access-control policies have associated rules that are not designed as per the discretion of the owner or user; instead they might be based on organization-specific rules. For example, history-based separation-of-duty (SoD) policy regulates the number of times a subject can access the same object.

35.4.5   Types of Access-Control Models

Access-control models should make sure that there is no leakage of access permissions to an unauthorized principal. Some of the access-control models used in the cloud computing paradigm are given below.

35.4.5.1   Role-Based Access-Control (RBAC) Model

This model is suitable for organizations where a static hierarchy is maintained and the members have defined roles such as manager, accountant or clerk, and specific access rights are associated with each role. Hence, the resources that the users are allowed to access are decided by their roles. Users are mapped to roles and roles are mapped to privileges (Ferraiolo *et al.*, 2001). The basic RBAC model is shown in the Figure 35.4. DBMS like SQL also use role- and rule-based access-control mechanisms to achieve distributed access control effectively.

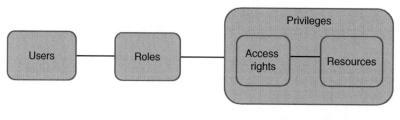

Figure 35.4   *Basic role-based access-control model*

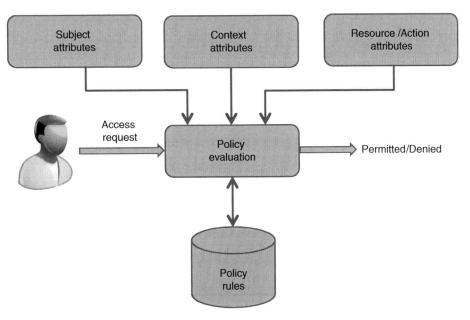

Figure 35.5   *Attribute-based access-control model*

35.4.5.2   *Attribute-Based Access-Control Model*

In this model, access to the resources is allowed or rejected depending on the attributes of the subject, context, resource and action. Subject attributes include user identity and group membership. Context attributes include time and location of access request made. Resource attributes include the resource identity, and the action represents the type of operation requested by the user, such as read or write. As shown in Figure 35.5, whenever an access request is initiated, the attributes pertaining to the access request are compared with the stored access-control policies and the policy evaluation module either accepts or rejects the access request.

35.4.5.3   *Risk-Based Access-Control Model*

In the risk-based model, real-time decisions are made to allow or reject a user request to access a particular resource, by calculating the risk involved in allowing the access against the perceived benefit of doing that. Figure 35.6 shows the risk-based access-control model in which the access requests are given to the policy enforcement point (PEP). The PEP contacts the policy decision point (PDP) for the access decision and the PDP takes the access decision by calculating the risk involved in the access request. In this case, calculation of the risk considers the attributes of the user, which also includes the trustworthiness of the requestor other than the identity information, context information of the access request such as time, location and the previous access history, and also the sensitivity level of the resource requested. Risk threshold levels are specified by the security policy from time to time, and the decision to grant or deny a request is taken based on that threshold value.

35.4.6   Selection Criteria of Access-Control Models

In cloud computing, which is a highly dynamic distributed computing system, the selection of a particular access-control model depends on factors such as the service delivery models, application architecture,

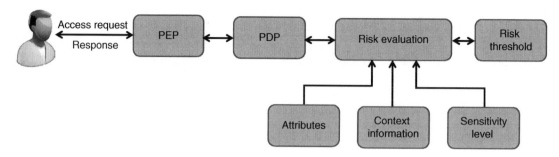

Figure 35.6   *Risk-based access-control model*

functionalities supported by the applications, sensitivity level of the application, the data stored, level of granularity of access control required, and so forth.

35.5   Identity Management in the Cloud

Identity management in cloud deals with the management of identities of various users in the cloud, which helps to provide better access control over the available resources in the system based on those identities.

35.5.1   Need for Identity Management (IdM) in the Cloud

In open service-oriented systems like the cloud, in many cases, the service providers and the service consumers are strangers. As they do not have pre-established trust between them, the service provider must be able to authenticate the unfamiliar users, and then determine whether the requestors have enough privileges to access the requested services. In cloud computing, a user or an organization may subscribe to services from multiple service providers. Proper identity management is the basis of stronger authentication, authorization, and availability features which are crucial in enforcing an effective access-control system in the cloud environment. The major functional requirements of IdM in the cloud environment are dynamic provisioning and deprovisioning, synchronization, entitlement and life-cycle management.

35.5.2   Identity Life-Cycle Management

Identity life-cycle management shows the various stages through which a user identity goes. The five stages in the identity life-cycle management (Mather *et al.*, 2009) could be provisioning and de-provisioning of user identities, authentication and authorization, self-service (how the user can maintain, update or reset credentials), password management (how the user password is stored in cloud), and compliance and audit (how the access is monitored for security purposes). Identity provisioning involves assigning the identity to a particular user by the CSP or the identity provider (IdP), and identity deprovisioning involves terminating the identity so that the specific identity is no longer valid for a given user.

35.5.3   Identity Providers

In cloud computing, instead of using separate credentials for different applications, cloud users can submit user-centric identity tokens to the CSPs. The identity providers are trusted entities in the cloud computing environment, which provide the identity tokens to the cloud users. These tokens could be used by the cloud users for services from the CSPs. Examples of IdPs include Ping Identity, and Symplified.

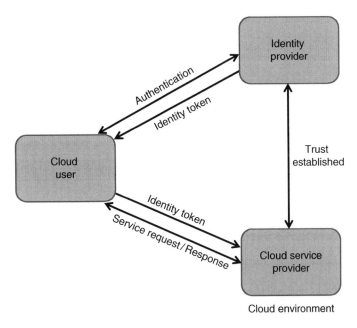

Figure 35.7   *Federated identity management*

35.5.4   Federated Identity Management

The users in a cloud federation do not need to use separate credentials for each cloud service provider or service they subscribe to; instead, they can use the identity token issued by the identity provider. The users can submit the security tokens (normally SAML assertions) issued by the identity provider to the service providers in the cloud federation. This is both efficient and secure, and relieves the users of the multiple credentials problem when accessing services from multiple cloud service providers. As shown in Figure 35.7, federated identity management in the cloud involves the federation between following entities: client, relying party or service provider (SP), and the identity provider (IdP). Before accessing the service from the SP, the service consumer has to be authenticated as a valid user by the IdP. The client contacts the identity provider listed in the trusted domain of the SP to obtain the identity token, and that token is submitted to the SP in order to obtain the requested services. As the SP and IdP are part of the federation and they have mutual trust, the user is allowed to access the services from the SP after successful authentication. Because of the federated identity management, service providers can concentrate more on their core services because the identity management operations are taken care of by the identity provider.

35.5.5   Single Sign-On (SSO)

This is the process of accessing more than one service from same or different service providers, by logging into the system only once. Identity federation supports single sign-on (SSO) as the users are able to access multiple services from the same or different CSPs using the same identity token issued by the identity provider. Figure 35.8 shows an overview of the SSO process. In this case, a user first accesses the services from service provider-1 by logging onto it. As shown in the figure, the user is authenticated by the identity provider. All the three service providers have mutual trust established with the identity provider. In this case, the user is not required to enter the identity credentials again to access the services from service provider-2 and service provider-3 as there is a SSO system established between the three service providers.

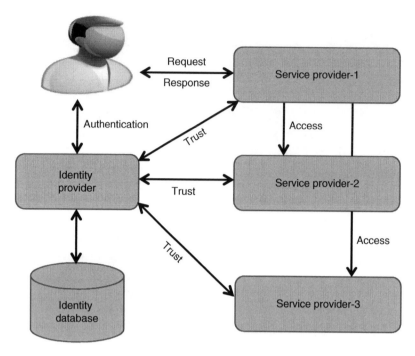

Figure 35.8   *Single sign-on overview*

35.5.6   Identity Management Standards and Protocols

The various identity-management protocols differ in features such as the data format supported and protocols to exchange credentials between the entities involved. Some of the well known identity-management protocols that help in establishing federation among the individual partners are given below.

35.5.6.1   Security Assertion Markup Language (SAML)

Security Assertion Markup Language includes a set of specifications for exchanging the authentication, authorization and attributes assertions across the federation. It uses an XML-based data format and this protocol is managed under the Organization for the Advancement of Structured Information Standards (OASIS).

35.5.6.2   Service Provisioning Markup Language (SPML)

This is an XML-based security framework developed by OASIS, and it helps in automating the provisioning and deprovisioning of user accounts with the CSP.

35.5.6.3   eXtensible Access-Control Markup Language (XACML)

In order to implement an access-control mechanism, this XML-based access-control language developed by OASIS could be used. This language provides the XML-schema, which could be used to protect the resources by making access decisions over these resources.

35.5.6.4   *Shibboleth*

Shibboleth is an open source identity management project, which helps to establish single sign-on solutions in the federation. Shibboleth uses the SAML specifications to achieve the authentication and authorization in the federated environment.

35.5.6.5   *OpenID*

OpenID provides a user-centric identity framework for the authentication purposes. OpenID 2.0 supports the identification of users through URL or XRI addresses.

35.5.6.6   *OAuth*

OAuth is an open-source identity management protocol used to provide the authorization of users' data across different applications, without disclosing the user's identity credentials. Identity tokens, issued by the identity provider, are used by the third-party applications to gain access to the user's protected data. Various applications have different tokens associated with them, and this helps to maintain the privacy of the data resources.

35.5.6.7   *OpenID Connect*

OpenID Connect is an open-source identity management protocol used to provide standardized authentication and authorization functions across federated applications. OpenID Connect combines the authentication and authorization processes of OpenID and OAuth. OpenID Connect protocol uses OpenID 2.0 and OAuth 2.0, and provides APIs to be used by third-party applications.

35.5.6.8   *WS-Federation*

WS-Federation is a part of the Web services security specification and it is meant for the federation of applications or Web services. WS-Federation specifications are extensions to WS-Trust protocol. This protocol can be used to share the identity information of various users across multiple security domains and organizations.

   As the different identity management protocols have dissimilar features, the selection of a particular protocol depends on the requirements of the applications and also on their architecture features.

35.5.7   Single Sign-On Products

There are many industry products offering SSO solutions to organizations. Some of them are given below.

35.5.7.1   *DACS*

Distributed Access-Control System (DACS) is a lightweight single sign-on and role-based access-control system for Web servers and server-based software, released under an open-source license (Distributed Access Control System, 2014). It ensures secure resource sharing and remote access via the Web. It is used to provide SSO functionality across organizational or departmental Web servers, and also to limit access to Web-based resources. It provides a secure, Web-based SSO and rule-based authorization process that can be applied selectively to any resource or activity (Web services, Web content, program features).

35.5.7.2   ESSO

Enterprise Single Sign-On (ESSO) from Dell Inc. enables an organization to achieve SSO (ESSO, 2014). It bases application logins on the existing active directory identities in the organization without the need for additional authentication methods. It supports the standard username/password logins and various strong authentication methods such as biometrics, smart cards or token-based two-factor authentication.

35.5.7.3   TrewIDM Cloud

TrewIDM Cloud's Identity Management engine supports various features of identity and access management such as user provisioning, deprovisioning and granting fine-grained access to specific applications. TrewIDM provides a seamless single sign-on feature for accessing both enterprise and software-as-a-service (SaaS) based applications.

35.6   Reputation and Trust

Nowadays, trust-based access control is gaining dominance in cloud computing, because the cloud is such a large distributed system, where it is not possible for a CSP to know in advance every other user or entity in the system. Trust represents confidence that somebody would behave exactly the way he is expected to behave. Mutual trust between the service providers and service consumers, and also between the providers of various services and the identity providers, is highly important in cloud computing to make the best use of it in a secure way.

35.6.1   Trust Types

In the cloud environment, trust is not static but is considered dynamic because it changes from time to time. Trust could also be classified into direct trust and indirect trust. Direct trust is the trust that an entity develops for another entity based on its past transactions or previous experience with the other entity. In the case of indirect trust, entity A may trust entity B, based on the recommendation from another entity, C, who is trusted by entity A. This form of trust is also known as recommended trust. In a huge distributed system like cloud, direct trust is often not possible. Methods or models should therefore be designed to use indirect trust effectively to achieve distributed access control in the cloud.

35.6.2   Reputation

The reputation of an entity, which is generated through successful transactions with other entities over a period of time, could result in an increased trust level for that entity. A more trusted person or entity may be given permission to carry out more privileged transactions or operations in the cloud-computing domain. At the same time, trust is hard to build and easy to lose. It takes time to build trust for an entity and a single instance of misbehavior or violation of the contract can reduce the trust level of the entities such as the CSP, CSC, or the IdP drastically in the cloud.

35.6.3   Dynamic Trust Management in the Cloud

In the cloud computing scenario, trust is dynamic because the same entity can have different trust values at different points of time. The trust value of an entity could be calculated based on various parameters such as past behavior and the history of previous transactions with the same entity and, also, by considering feedback from other trusted third parties in the cloud. Hence, in cloud computing, a trust model should be developed to calculate the correct trust value of various entities at different points of time, and this value could be used in access-control decisions. Generally, a decay function could be designed to represent the variation in trust values of the entities.

35.7　Data Security

In cloud computing, a proper access-control mechanism should be used to ensure the confidentiality, integrity, and availability of users' data. Data security concerns vary from one deployment model to another and the concerns are more serious when the public cloud model is used because of its inherent characteristics. Data security in the cloud model should deal with the data in transit, data at rest, and also data being processed. There should be suitable mechanisms adopted by the CSPs and the CSCs to protect the data while it is passing through each stage of the data life cycle. Encryption of the data and usage of secure standards such as SSL/TLS can be deployed to protect the security of the data in the cloud environment.

35.8　Conclusions and Research Directions

In this chapter, we have discussed the issue of distributed access control (DAC) in cloud computing systems. Important features and functions of DAC were discussed. Distributed access-control architecture for heterogeneous cloud computing environments was presented along with a discussion of the authorization requirements. Various access-control policies and models were also analyzed. The issues of identity management in cloud, effective identity management approaches, and various identity management standards and protocols were discussed. A few industry products offering SSO solution were also highlighted. Further research is needed on the establishment of a dynamic trust relationship between user domains and cloud domains, and between various cloud domains, in order to have a proper solution for distributed access control. DAC has enormous potential for further active research, in order to make the cloud computing paradigm secure, reliable, and scalable.

References

Almutairi, A., Sarfraz, M., Basalamah, S., *et al.* (2012) A distributed access control architecture for cloud computing. *Software* **29**(2), 36–44.

Distributed Access Control System (2014) *DACS Distributed System*, http://dacs.dss.ca/ (accessed December 28, 2014).

ESSO (2014) *Enterprise Single Sign-On*, http://software.dell.com/products/esso/ (accessed January 2, 2016).

Feng, F., Lin, C., Peng, D., and Li, J. (2008) *A Trust and Context Based Access Control Model for Distributed Systems*. Tenth IEEE International Conference on High Performance Computing and Communications (HPCC'08). IEEE, pp. 629–634.

Ferraiolo, D. F., Sandhu, R., Gavrila, S., *et al.* (2001) Proposed NIST standard for role-based access control. *ACM Transactions on Information and System Security (TISSEC)* **4**(3), 224–274.

Gunjan, K., Sahoo, G., and Tiwari, R. (2012) Identity management in cloud computing – A review. *International Journal of Engineering Research and Technology* **1**(4), 1–5.

Hu, V. C., Ferraiolo, D., and Kuhn, D. R. (2006) *Assessment of Access Control Systems*, National Institute of Standards and Technology, Gaithersburg, MD, http://csrc.nist.gov/publications/nistir/7316/NISTIR-7316.pdf (accessed January 2, 2016).

Lang, B., Wang, Z., and Wang, Q. (2007) *Trust Representation and Reasoning for Access Control in Large Scale Distributed Systems*. Second International Conference on Pervasive Computing and Applications (ICPCA 2007). IEEE, pp. 436–441.

Mather, T., Kumaraswamy, S., and Latif, S. (2009) *Cloud Security and Privacy: An Enterprise Perspective on Risks and Compliance*. O'Reilly Media, Inc., Sebastopol, CA.

Pereira, A. L. (2011) *RBAC for High Performance Computing Systems Integration in Grid Computing and Cloud Computing*. IEEE International Symposium on Parallel and Distributed Processing Workshops and PhD Forum, 2011 (IPDPSW). IEEE, pp. 914–921.

Turkmen, F., Jung, E., and Crispo, B. (2011) *Towards Run-Time Verification in Access Control*. International Symposium on Policies for Distributed Systems and Networks. IEEE, pp. 25–32.

Wei, Y., Shi, C., and Shao, W. (2010) *An Attribute and Role Based Access Control Model for Service-Oriented Environment*. Chinese Control and Decision Conference (CCDC). IEEE, pp. 4451–4455.

36

Cloud Service Level Agreement

Thomas J. Watson Research Center, USA

36.1 Introduction

Cloud-based services are becoming commonplace. Examples of these services include on-demand creation of virtual machines (VMs), backup of user data, and rapid deployment and automatic scaling of applications. These services are typically offered under three different service models: infrastructure as a service (IaaS), platform as a service (PaaS), and software as a service (SaaS) (Mell and Grance, 2011). Cloud providers, large and small, offer these services to individuals, small and medium businesses, and large enterprises. Each service is typically accompanied by a service-level agreement (SLA), which defines the service guarantees that a provider offers to its users.

While some cloud service deployment models are better understood than others (e.g., IaaS is arguably the most understood service model at the time of writing of this chapter), by and large cloud services are rapidly evolving and hence not standardized. As a consequence, the services offered by providers vary. Even for services deemed similar across providers, the SLAs can vary considerably. Service-level agreements also serve as a marketing tactic to attract new customers, either by bundling together different services, or by offering attractive guarantees with small print that may not ultimately benefit a cloud consumer. Such groupings of SLAs make comparison across cloud providers even more difficult. Moreover, a cloud provider may offer a canned SLA to all its consumers, irrespective of their requirements or size. An entity running mission-critical applications on a cloud would likely require a different SLA from an entity that is just beginning to evaluate the use of a cloud-based service.

In this chapter, we briefly explain the entities involved in cloud-based services (section 36.1.1). We describe elements of an SLA of a cloud-based service with detailed examples, which can help a cloud consumer in comparing SLAs of cloud-based services. We then give an overview of SLAs of two well known public cloud providers, namely, Amazon and Rackspace (section 36.3). Finally, we discuss future directions in cloud SLAs (section 36.4).

Encyclopedia of Cloud Computing, First Edition. Edited by San Murugesan and Irena Bojanova.
© 2016 John Wiley & Sons, Ltd. Published 2016 by John Wiley & Sons, Ltd.

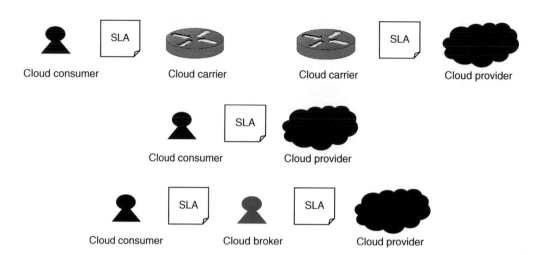

Figure 36.1   *Service level agreement relationship between cloud providers, cloud consumers, cloud carriers, and cloud brokers*

36.1.1   Entities involved in Cloud-Based Services

The National Institute of Standards and Technology (NIST) has defined a reference architecture for cloud services (Liu *et al.*, 2011). The reference architecture defines five unique actors involved in cloud-based services. They are:

- Cloud consumer. A person or entity that uses the services offered by one or more cloud providers.
- Cloud provider. An entity that offers services under an appropriate deployment model to cloud consumers.
- Cloud carrier. One or more entities that provides network services between cloud provider(s) and cloud consumers.
- Cloud broker. An entity that negotiates relationships between cloud provider(s) and cloud consumers.
- Cloud auditor. An entity that can conduct an independent verification of the offered cloud services and their usage.

Service-level agreements are typically defined between cloud providers, cloud consumers, and network carriers (cloud carrier according to NIST Reference Architecture – Liu *et al.*, 2011). Moreover, an SLA may also be defined between a cloud consumer and cloud broker(s). Figure 36.1 shows the entities involved in cloud-based services, and their SLA relationships.

36.2   Evaluating SLAs of Cloud-Based Services

Due to lack of standardization in cloud-based services, it is difficult for a cloud consumer to compare SLAs of different providers. In this section, we break down a cloud-based service SLA into easy-to-understand components and provide detailed examples for each component. By understanding the different elements of a cloud-based service SLA, a cloud consumer can make an informed choice when selecting a cloud provider.

36.2.1 Service Guarantee

A service guarantee specifies the metrics, which a provider strives to meet. Failure to achieve those metrics will result in a service credit to the customer. A service guarantee is sometimes also referred to as a service commitment.

Typically, service guarantees are specified over a time period (see section 36.2.2). Availability (e.g., 99.9% of a VM, response time (e.g., less than 50 ms for a storage request), disaster recovery (e.g., restoring data within 24 hours), ticket resolution (e.g., within 1 hour of reporting), and data backups (e.g., every day) are examples of service guarantees over a time period.

In this section, we give detailed examples of performance and data guarantees that may be offered by a cloud provider.

36.2.1.1 *Performance Guarantees*

A service guarantee may specify the performance objectives associated with one or more services. These performance guarantees bind a cloud provider to deliver acceptable performance for cloud services. Failure to meet these guarantees will result in a service credit to a cloud consumer. Typical examples of performance guarantees address availability, response time, and throughput. We briefly discuss these three guarantees here.

- *Availability,* informally, is the percentage of time for which a cloud system (or a subsystem) can serve requests. Availability can be specified on a per resource basis (e.g., VM or database), or on a large scale (e.g., an availability zone or a data center). A cloud provider offering an availability service guarantee may include a provision in an SLA to exclude any downtime due to maintenance from availability calculations. A cloud consumer must pay close attention to service guarantee exclusions from availability calculations. Availability can also be measured for a group of resources instead of a single resource. It can be computed over a group of servers or VMs, for example as the aggregate uptime of all servers (for more details, see section 36.2.3).
- *Response time,* informally, measures the processing time for a request. The response time calculations are subject to variations across cloud providers. A cloud consumer is interested in the end-to-end response time – the time elapsed from when a cloud consumer issues a request until the time a response is received. However, a cloud provider may choose to define response time as the duration between a request entering and leaving its cloud network. Any delay due to network issues beyond the control of a cloud provider may be excluded from response-time calculations. To provide bounded guarantees on the network delays of a request, some cloud providers are now offering a direct physical link between the cloud consumers' premises and their data centers.
- Throughput, informally, is defined as the total number of valid requests that are successfully served within a time unit. An SLA may define a maximum limit on the number of requests that can be served within a time unit, which will have an associated response time guarantee. If a cloud consumer were to initiate more requests than the limit defined in an SLA, it may be a violation of an acceptable usage policy and excluded from service guarantee calculations.

A cloud SLA needs to specify clearly how these metrics are collected and how often. Any exclusion from service guarantee calculations must be clearly specified in the SLA.

Currently, no standard benchmark exists for comparing the performance of cloud services. The cloud subcommittee (https://www.spec.org/osgcloud/, accessed December 26, 2015) of the operating system group (OSG) of Standard Performance Evaluation Corporation (SPEC) is working on standardizing a benchmark for comparing the performance of IaaS cloud services. Such a benchmark can help a cloud consumer greatly in selecting an appropriate cloud service for its usage.

36.2.1.2   *Data Guarantee*

Data service guarantees define which data is stored, where it physically resides, whether it is backed up, who can view the data, and what happens in case of lawful seizure of data or if a cloud provider or a consumer were to go out of business. Below, we discuss some of these issues.

Which Data to Store in Cloud?

A cloud consumer may choose to store some or all of its data in the cloud. Typically, a decision to store some or all data in the cloud is guided by the sensitive nature of data, and whether the cloud is hosted by a cloud provider or is setup on cloud-consumer premises, and any data protection guarantees offered by the cloud provider. A cloud consumer must communicate cloud data storage guidelines to its users (that is, employees of its organization). Moreover, to detect any violation of the data storage guidelines by its users or data protection guarantees offered by a cloud provider, a cloud consumer may choose to perform regular audits of the data being stored in a cloud. A cloud consumer may also entrust a third party for performing such an audit.

Where is Data Physically Stored?

While evaluating an SLA, a cloud consumer must evaluate where the data will be physically stored. It is important to be aware of the physical data storage location for several reasons, such as meeting regulatory compliance requirements such as the Health Insurance Portability and Accountability Act (HIPAA), country or area jurisdiction, or performance. The data can be stored on the premises of a cloud consumer, or in a data center managed by a cloud provider. A cloud consumer must evaluate whether a cloud provider can realistically deliver a data storage solution that spans multiple jurisdictions (e.g., legal or country).

What are Guarantees against Data Loss?

Not all data requires a backup. A cloud consumer should formulate a redundancy policy for various types of data, the size of the data that requires backup, and the access patterns. It can then evaluate data-redundancy solutions offered by cloud providers, and their costs. Cloud providers offer data redundancy in several ways. They can either replicate the data within a data center or replicate it across data centers that are geographically separate. Each redundancy solution has its merits and associated costs, which a cloud consumer must evaluate against its business objectives.

In the event of a disaster, the time to recovery may vary across cloud providers even if they provide similar data redundancies. It may also vary due to the physical location of data. An SLA may provide a service guarantee time to recovery in the event of a disaster. Obviously, this metric is of interest to cloud consumers who cannot afford any downtime due to disasters.

What are Data Privacy Policies?

A cloud consumer must evaluate the data privacy policies of a cloud provider. The public cloud providers only offer a standard SLA and a standard privacy policy. It is possible that the default privacy policies of a cloud provider may not meet the requirements of a cloud consumer. In such a case, a cloud consumer may choose to negotiate a new data privacy policy or consider a different cloud provider. An SLA should clearly define the responsibilities of a cloud provider for protecting data privacy, and any service credits to the cloud consumer due to advertent or inadvertent leakage of data by a cloud provider.

Throughout the world, there are laws and regulations that govern the privacy of personal data, and security of information systems. These laws vary from one country to another. Within a country, these regulations may vary across geographical boundaries (e.g., state) or jurisdictions (e.g., hospitals, municipalities). The preservation of data in the cloud, for instance for backup, requires that data be maintained for an extended period, even beyond its useful life. The responsibility for maintaining the privacy of this data typically rests with the cloud provider.

A cloud consumer must be aware that its data can be lawfully intercepted or seized. As part of an SLA, it may be the responsibility of a cloud provider to inform a cloud consumer of any lawful data intercept or seizure as appropriate.

A cloud consumer may also need to define privacy policies for its users (e.g., employees of its organization or users of its service). Its users must be made aware of the type of data that can be stored in a cloud. Moreover, a cloud consumer may need to set up appropriate authorization policies for users in its organization, just as it would do for noncloud storage.

Who Performs Data Audit?

As part of an SLA, a cloud consumer and a cloud provider may choose a third party to audit data storage to ensure compliance with regulations and SLA. Such an audit can be a part of a general audit of cloud services.

36.2.2   Service Guarantee Time Period

Service guarantee time period describes the duration over which a service guarantee is calculated. The time period can be small (in minutes or hours), span an entire billing month, the duration of service, or the duration of contract. As an example of a small service guarantee time period, consider the percentage of I/O operations (e.g., 99.9%) that must meet the quality-of-service threshold within a minute. A small time period of one minute leaves a provider with little wiggle room for erring. A service guarantee over a small period is likely going to command a higher premium than a similar guarantee over a longer period.

A service guarantee time period calculated over the entire billing month might, for example, cover the availability of a VM (e.g., 99.9%).

An example of a service guarantee over service life cycle might be a guarantee that physical copies of data are stored within a data center.

36.2.3   Service Guarantee Granularity

Service guarantee granularity describes the scale of the resource on which a provider specifies a service guarantee. For example, the granularity can be per service, per transaction basis, or per data center. As with the service guarantee time period, the service guarantee can be stringent if the granularity of service guarantee is fine-grained. An example of a fine-grained service guarantee is availability of a VM. In contrast, if service guarantee granularity is per data center, then VMs in the data center may not be available even if the data center is available.

Service guarantee granularity can also be calculated as an aggregate of the considered resources, such as VMs or requests. As an example of calculating service guarantee over aggregate resources, consider a cloud provider that provides an availability guarantee of 99.95% over all VMs of a cloud consumer. This guarantee is weaker than per-VM availability of 99.95%, as some VMs in the aggregate computation may have a lower availability than 99.95%.

36.2.4   Service Acceptable Usage Policy

Service acceptable usage policy defines the criteria for a cloud consumer to receive service guarantees. For example, in an IaaS cloud, an acceptable usage policy may be defined as the number of VM creation requests issued per minute. Another example of an acceptable usage policy is that the number of storage requests per minute must not exceed a certain threshold. Failure to comply with an acceptable usage policy will result into a violation of a service guarantee by a cloud consumer. As such, any service requests not meeting the acceptable usage policy are likely to be excluded from service guarantee calculations (see section 36.2.5) and may result in termination of services provided to a cloud consumer.

A cloud provider typically measures and records use of its services so that it can determine if there has been unacceptable use. It is typically up to a cloud provider to indicate to a cloud consumer whether acceptable usage is being violated before suspending the service. Such an indication can be done through API, or through an off-band channel such as e-mail.

36.2.5 Service Guarantee Exclusions

Service guarantee exclusions specify the instances of a cloud service that are excluded from service guarantee calculations. Typically, these exclusions include:

- an abuse or unacceptable use of the service by a cloud consumer;
- denial of service attacks;
- any downtime associated with scheduled maintenance;
- any downtime associated with a roll out of a new service version.

Scheduled maintenance can happen due to several reasons. For example, a hypervisor or a VM may need to be unavailable or may require a reboot in order to apply all operating system security updates. Or a database-as-a-service may be offline to perform maintenance on the underlying storage hardware. A cloud provider may design its services in a way to limit any downtime for a cloud consumer due to scheduled maintenance. Alternatively, a cloud provider may expose a maintenance schedule to a cloud consumer. In both cases, any potential disruption due to scheduled maintenance should be specified in an SLA. While examining service guarantees, a cloud consumer should pay careful attention to service-guarantee exclusions due to planned maintenance, and their disruption on its operations.

Service version roll out defines how a newer version of a cloud service is rolled out and when, and what impact it may have on a cloud consumer. In some cloud provider SLAs, service version rollout is considered part of scheduled maintenance. A cloud consumer needs to consider whether a roll out of a new service version may result in any service disruption. A cloud provider may specify any disruptions due to service version rollout as part of scheduled maintenance (and therefore, exclude it from service guarantee calculations).

36.2.6 Service Credit to a Cloud Consumer

A service credit is given to a cloud consumer if one or more service guarantees are not met. In some cases, a service credit may only be applied towards future service usage credit. In other cases, a service credit is real money refunded to a cloud consumer. A service credit can be a full or a partial credit to a cloud consumer for the affected service. Typically, a service credit never exceeds the service fees of the affected service. However, the affected fees may not reflect the cost incurred by a cloud consumer due to service disruption.

36.2.7 Service Violation Detection and Measurement, and Service Restoration

Service violation detection and measurement, and service restoration describes how and who detects, measures, and reports the violation of service guarantee (or SLA violation), and when a disrupted service is restored. A majority of public cloud providers surveyed at the time of writing of this book chapter place the burden of service violation detection on the cloud consumer.

For a cloud consumer, detecting and measuring SLA violations requires storage of request logs, service request metadata, and any other relevant data. When a cloud provider receives a request for verifying service violation, it must compare the evidence offered by a cloud consumer against its metrics and logs, and respond to cloud consumer within a reasonable time frame. An SLA of a cloud-based service typically states the time period during which a cloud consumer must file a service violation report. This time period determines how long a cloud provider must retain the relevant data for any potential claims of service violation.

Some cloud consumers may require an SLA conformance report from a cloud provider every service guarantee time period (e.g., one billing month). As part of SLA, cloud consumers may also require an independent evaluation of SLA conformance by a cloud auditor.

An SLA will stipulate how quickly a cloud provider restores a disrupted service. The time duration for service restoration may vary according to service problems. For example, (i) once the disruption in the connectivity to a service is established, it must be restored within an hour; (ii) if a physical machine's disk is faulty, it will be replaced within a day.

36.2.8 Service Renewals

Service renewals specify how a cloud consumer may renew the service. Typically, cloud usage is pay per use, and service renewal is implicitly embedded in it, that is pay and renew. However, some (large) enterprises may negotiate a contract that has a starting and ending date. At the end of a contract, a cloud consumer may choose to renew the contract under the same or a new SLA, or terminate it. An SLA should clearly specify the expectation for a cloud provider and consumer in case the cloud consumer chooses not to renew the contract.

36.2.9 Service Activation and Deactivation

Service activation and deactivation specifies the time when a service becomes active or is terminated. For example, in an IaaS cloud, a cloud consumer must explicitly request the creation of a VM, and a cloud consumer may only be charged when remote login to a VM succeeds after a VM is created. Similarly, service deactivation specifies the time when a cloud provider stops charging the cloud consumer for service usage. Service deactivation can be involved. It will likely include provisions on how cloud consumer data may be handled upon a service termination. The precise timing of service activation and deactivation has a direct bearing on the costs incurred by a cloud consumer. Such timing must be clearly specified in an SLA.

36.2.10 Service Excess Use

Service excess use defines how a cloud consumer may be charged for excess service use. Typically, cloud service is pay per use. However, a cloud consumer may also negotiate a long-term contract with a cloud provider, which will include provisions on any excess use of services by a cloud consumer.

36.2.11 Service Transferability

If a cloud consumer were to sell its business, or initiate a project with another cloud consumer and/or within its different business units, it will need to transfer its cloud resources to the seller or another cloud consumer. Such a transfer of resources may not always be possible due to business, technical, or legal reasons. A cloud consumer should understand whether a cloud provider limits any transfer of resources.

Moreover, a cloud provider's service may be terminated possibly in lieu of a new offering, or a cloud provider may also go out of business. A cloud consumer should pay attention to clauses that limit or prohibit transfer of data and resources in such cases, and any timelines on the transfer of data and resources.

36.2.12   Other Things to Consider

We briefly discuss some other aspects that should be considered when evaluating SLAs.

36.2.12.1   Non-Negotiable or Customized SLA

At the time of writing, all public cloud providers offer a non-negotiable SLA for their cloud services. These SLAs, generally, favor cloud providers. A cloud consumer should examine carefully any future impact on its operations when using cloud services under non-negotiable SLAs.

Depending on the size of cloud services being used, an enterprise may negotiate an SLA with a cloud provider to meet its business objectives. Such an SLA, if it differs from the standard SLA offered by a cloud provider, will likely command a higher price.

36.2.12.2   Cloud Service and Deployment Models

Infrastructure as a service, PaaS, and SaaS are different service-delivery models for cloud services. These services can be offered on a cloud hosted within a cloud consumer premise or in a data center managed by a cloud provider. For each service and deployment model combination, the precise definition of service guarantees can vary. For example, in an IaaS cloud, an availability service guarantee may be offered on a per VM basis whereas in a SaaS cloud, the availability service guarantee will likely be offered on a per application basis, which in in turn can be hosted on a IaaS and/or PaaS cloud.

Cloud consumers may have less control over where their data resides in PaaS and SaaS clouds and the format in which the data is stored. The lack of control and proprietary format leads to vendor lock in, which is a major concern for cloud consumers. Even if the cloud provider offers service guarantees against lock in, the size of cloud consumer data may make it difficult for a cloud consumer to change providers. An enterprise cloud consumer may require explicit guarantees on the data format and time to move data across cloud providers for PaaS and SaaS service delivery models.

36.2.12.3   Subcontracted Services

It is conceivable that a cloud provider may subcontract services to other vendors. For example, a cloud provider may in turn obtain services from multiple cloud providers and provide them as a bundled service to a cloud consumer. Software as a service cloud providers are more likely to use services of platform and infrastructure cloud providers to offer SaaS services to their consumers.

A cloud consumer may negotiate an SLA with a cloud provider providing bundled services or directly with each of the cloud providers. A cloud consumer should examine whether the SLA from a cloud provider offering bundled services from other providers meets its business requirements.

36.3   Service Level Agreements of Public Cloud Providers

We briefly give an overview of SLAs of two well known public cloud providers at the time of writing, namely Amazon (http://www.amazon.com, accessed December 26, 2015) and Rackspace (http://www.rackspace.com/, accessed December 26, 2015). These cloud providers offer IaaS and PaaS compute and storage services. The compute service comprises of a virtual machine (or instance) or CPU cycles that a customer can purchase on an hourly, monthly, or yearly basis. The storage service allows storage and retrieval of blob or structured data.

36.3.1   Amazon

Amazon (www.amazon.com, accessed December 26, 2015) is an IaaS provider and offers compute (Elastic Compute Cloud (EC2) – https://aws.amazon.com/ec2, accessed December 26, 2015), block storage (Elastic Block Store (EBS) – http://aws.amazon.com/ebs/, accessed December 26, 2015), and object storage (Simple Storage Service (S3) – https://aws.amazon.com/s3/, accessed December 26, 2015) services. In EC2, a customer can obtain virtual machines (instances in Amazon speak) by the hour or reserve them in advance for an entire year (https://aws.amazon.com/ec2/purchasing-options/reserved-instances/, accessed December 26, 2015). In addition, EC2 offers spot instances where a customer can bid for compute capacity. EC2 SLA (https://aws.amazon.com/ec2/sla/, accessed December 26, 2015) is applicable to hourly, spot, and reserved instances. The storage service S3 provides mechanism for storing and retrieving data objects using put(), get() operations. The data size of each object can be up to 5 TB.

36.3.1.1   EC2 SLAs

At the time of writing, EC2 and EBS SLA (https://aws.amazon.com/ec2/sla/, accessed December 26, 2015) promise a 99.95% region availability service guarantee (a data center is referred to as a region in Amazon EC2 service). A region comprises of one or more availability zones, which are groups of physical machines that share a minimal set of resources within a region (e.g., cooling of data center). EC2 does not provide any per VM availability service guarantee.

A region is considered unavailable if more than one availability zones are unavailable. A region is unavailable for EC2 if the virtual machines of a cloud consumer running in more than one availability zones within a region have no external connectivity. A region is unavailable for EBS if all attached volumes perform zero read or write I/O, with pending I/O in the queue. The monthly percentage uptime is calculated by subtracting from 100% the percentage of minutes during which an Amazon region was unavailable.

A cloud consumer must provide evidence of region unavailability to Amazon by the end of second billing cycle in which the incident in question occurred. A cloud consumer is eligible for a service credit exceeding one dollar if a region is unavailable for more than 0.05% of the time within a billing month. The service credit is up to 10% of a cloud consumer's bill (excluding any one-time costs) for the EC2 instances and EBS devices affected by the outage. Service credits are typically only given towards future EC2 or EBS payments but can also be issued to the credit card used to pay for services during the month in which the incident occurred.

Amazon's EC2 service does not provide any service credit for failures of individual virtual machines not attributable to region unavailability. This clause means that cloud consumers must implement appropriate reliability mechanisms for their applications. Further, Amazon does not provide any service credits if virtual machines suffer from any performance issues. A virtual machine can suffer performance degradation due to co-location or hardware differences with the underlying physical machine. Amazon EC2 SLA is vague in terms of the specifics of scheduled or unscheduled region maintenance, which is excluded from service guarantee calculations

36.3.1.2   S3 Service Level Agreements

Amazon Simple Storage Service (S3) SLA (https://aws.amazon.com/s3/sla/, accessed December 26, 2015) provides storage request completion guarantee of 99.9% over a billing month (service guarantee time period). A storage request is considered failed if an S3 server returns an "Internal Error" or "Service Unavailable" response to a request. These responses correspond to HTTP response codes 500 and 503. The burden of reporting request failure and providing evidence is on the cloud consumer.

In the S3 service, failed requests are calculated over a 5 min interval, which are then averaged over a billing month. The failed requests are calculated by dividing the number of requests generating an error response to the total number of requests in the 5 min interval. The percentage of completed transactions in the billing month is calculated by subtracting from 100% the average of failed request rates from each 5 min period.

The service credit is 10% of a cloud consumer's bill if completion rate is over 99.9%, and 25% of the cloud consumer's bill if completion rate is less than 99%. Similar to EC2, a service credit is typically applied towards future service usage but may also be issued to the credit card that was used to purchase the services affected. Amazon must receive the claim by the end of second billing month from the month in which the incident occurred. The S3 service does not specify any performance guarantees on the storage requests.

36.3.2 Rackspace

Rackspace provides a compute service, namely, Cloud Servers (http://www.rackspace.com/information/legal/cloud/sla#cloud_servers_next_gen_sla, accessed December 26, 2015), similar to EC2. It also provides a block storage service Cloud Block Storage (http://www.rackspace.com/cloud/block-storage/, accessed December 26, 2015) similar to EBS, and a file and object storage service for storing and retrieving files, namely, Cloud Files (http://www.rackspace.com/cloud/files/, accessed December 26, 2015). Cloud Servers is offered as first or next generation. We now discuss SLAs for the next generation of Cloud Servers.

36.3.2.1 Cloud Servers SLA

Rackspace breaks down its Cloud Servers (next generation) SLA into the following components:

- management stack used to create virtual machines (cloud servers in Rackspace speak);
- virtual machines.

The service guarantee for the management is computed over a billing month and is defined as the percentage of well formed success API requests to the management stack. If the percentage of successful API requests is below 99%, the service credit is 30% of the cumulative Cloud Servers fees within a billing month. An API request is considered failed if an HTTP 5xx response is received or if no response is received at all. Network issues outside the Rackspace data center may cause a response to be lost; however, such losses are excluded from service guarantee calculations.

For virtual machines, Rackspace provides a service guarantee that its data center network, HVAC, and power will be 100% available in a billing month, excluding the scheduled or emergency maintenance (service guarantee exclusions). Scheduled maintenance does not exceed 60 minutes in any calendar month and must be announced at least 10 days in advance to the customer. Emergency maintenance is to address critical issues such as security vulnerabilities or reliability issues.

If a physical server running the virtual machine fails, Rackspace gives a service guarantee that the failed VM will be repaired within an hour of problem identification. Further, if VMs need to be migrated due to server overload, a cloud consumer is notified 24 hours in advance. The SLA does not specify if Rackspace performs live or offline migration of a VM.

Rackspace computes service guarantee violations in increments of 30 min for data center network, HVAC, and power, and in 1 hour increments for downtime associated with physical servers or migration. If the data center network, HVAC, power, or physical servers are down, or if VMs need to be migrated, the service credit starts from 5% of a cloud consumer's bill up to 100% of the bill for affected VMs. The implication of Cloud Servers SLA is that Rackspace provides a service guarantee on a per virtual machine (instance) basis.

A customer must contact Rackspace within 30 days following the downtime of a virtual machine and provide evidence of the problem in order to receive a service credit. However, it is unclear how a customer can provide evidence for a specific problem such as HVAC, power, or network failure. Perhaps Rackspace maps the customer's evidence to specific problems and determine the service credit accordingly.

36.3.2.2 Cloud Files SLA

Rackspace provides a 99.9% request completion rate and Cloud Files server availability guarantee in a billing cycle (http://www.rackspace.com/cloud/legal/sla/, accessed December 26, 2015). The service is considered unavailable if a data center network is down, or if the service returns an error response (HTTP 500-599 status code) to a request within two or more consecutive 90s intervals, or if an average download time for a 1-byte document exceeds 0.3 s.

Unavailability due to scheduled maintenance is excluded from the availability calculations. As with Cloud Servers SLA, the scheduled maintenance period does not exceed 60 min and must be announced 10 days in advance.

A customer must contact Rackspace within 30 days following the down time and provide evidence of the problem in order to receive a service credit.

Tables 36.1 and 36.2 summarize our comparison of compute and storage service SLAs offered by Amazon and Rackspace. A detailed comparison of SLAs across cloud providers can be found in our earlier work (Baset, 2012).

36.4 Future of Cloud SLAs

In this section, we consider how a cloud provider may define SLAs for cloud services in the future.

Table 36.1 Compute SLA comparison of Amazon and Rackspace compute services

	Amazon EC2	Rackspace Cloud Servers
Service guarantee	Availability (99.95%)	Availability
Granularity	Data center	Per instance[a] and data center mgmt. stack
Scheduled maintenance	Unclear if excluded	Excluded
Patching	N/A	Excluded if managed
Guarantee time period	Per billing month	Per month
Service credit	10% if<99.95%	5% to 100%
Violation report respon.	Cloud consumer	Cloud consumer
Reporting time period	N/A	N/A
Claim filing timer period	By the end of second billing month from the month in which the incident occurred.	Within 30 days of downtime
Credit only for future payments	Yes (a dollar credit may be issued to a cloud consumer at Amazon's discretion)	No

Note: [a] Implied by SLA.

Table 36.2 *Storage SLA comparison of Amazon and Rackspace storage services*

	Amazon S3	Rackspace Cloud Files
Service guarantee	Completed transactions	Completed transactions
Granularity	Per transaction	Per transaction
Guarantee time period	Billing month	Billing month
Service credit	10% if<99.9% 25% if<99%	10% if<99% 100% if<96.5%
Violation report responsibility	Cloud consumer	Cloud consumer
Reporting time period	N/A	N/A
Claim filing timer period	By the end of second billing month from the month in which the incident occurred.	Within 30 days following unavailability
Credit only for future payments	Yes (a dollar credit may be issued to a cloud consumer at Amazon's discretion)	No

36.4.1   Service Guarantee

At the time of writing, the public cloud providers only offer uptime guarantees for IaaS compute services. The cloud providers may also want to offer other guarantees such as performance, security, and ticket resolution time. Providing a performance guarantee becomes necessary if cloud providers oversubscribe the resources of physical servers to decrease the number of physical servers used and increases their utilization. The oversubscription of the physical servers implies that performance of virtual machines running on physical servers may become a concern. Further, co-location of a virtual machine with other workloads may also impact the CPU, disk, network, and memory performance of a VM. Moreover, enterprises purchasing cloud-based services may demand a minimal level of performance guarantee. Therefore, it may be necessary for a cloud provider to offer performance based SLAs for its IaaS compute services with a tiered pricing model, and charge a premium for guaranteed performance.

36.4.2   Service Guarantee Time Period and Granularity

The service guarantee time period and granularity determine the stringency of the underlying service guarantee. A service guarantee is stringent if the guarantee's metric is performance based for a fine-grained resource over a small time period – for example, 99.9% of memory transactions in a 5-minute interval must complete within one microsecond. Such a stringent guarantee can be loosened by aggregating the service guarantee over a group of resources (e.g., aggregate uptime percentage of all instances must be greater than 99.5%) or by increasing the service guarantee time period (e.g., over an hour). Providers can use a combination of service guarantee granularity and service guarantee time period to price their services appropriately. For enterprise and mission critical workloads, a cloud provider may have no choice but to provide finer service guarantees.

36.4.3   Service Violation Detection and Credit

At the time of writing, none of the public cloud providers automatically detect SLA violation and they leave the burden of providing proof of the violation on the customer. This aspect may not be acceptable to cloud consumers with mission-critical or enterprise workloads. A cloud provider can differentiate the pricing of its offering if it automatically detects and credits the customer for SLA violation. However, the tooling cost to automatically measure, record, and audit SLA metrics can be a concern.

36.4.4   Standardization of SLAs

The lack of maturity in cloud service delivery (PaaS and SaaS) and lack of standardization in cloud SLAs makes it difficult for a cloud consumer to compare them effectively. As cloud services mature, and as the vision of utility computing is realized, the standardization of SLA is likely to take center stage. Structured representation of SLAs (e.g., in XML) may be a necessary step towards standardized SLAs.

36.5   Related Work

Patel *et al.* (2009) describe how a Web service legal agreement framework, developed in the context of service oriented architecture, can be applied to cloud SLAs. Truong *et al.* (2012) describe how to design contracts and exchanges for sharing and using data among cloud providers and consumers. Alhamad *et al.* (2010) describe nonfunctional requirements that cloud consumers need to consider when negotiating SLA with a cloud provider. The *Practical Guide to Cloud Service Level Agreements* (Cloud Standards Customer Council, 2012) provides a guide to cloud consumers on evaluating cloud SLAs.

36.6   Conclusion

In this chapter, we have provided an overview of entities involved in cloud-based services from an SLA perspective and have broken a cloud SLA into easy-to-understand components. We have discussed guidelines for consumers of cloud services to evaluate cloud providers, and then described SLAs of two well known cloud providers. Finally, we have discussed future of SLAs for cloud services.

The rapid evolution of cloud services makes the standardization of cloud services challenging. The lack of standardization of cloud-based services in turn makes it difficult to compare SLAs of different cloud providers. We hope that, as cloud services evolve, they will become more standardized as well as their SLAs.

References

Alhamad, M., Dillon, T., and Chang, E. (2010) *Conceptual SLA Framework for Cloud Computing*. Fourth IEEE International Conference on Digital Ecosystems and Technologies (DEST). IEEE, pp. 606–610.

Baset, S. (2012) Cloud SLAs: Present and future. *ACM SIGOPS Operating Systems Review* **46**(2), 57–66.

Cloud Standards Customer Council (2012) *Practical Guide to Cloud Service Level Agreements*, http://www.cloud-council.org/2012_Practical_Guide_to_Cloud_SLAs.pdf (accessed December 26, 2015).

Liu, F., Tong, J., Bohn, R. B., *et al*. NIST Cloud Computing Reference Architecture. Publication 500-292, NIST, Gaithersburg, MD.

Mell, P. M., and Grance, T. (2011) *The NIST Definition of Cloud Computing*. Special Publication 800-145. NIST, Gaithersburg, MD, http://www.nist.gov/customcf/get_pdf.cfm?pub_id=909616 (accessed November 25, 2015).

Patel, P., Ranabahu, A. H., and Sheth, A. P. (2009) *Service Level Agreement in Cloud Computing*. Wright State University Report, http://corescholar.libraries.wright.edu/cgi/viewcontent.cgi?article=1077&context=knoesis (accessed December 26, 2015).

Truong, H.-L., Comerio, M., De Paoli, F, *et al*. (2012) Data contracts for cloud-based data marketplaces. *International Journal of Computational Science and Engineering* **7**(4), 280–294.

37

Automatic Provisioning of Intercloud Resources driven by Nonfunctional Requirements of Applications

Jungmin Son, Diana Barreto, Rodrigo N. Calheiros, and Rajkumar Buyya

The University of Melbourne, Australia

37.1 Introduction

Cloud computing has emerged as a new computing paradigm offering subscription-oriented services in place of traditional in-house computing infrastructure. Through its utility computing concept, with a pay-as-you-go model, enterprises can avoid upfront investment for establishing infrastructures to provide computation power in the face of uncertain or fluctuating demand. For system administrators in an enterprise, instead of installing new servers and network equipment, they can easily acquire computing resources from one or more cloud providers with a few clicks on a Web page and pay only for their actual use. They only need to select one or more cloud providers and services that fit for their applications from Inter-cloud environment (Buyya *et al.*, 2010).

Throughout this chapter, "cloud provider" or "provider" refers to a company or an organization that provides public cloud computing services. "Customer" refers to a company, an organization or a person who uses the cloud computing service offered by cloud providers to deploy their application and serve to "end-users." A "system administrator" is a person who is in charge of managing computer resources and configuring infrastructures such as servers and network. In short, an enterprise will become a customer of the cloud provider when the system administrator in the enterprise decides to use the cloud service of the provider.

Cloud computing delivers its service to customers in three models: software as a service (SaaS), platform as a service (PaaS), and infrastructure as a service (IaaS). SaaS provides a complete stack of applications

Encyclopedia of Cloud Computing, First Edition. Edited by San Murugesan and Irena Bojanova.
© 2016 John Wiley & Sons, Ltd. Published 2016 by John Wiley & Sons, Ltd.

from the cloud provider, whereas PaaS provides software platforms that can be used by the customer's application. In both cases, the provider is in charge of managing and controlling the underlying environment of the services. In contrast, IaaS provides low-level computer resources – virtual machines (VMs) on which applications are deployed. Thus, the customer must configure and control the computing resources of VMs, whereas only the underlying physical infrastructures are managed by the provider.

When deploying a SaaS or PaaS service from multiple cloud providers, system administrators have less need to take infrastructural decisions because services are limited to specific applications or platforms, which have fewer options to be chosen, and cloud providers offer automatic features for their services such as automatic scaling.

On IaaS, however, system administrators are expected to make more decisions, because the parameters of VMs can be selected, which gives rise to several challenges. First of all, there are a vast number of options regarding the dimensions and quantities of VM resources from several possible providers. In most providers, the size of the VM is defined by the number and power of CPU cores, the amount of RAM, and the size of storage space, which comes with either a predefined amount set by the provider or is flexibly configurable by the customer. As each provider has its own set of resource types with different pricing policies, there is no common rule for definition of VM types. For example, Amazon EC2 (http://aws.amazon.com/ec2/, accessed December 26, 2015) offers 39 predefined types of VM depending on the sizes of resources, whereas Microsoft Azure (http://www.windowsazure.com/en-us/solutions/infrastructure/, accessed December 26, 2015) has 30 different VM types. Moreover, providers such as CloudSigma (http://www.cloudsigma.com/#features, accessed December 26, 2015) provide a more flexible configuration where a resource set is freely selectable by the customer without any fixed size. In addition to the size of the VM, several factors affect the decision on how the price for VM usage is determined, such as choice of operating system, location of the datacenter, and contract duration. In summary, system administrators are in charge of deciding the best option among various resource types, pricing schemes, and cloud providers.

Secondly, estimating the right amount of resources is important in order to determine the optimal size of VMs. Although cloud computing provides elastic scaling, which allows changes in the number and types of VMs after setup, determining the initial resource set is vital as it reduces the need to reconfigure, which demands time to be completed as it requires booting new VMs with the new configuration. Meanwhile, nonfunctional requirements, such as the expected number of end users, usage patterns and acceptable response time, are obtainable from accumulated statistics for existing application, or can be predicted from the application specification. Nonetheless, converting nonfunctional requirements to low-level VM resource requirements, such as number and computing power of CPU cores and amount of RAM, is difficult without comprehensive knowledge about the underlying infrastructure. By estimating proper resource sizes, an enterprise can avoid overprovision, leading to spending funds with unnecessary resources, or underprovision causing performance loss or failure of the application, which results in end-user dissatisfaction and consequent loss of revenue.

Finally, VMs must be allocated from the chosen provider with the preferred resource type within the set amount of time. As customers do not have control over the cloud infrastructure, there are no guarantees that resources will be allocated to the customer unless some type of reservation is made beforehand. When no reservation is in place, resource unavailability in one provider forces system administrators to find another provider that meets all the determined requirements and is able to fulfill the resources request. It requires extra effort and cost to the system administrator and, in the worst case, it may cause a service failure if it takes too long to be accomplished.

In this work, we propose an architecture to address the challenges emerging from the system administrator's perspective. Using our architecture, administrators can acquire the desired number of VMs from the best provider with proper resource size that covers their nonfunctional requirements. This will help system administrators to migrate their applications to the cloud by setting up the required IT infrastructure without being too concerned about calculating amounts of resources. Furthermore, enterprises can reduce the cost of cloud usage by selecting the optimal set of resources for their applications based on the supplied applications' nonfunctional requirements.

The rest of this chapter is organized as follows. We look at related work in section 37.2 and explain the background to this work in section 37.3. The architecture is described in section 37.4, with details of each component and its implementation. In section 37.5, we present the performance evaluation of a system prototype based on the proposed architecture. Finally, section 37.6 concludes the chapter and proposes future work.

37.2   Cloud Provisioning, Monitoring and Resource Selection

Several studies have been conducted by different groups in cloud provisioning, monitoring services, and resource selection. As our proposal integrates each of these features into one single framework, individual works are reviewed for each of these areas.

37.2.1   Cloud Provisioning

Resource provisioning in cloud computing refers to the decision about number, type, and location of resources to be deployed for a specific purpose. Definition of resources for provisioning may also include details on required processors, amount of storage, network bandwidth, and other relative resources from the cloud provider. The large number of variables related to resource definition makes it a complex problem. Nevertheless, if multiple cloud providers are to be used simultaneously, the problem becomes even more challenging.

Several approaches have been proposed for resource provisioning on multiple cloud providers. Grozev and Buyya (2014) reviewed and compared the architectures and brokering mechanisms of those intercloud systems. The authors proposed taxonomies for intercloud architectures and presented detailed surveys of each project.

In centralized federation architectures for interclouds, there exists a central component that aggregates the status of cloud providers and finds available resources from participating datacenters. For example, if one provider receives a request to provision resources from its client but cannot provide them, the request is redirected to another provider that can offer the desired resources.

Peer-to-peer federation architecture is similar to centralized federation approach except for the absence of the central component. In this architecture, cloud providers communicate and negotiate with each other directly, without a centralized server.

Independent intercloud approaches enable resource provisioning from multiple clouds without direct exchange between providers, as in the previous approach. This is achieved with an independent service or library that supports multiple cloud providers. For example, RightScale (http://www.rightscale.com/cloud-portfolio-management/benefits, accessed December 27, 2015) gives a single Dashboard and APIs to manage multiple clouds. They provide a configuration framework with templates to set up the VMs easily. They also provide an easy management tool on multiple cloud providers but do not perform the provider selection.

Independent approaches also include providing APIs for cloud application development and support deployment on multiple clouds, which allows developers to regard the heterogeneous clouds as a single platform with transparent access. Instead of developing an application using different APIs provided by each provider, these libraries include homogeneous controlling and provisioning functions supporting multiple providers. Apache Jclouds (http://jclouds.apache.org/, accessed December 27, 2015), for example, is a Java library for Java-based interaction with various providers, which supplies a provider-independent API for execution of operations regarding provisioning of computing resources and storages. While these libraries are helpful in developing an application able to execute on various providers, they just provide an alternative to provider-specific APIs, and therefore they do not offer cloud provider selection or automatic resource provisioning, which is still a task of system administrators' using such libraries.

37.2.2   Cloud Monitoring Services

Cloud monitoring is an important service to check the health of each datacenter and compare different types of VMs in different cloud providers. Infrastructure as a service cloud providers offer the computing resources in terms of VM, which is composed of the unit of CPU cores, the amount of memory, and the size of disk spaces. These terms are defined by the provider itself; thus, it is not straightforward to compare different types of VMs in different providers by just comparing the number of computing units they advertise. Periodical checks are necessary to determine the reliability and the availability of the provider. These metrics can be obtained by measuring the percentage uptime of the provider.

The importance of knowing the performance of different public cloud providers has encouraged the development of monitoring services that report metrics to a give a better picture of the real behavior of the different services.

CloudHarmony (http://cloudharmony.com/, accessed December 27, 2015) reports results of benchmarks in regard to performance, network, and uptime for a wide set of public cloud providers. To collect these metrics, monitoring services are located inside and outside of the cloud provider and some benchmark applications are also executed on behalf of CloudHarmony.

37.2.3   Cloud Providers Ranking and Selection

Before starting the provisioning process, a provider and the type of resources that will satisfy all the requirements should be chosen by the administrator. The selection criteria can vary depending on the requirements. Constraints are requirements that must be fulfilled by the cloud provider. For example, some government applications may be restricted to run within their national territory due to legislation. In such cases, geographical constraints should be applied to choose the provider, so that providers who have no datacenter in such a nation will be excluded from the choice. Preferences are the criteria for ordering the providers. A particular percentage of uptime from the datacenter can be one of the preferences when an application needs higher reliability. Price of the service can be another preference if the system administrator is more concerned about the cost, resulting in the selection of the most economic provider.

Some research in this area has already been conducted by several scholars. Some of it is introduced in the following paragraphs.

Li *et al.* developed CloudCmp (Li *et al.*, 2010), which compares different cloud providers by using a tool to perform systematic benchmarking. It evaluates the performance of elastic computing, persistent storage and intracloud and wide-area networking for each provider, and compares providers using unified metrics in each service. CloudProphet (Li *et al.*, 2011) estimates applications' resources and performance in the cloud. This uses the trace-and-replay method, which records the workload of an application from a traditional infrastructure, and measures performance when the recorded workload is replayed in a cloud environment.

SMICloud (Garg *et al.*, 2011) is a framework to rank cloud providers for a given application considering the service measure indexes (SMI): accountability, agility, assurance of services, cost, performance, security, privacy, and usability. It operates by assigning different key performance indicators (KPI) to evaluate these indexes in different cloud providers.

Zhang *et al.* (2012) proposed a declarative recommender system to select a cloud provider. The system receives as input requirement parameters from system administrators and determines the best provider that satisfies the requirements. However, the types of input parameters are resource sizes, which should be transformed from nonfunctional high-level requirements.

Rak *et al.* (2013) presented a cost/performance evaluation tool on top of the mOSAIC platform (http://www.mosaic-cloud.eu/, accessed December 27, 2015). Evaluation is performed by simulating and estimating resources, cost and response time of an application. The authors propose the use of nonfunctional requirements to create a system that suggests the best option among a set of cloud providers.

37.3   Resource Provisioning driven by Nonfunctional Requirements

Resource provisioning in an organization is usually performed by system administrators. They are in charge of the IT infrastructure and make decisions about where to deploy applications. The system administrators also fix problems related to failures of the hardware and software that support this infrastructure. Although the move to the cloud frees system administrators from managing underlying IT infrastructure, and the resulting hardware and software issues, it also brings new challenges.

Before deploying an application in the cloud, system administrators need to consider the quantity of resources this application will consume in terms of resource capacity as defined by different cloud providers. It is a challenging task because this estimation may differ from the one for an in-house IT infrastructure, and from one cloud provider to the other. The constraints for cloud selection must also be considered as the selected provider should fulfill every requirement.

After determining resources and other requirements that the application needs to execute satisfactorily in the cloud, the next task to be performed by system administrators is to decide which cloud providers can supply the estimated resources and which of them are more appropriate to host the application. This selection of candidate cloud providers is a laborious task as there are a large number of available providers, each offering varieties of services, which cannot be directly compared with the services provided for others.

More than 95 public cloud providers are registered by the monitoring service CloudHarmony. Each offers a diverse range of services. For example, GoGrid (http://www.gogrid.com/, accessed December 27, 2015) offers just one type of machine x-Large, that is configured with eight cores of CPU and 8GB of memory, while Amazon offer m1.xlarge, m2.xlarge, m3.xlarge, and c1.xlarge configurations, all of them with different quantities of resources. Moreover, other cloud providers, such as CloudSigma, do not have a default configuration but allow flexible configuration of resources.

After selecting the candidate providers that can supply the services to deploy the application, the next concern of system administrators is to obtain the required resources from cloud providers. This is a challenging task because APIs and interfaces to communicate with cloud providers are diverse and nonstandardized. Furthermore, latency in communication, provider's outage, and eventually lack of resources can prevent the administrator's instructions from being carried out. In this case, another provider would need to be contacted to obtain resources.

As was discussed previously, these activities of estimating resources, selecting cloud providers, and allocating resources show that the work performed by system administrators is still complex. Solutions should therefore be developed for support system administrators to reduce this complexity and to motivate organizations to move their applications to the cloud.

37.3.1   Nonfunctional Requirements

Currently, when system administrators want to acquire resources from IaaS cloud providers, they need to know the details of the resources they need. For example, consider a situation where the local infrastructure has extra resources to execute applications successfully. In this environment, administrators may know about the expected behavior and performance of their applications and this knowledge can be represented using nonfunctional requirements.

Nonfunctional requirements, according to the definition provided by (Glinz 2007), are attributes and constraints of an application created to achieve some level of quality and performance. Therefore they are not related to functions that the systems should carry out, but to properties that the systems should have, including availability, reliability, portability, cost, efficiency, usability, and testability. For example, in a shopping application, nonfunctional requirements can include the number of online transactions it can support; however, shipment tracking functionality is not included in the nonfunctional requirements.

A nonfunctional requirement can be described with features associated with it. For example, to describe the cost of the local IT infrastructure, the related features are the maximum price to spend in the new hardware acquisition and the price for maintenance of the infrastructure. In the specific context of clouds, the related feature could be the maximum price for using the resources during a time period.

Some of the nonfunctional requirements that can be used to evaluate cloud providers are described below:

- *Portability* is related to the ease with which an application can be executed on various platforms. It includes a type of operating system or image running on VMs.
- *Reliability* includes mean time to repair (MMTR) and mean time to fail (MMTF) of cloud providers, which affect the probability of system failure. The type of environment, either on development, on test, or in production, will decide the level of reliability. It also relates whether an application is acceptable to use Amazon's spot price model, since the spot model sacrifices reliability while low cost can be achieved. Using this model, users can bid for resources that will be delivered only if the bid exceeds the market price of the service but once the bid becomes less than the market price, the computing service can be removed without any notice.
- *Availability* corresponds to the system's ability to respond to user requests, including uptime percentage of the cloud provider and locations of end-users. When multiple machines are required to balance the overload, load balancing is crucial feature to achieve high availability.
- *Efficiency* is a requirement related with the application performance such as expected throughput and response time of an application. Also, it is important to know the type and the amount of workload that applications should support, since optimal scheduling or provisioning techniques that maximize the performance can be chosen based on that information.
- *Cost* is related to how much the customer is willing to pay for a service that satisfies all the other nonfunctional requirements.

Features associated with each nonfunctional requirement are presented in Table 37.1. These nonfunctional requirements are considered to design the architecture for resource provisioning described in the following section.

Table 37.1 *Initial nonfunctional requirements to consider in the system*

Nonfunctional requirement	Relevant features
Portability	Virtual machine's operating system or image
	64 or 32 bit architecture of the H/W and O/S
Reliability	Mean time to repair (MMTF)
	Mean time to fail (MMTR)
	Requires application backup
	Type of environment (development, test, or production)
	Allows spot price model
Availability	Uptime percentage
	End-user locations
	Requires load balancing for high availability
Efficiency	Expected throughput
	Response time
	Type of workload, amount of workload
Cost	Maximum price to pay for a window of time

37.4   System Architecture

The architecture is composed of three independent modules: high-level nonfunctional requirements (NFR), translator, cloud service selector and resource allocator. Figure 37.1 shows the general view of the system. Each of its components is detailed next.

The high-level NFR translator translates the nonfunctional requirements provided by the system administrator into the technical specification for a cloud infrastructure. It receives nonfunctional requirements, such as efficiency, availability, and reliability of a specific application, and estimates the amount of resources, such as number of CPU cores, memory, and storage requirements. The output also includes specific constraints that cloud providers should take into account, such as locations or contract periods, which are evaluated from the given nonfunctional requirements. When it calculates the estimated resources, it interacts with the application profile database to store and retrieve the profile for the application. Furthermore, this module also interacts with the cloud information database to generate the selectable input options, such as available data-center locations and contract periods.

The cloud service selector is responsible for recommending the cloud providers that are more adequate for hosting an application. For this purpose, it receives the estimated size of resources and other parameters such as the constraints and the prioritization of the nonfunctional requirements, and then builds a list of suggested cloud configurations, including the recommended provider. It interacts with the cloud information database, which contains all the information about various resource types in each provider and their pricing information, to apply the constraints and calculate the total cost incurred by the recommended configuration.

The resource allocator searches for available resources based on the recommended providers and acquire resources directly from cloud providers. For this task, it uses the output list given by the cloud service selector and tries to obtain VMs from the most suitable provider. If the requested resources are not available from the top-listed provider, it tries to allocate the resources from the second best provider. The process continues until either all resources are allocated or no more providers are left in the list. It interacts with the provider credential database in order to obtain login credential information for each provider.

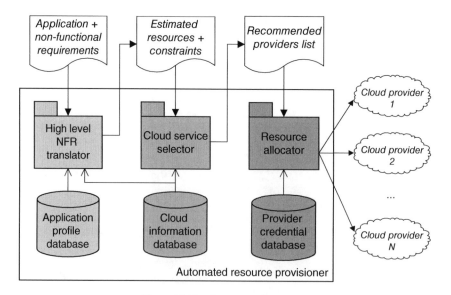

Figure 37.1   *System architecture*

As each component is designed to work independently, one component can be substituted by another program or module with better performance. For example, the declarative recommender system (Zhang *et al.*, 2012) can be used to produce the recommended providers list that is fed into the resource allocator. As their system includes blob storage and network usage costs, it might produce better results for those services, while our cloud service selector focuses more on computing services. In such case, the resource allocator could read the output of the declarative recommender system and try to allocate resources.

37.4.1   System Components

37.4.1.1   *High-Level NFR Translator*

The high-level NFR translator evaluates the nonfunctional requirements and translates them into the technical parameters for cloud providers. The architecture proposed for the cloud resources estimator follows a client-server model of three tiers composed of database, business, and presentation tier.

The presentation tier corresponds to the graphic interface in order to access the application functionality. It allows system administrators to select requirements, change their orders and enter the detailed parameters of each requirement.

The business tier, containing the main logic, retrieves data from the database, creates a workflow to evaluate nonfunctional requirements and performs estimating the amount of resources and determining the constraints. Figure 37.2 presents the class diagram of the business tier.

When the CloudEstimator receives the application and non-functional requirements from the presentation tier, it begins the nonfunctional requirements evaluation process using the WorkflowEvaluator and builds up

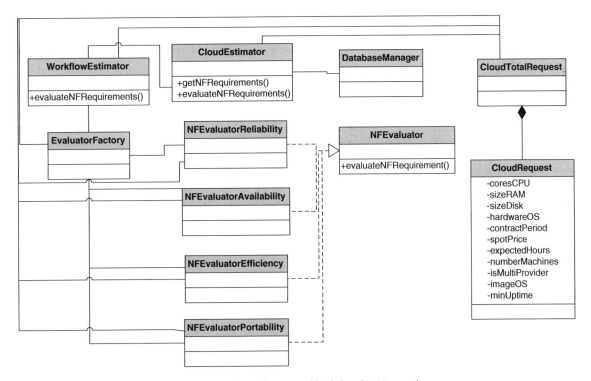

Figure 37.2   *Class diagram of high-level NFR translator*

the CloudTotalRequest object that holds information of required resources. The order of evaluation for each requirement is determined according to its priority, given by system administrators. Also, for each nonfunctional requirement, the associated evaluator class is used to update CloudTotalRequest. For example, NFEvaluatorAvailability is used to evaluate availability requirement and change value in CloudTotalRequest that adds a constraint to meet the availability requirement.

Among various evaluators, the efficiency of the evaluator is a key to estimating the size of resources. In order to evaluate the efficiency and estimate resources, we propose the use of an application profile technique that runs the application components with a given workload. This technique has been widely adopted in the literature (Shimizu *et al.*, 2009; Li *et al.*, 2010, 2011). The results of the execution, containing information about resources consumed and performance achieved, are stored in the application profile database. When the component can be profiled in different infrastructures, better estimations can be achieved (Shimizu *et al.*, 2009).

Once the profile is obtained, the information is fed into an estimator model. The new data obtained using the model should also be stored in the database and possibly updated with the values obtained after the application is in production in the cloud infrastructure. The data can also be used to update the model itself to increase its accuracy. After profiling and modeling the application and estimating the required resources, the output is generated by updating the existent CloudTotalRequest with the result of the estimation. Figure 37.3 summarizes the process.

In the database tier, details of the components that compose the applications are stored along with their profiles when executing on different infrastructures. An application comprises several components, and the throughput of the entire application can be predicted based on the performance of individual components and the cost of communication among them, as discussed by Stewart and Shen (2005). This approach has the advantage of allowing the estimation of the application performance with components deployed in different infrastructures and possibly in different cloud providers. It also retrieves entries from the cloud information database to show a list of requirement options for input parameters when the initial view is available.

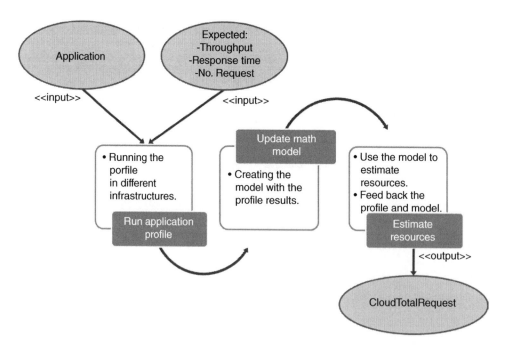

Figure 37.3   *Steps executed by the efficiency evaluator in high-level NFR translator*

37.4.1.2 *Cloud Service Selector*

The cloud service selector lists the recommended providers and their services by analyzing the estimated resources and constraints. The components are designed following the Model-View-Controller model, as it can easily separate each component by reducing dependencies on other components. Figure 37.4 shows the architecture of the cloud service selector.

The component receives the estimated resources and constraints as input from the high-level NFR translator through its view subsystem. Once the information is provided to the controller, it retrieves records satisfying the requirements from the model through a query to the cloud information database. The database stores information about providers and their services including resource configurations, geographical locations, operating systems, uptime percentages, contract periods, prices, and other parameters to evaluate the constraints and the cost.

The requirements are used as constraints for the model – for example, the retrieved records should have more resources than the requirements, matched location and operating system, and a shorter contract period. When three months' contract is specified, for example, a service with a 1 month contract can be retrieved but a service with a 6 months contract cannot.

Once it gets all candidate providers and their service types, the model calculates the expected price based on the use period. For some providers, such as GoGrid, only the contract period affects the total price. However, in other cases, such as Amazon's Reserved, the actual running time of the instance influences charges, in addition to the upfront fee for the contract establishment. Hence, the actual use period should be included to calculate the correct price. Once the contract and the actual use period are input by the system administrator, total cost can be simply calculated using pricing information stored in the cloud information database.

After applying constraints and calculating the actual price for each provider, a list of providers is generated and prioritized according to the order. If the system administrator selected price as the most concerning aspect, the cheapest provider will be the first entry of the output list. In other cases, the one with greater percentage of uptime will be the first if the administrator weights availability more than the cost.

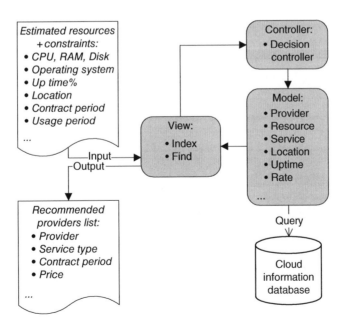

Figure 37.4 *Component architecture of cloud service selector*

37.4.1.3   *Resource Allocator*

The resource allocator interacts with providers and requests allocation of the resources defined in the list of specifications. The component requires a cloud configuration list as its input, which is the output from the selector. Each row in the list contains information about the provider, resource set, location, image information, and the number of machines to be allocated.

Figure 37.5 describes the flow of the resource allocator. It starts parsing the input configuration list and building a collection of cloud configuration objects storing the parsed information. Once all elements are

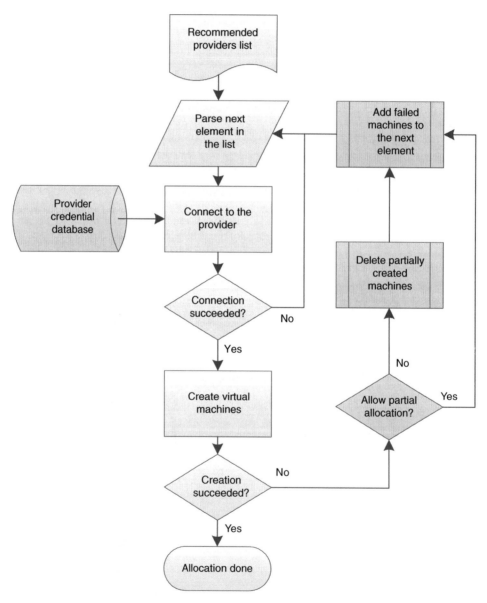

Figure 37.5   *Flowchart of resource allocator*

parsed, it tries to create VMs for the provider using those objects. When the system connects to the provider, it authenticates the identity using credentials stored in provider credential database. If the connection is successfully established, it requests the number of VMs with the specific information of resource, location, and image to the cloud provider with a timeout. Once either the allocation succeeds or the timeout is exceeded, the allocator analyses the results and proceeds to the next option.

Additionally, partial allocation on different providers is supported by the resource allocator. Depending on the selection of whether the allocation needs to occur on a single provider or can be split among multiple providers, there is an additional process to be carried out. If the partial allocation to several providers is allowed, the remaining number of VMs is added to the next option in the list. If the input specifies that only one provider can be used, it does not proceed with the allocation and repeats the process for the next entry.

37.4.2   Implementation

In order to illustrate and evaluate the functionality of the proposed architecture, we implemented a prototype for the high-level NFR translator, cloud service selector, and resource allocator. Each component is developed separately in order to ensure the independence of the component.

The high-level NFR translator is implemented using JavaEE, JavaEE GlassFish and MySQL, and is developed on a three-tier basis including the application profile database. The component's prototype is configured with four nonfunctional requirements: portability, availability, reliability, and efficiency. The requirements are selected with the subfeatures described in Table 37.1, which are necessary to create a resource estimation request.

The cloud service selector is implemented with conventional model-view-controller architecture using Ruby on Rails. The cloud information database is also included to store different resource types, pricing policies, and datacenter locations of each provider. By enabling the input of resource requirements, it displays the prioritized providers' list as an output.

The resource allocator is implemented with Java on console interface. In order to cover various providers in a single program, we use Jclouds, multicloud supporting library, written in Java. Jclouds allows developers to use homogeneous APIs to connect to different clouds, thus developers do not need to use different APIs for different providers. Instead, Jclouds provides a single interface to connect, create, and destroy VMs. Also, in the prototype, the provider credential database is implemented as a list of provider name, user name, and password, and parsed by the resource allocator.

37.5   Performance Evaluation

Evaluation is performed independently for each module of the system, as they are designed and developed independently. We evaluate the high-level NFR translator in order to have its performance measure. The cloud service selector is assessed in order to evaluate the functionality of selecting the cloud provider candidates that meet requirements, and to evaluate the benefits of prioritizing the candidate providers according to nonfunctional requirements. Finally, the performance of the resource allocator is also evaluated.

37.5.1   High-Level NFR Translator

The goal of this experiment is to evaluate whether the proposed architecture is able to scale dynamically when the number of requirements and requirements' features increase.

The platform used to test the application is the Australia Research Cloud NeCTAR (http://www.nectar.org.au/research-cloud, accessed December 27, 2015). This experiment uses an instance of type m1.small. Instances of

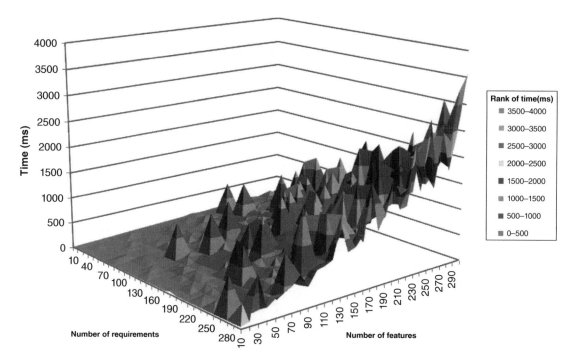

Figure 37.6   *High-level NFR translator performance*

this type have 4 GB of RAM, 1 CPU core and 10 GB of disk. The Operating System used in the VM is Ubuntu 13.04.

Requests with a varying number of requirements and a varying number of features per requirement are generated. The number of nonfunctional requirements varies from 10 to 300, in steps of 10. For each increment of requirements and their features, we measure the time taken to evaluate nonfunctional requirements and to transform them into technical parameters. The result, shown in Figure 37.6, demonstrates that the prototype scales satisfactorily when the number of requirements increases. It takes less than 4 s with 300 requirements and 300 features input, which is an acceptable time to process such a number of requirements.

37.5.2   Cloud Service Selector

The cloud service selector should be able to select a valid service and calculate the precise price based on the given requirements. It also should prioritize the provider candidates according to the priority of the nonfunctional requirements as defined by the system administrator.

For the evaluation of the selector, the contents in the cloud information database are crucial to the result, as the quality of the result depends on the database information. For the purpose of this this experiment, we create a static database with synthetic data that allowed us to test the functionality. Table 37.2 shows a part of the database elements used for the experiment, with various instance types and pricing schemes of different providers. In order to validate the system, we input the parameters of the user case described in Table 37.3 into the cloud service selector, and obtain the result of the ordered list.

The results (Table 37.4) show that the system suggests only services whose resources are greater than required, with any operating systems, in any locations, and for any contract periods less than 3 months. For example, the service from Provider 1 with 12 months contract is excluded, and the lowest cost service from

Table 37.2 *Database information used for cloud service selector experiment*

Provider	Resource type	CPU (cores)	RAM (MB)	Disk (GB)	OS	Datacenter location	Contract period	Resp. time (ms)
Provider 1	m1.small	1	1700	160	Windows	Oceania	None	10.39
Provider 1	m1.small	1	1700	160	Windows	Oceania	12 month	10.39
...								
Provider 2	small	1	1000	50	Linux	Europe	1 month	25.40
...								
Provider 3	custom	(Max) 20	(Max) 32 000	(Max) 1024	Linux	Europe	1 month	31.60
Provider 3	custom	(Max) 20	(Max) 32 000	(Max) 1024	Linux	N. America	1 month	18.02
Provider 3	custom	(Max) 20	(Max) 32 000	(Max) 1024	Linux	N. America	3 month	18.02
...								

Table 37.3 *User case for cloud service selector validation*

Field	Value
Resource	CPU = 1 RAM = 512 MB Disk = 20 GB
Operating system	Any
Location	Any
Contract period	3 months
Allow unreliable services?	No
Usage period	3 months
Order by	Cost

Table 37.4 *Results from cloud service selector*

Rank	Provider	Resource type	OS	Datacenter Location	Contract period	Resp. Time	Price
1	Provider 3	custom	Linux	N. America	3 months	18.02	81.11
2	Provider 3	custom	Linux	N. America	1 month	18.02	83.57
3	Provider 3	custom	Linux	Europe	1 month	31.60	91.80
4	Provider 2	small	Linux	Europe	1 month	25.40	108.75
...							
7	Provider 1	m1.small	Linux	Oceania	None	10.39	172.80
...							

Provider 3 is placed in the first rank. In addition, we evaluate the benefits of prioritizing by another nonfunctional requirement: response time. When an enterprise can make more profit with faster response time by providing better user experience, it will be willing to pay more for it. The system works precisely as it prioritizes the provider with fastest response time to the top of the list although its price is higher than others. As a result, the services of Provider 1, previously ranked number 7 in order of price, become the most suitable service with the fastest response time.

37.5.3 Resource Allocator

The resource allocator is tested using the list of cloud configurations that is obtained from the selector. Although the system supports every provider supported by Jclouds, we evaluate it with "t1.micro" instances, cost-free VMs provided by Amazon's EC2 service.

We build a cloud configuration list with two candidate configurations with Amazon-EC2, two t1.micro instances, Ubuntu 12.04 and Australia (ap-southeast-2) location, and another with an east United States (us-east-1) location. The allocator firstly tries to acquire VMs described on the first element, the Australia one. If the first request fails due to lack of resources in Australian datacenters, the allocator attempts to the next candidate. When the system was executed, it succeeded in allocating two machines in Sydney, and we can find the successfully created machines on the control panel of Amazon's web site.

37.6 Conclusions and Future Directions

Cloud computing enables a major paradigm shift in the way that computing resources are acquired. Without any hardware acquisition, system administrators are able to obtain computing power to deploy their services within minutes using cloud computing. It also makes it possible to pay only for resources consumed with no minimum contract and upfront costs.

However, deploying applications in the cloud is still a complex task for system administrators. They are expected to estimate resources required by their applications, which may be difficult because they frequently do not know exactly how many resources are actually necessary. Furthermore, administrators have to select the best cloud service amongst various providers and different types of services, and acquire them for applications to deliver the expected performance.

In this chapter, we proposed an architecture supporting system administrators in the arduous task of deploying applications on the clouds in three ways. Firstly, it translates nonfunctional requirements from administrators into cloud resource parameters. Secondly, it selects the most convenient provider among different candidates, which satisfies every requirement. Finally, the actual VMs are allocated automatically from the selected provider.

The proposed architecture is also verified through evaluation and validation. Each component is validated in performance and scalability for various sets of nonfunctional requirements. Also, we show that the number of VMs with adequate resources is actually allocated from the selected cloud provider at the end of the process.

We recommend a few areas for further work. The proposed architecture can be applied to measure the performance of various techniques in each module. Several resource estimation techniques can be applied to the estimator in the architecture, which can lead to the most accurate methodology to estimate resources in the clouds being found. Similarly, different approaches to select the best provider can be used for the selector module. In addition, the cloud information database can be improved by applying dynamic updates; thus, it will keep the consistency between the system and the providers, and will provide more accurate selection by including real-time metrics measured by monitoring services. Finally, dynamic resource provisioning can be applied to the system, which can dynamically perform the whole provisioning process depending on the real-time workload measured from the running application.

References

Buyya, R., Ranjan, R. and Calheiros, R. N. (2010) *InterCloud: Utility-Oriented Federation of Cloud Computing Environments for Scaling of Application Services*. Proceedings of the Tenth International Conference on Algorithms and Architectures for Parallel Processing. Springer-Verlag, Berlin, p. 13.

Garg, S. K., Versteeg, S. and Buyya, R. (2011) SMICloud: A framework for comparing and ranking cloud services. Proceedings of the Fourth IEEE/ACM International Conference on Utility and Cloud Computing (UCC'11), pp. 210–218.

Glinz, M. (2007) On non-functional requirements. Proceedings of 15th IEEE International Requirements Engineering Conference (RE '07), pp. 21–26.

Grozev, N. and Buyya, R. (2014) Inter-Cloud architectures and application brokering: taxonomy and survey. *Software: Practice and Experience* **44**(3), 369–390.

Li, A., Yang, X., Kandula, S. and Zhang, M. (2010) *CloudCmp: Comparing Public Cloud Providers*. Proceedings of the Tenth ACM SIGCOMM Conference on Internet Measurement (IMC '10). ACM, New York, NY, pp. 1–14.

Li, A., Zong, X., Kandula, S., *et al.* (2011) *CloudProphet: Towards Application Performance Prediction in Cloud*. Proceedings of the ACM SIGCOMM 2011 Conference (SIGCOMM'11). ACM, New York, p. 426.

Rak, M., Cuomo, A. and Villano, U. (2013) Cost/performance evaluation for cloud applications using simulation. Proceedings of 2013 IEEE 22nd International Workshop on Enabling Technologies: Infrastructure for Collaborative Enterprises (WETICE 2013), pp. 152–157.

Shimizu, S., Rangaswami, R., Duran-Limon, H. A., and Corona-Perez, M. (2009) Platform-independent modeling and prediction of application resource usage characteristics. *Journal of Systems and Software* **82**(12), pp. 2117–2127.

Stewart, C. and Shen, K. (2005) Performance modeling and system management for multi-component online services. *Proceedings of the Second Conference on Symposium on Networked Systems Design and Implementation*, Berkeley, CA, p. 71.

Zhang, M., Ranjan, R., Nepal, S., *et al.* (2012) *A Declarative Recommender System for Cloud Infrastructure Services Selection*. Proceedings of the Ninth International Conference on Economics of Grids, Clouds, Systems, and Services (GECON 2012) (eds. K. Vanmechelen, J. Altmann and O. Rana). Springer-Verlag, Berlin, Heidelberg, pp. 102–113.

38

Legal Aspects of Cloud Computing

David G. Gordon

Carnegie Mellon University, USA

38.1 Introduction

Prior to their widespread adoption, it was clear that cloud technologies would pose problems for current legal systems (Armbrust *et al.*, 2009). Many laws and regulations were written under the pretense that organizations handled all of their IT needs themselves, and any that were handled by another party could be done so "cleanly" (i.e. with a clear division of duties), albeit with significant time and effort. Thus, while laws that affected information technology were not without ambiguity, the line between a client and entity that provided a service to or on behalf of that client was fairly distinct.

Cloud computing challenged this assumption in many ways. First, the division of duties became less clear, as clients could choose to outsource *degrees* of their IT needs as delineated by the three types of cloud services: infrastructure as a service (IaaS), platform as a service (PaaS), and software as a service (SaaS), as addressed in previous chapters. Second, increasing bandwidth, storage, and processing capacities enabled this provisioning of services to be accomplished faster, easier, and more seamlessly than ever before, sometimes with the click of a button. Current laws were not designed for such rapid change or the increasing probability that provider and client would be governed by different legal systems, sometimes without either having knowledge of this occurring (Mather *et al.*, 2009; Mowbray, 2009). Third, laws were already contending with the growing use of social networking and other uses of personal information in online transactions, and cloud computing only served to complicate these matters further.

This chapter is intended to serve as an introduction to law and the cloud, and will provide basic information regarding the legal landscape in which the cloud exists; a survey of major laws across the United States, Europe, and elsewhere that affect cloud operations (Sotto *et al.*, 2010); the complex issue of transborder data flows, and the variety of mechanisms available to organizations to achieve compliance; and the usage of contracts between cloud service providers and their clients. The chapter will focus on legal concerns of cloud service providers, their clients, and the relationship between the two, based on

Encyclopedia of Cloud Computing, First Edition. Edited by San Murugesan and Irena Bojanova.
© 2016 John Wiley & Sons, Ltd. Published 2016 by John Wiley & Sons, Ltd.

US and European jurisprudence. The laws spoken to herein are frequently reinterpreted, supplemented, or modified, and there remains ambiguity in their application. As such, only the most basic, fundamental components of these laws will be considered, and under no condition should any content be used as a substitute for legal counsel.

This chapter will use the following terms and meanings frequently for convenience; they are *similar, but not identical to*, those found in many major data protection laws worldwide. Additionally, despite differences in authority and origin, for the purposes of this chapter, the term "law" will refer to laws, regulations, statutes, and directives unless otherwise stated.

- *Cloud Service Provider* (*CSP*): an entity that provides cloud services to a cloud client, e.g. IaaS, PaaS, and SaaS.
- *Cloud Client* (*CCL*): an entity that obtains cloud services from a cloud service provider.
- *Data*: most laws affecting the cloud described in this chapter are concerned with the security and privacy of personal information (Hon *et al.*, 2011), so *in the context of this chapter*, "data" will refer to personal information about an individual, such as first name and last name, address, telephone number, and so forth, unless otherwise indicated.
- *Jurisdiction*: a geographic area with an established legal system or body of law and means of enforcement, for example California, United States, or European Union. Note that a single location can be considered to fall under multiple jurisdictions; for example, an entity located in the state of California is subject to both California State law and Federal law.
- *Trans-border data flow*: the action of data being transferred from one jurisdiction into another.

38.2   Law, Technology, and the Cloud

Previously issued laws and regulations governing data handling and usage – even those enacted as recently as the late 1990s and early 2000s – were largely created under the assumption that the IT infrastructure of their time had neither the bandwidth, storage, controls, nor architecture for data to change hands or jurisdiction without considerable effort. The emergence of cloud computing reduced this effort, as cloud service providers enabled businesses and organizations to provision computational resources instantly across the globe in an adaptable, portable, and highly redundant environment. In so doing, however, this technology raised a number of complex legal issues that previous regulations had not anticipated. To address these matters, past laws would need to be reinterpreted or even amended, or new laws would need to be introduced.

Broadly, laws that affect the cloud span both public and private law, which serves as a natural divide for their discussion. The first of these, public law, pertains to the relationship between individuals and the government, or relationships between individuals that are relevant for the rest of society. With regards to the cloud, this area primarily concerns the hundreds of regulations issued by governments across the world designed to ensure the privacy and security of their citizens' information, such as the Health Insurance Portability and Accountability Act (HIPAA) in the United States or the Data Protection Directive (EU 95/46/EC) in the European Union (Jaeger *et al.*, 2008). Private law, in general, covers relationships between private individuals, such as the law of contracts or torts. In the cloud, this includes contracts between parties – namely, cloud service providers and their clients – which define their relationship and the expected level of service between them. Often these contracts may include obligations prescribed by public law. Both types of law will be discussed later in this chapter. However, before doing so, we will first provide an outline for the basic structure of these laws.

38.2.1 Structure of Laws

Many laws are structured similarly, including the data privacy and security laws addressed in the next subsection. Understanding aspects common to these laws will be helpful not only in understanding each law as it stands alone but also how they compare with one another. For the purposes of this chapter, we are chiefly concerned with:

- *Definitions*: terms found in the law that are given explicit meaning with a certain scope, such as throughout a statute or chapter. Definitions, which are sometimes technical in nature, can be used for a variety of concepts, such as stakeholders ("healthcare facilities"), objects or concepts ("computer", "encryption") or actions ("unauthorized access"). Definitions serve to give the law a common vocabulary and explicitly set forth the meaning of significant terms. They are seldom considered "complete," in that they may apply to an object or concept that is not stated verbatim, such as a cell phone being classified as an "electronic device that is capable of performing high-speed data processing."
- *Stakeholders*: types of entities that the law affects, and they can include individual persons, businesses, governments or government agencies, or the like. As mentioned, stakeholders are often explicitly defined early in the text.
- *Requirements*: actions the law imposes on stakeholders, including obligations, permissions, and prohibitions, which are actions a stakeholder must, may, or may not do, respectively.
- *Conditions*: criteria that appear before a requirement, or otherwise qualify a stakeholder or action; for example, a cloud service provider who "processes data on behalf of a client" (condition) "shall keep that data secure" (requirement).
- *Penalties / right to action:* the repercussions that can occur when an aspect of the law is not satisfied, for example when a cloud service provider neglects to secure data when the law obligates them to do so. Penalties can take numerous forms, such as monetary fees or restrictions on the operations of an organization, and may be brought about by different parties specified in the law.
- *Enforcement:* a legal body that is responsible for ensuring compliance with the law or affecting penalties for violations, and the means through which they may do so.

38.3 Major Laws and Regulations

Cloud service providers and their clients should be aware of the large number of laws that affect their cloud operations, the most significant of which are those enacted to protect the privacy and security citizens' personal information. These laws, which began to take shape in the early 1970s and have continued to be introduced and refined in the present day, prescribe practices, guidelines, and restrictions on the handling and uses of personal information, such as social security numbers or financial data. Many of the topics addressed by these laws relate to one or more of the recommendations for the protection of personal data issued by the Organization for Economic Cooperation and Development (OECD) in 1980, including:

- *Notice*: individuals should know when their information is collected or used.
- *Purpose*: an individual's information should be used for its stated purpose and no other.
- *Consent*: an individual's information should not be collected or used without their advance consent.
- *Security*: information should be protected from threats, both internal and external.
- *Disclosure*: individuals should know who possesses or uses their data.
- *Access*: individuals should be able to access their information, make corrections and alterations within reason, and control access to that information.
- *Accountability*: individuals should have means to hold organizations accountable for these principles.

The style of regulation adopted varies by jurisdiction, and often reflects correspondingly varied underlying intents. The United States, for example, lacks an all-encompassing data-protection law, relying instead on a patchwork of regulations divided by industry or function; for example, the HIPAA for healthcare, or the Gramm–Leach–Bliley Act for the finance sector. In contrast to this, most other nations instead rely on omnibus-style data-protection laws: that is, laws that apply to all data regardless of the industry or circumstance in which it is used. Prominent examples of these laws include the Personal Information Protect and Electronic Documents Act (PIPA) of Canada; the European Data Protection Directive 95/46/EC, its successor, the General Data Protection Regulation, and the Information Technology (Reasonable Security Practices and Procedures and Sensitive Personal Data or Information) Rules from India. In addition to protecting citizens' personal information (Ruiter and Warnier, 2011; Pearson, 2013), these laws serve other purposes, such as creating an attractive legal environment for data outsourcing. This chapter will focus on US and European law.

This section serves as the barest of introductions to some of the more prominent laws affecting the cloud environment for both providers and consumers of cloud services. As many of the laws themselves exceed the length of this chapter several times over and are subject to frequent updates and amendments, details contained herein should be used for preliminary guidance purposes only and are by no means comprehensive nor a substitute for legal advice.

38.3.1   United States of America

The United States has many laws that affect the cloud. As previously mentioned, the US lacks an omnibus data privacy and security law, and instead relies on laws to address each industrial sector individually. This allows each sector to create rules and requirements specifically relevant to their operations but it can also lead to difficulties and confusion over legal coverage as well as incompatibility with the data-protection efforts of foreign countries, nearly all of which rely on laws that apply to any body handling personal information. While laws have been introduced explicitly to regulate the cloud, e.g., the Cloud Computing Act of 2012 (S.3569), they have not gained sufficient traction in the United States Congress. This section will focus primarily on laws governing healthcare, financial, and government organizations, with additional notes on laws addressing specific issues, such as the Children's Online Privacy Protection Act.

38.3.1.1   Healthcare

Cloud operators serving the healthcare domain or healthcare organizations wishing to move their operations to the cloud are subject to the HIPAA and amendments made to it through the Health Information Technology for Economic and Clinical Health Act. Early on, the requirements contained in HIPAA's Privacy and Security Rules (45 C.F.R. 160, 162, and 164) did not necessarily apply to third parties who performed data storage or processing on behalf of a healthcare organization. As the law has matured and changed, however, cloud service providers are now considered "business associates" under HIPAA and through mandated "business associate agreements" are subject to many of the same requirements as healthcare organizations themselves. However, if a CSP acts as a "mere conduit" for data (like the United States Postal Service delivering health information) and does not have "routine access" to the information, it is not considered a business associate (Hall, 2013).

Additionally, all states and territories in the United States have their own laws regarding medical records or electronic healthcare records specifically. It has been established, however, that HIPAA provides a legal floor, in that when state requirements fall below the standards required by HIPAA, those prescribed by HIPAA preempt those issued at the state level. Some CSPs specifically provide services to the healthcare industry and claim to be HIPAA compliant. This title is not official, but often means that the organization has undergone an independent audit by a third party who will measure their practices against a set of testing criteria, such as the HIPAA audit protocol created by the Office for Civil Rights (Delgado, 2011).

HIPAA prescribes a number of safeguards divided into three categories: physical, technical, and administrative. Under a business associate agreement, cloud providers take on responsibility for some of these controls and share responsibility for others. While the CSP may assist the healthcare organization in achieving these controls, it is ultimately the organization itself that is responsible for ensuring that they are accomplished. Examples of each safeguard type are as follows:

- *Administrative*: periodic risk assessments and ongoing risk management, workforce security and privacy training, password management, security incident response procedures, data backup and recovery, login monitoring, workforce authentication and authorization.
- *Physical*: media disposal or reuse policies and procedures, facility access controls, workstation authorization, data backups and storage, maintenance records, access control and validation.
- *Technical*: automatic logoff, software auditing controls, encryption, means for verifying data integrity, transmission security, unique user identification.

Following the implementation of the Health Information Technology for Economic and Clinical Health (HITECH) Act, HIPAA also includes policies and procedures to be followed in the case of a data breach. Significantly, breaches of deidentified or encrypted data (when the encryption keys have not also been breached) are not considered breaches.

38.3.1.2   Corporate and Finance

The Gramm–Leach–Bliley Act (GLBA) or Financial Services Modernization Act of 1999 primarily served to remove market barriers between banking, security, and insurance companies. Included in the Act, however, were notable provisions for consumer privacy and protection of consumer data possessed by these and other financial institutions in the form of the Financial Privacy Rule and the Safeguards Rule (15 USC 6801–6809). The Financial Privacy rule contains many similar provisions to other privacy laws, ensuring that consumers are informed through a privacy policy about how their data is collected, used, shared, and protected. Customers must also be notified of their right to opt out of having their data shared with "unaffiliated parties" as granted under the Fair Credit Reporting Act. The Safeguards Rule mandates that financial institutions must have a documented information security program that ensures security and confidentiality for consumer data, protects against threats to those records, and prevents unauthorized access to the records that could harm the customer. This includes having an employee or department dedicated to the purpose, conducting periodic risk assessment of departments that handle private information, ongoing monitoring and assessment of security programs, and ensuring that safeguards are kept relevant for any changes in information collection or usage. The rules require that data be encrypted; however, organizations may choose not to encrypt data provided they can justify their position to the Federal Financial Institutions Examination Council (FFIEC). Organizations that handle common forms of payment data, such as credit card information or point-of-sale (POS) cards should consider adhering to the Payment Card Industry Data Security Standard (PCI-DSS).

Financial institutions looking to move portions of their services to the cloud should explicitly look for providers that have been proven through third-party audits to be compliant with the GLBA. Because the act requires that consumers be thoroughly informed with regards to the movements of their data, the CSP must be able to provide information about its practices to the client. Further, depending on the nature of the agreement between the CSP and their client, the CSP may be considered a nonaffiliated third party, meaning that the client would be prohibited from sharing information with the CSP without first obtaining consumer consent.

The Sarbanes–Oxley Act, or SOX, sets a number of standards for public companies and accounting firms, some of which can significantly affect IT architectures. Many of these standards relate to accountability, which can involve frequent and detailed logging of activities as well as widespread user authentication. Cloud service providers looking to specialize in this area may wish to consider going through a Statement on Auditing Standards (SAS) 70 audit developed by the American Institute of Certified Public Accountants (AICPA).

38.3.1.3   *Government and More*

Businesses aren't the only organizations that seek to capitalize on the advantages of cloud computing. In 2010, Vivek Kundra (then serving as the first Chief Information Officer of the United States) issued a report that included the "Cloud First" initiative, which called for increased use of cloud computing technologies in government operations. In order to supply cloud services to federal agencies, however, providers must be compliant with the Federal Information Security Management Act (FISMA) of 2002 – most importantly the FIPS 200 standard. Compliance with FISMA is not easily achieved and involves significant risk assessment regarding agencies' operations and information systems. The minimum security and assurance requirements, as well as controls that may be used to satisfy them, can be found in NIST Special Publication 800-53.

In order to facilitate compliance with FISMA, the Office of Management and Budget developed the Federal Risk and Authorization Management Program (FedRAMP). This program can assist government organizations in determining if CSPs meet compliance, authorizes third-party assessment organizations (3PAOs) to conduct independent audits of CSPs prior to approval or rejection by FedRAMP, and maintains records of CSPs that have been reviewed.

There are a number of other regulations that affect both federal and private institutions, including the Children's Online Privacy Protection Act (COPPA), which includes specific requirements for obtaining and handling data on persons under the age of 13; the multitude of data breach notification laws enacted by each state, which mandate that organizations notify affected individuals (be they data subjects or clients) about data breaches in their systems; and the Family Education Rights and Privacy Act (FERPA), which provides various protections and rules regarding storage of and access to adult educational records.

When the Uniting and Strengthening America by Providing Appropriate Tools Required to Intercept and Obstruct Terrorism (PATRIOT) Act of 2001 was passed following the events of September 11, 2001, it was uncertain how it would affect the emergent field of cloud computing. In short, the PATRIOT Act permits the United States government to obtain records and information that pertain to foreign intelligence or international terrorism using less accessible venues than required for investigation of domestic crimes. Despite claims that warrantless seizures of data are minimal, this act has been *construed* by many – particularly other countries – to permit the United States government unfettered access to any data kept in the cloud. While the chilling affect this has had on adoption of American cloud services by non-American organizations has not been measured, a number of high-ranking officials, such as the European Vice Commissioner Viviane Redding, have openly criticized the Act's effects on the industry and discouraged use of American cloud services for this very reason.

38.3.2   European Union

Privacy is viewed as a human right in Europe, having been codified under Article 8 in the European Convention on Human Rights treaty in the 1950s. With regards to data, this right has been protected under a number of regulations, including the EU Data Protection Directive 95/46/EC, its proposed revisions in the form of the General Data Protection Regulation issued in 2012, and the various regulations enacted by member states, such as Germany's Bundesdatenschutzgesehzt or the United Kingdom's Data Protection Act. The broad applicability of these regulations, as well as the strictness with which they are enforced, have resulted in them being the standard by which other jurisdictions' regulations are compared. Cloud Service Providers operating in the European Union, on behalf of organizations located in the European Union, or with data from European citizens, must ensure their practices are in line with these regulations.

38.3.2.1   Data Protection Directive (95/46/EC)

Unlike the United States, which governs data by the industrial sector in which it falls (e.g. healthcare, finance, etc.) most of the EU's data-protection laws are considered nonsectoral, in that they apply to all personal information, regardless of its nature. Significantly, the definition of personal information used in these laws is extremely broad, going well beyond common attributes such as address or phone number to include "any information relating to an identified or identifiable person". The law also makes a distinction between data controllers and data processors, in that processors store, operate on, or otherwise process data on behalf of controllers. In most cases, cloud service providers are considered data processors. However, under certain circumstances they may be considered data controllers: for example, when they process data stored for another organization for their own purposes. While controllers are considered responsible for meeting the criteria set forth in the directive, they are permitted to delegate these responsibilities onto data processors, so regardless of their status, CSPs should be aware of the requirements contained in these regulations.

If a CSP uses subprocessors, the CSP must ensure that the subprocessors are made aware of the legal requirements attached to the data and that the original data controller is aware of and has consented to the transfer. Cloud service providers are also responsible for assisting their clients in complying with access requests by data subjects. The Article 29 Working Party recommends that the requirements should not be "dispersed throughout the chain of outsourcing or subcontracting" so that responsibility is clearly allocated.

In principle, the directive itself is not legally binding, and member states are responsible for implementing and enforcing it through their internal law. All member states have enacted their own data protection regulations, which are enforced to varying degrees. The directive also contains provisions pertaining specifically to transborder data flows. These special circumstances are discussed in detail in the next subsection of this chapter.

The Data Protection Directive was followed years later by the Directive on Privacy and Electronic Communications or E-Privacy Directive (2002/58/EC), which establishes the right to privacy for the electronic communication sector. Cloud service providers that contribute to these services in the public communications sector should also consider the provisions of this regulation as well as those specified in its revisions (2009/136/EC).

38.3.2.2   General Data Protection Regulation

In 2012, the European Commission released a draft of the General Data Protection Regulation (GDPR), which was intended to harmonize disparate laws among member states as well as address concerns brought about by new technologies, which include cloud computing (Schellekens, 2013). Unlike Directive 95/46/EC, the GDPR is a regulation and thus has direct effect without being implemented by member states. The regulation is expected to go into effect in 2016, and at the time of this publication its contents are still under discussion.

In general, the Data Protection Regulation can be thought of as revising and extending EU Directive 95/46/EC; that is, most of the requirements contained in the directive are retained and strengthened in the regulation. Among the potential changes that are of concern for CSPs and their clients are an increase in noncompliance fees (up to 1 000 000 euros, or up to 2% of global annual company turnover), notification obligations similar to those found in the United States, prescription of privacy by design practices, the creation of data protection officers within each organization, and a "right to be forgotten." This last item requires that individuals are able to have their data deleted if there are no legitimate grounds for its retention. Such a right, if preserved, could have significant implications for data backups maintained in the cloud.

38.3.2.3   *Data Protection Authorities*

Each member state in the European Union has a data-protection authority that cloud service providers and their clients are likely to interact with in some capacity (EU Agency for Fundamental Rights, 2010). Essentially, each authority is a legal body that is responsible for enforcing data-protection laws within that jurisdiction. If a CSP or client operates out of a certain state or works with a significant amount of personal information on that state's citizens, they should be aware of the practices of the authority for that state. Unfortunately, the authorities vary widely from state to state; some are sectorally based, some are not; some are federal agencies, others exist as substate agencies. Of primary concern should be the authority's preferred method of enforcement (preventative, or after the fact), how aggressively the authority pursues violators, and whether or not the authority needs a judicial warrant prior to accessing cloud resources.

38.3.3   **Here, There, Everywhere**

While the United States and European Union may lead the way with regards to regulating the cloud, many other jurisdictions have made similar efforts. In addition to the primary objective of protecting their citizens and businesses, some of these laws have been issued in an effort to meet or exceed the standards issued in other countries (such as India's enactment of the 2011 Information Technology Rules), thereby creating attractive outsourcing markets. As it has become easier for businesses to reach new markets or take advantage of overseas technology services, CSPs and their clients should maintain awareness of these laws so should opportunities arise they will be able to act quickly. Cloud service providers should also be informed should they want to take advantage of the ability to subcontract some of their work to another CSP in one of these jurisdictions. In this section, we briefly describe recent laws enacted in countries outside of the United States and European Union, including Japan, India, and South Korea. Only the most significant features of laws in each country are discussed, for reasons of space.

38.3.3.1   *Japan*

Japan's Personal Information Protection Act, enacted in 2005, shares many similarities with the EU Directive as well as US laws. As the law itself was based on the EU Directive, it uses a similar (broad) definition of personal information and prescribes requirements in line with those for the European Union. However, the Act itself is interpreted by sectorally based government ministries who then issue administrative guidelines for their sector. Thus, clients from different industries may have different requirements, as in the United States. Its greatest difference from these other laws is that it provides exemptions for small data sets (5000 individuals or less). It should be noted that the Personal Information Protection Act does not apply to public agencies, who are instead affected by a similar law intended for "Administrative Organs."

38.3.3.2   *India*

India has a number of laws for the information technology sector, but of greatest concern is the Information Technology (Reasonable Security Practices and Procedures and Sensitive Personal Data or Information) Rules issued in 2011. As with Japan's Personal Information Protection Act, this law closely mimics the EU Directive in most aspects. However, it should still be viewed with great care, as the law has been subject to frequent reinterpretation since its creation. For example, its consent requirements (considered extremely strict) were first applicable to third parties, meaning that in order for an Indian organization to work with personal information on behalf of another firm (native or foreign) the Indian organization would first have to obtain consent from the data subject, and do so again upon any major change in practices. This was later clarified to not apply to organizations working with firms under a contractual obligation. Due to the uncertain stability of the law, CSPs and their clients should make sure their practices are in line with the latest interpretations before determining or claiming compliance.

38.3.3.3 South Korea

South Korea's Personal Information Protection Act (PIPA), which came into force in 2012, is considered one of the strictest privacy laws in existence. Given the country's strong track record for enforcement in this area, it is possible that it will be the precursor to similar laws in other countries. As with the previously mentioned laws, the PIPA is similar to the EU Directive. However, it differs significantly in that it mandates minimal collection of data as well as prescribing anonymization of data (if possible), creation of the position of privacy compliance officers within each organization, and requires that data subjects be notified of any processing performed by subprocessors. This latter requirement is particularly important for cloud service providers, who occasionally use subprocessors to provide their services. This is also discussed briefly in section 38.5.

38.4 Handling Transborder Dataflows

Many CSPs operate within the same jurisdiction as the clients that they serve (Narayanan, 2011). However, some offer their services to those beyond their borders in order to take advantage of lucrative foreign markets or more affordable operating costs. When the client's data leaves its municipality of origin for storage and processing elsewhere, a transborder dataflow is created. This introduces new legal complexities to the situation as the different parties operate under different legal regimes. In this context, which laws apply, to whom, and when? If a cloud service provider in India receives medical information regarding California patients in the United States, is it governed by the Health Insurance Portability and Accountability Act? What if the HIPAA prescribes data privacy requirements fundamentally incompatible with those found in India? If the Indian organization experiences unanticipated downtime or suffers a data breach, what responsibilities do the parties have to one another as well as to the patients themselves?

Fortunately, there are a variety of options to address these challenges, including the use of Binding Corporate Rules (BCR), Safe Harbor Certification, or Model Contract Clauses (MCC). Although these methods may only be legally mandated under specific circumstances, the standards they prescribe (owing to EU's strong data protection laws) make them suitable for other relationships involving transborder data flows. Each has different effects, and is suitable for different contexts: BCRs are for international transfers of data within an organization (e.g. multinational companies), Safe Harbor certification is for US organizations who anticipate working with data regarding EU citizens, and model contract clauses allow EU member nations to export data to a jurisdiction that has not been deemed to allow an adequate level of protection. In the following paragraphs, we describe the background behind each of these mechanisms, the parties they affect, and what they entail.

38.4.1 Binding Corporate Rules

Binding corporate rules, or BCRs, allow international organizations to transfer personal information outside of the European Union but remain in compliance with EU Directive 95/46/EC. Essentially, BCRs take the form of internal data privacy and security standards and practices that an organization is bound to for all data transfers. Once an organization develops a set of BCRs, they are submitted to a national data protection authority (DPA), which approves or disapproves the rules for their country. Significantly, the following countries have formed a mutual recognition system wherein one DPA (the "lead") can approve BCRs on behalf of the rest. Popular lead DPAs include the United Kingdom's Information Comissioner's Office (ICO) or France's Commission Nationale de l'Informatique et des Libertés:

Austria	France	Liechtenstein	Norway
Belgium	Iceland	Luxembourg	Slovenia
Bulgaria	Ireland	Germany	Spain
Cyprus	Italy	Malta	United Kingdom
Czech Republic	Latvia	The Netherlands	

Originally available only to data controllers, BCRs became available to data processors in January of 2013, and allow a data processor – such as a cloud service provider – to transmit its client's data outside of the EU. Most significantly, many data controllers who have not made other arrangements to have data transferred outside of the EU may rely on the BCRs of their processors, including in order to demonstrate compliance. Although BCRs vary from organization to organization, there are certain core components that must be present. These include assuming responsibility for all transfers of personal information within the organization, which includes data breaches; granting third-party beneficiary rights to data subjects, so individuals can enforce the BCRs directly against the organization; and limiting the transfer of information to other third parties.

In general, BCRs are valuable in that, if drafted broadly, they allow the organization to adapt to changes in their own corporate structure or dataflows. Further, they serve to strengthen the organization's privacy practices as a whole, and avoid specific negotiation of practices on a case-by-case basis. It should be noted, however, that upon submission, approval of an organization's BCRs may take a year or longer. Many organizations have made portions or summaries of their binding corporate rules available to the public.

38.4.2 Safe-Harbor Certification

As mentioned, EU Directive 95/46/EC prohibits the transfer of European citizens' data to countries outside of the European Economic Area (EEA) that did not meet the "adequacy" standard for data protection. To address this, the United States Department of Commerce consulted with the European Comission and developed the US-EU Safe Harbor framework, which simplified the process for US organizations to comply with the EU directive. Certifying that they comply with the US-EU safe harbor framework establishes that an organization within the United States provides adequate protection for personal data originating in the European Union, waives or grants automatic approval for member state requirements regarding data transfers, and that claims brought against US organizations by European citizens will be heard in the United States.

Participation in the US-EU safe harbor framework is voluntary. To participate, the organization self-certifies with the US Department of Commerce that it meets the requirements set forth in the framework, which are based on the seven principles of notice, choice, third-party transfer, access, security, data integrity, and enforcement. To demonstrate compliance, an organization may self-certify – which is discussed below – or have an audit conducted by an independent third party. Some organizations offer a safe harbor seal program, which can be used to demonstrate to clientele that an organization is compliant with some or all of the framework.

Membership in the US-EU safe harbor framework is valuable for cloud service providers, as European clients (or clients who have European data) will need assurance that adequate protections are in place. It should be noted, however, that the safe harbor framework has met with some criticism and may not be sufficient for certain clients or very sensitive data. Note also that the Safe Harbor framework is not available to all sectors, as telecommunications companies and financial organizations are not covered. Penalties, which are often enforced by the FTC, can include monetary sanctions or removal from the member list, which may prohibit an organization from using or operating with EU member data.

38.4.3 Model Contract Clauses

Model contract clauses are another means by which organizations may transfer EU data outside of the EEA. In the EU Directive, they take the form of clauses that may be included in contracts between data controllers and data processors, which, if adhered to, are sufficient to allow transfer of information outside the EEA. Note that the clauses vary depending on the relationship between parties; transfers of data between a data controller and data processor use different clauses than transfers between data controllers. Each relationship (controller to controller, and controller to processor) has two sets of clauses to choose from; however, Set I for controllers and processors is only valid for contracts set in place prior to March 2010; all current contracts between processors and controllers must use the newer clauses.

The two sets of clauses for data controllers differ primarily with regards to where liability is placed. In Set I, responsibility is jointly shared between the controllers; in Set II, a data subject can only enforce its rights against the party that is at fault in the breach of contract. It is important to note that should data subjects be unable to enforce their rights against the data importer, they still may be able to take action against the data importer for failing to ensure that the data importer has adhered to their requirements. The model clauses available for transfers between data controllers and processors also place liability on the party at fault. Note that should a data processor transfer data further to a data subprocessor, that data must be transferred under the same protections originally established by the model clauses, and allow the data subprocessor to be held responsible should the fault occur in its domain. The wording of the model clauses cannot be changed without forgoing the guarantee made by the European Commission that the data is sufficiently protected. Although technically it is possible for adequate protection to be assured under altered clauses, the parties must have evidence supporting these claims. The data importer cannot subcontract out without prior consent from the data exporter.

In addition to limiting data processing to reasons explicitly provided in the contract itself, model contract clauses also require the data importer to have appropriate levels of security, identify staff that will be trained under proper data protection practices, and ensure that the data exporter is aware of laws in the importer's country that allow authorities to access the exporter's data. Model clauses can be beneficial in that they allow selective guarantees of data protections on a case-by-case basis, and do not require the organization to maintain protection for all data received. However, they require frequent monitoring for updates, can result in dissimilarity in operations between clients, and like all contracts, are subject to prolonged periods of negotiation. Note that very few cloud providers offer model contract clauses to their clients.

38.5 Contracts and Terms of Service

Successful relationships between CSPs and their clients are founded on clear and comprehensive contracts that formally establish the functional and nonfunctional conditions under which the service is provided (Wieder *et al.*, 2011; Myerson, 2013). These legally binding documents vary considerably based on the size of the providers and their clients, the import and scope of the service provided, provider positioning (e.g. a provider who explicitly claims compliance with HIPAA provisions), and customization of services provided (Denny, 2010; Bradshaw *et al.*, 2011; Chief Information Officer Council and Chief Acquisition Officers

Council, 2012). In this section, we describe major components that both providers and clients should consider when engaging in a contractual relationship. For the purposes of this section, we use the term contract to refer to a legally binding agreement between parties that is enforceable by law. This may include traditionally conceived paper-based contracts as well as digital terms of service agreements (Patel *et al.*, 2009).

Although no cloud computing contracts are identical, they often contain similar components and provisions. Some of these components are common to contracts in all domains, such as specifying the applicable legal system and jurisdiction; others are more unique to technology fields or cloud computing in particular, such as enumerating the conditions under which data may be made accessible to thirrd parties, or the strength of encryption used for data in transit. Knowing these components as well as the range of values they may take can be valuable both for providers and clients: providers so that they may position themselves effectively and be aware of their offerings relative to their competitors, and clients so they can determine what values are most important to their organization in addition to as what they are (un)willing to compromise on during negotiations. The following represent a small sampling of some of the most relevant components found in cloud contracts organized around three categories: (i) legal components, which cover general legal issues like dispute resolution, (ii) data components, which cover practices specifically regarding cloud data (ownership, security), and (iii) service components, which cover aspects of how the service will be provided, such as uptime.

38.5.1   Legal Components

One of the first components seen in many contracts sets forth both the legal system through which the contract is enforceable and a specific jurisdiction where claims may be brought. Because cloud service providers will often be in a different state or country from their clients, these components are particularly important. Unless there are specific reasons to the contrary, both clients and providers should engage in contracts that are enforceable under their native legal systems, for many reasons: it minimizes the possibility of differences in legal process, it allows each party to draw on readily available legal resources who will be most familiar with that body of law, and reduces the chances of one party being unable to enforce a provision against the other. Relatedly, the contract may specify a specific venue through which disputes may be settled; for example, clients of Amazon's EC2 service consent to most disputes being heard in the state or federal court of King County, Washington. If the contract contains a choice of venue provision, it will often benefit one party or the other.

Most contracts will also provide alternative means of dispute resolution, such as arbitration (the parties agree to a resolution determined by an independent third party) or mediation (the parties meet with an independent third party who guides the development of a resolution, but they are not obligated to agree to that resolution). To take this a step further, nearly all contracts – particularly non-negotiable contracts – will partially or fully indemnify the parties against specific types of losses: for example, if a provider's services experience difficulties that lead to a loss of profit on the part of a client, the client will be unable to hold the provider accountable for those losses.

38.5.2   Data Components

Perhaps most relevant for this chapter are components that describe how data will be kept secure and confidential during storage and transmission, as well as how long it will be retained. Clients should look for text that explicitly states that a provider meets the security or confidentiality requirements of relevant laws. Unfortunately, many providers will claim that the responsibility for keeping data secure and confidential falls to the client. Relatedly, providers should ensure that their providers do not engage in advertising and / or data mining driven by client data, which may be addressed in a different section of the contract. Note that providers will be obligated to provide access to data under court orders or for law enforcement purposes.

Provided these matters are up for negotiation, both parties should expect explicit details. That is, a contract should avoid using ambiguous phrases, such as "the data will be kept secure" or "will be encrypted." When terms like these are used, both parties will be apt to interpret them in their favor, which can be problematic. Instead of these terms, parties should provide details regarding how data will be kept secure (e.g. cloud facilities will require an RFID badge for entry), or the strength of encryption used (e.g. Triple DES, AES). Clients should also look for text that describes how and how often these claims are certified or tested.

38.5.3   Service Components

The majority of space in cloud contracts is devoted to the conditions under which the service will be provided. Many clients are most concerned with guaranteed uptime, or the amount of time the provider promises to have the service available, and the type of acts that are not considered in this metric, such as utility failures or natural disasters. It is important to note that, even if an act does count towards affecting the metric, the contract may include language that protects the provider from losses as a result of that act.

Should providers experience problems that would affect their ability to provide their services, they may subcontract the client's work to another cloud service provider in an effort to maintain consistent service. Providers may also subcontract in the ordinary course of business, without the client being aware that any changes have occurred. Although consistent, reliable service is obviously desirable, clients should be very concerned that it does not come at the expense of the security or privacy of their data, and that any subcontractors used are guaranteed to provide security and confidentiality at the same level as can be expected from the original provider.

Clearly, communication between provider and client is important, and many contracts will contain details specifying what will be communicated, and how such communication will be handled. For example, many of the laws discussed previously obligate the provider to notify the client in the event of security breaches of their systems. Contracts will often, but not always, detail how this notification will be provided, which includes specifying to whom it will be delivered, the medium used, the content of the notification (e.g. amount of data exposed, exact time of breach), and how quickly the notification will be made after the provider discovers or is made aware of the breach. Contracts that may be changed at the discretion of the provider will often contain provisions that allow them to change the contract at any time and notify affected clients by posting an update to the provider's main web site.

38.6   Conclusion

New technologies extend the legal frontier, exposing new issues for lawyers, engineers, and policy makers to explore together – and those issues should not be underestimated. Fortunately, there are many strategies, resources, and organizations available to help both CSPs and their clients address their legal concerns; the various mechanisms and supporting organizations for transborder data flows discussed in this chapter are one such example. However, as laws have changed to adapt to the cloud, it is inevitable that with further advances they will change again. This, as well as the fact that cloud service providers are often bound by the legal requirements applicable to the clients they serve, means that the two must work collaboratively in determining what needs to be done, by whom, and when. At the same time both must recognize that any measures taken to protect data, arguably the biggest concern of most laws and contracts, are not foolproof, and that the risk of failure is known and mitigated to the fullest extent possible.

References

Armbrust, M., Fox, A., Griffith, R. *et al.* (2009) *Above the Clouds: A Berkeley View of Cloud Computing*. Electrical Engineering and Computer Sciences University of California at Berkeley, https://www.eecs.berkeley.edu/Pubs/TechRpts/2009/EECS-2009-28.pdf (accessed December 29, 2015).

Bradshaw, S., Millard, C., and Walden, I. (2011) Contracts for clouds: Comparison and analysis of the terms and conditions of cloud computing services. *International Journal of Law and Information Technology* **19**(3), 187–223.

Chief Information Officer Council and Chief Acquisition Officers Council (2012) *Creating Effective Cloud Computing Contracts for the Federal Government*. Federal Cloud Compliance Committee, https://cio.gov/wp-content/uploads/downloads/2012/09/cloudbestpractices.pdf (accessed December 28, 2015).

Delgado, M. (2011) *The Evolution of Health Care IT: Are Current U.S. Privacy Policies Ready for the Clouds?* Paper presented at the IEEE World Congress on Services, 2011. IEEE, pp. 371–378.

Denny, W. R. (2010) Survey of recent developments in the law of cloud computing and software as a service agreement. *Business Lawyer* **66**, 237.

EU Agency for Fundamental Rights (2010) *Data Protection in the European Union: The Role of National Data Protection Authorities*, European Agency for Fundamental Rights, Luxembourg, http://fra.europa.eu/sites/default/files/fra_uploads/815-Data-protection_en.pdf (accessed December 28, 2015).

Hall, J. L. (2013) *Final Rule Confirms that ISPs Transmitting PHI are not Business Associates" Center for Democracy and Technology*, https://cdt.org/blog/hipaa-final-rule-confirms-that-isps-transmitting-phi-are-not-business-associates/ (accessed December 29, 2015).

Hon, W. K., Millard, C., and Walden, I. (2011) The problem of "personal data" in cloud computing: What information is regulated? The cloud of unknowing. *International Data Privacy Law* **1**(4), 211–228.

Jaeger, P. T., Lin, J., and Grimes, J. M. (2008) Cloud computing and information policy: Computing in a policy cloud? *Journal of Information Technology and Politics* **5**(3), 269–283.

Mather, T., Kumaraswamy, S., and Latif, S. (2009) *Cloud Security and Privacy: An Enterprise Perspective on Risks and Compliance*, O'Reilly Media Inc., Sebastopol, CA.

Mowbray, M. (2009) *The Fog Over the Grimpen Mire: Cloud Computing and the Law*. HP Laboratories, http://www.hpl.hp.com/techreports/2009/HPL-2009-99.pdf (accessed December 29, 2015).

Myerson, J. (2013) *Best Practices to Develop SLAs for Cloud Computing*, IBM Developer Works, http://www.ibm.com/developerworks/cloud/library/cl-slastandards/ (accessed December 29, 2015).

Narayanan, V. (2011) Harnessing the cloud: International law implications of cloud-computing. *Chicago Journal of International Law* **12**, 783.

Patel, P., Ranabahu, A., and Sheth, A. (2009) *Service Level Agreement in Cloud Computing*, Ohio Center of Excellence in Knowledge-Enabled Computing (Kno.e.sis), http://corescholar.libraries.wright.edu/cgi/viewcontent.cgi?article=1077&context=knoesis (accessed December 29, 2015).

Pearson, S. (2013) *Privacy, Security, and Trust in Cloud Computing*. HP Laboratories, http://www.hpl.hp.com/techreports/2012/HPL-2012-80R1.pdf (accessed December 29, 2015).

Ruiter, J. and Warnier, M. (2011) Privacy regulations for cloud computing: Compliance and implementation in theory and practice, in *Computers, Privacy, and Data Protection: An Element of Choice* (eds S. Gutwirth, Y. Poullet, P. De Hert, and R. Leenes). Springer, Berlin, pp. 361–376.

Schellekens, B. J. A. (2013) The European Data Protection Reform in the Light of Cloud Computing. Master's thesis. Tilburg University.

Sotto, L. J., Treacy, B. C., and McLellan, M. L. (2010) Privacy and data security risks in cloud computing. *Electronic Commerce and Law Report* **5**(2), 38.

Wieder, P., Butler, J. M., Theilmann, W., and Yahyapour, R. (2011) *Service Level Agreements for Cloud Computing*, Springer, Berlin.

39

Cloud Economics

Sowmya Karunakaran

IIT Madras, India

39.1 Introduction

There are many speculations about how cloud computing will progress in the coming years. Information technology pioneers and leaders have a critical need for a clear vision of the industry's future. There is some indication that cloud computing will reach the plateau of productivity very soon, increasing the urgency to find this vision. The appropriate way to formulate such a vision is to understand the underlying economics, which drive long-term trends. In this chapter, we will assess the economics of cloud computing.

Economics research themes will broadly include aspects such as pricing and markets. Most research on cloud computing focuses on technological aspects and it is vital to look at the economic aspects to bring a holistic perspective to discussions about cloud computing. Studies of these aspects in the context of cloud computing are at a nascent stage. In terms of review studies, there are several works that provide a review of themes pertaining to cloud. While most reviews deal with taxonomy, few deal in detail with specific focus areas such as pricing or adoption.

This chapter will serve as an enchiridion of topics pertaining to economic aspects of cloud computing. Section 39.2 provides the background on cloud computing economics and discusses metrics to quantify the economic benefits. Section 39.3 provides an overview of cloud pricing. Section 39.4 discusses the popular market structures for cloud computing services. Section 39.5 gives a snapshot of the cloud supply chain. Section 39.6 discusses the role of enablers in adding economic value. Section 39.7 describes the terms "demand aggregation" and "network effects" in the context of cloud computing. Finally, Section 39.8 discusses future research directions.

39.2   Economics of Cloud Computing

39.2.1   History, Developments, Trends

The advent of parallel computing dates back several decades. Since then researchers have been constantly looking for ways to harness idle CPU cycles and to make network systems look like a single large computer. Advances in virtualization and software-oriented architecture (SOA), and in Internet technology, have made utility computing a reality. Cloud computing offers many infrastructure, platform, and software services as a utility (Buyya *et al.*, 2009). This growing supply and adoption of cloud, which is perceived as the fifth utility, has triggered the commoditization of IT. These services have transformed the way IT delivery happens in an organization. Perhaps, the most significant transformation is a shift from viewing IT as a capital expenditure, seeing it instead as an operational expenditure (Marston *et al.*, 2011). Consumers are aggressively pursuing this shift and this is evident from Forrester's view that the cloud-computing market is expected to exceed $241 billion by the year 2020.

39.2.2   Quantifying Economic Benefits

Quantifying economic benefits is critical for cloud adoption decisions. For example, consider the case of cloud computing in a resource-intensive industry such as pharmaceutics. Reports indicate that the adoption of a virtual screening process resulted in average savings of $130 million and 8 months per drug in its development cycle (Seifert *et al.*, 2003). In another case, a top-tier pharmaceutical company handled a molecular modeling problem in approximately 8 hours at a fee of just $1279/hour using 30 472 instances, which was equivalent to 95 078 hours of computer work (Amazon, 2011). The benefits discussed above are *ex post facto* but often the challenge facing cloud adoption decision makers is in predicting these benefits of transitioning to cloud. The first step in quantifying economic benefits is to look at the sources of various costs. A basic list of costs to consider for an in-house setup includes costs of the server, software, networking, bandwidth, cooling, power, facilities, real estate, space, and human resources. Comparing and contrasting the in-house costs with the cost of moving to cloud is an important step in understanding the benefits from cloud adoption. Cost considerations for the cloud typically include software costs, hardware costs, and migration costs. Researchers have studied cloud application economics at the transaction level and have arrived at four constituents for transaction cost: processing cost, storage cost, bandwidth cost, and services cost. In addition to cost considerations, various other factors can play a role in cloud adoption decisions. For example, security and privacy, reliability of the cloud service, performance, interoperability, and operational factors (which include governance, control and change management) have been widely studied. Cultural impact, elasticity and size of demand, support provided by cloud service provider, and legal aspects are other decision factors that can influence cloud adoption. The following section provides a discussion on metrics, which can guide a decision maker both from a provider and from user perspective to derive the economic value of the decision.

39.2.3   Metrics

39.2.3.1   Cloud Provider Metrics

Managing and providing computer resources to cater for user requests remains one of the biggest challenges for the cloud-computing service-provider community. Current solutions depend on job abstraction for resource control in order to manage resources. Users typically submit their computation tasks as batch jobs. The resource management system handles these requests and takes care of job scheduling and resource allocation. Although a large number of users and various scientific applications use this model, it requires the user to know the environment in which the application will be executed. This calls for increased transparency at the provider's end. It is inevitable for the provider to track and display some key metrics that will help users

Table 39.1   *Cloud provider metrics*

Metric	Description
Cost effectiveness (time based)	Cost over a fixed time (e.g. $/month)
Cost effectiveness (performance based)	Ratio of price to performance
Incremental cost effectiveness	Change in cost per unit speedup
Cost efficiency	Number of users supported on a given budget
Average revenue per server	Ratio of revenue to the number of servers
Average revenue per user	Ratio of revenue to the number of users
Cost per benchmark	Ratio of benchmarked task's finishing time to the published per hour price.

Table 39.2   *Cloud consumer metrics*

Metrics	Description
Savings	
Supply-chain cost	Change in cost to deliver due to flexibility and choice offered by cloud.
License cost	Change in license cost due to cloud adoption.
Green cost	Change in cost of maintaining green IT.
Financial ratios	
Net present value (NPV)	Net present value is calculated as net benefit from cloud adoption minus the cloud's investment cost. It is an absolute metric and a positive value indicates an economic benefit.
Benefit-to-cost ratio (BCR)	This is calculated as discounted net benefit from cloud adoption divided by the investment costs. Unlike the NPV, BCR is a relative economic metric. A number greater than 1 indicates a positive economic benefit.
Discounted payback period (DPP)	The DPP is calculated as the number of years it takes for benefits from cloud adoption to equal total investment costs.
Operation and migration	
Migration cost	Initial deployment cost of a new IT system on the cloud. Future cost of migrating an IT system from one cloud to another.
Operational cost	Costs for managing the cloud deployments, procuring / sourcing instances, monitoring, backup, and recovery.

to manage their computing jobs. More importantly, providers have to track metrics that can help them assess the economics of their cloud services. For example, cost-effectiveness metrics focus on cost per unit of outcome and incremental metrics focus on change (Li *et al.*, 2012). Table 39.1 provides a sample list of metrics pertinent to economic evaluation for cloud providers.

39.2.3.2   *Cloud Consumer Metrics*

While many cloud providers offer cloud computing services, their varying pricing schemes and their approaches to infrastructure, virtualization, and software services leaves the cloud consumer with a problem. A much bigger challenge for a potential cloud consumer is whether to move to the cloud or to continue with in-house facilities. Analysis of the current IT landscape benefit ratios and associated migration costs is crucial to address this challenge. Analysis of metrics that show savings is required to assess the benefits after cloud adoption. Table 39.2 provides a sample list of these metrics. In addition to these metrics there are other intangible factors that need to be analyzed during a cloud adoption. Examples of such factors include stakeholder impact, socio-political feasibility, operational viability, technological suitability, and regulatory and privacy requirements.

39.3 Pricing Cloud Services

Different cloud service providers have adopted varied pricing schemes. The providers base their tariffs and charging models on their business objectives. The subsequent sections discuss the pricing structures set by different cloud providers and constituent elements.

39.3.1 Pricing Elements

From the provider's perspective, key considerations for developing a pricing structure include the costs of servers, software, networking, cooling, power, facilities, real estate, space and utility. From the consumer's perspective it is imperative to determine the valuation for a cloud service. What is a good price to pay? In situations where the consumer's valuation is unknown, service providers elicit the consumer's valuation through dynamic pricing schemes and auction formats.

39.3.2 Pricing Schemes

The most widespread method of pricing in cloud is pay per use. An example is Amazon's EC2 on-demand model, which is based on units with a constant price (Weinhardt *et al.*, 2009). Another frequently used pricing model is subscription-based pricing, whereby users enter a contract for few months or a year. An example is Amazon's EC2 reserved instance model. Obviously, customers and providers would like to use static and simple pricing models in order to make it easier to predict costs. This explains the prevalence of the above models. A newer pricing scheme called dynamic pricing has also recently become available. An example is Amazon EC2 Spot Instances. Spot Instances enable cloud consumers to place bids for unused capacity – the excess capacity that remains with the cloud provider after fulfilling the pay per use and subscription demands. The service provider sets the spot price and users are charged the spot price. The spot price fluctuates periodically, depending on supply and demand.

In general, there are two broad pricing schemes: static and dynamic. In case of static schemes, the price does not fluctuate frequently. The price is listed and remains constant for weeks or months. By contrast, in the case of dynamic pricing schemes, the price fluctuates frequently. Different pricing models under these schemes are discussed in Table 39.3.

Most IaaS vendors provide a base plan consisting of 512 MB RAM and about 10 GB storage. There are also no costs associated with inbound data transfer to the servers. Outbound data transfers are charged. Table 39.4 provides a brief review of cloud service providers and their pricing schemes along with the details of the base plan.

39.3.3 Collaborative Pricing

Market structures where a vendor can collaborate with other vendors and deliver services to a consumer are becoming popular. Researchers have also started looking at collaborative pricing schemes to suit business models where the infrastructure resources are derived from multiple providers. For example, each user can bid a single price value for different composite/collaborative services provided by cloud providers. Similarly, collaborating providers can set a common price for their collaborative services (Hassan and Huh, 2010). Figure 39.1 is a schematic of a collaborative services model.

Table 39.3   *Cloud pricing schemes*

Model	Description	Example
Static		
Pay as you go (PAYG)	Users are billed per unit of time usage	Rackspace Cloud Servers Amazon EC2 On-Demand
Subscription based	Users subscribe in advance for computing resources	IBM Smart Cloud Reserved Capacity
Prepaid per use	Users are billed hourly from a prepaid credit	GoGrid cloud servers – hourly basis
Subscription per use	Combination of subscription and PAYG models	Joyent monthly SmartMachines and daily usage
Dynamic		
Spot	Users bid for instances and bids, successful bidders pay the spot price	Amazon EC2 Spot

Table 39.4   *Review of service providers and their pricing schemes*

Service provider	Pricing options	Primary services	Starting / base plan
Amazon EC2	PAYG, subscription, spot pricing	IaaS	0.6 GB RAM, 1 vCPU, elastic block storage
Amazon RDS	PAYG, subscription	DBaaS	630 MB memory, 64-bit, low I/O capacity, backup
DropBox	Per user / year	STaaS	100 GB storage
GoGrid	PAYG hourly plan monthly / yearly prepaid plan	IaaS /STaaS	0.5 GB RAM, 10 GB storage
Joyent	PAYG	IaaS	Extra small 0.5 GB smart OS/Linux
LunaCloud	PAYG	IaaS	512 MB RAM, 1vCPU, 10GB storage
Microsoft	PAYG, 6 months, yearly	IaaS/PaaS	1.6 Ghz CPU, 1.75 GB RAM, 225GB instance storage
Rackspace	PAYG	IaaS	512 MB RAM, 1 vCPU, 2GB local storage
Salesforce.com	subscription plans (per user/month)	SaaS	–
Verizon/Terremark	PAYG	IaaS	0.5 GB RAM

39.3.4 Pricing Fairness

With more and more players entering the cloud ecosystem, competition among cloud services providers is increasing. Cloud users constantly evaluate these providers on a wide range of parameters. One of the key elements is pricing fairness. In economics, pricing fairness includes personal and social fairness. Personal fairness is subjective in nature and varies from person to person. It typically means that the pricing should be low enough. Social fairness investigates whether users have the same financial cost for the same set of tasks. One instance of pricing fairness can arise in resource allocation – identifying bottleneck resources to understand the allocation profile of users to ensure fairness. This can be an important

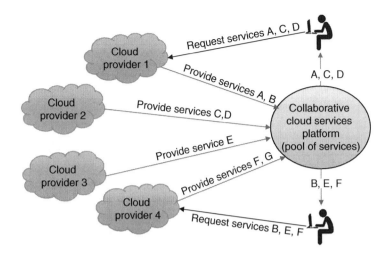

Figure 39.1 *Collaborative cloud services*

consideration for providers while developing charging models. Another instance of pricing fairness can arise due to the impact of different levels of interruptions faced by the cloud users for the same service (particularly in a dynamic price market).

39.4 Market Structures

39.4.1 Standard Vendor-Customer Structures

The traditional or standard model involves direct one-one interaction between the vendor and the customer. Each vendor publishes the list of services offered and the price of each service. The customer chooses the vendor based on current needs (the service required) and places a request for service. The vendor caters to the request based on availability and other relevant factors. If there is no single vendor who caters for all of the services required by the customer, the customer has to choose two or more service providers manually to fulfil the requests. Here, vendors need to devise and publish certain metrics that can help consumers in their adoption decision.

39.4.2 Cloud Markets such as Cloud Bank Model

Federated cloud is a mechanism for sharing resources thereby increasing scalability. In federated clouds, users request more than one type of resource from different providers. The biggest advantage of using a federated cloud service is that users who generally purchase different resources from many cloud providers for their different compute and storage needs can obtain all the services seamlessly from a single federated cloud market. The challenge with federated cloud markets is that the users need information about all service providers and the status of each provider. To overcome this challenge, researchers have proposed various platforms and market models of cloud services such as Mundi and Cloud Bank. The market exchange would allow autonomous agents representing providers, consumers and brokers to manage and distribute resources through economic models such as posted pricing, auctions and negotiations. For example, Spotcloud launched a capacity-clearing market where buyers select available capacity based on cost and location.

39.4.3   Review of Auction Formats Employed in Cloud Markets

A variety of auction formats for the cloud have been suggested in the research literature. Advances in cloud economics have led to the evolution of the market infrastructure in the form of a market exchange that facilitates trading between consumers and cloud providers, for example Mandi (Garg *et al.*, 2013). Researchers have proposed the adoption of open markets for trading IaaS resources using a continuous reverse auction mechanism (Roovers *et al.*, 2012). Research has also indicated that use of combinatorial auction-based allocation mechanisms can improve allocation efficiency (Zaman and Grosu, 2013). A real-time group auction system for efficient allocation in cloud instance market has been developed based on a combinatorial double auction (Lee *et al.*, 2013). A combinatorial auction-based cloud market model that facilitates dynamic collaboration among cloud providers to provide composite/collaborative cloud services to consumers has been developed (Hassan and Huh, 2010). Research also indicates the use of (n + 1)th price auction of multiple goods to maximize provider's revenue, with an *ex post facto* supply limitation – after the users place their bids. In an (n + 1)th price auction of multiple goods, each client bids for a single good. The provider chooses the top N bidders. The provider may set "N" up front on the basis of available capacity or in case of revenue maximization choose to set N after receiving the bids. In both cases, N is constrained by the available spare capacity and cannot exceed that. The top N winning bidders pay the published price and are allocate the instances (Agmon Ben-Yehuda *et al.*, 2011).

The current form of dynamic pricing adopted and practiced by cloud vendors like Amazon is closer to a uniform price auction in which all bidders who win the auction pay the same price. It can also be categorized as a "sealed bid" because the bids are unknown to other bidders. Amazon terms its dynamic pricing scheme "spot." It is market-driven, since the spot price is set according to the clients' bids. In a spot pricing scheme, if the bid price exceeds the current spot price, the instance is allocated until either the user chooses to terminate upon task completion or the vendor initiates the termination once the spot price increases above the bid price.

39.5   Cloud Supply Chain

A Cloud Supply Chain is two or more parties linked by the provision of cloud services, related information and funds (Lindner *et al.*, 2010). Figure 39.2 gives a macro level view of the cloud supply chain and its entities. At the top of the chain are hardware and software manufacturers that provide the requisite infrastructure and software support to the cloud vendors. The cloud vendors form the heart of the cloud supply chain and are responsible for running datacenters and providing the first level of cloud services. The cloud providers can take many roles within the cloud supply chain. They might act as infrastructure (IaaS), platform (PaaS) or software providers (SaaS) and could be directly in contact with the end customer but they may also act as agents/brokers or business partners. They could also be service aggregators and value-added resellers that use the service provided and combine or enrich it with another service or a new functionality to create a composite service. The end customers usually consume a product that is either a simple or a composite service provided directly by a service provider over the cloud supply chain or via an agent or a value-added reseller based on the complexity and nature of their need.

39.6   Role of Enablers in Adding Economic Value

39.6.1   Agents and Brokers

Today, a wide range of brokering mechanisms exists. For example, the utility adaptive personal cloud service brokering mechanism matches a service user (consumer) to a specific service to maximize the user's utility; the agent-based online measure infrastructure, which can be used by consumers to evaluate the quality of

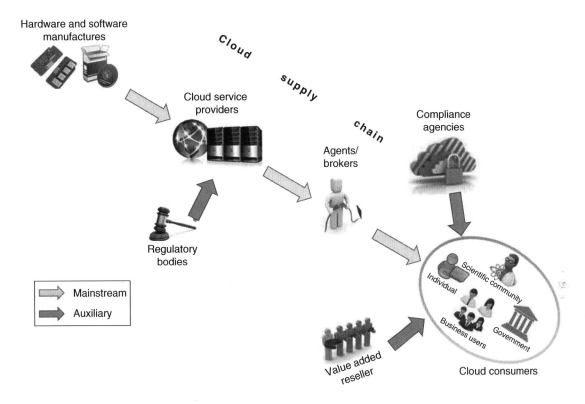

Hardware and software
manufactures

Cloud service
providers

Cloud supply chain

Compliance
agencies

Agents/
brokers

Regulatory
bodies

Mainstream

Auxiliary

Value added
reseller

Individual

Scientific community

Business users

Government

Cloud consumers

Figure 39.2   *Cloud supply chain*

service of the computing environment; the Cloud Service Measurement Index Consortium (CSMIC) – a framework based on common characteristics of cloud services, which can be used by consumers to compute a relative index for comparing different cloud services.

Architectural frameworks capable of powering the brokerage-based cloud services are available and provide a list of desirable features for a broker. Broker cloud management systems have been developed with a view to minimizing the cost of the link between the broker and the cloud. An efficient negotiation between brokers requires the usage of resource-level information to increase the accuracy of negotiated SLAs and facilitate achievement of business and performance goals. Researchers have proposed a nonblocking resource broker for a private cloud, which provides the choice of dynamically updating the resource details in the broker.

Figure 39.3, illustrates a simple cloud brokering service. Consumers place requests to a broker, specifying a reservation price and a deadline. For example, a user may request a service and be willing to pay a maximum price of $10 for the job and expect a completion time no later than 8 hours. Using this information, the broker then negotiates with service provider and offers several proposals that vary on price and time such as: ($8, 6 hours), ($6, 7 hours), ($4, 8 hours) and ($2, 9 hours).

39.6.2   Government

It has been widely perceived that government departments and agencies can apply cloud-based services to improve transparency while addressing their administrative goals of scalability and interactive citizen web sites and portals. The cloud can also help government agencies to increase collaboration across different departments, deliver volumes of data to citizens in simple, effective and cost-efficient ways. Unlike other

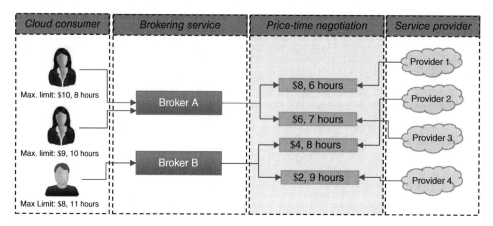

Figure 39.3   *Example of a cloud brokering service*

enterprises, government and public-sector firms need to operate even during times of crisis. Cloud computing can continue to provide operational efficiency even during times of crisis as the portals, applications, and servers are not completely dependent on in-house servers or on-site staff.

A multitude of e-Governance systems are currently operational, which provide a diverse range of government-to-consumer (G2C), government-to-business (G2B) and government-to-government (G2G) services. The various stakeholders within the realm of e-Governance systems have started to recognize the need to develop interoperable and reusable systems due to the increasing maturity and complexity of these systems. In terms of scalability of functionality and spread, interoperability and reusability together enable rapid scalability. However, to assimilate these benefits e-Governance systems need to be designed and developed by various government agencies and departments in unison. E-Governance projects can easily lose track of these benefits if they set their focus on only attaining their departmental goals.

Many e-Governance projects have embraced the enterprise architecture approach with this insight in mind. IEEE 1471 is a standard for describing architecture of a software system commonly known as enterprise architecture. Various governments and government agencies have created architectural frameworks for use within their domain. Some examples include the European Interoperability Framework and the Federal Enterprise Architecture Framework. Many other e-governance projects are also following this trend and building their own frameworks. With the rise in Cloud adoption rates and the various benefits it offers, eGovernance projects are beginning to take advantage of cloud computing. Enterprise architecture techniques help to describe the features of a cloud computing system. IEEE 1471 supports the indispensable association between the business needs of an e-governance project and the allied computing resources needed to address them.

39.6.3   Standards and Regulations Bodies

A lack of well defined standards could limit the usefulness of cloud computing. It could impose severe restrictions on cloud implementation and adoption by limiting interoperability among various cloud platforms and raising doubts about security. In addition to interoperability and security, other key issues for firms that are moving into cloud is to provide governance for data that it can no longer directly control. Such organizations need to understand how its business practices will continue to guarantee compliance with several other existing industry standards, for example HIPAA, PCI, and ITIL. Service-level agreements, confidentiality agreements, compliance audits, and so forth, need to be extended or retuned to combine issues arising from hosting data in the cloud (Cochran and Witman, 2011). Most importantly, Standards and regulations for the cloud could have an impact on public finances. The Federal Cloud Computing Initiative (FCCI) was constituted with the goal of supporting standards and rules for the adoption of cloud services, which in turn can reduce the US government's IT spending.

39.7 Economic Terms in the Context of Cloud Computing

39.7.1 Demand Aggregation

Demand aggregation assists enterprises to become aware of the purchasing requests that arise from different departments within the company, and makes these requests visible. This information helps companies to increase their purchasing power and obtain greater savings on the costs of goods and services. Enterprises that can design their systems so that their demand for computer power is flexible should receive better pricing as the economics work out. For example, batch jobs that are not time sensitive can be run at cheaper prices at nonpeak hours than if they are processed constantly through the day. On the other hand, demand aggregation is beneficial to cloud providers too. A substantial portion of the demand for computing services will have constant demand profiles without the spikes and troughs characteristic of business users – for example, bioinformatics, pharmaceutical modelling and other scientific simulations that run around the clock. Sellers of utility computing services would do well to segment these customers out and offer them lower prices. This technique can greatly improve asset utilization and has been widely used by telecommunications providers for years and is currently also employed in the cloud community. Many third-party providers enable demand aggregation through their tools and services. Table 39.5 provides a snapshot of some of these tools.

39.7.2 Network Effects

39.7.2.1 Data

Handling data-intensive tasks automatically leads to network effects because data processing generates new data. A company that stores its data on the cloud may also decide to process it. If both the storage and processing were to use the services of a common service provider, issues such as portability, and

Table 39.5 *Snapshot of various tools that enable demand aggregation*

Tool/solution	Description
Gravitant cloudMatrix	Delivers comprehensive cloud planning services for enterprise
SnapLogic	Connects private and public cloud applications.
Talend	Open-source-based platform enabling enterprises to build their own platform to connect private and public cloud applications.
OneSaaS	Connects public cloud applications and synchronizes contacts, leads, product details, invoices, sales data, and financial information.
HP Cloud Aggregation Platform	Facilitates and standardizes the processes for CSPs to enable a SaaS business model.
Mulesoft	Synchronizes data between an on-premises and cloud applications and automates business processes between cloud services.
Elastic.io	Cloud integration platform, taking care of reading data and API formats, data mapping and data transformation.

interoperability would not arise. For example, if a department within a company decides to place data in Amazon Simple Storage Service and to place the code chunk to process it on Amazon EC2, it would be simpler for a different department within the company, which needs to run another code on the data, to move the processing to Amazon EC2. Each of these processing steps generates even more data. This process develops into a cycle. When all of a company's data-processing tasks move into the same cloud, to take advantage of the co-location and minimize data-transfer latency and overall costs, the company reaches a steady state.

The network effect can extend well beyond the boundary of a single enterprise. In reality, for example, Firm Y often consumes data created by Firm X. When the volume of this data interchange is huge, it makes economic sense for Firm Y to move into the same cloud as Firm X. Clearly service providers can exploit this network effect, particularly by identifying complementary firms that share huge volumes of data. One such instance is AppNexus, a utility cloud that is optimized for the use of ad networks and their ecosystem, which consists of publishers and advertisers, who use the data predominantly for predictive analytics.

39.7.2.2 Platform Switching

Most application software programs are typically compatible with a specific set of operating systems. This means that network effects play a major role in determining the market dynamics. Fershtman and Gandal (2012) classified these network effects into direct and indirect. There were strong network effects in the operating systems market. The direct network effects occurred when consumers increasingly used operating systems that were available on many computers. The indirect network effects can be attributed to more and more users adopting operating systems that offer a large variety of application software. Application developers also preferred to develop applications for operating system with a large user base. The transition to cloud will weaken both the direct and indirect network effects due to wider access to technologies like virtualization, and will therefore impact the pattern of competition in the operating system arena. In particular, indirect network effects may diminish for the operating systems in the cloud environment. It is also likely that multiple platforms can co-exist in the cloud (Fershtman and Gandal, 2012).

39.8 Research Directions

39.8.1 Cloud Providers

With the emergence of collaborative cloud services, demand for algorithms for resource sharing and revenue sharing, with various cloud providers working together, is bound to increase. The introduction of dynamic pricing schemes calls for studies related to factors that influence user buying behavior and related concepts from a microeconomic and macroeconomic point of view. Methods that can reduce the time required to reach the maximum profit point and enrich relevant approaches from the field of autonomic systems can form a valuable research area. Rising cloud adoption rates make it imperative for cloud providers to study the differences among their users and their corresponding demands, workloads, and usage patterns. Evaluation of service offering tradeoffs is another valuable area of research. For example, on one hand the diverse needs of growing cloud demand may tempt a vendor to offer diversified services, and, on the other hand, considering economies of scale, the vendor may only focus on a single service offering. Providers need to consider the impact of their QoS decisions on consumer satisfaction and revenue management. For the success of any industry, human resources are a key factor. Studies of human resource factors need to be conducted in the context of cloud – for example, a research study focusing on identifying the skill sets required for running and managing large-scale datacenters.

39.8.2 Cloud Consumer

Cloud consumer organizations, in addition to cost modeling, need to investigate project management and software cost estimation techniques to unearth the full costs of cloud migration. Study and analysis of long-term workloads can also provide the user with different possible budgets on a monthly or yearly basis. Prediction methods can be developed to minimize costs and completion times by identifying correlations between past and current prices and between instance types. Firms that have transitioned to the cloud need to develop systematic tracking mechanisms to monitor use of cloud services. Without this internal monitoring users could spawn instances unnecessarily and cloud costs could turn out to be much higher than predicted.

Software development firms need to re-examine current software engineering practices and tailor them to complement development in the cloud. Research on performing application discovery to obtain essential input parameters such as application dependencies, component response times, and traffic exchanged between components can be valuable. Researchers and businesses need to explore the impact of vendor lock in and develop models to estimate the potential loss due to this.

From a human resource perspective, studies need to be carried out to identify the competencies that traditional software developers need to learn in order to build and run applications suited for cloud computing environments.

39.8.3 Others

The role of agents and brokers is bound to increase with the expansion of cloud market share. Agents need to develop methods that allow them to meet the requirements (such as SLA and security), matching user needs with what the cloud provider can offer. Third-party brokers need to study brokerage-based cloud services, cloud aggregation, and cloud bursting. Value-added resellers need to examine what kind of standards need to be established for a service to become truly valuable to a large community. In the interest of all the stakeholders, researchers need to analyze the impact of storing data across multiple geographies. This research can address many open questions. For example, one question that remains even today is whether the laws of all the countries where the servers reside be analyzed and understood in case litigation occurs.

39.9   Conclusion

Amazon has started offering Elastic Computing Cloud at prices close to 10 cents per hour. Microsoft is investing billions and adding up to 35 000 servers a month to build cloud computing. Google applied for a patent for ship-based datacenters generating tidal power to run cooling pumps. Cisco is researching "intercloud," a federation of clouds, in the same way that the Internet is a network of networks. With all of these happening, the significance of moving to cloud needs no emphasis.

However, industry reports indicate that cloud spending is yet to make a dent, although it is on the rise. This can be attributed to the dilemma faced by adopters due to lack of insight into the economic aspects of moving to a cloud. In this chapter, we saw an overview of various terms and concepts pertaining to the economic considerations of cloud. This could typically benefit managers in an organizational setting and researchers in the IS space. In the course of decision making, organizations need to consider economic implications pertinent to the decision and use appropriate models and techniques to support their decision making. Ideas stemming from fields such as economics, management, and decision theory are increasingly becoming relevant to this new domain of computing. Researchers need to understand the gaps and the urgency in this emerging area of study, identifying opportunities for research and addressing them appropriately.

References

Agmon Ben-Yehuda, O., Ben-Yehuda, M., Schuster, A., and Tsafrir, D. (2011) Deconstructing Amazon EC2 Spot Instance Pricing. IEEE Third International Conference on Cloud Computing Technology and Science (CloudCom). IEEE, pp. 304–311.

Amazon (2011) Amazon Elastic Compute Cloud (Amazon EC2), http://aws.amazon.com/ec2 (accessed December 29, 2015).

Buyya, R., Yeo, C. S., Venugopal, S., *et al.* (2009) Cloud computing and emerging IT platforms: Vision, hype, and reality for delivering computing as the fifth utility. *Future Generation Computer Systems* **25**(6), 599–616.

Cochran, M. and Witman, P. D. (2011) Governance and service level agreement issues in a cloud computing environment. *Journal of Information Technology Management* **22**(2), 41–55.

Fershtman, C. and Gandal, N. (2012) Migration to the cloud ecosystem: Ushering in a new generation of platform competition. *Communications and Strategies* **85**(1), 109–123.

Garg, S. K., Vecchiola, C., and Buyya, R. (2013) Mandi: A market exchange for trading utility and cloud computing services. *Journal of Supercomputing* **64**(3), 1153–1174.

Hassan, M. and Huh, E.-N. (2010) A novel market oriented dynamic collaborative cloud service infrastructure, in *Handbook of Cloud Computing* (eds. Furht, B and Escalante, A.). Springer, Berlin.

Lee, C., Wang, P., and Niyato, D. (2013) A real-time group auction system for efficient allocation of cloud internet applications. *IEEE Transactions on Services Computing* **8**(2), 251–268.

Li, Z., O'Brien, L., Zhang, H., and Cai, R. (2012) On a Catalogue of Metrics for Evaluating Commercial Cloud Services. Proceedings of the 2012 ACM/IEEE Thirteenth International Conference on Grid Computing (GRID '12). ACM/IEEE, pp. 164–173.

Lindner, M., Galan, F., Chapman, C. *et al.* (2010) The cloud supply chain: A framework for information, monitoring, accounting and billing. Second International ICST Conference on Cloud Computing (CloudComp 2010).

Marston, S., Li, Z., Bandyopadhyay, S., *et al.* (2011) Cloud computing – the business perspective. *Decision Support Systems* **51**(1), 176–189.

Roovers, J., Vanmechelen, K., and Broeckhove, J. (2012) A reverse auction market for cloud resources. Lecture Notes in Computer Science 7150, 32–45.

Seifert, M., Wolf, K., and Vitt, D. (2003) Virtual high-throughput in silico screening. *Biosilico* **1**, 143–149.

Weinhardt, C., Anandasivam, A., Blau, B., and Stößer, J. (2009) Business models in the service world. *IT Professional* **2**, 28–33.

Zaman, S. and Grosu, D. (2013) A combinatorial auction-based mechanism for dynamic VM provisioning and allocation in clouds. *IEEE Transactions on Cloud Computing* **1**(2), 129–141.

Part VIII

Cloud Applications and Case Studies

40

Engineering Applications of the Cloud

Kincho H. Law,[1] Jack C. P. Cheng,[2] Renate Fruchter,[1] and Ram D. Sriram[3]

[1] *Stanford University, USA*
[2] *Hong Kong University of Science and Technology, China*
[3] *National Institute of Standards and Technology, USA*

40.1 Introduction

Cloud computing has emerged as a new computing paradigm that promises to deliver significant value to businesses as well as engineering enterprises. The cloud computing concept grew out of recent advances in service-oriented architecture (SOA), distributed and network computing, and virtualization. Cloud computing has huge potential beyond the recent technological advances as business models have rapidly evolved to harness the network infrastructure, hardware resources, massive data storage, and programming platforms provided by cloud service providers. Large corporations can now move from owning their own computing resources and internal support to use powerful hardware and software resources over a communication network such as the Internet. Small businesses can focus on creating valuable applications without spending significant capital on installing and maintaining complex computer resources. As success stories about the deployment of cloud services – from Internet companies, such as Snapchat, to traditional consumer companies, such as 3 M – have demonstrated, the cloud service model will continue to grow, not only in the Web-based and business industries, and will have a significant impact on science and engineering.

As engineering software applications become Web-based services, engineers and designers can easily access the services as if the tools are running on their desktop. With storage service provided on the cloud, engineers can share their designs, drawings, models, data, and documents with project members. Furthermore, with trust as well as appropriate access control mechanisms, project partners can collaborate across the supply chain. As engineering projects are increasingly globalized and becoming more complex, the cloud service environment can be deployed to allow customers, designers, subcontractors, manufacturers, and company owners to share information quickly and to develop engineering solutions. In other words, in addition to providing engineering software as a service, the cloud computing environment can be deployed as a facilitator for collaboration.

Encyclopedia of Cloud Computing, First Edition. Edited by San Murugesan and Irena Bojanova.

For cloud to truly deliver value to engineering enterprises beyond serving application services, which are now implemented in a similar way to the traditional service-oriented architecture, the services in the cloud need to be interoperable (preferably among services by different cloud providers). Cloud infrastructure standards have been actively pursued by industry consortia with participation by several large cloud service providers. To ensure interoperability among engineering services, engineering information models and standards are needed to break down silos and islands of automation, particularly as applications are rapidly transiting from desktop "close-box" tools to cloud-based open software services. Standard industry ontology can play an important role in supporting interoperable solutions for engineering collaborations in cloud computing (Fenves *et al.*, 2005).

Cloud computing holds out the promise of providing new ways to support collaboration everywhere, at any time, through a multitude of platforms, from desktops and tablets to mobile phones. Design changes and discussions can be tracked, monitored, and shared in real time by project members, in a way similar to popular social apps. Collaborative technologies, with appropriate access control, are needed to support synchronous (real-time) and asynchronous collaboration in an engineering environment and in global project development (Sriram, 2002). The cloud services model provides an appropriate infrastructure to implement the technologies as outlined by Sriram (2002).

This chapter is organized as follows. We first review the basic definitions that are commonly used in cloud computing and discuss the approaches by which engineering enterprises deliver their applications as cloud services. We then discuss service and information interoperability in the cloud environment, review some current standardization efforts, and introduce the deployment of engineering information models and exchange standards in the cloud service environment. In addition to service and information interoperability, the cloud environment has great potential in supporting collaboration among project participants. We discuss an example application: deploying cloud services to facilitate virtual brainstorming and team engagement in collaborative engineering design. The chapter concludes with a brief discussion on future directions.

40.2   Cloud-Based Engineering Services

Fundamentally, cloud computing is a utility over a network model that enables on-demand access to computing resources such as servers, storages, applications, and services. To enhance understanding of the technology, taxonomies and models on cloud services have been defined to provide a classification of cloud systems. The National Institute of Standards and Technology (NIST) has defined three broadly adopted service models, namely software as a service (SaaS), platform as a service (PaaS), and infrastructure as a service (IaaS) (Mell and Grance, 2011). The service model definitions have also been expanded to include many service utilities and acronyms. For instance, Youseff *et al.* (2008) suggest expanding the infrastructure definition by separating data and storage (DaaS), communications (CaaS), and computational resources (IaaS) as services, and to include software kernels (such as operating systems, virtual machines, and grid tools) and firmware and hardware (HaaS) as additional services. Other suggested cloud terminologies from various vendors include: business process as a service (BPaaS) and, in essence, everything as a service (EaaS or XaaS). On the other hand, Armbrust *et al.* (2010) view the cloud platform and infrastructure merely as the basic utility to deliver applications over the Internet. Generally, users access cloud-based applications through thin client interfaces, such as a Web browser or a mobile app; the application software and the data are stored on servers at a remote location hosted by the service provider.

Major manufacturers and engineering software vendors in electronic design automation (EDA) and (computer-aided design / computer-aided manufacturing / computer-aided engineering) (CAD/CAM/CAE), such as Synopsys, Cadence, Mentor Graphics, SiCAD, Nimbic, Tabula, Autodesk, MSC, Dassault Systemes, Bentley, Belmont, Intergraph, ANSYS, Siemens, and many others, have started or are starting to deliver a wide variety of engineering applications as cloud-based services. The four basic deployment models, namely public, private, community, and hybrid cloud, as defined by NIST (Liu *et al.*, 2011; Mell and Grance, 2011) can be used to describe how engineering services are beneficially deployed in the cloud-computing environment.

- Public cloud is probably the most prevalent model for service deployment being reported. Cloud service providers such as Google, Microsoft, Amazon, IBM, HP, and Salesforce, provide computing resources, platform, and infrastructure that clients can use to deliver services to customers. Engineering software vendors, such as Mentor Graphics and others, have deployed public cloud services to house their applications for use over the Internet.
- Private cloud is provisioned for secured use internally by an organization or private group. One example is Futjisu's Engineering Cloud, which is designed to serve the internal engineering divisions in the company and is planned to serve companies partnering with Fujitsu (Yasuda, 2012). Software vendors, such as Tabula and others, have begun to support their platforms in a private cloud computing environment.
- Community cloud is referred to infrastructure and services provided for and used exclusively by a community or organizations. In engineering, the community-cloud model can be deployed to support (lifecycle) project-based development. A community or project-based cloud environment with services and resources made available through public and private IT clouds could be an ideal resource for supporting collaboration. For example, cloud-based project-based collaborations, where project participants can share design information and engineering models, are now supported by Autodesk 360, Integraph SmartPlant, and other vendors.
- Hybrid cloud infrastructure and services combine the utilities and services offered by the public, community, and private cloud providers. Many engineering companies are interested in seeking ways to deploy the hybrid model, in that security (not knowing where the data is and who accesses the data) and legality issues (such as ownership of the data and change of contractual agreement) can be protected (Redmond *et al.*, 2012). The hybrid cloud model is widely supported by software vendors, such as IBM and others, to allow clients to integrate services between cloud service providers and in-house applications.

Although current cloud-based engineering services are mostly in design automation (and business enterprises), cloud computing models have been proposed in the manufacturing and biomedical information domains (Rosenthal *et al.*, 2010; Xu, 2012).

Irrespective of the cloud deployment models, from the economy of scale perspective the major benefits of cloud utilities are the shared resources and application services. Using the cloud services hosted by a service provider, consumers are relieved of the costly maintenance and upgrade of software, hardware, and infrastructure. The pay-per-use model for cloud services and resources could potentially reduce the burden of IT expenditure for companies. Users can also access high-end computing resources, such as supercomputers and computing clusters, for large-scale simulations on an as-needed, pay-as-you-go basis. In contrast to desktop-based software development, by deploying computing cloud infrastructure, application developers can continue to update software services and to incorporate new features without interrupting users. Furthermore, as cloud service providers continue to incorporate enhancements on existing applications and to develop new tools, users can beneficially deploy these services faster. Irrespective of the pay-per-use or subscription cost model, the potential benefit of reduced cost and easy access to shared computing resources and application services is one of the driving forces for the adoption of cloud computing by companies, particularly for small and medium enterprises (SMEs). Last but not least, an application residing in a cloud can be used by multiple users from around the world. For instance, a company may deploy a data-analytic tool from a cloud service to analyze data collected from different facilities and manufacturing plants in geographically separated locations. As a growing number of engineering services are now available as cloud-based services, interoperability among the services is of interest to users (Redmond *et al.*, 2012).

40.3   Interoperability and Cloud Services

Interoperability refers to the ability of systems and applications to communicate, query and exchange information, and to work together. The purpose of interoperation is to increase the value of information when information from multiple and, likely, heterogeneous sources is accessed, related, and combined. Interoperation among systems increases the value of each individual, isolated engineering service and enhances efficiency and productivity in the engineering product-development life cycle and business supply chain. Companies around the world are trying to take advantage of information and communication technologies (ICT) to create virtual engineering services and supply chains where customers, suppliers, and business partners collaborate with each other.

There are four basic categories for service interoperability in the cloud computing environments, as illustrated in Figure 40.1:

- within a cloud environment (offered by a service provider), service invocation, service integration and information exchange between services should be supported;
- across different cloud environments, services should be able to exchange information and invoke operations;
- the local application service should be able to connect and integrate with services residing in the cloud environment;
- cloud services should be portable and be able to migrate from one cloud environment to another.

Standardized interfaces and protocols should be established to enable: (i) service (or system) portability and migration, (ii) service invocation / integration, and (iii) information sharing and exchange among services within a single cloud environment and across different cloud platforms.

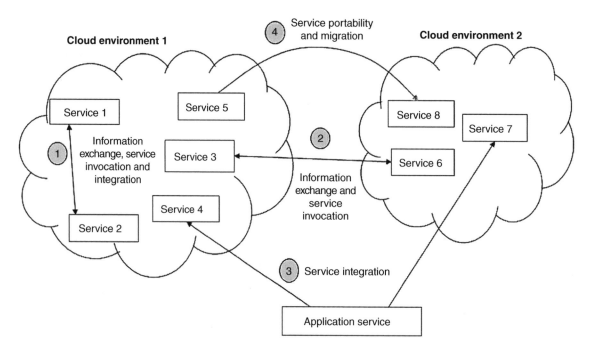

Figure 40.1   *Categories of interoperability in cloud computing environments*

Portability and migration of applications (from one service provider to another) and invocation and integration of services (within and across service providers) are issues that are important not only for the users of the cloud service but also the software service developers and the companies that market cloud platform and infrastructure resources. To support service portability and service invocation, many organizations and consortia have initiated efforts to define standard interfaces or to provide open source interface and protocols to address the platform and infrastructure interoperability issue. For instance, OpenStack (see http://www.openstack.org/, accessed January 3, 2016) aims to provide "ubiquitous open source cloud computing platform for public and private clouds." The Open Virtualization Format (OVF) (see http://www.dmtf.org/standards/ovf, accessed December 30, 2015) proposed by the Distributed Management Task Force (DMTF), Inc., aims to define the semantics and implementation details to address the portability and deployment of virtual applications and to achieve interoperable cloud environment. A unified code interface (see http://code.google.com/p/unifiedcloud/, accessed December 30, 2015), which uses the Resource Description Framework (RDF) to describe a semantic cloud data model (taxonomy and ontology), has been proposed as an open and standardized cloud interface for unification of various cloud APIs. The Organization for the Advancement of Structured Information Standards (OASIS) has developed the Topology and Orchestration Specification for Cloud Applications (TOSCA) (see http://docs.oasis-open.org/tosca/TOSCA/v1.0/TOSCA-v1.0.html, accessed December 30, 2015), which aims to simplify the migration of applications from one cloud to another and orchestrates them across multiple cloud environments. Efforts have been reported to enhance Web service standards, such as Web Service Definition Language (WSDL) and Business Process Execution Language (BPEL), commonly deployed for describing service interfaces and orchestrating services, to support service invocation and service integration in the cloud service environment (Anstett *et al.*, 2009). Enterprise software companies, such as WSO2, have incorporated BPEL in their cloud service platforms, whereas others, such as IBM, offer support for delivering service integration with other open standards.

Engineering information models and interoperability standards play an important role in supporting information sharing and exchange and service integration (Fenves *et al.*, 2005). As engineering designs become more complex, the variety and volume of digital information and the demand on easy access to the data in engineering projects has dramatically increased. Consider building engineering as an example. The design and construction of the Beijing National Aquatics Center in China have generated over 1.2 million drawings, which required over 200 GB of storage. The amount of building information increases along project life cycle, from planning and design, to construction, operations, and maintenance. A cloud-based storage service is desirable for managing building information over the facility life cycle and allows project partners to access and share information everywhere and anytime. One approach is to extend storage (infrastructure) as a service and to adopt industrywide standard information and ontology models to facilitate information sharing and service integration within and across cloud environments.

In the building and construction industry, building information modeling (BIM) applications have begun to emerge as a vehicle to support integrated project-delivery process through exchange of information with open standards. By adhering to open standard data modeling and format, such as the Industry Foundation Classes (IFC) developed by buildingSMART (formerly called the International Alliance for Interoperability, IAI), information exchange and interoperability are supported by BIM application software. Redmond *et al.* (2012) conducted a semistructured interview of 11 BIM experts and concluded that Web-based BIM exchanges on a cloud platform can lead to enhanced information and service interoperability between different construction applications.

The BIM-PDE server is a prototype cloud-based framework for storing and retrieving BIM information (Cheng and Das, 2013). Unlike in the traditional approach where every end user has his / her own version of BIM data, the cloud-based framework provides a centralized server where the information of an integrated building model is stored and the data of customized models can be retrieved. The end users could download or upload partial models by executing updates on the integrated building model in the server. This enhances information-sharing efficiency as only the key information is shared instead of the entire model.

Apache Cassandra (see http://cassandra.apache.org/, accessed December 30, 2015), an open-source distributed database management system, is used to implement the BIM-PDE server. The architecture of a Cassandra instance consists of a set of independent nodes (computers/ disk space) configured together in a cluster. An instance of Cassandra may run on one or more virtual machines called nodes located in the cloud. The system architecture of Cassandra is compatible with popular cloud infrastructure providers, such as Amazon Elastic Compute Cloud (EC2). The data stored in the BIM-PDE server is partitioned and distributed among the nodes in the Casandra distributed database system to facilitate efficient data retrieval and update using parallel threads.

A simple example scenario is presented to demonstrate the retrieval and updating of building models in a storage (or infrastructure) as a service environment. The BIM-PDE server is invoked via a user interface as shown in Figure 40.2. The welcome page checks credentials of the users and authenticates the authorized user to the "BIM Partial Data Exchange" Web page. As shown in Figure 40.2, three data-exchange functions are supported by the BIM-PDE server – (i) uploading a new building model, (ii) downloading a partial model, and (iii) uploading a partial model.

Figure 40.3 shows the new building model (IFC1) that the designer creates using Autodesk Revit Architecture. The architect imports the partial model (IFC2) to the BIM software, Graphisoft ArchiCAD, and makes changes to the partial building model. With the partial model, the architect deletes the highlighted wall, adds a new wall, and exports an IFC file (IFC3) to the server. The architect updates these changes to the integrated building model stored in the BIM-PDE server without affecting the other parts of the same model. Internal functions leveraging the IFC standard data model are implemented to update the key information (including geometric and material properties) read from the partial building model file (IFC3) to the

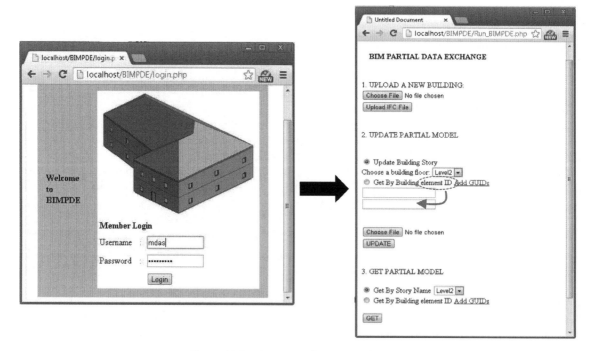

Figure 40.2   *Access to the BIM-PDE server*

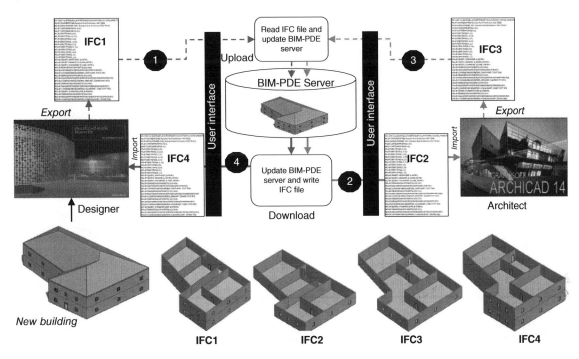

Figure 40.3 *Example scenario describing partial BIM building model transfer (the roof is hidden in IFC1 and IFC4)*

BIM-PDE server. As only the partial model is exchanged, data exchange and model updating can be performed efficiently. The whole building model, IFC4, can now be downloaded by other services, with relevant changes reflected in the new model. While SaaS is a common paradigm currently deployed by engineering application services, the use of IaaS or DaaS using standard ontology as a vehicle for information and software interoperability in the engineering domain could prove beneficial to provide added value to individual cloud applications and enhance collaboration. It should be emphasized that although we have provided an example in the building engineering domain, similar strategies can be achieved using various information models and ontologies developed in the manufacturing and other engineering domains (Fenves *et al.*, 2005).

40.4 Collaborative Technologies for Cloud Platform

Design of complex engineering systems is a collaborative task among designers or design teams that are physically, geographically, and temporally distributed (Sriram, 2002). Ensuring comprehensive technical proficiency in a world where trends are toward more multidisciplinary design can be a costly undertaking for a company. Effective collaborative design environment is important for realization of multidisciplinary design.

Perhaps the most important collaborative activities in the product development process lie in the conceptual design phase. Meetings are commonly convened to explore and brainstorm design concepts and develop design schemas. Creativity, however, is not necessarily restricted within meetings. Many creative or "light-bulb" design ideas are instantaneous. Cloud computing, with its capability to support devices, ranging from personal and mobile devices to HPC resources, can have a significant impact on creative and collaborative design. In addition to facilitating information capturing, sharing and access, the cloud service environment offers a new social network paradigm that allows communications, interactions and collaborations among collocated and distributed participants and their devices.

The Project-Based Learning (PBL) laboratory at Stanford University has been focused on the development, deployment, and assessment of collaboration technologies in support of synchronous and asynchronous collaborative interaction in cross-disciplinary geographically distributed project teamwork. Collaboration software applications and virtualized storage have been deployed on Microsoft's Azure cloud platform. Two examples of such cloud applications are:

- *BrainMerge:* a virtual brainstorming application service that facilitates crossdisciplinary, geographically distributed teams. The purpose of BrainMerge is to: (i) make individual "local conditions" visible in real time during a specific collaborative event, and (ii) provide a mechanism to harvest the team members' creativity.
- *eMoC* (engagement Matrix of Choices): a mobile application and service that allows project members to make their "local conditions" transparent, update them in real time, and track them over time. The purpose is to provide feedback on the status of distributed project members towards establishing mutual understanding, thereby enhancing their degree of engagement, and improving work productivity (Fruchter and Medlock, 2013).

The "local conditions" in BrainMerge and eMoc include not only location awareness – time and geographic location information – but also type of location, for example home, café, lab; resource information such as available networks, devices, and tools; workload; personal knowledge profile, and intelligent feedback regarding potential degree of engagement as a function of the existing conditions of all team members. These local conditions include not only the work conditions but also the social and cultural attitudes that are important to assess and measure the collaboration and performance of a project team.

The virtualization of both BrainMerge and eMoc cloud applications and services led to similar architecture implementations as shown in Figures 40.4a and 40.4b, with a MySQL database cloud storage and the respective virtual machine application and service running on a Microsoft Windows Azure platform. The participants interact with the BrainMerge and eMoc cloud applications and services through the corresponding BrainMerge web site and eMoc web site. Distributed participants notify the cloud applications and services,

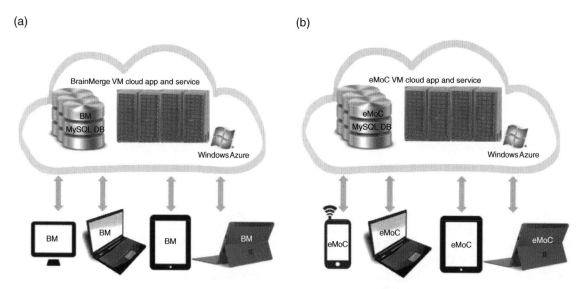

Figure 40.4 *Cloud applications and services: (a) BrainMerge; (b) eMoC*

i.e., BrainMerge and eMoc, of their local conditions and ideas. This information is collected, updated, and tracked in real time, then correlated, integrated, and fed back to the end-point mobile and desktop devices of the participants. This allows teams to best leverage their knowledge and resources during synchronous and asynchronous interactions.

40.4.1 BrainMerge – A Cloud Service Supporting Brainstorming Sessions

To harvest knowledge and foster creativity in collaborative brainstorming sessions, BrainMerge is designed to allow all the geographically distributed participants to make their local conditions visible in real time from their mobile or desktop devices. Figure 40.5 illustrates a digital badge example for Dr. Renate Fruchter, showing her geographic location (sun icon), day time – engaging in a brainstorming session during work hours (hammer icon), feelings (smiley face), availability – coming from a meeting and going to a meeting (running icons left and right the image), workload – number of projects (left side bar), type of network connectivity (broadband icon). Furthermore, as shown in Figure 40.6, BrainMerge offers a meshup with Google Maps to show the participants' locations especially for brainstorming sessions where team members have significant time zone differences.

BrainMerge supports concurrent, real-time, synchronous ideation by connecting the participants' end-point devices with the cloud application and service. It captures, tracks, compiles, and feeds all the participants' inputs. Most importantly, it gives all participants a voice, allowing every participant to review everyone else's ideas and expand or build on others' ideas, as well as vote, prioritize, and cluster ideas. Figure 40.6 illustrates a building solution (Figure 40.6a), and a brainstorming session (Figure 40.6b and 40.6c) from a global project team of eight participants – architect at University of Ljubljana in Slovenia, structural engineers at Stanford University, construction managers at KTH in Stockholm Sweden and Stanford University,

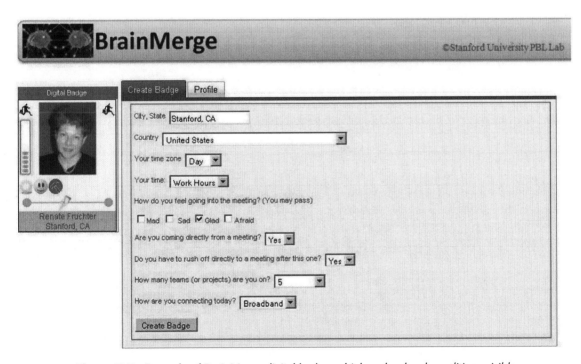

Figure 40.5   *Example of BrainMerge digital badge, which makes local conditions visible*

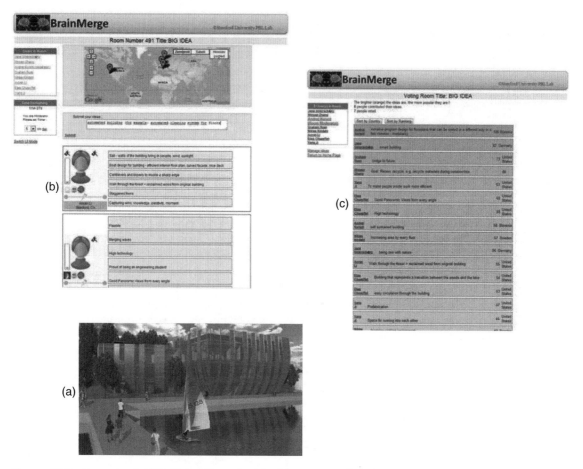

Figure 40.6 *Example of AEC Global Team final building proposal and BrainMerge virtual brainstorming and voting results*

MEP designers at Stanford University, and LCFM at Bauhaus University in Germany. As shown in the Google map (Figure 40.6b), the brainstorming title is the "*Big Idea*" that the team members submitted their overall objectives or ideas about the building (such as "Boat design for (the) building"). The BrainMerge screenshots from a participant's device illustrate: (i) a moment during the brainstorming session as a participant reads team members' current ideas and enters a new idea in the text box (Figure 40.6b); and (ii) a snapshot of the voting room (Figure 40.6c) for the results compiled by the BrainMerge cloud service from the votes casted by seven out of eight team members and fed back in real time to the participants on their devices.

40.4.2 eMoC – A Cloud Service Supporting Team Engagement in Collaborative Environment

One of the key challenges for collaborative engineering among distributed project teams is to make their local conditions visible and, possibly, make other team members aware of them. Transparency, visibility, and alignment of local conditions are critical for achieving high-performance teamwork and are fundamentally important in building a sense of team identity and belonging (Fruchter *et al.*, 2010). The collaborative tool, engagement

Matrix of Choices (eMoC) is developed to assist global team workers to achieve mutual understanding of their local work environment conditions (Fruchter and Medlock, 2013). eMoC enables teams to assess and make explicit choices continuously related to the physical, digital, and interaction as available options in their work environment.

Engagement involves people, and the content and equipment they create, manipulate, and operate. From the physical, digital, and interaction context, we define four degrees of engagement, namely Awareness, Attention, Participation and Engagement (AAPE):

- Awareness – all participants can only hear each other.
- Attention – all participants can only hear and see each other.
- Participation – all participants can hear and see each other, and co-control shared content.
- Engagement – all participants can co-create the collaboration space, content, products, in addition to being able to hear and see each other, and co-control shared content.

Additionally, eMoC assists individuals (i) to assess, reassess, and realign their local work environment and build awareness of their current conditions, and (ii) to progressively move from the individual level to the dyad and to the team, followed by commitment towards full engagement in the project. Tool kits allow individuals and project teams to report and identify their location (including place, network, and devices employed), skill levels, workload, and so forth. By formalizing and implementing eMoC as a cloud service, the tool allows participants flexible engagements in which they can use their mobile and desktop devices through a Web-based user interface.

The eMoc cloud service captures, stores, tracks, and updates the local conditions of all participants in real time. Figure 40.7 illustrates an eMoC individual profile made transparent to the team – vSpeed (displaying reported versus actual workload) and ccKit (showing available networks, devices, ICT tools and potential AAPE engagement capacity with the e-mail tool as an example). The eMoc cloud service also compiles, and feeds back to the team members, the current level of ICT alignment / misalignment, for example how many team members have access to a WiFi network connection, a device like a Smartboard, or tool like Gotomeeting Web conferencing. This allows teams to plan their synchronous and asynchronous interactions accordingly, to achieve the highest potential degree of engagement.

BrainMerge and eMoC cloud applications and services have been deployed and used in global course projects in education testbeds (see http://pbl.stanford.edu/AEC%20projects/projpage.htm, accessed December 30, 2015) as well as corporate pilot case studies in Fortune 500 high tech and manufacturing enterprises. These education and industry testbeds enabled us to assess the user experience and impact on team dynamics and performance (Fruchter *et al.*, 2010; Fruchter and Medlock, 2013). BrainMerge has enabled deliberations and collaborative development of design concepts. Industry users have commented that eMoC have made them aware of their team members' local conditions and workload as well as how important it is to have this information in order to have more realistic deadlines and work plans. Deploying the collaborative SaaS tools on a cloud platform, such as Windows Azure, has significantly reduced the infrastructure and resource cost as well as increased the reliability, scalability, storage capacity, and work productivity in the PBL Lab.

40.5   Summary and Discussion

Cloud computing has the potential to transform the practice of engineering by providing services over a network to support the entire product life cycle in a transparent manner. In this chapter, we have briefly reviewed the basic cloud service and deployment models and discussed how engineering enterprises and engineering service providers deliver their applications as cloud services. The business models, such as pay-per-use on an

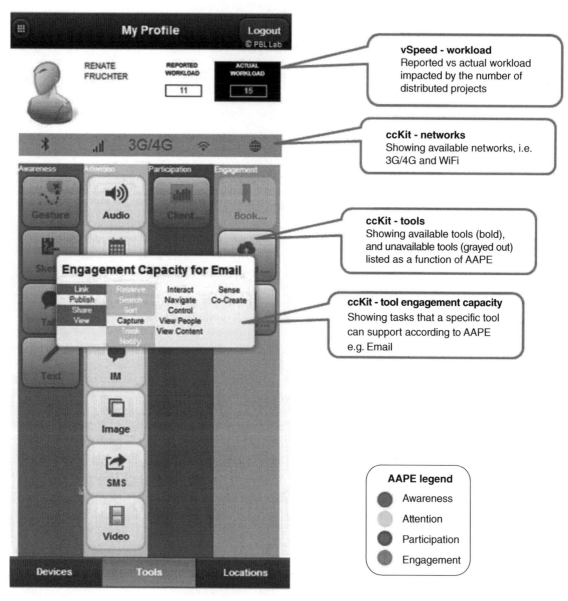

Figure 40.7   *eMoC mobile example of an individual profile made transparent to project team*

as-needed basis, offered by cloud service providers can significantly reduce the IT expenditures of engineering companies. As advanced CAD/CAE/CAM tools become available in the cloud, small and medium enterprises (SME) in engineering and manufacturing will be able to utilize the most appropriate, up-to-date computational tools (which may only be available to large corporations in the past) to develop high-quality products.

In order to deliver services effectively to support the entire engineering product life cycle, information sharing and service interoperability between various clouds must be adequately addressed. Furthermore, synchronous and asynchronous cloud-based collaborative tools can greatly enhance collaboration and the productivity of project teams in a geographically distributed setting. In this chapter, we have described two case studies of cloud-based computing in engineering – one in product information sharing and the other in distributed collaboration. Continuing research and development efforts in supporting information and service interoperability and distributed collaboration are needed to realize the full potential of cloud computing in engineering. Last but not least, as sensors and actuators are increasing being deployed to link computational systems to the physical world, a cyber-physical cloud computing environment, where sensors and actuators are (remote) services and the information generated is effectively stored, processed, shared in the cloud environment, represents the first step towards smart networked engineering systems (Simmon *et al.*, 2013).

Acknowledgements

The authors would like to acknowledge support from the Information Technology Laboratory and the Enterprise Systems Group at the National Institute of Standards of Technology (NIST), the SAVI EAGER program of the US National Science Foundation (NSF), Grant No. 1265953, and the Hong Kong Research Grants Council (RGC), Grant No. DAG09/10.EG02. The Project-Based Learning Lab research team at Stanford University would like to thank Microsoft, Inc., for the grant from Azure for Academic Institutions. Any opinions, findings, and conclusions or recommendations expressed in this material are those of the authors and do not necessarily reflect the views of NIST, NSF, and the Hong Kong RGC.

Disclaimer

Certain commercial software systems are identified in this paper. Such identification does not imply recommendation or endorsement by the National Institute of Standards and Technology (NIST); nor does it imply that the products identified are necessarily the best available for the purpose. The software tools and the vendors mentioned in this article are not comprehensive and are only provided to show the growing popularity of the cloud model. Any omission of tools is not intentional. Further, any opinions, findings, conclusions or recommendations expressed in this material are those of the authors and do not necessarily reflect the views of NIST or any other supporting US government or corporate organizations.

References

Anstett, T., Leymann, F., Mietzner, R., and Strauch, S. (2009) Towards BPEL in the cloud: Exploring different delivery models for the execution of business processes. Proceedings of the International Workshop on Cloud Services (IWCS 2009) in conjunction with the 7th IEEE International Conference on Web Services (ICWS 2009), pp. 670–677.

Armbrust, M., Fox, A., Griffith, R., *et al.* (2010) A view of cloud computing. *Communications of the ACM* **53**, 50–58.

Cheng, J. C. P. and Das, M. (2013) A cloud computing approach to partial exchange of BIM models. Proceedings of the 30th International Conference on Applications of IT in the AEC Industry, Beijing, China.

Fenves, S., Sriram, R. D., Subrahmanian, E., and Rachuri, S. (2005) Product information exchange: Practices and standards. *Transactions of the ASME Journal of Computing and Information Science in Engineering* **5**(3), 238–246.

Fruchter, R., Bosch-Sijtsema, P., and Ruohomaki, V. (2010) Tension between perceived collocation and actual geographical distribution in project teams. *International Journal of AI and Society* **25**, 183–192.

Fruchter, R. and Medlock, L. (2013) The journey from island of knowledge to mutual understanding in global business meetings. In *Shikakeology: Designing Triggers for Behavior Change* (eds. N. Matsumura and R. Fruchter). AAAI, Palo Alto, CA.

Liu, F., Tong, J., Mao, J., *et al.* (2011) NIST Cloud Computing Reference Architecture – Recommendations of the National Institute of Standards and Technology. Special Publication 500-292. National Institute of Standards and Technology, Gaithersburg, MD.

Mell, P. M., and Grance, T. (2011) *The NIST Definition of Cloud Computing*. Special Publication 800-145. NIST, Gaithersburg, MD, http://www.nist.gov/customcf/get_pdf.cfm?pub_id=909616 (accessed November 25, 2015).

Redmond, A., Hore, A., Alshawi, M., and West, R. (2012) Exploring how information exchanges can be enhanced through Cloud BIM. *Automation in Construction* **24**, 175–183.

Rosenthal, A., Mork, P., Li, M. H., *et al.* (2010) Cloud computing: A new business paradigm for biomedical information sharing. *Journal of Biomedical Informatics* **43**, 342–353.

Simmon, E., Kim, K.-S., Subrahmanian, E., *et al.* (2013) *A Vision of Cyber-Physical Cloud Computing for Smart Networked Systems*. Report No. NISTIR 7951. National Institute of Standards and Technology, Gaithersburg, MD.

Sriram, R. D. (2002) *Distributed and Integrated Collaborative Engineering Design*, Sarven Publishers, Glenwood, MD.

Xu, X. (2012) From cloud computing to cloud manufacturing. *Robotics and Computer Integrated Manufacturing* **28**, 75–86.

Yasuda, M. (2012) Fujitsu's engineering cloud. *Fujitsu Scientific and Technical Journal* **48**, 404–412.

Youseff, L., Butrico, M., and Da Silva, D. (2008) Toward a unified ontology of cloud computing. Grid Computing Environments Workshop, Austin, TX, pp. 1–10.

41

Educational Applications of the Cloud

V. K. Cody Bumgardner, Victor Marek, and Doyle Friskney

University of Kentucky, USA

41.1   Introduction

Organizations around the world are adopting cloud computing and the education sector is no exception. Education providers, from primary and secondary education to higher education, hope to solve many of their technical challenges with cloud technologies. Cloud computing provides an educational institution with access to resources that were once only available to wealthy institutions. A perfect example of this is Khan's Academy, which offers individual students, schools, and universities resources to improve their success in the classroom. Resources available from cloud computing will challenge traditional methods of education. Today, students can graduate from high school or college without having to sit in a traditional classroom. Students in classrooms have access to outstanding educational resources previously unaffordable to them. The consumerization of technology and emergence of mobile computing pose unique challenges for educators. Consumers expect to be able to interact with traditional education institutions in much the same way they use public cloud applications (YouTube, Khan Academy, etc.). Access to academic resources from any device at any time is not just a benefit – it is expected. This challenge is not just for traditional institutions, as licensing restrictions around software or even content can limit the most technically proficient online institutions.

This chapter discusses the adoption of cloud technology, applications related to instruction, front-office interaction, and back-office operations. The use of cloud computing in academic research will also be covered. The chapter provides a broad overview broad overview of cloud computing in education and outlines recommendations and considerations for selecting cloud services.

Encyclopedia of Cloud Computing, First Edition. Edited by San Murugesan and Irena Bojanova.
© 2016 John Wiley & Sons, Ltd. Published 2016 by John Wiley & Sons, Ltd.

41.2   Adoption of Cloud Technologies in Education

From a technology standpoint, educational institutions are in a uniquely difficult position when it comes to technology. Education providers experience many of the same problems as producers of consumer products. How can all possible technology platforms be supported that might be used by their consumers? As with consumer products, there is a careful balance to be made between supporting a subset of technologies effectively, and isolating potential/actual customers. To deal with these challenges, some education providers, like some consumer technology companies, only support a very specific platform (for instance IOS 6.x on iPad2 or greater). Some education providers are in a position to provide a computing platform as part of the cost of education, often referred to as 1:1 programs but, mostly, devices are provided by students and their families. This is often referred to as a "bring your own device" (BYOD) scenario.

Consumerizaton has changed how corporations and educational organizations support today's computing environment. Cloud computing resources combined with personally owned technology minimize the impact that IT organizations have on mandating standards. This is both a blessing and a curse. The adoption of cloud computing by school organizations will continue to transform how IT organizations support their customers (students). In addition to the BYOD problem, education providers also have to overcome all the technical challenges experienced by corporations. A large K-12 school district can have over 100 000 users, spread out over a thousand locations. A single large university, from an administrative computing prospective, can resemble a Fortune 500 company. Compliance requirements related to regulatory agencies require educational organizations to address security issues such as PCI (credit card processing), and comply with the Family Educational Rights and Privacy Act (FERPA), which defines rules for the protection of student data, and the Health Insurance Portability and Accounting Act (HIPAA), which defines rules for privacy protection and end use, resulting in challenges for education providers.

41.2.1   Private and Public IaaS Adoption

In recent years, Amazon Web Services (EC2) (http://aws.amazon.com/ec2/, accessed December 30, 2015) has been the leader in public infrastructure as a service (IaaS) cloud. During the 2012 Amazon Web Services (AWS) Public Sector Summit in Washington, DC, Amazon announced that more than 1500 education institutions were leveraging AWS for a wide range of uses including big data analytics, high-performance computing, Web and collaboration applications, archiving and storage, and disaster relief. Despite this, the majority of IaaS resources remain centralized at the institutional level. For many institutions a great deal of cost avoidance has already been achieved through server virtualization. These centrally managed pools of virtual resources are called "private clouds."

41.2.2   PaaS Adoption

In comparison to IaaS, if a platform-as-a-service (PaaS) application is not being used, the cost decreases. While technically attractive, very few education providers take direct advantage of these services. In part, this is due to limited in-house software development, outside of traditional application frameworks. As previously mentioned, existing architectures can't take advantage of PaaS, and when they are rewritten to do so, software vendors often force customers to make the direct transition to software as a service (SaaS). It is also noteworthy that many educational technology providers use IaaS as their primary means to provide products to the educational community.

41.2.3   SaaS Adoption

The highest growth area is SaaS because it allows the student to select the resource needed with intervention by an IT organization. Software as a service is often the choice of school teachers, administrators, and parents.

A commonly used example of SaaS by students is YouTube, offered by Google. It is rare to find software being introduced to education providers by their vendors that would be considered anything but SaaS. These SaaS offerings include: Facebook (http://www.facebook.com), Twitter (http://www.twitter.com), Google Apps (https://www.google.com/enterprise/apps/education/, accessed December 30, 2015), and Microsoft Office 365 (http://office.microsoft.com/en-us/academic/, accessed December 30, 2015). These applications are highly available, globally redundant, and fully distributed. Unfortunately, even when software is hosted on a single server by the software vendor, the solution is sold as a "cloud" application. Technically superior software solutions take advantage of PaaS or build their own distributed cloud frameworks. While often software vendors do not disclose their internal architecture, those who have invested in cloud architectures will use this as a selling point. From the standpoint of adopting new software solutions there is often no choice but to accept a hosted solution if one wants to use the software. Unlike IaaS or even PaaS solutions, software vendors providing SaaS will often take on the responsibility and liability of securing application data.

In terms of existing software, organizations are being faced with the choice of upgrading vendor-provided software for local deployment or migrating to a hosted solution. Often this is not a question about using cloud computing; it is a question of licensing model. However, the vast majority of SaaS offerings are based on "named-user" licensing. This is particularly problematic for education providers, where perhaps only a small section of their population will use a particular software package. To illustrate, consider a statistics package locally hosted by a university department. In the past, the licensing for the software might be related to a percentage of expected users, mostly from a specific department, but the software was available campus-wide. In a named-user model, anyone that used the software package would have to be designated to do so. This change in licensing forces institutions to choose between maintaining software locally, restricting software availability, or paying higher prices to include all users. While these are not technically cloud problems, they have become barriers to adoption.

41.2.4 Platform Adoption Summary

It is nearly a foregone conclusion that new software introduced by vendors to education institutions will be delivered through SaaS. The efficiency with which SaaS vendors can provide services will be rooted in their own cloud-adoption abilities. There is an active debate concerning security around cloud services. There are those who think that by giving up control of the infrastructure, platform, and software layers, they are giving up their ability to secure their data. One could also argue that the baseline security measures provided by cloud providers might not meet the requirements of all organizations. It is also conceivable that large service providers are bigger targets for cyberattacks. The other side of this argument is that a fully staffed baseline security service is superior to what many institutions are currently able to provide. By migrating the software to the cloud, we not only remove risk associated with providing infrastructure but also remove a great deal of security risk related to internal threats.

41.3 Cloud Applications in Education

We often associate cloud technology with ubiquitous computing (where an application appears to be accessible from any device). This is reasonable considering that the majority of the consumer-focused cloud applications (Facebook, Twitter, etc.) have multiple access methods, including mobile, desktop, and Web applications. Educational publishers have embraced SaaS to extend their textbook offerings to include interactive personalized learning experiences. Users have come to expect that applications will be made available in a wide variety of ways from anywhere at any time, and cloud solutions provide this capability. However, most organizations simply can't afford to meet this expectation with applications developed in house.

The majority of new applications are delivered as SaaS, which eliminates the need for institutions to maintain expensive datacenters and information technology staff with dated skills. Cloud-based education technology companies compete with the newest ideas to enhance the learning environment. As applications become successful they are integrated into the product offerings of mature educational vendors. Integration and agility to meet the needs of learners become critical. Most organizations struggle to calculate the savings related to infrastructure related to cloud migrations; often the savings are not as important as meeting the needs of the students. Cloud vendors provide innovative alternatives to traditional educational solutions. Cloud migration strategies must include ways to quantify reasonably the value used by eliminating traditional datacenter requirements. While one can argue that cost of cloud communication will increase, the benefit of eliminating traditional datacenters should more than offset communication cost.

In the following subsections, we will discuss specific areas impacted by cloud adoption.

41.3.1   Cloud in Instruction

The technology cornerstone of most educational institutions is the learning management system (LMS). The basic task of a LMS is to provide a framework to manage all aspects (content, instruction, assignments, testing, grading, etc.) of the learning process. Usage varies greatly based on online requirements and institutional guidelines for course standardization. Learning management systems have evolved to include content management systems (CMS), collaboration tools, and a host of other features. The LMS has been around for decades and, like many systems of this lineage, they are designed to be hosted (installed locally) by the institution. It is common to have many pieces of software, from many vendors, work in conjunction with an institutional LMS. Traditional campus LMS integration is shown in Figure 41.1. Components are tightly

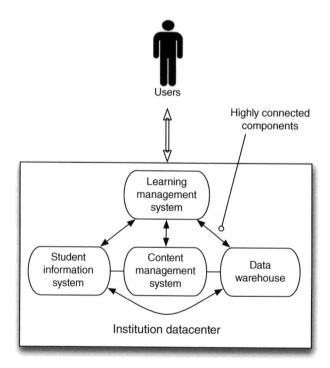

Figure 41.1   *Traditional "highly connected" model for LMS campus integration*

integrated with each other using many integration points. This tight integration, often constructed using custom developed interfaces, is an impediment when searching for cloud-based alternatives.

Emerging LMS systems, like most new software, are being delivered in the SaaS model. A good example
of this is Instructure's Canvas (http://www.instructure.com, accessed December 30, 2015). One of the key
differences between legacy LMS systems and emerging cloud offerings is the openness of the ecosystem.
Traditional systems are highly proprietary whereas emerging systems are created on open standards and
typically provide APIs that encourage interoperability. Even if a competing LMS could somehow integrate
with remaining local services, the overall user experience could be limited by locally hosted dependencies.
The problem becomes harder when components related to an LMS are available from various cloud vendors.
When everything was hosted locally, one conceivably had low-level access to components of the system. This
allowed for various levels of product integration, based on the desired user experience level. In Figure 41.2,
we show three separate SaaS cloud providers with the same application features as those shown in Figure 41.1.
There is an important difference between these figures, related to their level of integration. The components
in Figure 41.1 are highly connected, allowing application data in the various components to be bidirectionally
shared. The components in Figure 41.2 don't have the same bidirectional communication capabilities as those

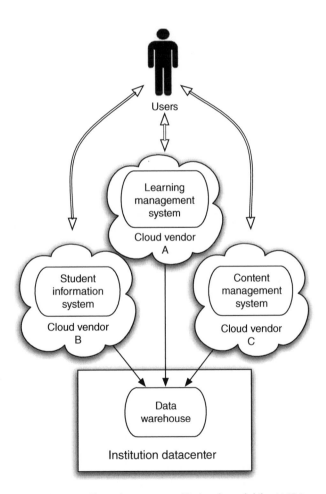

Figure 41.2 *Current "loosely-connected" cloud model for LMS integration*

in Figure 41.1. These components are limited to both the level of API access provided by the respective SaaS providers, and the data exposed by the APIs.

In response to these problems, institutions have formed consortiums to develop open mashups of learning platforms (Mikroyannidis *et al.*, 2012). One such consortium is Unizin (http://unizin.org, accessed December 30, 2015). The goal of Unizin is to provide a common platform focused on digital content development and data analytics. Open access to interaction data across clouds (Rizzardini and Amado, 2012) is necessary for recommendation engines (Leony *et al.*, 2012) used in adaptive (Chaabouni and Laroussi, 2012), and personalized (Gillet and Bogdanov, 2012) learning.

No discussion about cloud and education would be complete without mentioning the so-called massive open online courses (MOOCs). While the idea of a MOOC is really more of a business model based on universal (possibly free) access to education than a specific technology, it provides an edge case for technology problems. If instruction were open to everyone in the world and content were made freely available everywhere, one would need to develop a unified platform that operates very efficiently.

41.3.2 The Use of Cloud in the Front Office

The term "front office" is used to describe the departments or services of an institution that come into direct contact with the end user. Most school organizations use cloud services to support library systems, administrative, and academic system. The use of cloud computing was introduced by offerings from companies like Google and Microsoft with e-mail and storage applications. From an education prospective we could certainly consider the user-facing component of an LMS, a front-office service. However, there are many other services beyond those used in instruction that institutions have traditionally provided. Collaboration services (online messaging / meeting software), e-mail, and personal storage all fall into this category. Even more recently, these offers have been extended to faculty and staff. The term "free" in this respect is really "with no additional charges," because often existing support agreements must be maintained. Nevertheless, organizations have the option to migrate cost and risk associated with maintaining local infrastructure to another organization. As with the LMS we discussed in the previous subsection, the migration of local services to the cloud is not always a clean one. The "free" option might not include all of the capabilities that are provided with local hosting. Many vendors have options to designate paid and free accounts within the same institution. If institutions can be flexible, not only can services be migrated to the cloud, but the resources provided to their users are often much greater (bigger mailbox, storage, etc.) than they themselves can provide.

41.3.3 The Use of Cloud in the Back Office

The term "back office" is used to describe the departments and services of institutions that are dedicated to running the institution itself. In subsection 41.3.1 we discussed how LMS systems were used in instruction. An LMS has components of both front office and back office. Back office functions of an LMS include grading, assessment, and other administrative functions. Many institutions maintain an enterprise resource planning (ERP) system in addition to their LMS system. The ERP system might also have front-office functions in the form of student, faculty, and staff self-service functions. However, the majority of ERP functions are dedicated to the administrative aspects of running the organization. Enterprise resource planning and LMS systems share many of the same cloud migration challenges. Considering that ERP systems have been around longer than LMS systems, it is reasonable to think that decoupling an ERP system from institutional dependencies could be even more challenging. Much of the risk associated with providing ERP infrastructure can be mitigated by remotely hosting private cloud services. Figure 41.3 shows a multisite IaaS cloud of traditional services. All of the existing integration benefits are maintained, while risks related

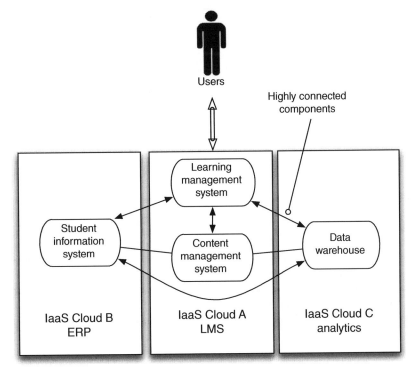

Figure 41.3 Hybrid "highly connected" IaaS clouds

to maintaining infrastructures are eliminated. In this model the vendor provides remote IaaS services but, unlike public cloud, these resources are isolated from other customers and generally have a fixed size and cost. This arrangement can also be expanded to include additional dependencies, including custom developed solutions.

Compared to front-office applications there is less of an expectation for back-office applications to provide ubiquitous client interfaces. Back-office applications are typically developed based on institutional business practices, which vary between education institutions. For this reason flexibility is valued over "canned" application functionality. In the short term one can expect institutions to adopt remotely hosted private cloud solutions for back-office applications.

41.3.4 Social Networks, Gamification, and Student Analytics

Institutions can mandate that students utilize online systems for instruction. Individual courses might even require student interactions through blog post or online meetings. However, it is something else entirely to make students want to use these services. There has been limited success in building online communities on an institutional level. Social media sites maintained on an institutional level can be conceived as "big brother," which is a deterrent to participation.

The practice of providing learners with feedback related to academic performance and engagement is sometimes referred to as the "gamification" of learning. Most often this data is presented to students as a scorecard, which shows achievements as if they were participating in a game. The objective of this practice is to entice the student to engage and to identify themselves as part of their institution. Increases in institutional engagement are linked to everything from student retention to alumni endowment. There have been many cloud applications developed for student engagement from many education software vendors. Many of these applications provide the ubiquitous user experience expected of cloud software. Gamification has opened the door for publishers to create personalized learning solutions. The principles of adaptive learning and gamification complement one another and create a platform for creating exciting interactive learning environments. An example of this is the partnership between Arizona State University, Pearson's and Knewton (Upbin, 2012) to develop an adaptive tutoring system.

41.3.5   Education Application Summary

Student analytic services allow institutions to compare and analyze the needs of their students against state and national standards. Student sentiments, the development of their social networks, and interaction with the institution all play a role in this analysis. Vendor control of the application's API can be used as a competitive advantage for services that the vendor might also provide. Figure 41.4 shows two possible options for CMS systems: CMS A and CMS B. CMS A is more expensive and provides fewer features than CMS B. However, CMS A is provided by the same cloud vendor as our LMS system and the features of both applications are tightly integrated. In this case the vendor with the integrated solution has no incentive to provide a robust API that could be used to integrate other options. Care must be taken to evaluate the APIs provided by SaaS vendors, to ensure appropriate access to data can be obtained. As previously mentioned in subsection 41.3.1, if an instructional cloud standard is adopted by a large number of institutions, a common API will likely be defined to overcome data integration problems. Cloud-based companies like StarFish (http://www.starfishsolutions.com, accessed December 30, 2015) and Civitas (http://www.civitaslearning.com, accessed December 30, 2015) are examples of K-20 focused student analytic applications.

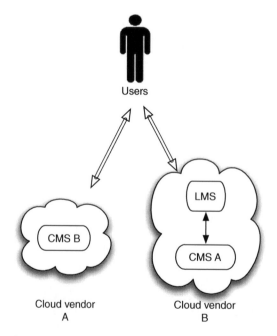

Figure 41.4   *Competitive advantage through the API*

41.4   Cloud Computing in Research

While purpose-built supercomputers have existed for several decades, much of the high-performance computing that exists in research institutions today has evolved from augmentations of more general-purpose computing services based on Intel/AMD processors, with low-latency interconnects between each server. The use of research computing throughout its relatively short history has been dominated by the "hard sciences." Specifically, various areas of theoretical physics, computational chemistry, computational biology, and materials engineering are often primary consumers of computational resources. One of the primary reasons for this is that in theoretical pursuits, the problems are rarely solved and computational models can generally be expanded to match all available resources. In recent years, bioinformatics, linguistics, and other disciplines are challenging the status quo in terms of research computing. New needs do not always fit the old paradigm, and new architectures are needed to satisfy these new needs. Perhaps the biggest challenge in bringing on board new users will be in adapting their needs to existing resources. Learning ways to take advantage of a supercomputer is not yet generally part of academic training in most areas. We conclude that, in research computing, one size does not fit all. The next phase of computing for research institutions might very well be the era of cloud computing. National and regional research laboratories connected through high-speed research networks, along with institutionally deployed purpose-built computers, might fulfill the needs of users that are traditionally consumers of significant amounts of computation. Others might need "fast data," "big data," and throughput-intensive resources.

41.4.1   Research Computing, Supercomputing, and Cloud Orchestration

Far too many researchers use their own personal computers as their primary computational tool. Much of the reason for this is the barrier to entry for traditional research computing and the inflexibility of most institutional offerings. Most supercomputer environments are multiuser but not multitenancy. This means that people sharing an environment must follow the same rules, and make provisions not to interfere with the jobs of others. The idea of "less than HPC" has gained some following in recent years. For many workloads researchers are willing to trade the speed of traditional supercomputers for the flexibility of virtual multitenant environments. In addition, virtual environments can be used for HPC training (Gomez-Folgar *et al.*, 2012). Resources might be provided in the form of public cloud such as Amazon EC2 or Rackspace (https://www.rackspace.com, accessed February 4, 2016), or private cloud, such as OpenStack or VMware (http://www.vmware.com). Care must be taken by organizations to leverage cloud resources with open standards, so as not to be caught by vendor lock in on the cloud API level.

In many ways, simply providing infrastructure is a step back from traditional research computing services. For these reasons, cloud application orchestration is an area of great interest and development. Ubuntu Juju and OpenStack Heat, discussed above, are examples of application orchestration packages, which enable users to deploy complex application suites across clouds. In such arrangement, a student or researcher deploys a large storage cluster or full multinode Hadoop cloud at the click of an icon. While generally limited to single-sites, cloud orchestration will no doubt evolve into a multisite control model, allowing for deployments and migrations on multiple IaaS offerings simultaneously. When this occurs, the applications or the platforms they run on can make *active* decisions related to the best fit resource providers for application resources.

41.4.2   Regional Network and National Laboratories

In general, since the early 1990s, institutional research computing and research networks have kept pace with each other. Often these resources are funded from recurring government sources, which are relatively stable. It should be observed, though, that although stable, these funds have become stagnant over the past decade, leaving many institutions unable to compete. National laboratories, such as Los Alamos National Laboratory, have been at the forefront of supercomputing since its inception, while many individual research institutions (universities, small laboratories, commercial research, etc.) that have touted "top 500" ranked supercomputers for decades now find themselves unable to compete. National laboratories and consortium-funded supercomputers dominate the top supercomputers in the world. As institutions that once had their own central resources now look to national laboratories and consortiums for resources, network connectivity will be critical. Research networks are used relatively lightly, compared to their institutional research computing counterparts. One reason for this is that research networks connecting institutions and national labs are generally used for applied purposes such as to transfer data from point X to Y, and then they use the resources at a remote facility. Research computing, on the other hand, is often very theoretical and ongoing. In addition, even very large amounts of data related to theoretical work can often be regenerated with relatively small amount of input data. Most often, even in the case of national laboratories, computational results remain on the supercomputers that generated them. This is not to say that the research networks are devoid of experimentation – in fact one such NSF-funded project that supports research network experimentation is the Global Environment for Network Innovations (GENI) (one of the authors is involved in this research). Still, GENI experimentation is more applied in focus than theoretical research computing.

At present, the financial support for individual institutions to operate on a world-class computational level is waning. On the other hand support for regional and superregional consortium-based supercomputers is growing. Initially, underutilized research networks will be used more but as means to transport preprocessed and post-processed data. The idea of *active* "cloud computing," as described in section 41.4.1, will fully utilize research networks with active computational data. This type of resource optimization will universally benefit researchers.

41.4.3   "Big Data" versus "Fast Data"

The term "big data," while fashionable like "The Cloud," does actually describe a serious challenge for computer scientists and engineers. Industry experts (Gantz and Reinsel, 2012) predict a 50-fold increase in digital data from 2010 to 2020. The same experts estimate that as of 2012, only 3% of data that is generated was "tagged" with useful metadata. The majority of digital data today is created and used by consumers, not scientists. Much of this new data is generated from social media, sensors, and multimedia content. While large sources (Lynch, 2008) of data in single locations, like the Hadron Collider (LHC) exist, most of this "big data" is distributed. Not only is the source of the data distributed, but the analysis and use may be distributed as well (see Figure 41.5). In some cases, distributed data sources can be combined and processed in batches, whereas others need real-time distributed and concurrent processing. The term "fast data" is used to describe data that, while large in volume, must also be processed as it arrives. This is to say that we cannot wait until all data is transferred before starting processing. Often, data never stops, and one must react to the characteristics of the stream.

This is just one such example of many related to the processing of data "in movement." In the future this type of data processing will be necessary for everything from dynamic billing of cloud resources, algorithms for traffic routing, or even personal health services.

41.4.4   Research Cloud Summary

Let us imagine, for a moment, pockets of standards-based infrastructure in a highly connected global network. This is an accepted form of cloud computing, which we often designated as IaaS. Based on this definition, one could argue that the connected research and defense computers in the early days, of what would become the Internet was a form of cloud computing. Add to this collection of resources applications that can self-distribute and *actively* react to changes in both the connectivity and health of available resources. While many applications could run on this *active* cloud, perhaps the most benefit can be extracted in the areas of

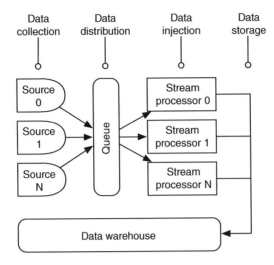

Figure 41.5   *"Big" and "fast" data processing"*

"big data" and "fast data." If one extends this type of thinking far enough into the future, a vision of the "Internet of things" comes to mind.

41.5   Conclusions

There is much to be gained through the use of cloud computing in education. Many of the unique challenges related to providing an engaging educational experience can be addressed with cloud technologies. The ubiquitous experience expected by consumers can be achieved through standardized platforms with a global reach. However, this utopian user experience will not occur overnight. While most new applications are being delivered through SaaS, the majority of existing applications will be limited to single-site deployments. Luckily, even single-site deployments can take advantage of private and public IaaS. The adoption of cloud, even on this level, removes the risk and cost of supporting infrastructure. Further benefits can be obtained through the use of cloud orchestration technologies, which in time can help users to transition traditional applications to PaaS or SaaS. The evaluation of SaaS APIs for both access method and data availability will be critical. Cloud software vendors can create a competitive advantage for their service by controlling the API layer. Institutions must develop long-term integration strategies and architectural guidelines before migrating core business services to SaaS. In the education industry, if common APIs can be adopted, vendor lock in can be avoided.

Common cloud applications used in academic settings are given in Table 41.1.

Table 41.1   *Common cloud applications used in academic settings*

Provides	Service	Provider
General Digital Media	SaaS	YouTube
Instructional Digital Media	SaaS	Khan Academy
Virtual Machines (System)	IaaS	Amazon EC2
Virtual Machines (Platform)	PaaS	Microsoft Azure
Learning Management System (Instructure)	SaaS	Canvas
Student Analytics	SaaS	StarFish

References

Chaabouni, M. and Laroussi, M. (2012) *A Cloud Computing for the Learner's Usage Tracks Analysis*. Proceedings of the First International Workshop on Cloud Education Environments (WCLOUD 2012), Antigua, Guatemala, http://ceur-ws.org/Vol-945/paper3.pdf (accessed December 31, 2015).

Gantz, J. and Reinsel, D. (2012) *The Digital Universe in 2020: Big Data, Bigger Digital Shadows, and Biggest Growth in the Far East – United States*. IDC iView: IDC Analyze the Future, https://www.emc.com/collateral/analyst-reports/idc-digital-universe-united-states.pdf (accessed December 30, 2015).

Gillet, D. and Bogdanov, E. (2012) *Personal Learning Environments and Embedded Contextual Spaces as Aggregator of Cloud Resources*. Proceedings of the First International Workshop on Cloud Education Environments (WCLOUD 2012), Antigua, Guatemala, http://ceur-ws.org/Vol-945/paper8.pdf (accessed December 31, 2015).

Leony, D., Pardo, A., Gélvez, H., and Kloos, C. D. (2012) A Cloud-based Architecture for an Affective Recommender System of Learning Resources. Proceedings of the First International Workshop on Cloud Education Environments (WCLOUD 2012), Antigua, Guatemala, http://orff.uc3m.es/handle/10016/18712 (accessed December 31, 2015).

Lynch, C. (2008) Big data: How do your data grow? *Nature* **455**(7209), 28–29.

Mikroyannidis, A., Okada, A., Scott, P. *et al.* (2012) *weSPOT: A cloud-based approach for personal and social inquiry*. First International Workshop on Cloud Education Environments (WCLOUD 2012), Antigua, Guatemala, http://ceur-ws.org/Vol-945/paper2.pdf (accessed December 31, 2015).

Rizzardini, R. H. and Amado, H. (2012) *Measuring emotional responses to experiences with cloud-based learning activities*. Proceedings of the First International Workshop on Cloud Education Environments (WCLOUD 2012), Antigua, Guatemala, http://citeseerx.ist.psu.edu/viewdoc/download?doi=10.1.1.416.7864&rep=rep1&type=pdf (accessed December 31, 2015).

Upbin, B. (2012) Knewton is building the world's smartest tutor. *Forbes* (February 22), http://www.forbes.com/sites/bruceupbin/2012/02/22/knewton-is-building-the-worlds-smartest-tutor/ (accessed December 31, 2015).

Gomez-Folgar, F., Garcia-Loureiro, A., Fernandez Pena, T., *et al.* (2012) *Cloud Computing for Teaching and Learning MPI with Improved Network Communications*. Proceedings of the First International Workshop on Cloud Education Environments (WCLOUD 2012), Antigua, Guatemala, http://citeseerx.ist.psu.edu/viewdoc/download?doi=10.1.1.416.8266&rep=rep1&type=pdf#page=26 (accessed December 31, 2015).

42

Personal Applications of Clouds

Cameron Seay,[1] **Montressa Washington,**[2] **and Rudy J. Watson**[3]

[1] *North Carolina A&T University, USA*
[2] *Case Western Reserve University, USA*
[3] *University of Maryland University College, USA*

42.1 Introduction

Personal cloud applications can be defined as public cloud services that focus on individual or personal use as opposed to business use. "The public cloud is used by the general public cloud consumers and the cloud service provider has the full ownership of the public cloud with its own policy, value, and profit, costing, and charging model" (Singh *et al.*, 2011). Consumers are using personal cloud applications for shared calendars, shopping lists, social networking, and location-based services. With respect to personal cloud applications, our model views individuals as consumers who, from a vast number of alternatives, consolidate a coherent solution that suits their individual purposes. What emerges is the following: (i) cloud users carry around with them a suite of applications ideally suited for their purposes; and (ii) because selection and usage is easily tracked by marketers and developers, this behavior drives new offerings toward increasingly more pertinent functionality. Increasingly irrelevant are the issues of architecture and operating system. For an application to be appropriately labeled "for the cloud" it must operate anywhere (provided there is connectivity to the cloud) and on virtually anything.

Some of the advantages of using personal cloud applications include ease of collaboration, synchronization of multiple devices, and automated backup of data. In addition to simply sharing information, some cloud applications allow for simultaneous entry and updating of data while keeping track of the specific user and facilitating version control. Since some users may now use desktops at work, laptops and tablets at home, and smartphones on the go, the almost effortless synchronization of information on the various devices is a desired feature. Whether one is using multiple devices or a single device for access, automated backup means that no longer will a hard-drive crash or some physical damage to a device mean that data will be lost or unrecoverable. As the pricing for cloud services is based upon economies of scale, collectively, the users have technology and function available to the individual that ordinarily would not be cost justified.

We take a two-dimensional view of using cloud applications: a generic type of application and usage of specialized or customized implementations. Generic applications include e-mail, collaboration tools, and shared data storage. Specialized implementations include applications that require geospatial location data and applications that are customized to the individual's specific detailed requirements. This includes applications that provide information based upon location and applications that are configured based upon the user's perceptions, which may be overtly stated or deduced.

E-mail is a generic application that has been around for over 40 years, but initially users had to be on a specific network to access it. The advent of the Web has eliminated this constraint. Likewise, for document collaboration, backup and recovery services, the only true requirement is access to the Internet. The applications run in the cloud and the user is unconcerned about where the application is or exactly how they will connect to it.

Access is one of the key drivers of personal cloud applications, specifically ubiquitous access from any device and from practically any place. This applies to data and information related to one's work as well as to personal data such as contacts, e-books, music, and social media. One might think that the future direction is only limited by the user's imagination and desires. However, there are more forces at work than market demand. An interesting question to analyze is whether the market is user driven or provider driven. McKinsey Global Institute ranks cloud technology as one of the top four most potentially disruptive technologies that will most radically transform human life between now and the year 2025 (Manyika *et al.*, 2013). They ranked mobile Internet as number one – and it is the foundation for ubiquitous personal cloud application usage.

Personal cloud applications are being developed and marketed at an increasing rate. In the United States, 77% of online adults use a cloud storage service (Gillett, 2013). This poses potential issues for businesses that desire employees to keep their business and personal use separate. Often, early adopters of personal cloud applications lead their IT suppliers to the applications that become the business standard. There are many personal cloud application services for consumers to choose from ranging from file and document sharing to sophisticated personal finance applications. Examples are Google Drive and Drop Box. The most widely used personal cloud application providers offer some form of free space for personal use, which ranges from 2 GB to 50 GB depending on the provider (Drago *et al.*, 2012). Users wanting more space or business accounts can expect to pay fees ranging from $3.95 per month to upwards of $600.00.

42.2   Personal Cloud Application Major Providers

Table 42.1 shows the major providers of personal cloud applications. It depicts the key features of the offerings by the various providers. The table is, however, not all inclusive as new companies and offerings are added on a monthly basis; however, it does identify the current major players (Vaughan-Nichols, 2013).

42.3   Issues and Limitations

Consolidation of services by cloud computing providers for the enterprise will continue as most providers offer personal cloud application services for the consumer in some form. Security of personal cloud applications is not the major concern that one may suspect. In August 2013, SafeNet Labs asked with hundreds of business professionals worldwide and reported that 52% of them were concerned about the security of cloud applications. However, 64% of the total respondents revealed that they frequently use cloud based applications

Table 42.1 *Personal cloud application providers*

Application	Key features
Amazon Cloud Drive	Store photos, videos, documents, and other digital files
	Cloud Drive app for Windows and Mac. You can also install Cloud Drive Photos for Android and iPhone
	Quick access from any Web browser
Apple iCloud	Music, apps, books, and TV shows purchased from the iTunes store, Photo Stream, can also be stored and streamed from it, and none of the purchased media counts against your storage quota
	Access to Apple's wireless service
Box	View and access files on demand
	Share a whole folder of files; create a new folder, upload files, then invite others to join.
	Access your files anywhere, anytime, on any device: desktop, laptop, iPhone, iPad, Android phones and tablets
	Online collaboration – post comments and assign tasks
Dropbox	Does not need a Web-browser interface. It will run natively on almost any PC, including Linux computers or devices running Android or iOS
	Access any files stored because, by default, it syncs with all of your local devices
Google Drive	Automatically syncs with the cloud so that everything is consistent across all of your devices
	Integrates with Windows and Mac file systems, does not natively support Linux. Supports Google's own Chrome OS, Android, and Apple's iOS
	Enables you to share and collaborate on any kind of file, including documents, music, images, and videos
	Any content you create in Google Drive does not count against your storage quota
JustCloud (UK)	Sync multiple computers
	Unlimited storage
	Mobile access
Kanbox (China)	Cloud storage and sharing platform
	Allows users to store photos, videos, and documents across multiple locations and devices
	Offers mobile access
Media Fire	MediaFire's free version includes ads
	Only download files from a folder one at a time
	Supports Linux, Mac OS X, and Windows on the PC, and Android and iOS on devices
SpiderOak	Client software, which supports Linux, Windows, and Mac OS X in PCs, and Android and iOS, encrypts everything before it hits SpiderOak's servers
	The service can only be accessed through its secure client software – you cannot use it via a Web browser or by your native operating system. SpiderOak is designed for security first and foremost;
Microsoft SkyDrive	SkyDrive is free online storage for your files, which you can access from anywhere
	SkyDrive desktop app allows you to sync your files to your devices automatically;
	All you need is a Microsoft account
	If you have used Microsoft services in the past – like Xbox, Hotmail, Skype, or Outlook.com – you already have one.
	SkyDrive will let you grab files from any PC that is associated with your account and pull them into the cloud remotely
	Works natively with Windows phones
MyCloud	Content stored at home via device
	Shared Storage and backup
	Anywhere access
Ubuntu One	Linux offers storage and music streaming
	Available on Linux, Windows XP or higher and Mac OS X 10.6 or higher. Ubuntu One is also available on both Android and iOS.
ZipCloud (UK)	Backup for Macs
	File sharing
	Remote access

to store their personal and work related data (SafeMonk, n.d.). Consumers have accepted that personal cloud applications are secure enough for the storage of personal information considering that there have been few breaches. Some businesses fought the trend until they realized the productivity gains. Employees recognized the file-sharing efficiencies and pushed their IT suppliers to respond. What may lag behind are the corporate policies providing direction for use.

The intermix of work-related generic cloud applications and personal applications has potential pitfalls. For instance, when an employee forwards all of his or her individual e-mail accounts to a single account, there may be unintended consequences. Sensitive corporate data may end up in an employee's personal e-mail account. Conversely, an employee's personal communications may eventually be stored in the business e-mail cloud server and thus be available for viewing by the corporation. In the case of government or classified information, e-mails ending up in personal accounts can lead to legal liabilities.

There have already been legal cases involving the ownership of data after work data and personal data have been inter-mixed (Milligan and Salinas, 2013). PhoneDog is a company that sued a former employee over continued use of a Twitter account that was associated with the company. The company argued that the Twitter account and password constituted a trade secret. The case was eventually settled out of court but, prior to a settlement, the company was initially allowed to proceed with the case (United States District Court, Northern District of California, PhoneDog versus Noah Kravitz, No. C 11-03474 MEJ, Order on Defendant's Motion to Dismiss Pursuant to FRCP, 12(B)(1) AND 12(B)(6), http://www.tradesecretslaw.com/uploads/file/phonedog%281%29.pdf, accessed December 31, 2015). An example of the intermix of work and personal data involves social media as a personal cloud application. Many people are beginning to rely on applications such as LinkedIn to be the repository for their contact information. Another example of a potential issue is a situation where individuals link to their employer's client list. If they leave employment, should they still have access to the clients who are now a part of their personal cloud contact application?

It may be apparent that, when an individual has data in a personal cloud application, ownership of the data resides with the individual. What happens if personal data is modified or captured by the service provider? Providers can sell information for marketing studies and behavioral targeting (Milligan and Salinas, 2013). For example, when using certain Google applications, information may be captured such as location information to facilitate customized marketing. A potential issue is whether it is an infringement of privacy or should it be an expectation when procuring the service.

When a personal cloud application is used to store copyrighted or licensed material such as music, it raises major issues such as who has the responsibility to enforce the applicable laws as it relates to sharing, duplicating, and use of the material. Depending upon the jurisdiction, the user, the application owner, or even the cloud service provider may be liable. For example, Trout (2013) surmises "if music locker services continue to operate as they are now under the existing Digital Millennium Copyright Act, it is likely that a court could find them secondarily liable for copyright infringement." This can lead to ineffective enforcement of the various laws. It raises the question of whether inconsistent application of the law has hindered the proliferation of cloud applications through uncertainty or whether this uncertainty has created a more fertile environment.

Another disruptive technology that may have an influence on the use of personal cloud applications is the Internet of things. Through the increased use of embedded sensors, the way information is shared is changing. In addition to an individual using a personal cloud application to perform a particular function, information serving as a trigger may be obtained from an information network enabled by the Internet of things. An example is being awakened by a particular song depending upon the weather. People may be able to set their alarm for a particular time. Instead of a buzzer, the device goes to a cloud application and plays music. Prior to selecting a particular song, the device will detect information from a sensor that registers temperature and precipitation, and obtains forecast data. If it is going be sunny and warm as opposed to cold or raining, the music played to awaken the individual would differ.

42.4   Personal Application Categories

Although there are other various types of personal cloud applications, three of the most common are mobile-cloud applications, location-based applications, and personal cloud storage.

42.4.1   Mobile-Cloud Personal Applications

Fan *et al.* (2011) state that definitions of mobile cloud computing can be divided into two categories. The first category refers to situations where data storage and processing take place outside of the mobile device. This limits the computing capability of the device and allows for more security through centralization of the software. The second category refers to situations where the data storage and processing take place within the mobile device. This minimizes the network traffic and provides for more efficient access to data stored on other mobile devices or sensors.

The growth of personal cloud applications is largely due to the smartphone. More consumers are using personal cloud applications primarily driven by the rapid growth of mobile computing. This growth is fueled by the need for converged collaborative services, the widespread adoption of mobile broadband service and the deployment of key technological enables such as HTML5 and the Open Mobile Appliance Smart Card Wed Server (SCWS).

Cloud-mobile applications are envisioned to minimize smartphones' resource consumption by leveraging rich cloud resources with no quality degradation. A major problem with mobile phone applications is the amount of energy and resources required to run them. Based on research by Abolfazli, *et al.* (2012), mobile cloud applications are more efficient and are less resource intensive. Hung *et al.* (2011), proposed a framework for offloading the demanding jobs from the mobile device to a virtual server in the cloud. This would reduce the mobile device's power consumption, and assist with computational speed, memory size, and wireless bandwidth constraints.

In a comparison of mobile augmentation approaches, Abolfazli *et al.* (2012) identified the required characteristics of a mobile cloud application that would make mobile devices more efficient in the use of cloud applications. The metrics were separated into low, medium, and high. The characteristics identified as high were: quality of experience and data safety. The characteristics identified as medium were: implementation cost, implementation complexity, and network delay and security. The characteristics identified as low were local resource consumption, device side maintenance, and execution.

Stanford University researchers led an NSF-funded project investigating personal cloud computing infrastructures (Kim *et al.*, 2010). The intent was to create a federated storage system out of existing free applications. This would allow a user, in effect, to create a single indexed view of a virtual personal cloud environment consisting of components from various Web services. A higher level management system would create the index of current data. It avoids the need for users to migrating existing data that they may currently have stored.

42.4.2   Location-Based Services

Applications with location-based services (LBS) are increasingly becoming more valuable and popular mobile cloud applications. When traveling to a new place, mobile users often turn to services like Yelp or TripAdvisor to get an idea of what restaurants, bars and/or shopping available nearby.

An April, 2012 report by TNS found that almost one-fifth (19%) of the world's six billion mobile users are already using LBS, with more than three times this number (62%) aspiring to do so in the future:

> Navigation with maps and GPS is currently the most popular motivation behind LBS uptake (46 per cent), but there is growing interest in more diverse activities, with 13 per cent of current social network users "checking-in"

through platforms like Foursquare – a 50 per cent uplift on 2011. LBS users are increasingly using services to enrich their social lives, with one in five (22 per cent) using it to find their friends nearby. Around a quarter use the technology to find restaurants and entertainment venues (26 per cent) or check public transport schedules (19 per cent) and 8 per cent to book a taxi. (TNS, 2012)

Another location-based service is the use of mobile phones to navigate golf courses and maintain records on strokes, putting and over all scores. The applications connect to satellites to obtain GPS data and may also connect to cloud servers to obtain and store performance data (Zahradnik, 2013).

42.4.3   Personal Cloud Storage

Personal cloud-storage devices and applications are feasible alternatives to using a cloud service provider. This involves a user setting up one's own storage device and configuring secure remote access. The primary advantages include the elimination of monthly or annual fees and the provision of essentially unlimited storage. When connected on the same network, access speed can be much faster than a solution that is only available remotely.

Although the setup is becoming easier, especially with help of the vendor, using a personal cloud requires the user to have the capability to address individual devices behind the router that is connected to the Internet. Technically, it requires dynamic name-resolution services. One will also need a domain name from DDNS to avoid issues with the changing IP addresses received from one's Internet provider.

42.5   Lessons Learned

The use of personal cloud applications is an emerging area with considerable opportunities for future research, which include issues related to privacy and security, robustness of storage space, customization, upload / download times, and bandwidth and the various types of uses that are in demand by consumers.

The growth in the Internet of Things will have a profound effect on the use of personal cloud applications. Tektonidis *et al.* suggests various scenarios on the use of content services. One scenario is that of a blind person living alone, making use of the feedback from sensors in the environment, providing data to personal cloud applications, and mobile devices helping that person to navigate on foot, access public transportation, or make online purchase decisions (Tektonidis and Koumpis, 2012).

The proliferation of personal applications that are similar to each other initially saturates the market until market forces drives usage towards a balance. An example is how the decisions of whether to store one's music in iTunes or Amazon or alternatively use Spotify or Pandora will change the basic concept of music ownership. Consumer decisions on the use of personal applications have far-reaching implications for industry and society. Music, movies and the arts are examples but it extends to any type of creative enterprise even knowledge creation. In fact, according to the McKinsey Global Institute study, automation of knowledge work will be the second most disruptive technology in the coming years (Manyika *et al.*, 2013). In the near future, understanding the art of the possible will become crucial to the supplier of services as well as to the users of personal cloud applications.

References

Abolfazli, S., Sanaei, Z., and Gani, A. (2012) Mobile cloud computing: A review on Smartphone augmentation approaches. First International Conference on Computing, Information Systems, and Communications, Singapore.

Drago, I., Mellia, M., Munafo, M., *et al.* (2012) Inside dropbox: Understanding personal cloud storage services. Proceedings of the 2012 ACM Conference on Internet Measurement, November, pp. 481–494.

Fan, X., Cao, J. and Mao, H. (2011) *A Survey of Mobile Cloud Computing*. ZTE Communications. No. 1, http://wwwen. zte.com.cn/endata/magazine/ztecommunications/2011Year/no1/articles/201103/t20110318_224532.html (accessed December 31, 2015).

Gillett, F., (2013) *Stay Alert: Merging Personal and Work Data: As Consumers and Employees Embrace Their Digital Selves CIOs Will Face New Challenges*, http://www.cio.com/article/741994/Stay_Alert_Personal_and_Work_Clouds_ Are_Merging (accessed December 31, 2015).

Hung, S., Shieh, J., and Lee, C. (2011) Virtualizing smartphone applications to the cloud. *Computing and Informatics* **30**, 1083–1097.

Kim, P., Ng, C., and Lim. G., (2010) When cloud computing meets with semantic web: A new design for e-portfolio systems in the social media era. *British Journal of Educational Technology* **41**(6), 1018–1028.

Manyika, J., Chui, M., Bughin, J., *et al.* (2013) *Disruptive Technologies: Advances that Will Transform Life, Business, and the Global Economy*. McKinsey Global Institute, 4, http://www.mckinsey.com/insights/business_technology/ disruptive_technologies (accessed December 31, 2015).

Milligan. R., and Salinas. D. (2013) Keeping trade secrets in social media and cloud computing. *The Licensing Journal* **33**(6), 29–38.

SafeMonk (n.d.) *Cloud App Usage vs Data Privacy Survey: Despite Concerns Over Data Security, Cloud-based App Use Prevalent*, http://www2.safenet-inc.com/survey/SafeMonk-Survey-Results-v6.pdf (accessed December 31, 2015).

Singh, B., Khanna, R. Gujral, D. (2011) Cloud computing: A need for a regulatory body. Proceedings of the International Conference on High Performance Architecture and Grid Computing (HPAGC), Chandigarh, India, p. 120.

Tektonidis, D. and Koumpis, A. (2012) Accessible Internet-of-things and Internet-of-content services for all in the home or on the move. *International Journal of Interactive Mobile Technologies* **6**(4), 25–33.

TNS (2012) *Two Thirds of World's Mobile Users Signal They Want to be Found*, http://www.tnsglobal.com/press-release/ two-thirds-world%E2%80%99s-mobile-users-signal-they-want-be-found (accessed December 31, 2015).

Trout, B. (2013) Infringers or innovators? Examining copyright liability for cloud-based music locker services. *Vanderbilt Journal of Entertainment and Technology Law* **14**(3), 729–757, http://www.jetlaw.org/wp-content/journal-pdfs/Trout. pdf (accessed December 31, 2015).

Vaughan-Nichols, S. (2013) *The Top 10 Personal Cloud-Storage Services*. ZDNET, http://www.zdnet.com/the-top-10- personal-cloud-storage-services-7000011729/ (accessed December 31, 2015).

Zahradnik, F. (2013) *Five Best iPhone Golf GPS Apps*. About.com, http://gps.about.com/od/sportsandfitness/tp/Best_ iPhone_Golf_GPS_Apps.htm (accessed December 31, 2015).

43

Cloud Gaming

Wei Cai, Fangyuan Chi, and Victor C. M. Leung

The University of British Columbia, Canada

43.1 Introduction

Video games are among the most profitable products in the software marketplace. Realizing the cloud's virtually infinite processing power, game companies have started to seize the opportunities to host game applications on the cloud (Ross, 2009), and the term "cloud games" has appeared. Cloud games offer a thin-client approach to computer games by having all game data stored in the cloud's datacenters and enabling computation-intensive tasks to be offloaded to the cloud, and enable players to gain full access to their personalized game environment from any mobile device and virtually anywhere. This is a relatively new trend in the gaming industry, which started in recent years when mobile devices' technological advancements rapidly accelerated. We believe cloud games are the future of the gaming industry and the reason is twofold. From the game providers' perspective, cloud computing services such as Amazon Web Services, which feature a "pay-as-you-go" model, allow game providers to pay only for the required computing resources, which means that the game providers can focus almost exclusively on developing the games themselves. From the players' perspective, cloud gaming frees them from upgrading their hardware frequently, and eliminates physically purchasing software, as only a thin client is needed to access the game via the Internet.

Cloud gaming can be extended to game players in the form of gaming as a service (GaaS) (Cai *et al.*, 2014a), which exhibits many attractive and unique features including:

- An effective anti-piracy solution: as binary code is hosted in a secured cloud server, the cloud gaming model is a potential solution to the ever-present issue and concern of gaming software piracy.
- A flexible business model: transforming from gaming software retailing to gaming service provisioning brings in more attractive and flexible business models such as "pay per play," prepaid, postpaid, and monthly subscription.

Encyclopedia of Cloud Computing, First Edition. Edited by San Murugesan and Irena Bojanova.
© 2016 John Wiley & Sons, Ltd. Published 2016 by John Wiley & Sons, Ltd.

- Click-and-play. The game copies need not be installed in the game terminals, which reduces the time cost for players. Nowadays, the storage requirements of gaming programs have significantly increased. However, many games are seldom played after installation. Therefore, "click-and-play" can attract more potential players' attention, thus increasing the popularity of the game.

In this chapter, we categorize cloud gaming services into four models: video-based cloud gaming, instruction-based cloud gaming, file-based cloud gaming, and component-based cloud gaming. We describe architecture and system design of each of them. We also discuss selection among cloud gaming models.

43.2 Video-based Cloud Gaming

Video-based cloud gaming, also known as gaming on demand, is the most popular service model in the market.

43.2.1 Offloading Everything

The basic idea of video-based cloud gaming is to offload everything of a game into the cloud, including the game engine, artificial intelligence (AI) processing, and rendering modules. User inputs and the encoded video frames are transmitted between the cloud game server and game players. This model allows direct and on-demand streaming of game videos onto computers, consoles, and mobile devices, in a similar way to video on demand, through the use of a thin client. The actual game is stored in the operator's or game company's server and is streamed directly to clients accessing the server. This allows access to games without the need for a console and largely makes the capability of the user's client computer unimportant, as all the processing needs are satisfied by the server.

43.2.2 Architecture

Figure 43.1 shows the architectural framework for video-based cloud games. In this model, the cloud virtualizes an execution environment and initiates a game instance once it receives a connection request from a player. Depicted as a data flow, the real-time gaming video is rendered by the game instance in the cloud and then recorded by the video capture module, frame by frame. Afterwards, the video frames are encoded and transmitted to the player's terminal over the Internet. These encoded frames are reconstructed at the video decoder in the game terminal and displayed by the video player on the player's screen. In reverse, as indicated by the control flow, the player's inputs are recorded by the user controller and transmitted to the cloud server

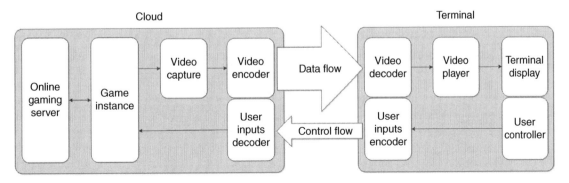

Figure 43.1 *Video-based cloud gaming architecture*

after encoded by the input encoder. The cloud server receives these encoded signals and decodes them into control inputs for the game instances, to reproduce the player's interactions in the cloud.

43.2.3   Industry

This revolutionary new model of gaming has its roots in Internet Protocol television (IPTV) from 2004. However, the model did not see many signs of adoption until 2009, when two of the biggest brands in the market, Gaikai and OnLive, were launched. Currently, companies such as G-Cluster, StreamingMyGame, OTOY, CiiNOW, Sony, and Microsoft have also started to provide such commercialized game services to the public. OnLive, Gaikai and G-Cluster have significantly different infrastructures. They all provide the service on multiple platforms but OnLive requires users to install a software client, whereas Gaikai only requires a modern browser to run, and G-Cluster requires a TV-set-top box. All service providers require users to have a decent Internet connection speed; in particular, OnLive requires users to have at least a 2 mbps connection, and Gaikai 3 mbps, but both have a recommended connection speed of 5 mbps for optimal performance and experience. One major difference between the three companies is that OnLive requires users to be located within 1000 miles (1600 km) of one of its five datacenters across the United States; Gaikai has 300 datacenters, to date, across the United States, and has also signed deals with local broadband providers to install servers at another 900 peering locations; and G-Cluster provides white-labeled services that are sold to Internet service providers. The difference in the number of datacenters operated by OnLive and Gaikai means that users in more isolated areas will have significantly different experiences on the these two platforms, and Gaikai likely achieves lower latency for all users. GaiKai was acquired by Sony in 2012 at the value of 280 million USD, which led to PS NOW's launch in 2014. In 2015, OnLive ceased operations after 3 years of financial difficulty and sold their patents to Sony. With PS Now, Sony allows their customers to play hundreds of PlayStation 3 (PS3) games on the latest PlayStation 4 (PS4) consoles without porting the games.

43.2.4   Ongoing Research

Cloud gaming is also a hot topic in academia. Researchers have been studying the measurement mechanism to analyze existing commercial cloud gaming frameworks, proposing and implementing their own cloud gaming systems and finding ways to optimize the system's performance.

43.2.4.1   Measurements

Several models have been proposed to measure cloud gaming performance in terms of system parameters representing quality of service (QoS) – interaction latency. An empirical study (Claypool *et al.*, 2012) on OnLive has performed a detailed analysis. The authors selected three types of game genres, including first person, third person and omnipresent, to analyze the network turbulence of upstream and downstream, including bit rate, package size and interpackage time. Similarly, Shea *et al.* (2013) measured the interaction delay and image quality of the OnLive system under diverse game, computer, and network configurations. Experimental results indicate that cloud processing introduces an additional latency of 100 to 120 ms to the overall system, hence more efficient designs in terms of video encoders and streaming software are required. Another measurement methodology is proposed in Chen *et al.* (2011), which assessed the system response delay (RD) of a cloud gaming system. In that paper, RD was segmented into three components, namely network delay, processing delay (elapsed time between when the server receives user actions and responds with a corresponding encoded frames), and playout delay (elapsed time between when the client receives the video frame and presents it on screen after decoding the frame). A series of mechanisms is developed to measure and analyze the total RD as well as the delay components.

For the game players, quality of experience (QoE) is the determining factor for the success of a game. Hence, the ability to model and evaluate the QoE of a cloud gaming service is important to game providers, so that they can provision for the appropriate QoE levels, monitor the QoE achieved, and take steps to improve the service as needed. In Jarschel *et al.* (2011), a measurement study based on subjective tests was used to evaluate QoE. The study first developed a test bed that emulated the cloud gaming services, then a series of tests was developed to gauge a user's reactions to varying settings of network delay and packet loss. Finally a survey was conducted with each individual test person. The collected survey data was then analyzed to determine the impact of network delay and package loss on QoE.

Similarly, in Wang and Dey (2012), a measurement metric, which is similar to the subjective quality assessment methodology termed Game Mean Opinion Score (GMOS) was introduced and developed to quantitatively measure QoE in a real-time gaming session. Lee *et al.* (2012) considered MOS to be a comparatively costly approach, due to its fewer and coarser responses. Therefore, they propose a more objective method: they use the fEMG potentials measured at the corrugator supercilii muscle, known to be associated with negative human emotions, to indicate how much a player is annoyed by the latency in a game's input-response loop. According to their results, not all games are equally friendly to cloud gaming.

43.2.4.2 *System Design*

A system similar to OnLive, called GameOn, is implemented to gain an understanding of the OnLive system. The prototype includes both the GameOn server and its thin client. Communication between the server and the thin client is via the User Datagram Protocol. The client runs two threads; one thread captures the user inputs while another one receives and decodes video frames transmitted by the server. The server, in the meantime, runs three threads, which are responsible for streaming game video, receiving inputs from clients, and providing the game menu.

Another well known cloud gaming system is called GamingAnyWhere (Huang *et al.*, 2013), which is an open system that allows game developers to implement their own algorithms and protocols to extend the capabilities of the system. It provides diverse system parameters so that researchers could tune the system for different experimental purposes. Two network flows, namely data flow and control flow, are defined, where the data flow is used to stream audio and video (A/V) frames from the server to the client, and the control flow is used to send the user's actions back to the server. Associated with each selected game running on the same server is an agent that behaves as the user and interacts with the server by replaying the received user inputs. The client is basically a customized game console implemented by combining a real-time streaming protocol / real-time transport protocol (RTSP/RTP) multimedia player and a keyboard / mouse logger. The architecture also allows observers. As the server delivers encoded A/V frames using the standard RTSP and RTP protocols, observers could access and watch a game using multimedia players.

43.2.4.3 *Optimization*

To meet constantly changing network communication and cloud computation constraints, in Hemmati *et al.* (2013), a selective object-encoding method is proposed to reduce the required network bandwidth and processing power without much impact on user-perceived QoE. The key idea is to add fewer objects to the scene so that the game processing and image generating time is reduced. On the other hand, Shi *et al.* (2011) introduced a video encoder that selects a set of key frames in the video sequence, and uses the three-dimensional (3D) image-warping algorithm to interpolate other nonkey frames. This approach takes advantage of the pixel depth, rendering viewpoints, camera motion pattern, and even the auxiliary frames that do not actually exist in the video sequence to assist video coding. Furthermore, in Wang and Dey (2010), rendering parameters, such as realistic effect, view distance, texture detail, environment detail, and rendering frame rate, are studied and their effects on communication and computation costs are characterized. Then, an adaptive rendering technique is proposed, which dynamically varies the graphic rendering parameters in the cloud servers in response to the constantly changing communication and computation constraints. The adaption process includes both offline and online steps, where the offline step is used to derive optimal rendering settings for different adaption levels (determined by the communication and computation costs), and a run-time adaptation scheme that can select the optimal adaptation level depending on the current network and computation environment.

43.2.5   Challenges

Video-based cloud gaming takes advantage of broadband connections; however, this model encounters practical issues in existing network provisioning. First, computer games are response-sensitive applications. For video based cloud gaming, the cloud server will have to respond to and process user inputs, render and encode video frames, and transmit the results back to user in near real time. Also, in order for users to receive the video frame as well as transmit user inputs, a reliable Internet connection is required. However, video transmissions consume a great deal of network bandwidth and hence will be very costly for both the user and the game provider. Therefore, bandwidth saving is another problem for researchers to solve.

43.3   Instruction-Based Cloud Gaming

43.3.1   Offloading Game Logics

With the improvement of hardware capability, most gaming terminals, including mobile devices, are capable of performing complicated rendering for game scenes. Under this circumstance, a mechanism for on-device rendering instruction execution (also known as instruction-based rendering) could be adopted, where only instructions needed to render the image are sent to the client instead of the encoded video frames. Such a mechanism is used in Baratto *et al.* (2005), which proposes a virtual display architecture for thin-client computing. In this architecture, applications are hosted on a remote server, with a virtual device driver that simulates users' inputs received from client devices, and a virtual display driver that intercepts and transmits drawing commands of screen updates over the network to client devices to display. Applying instruction-based rendering to cloud gaming, an instruction-based cloud gaming model is proposed to encode the game presentations, including characters and scenes, as a set of instructions, which are rendered afterwards in the gaming terminal to eliminate the high burden of real-time video transmissions on the network. Essentially, this cloud gaming model offloads only the game logics to the cloud.

   The most highlighted benefit for the instruction-based cloud gaming system is that the cloud server no longer needs to transmit real-time gaming video frames to the terminals through the Internet, which significantly reduces the network workload. In the meantime, the instruction-based cloud gaming system still has the benefits that video-based cloud gaming has, including "click-and-play," anti-piracy, development cost reduction, and so forth. Moreover, this model also supports a cross-platform gaming experience, given that the corresponding implementations of interface renderers for multiple platforms are available.

43.3.2   Architecture

Figure 43.2 depicts the architectural framework for instruction-based cloud gaming. As we can see from the instruction flow, the gaming video is no longer rendered in the cloud server. Instead, the gaming logic generates a set of display instructions to represent the gaming contents, and sends them to the gaming terminal via the Internet. The terminal then interprets the display instruction with a designated instruction set and renders the gaming video locally at the terminal. On the other hand, the reverse control flow is similar to that in the video-based cloud-gaming model.

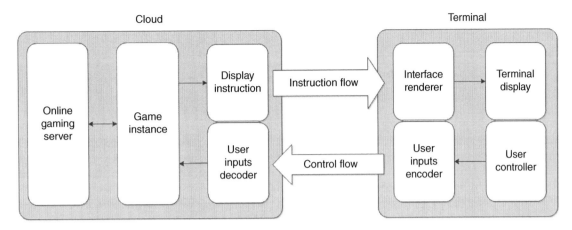

Figure 43.2   *Instruction-based cloud gaming architecture*

43.3.3   Industry

Browser games (Vanhatupa, 2011), also known as "web games", are examples of instruction-based cloud gaming, which have great commercial potential. The web browser is the most accessible thin client because it is the most commonly installed application in both computers and mobile devices. Also, the widespread of the latest version of hypertext markup language (HTML), HTML5, and JavaScript library provides the Web game developers opportunities to support the latest multimedia in their games, which makes the browser an ideal candidate to be the client of the instruction-based cloud gaming model. Browser games have existed long before cloud gaming and it is important for us to know its evolution, which can be categorized as follows.

- The script phase. This is the initial phase, which supports games with low image-quality requirements and no frequent interactions (e.g., chess games). The supported games are easy to play and implement as they are usually written as a script that manipulate the static HTML elements with the document object model (DOM).
- The plugin phase. This is also the phase when the graphic quality of browser games is enriched by browser plugins such as Adobe Flash and Java Applet, and large-scale video games are implemented and deployed using integrated development environment (IDE). In this phase, browser games have started to be rendered as three-dimensional (3D) graphics.
- The HTML5 phase. This is the time when the industry has started to turn its attention towards plugin-free browser games as plugin-powered browser games fail to provide cross-platform services. With the emergence of the HTML5 standard, along with related Web technologies, HTML5 games have become the new trend in the gaming industry. Currently, many HTML5 game engines (e.g., Akihabara and ammo. js) already exist, which provide good support for the development of such browser games.

43.3.4   Ongoing Research

Recent studies have focused on utilizing the power of cloud computing while considering the advantages of instruction-based architecture. Such architecture is proposed in Gorlach *et al.* (2014), where a cloud proxy client is introduced to act as an intermediary between the cloud server and its mobile clients. Running on the cloud, the proxy client is responsible not only for computing user actions based on received user inputs and forwarding them to the cloud game server but also applying game logic to the latest received game states and creating the rendering instructions for updating game scene. More importantly, the proxy client intercepts the execution of rendering instructions and transmits them to the mobile device for local execution. By decoupling the creation of rendering instructions from its execution and transmitting only small-size rendering instructions over the Internet, the communication burden caused by video transmission is eased, and hence addresses the challenges caused by the limitations of the mobile networks.

43.3.5   Challenges

For the instruction-based cloud gaming platform, the most critical challenge is to design an instruction set that (i) can represent all gaming images for various games, and (ii) can be efficiently transmitted via the Internet. On the other hand, interpreting the display instruction and rendering the gaming video in the terminal efficiently and accurately is also an open research issue.

43.4   File-based Cloud Gaming

43.4.1   Progressive Downloading

Cloud gaming is distributed to various terminals with diverse capacities, some of which are even capable of executing the game locally. In addition, instruction-based rendering cannot render high-quality images, and hence the games using this method cannot be very complicated and attractive. Thus, downloading the game logic for local execution appears to be a better solution.

File-based cloud gaming, also known as progressive downloading, initially downloads only a small part of the game onto the user's device. As the game proceeds, more game fragments are downloaded and the game is executed on the user's device. This technology makes it possible to download the data in the form of fragments. A small part of a game is downloaded initially so that the player can start playing the game quickly. The remaining game content is downloaded to the end user's device while playing. This allows instant access to games with low-bandwidth Internet connections with minimal lag. The cloud is used for providing a scalable way of streaming the game content and for big data analysis. Compared to video-based cloud gaming, file-based gaming services can reduce distribution costs and enhance user experience – not only by offering instant playability but also by reducing the number of steps to start playing.

43.4.2   Architecture

Figure 43.3 illustrates the architectural framework for the file-based cloud gaming. Game contents, including both fragmented game data and the binary code, are progressively downloaded to the terminal, and the game instance is then executed locally. The locally executing code fragments directly process both user inputs and game scene rendering. Game states, as well as user information, are synchronized back to the online gaming server residing in the cloud, for the purposes of inter-player message exchange, storage and statistical analysis.

43.4.3   Industry

Companies such as Kalydo, Approxy, and SpawnApps are providing file-based cloud gaming. Kalydo is now supplying the service with over 150 million sessions served in over 15 countries worldwide. Its file-streaming service sends data to players only when needed, via the proprietary Kalydo Player plugin, which can run in a browser, or from the PC desktop. Many games that have been launched using Kalydo are subsequently serviced through social networks and game portals such as Facebook, Kongregate, and Mini Clip. Approxy is trying to apply the same progressive download approach to cloud gaming via a technology it calls "cloudpaging." Cloudpaging works by breaking applications up into 32-Kbyte "pages," which are then fetched on demand over a secure

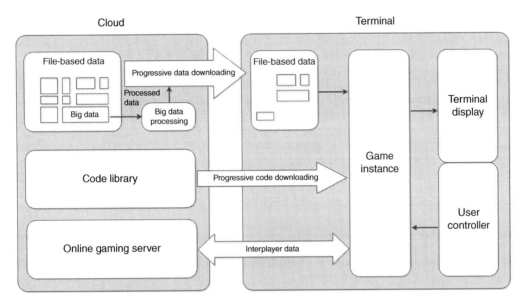

Figure 43.3 *File-based cloud gaming architecture*

hypertext transfer protocol connection by a virtual memory management unit on the client machine, and this allows the game to immediately start executing inside a virtual console, without any installation. The full game may then be downloaded in the background while the user plays it. With cloudpaging, a single Approxy server can serve 10 000 customers, far cheaper than a dedicated "pixel pusher" server that must dedicate racks of graphic processing units to rendering a game. More importantly, Approxy claims that games can be played offline, even if an Internet connection is unavailable, which is something that OnLive cannot offer. On the other hand, Approxy includes its own "pixel-pushing" operations to other machines on the network, like OnLive. It is a mixed mechanism of the file-based and video-based cloud gaming: a local server residing in a local area network has the game downloaded progressively, and actually renders and pushes the gaming video frames to the player's mobile device, which accesses the local server over the local area network. This two-layered architecture works well because the latency within a local area network is usually very low.

43.4.4 Challenges

However, with file-based cloud gaming the games must still be downloaded, and users must also invest in their own top-of-the-line graphics hardware to play the most advanced games, which is a disadvantage that OnLive and similar cloud gaming services do not have. In fact, file-based cloud gaming does not reduce the minimum specifications of the processing and graphic hardware that users must have in their machines. Hence, file-based cloud gaming does not fully utilize the computing resources in the cloud.

43.5   Component-Based Cloud Gaming

43.5.1   Dynamic Partitioning

To overcome the problem in file-based cloud gaming system, a more flexible solution can be explored. In fact, computer games are software applications that could be written using many different programming languages. In general, the core of any game is the game loop, which could be described in three steps: read user inputs, run AI, and render results. This particular procedure might contain different input / output methods and involve information exchange between multiple players. However, in general a game program may be considered as a series of interconnected modules with distinct functionalities. From this perspective, file-based cloud gaming means all game modules are implemented at the terminal, while the networking module provides the interface between game clients and the online gaming server. In this case, the cloud is only used as an information exchange server, which is the traditional design for online games. Video-based cloud gaming goes to the other end of the spectrum: the terminal only contains the input module, while the cloud hosts all of the remaining modules / components. In this case, the rendered real-time gaming videos are transmitted to the player via the Internet. In contrast, instruction-based cloud gaming illustrates another design idea: the input module and the rendering module are executed in the terminal, while the other modules are running in the cloud. According to the three schemes, we can see that the essence of cloud gaming is to leverage the cloud resources to execute a number of gaming modules for the purpose of reducing terminal workload and achieving better efficiency.

This observation gives rise to the following questions. Is there a more flexible solution adapting the cloud gaming service to the varying circumstances, for example the unstable quality of network connectivity? To this end, a new design methodology termed component-based cloud gaming becomes promising, whereby a cloud-based game is constructed by a set of interdependent components that are executed either in the cloud server or the player's terminal, as determined by the current conditions of the terminal and its network connectivity. Such cognitive capabilities, named dynamic partitioning, enable optimal component allocations that can potentially provide a high efficiency to the cloud gaming model. As a flexible and intelligent platform, component-based cloud gaming monitors the real-time environment status, sets an optimization target (e.g., the overall interaction latency, computational resource upper bound, or bandwidth ceiling), and achieves the target through dynamic partitioning. Despite the cognitive partitioning, video-based, instruction-based and file-based cloud gaming, which are based on a static partition of the game modules, are all special cases of component-based cloud gaming.

43.5.2   Architecture

Figure 43.4 illustrates the architectural framework for the component-based cloud gaming. In this category, the game programs are modularized as components, which are able to migrate from the cloud to the terminal during gaming session and dynamically concatenate with each other to form a complete game. In other words, the terminal can fetch and execute a set of redundant game components from the cloud to reduce the burden on the cloud. As shown in the figure, a partitioning coordinator layer incorporating cognitive capabilities is designed to manage the execution of game components. Once a control instruction from the player is transmitted, the coordinator intelligently assigns the responsive components, whether in the cloud or in the terminal. The same mechanism applies to the invocation between components: all "invoke" messages are scheduled by the partitioning coordinator in order to achieve dynamic partitioning.

43.5.3   Ongoing Research

Cai *et al.* (2013) designed and implemented a cognitive platform that both supports "click-and-play" and cognitive resource allocation for decomposed cloud games. Consisting of a set of controllers and coordinators, the cognitive platform is able to dispatch selected gaming components from the cloud to a player's terminal and later use dynamic partitioning to adapt its service QoS to the real-time system environment. As a development-friendly environment, the platform also provides a set of application programming interfaces, so that game developers need not be concerned with the lower layer resource management details but only focus on the design of game programs.

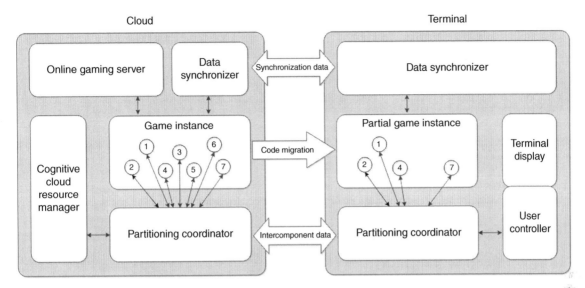

Figure 43.4 *Component-based cloud gaming architecture*

From the prospective of cloud resource management and scheduling, Cai *et al.* (2014b) modeled the component-based game and investigated the capacity of intelligent resource management for different optimization targets, including cloud resource minimization and throughput-oriented optimization. Experimental results show that, with cognitive resource management, the cloud system can adapt to various service requirements, such as increasing the quantity of supported devices and reducing the network throughput of user terminals, while satisfying players' QoE expectations.

43.5.4 Challenges

As a cognitive system, to achieve adaptive system optimization in real-time is the most critical challenge. Note that, as a component may invoke another remote component as scheduled by the partitioning coordinator, the communication latency can be controlled explicitly to provide an acceptable QoE level. In addition, as identical components may reside in both the cloud and terminal, developing an efficient mechanism to synchronize the data in the components is also a challenge. Besides, in order to support "click-and-play," no gaming component resides in the terminal at the beginning of the gaming session, so how to efficiently dispatch the necessary components to the terminal without interrupting the gaming experience is also an open issue.

43.6 Selection Among Cloud Gaming Models

Given that the various cloud gaming models are quite different and have different advantages and disadvantages for different game genres, the game developers and service providers need to weigh the advantages and disadvantages of each to decide which cloud gaming model best fits their needs.

43.6.1 Game Genres

We classify game genres on the basis of two elements. Scene observation is how a player's observation of the game scene determines the variety of output images on the screen. In general, the most common scene

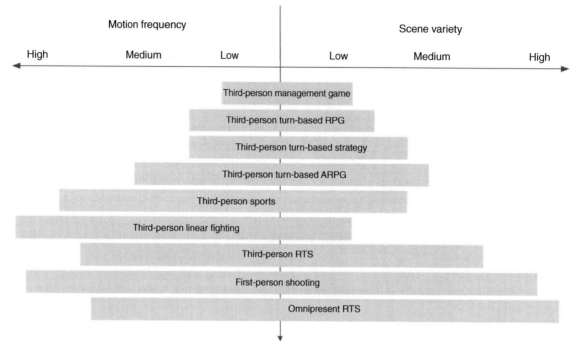

Figure 43.5   *Scene variety and motion frequency*

observation styles are categorized as first-person, third-person, and omnipresent. Game topic refers to the game content provided, which determines the interaction behavior between players and the game. Game topics include shooting, fighting, sports, turn-based role-playing (RPG), action role-playing (ARPG), turn-based strategy, real-time strategy (RTS), and management.

Figure 43.5 shows the relationships between the scene variety and motion frequency for the most common game genres. The figure indicates that these features vary among distinct game genres, so we must consider each architectural framework's advantages and disadvantages and use the best solution for the target gaming services.

43.6.2   Solution Comparison

For video-based cloud gaming solutions, simplicity is one factor that attracts traditional gaming companies: with a well-designed video-streaming server and an efficient virtual machine in the cloud, all of the existing video games can easily be converted to and steamed as cloud games. The virtually infinite resources in the cloud mean that the overall system's QoS will not be affected by the systemic complexity of game design. Having all virtual machines running on the cloud and connected to each other via internal network links, video-based cloud gaming also makes multiplayer game more efficient as, by default, the cloud is assumed to have high-quality internal network connectivity.

In contrast, instruction-based cloud gaming is still limited by the capacity of the local renderer and the efficiency of the instruction set. However, as the gaming logic and rendering modules are decoupled, it is a perfect choice for those gaming services that focus on strategy management instead of frequent real-time interactions with fascinating graphics. For instance, most of the Web-page games integrated in online social networks fall into this category.

On the other hand, file-based cloud games are intrinsically executed locally as most of the conventional game software only uses the cloud as a progressive downloading server for data and compiled codes. This feature makes it a good candidate for games with specific storylines: a minimal proportion of the game is sufficient for players to get their games up and running, while the forthcoming data and codes are streamed to the terminal during their gaming session.

Representing a future solution, component-based cloud gaming intends to address existing issues in the above three solutions. The dynamic partitioning solution enables the mobile devices to extend their gaming functionalities in a flexible fashion. With intelligent component migration, the platform can adaptively balance the workloads for the cloud and terminals to optimize the whole gaming system while satisfying the players' perceived QoE expectations. This adaption to the environment, including network quality and terminal capacities, makes it an ideal solution for providing ubiquitous cloud gaming services for various devices, ranging from powerful desk top computers to mobile terminals with constraints on various resources.

43.7 Conclusion

Cloud gaming provisioning is considered as the next generation modality of the gaming industry. This chapter summarized the platforms aiming to provide cloud-based gaming services. We have compared differences between the existing cloud gaming models and architectural frameworks and have examined their benefits, challenges, and research status. Based on these studies, we have analyzed a selection of cloud gaming provisions based on game characteristics and a variety of terminals. We have identified the component-based cloud gaming model as a future solution that can address most of the issues concerning other cloud gaming models.

References

Baratto, R., Kim, L. and Nieh, J. (2005) THINC: A virtual display architecture for thin-client computing. In Proceedings of the 20th ACM Symposium on Operating Systems Principles (SOSP 2005), Brighton, UK, pp. 277–290.

Cai, W., Chen, M. and Leung, V. (2014a) Toward Gaming as a Service. *IEEE Internet Computing* **18**(3), 12–18.

Cai, W., Chen, M., Zhou, C. *et al.* (2014b) Resource management for cognitive cloud gaming. Proceedings of the 2014 IEEE International Conference on Communications (ICC2014), pp. 3462–3467.

Cai, W., Zhou, C., Leung V. and Chen, M. (2013) A cognitive platform for mobile cloud gaming. Proceedings of the IEEE Fifth International Conference on Cloud Computing Technology and Science (CloudCom2013), pp. 72–79.

Chen, K., Chang, Y., Tseng, P. *et al.* (2011) Measuring the latency of cloud gaming systems. Proceedings of the 19th ACM International Conference on Multimedia (MM '11), pp. 1269–1272.

Claypool, M., Finkel, D., Grant, A., and Solano, M. (2012) Thin to win? Network performance analysis of the OnLive thin client game system. Proceedings of the 11th Annual Workshop on Network and Systems Support for Games (NetGames), pp. 1–6.

Gorlatch, S., Meilaender, D., Glinka, F., *et al.* (2014) Bringing mobile online games to clouds. Proceedings of the 33rd IEEE International Conference on Computer Communications (INFOCOM).

Hemmati, M., Javadtalab, A., Shirehjini, A., *et al.* (2013) Game as video: bit rate reduction through adaptive object encoding. Proceedings of the 23rd ACM Workshop on Network and Operating Systems Support for Digital Audio and Video (NOSSDAV '13), pp. 7-12.

Huang, C., Hsu, C., Chang, Y., and Chen, K. (2013) GamingAnywhere: An open cloud gaming system. Proceedings of the Fourth ACM Multimedia Systems Conference, Oslo, Norway, February / March, pp. 36–47.

Jarschel, M., Schlosser, D., Scheuring, S., and Hoβfeld, T. (2011) *An Evaluation of QoE in Cloud Gaming based on Subjective Tests.* In Proceedings of the 2011 Fifth International Conference on Innovative Mobile and Internet Services in Ubiquitous Computing (IMIS '11). IEEE Computer Society, Washington, DC, pp. 330–335.

Lee Y., Chen K., Su H., and Lei C. (2012) Are all games equally cloud-gaming-friendly? An electromyographic approach. Proceedings of the 11th Annual Workshop on Network and Systems Support for Games (NetGames), pp. 1–6.

Ross, P. E. (2009) Cloud computing's killer app: Gamin. *IEEE Spectrum* **46**(3), 14.

Shea, R., Liu, J., Ngai, E., and Cui, Y. (2013) Cloud gaming: architecture and performance. *IEEE Network* **27**(4), 16–21.

Shi, S., Hsu, C., Nahrstedt, K. and Campbell, R. (2011) Using graphics rendering contexts to enhance the real-time video coding for mobile cloud gaming. Proceeding of the Ninth ACM International Conference on Multimedia, pp. 103–112.

Vanhatupa, J. (2011) On the development of browser games – technologies of an emerging genre. Proceedings of the Seventh International Conference on Next Generation Web Services Practices (NWeSP), pp. 363–368.

Wang, S. and Dey, S. (2010) Rendering adaptation to address communication and computation constraints in cloud mobile gaming. Proceedings of the IEEE Global Telecommunications Conference, pp. 1–6.

Wang, S. and Dey, S. (2012) Cloud mobile gaming: Modeling and measuring user experience in mobile wireless networks. *ACM SIGMOBILE Mobile Computing Community* **16**(1), 10–21.

Part IX
Big Data and Analytics in Clouds

44

An Introduction to Big Data

Mark Smiley

The MITRE Corporation, * *USA*

44.1 Introduction

Our world is awash in a rising ocean of data. Modern modes of transportation – planes, cars, and ships – contain thousands of sensors that constantly generate data (Hemsoth, 2013). Medical researchers sort through thousands of genes in millions of patients, attempting to find genes that lead to diseases such as cancer (Hurd, 2013; Kinney, 2013). Physicists use enormous colliders to smash particles and improve our understanding of the universe; the sensors in these colliders generate more raw data than the system can store (O'Luanaigh, 2013). Online businesses such as Amazon and Netflix track their customers' online behavior and use that information to suggest books or movies for them to purchase. Smartphones record audio, video, and images, as well as metadata about when and where the video and images were recorded, and their owners post them to social media sites. All these are examples of big data. The age of big data has begun, and exploiting big data is changing our world.

The rise of sensors, search engines, smartphones, online retail, and social media have all led to a tremendous increase in the amount of data collected about each of us. In fact so much data is collected that it may seem to be too much to process. Data scientists have had to devise new techniques to analyze and process all this data – transforming it from data into useful information. Advances in genomics have led to the need to process huge quantities of DNA information in a short period of time (Kinney, 2013). Numerous oceanographic sensors collect data about the ocean (OOI, 2012). Radio telescopes collect large amounts of data for astronomers to pore over (McKenna, 2013). In so many ways, the analysis of big data touches everyone's lives.

---

* This chapter is © 2013 – The MITRE Corporation. All rights reserved. The author's affiliation with The MITRE Corporation is provided for identification purposes only, and is not intended to convey or imply MITRE's concurrence with, or support for, the positions, opinions or viewpoints expressed by the author.

Encyclopedia of Cloud Computing, First Edition. Edited by San Murugesan and Irena Bojanova.
© 2016 John Wiley & Sons, Ltd. Published 2016 by John Wiley & Sons, Ltd.

44.2   What is Big Data?

44.2.1   Preliminary Concepts

Before delving into big data, there are a few concepts that we should understand.

44.2.1.1   *Types of Data*

Data is structured, semistructured, or unstructured. Structured data is data that is organized in a structure. This structure can either be fixed fields inside a record (as in a relational database), or in a well formed format such as XML or JavaScript Object Notation (JSON). Semistructured data has some structure, but the data isn't expressed in terms of rows and columns or a formal structure such as XML with a schema. An HTML page is an example of semistructured data. Unstructured data does not have fields in fixed locations, nor does it follow a standard format such as XML or JSON. Examples of unstructured data include raw text files such as a server log, a Microsoft® Word document, a Portable Document Format (PDF) file, or an e-mail.

44.2.1.2   *Sources of Data*

There are many processes that produce big data. Sources of big data include business operational data, scientific data, social networking, web logs, video streaming, sensor data, smartphone data, and more. Here are some examples:

- Imagery. Google Maps offers over 20 petabytes (PB) of imagery (McKenna, 2013).
- Video streaming. Netflix has over 3.14 PB of video in the master copies alone (Vance, 2013, p. 4).
- Social networks. By February of 2012, Facebook had stored over 100 PB of data (McKenna, 2013).
- Scientific sensor data. As of February 2013, the CERN Data Centre had recorded over 100 PB of physics data over a period of 20 years. The Large Hadron Collider (LHC) at CERN in Switzerland and France generated about 75 PB of that data in only 3 years. One hundred PB is about the same size as 700 years of HD movies (O'Luanaigh, 2013). IBM estimates that the Square Kilometer Array (SKA), a radio telescope due to be completed in 2024, is expected to produce 1376 PB per day (McKenna, 2013).
- The Internet. In 2007, Google was already processing over 403 PB of data per month (Dean and Ghemawat, 2008). Cisco estimated that all global Internet traffic in 2012 was about 400 Exabytes (EB) (http://www.cisco.com/en/US/solutions/collateral/ns341/ns525/ns537/ns705/ns827/images/qa_c67-482177-1.jpg, accessed January 1, 2016).

These are just some of the examples of big data. There are many more.

44.2.1.3   *Streams*

Some data is produced in streams. A data stream is "a sequence of digitally encoded signals used to represent information in transmission" (Federal Standard 1037C, 1996). Some examples include click streams, packet streams, sensor data, satellite data, a video stream produced by an online video camera, and financial data such as stock-market data.

44.2.2   Big Data Defined: Volume, Velocity, and Variety

What is big data? One proposed definition of big data is data that has grown to a size that requires new techniques to store, organize, and analyze the data (Butler, 2013). This seems to be a reasonable definition at the moment but this definition will inevitably break down over time as data scientists develop ways of handling what is now considered big data. The data will still be big but it may no longer be necessary to innovate in order to handle it. Thus this book will use another popular definition of big data involving the three "Vs": volume, velocity, and variety.

Big data has one or more of three key features: a large volume of data, a high velocity with which the data is created, or a high degree of variety in the data (Gartner, 2013). Volume simply means the amount of data. Velocity means that the data is being created rapidly (e.g., hundreds of messages per second); velocity is typically associated with streams. High variety is usually associated with unstructured data such as server logs. So big data is associated with three "Vs": volume, velocity, and variety.

Consider velocity. How can data be created at a high velocity? It is typical for a sensor to produce high-velocity data. The GPS in a smartphone, for example, can emit location data every second. Similarly, Airbus has over 100 000 sensors in each of their planes, each sensor generating a stream of data (Hurd, 2013). Airbus processes this data in real time to feed information to the pilot. Later, they mine the data to help improve future aircraft designs. It is also possible for a large group of people to create high-velocity data – for example, the set of all links clicked on by everyone on the Internet represents a high velocity of data. Another example is the stream of all stock trades each day that the market is open.

Big data needs big data analytics. In its raw state, big data is too large to be very useful – it takes too long for a human to consume it and make sense of it. To become useful, it must be processed into something smaller. Big data analytics involve processing big data into useful information.

44.2.3   How Big is Big?

Big data can be relative to the size of data that a business is accustomed to handling. For example, a business may have tracked inventory and accounting before but now it may be tracking what its customers are buying, and analyzing their buying habits (e.g. what things they buy together, when they shop, how often they buy a particular product, etc.). That's a different level of data collection that a business may consider to be big data.

But generally the term "big data" is used to indicate data above the 100 GB range. Often it deals with hundreds of terabytes (TBs), and possibly a PB or more. This threshold will no doubt move upwards as time passes.

44.2.4   Applications of Big Data

As noted above, sources of big data include search engines, social media, online retail, sensors, and smartphones. Some very successful companies rely on big data generated by the Internet. Google, Amazon, Yahoo, and Facebook are well known examples of such companies. Netflix uses data collected from customer behavior to offer its customers movies that they might enjoy; Amazon does the same for books, music, movies, and other items that they purvey.

Weather-prediction systems consume data taken from weather sensors around the country and in many parts of the world. These sensors collect data on temperature, wind speed, pressure, and humidity. Satellites also offer more data about the weather. All this data is consumed by weather prediction algorithms to help predict the weather each day (Hardy, 2012).

The National Cancer Institute is investigating gene-to-cancer interaction, correlating 17 000 genes in 60 million patients with five major cancer types, and handling 20 million medical publications (Hurd, 2013). Similarly, Georgetown University is using genomics to help understand the causes of premature birth (Kinney, 2013).

NTT Docomo is the largest Japanese mobile phone operator. It has 13 million smartphone users, generating 700 000 events per second. The company is performing real-time mobile traffic processing and using the resulting analysis to optimize its network (Hurd, 2013).

The New York Stock Exchange generates 185 000 messages per second, with 2 TB per day of data. It has to be reliable as this data describes about $2.5 trillion in transactions per day. It all has to be fast, measuring latency in milliseconds (Hurd, 2013).

The United States Securities and Exchange Commission (SEC) worked with Tradeworx to develop a system that runs on Amazon's cloud-computing services to perform real-time analysis of 20 billion messages per day to allow them to reconstruct any market, on any given day in history (Kinney, 2013).

Ford collects data on its new cars. Ford's Fusion hybrid model can create 25 GB of data per hour. This data can be used to fine-tune the car. Ford's use of big data analytics includes analyzing computer-aided design (CAD) and computer-aided engineering (CAE) models, along with running manufacturing simulations (Hemsoth, 2013).

The Ocean Observatories Initiative (OOI) is collecting data from various sensors such as those on buoys, underwater cameras on submersibles, and undersea robots. These sensors can measure the depth, salinity, and temperature of the water. Some can sample the phytoplankton to count how numerous they are in a particular area. The goal of the project is to collect and analyze this data to better understand a number of important oceanographic related issues, including climate, the ecosystem, and sea-floor dynamics. The data will be shared with the public in real-time (OOI, 2012).

Ancestry.com® has over 10 PB of data. This data is related to genealogy, and includes DNA data, pictures, and a variety of records (Zhukov, 2013).

There are many other sources of big data, but the above examples offer a taste of some of the big data that organizations are collecting and processing.

44.2.5   Preconditions

There are three main preconditions for the rise of big data:

- The ability to store large volumes of data in a form that is accessible (i.e., on hard drives or solid-state drives, not tape drives).
- The ability to process big data rapidly at a reasonable cost. This means inexpensive computing power, particularly in the form of cloud computing.
- The existence of producers of big data.

The first of these is the ability to store large volumes of data in a format, such as a disk array, that makes the data accessible. This storage needs to be at a reasonable cost. In November 2014, an internal consumer-grade 1 TB drive sold for less than $65, and the price per TB continues to drop. Typical cloud computing storage costs more than that as the data is normally replicated to at least three drives for redundancy.

Second is the ability to process that data into something that is understandable to humans in a reasonable amount of time at a reasonable cost. Two main things that have made this possible are inexpensive computing power in the form of faster and less expensive central processing units (CPUs) and the rise of cloud computing. Cloud computing services, such as those offered by Amazon Web Services (AWS), put massive and inexpensive compute power in the hands of anyone who can afford it. There is no longer a need to purchase an expensive supercomputer and lease a building to house it. Now big data can be crunched on a set of Amazon Elastic Compute Cloud (Amazon EC2) instances and stored using Amazon Simple Storage Service (Amazon S3).

Third, for big data to exist, copious data producers must exist. Important data producers that generate all this data include sensors, mobile devices, social media such as Facebook and Twitter, and user interaction with web pages. Examples of sensors include a weather sensor, a tsunami sensor in a buoy, a pressure sensor on a plane, the GPS in a smartphone, the microphone in a phone, a digital camera, and the sensors in the

Large Hadron Collider. Each day millions of people visit the Internet and click on various links; that is another source of big data. Google and Facebook, for example, use such data to offer targeted advertisements to their users.

A modern smartphone contains several sensors: a microphone, a camera, and a GPS. So a smartphone can collect location data, audio, pictures, and videos, as well as data on which applications (apps) the owner runs and which links they click on in Web pages accessed from that phone.

44.2.6   Why Big Data?

Some of the reasons that have driven businesses to implement big data analytics are: to aid in making critical business decisions, improve customer response, increase revenue, and carve out new lucrative businesses. Some of these businesses are quite well known, and completely dependent on their ability to consume, store, and process big data. These are companies that operate at internet scale, companies like: Google, Amazon, Facebook, Netflix, Yahoo, and Twitter.

Consider retail data. When a consumer purchases items at a store, the cashier scans the bar code of the item. If that consumer also has a rewards (or membership) card for that store, the store can track which items were purchased, when, and how frequently. This data can help the store manager know how much of an item to stock in the store, or when to offer a coupon for a particular item to that consumer. Similarly, a rewards card helps restaurants track what items customers order and when to offer them an incentive to dine there.

Online retailers don't need to offer a membership card – each customer has an account. Amazon, for example, tracks purchases and suggests other items that customers might like, based on their purchase history. A recommendation engine then analyzes the data and decides what to suggest. Netflix is another good example of a company that uses a recommendation engine.

44.3   What Can You Do With It? Big Data Analytics

If big data existed but there was no way to analyze it, it would be of little use. It is partly the ability to store such large volumes of data that has allowed the rise of big data but it is also the ability to process that data into something that is understandable by humans in a reasonable amount of time that has allowed the exploitation of that data. Successful big data analytics must be able to process the data into something smaller than the raw data, allowing an application to present the results in a way that makes sense to a human.

Because big data is so large, it normally cannot be processed sequentially in a reasonable amount of time. So the data is broken up into chunks, which are analyzed by a set of processes running in parallel. These processes typically run one per CPU core in a server, or one per virtual machine (VM). The results of the parallel analysis are then joined together to create the result.

To help understand how some typical big data analytics work at a high level, consider the following example. Suppose a company offers a web site and collects the click logs from all those who visit the web site. These logs may be several TB in size. For example, the company may want to know the top paths that people follow through the web site to make a purchase.

To apply big data analytics, first chop up the logs into multiple pieces. Then create a cluster of machines to work on the problem. Feed a piece of the log to each machine in the cluster, and let it work on that piece. The result of the processing must be smaller than the original piece. When the machine has completed processing that piece, it reports back its result and accepts another piece to process. Other machines work on their respective pieces and report their results. Once all the pieces have been processed, the results are aggregated and reported. This is illustrated in Figure 44.1. Note that this is a simple case. In other cases, the results are fed into other sets of parallel processes and reduced again and again before the final result.

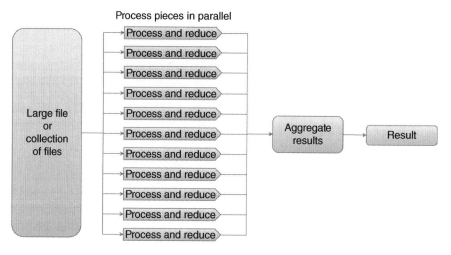

Figure 44.1 *High-level big data analytics*

Thus one of the key features of big data analytics is that they typically use parallel processing. This need for parallelism to process big data has led to the need for frameworks that enable parallel processing of various kinds. An example of this is reactive steams, discussed in section 44.7.3.2.

44.4   Limitations of Big Data Analytics

Parallel processing of big data is crucial yet there are limits to how much an algorithm can be sped up by parallelizing most of it. There is usually a part that cannot be parallelized. This issue leads to Amdahl's law (Amdahl, 1967).

44.4.1   Amdahl's Law

Amdahl's Law states that if p is the number of times part of the algorithm can be sped up, f represents the fraction of the computational load that was not improved by parallelizing it, and m is the maximum speed increase possible, then

$$m \leq \frac{p}{1 + f(p-1)}$$

For example, if an algorithm has a part that can be done in parallel on $p = 100$ machines, but 1/10 of the original algorithm cannot be parallelized, then $m \leq 100 / 10.9 \approx 9.17$. So a ninefold increase is about the best a new algorithm could do in that case. If $p = 1000$ machines, then $m \leq 1000 / 100.9 \approx 9.91$, so the improvement does not change much if more machines are allocated to the problem. Thus, parallelizing will have the greatest effect if it is accompanied by a corresponding improvement in the sequential portion of the algorithm, or if the sequential part can be minimized.

44.5   The Big Elephant in the Room: MapReduce and Hadoop

The seminal paper in the field of big data analytics (Dean and Ghemawat, 2004) came from Google and introduced MapReduce. Then Yahoo implemented a version of MapReduce, along with a supporting environment, and donated it to the open-source community. This implementation became Apache™ Hadoop®. Apache Hadoop is currently the most popular tool for running big data analytics. Apache Hadoop is discussed in detail in Chapter 48, so it is not covered here.

Later, Google found a way to improve the performance of their web-page indexing using Percolator, a tool created at Google. They found that Percolator improved performance over MapReduce by a factor of 100 (Peng and Dabek, 2010). However, this technology is currently only available at Google.

44.6   Everything is a Network: Big Data Graph Analytics

There are other big data analytics, besides those based on Hadoop MapReduce. One important type involves big data graph analytics.

A *graph* is a set of vertices (or nodes) along with a set of edges connecting the vertices. Graphs are sometimes called network diagrams. Examples of graphs include the flights between airports, the servers on the web along with their connections, and social graphs. Figure 44.2 depicts a small sample social graph.

In this example, the disks indicate vertices and represent people while the edges are the lines connecting the vertices, indicating friendship. Here Bob, Carol and Alice all know each other but Ted knows only Carol.

Imagine a graph of all the one billion Facebook users connected to each of their friends. That would be a very large graph indeed. Back in September 2012, when Facebook had one billion users, they had 140.3 billion connections (Taylor, 2012). An intern at Facebook, Paul Butler, created such a graph back in 2010, when Facebook had only 500 million users. He plotted a point for each city and drew a great circle between each pair of cities that had at least two connected people, shading the curve based on the number of connections the great circle represented. The result is a stunning image of the free world (Butler, 2010).

What can be learned from studying such social graphs? It is possible to study the structure of social connections, as well as social dynamics: how the social connections evolve over time. For example, it is possible to look at the degrees of separation between people. This is the average distance between users in a social graph, where distance is measured in the minimum number of vertices that must be traversed to follow the graph from one user to another. In 1969 it was thought that this number was 5.2 (Travers and Milgram, 1969), but that was based on a small study. For Facebook in 2011, the average distance between users was computed to be 4.74. Note that this latter number was computed by looking at all the users and their connections on Facebook, not just a sample (Markoff and Sengupta, 2011).

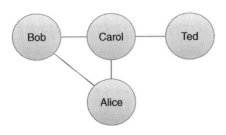

Figure 44.2   *Small social graph*

This example illustrates one of the benefits of big data analytics: moving from small statistical studies to actually crunching huge amounts of raw data to find more precise answers. When big data is available, the old style of working with a small statistical sample and projecting an answer within a margin of error can often be replaced simply by computing results for all the data – not just a sample – with no margin of error.

For another example of a large graph, let each web page in the world be a vertex and each hyperlink an edge. As of September 2013, the number of web sites was estimated at 739,032,236 (*Netcraft*'s September 2013 *Web Server Survey*). Other examples include the flow of goods and the flow of water.

Big data Web analytics (Web-scale graph processing) are one of the keys to Google's success. Consider, for example PageRank, Google's algorithm for ranking pages and determining the order in which to return search results (Brin and Page, 1998).

44.6.1   Drawbacks to MapReduce for Graph Problems

MapReduce is very useful, but there are drawbacks to applying it to graph problems. The speed is slower for two main reasons. The first is that each iteration is a separate MapReduce job, with all the overhead that entails. The second speed issue is that there is too much disk access. With MapReduce, the graph data is read from disk, and the intermediate results are written to the Hadoop File System (HDFS). Another drawback is that joins need to be implemented by hand and are data dependent.

44.6.2   Pregel

The above drawbacks led Google to develop a big data graph analytic engine called Pregel (Malewicz, 2010). Pregel is designed for processing large-scale graphs, such as a set of Web pages with connecting hyperlinks. Pregel uses bulk synchronous parallel (BSP) execution.

The main computational component in Pregel is a *superstep*. Within each superstep, each vertex in the graph can call a user-defined function. The function can modify the vertex state, and functions can vary by vertex and by superstep. Functions can read messages sent to the vertex during the previous superstep and can send messages to other vertices, which will receive them in the following superstep. Messages typically propagate along edges, but a message can be sent to any vertex known to the sending vertex (Malewicz, 2010). Figure 44.3 depicts this process conceptually.

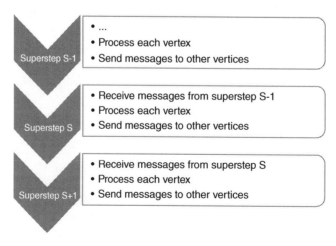

Figure 44.3   *Pregel process*

Pregel is vertex-oriented. Each vertex has an ID, a list of adjacent vertex IDs, and their corresponding edge values. Each vertex is invoked in each superstep, can do some processing on its information, and can send messages to other vertices that are delivered over superstep barriers.

Pregel uses a master-slave design. Vertices are partitioned and assigned to workers, while the master assigns and coordinates tasks. Workers execute vertices and communicate with one another. It can use termination votes, which means that vertices vote on whether or not to terminate the job. It also uses checkpointing to improve fault tolerance.

Unfortunately, Pregel is not available outside Google. However, there is an open source project that endeavors to emulate Pregel: Apache Giraph.

44.6.3 Apache Giraph

Apache Giraph is an open-source version of Pregel. Yahoo! developed the original source code, then gave it to Apache to open source it. Facebook, LinkedIn, and Twitter have all contributed to its development.

Giraph runs on the Hadoop infrastructure. Computations are executed in RAM. It is fault tolerant, and it can run BSP operations on big data that can be represented as a graph. It can run in a Hadoop job pipeline as a normal MapReduce job, and it uses Apache Zookeeper for synchronization.

Some other big data graph analytic tools include: Hama, GoldenOrb, and Signal/Collect.

44.7 Data Streams

We have seen that some data is produced in streams. A key feature of streams is that the data is read and processed as it arrives. Streams pour data into a system. Contrast this with data in a standard relational database, in which relationships between data elements are persistent.

44.7.1 Event Processing

Data streams can be thought of as a series of events. Each event contains data collected or generated over some time interval. New events are constantly being generated. The systems that process the stream of events are typically set up to use continuous queries – queries that never finish. That's quite different from a standard relational database. Event processing operates on a stream of events and converts it into useful information. Complex event processing (CEP) combines information from multiple sources and attempts to discern actionable patterns in the information.

44.7.2 Data Stream Management Systems

A data-stream management system (DSMS) is a system that manages continuous streams of data. Typically, a DSMS has one or more standing queries which operate on the data stream as the data is ingested into the system.

44.7.3 Stream Solutions

There are a number of existing products in the stream solutions space. This section discusses a few of these.

44.7.3.1 Streams in Java

Starting with Java 8, Java offers lambda expressions and streams. Lambdas make it easier to operate in parallel on streams. Converting lambda expressions from sequential to concurrent processing is as simple as changing `stream` to `parallelStream`. Hearkening back to Amdahl's law, note that there is some overhead to the new parallel methods in Java. So it is important to carefully decide what should be parallelized and what should not.

44.7.3.2 Reactive Streams

Reactive Streams (http://www.reactive-streams.org/, accessed January 1, 2016) is an interesting initiative to devise an asynchronous stream-processing standard when there is nonblocking back pressure on the Java virtual machine. Back pressure means that the receiver can assume that it only needs to buffer a preset maximum amount of data. This allows each queue on the receiving side to be bounded while the number of queues can increase as needed.

The related Reactive Extensions (Rx) offer an API for asynchronous programming with observable streams (see http://reactivex.io/, accessed January 1, 2016). Rx has implementations in multiple languages, including Java, JavaScript, C#, Scala, Clojure, C++, Ruby, Python, and Groovy. Organizations such as Netflix, Microsoft and GitHub use Rx.

A related document is the Reactive Manifesto (http://www.reactivemanifesto.org/, accessed January 1, 2016), which describes Reactive Systems as:

- responsive;
- resilient;
- elastic;
- message driven – using asynchronous messaging.

44.7.3.3 Odysseus

Odysseus is an open source in-memory DSMS and CEP system written in Java. It is designed to process a continuous stream of events in near real time. It offers a number of types of processing steps, including the ability to filter and to correlate events.

44.7.3.4 Apache Storm

Apache Storm is an open source system for doing computation on streams in real-time. It is discussed in Chapter 48.

44.7.3.5 Oracle Data Stream Solution

Oracle Corporation has a data stream solution called Oracle® Streams. Each unit of information is called a message. Oracle® Streams can send messages within a database or from one database to another, even if the second database is not an Oracle database. Messages are captured, staged, and consumed. The database initiating the message is the source database, whereas the database consuming the messages is the downstream database. Messages can be staged in a queue and consumed when the downstream database is able to process them. Oracle Streams can also transform messages based on a set of rules. Oracle Streams were designed for data replication; they can capture changes to a database and replicate them to other databases.

44.7.3.6 IBM Data Stream Solution

IBM® InfoSphere® Streams is a streaming product from IBM that can run either virtualized or on bare metal (in other words, not in a virtual machine). It can scale out (horizontally), so it is possible to add more nodes that operate in parallel. Applications written for InfoSphere® Streams use IBM® Streams Processing Language (SPL), which works with data streams, tuples, operators, processing elements and jobs. IBM® InfoSphere® Streams include data-visualization tools, some analytics, including some geospatial analytics, and support for the statistics packages R and SPSS.

44.8 Conclusion

The rise of big data producers such as sensors, mobile devices, social media, and user interaction with Web pages has led to an unprecedented amount of data becoming available. Processing such data requires new paradigms, leading to big-data analytics.

Some big-data tools, such as Pregel and Apache Giraph, operate on graphs, and they work well on problems that can be expressed in that space, such as the analysis of social networks.

On the other hand, some data is generated in streams, and companies have created various systems and technologies to process streams, such as Reactive Streams, Apache Storm, and others. These and other big data analytics help companies run at Internet scale.

The world of big data continues to expand. The current state of data science offers considerable room for new ideas and new tools to pave the way to assimilating the enormous quantity of data in this new world of big data.

References

Amdahl, G. (1967) Validity of the single processor approach to achieving large-scale computing capabilities. *AFIPS Conference Proceedings*, pp. 483–485. http://www-inst.eecs.berkeley.edu/~n252/paper/Amdahl.pdf (accessed January 1, 2016).

Brin, S., and Page, L. (1998) The anatomy of a large-scale hypertextual Web search engine. *Computer Networks and ISDN Systems* **30**, 107–117, http://infolab.stanford.edu/pub/papers/google.pdf (accessed January 1, 2016).

Butler, B. (2013) EMR for fun and for profit, Amazon Web Services Worldwide Public Sector Summit, Washington, DC, September 10, 2013. http://d36cz9buwru1tt.cloudfront.net/145AB-130-EMR-for-Fun-and-Profit-final.pdf (accessed January 1, 2016).

Butler, P. (2010) *Vizualizing Friendships*, https://www.facebook.com/notes/facebook-engineering/visualizing-friendships/469716398919 December 13, 2010 (accessed January 1, 2016).

Dean, J. and Ghemawat, S. (2004) MapReduce: Simplified Data Processing on Large Clusters. Sixth Symposium on Operating System Design and Implementation (OSDI'04), San Francisco, CA, December, 2004, http://research.google.com/archive/mapreduce.html (accessed January 1, 2016).

Dean, J. and Ghemawat, S. (2008) MapReduce: Simplified data processing on large clusters. *Communications of the ACM* **51**(1), 107–113.

Federal Standard 1037C (1996) Telecommunications: Glossary of Telecommunication Terms, National Communications System Technology and Standards Division, General Services Administration Information Technology Service, August 7, 1996, http://everyspec.com/FED-STD/FED-STD-1037C_4685/ (accessed January 1, 2016).

Gartner (2013) Big Data, *IT Glossary*, http://www.gartner.com/it-glossary/big-data/ (accessed January 1, 2016).

Hardy, Q. (2012) Active in Cloud, Amazon Reshapes Computing, *The New York Times* (August 27), http://www.nytimes.com/2012/08/28/technology/active-in-cloud-amazon-reshapes-computing.html?_r=2& (accessed January 1, 2016).

Hemsoth, N. (2013) How Ford is Putting Hadoop Pedal to the Metal, *Datanami* (March 16). http://www.datanami.com/datanami/2013-03-16/how_ford_is_putting_hadoop_pedal_to_the_metal.html (accessed September 8, 2013).

Hurd, M. (2013) *Oracle OpenWorld Keynote* (September 23), http://medianetwork.oracle.com/video/player/2686420812001 (accessed January 1, 2016).

Kinney, J. (2013) Big data in the cloud: Accelerating innovation in the public sector, 2013 Amazon Web Services Worldwide Public Sector Summit, Washington, DC, September 10, http://d36cz9buwru1tt.cloudfront.net/146CB-300-Big-Data-in-the-Cloud-Accelerating-Innovation-in-the-Public-Sector-final.pdf (accessed January 1, 2016).

McKenna, B. (2013) What does a petabyte look like? *Computer Weekly* (March), http://www.computerweekly.com/feature/What-does-a-petabyte-look-like (accessed January 1, 2016).

Malewicz, G., Austern, M. H., Bik, A. *et al.* (2010) Pregel: A system for large-scale graph processing. Proceedings of the 2010 ACM SIGMOD International Conference on Management of Data (SIGMOD'10), June 6–11, 2010, Indianapolis, Indiana, http://dl.acm.org/citation.cfm?doid=1807167.1807184 (accessed January 1, 2016).

Markoff, J. and Sengupta, S. (2011) Separating You and Me? 4.74 Degrees. *New York Times* (November 21) http://www.nytimes.com/2011/11/22/technology/between-you-and-me-4-74-degrees.html (accessed January 1, 2016).

O'Luanaigh, C. (2013) *CERN Data Centre Passes 100 Petabytes*. CERN, http://home.web.cern.ch/about/updates/2013/02/cern-data-centre-passes-100-petabytes. (accessed January 1, 2016).

OOI (2012) *Ocean Observatories Initiative* [brochure], http://oceanobservatories.org/wp-content/uploads/2010/05/2012_OOI_Brochure_Pages_Web.pdf (accessed January 1, 2016).

Peng, D., and Dabek, F. (2010) Large-scale incremental processing using distributed transactions and notifications. Proceedings of the 9th USENIX Symposium on Operating Systems Design and Implementation (USENIX 2010), http://research.google.com/pubs/pub36726.html (accessed January 1, 2016).

Taylor, C. (2012) The Most Important Facebook Number: 140.3 Billion. *Mashable,* October 5, http://mashable.com/2012/10/05/the-most-important-facebook-number-140-billion/ (accessed January 1, 2016).

Travers, J. and Milgram, S. (1969) An experimental study of the small world problem. *Sociometry* **32**(4), 425–443, http://www.cis.upenn.edu/~mkearns/teaching/NetworkedLife/travers_milgram.pdf (accessed January 1, 2016).

Vance, A. (2013) Netflix, Reed Hastings Survive Missteps to Join Silicon Valley's Elite. *Bloomberg Businessweek* (May 9), http://www.businessweek.com/articles/2013-05-09/netflix-reed-hastings-survive-missteps-to-join-silicon-valleys-elite#p4 (accessed January 1, 2016).

Zhukov, L. (2013) Data Science with Billions of Historical Records. *Ancestry.com* (June 14), http://blogs.ancestry.com/techroots/data-science-with-billions-of-family-trees-and-historical-documents/ (accessed January 1, 2016).

45

Big Data in a Cloud

Mark Smiley

The MITRE Corporation, * *USA*

45.1   Introduction

Cloud computing is ascendant; it offers virtually limitless compute power to consume and process previously unimaginably large amounts of data. Cloud computing enables people and organizations to access this power with little upfront cost. The emergence of big data and cloud computing, along with the ability to process that data so that humans can consume and make sense of it, is leading to revolutions in many fields of human endeavor.

It is not a coincidence that the rise of cloud computing has helped enable the ability to process larger and larger amounts of data. The ability to run large-scale big data analytics in a cloud has given some businesses the capability to analyze and process data that they could not handle on their own systems. For example, the TimesMachine discussed in subsection 45.3.5 illustrates how the *New York Times* used the cloud to process its archives.

Running big data analytics on the cloud is superb for occasional use, obviating the need to purchase, house, and maintain the hardware. On the other hand, for those businesses and organizations that run big data analytics constantly, there are some good reasons for running those analytics on bare metal, perhaps on premises. This chapter discusses advantages and disadvantages of running big data analytics on a cloud.

Often, it is not enough to store raw data; instead the data must be stored in a structure so that it can be queried and retrieved. Traditional relational databases, which use Structured Query Language (SQL), tend to focus on

* This chapter is © 2013 – The MITRE Corporation. All rights reserved. The author's affiliation with The MITRE Corporation is provided for identification purposes only, and is not intended to convey or imply MITRE's concurrence with, or support for, the positions, opinions or viewpoints expressed by the author.

a single large database server. In an enterprise setup, of course, there is usually a failover database; sometimes there is a master database for creating, updating, or deleting records, and a few slave databases from which to read records. But these relational databases generally do not perform well when working with big data, and when they are distributed across dozens of nodes. Thus new databases have arisen to meet the needs of working with large distributed data. These new databases are often referred to as Not Only SQL (NoSQL) or occasionally as NewSQL databases. These are typically distributed databases that can scale out. This chapter starts with a discussion of NoSQL databases, then moves into a discussion of running big data analytics on a cloud.

45.2   Databases for Big Data

This section considers how to organize large amounts of data so that it can be found. Organizing such data requires a database that can handle big data.

45.2.1   SQL and ACID

Modern relational database management systems (RDBMS) support the use of Structured Query Language (SQL). At a high level, the contents of an RDBMS may be thought of as a set of interrelated tables, each row of which may be considered a record. For example, a row might contain information about a customer; in this case the columns (or fields) might be: name, address, phone number, e-mail, and so forth.

A key advantage of SQL databases is that they support ACID transactions. ACID stands for: Atomicity, Consistency, Isolation, and Durability (Haerder and Reuter, 1983). Atomicity means that either the whole transaction is processed and saved or nothing is saved. Consistency means that data moves from one consistent state to another; this includes the notion that all database nodes see the same data at the same time. Isolation means that concurrent transactions are isolated from each other, and durability means that once transactions succeed, the data is never lost. There are several advantages associated with SQL: it has been around a long time, many people have experience with SQL, and there is a rich set of tools that work well with SQL.

Relational database management systems are great for transactional data, which requires ACID. But when the data is too large to fit into a single RDBMS, difficulties arise. Suppose the database load is split between two databases. What happens if two people try to write to the same record? An RDBMS wants to keep the data consistent, so it must lock the record when one person asks to edit it, and not allow the second person to edit it. What if the first person doesn't finish editing the record for several hours? Then the second person has to wait. This is just a simple example of various locking issues associated with RDBMSs.

There are other partial solutions to the problem, such as setting up a master and a slave database. It is then possible to allow writes only to the master, while reads can access the slave. But if the data is too large for the master, this will not work either. Another approach is to use sharding. Sharding splits up database tables by rows, based on some rule. For example, the system might keep all rows representing people whose address is west of the Mississippi in one database and those who live east of the Mississippi in another. But then what if each of those databases grows too large? Large online search engines, such as Google's, for example, contain too much data for any RDBMS. Thus Google and others have had to invent new techniques to store and query such large amounts of data, as described in the next section.

45.2.2   NoSQL

The rise of data that is too large to fit into a single RDBMS has led to the creation of other kinds of database systems. These systems are called Not Only SQL (NoSQL) databases. Before delving into the different kinds of NoSQL databases, it is important to discuss a few other concepts to set the stage.

45.2.2.1 *Brewers CAP Theorem*

Three main goals of database design are consistency (defined above), availability and partition tolerance. Availability means that database node failures do not prevent surviving nodes from continuing to operate. Partition tolerance means that the system continues to operate despite arbitrary message loss (but no data loss) between two sets of servers.

The consistency, availability and partition tolerance (CAP) theorem (also known as Brewer's theorem), proves that it is not possible to achieve all three of these database goals, though it is possible to achieve any two of the three goals, (Gilbert and Lynch, 2002). Figure 45.1 illustrates the CAP theorem. NoSQL databases fit into several different regions in Figure 45.1.

Databases that have consistency and availability, but not partition tolerance, include: MySQL, MongoDB and Redis. Examples of those with availability and partition tolerance are Amazon's Dynamo, Cassandra, and CouchDB. Examples with consistency and partition tolerance are: Apache HBase™, Accumulo, and Paxos. Many of these are offered on the Amazon cloud either directly from Amazon (e.g., Dynamo), or through the Amazon Marketplace (e.g., MongoDB and CouchDB).

45.2.2.2 *Eventual Consistency and BASE*

In contrast to ACID, there is the concept of "basically available, soft state, eventual" (BASE) consistency. The key idea is that, for some systems, it is acceptable to have eventual consistency, rather than enforcing immediate consistency, as in a standard RDBMS. For example, suppose two people do a Web search for the same topic at about the same time. Does it matter if the top 10 results are in a slightly different order? In this case, it is more important to respond to both users as quickly as possible, rather than to have consistency. That's a natural tradeoff that Google and others have made. Vogels (2008a, b) discusses eventual consistency in some detail, and states that in order to create complex distributed systems to operate at the Internet scale it is sometimes necessary to choose availability over consistency. This is not surprising but it is important to remember that although BASE is an excellent solution for large distributed systems, some use cases (such as monetary transactions) require ACID.

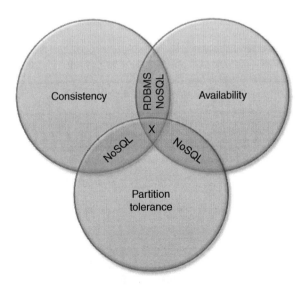

Figure 45.1 *Brewer's CAP theorem*

45.2.2.3   *Types of NoSQL Databases*

There are four main types of NoSQL data stores: key value, document, column, and graph. Key-value stores are similar to a large persistent HashMap (or HashTable). Each item has a key and a value. The value can be something large, such as a video. The key may just be a string, or it may be more complex. Amazon's DynamoDB is an example of a key-value store (DeCandia *et al.*, 2007; Wood, 2012); Redis is another example.

Document databases also use a key to access a document but each document also has a structure that can be queried. The documents typically contain other sets of key-value pairs, or other kinds of documents, such as JSON documents (which contain key-value pairs as well as arrays). Queries work better with document databases than with simple key-value stores. Good examples of document databases include MongoDB (which uses Binary JSON) and CouchDB.

There are also NoSQL databases designed to store and query data that is organized into graphs; Chapter 44 discussed graph analytics. Graphs consist of nodes and edges (relationships). Social networks are expressed well as graphs. Neo4J is a good example of a graph database.

Columnar databases store columns of data together, rather than rows. The father of columnar databases is Google's BigTable, which is a distributed NoSQL database designed to scale to handle petabytes of information, (Chang *et al.*, 2006). BigTable is a sparse, sorted, columnar database. Google uses BigTable for Web indexing, Google Earth, and Google Finance (Chang *et al.*, 2006). Apache HBase™ was modeled on BigTable. Another example of a columnar database is Apache Cassandra, which scales linearly (Cockcroft and Sheahan, 2011); Netflix uses Cassandra on Amazon Web Services (AWS) as part of its video streaming service (Cockcroft and Sheahan, 2011).

Apache Accumulo™ is another columnar database that is based on HBase, but which includes cell-based access control. In Accumulo, a *key* is multidimensional and consists of several parts:

1.   Row ID
2.   Column, which has three parts:
     • Family
     • Qualifier
     • Visibility (access control)
3.   Timestamp

Each of these parts is a byte array, with the exception of the timestamp, which is a long. Accumulo is particularly popular in areas in which cell-based access control is important.

45.2.3   NewSQL

NewSQL is another type of database, related to NoSQL databases. The idea is to offer the scalability of a NoSQL database, while retaining ACID properties and offering an SQL-like query language. Examples include VoltDB and NuoDB.

45.3   Big Data on a Cloud

This section discusses running big-data applications in a cloud. Big data can be hosted in a cloud, and big data analytics can be executed on the cloud. It does not require cloud computing but the rise of cloud computing has opened up the exploration and analysis of big data to many more people, due to its low cost and ready availability.

45.3.1 Analytic Clouds and Utility Clouds

An environment that is set up to handle big data and run big data analytics, including a set of machines to run the parallel analytic processes, is sometimes called an *analytic cloud.* An example of an analytic cloud is a large cluster of commodity servers running Hadoop, as discussed in Chapter 48.

An analytic cloud is quite different from the standard NIST definition of a cloud (Mell and Grance, 2011). That definition includes the five essential characteristics discussed in more detail elsewhere in this book:

- on-demand self-service;
- broad network access;
- resource pooling;
- rapid elasticity;
- measured service.

A cloud meeting the NIST definition of cloud computing is sometimes called a *utility cloud,* to distinguish it from an analytic cloud. A utility cloud uses an infrastructure as a service (IaaS) or platform as a service (PaaS) service model, or both. Classic public utility clouds include those offered by Amazon and Microsoft, among others, as described elsewhere in this book. Unless otherwise specified, the term "cloud" means "utility cloud."

An analytic cloud may run natively (i.e., without the use of virtualization) on a set of physical machines running an environment such as Hadoop, along with its ecosystem of utilities. On the other hand, it may be a set of virtual machines running in a large datacenter with a similar ecosystem of tools. In the latter case, these machines may be running in a utility cloud.

Virtualization is a key component of a utility cloud. On the other hand, an analytic cloud may or may not use virtualization. It is possible to run an analytic cloud on a utility cloud, but not the converse. Indeed, part of the reason for the growth of big data analytics has been the existence of utility clouds, and the ability to use them to host analytic clouds.

45.3.2 Running Big Data Analytics on a Utility Cloud

There are advantages and disadvantages to running big data analytics on a utility cloud. Advantages of running big data analytics in a utility cloud include the fact that it is convenient, readily available, and it can be affordable. Naturally, the size and length of the jobs have a large effect on the price. Yet the availability of public clouds means that it is no longer necessary to purchase a large number of servers to run the analytics or find a suitable room to house the servers, the funding to pay for the power and cooling, the systems administrators to maintain the servers, and so forth. Instead, it is possible to run big data analytics on a public cloud.

On the other hand, if an organization sets up its big data analytics to run on bare metal – that is, on hardware, using no virtualization – that organization will be rewarded with a speed improvement in the analytics. This speed increase is partly due to the input / output (I/O) penalty paid for virtualization.

Another factor is the noisy neighbor problem. In this case, one of the virtual machines (VMs) may be running on the same server as someone else's VM. Moreover, their VM may use a lot of I/O. As the company is pumping big data through the VM, it also requires a substantial amount of I/O to move the big data into the VM. The two VMs then vie for the same physical amount of I/O bandwidth, slowing down both VMs, or the one with the lowest priority. In Figure 45.2, VM 2 is the noisy neighbor for VM 1; it consumes so much bandwidth that VM 1 cannot attain the bandwidth that it desires.

There are some ways to mitigate the noisy neighbor problem:

- Move loads that perform poorly to other physical machines. Or: "If your neighbor is noisy, you can move."
- Reserve specific machines for use, so that their workloads can be tuned to avoid the issue.

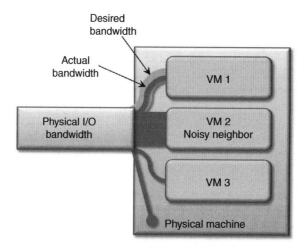

Figure 45.2   *Noisy neighbor*

- If necessary, use high I/O instances, though these will typically be more costly than normal instances.
- If the data is streaming data, store it in place, distributed on the nodes, so less I/O is needed.
- In a private cloud, use more NICs per machine, or use NICs with enough bandwidth.

Although big data analytics run faster on bare metal, the mean time to job completion (MTTJC) may be shorter on a utility cloud. There are two main reasons for this improvement in the MTTJC. The first reason is because it is possible to set up and run jobs on a cloud, without having to purchase the hardware that a bare metal setup requires. The second reason is because it is possible to scale up the number of virtual machines as needed to process the data. In fact this scaling can be automated, based on a predefined policy. This feature of a utility cloud is called *rapid elasticity* (Mell and Grance, 2011).

45.3.3   Running Big Data Analytics on Bare Metal

With all the advantages of hosting big data on a utility cloud, why would anyone want to use bare metal? The main reason is performance. As mentioned above, using virtualization (a technology that underlies utility clouds) causes a performance hit over running an analytic cloud on bare metal. Also, a bare metal analytic cloud can be designed with high I/O characteristics. Dedicated physical machines, along with a proper design, can avoid the noisy neighbor problem.

Disadvantages include the fact that a bare metal analytic cloud is expensive and time consuming to set up and maintain. An organization must purchase all that hardware, house it in a building, run cables, and so forth. It must also keep a staff of systems administrators on call to keep the servers running. All this makes a bare metal analytic cloud quite impractical either for occasional use, or when a job requires rapid elasticity of computing resources. If an organization needs to run analytics constantly, however, then running them on bare metal may make sense. Google is a good example of a company that runs its own bare metal analytic cloud.

For those who wish to purchase a bare metal analytic cloud, Oracle® offers a big data cloud in a box, called the Oracle Big Data Appliance, which includes: the Cloudera® distribution of Apache Hadoop, R (an open source statistics tool), and the Oracle NoSQL database. It can be connected to a standard Oracle database, and it can stream 15 TB per hour from Hadoop to the Oracle database, (Oracle Corporation, 2013).

45.3.4   Which Big Data Analytics Belong on a Utility Cloud?

Each situation is different but here are some guidelines for deciding when to use a utility cloud for big data analytics.

- Some analytics may only need to run occasionally, such as for research purposes, to prototype a new application, or for one-time use. This is a natural reason to use a utility cloud.
- If the data to be analyzed is already in a utility cloud, that is another good reason to use a utility cloud.
- Another reason to use a utility cloud is if the nature of the big data analytic problem being solved involves rapid changes in resource requirements. In a bare-metal environment, the amount of resources for any given job remains fixed, so if the job size varies, it causes the MTTJC to vary. On the other hand, a utility cloud is elastic, so it can scale rapidly to complete the job in a shorter time.
- Using a utility cloud is also convenient, available, and affordable. All these are more good reasons to use a utility cloud.

45.3.5   TimesMachine: An Example of Using a Utility Cloud for Big Data

To illustrate an example of a corporation using a utility cloud to process big data, the *New York Times* decided to digitize all their newspapers from 1851–1922. This project was called TimesMachine, and the process to create it is described in Gottfrid (2008). The *Times* started by scanning in the newspapers and creating large tagged image file format (TIFF) images, along with some metadata and the text of the articles acquired via Optical Character Recognition (OCR). That yielded several TB of data, which was not easily consumable via the Web. So Gottfrid used AWS EC2 virtual servers and S3 virtual storage along with Hadoop to:

- ingest 405 000 TIFF images with an equal number of XML metadata files to map 3.3 million SGML articles to rectangular areas in the TIFF images;
- convert the above data into 810 000 PNG images, both thumbnails and images, along with 405 000 JavaScript files, composing content for consumption on the Web.

Gottfrid's team used hundreds of virtual machines in the cloud to process the data in under 36 hours, (Gottfrid, 2008). The *Times* certainly did not need to create a bare metal analytic cloud to process their data. The utility cloud offered a perfect solution for its needs.

45.3.6   Running Hadoop in a Cloud

As Hadoop is the most popular big data ecosystem, it is worth discussing running Hadoop in a cloud in more detail. Chapter 48 discusses some implementations of Hadoop on a utility cloud. To round out the picture, this section deals with some general issues of running Hadoop on a utility cloud.

The main issues are:

- Hadoop is rack and host aware, while cloud providers do not offer that information;
- virtualization reduces performance.

Hadoop is rack-aware and host-aware. That is, Hadoop has knowledge of the underlying physical topology. By default, Hadoop stores each piece of data on three nodes. The Hadoop NameNode keeps track of which nodes contain what data. Thus it is important for it to know on which rack each node resides. If it lacks this knowledge, it might put all three copies of the data on the same rack (or even the same physical machine). Yet on a typical IaaS or PaaS utility cloud, cloud consumers have no knowledge of which physical machines contain their virtual machines. This is a potential problem with running Hadoop on a utility cloud.

However, there are some ways to mitigate this problem. One is for the cloud service provider (CSP) to offer a Hadoop service. The CSP knows the physical topology, so it can build that into its offering. Most of the major CSPs now offer a Hadoop service. For example, the next section discusses Amazon's offering: Elastic MapReduce (EMR). Google, Microsoft, and others also offer similar services.

VMware continues to work on addressing the issue of running Hadoop on virtual machines. To this end, it started the open source Project Serengeti. The goal of Serengeti is to facilitate deploying Hadoop in a virtual environment. Serengeti is able to deploy the Hadoop ecosystem on top of VMware's vSphere virtualization platform. A number of popular Hadoop distributions have endorsed Serengeti.

Conventional wisdom is that virtualization reduces performance. After all, the hypervisor adds some processing and memory overhead. However, a recent paper offered by VMware (Buell, 2013) ran Hadoop virtualized on VMware vSphere. The paper suggests that the deleterious performance characteristics may be minimized by properly sizing and configuring the VMs. Independent tests will naturally need to verify this finding but it is an encouraging prospect.

It is not possible to cover all the issues with virtualizing Hadoop here. An excellent site with considerable detailed information on running virtualized Hadoop is http://wiki.apache.org/hadoop/Virtual%20Hadoop (accessed January 2, 2016).

An experiment by Wendt (2014) found that, in their case, running Hadoop in a cloud was less expensive than running it internally on bare metal.

45.3.7 Amazon Elastic MapReduce

It takes considerable time and thought to tune an analytic cloud to run on a utility cloud. For this reason, Amazon offers an analytic cloud service called Amazon Elastic MapReduce (EMR). The content in this section is based on a presentation by Ben Butler, Senior Manager of Big Data at Amazon (Butler, 2013). It is helpful to go through this information in detail, as it is an excellent exemplar of how to process big data in a utility cloud. As of 2013, there were 5.5 million EMR clusters, so it is currently the most popular way to run big data analytics on a utility cloud.

45.3.7.1 Getting the Data into the Cloud

The data must first get into the cloud. Amazon offers four main ways to do that:

- AWS Direct Connect, which provides dedicated low-latency bandwidth.
- AWS Import/Export, which means physically shipping the media (e.g., hard drives) to AWS.
- Queuing, such as the Amazon Simple Message Service (SMS), which offers highly scalable event buffering. This would be one way to feed in high velocity data.
- Amazon Storage Gateway, which allows synchronization of local storage to the cloud.

45.3.7.2 Storing the Data

Of course there must be a way to store the data in the cloud. Amazon offers several suitable types of storage, depending on the data and analytic needs.

- Amazon Relational Database Service (Amazon RDS). This is a relational database management system (RDBMS), which offers a variety of engines, including: MySQL, Oracle, and Microsoft SQL Server®. It can handle up to 3 TB.
- Amazon SimpleDB. This is an example of a NoSQL database, so it is a nonrelational database that is schema-less. It can be appropriate for smaller datasets.

- Amazon DynamoDB. This is another NoSQL database, and it is also schema-less. Furthermore it is designed for high throughput, storing data on Solid State Drives (SSDs), and spread across geographic zones for reliability. It currently offers speeds of 5 ms for a read, and 10 ms for a write (Wood, 2012).
- Amazon S3. This is Amazon's popular key-value object data storage. It is highly durable, with 99.999999999% durability, and each object can contain up to 5 TB. Access time is in the range of tens to hundreds of ms (Butler, 2013).
- Amazon Glacier. This is a service designed for long-term cold storage, to replace older tape storage. It is also highly durable but it takes longer to access data, so it is not appropriate to run data analytics against it. An item can be accessed within 5 hours of the request (Butler, 2013).

Amazon also offers Amazon Redshift, a data warehouse service designed for sizes from 2 TB to 1.6 PB. This may be more appropriate once the data has been processed by EMR and the results stored in one of the above.

45.3.7.3 Typical Usage

A typical way to use EMR is to follow these steps:

1. Put the data into S3 or HDFS. (Both copy the data to three machines.)
2. Launch the EMR cluster and process the data, storing the results in S3.
3. Aggregate the results from all the nodes and store that in S3.
4. Retrieve the results from S3.
5. Terminate the cluster when complete.

It is possible to run multiple clusters against the same data, each researching a different question. To improve security, the EMR cluster can launch in a Virtual Private Cloud (VPC).

45.3.7.4 Best Practices

There are a couple of best practices for using EMR. The first is not unique to EMR but is true of all big data applications.

- Move applications to the data, rather than the other way around. It takes too long to send the data elsewhere.
- Use the Amazon EC2 Spot market to reduce costs. Spot instances are EC2 instances that are unused, so Amazon offers them at prices that vary from moment to moment. You can specify a limit on how much you would like to pay for an instance but if the price goes over your limit you could lose your instance within minutes.

Deyhim (2013) contains a more detailed set of best practices for EMR.

45.3.7.5 Amazon's Public Data Sets

Amazon offers some public data sets for use with EMR or other big data analytics. One example is the 1000 Genomes Project, containing over 200 TB of data – see http://aws.amazon.com/1000genomes/ (accessed January 2, 2016), for more information.

45.3.8 Windows Azure HDInsight Service

Microsoft also offers a big data service that runs in their Azure cloud, called Windows Azure HDInsight Service, which deploys and provisions Apache™ Hadoop® clusters in the Azure cloud, (Microsoft, 2013).

This offering supports an entire Hadoop ecosystem of tools, such as Apache Pig, Apache Hive, and Apache Sqoop. Those tools are discussed in Chapter 48.

45.3.9   Other Cloud Implementations

AppEngine-MapReduce is an open-source library for doing MapReduce-style computations on the Google App Engine platform with pricing that is completive with Amazon EMR. See https://code.google.com/p/appengine-mapreduce/.

The Sahara project is working on supporting Hadoop on top of OpenStack. OpenStack is a popular IaaS implementation that may be instantiated in a private cloud. For more information on Sahara, see https://wiki.openstack.org/wiki/Sahara (accessed January 2, 2016).

45.4   Conclusion

By taking a different approach to the CAP theorem, many NoSQL databases have been designed to run in a distributed fashion, so that they may run at Internet scale, unlike traditional RDBMS tools. Thus they fit nicely onto a utility cloud.

The existence of utility clouds has allowed more people to become involved with big data analytics. Because of these readily available public clouds, it is no longer necessary to purchase, house, run and maintain a room full of servers to analyze big data.

Disadvantages of running big data analytics in a utility cloud include reduction in speed due to the I/O virtualization penalty, and the noisy neighbor problem (which can be mitigated). An additional disadvantage for Hadoop is that it is rack aware, while public clouds do not provide that information. Fortunately cloud providers and virtualization vendors are aware of the issue and most offer a hosted Hadoop solution, such as Amazon's EMR or Project Serengeti.

Cloud computing lowers the barrier to enter the world of big data. Thus it allows researchers to try new ideas without imposing on them a large financial burden and a time consuming setup process.

References

Buell, J. (2013) *Virtualized Hadoop Performance with VMware vSphere® 5.1,* http://www.vmware.com/files/pdf/vmware-virtualizing-apache-hadoop.pdf (accessed January 2, 2016).

Butler, B. (2013) EMR for fun and for profit, 2013 Amazon Web Services Worldwide Public Sector Summit, Washington, DC, September 10, 2013. http://d36cz9buwru1tt.cloudfront.net/145AB-130-EMR-for-Fun-and-Profit-final.pdf (accessed January 2, 2016).

Chang, F., Dean, J., Ghemawat, S., *et al.* (2006) BigTable: A distributed storage system for structured data, OSDI'06: Seventh Symposium on Operating System Design and Implementation, Seattle, WA, http://research.google.com/archive/bigtable.html (accessed January 2, 2016).

Cockcroft, A. and Sheahan, D. (2011) Benchmarking Cassandra Scalability on AWS – Over a million writes per second. *The Netflix Tech Blog.* November 2, http://techblog.netflix.com/2011/11/benchmarking-cassandra-scalability-on.html (accessed January 2, 2016).

DeCandia, G., Hastorun, D., Jampani, H., *et al.* (2007) Dynamo: Amazon's highly available key-value store. Proceedings of the 21st ACM Symposium on Operating Systems Principles, Stevenson, WA, October 2007, pp. 205–220. http://www.allthingsdistributed.com/files/amazon-dynamo-sosp2007.pdf (accessed January 2, 2016).

Deyhim, P. (2013) *Best Practices for Amazon EMR,* http://media.amazonwebservices.com/AWS_Amazon_EMR_Best_Practices.pdf (accessed December 2, 2016).

Gilbert, S. and Lynch, N. (2002) Brewer's conjecture and the feasibility of consistent, available, partition-tolerant web services. *ACM SIGACT News* **33**(2), 59.

Gottfrid, D. (2008) *The New York Times Archives + Amazon Web Services = TimesMachine*, http://open.blogs.nytimes.com/2008/05/21/the-new-york-times-archives-amazon-web-services-timesmachine/ May 21, 2008 (accessed January 2, 2016).

Haerder, T. and Reuter, A. (1983) Principles of transaction-oriented database recovery. *Computing Surveys* **10**(4), 287–317.

Mell, P. M., and Grance, T. (2011) *The NIST Definition of Cloud Computing*. Special Publication 800-145. NIST, Gaithersburg, MD, http://www.nist.gov/customcf/get_pdf.cfm?pub_id=909616 (accessed November 25, 2015).

Microsoft (2013) Introduction to Windows Azure HDInsight Service, https://azure.microsoft.com/en-gb/documentation/videos/introduction-to-hdinsight-service/ (accessed January 2, 2016).

Oracle Corporation (2013) *Oracle Big Data Appliance X4-2*. Oracle Data Sheet, http://www.oracle.com/technetwork/server-storage/engineered-systems/bigdata-appliance/overview/bigdataappliance-datasheet-1883358.pdf (accessed January 2, 2016).

Vogels, W. (2008a) Eventually consistent. *ACM Queue* **6** (6), http://queue.acm.org/detail.cfm?id=1466448 (accessed January 2, 2016).

Vogels, W. (2008b) Eventually consistent – revisited. *All Things Distributed,* http://www.allthingsdistributed.com/2008/12/eventually_consistent.html (accessed January 2, 2016).

Wendt, M. (2014) *Cloud-Based Hadoop Deployments: Benefits and Considerations*. Accenture Technology Labs, https://www.accenture.com/t00010101T000000__w__/jp-ja/_acnmedia/Accenture/Conversion-Assets/DotCom/Documents/Local/ja-jp/PDF_2/Accenture-Cloud-Based-Hadoop-Deployments-Benefits-and-Considerations.pdf (accessed January 2, 2016).

Wood, M. (2012) *Introducing DynamoDB,* March 20, http://www.slideshare.net/AmazonWebServices/webinar-introduction-to-amazon-dynamodb (accessed January 2, 2016).

46

Cloud-Hosted Databases

Sherif Sakr

University of New South Wales, Australia

46.1   Introduction

Cloud computing technology represents a new paradigm for hosting software applications. This paradigm simplifies the time-consuming processes of hardware provisioning, hardware purchasing, and software deployment. It has revolutionized the way computational resources and services are commercialized and delivered to customers. In particular, it shifts the location of its computing infrastructure to the network to reduce the costs associated with the management of hardware and software resources. It, therefore, fulfils the long-held dream of envisioning computing as a utility where economies of scale help to drive down the cost of computing infrastructure effectively (Armbrust *et al.*, 2009). It offers a number of advantages for the deployment of software applications such as a pay-per-use cost model, low time to market, and the perception of (virtually) unlimited resources and infinite scalability. In practice, the advantages of the cloud-computing paradigm open up new avenues for deploying novel applications that are not economically feasible in a traditional enterprise infrastructure setting. The cloud has become an increasingly popular platform for hosting software applications in a variety of domains such as e-retail, finance, news, and social networking. We are witnessing a proliferation in the number of applications with a tremendous increase in the scale of the data generated and being consumed by such applications. Cloud-hosted database systems powering these applications form a critical component in the software stack of these applications.

In general, data-intensive applications are classified into two main types:

- online transaction processing (OLTP) systems that deal with operational databases up to a few terabytes in size with write-intensive workloads that require ACID (Atomicity, Consistency, Isolation, Durability) transactional support and response-time guarantees;

Encyclopedia of Cloud Computing, First Edition. Edited by San Murugesan and Irena Bojanova.
© 2016 John Wiley & Sons, Ltd. Published 2016 by John Wiley & Sons, Ltd.

- online analytical processing (OLAP) systems that deal with historical databases of very large sizes, up to petabytes, with read-intensive workloads that are more tolerant to relaxed ACID properties.

In this chapter, we focus on cloud-hosted database solutions for OLTP systems. A successful cloud-hosted database tier of an OLTP system should sustain a number of goals:

- *Availability*. They must always be accessible, even during a network failure or when a whole datacenter has gone offline.
- *Scalability*. They must be able to support very large databases with very high request rates at very low latency. In particular, the system must be able to replicate and redistribute data automatically to take advantage of the new hardware. They must be also able to move load between servers (replicas) automatically.
- *Elasticity*. They must cope with changing application needs in both directions (scaling up / out or scaling down / in). Moreover, the system must be able to respond gracefully to these changing requirements and recover quickly to its steady state.
- *Performance*. On public cloud computing platforms, pricing is structured in such a way that one pays only for what one uses, so the vendor price increases linearly with the requisite storage, network bandwidth, and compute power. Hence, the system performance has a direct effect on its costs. Thus, efficient system performance is a crucial requirement to save money.

Arguably, one of the main goals of cloud-hosted database system is to facilitate the job of implementing every application as a *distributed, scalable,* and *widely accessible* service on the Web. The Amazon online retailer, eBay, Facebook, Twitter, Flickr, YouTube, and LinkedIn are just examples of online services that are currently able to achieve this goal successfully. Such services have two main characteristics: they are *data intensive* and *very interactive*. Currently, a common goal is to make it easy for every application to achieve high scalability, availability, and performance targets with minimum effort.

The quest to conquer the challenges posed by hosting databases on cloud computing environments has led to a plethora of systems and approaches. In practice, there are three main technologies that are commonly used for deploying the database tier of software applications in cloud platforms, namely, the services of NoSQL storage systems, database-as-a-service (DaaS) platforms, and virtualized database servers. This chapter aims to discuss the basic characteristics and the recent advancements of each of these technologies, illustrating the strengths and weaknesses of each technology.

46.2 NoSQL Database Systems

For decades, relational database management systems (e.g. MySQL, PostgreSQL, SQL Server, Oracle) have been considered as the *one-size-fits-all* solution for providing data persistence and its retrieval. These systems have matured after extensive research and development efforts and have very successfully created a large market of solutions in different business domains. However, the ever increasing need for scalability, and new application requirements, have created new challenges for traditional relational database management systems (RDBMS). There has therefore been some dissatisfaction with this one-size-fits-all approach in deploying the data storage tier for large-scale online Web services (Stonebraker, 2008), which resulted in the emergence of a new generation of low-cost, high-performance database software that challenges the dominance of relational database management systems. A big reason for this movement, known as NoSQL (not only SQL), is that different implementations of Web, enterprise, and cloud computing applications, which have different requirements from their data-management tiers (e.g. not every application requires rigid data consistency), have opened up various possibilities in the design space. For example, for high-volume web sites (e.g. eBay, Amazon, Twitter, Facebook), scalability and high availability are essential requirements that cannot be compromised. For these applications, even the slightest outage can have significant financial consequences and impacts customer trust. The CAP theorem (Brewer, 2000) has shown that a distributed database system can only choose at most two properties from consistency, availability and tolerance to partitions. Therefore, most of these systems decide to compromise the strict consistency requirement. In particular, they apply a relaxed consistency policy called *eventual consistency* (Vogels, 2008), which guarantees that if no new updates are made to a replicated object, eventually all accesses will return the last updated value (Vogels, 2008). If no failures occur, the maximum size of the inconsistency window can be determined based on factors such as communication delays, the load on the system, and the number of replicas involved in the replication scheme.

Google's BigTable (Chang *et al.*, 2008) and Amazon's Dynamo (2007) (presented by Amazon) have provided a proof of concept that inspired and triggered the development of a new wave of NoSQL systems. In particular, BigTable has demonstrated that persistent record storage could be scaled to thousands of nodes while Dynamo has pioneered the idea of eventual consistency as a way to achieve higher availability and scalability. In principle, the implementations of NoSQL systems have a number of common design features such as:

- Supporting flexible data models with the ability to dynamically define new attributes or data schema.
- A simple call-level interface or protocol (in contrast to a SQL binding), which does not support join operations.
- Supporting weaker consistency models than ACID transactions in most traditional RDBMS. These models are usually referred to as Basically Available, Soft-State, Eventually Consistent (BASE) models (Pritchett, 2008).
- The ability to horizontally scale out throughput over many servers.
- Efficient use of distributed indexes and RAM for data storage.

Commercial cloud offerings taking this approach include Amazon S3, Amazon SimpleDB and Microsoft Azure Table Storage. There is also a large number of open-source projects that have been introduced, which follow the same NoSQL principles, such as HBase, Cassandra, Voldemort, Dynomite, Riak, and MongoDB. In general, these NoSQL systems can be classified with respect to different characteristics. For example, based on their supported data model, they can be classified into the following categories:

- *Key-value stores.* These systems use the simplest data model, which is a collection of objects where each object has a unique key and a set of attribute / value pairs.
- *Extensible record stores.* They provide variable width tables (column families) that can be partitioned vertically and horizontally across multiple servers.
- *Document stores.* The data model of these systems consists of objects with a variable number of attributes with a possibility of having nested objects.

In addition, the systems can be classified into three categories based on their support of the properties of the CAP theorem:

- *CA systems.* Consistent and highly available but not partition tolerant.
- *CP systems.* Consistent and partition-tolerant but not highly available.
- *AP systems.* Highly available and partition tolerant but not consistent.

Table 46.1 *Design decisions of sample NoSQL system*

System	Data model	Consistency guarantee	CAP option	License
BigTable	Column families	Eventually consistent	CP	Internal at Google
PNUTS	Key-value store	Timeline consistent	AP	Internal at Yahoo!
Dynamo	Key-value store	Eventually consistent	AP	Internal at Amazon
S3	Document store	Eventually consistent	AP	Commercialized By Amazon
Simple DB	Key-value store	Eventually consistent	AP	Commercialized By Amazon
HBase	Column families	Strict consistent	CP	Open Source – Apache
Cassandra	Column families	Eventually consistent	AP	Open Source – Apache
MongoDB	Document store	Eventually consistent	AP	Open Source – GPL

In practice, choosing the appropriate NoSQL system (from the very wide available spectrum of choices) with design decisions that best fit with the requirements of a software application is not a trivial task and requires careful consideration. Table 46.1 provides an overview of different design decisions for a sample of NoSQL systems. For comprehensive survey of the NoSQL system and their design decisions, we refer the reader to Cattell (2010) and Sakr *et al.* (2011).

In general, the capabilities of the NoSQL systems have attracted a lot of attention. However, many obstacles still need to be overcome before these systems can appeal to mainstream enterprises such as:

- *Programming model*. NoSQL databases offer few facilities for ad-hoc query and analysis. Even a simple query requires significant programming expertise. The inability of such systems to express an important join operation declaratively has been always considered one of the main limitations of these systems.
- *Transaction support*. Transaction management is one of the powerful features of RDBMS. The current limited support (if any) of the transaction notion from NoSQL database systems is considered to be a big obstacle to their acceptance for implementing mission-critical systems. In principle, developing applications on top of an eventually consistent NoSQL datastore requires a greater effort than traditional databases because they hinder the ability to support key features such as data independence, reliable transactions, and other crucial characteristics often required by applications that are fundamental to the database industry.

Migration. Migrating existing software application that uses relational database to NoSQL offerings would require substantial changes in the software code due to the differences in the data model, query interface and transaction management support. In practice, it might require a complete rewrite of any source code that needs to interact with the data management tier of the software application.

Maturity. relational database management systems are well known for their high stability and rich functionality. In comparison, most NoSQL alternatives are still preproduction versions with many key features being either not stable enough or yet to be implemented, so enterprises are still approaching this new wave of data management with extreme caution.

There is thus still a big debate between the proponents of NoSQL and RDBMS, which is centered around the right choice for implementing online transaction processing systems. Relational database management systems proponents think that the NoSQL camp has not taken sufficient time to understand the theoretical foundation of the transaction processing model. For example, the eventual consistency model is still not well defined and different implementations may differ significantly from each other. Figuring out all this inconsistent behavior adds to the application developers' responsibilities and make their lives very much harder. They also believe that NoSQL systems could be more suitable for OLAP applications rather than for OLTP applications (Abadi, 2009). On the other hand, the NoSQL camp argues that the domain-specific optimization opportunities of

NoSQL systems give back more flexibility to the application developers, who are now no longer constrained by a one-size-fits-all model. However, they admit that making such optimization decision requires a lot of experience and the process can be very error prone and dangerous if not done by experts.

46.3 Database-as-a-Service (DaaS)

Datacenters are often underutilized due to overprovisioning as well as the time-varying resource demands of typical enterprise applications. *Multitenancy*, a technique that is pioneered by salesforce.com, is an optimization mechanism for hosted services in which multiple customers are consolidated onto the same operational system and thus economies of scale help to drive down the cost of computing infrastructure effectively. In particular, multitenancy allows pooling of resources, which improves utilization by eliminating the need to provision each tenant for their maximum load. Multi-tenancy is therefore an attractive mechanism for both of the cloud providers, who are able to serve more customers with a smaller set of machines, and also to customers of cloud services who do not need to pay the price of renting the full capacity of a server.

In practice, there are three main approaches for the implementation of multitenant database systems:

- *shared server*, where each tenant is offered a separate database in the same database server;
- *shared process*, where each tenant is offered its own tables while multiple tenants can share the same database;
- *shared table*, where the data of all tenants is stored in the same tables and each tuple has an additional column with the tenant identifier.

Database as a service is a technology where a third-party service provider hosts a relational database as a service (Agrawal *et al.*, 2009). Such services alleviate the need for their users to purchase expensive hardware and software, deal with software upgrades, and hire professionals for administrative and maintenance tasks. Cloud offerings taking this approach include *Amazon RDS, Microsoft SQL Azure, Google Cloud SQL,* and *Heroku Postgres*. While the shared table multitenancy model can be used by SaaS providers (e.g. Salesforce. com) because all tenants share the same database structure for their application, the shared server multitenancy model is the most commonly used by most commercial DaaS providers as it is considered to be the most effective approach to secure the isolation of each tenant's data and allocated computing resources.

Amazon RDS is an example of a relational database service that gives its users access to the full capabilities of a familiar MySQL database or Oracle. Hence, the code, applications, and tools that are already designed on existing MySQL or Oracle databases can work seamlessly with Amazon RDS. Once the database instance is running, Amazon RDS can automate common administrative tasks such as performing backups or patching the database software. Amazon RDS can also conduct automatic failover management. Google Cloud SQL is another service that provides the capabilities and functionality of MySQL database servers, which are hosted in Google's cloud. Although there is tight integration of the services with Google App Engine (Google's platform-as-a-service software development environment), in contrast to the original built-in data store of Google App Engine, Google Cloud SQL allows software applications to move their data in and out of Google's cloud easily without any obstacles. Microsoft has released the Microsoft SQL Azure database system as a cloud-based relational database service, which has been built on Microsoft SQL server technologies. It provides a highly available, multitenant database service hosted by Microsoft in the cloud. Applications can create, access, and manipulate tables, views, indexes, roles, stored procedures, triggers, and functions. It can execute complex queries and joins across multiple tables. It also supports Transact-SQL (TSQL), native ODBC, and ADO.NET data access. In particular, the SQL Azure service can be seen as running an instance of an SQL server in a cloud hosted server, which is automatically managed by Microsoft instead of running on an on-premises managed server. Similarly, Heroku Postgres provides a Web service that provides the functionalities of the SQL-compliant database, PostgreSQL.

Relational cloud (Curino *et al.,* 2011) represents a research effort for developing a system that hosts multiple databases on a pool of commodity servers inside one datacenter. In order to allow workloads to scale across multiple servers, the system relies on a graph-based data partitioning algorithm that groups data items according to their frequency of co-access within transactions / queries. The main goal of this partitioning process is to minimize the probability that a given transaction has to access multiple nodes to complete its execution. In addition, in order to manage and allocate the available computing resources to the different tenants effectively, the system monitors the access patterns induced by the tenants' workloads and the load of each database server, and uses this information to determine periodically the best way to place the database partitions on the back-end machines. The goal of this monitoring process is to minimize the number of machines used and balance the load on the different servers.

In practice, the migration of the database tier of any software application to a relational database service is expected to require minimal effort if the underlying RDBMS of the existing software application is compatible with the service offered. This helps the software applications to achieve faster "time to market" because they can host the database tier of their application in cloud platforms quickly, and use their features and advantages. However, many relational database systems are, as yet, not supported by the DaaS paradigm (e.g. IBM DB2, Informix, Sybase). In addition, some limitations or restrictions might be introduced by the service provider for different reasons (e.g. the maximum size of the hosted database, the maximum number of possible concurrent connections). Moreover, software applications do not have sufficient flexibility in being able to control the allocated resources of their applications (e.g. dynamically allocating more resources for dealing with increasing workload or dynamically reducing the allocated resources in order to reduce the operational cost). The whole resource management and allocation process is controlled at the provider side, which requires an accurate planning for the allocated computing resources for the database tier and limits the ability of the consumer applications to maximize their benefits by leveraging the elasticity and scalability features of the cloud environment.

46.4   Virtualized Database Servers

Virtualization is a key technology of the cloud computing paradigm that abstracts away the details of physical hardware and provides virtualized resources for high-level applications. A virtualized server is commonly called a *virtual machine* (VM). Virtual machines allow the isolation of applications from the underlying hardware and other VMs. Ideally, each VM is both unaware of and unaffected by other VMs that could be operating on the same physical machine. In principle, resource virtualization technologies add a flexible and programmable layer of software between applications and the resources used by these applications. The approach of *virtualized database server* makes use of these advantages where an existing database tier of a software application that has been designed to be used in a conventional datacenter can be directly ported to virtual machines in the public cloud. Such a migration process usually requires minimal changes in the architecture or the code of the deployed application. In this approach, database servers, like other software components, are migrated to run in virtual machines. While the provisioning of a virtual machine for each database replica imposes a performance overhead, this overhead is estimated to be less than 10% (Minhas *et al.,* 2008). In principle, this approach represents a different model of multitenancy – *shared physical machine* – where a VM of a virtualized database server can be running on the same physical machine with other VMs, which are not necessarily running database operations.

Dynamic provisioning is a well known process of increasing or decreasing the allocated computing resources (e.g. number of virtualized database servers) to an application in response to workload changes. In practice, one of the major advantages of the virtualized database server approach is that the application can have full control in dynamically allocating and configuring the physical resources of the database tier (database servers) as needed (Soror *et al.,* 2008; Cecchet *et al.,* 2011; Sakr and Liu, 2012). Hence, software applications can fully use the elasticity feature of the cloud environment to achieve their defined and customized scalability or cost reduction goals. However, achieving these goals requires the existence of an *admission control* component, which is responsible for monitoring the system's state and taking the corresponding actions (e.g. allocating more / less computing resources) according to the defined application requirements and strategies. Therefore, one of the main responsibilities of this admission control component is to decide when to trigger an increase or decrease in the number of the virtualized database servers, which are allocated to the software application.

In general, the decision of *when* to increase or decrease the allocated computed resources is made in a lazy fashion for the Web and the application tiers of the software application, in response to an actual or anticipated significant workload change. Such lazy triggers are appropriate for these tiers because new capacity can be added in a quick manner, whenever required, as the only resulting latency is for the virtual machine startup. However, provisioning of a new database replica involves copying and restoring a new replica, which can take minutes or hours depending on the database size. The *Dolly* system (Cecchet *et al.,* 2011) has presented an approach that takes the latency of provisioning new database replicas into account when triggering provisioning decisions. In particular, Dolly incorporates a model to estimate the latency to create a replica, based on the virtual machine snapshot size and the database resynchronization latency, and uses this model to trigger the replica spawning process well in advance of the anticipated workload increase.

A service-level agreement (SLA) is a contract between a service provider and its customers. In principle, SLAs capture the agreed guarantees between a service provider and its customer. They define the characteristics of the service provided, including service-level objectives (SLOs) (e.g. maximum response times), and define penalties if these objectives are not met by the service provider. In practice, flexible and reliable management of SLA agreements is of paramount importance both for cloud service providers and consumers. The *CloudDB AutoAdmin* framework (Sakr and Liu, 2012; Zhao *et al.,* 2014) has presented another approach for the SLA-based dynamic provisioning of the database tier of the software applications based on application-defined policies for satisfying their own SLA requirements. In this framework, the SLAs of the consumer applications are defined in terms of rules and goals that are subjected to a number of constraints that are specific to the applications' requirements. The framework continuously monitors the application-defined SLA and automatically triggers the execution of necessary provisioning actions when the conditions of the rules are met. Hence, the software applications have more flexibility in defining their own lazy or eager provisioning rules.

46.5 Discussion and Conclusion

This chapter presented an overview of the state of the art of existing technologies hosting the database tier of software applications in cloud environments. Table 46.2 summarizes some of the tradeoffs between the different cloud-hosted database technologies. Based on our discussion, we can make the following recommendations:

- NoSQL systems are viable solutions for applications that require scalable data stores, which can easily scale out over multiple servers and support flexible data model and storage scheme. However, the access pattern of these applications should not require many join operations and can work with limited transaction support and weaker consistency guarantees. In general, NoSQL systems are recommended for newly developed applications but not for migrating existing applications which are written on top of traditional relational database systems. For example, Amazon Web Services describe the antipatterns for using its

Table 46.2 *Tradeoffs of different cloud-hosted database technologies*

Technology / criteria	NoSQL	DaaS	Virtualized database servers
Ease of adoption / migration for existing databases	Low (more appropriate for developing new applications)	Moderate (migration process can be straightforward if a cloud service of the underlying database system is available)	High (database server can be migrated easily to a virtual machine like any other software component)
Ease of management for the hosted data	High (the service provider takes care of all the management of the hosted data)	High (the service provider takes care of all the management of the hosted data)	Low (user needs to take care of all the management of the hosted database in a virtualized environment)
User control of the elasticity and scalability of the allocated computing resources	Low (the user has no influence on the behavior of the underlying system / service)	Moderate (the user has limited control in configuring the service according to the requirements of his application)	High (the user has full control in configuring the service according to the requirement of his application)

cloud-hosted NoSQL solution, SimpleDB, to include predeveloped software applications, which are tied to traditional relational database or applications that may require many join operations and complex transactions. In addition, with the wide variety of currently available NoSQL systems, software developers need to understand thoroughly the requirements of their application in order to choose the NoSQL system with the appropriate design decisions for their applications.

- Database-as-a-service solutions are recommended for software applications that are built on top of relational databases. They can be migrated easily to cloud servers and alleviate the need to purchase expensive hardware, deal with software upgrades and hire professionals for administrative and maintenance tasks. However, these applications should have the ability to accurately predict their application workloads and provision the appropriate computing resources so that they can achieve their performance requirements. Unfortunately, these applications should be ready not to automatically leverage the elasticity and scalability promises of cloud services.
- Virtualized database servers are recommended for software applications that need to leverage the full elasticity and scalability promises of cloud services and need to have full control of the performance of their applications. However, these application need to build and configure their admission control for managing the database tier of their applications.

In practice, users of cloud database services often have the challenge of choosing the appropriate technology and system that can satisfy their specific set of application requirements. A thorough understanding of current cloud database technologies is therefore essential for dealing with this situation. In addition, a set of research challenges has been introduced by the cloud computing paradigm; these need to be addressed in order to ensure that the vision of designing and implementing successful management solutions in the cloud environment can be achieved. We shed the light on some of these challenges as follows (Sakr, 2014):

- *True elasticity*. To unleash the power of the cloud computing paradigm, cloud database systems should be able to manage and utilize the elastic computing resources transparently to deal with fluctuating workloads. The commercial NoSQL cloud offerings (e.g. Amazon SimpleDB) and commercial DaaS offerings (e.g. Amazon RDS, Microsoft SQL Azure) do not provide their users with any flexibility to dynamically increase or decrease the allocated computing resources of their applications. While NoSQL offerings claim to provide elastic services for their tenants, they do not provide any guarantee that their provider-side elasticity management will provide scalable performance with increasing workloads. Moreover, commercial DaaS pricing models require their users to predetermine the computing capacity that will be allocated to their database instance as they provide standard packages of computing resources (e.g. Micro, Small, Large and Extra Large DB Instances). In practice, predicting workload behavior (e.g. arrival pattern, I/O behavior, service-time distribution) and consequently accurate planning of the computing resource requirements with consideration of their monetary costs are very challenging tasks. Users might therefore still tend to overprovision the allocated computing resources for the database tier of their application in order to ensure satisfactory performance for their workloads. As a result of this, the software application is unable to fully utilize the elastic feature of the cloud environment.

- *Live migration.* Live migration is an important component of the emerging cloud computing paradigm, which provides extreme versatility for management of cloud resources by allowing applications to be transparently moved across physical machines with a consistent state. It represents an important tool for achieving elasticity and dynamic provisioning. It is also used to ensure availability by tenants migrating to other servers when the host server is planned to go down for maintenance. Moreover, it can be used to consolidate multiple tenants onto a relatively idle server, which alleviates the need for extra servers, which can be shut down, thus reducing operating costs. However, live migration is a resource-intensive operation and can come at the cost of degraded service performance during migration due the overhead caused by the extra CPU cycles, which are consumed on both the source and the destination servers in addition to the extra amount of network bandwidth, which is consumed for the transmission process. In principle, live migration of databases in a timely fashion is a challenging task. In addition, there is a tradeoff between the migration time, the size of the database, and the number of update transactions in the workload that are executed during the migration process. Furthermore, in a multitenancy environment, the challenges of deciding which tenant to migrate and where (to which server) this tenant should be migrated to remain open issues for further investigation and careful consideration.

- *SLA management.* In general, SLA management is a common general problem for the different types of software systems that are hosted in cloud environments for different reasons, such as the unpredictable and bursty workloads from various users in addition to the performance variability in the underlying cloud resources (Cooper *et al.*, 2010). The state-of-the-art cloud databases do not allow the specification of SLA metrics at the application or at the end-user level. In practice, cloud service providers guarantee only the availability (uptime guarantees), but not the performance, of their services. In general, adequate SLA monitoring strategies and timely detection of SLA violations represent challenging research issues in the cloud computing environments. In practice, traditional cloud monitoring technologies (e.g. *Amazon CloudWatch*) focus on low-level computing resources (e.g., CPU speed, CPU utilization, I/O disk speed). In general, translating the SLO of software applications to the thresholds of utilization for low-level computing resources is a very challenging task and is usually done in an ad hoc manner due to the complexity and dynamism inherent in the interaction between the different tiers and components of the system. Furthermore, cloud service providers do not automatically detect SLA violations and leave the burden of providing the violation proof on the customer. It therefore becomes a significant issue for the cloud consumers to be able to monitor and adjust the deployment of their systems if they intend to offer viable service-level agreements (SLAs) to their customers (end users). It is an important requirement for cloud service providers to provide cloud consumers with a set of facilities, tools and framework that ease their job of achieving this goal effectively.

- *Benchmarking.* In principle, benchmarks need to play an effective role in empowering cloud users to make better choices regarding the systems and technologies that suit their application's requirements. In general, designing a good benchmark is a challenging task due to the many aspects that should be considered and can influence the adoption and the usage scenarios of the benchmark. We believe that it is important that cloud users become able to paint a comprehensive picture of the relationship between the capabilities of the different type of cloud database services, the application characteristics and workloads, and the geographical distribution of the application clients and the underlying database replicas. As yet, the literature does not contain any comprehensive assessments and measurements of the performance, scalability, elasticity or consistency guarantees of the different categories of cloud database services. This is a clear gap that should attract more attention from the research community.

References

Abadi, D. J. (2009) Data management in the cloud: Limitations and opportunities. *Bulletin of the IEEE Computer Society Technical Committee on Data Engineering* **32**(1), 3–12.

Agrawal, D., El Abbadi, A., Emekci, F., and Metwally, A. (2009) Database management as a service: Challenges and opportunities. IEEE 25th International Conference on Data Engineering (ICDE), pp. 1709–1716.

Armbrust, M., Fox, A., Griffith, R., *et al.* (2009) *Above the Clouds: A Berkeley View of Cloud Computing.* Technical report UCB/EECS-2009-28, University of California, Berkeley, CA.

Brewer, E. A. (2000) Towards robust distributed systems. Proceedings of the 19th Annual ACM Symposium on Principles of Distributed Computing, July 16–19, Portland, Oregon, p. 7.

Cattell, R. (2010) Scalable SQL and NoSQL data stores. *SIGMOD Record* **39**(4), 12–27.

Cecchet, E., Singh, R., Sharma, U., and Shenoy, P. J. (2011) *Dolly: Virtualization-Driven Database Provisioning for the Cloud.* Proceedings of the International Conference on Virtual Execution Environments (VEE). ACM, pp. 51–62.

Chang, F., Dean, J., Ghemawat, S., *et al.* (2008) Bigtable: A distributed storage system for structured data. *ACM Transactions on Computer Systems* **26**(2), 1–26.

Cooper, B. F., Silberstein, A., Tam, E., *et al.* (2010) *Benchmarking Cloud Serving Systems with YCSB.* Proceedings of the First Symposium on Cloud Computing (SOCC 10). ACM, pp. 143–154.

Curino, C., Jones, E. P. C., Popa, R. A., *et al.* (2011) Relational cloud: A database service for the cloud. Proceedings of the Fifth Biennial Conference on Innovative Data Systems Research (CIDR), pp. 235–240.

DeCandia, G., Hastorun, D., Jampani, M., *et al.* (2007) Dynamo: Amazon's highly available key-value store. Proceedings of 21st ACM SIGOPS Symposium on Operating Systems Principles (SOSP '07), pp. 205–220.

Minhas, U. F., Yadav, J., Aboulnaga, A., and Salem, K. (2008) Database systems on virtual machines: How much do you lose? Proceedings of the IEEE 24th International Conference on Data Engineering Workshop (ICDE'08), Cancun, Mexico, pp. 35–41.

Pritchett, D. (2008) BASE: An ACID alternative. *ACM Queue* **6**(3), 48–55.

Sakr, S. (2014) Cloud-hosted databases: Technologies, challenges and opportunities. *Cluster Computing* **17**(2), 487–502.

Sakr, S. and Liu, A. (2012) SLA-based and consumer-centric dynamic provisioning for cloud databases. Proceedings of the IEEE Fifth International Conference on Cloud Computing (CLOUD), pp. 360–367.

Sakr, S., Liu, A., Batista, D. M., and Alomari, M. (2011) A survey of large scale data management approaches in cloud environments. *IEEE Communications Surveys and Tutorials* **13**(3), 311–336.

Soror, A. A., Minhas, U. F., Aboulnaga, A., *et al.* (2008) *Automatic virtual machine configuration for database workloads.* Proceedings of the 2008 ACM SIGMOD International Conference on Management of Data. ACM, pp. 953–966.

Stonebraker, M. (2008) One size fits all: An idea whose time has come and gone. *Communications of the ACM* **51**(12), 76.

Vogels, W. (2008) Eventually consistent. *ACM Queue* **6** (6), http://queue.acm.org/detail.cfm?id=1466448 (accessed January 2, 2016).

Zhao, L., Sakr, S., and Liu, A. (2014) A framework for consumer-centric SLA management of cloud-hosted databases. *IEEE Transactions on Service Computing* **1**, 1.

47

Cloud Data Management

Lingfang Zeng,[1] Bharadwaj Veeravalli,[2] and Yang Wang[3]

[1] *Huazhong University of Science and Technology, China*
[2] *National University of Singapore (NUS), Singapore*
[3] *Shenzhen Institute of Advanced Technology, China*

47.1 Introduction

Cloud data management (CDM) offers significant advantages and is already favored by many enterprises as a first choice. The world's major IT companies, such as IBM, HP, Cisco, Microsoft, and Google, have set up their own datacenters to provide CDM solutions. Several CDM systems such as GFS (Ghemawat *et al.*, 2003), BigTable (Chang *et al.*, 2006), HDFS (Borthakur, 2013), HBase (http://hbase.apache.org/, accessed January 18, 2016), Dynamo (DeCandia *et al.*, 2007), SimpleDB (http://aws.amazon.com/simpledb/, accessed January 18, 2016), S3 (http://aws.amazon.com/s3/, accessed January 18, 2016), Cassandra (http://cassandra.apache.org/, accessed January 18, 2016), Azure (http://www.windowsazure.com/en-us/, accessed January 18, 2016), and PNUTS (Cooper *et al.*, 2008) are currently being used. Energy-saving measures, virtualization, secure data handling, server consolidation, intelligent control and management, and new datacenter concepts will have a profound impact on the future of CDM. Some of the key characteristics of CDM are parallelism and high-performance computing, data backup and data protection technology, virtualization, and effective data organization and datacenter management.

47.1.1 Parallelism and High-Performance Computing

Cluster computing and multicore processors, which accelerated the development of parallel applications, are widely used in large datacenters. Although parallel computing has solved many bottlenecks, the slow memory I/O system leads to a performance bottleneck. As datacenters need huge amounts of storage, the storage subsystem should employ the parallel technology. Network File System (NFS) (Tykhomyrov and Tonkonog, 2002) has a new standard (NFS version 4.1) including an expansion of Parallel NFS (pNFS) (Shepler *et al.,* 2003), which has a high transmission rate.

47.1.2 Data Backup and Data Protection against Failure

Redundant Array of Independent Disks, or Redundant Array of Inexpensive Disks (RAID) (Patterson *et al.,* 1988) was the first technology to protect data storage systems against failure. RAID levels 1, 3, and 5 use different architecture for the protection of data on the hard drives. The latest RAID6 technology can tolerate two disk failures (data corruption), further increasing data protection. Data snapshots are the most commonly used data backup and recovery technology. For example, Copy-On-Write (COW) and Continued Data Protection (CDP) provide block-level or file-level data protection. The latest file system technology consists of the log-based file systems (for example, Log-Structured File System, Ext3, ReiserFS, XFS, JFS, WAFL), which track content change. If the entire write operation is interrupted for some reason (such as system power down), the system is restarted according to the log – the system resumes from the interrupted operation.

In the industrial sector, data disaster recovery has also been widely used; products for data protection include IBM's FlashCopy / Metro and Mirror / Global Copy / Global Mirror / PPRC (http://www-03.ibm. com/systems/storage/software/, accessed January 18, 2016), HP's OpenView/CASA/XP CA/EVA CA (http:// www8.hp.com/us/en/software/enterprise-software.html, accessed January 18, 2016), and EMC's TimeFinder and SRDF (http://www.emc.com/index.htm, accessed January 18, 2016). Data deduplication, a lossless compression technology, can reduce the amount of storage space required and can reduce storage costs significantly. Deduplication stores data by calculating the dataset repetition rate for duplicate data in a storage system; it stores only one copy of the data through the data's reference identification count.

47.1.3 Virtualization

Traditionally, the concept of virtualization only involved a server and storage but now it has extended to include I/O, desktops, and unified communications. Storage virtualization is the abstraction of storage services from the underlying hardware resources. Virtualization allows a large number of storage resources to be consolidated into a single storage pool; the end users will not see a specific storage disk or tape, and will not care about where their data is actually stored (or how data travels through to a specific storage resource). The virtual storage pool is a centralized management point where resources are allocated dynamically according to an application's demands. It is difficult to integrate a datacenter's applications into other applications without virtualization.

An example of virtualization technology is the tape library. It can be used like a disk array with low latency, and this is widely applied in today's virtual tape library (VTL) where it plays an increasingly important role. Putting scattered storage resources together into a virtual "storage pool" can improve the system's overall efficiency while potentially reducing the system's administration cost. Storage virtualization can also support resource allocation functions with resource partitioning and distribution capacity. Based on service-level agreements (SLAs), it can integrate the requirements of the storage pool to be divided among different stakeholders to fulfill the applications' performance and capacity needs. Many storage vendors use different ways to achieve storage virtualization capabilities: host-based virtual memory, virtual memory-based storage devices, and network-based virtual storage.

47.1.4 Effective Data Organization and Datacenter Management

Datacenters have the capacity to handle a huge amount of data. Storing, organizing, and searching in a huge data set needs an efficient data-management system. A distributed file system (such as GFS) is a large-scale data-management system that can store and organize data in an effective way. An object-oriented storage system (OBS) stores data as an abstract object; the data attributes and operations are bound together within the target range storage. This approach greatly simplifies and improves data management and content-based storage such as content-addressable memory (CAS) in object storage systems. In the CAS, the representation of the data object is a globally unique numeric identifier referred to as a digital fingerprint. A common approach is based on the data contents of the fixed-length hash, which is calculated to replace the file name. For the network storage system, clients simply use the numeric identifier to access the content. Well known content-based storage (prototype) systems are the Venti network storage system (Quinlan and Dorward, 2002), developed by Bell Labs; Deep Store archival storage systems, developed by the University of California and the Intel development center, and the CASPER distributed file system (Tolia *et al.*, 2003).

One of the current issues faced by datacenters is the management of unstructured data such as video, audio, images, XML documents, and Web pages. Attributes of unstructured data are different from traditional structured data. Unified storage and content management systems, such as HP's unified storage systems, provide a better solution for the management of unstructured data. Structured data can be managed by the Tsinghua Tongfang content management system (CMS).

47.1.5   Green Datacenters

The energy consumption of datacenters accounts for 40% of the energy consumption of the entire IT industry. A datacenter's power consumption depends on various aspects of the datacenter. At the macro level, it depends on the datacenter's location, construction, internal structure, wiring, and cooling system. At the micro level, power consumption depends on the servers and storage system. Energy-saving techniques such as virtualization may be applied to reduce their energy consumption.

47.2   Cloud Data Management Techniques: An Overview

In this section, we outline key cloud data management technologies such as unified storage resource management, data organization, and cloud storage gateway, and also highlight technical difficulties in cloud data management.

47.2.1   Data Management Techniques

47.2.1.1   Unified Storage Resource Management

Unified storage resource management can be divided into two major categories: storage resource virtualization and tiered storage management.

Storage resource virtualization of heterogeneous storage devices and subsystems can achieve block-level storage virtualization. Hierarchical management of heterogeneous storage devices can be used to manage data properly. Such systems are characterized by frequency-related acquisition, data management strategies, intelligent data migration, and dynamic adaptive rights management technology.

47.2.1.2   Data Organization

Currently, popular CDM systems are GFS and BigTable, HDFS and HBase, Dynamo and SimpleDB, S3, Cassandra, Azure, and PNUTS. These systems focus on the following five aspects: object-based massively parallel file-system architecture, high performance and high reliability of the metadata server cluster, rapid retrieval of metadata technology, high-performance local object file systems, and distributed object locking mechanisms:

- Object-based massively parallel file systems relate to the methodology of building an efficient PB-class storage system, storage system self-organization, self-management, self-testing and self-providing system security architecture, and a protection implementation mechanism.
- Metadata clustering technology relates to how the rational allocation of preserving metadata storage hierarchy at all levels (quantity, type, and scheduling policy) is conducted. An efficient cache algorithm of metadata and metadata access speed can improve user access performance.
- Metadata retrieval technology is mainly used to save storage space; data structures are used to organize and manage large volumes of information, metadata, and multidimensional attribute-based information, supporting quick search functions.
- A high-performance local object file system's main research fields are: object data indexing and storage, searching and managing free space, managing object metadata, and storing and retrieving of object properties to achieve long-term performance.
- A distributed object locking the storage system's main mechanism is the concurrency control mechanism. Using the metadata server, the object storage device and clients can work together to achieve tripartite concurrency control.

47.2.1.3 *Cloud Storage Gateway*

Application of cloud storage services is relatively limited as cloud storage has limitations in terms of its interface. Enterprises have many mature applications such as file sharing and data backup (built on storage devices). These systems only support the traditional storage interface (block interface such as iSCSI) or file interface (such as NFS and CIFS). Cloud-storage service providers often do not provide support for the traditional storage interface, but they use and support common REST (Fielding, 2000) or SOAP (Zur Muehlen, 2005) interfaces and protocols. The cloud storage gateway solves the interface's matching problem. A cloud storage gateway in the client side of cloud storage service is a special gateway device mainly used for the transformation between cloud storage protocols and traditional storage protocols.

47.2.2 Technical Difficulties

Cloud data storage presents the following technical difficulties:

- *Unified description of heterogeneous storage resources.* The practical application of the mass storage system function zoning of complex equipment (and a wide range of agreements) needs effective storage resource management. The uniform storage resource description, however, is difficult to achieve. There is therefore a need to adopt actively and absorb current international standards such as SNIA (http://www. snia.org/, accessed January 18, 2016), which is endorsed by the International Organization for Standardization.
- *Reliable preservation and fast retrieval of mass storage resource management information.* A lot of storage resource management information is described by a large number of storage nodes producing a unified model. The information includes the creation and expansion of virtual volumes. The information's accuracy and timely access are important. The mass storage resource management information helps to achieve reliable and rapid retrieval preservation.

- *Storage virtualization performance overheads.* In a virtualized environment, the system contains a number of different hardware and software characteristics. Different access methods are used to implement storage virtualization. This process can increase the complexity of storage virtualization software. On the other hand, various types of applications can access a highly complex mixed load concurrently. To ensure the system's storage efficiency, appropriate tools are needed to manage storage resources on demand; it can increase the system's operational overheads. Careful research and development of storage virtualization software is required to provide on-demand storage services with minimum system performance penalty.
- *Structured data and unstructured data integration.* While structured data is stored in the database, unstructured data is stored in the file system. In general, creating, retrieving, updating, and deleting a data resource is very complex, as is maintenance and the transactional consistency associated with the corresponding heterogeneous data sources.
- *Backup and recovery of deduplication-based systems.* Data deduplication has been a hot research topic in recent years. It has a broad and effective application setup in the backup and archive storage system. Deduplication can effectively remove duplicate data in the data flow, improve the utilization rate of storage, and save network bandwidth. However, there are still many problems to be solved in order to build a high-performance data-backup system using deduplication.
- *Key-value storage.* The most notable features of the big data era are that the amount of data is huge and there are many types of data, like structured and unstructured data. The traditional relational database does not have the capability to deal with such data. The massive growth gave birth to the development of NoSQL database, and the key value databases such as MongoDB (Membrey *et al.*, 2010), Redis (http://redis.io, accessed January 18, 2016), Tokyo Tyrand / Cabinet (Hirabayashi, 2010), and CouchDB (Anderson *et al.*, 2010). Key-value databases are generally for specific applications. However, maintaining the high performance-database (in the case of using a few system resources) is still a problem.

47.3 Design and Implementation

The objective of the hierarchical management approach for CDM is to achieve a unified resource management system for various heterogeneous devices. A smart datacenter supports many heterogeneous storage devices such as the self-development of high-end disk arrays, cost-effective heterogeneous disk arrays, tape libraries, and other low power-consumption storage devices. The object-based distributed file system can achieve effective organization of mass data. A smart datacenter, shown in Figure 47.1, is divided into five basic technical architecture levels: the device layer, the device interface layer, the base software application layer, the application programming interface (API) layer, and the application layer.

Key CDM technologies that support a datacenter are unified resource management, massive data storage organization, data security, and data resource integration, which are described below.

47.3.1 Uniform Resource Management

Under the guidance of the management strategy, storage virtualization can achieve data migration, replication, compression, and other management functions, and optimize resource utilization. Depending on the access frequency and feature mapping between the storage characteristics, the formation of intelligent storage resource-management strategies (based on the access frequency of stored data) can reduce unnecessary migration bumps. We propose an automatic data-migration program to ensure appropriate data at the right time. Administrators can manage the data in a timely manner to achieve rational and efficient use of storage resources.

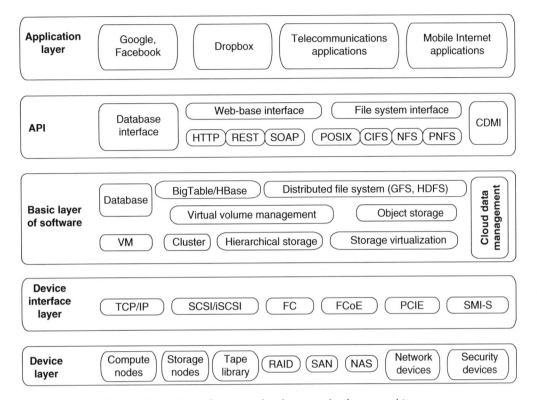

Figure 47.1   *Smart datacenter hardware and software architecture*

Mass storage systems are often deployed from different vendors' storage devices and subsystems; incompatibility is widespread in such cases. Although the industry has defined standards, the integrated management of a heterogeneous storage environment is still far from application in practice. Incompatibility can result in a tremendous waste of storage resources and excessive duplication of investment. It can also affect consistent data access and backup disaster recovery efforts.

47.3.2   Large-Scale Data Storage Organization

Mass data storage architecture, shown in Figure 47.2, is a typical parallel file-system structure consisting of massive parallel file system clients, metadata storage node clusters, and cluster components. Traditional parallel file systems are not well organized mass storage systems to meet large-scale, high-performance, scalable, highly available, and easy management on demand services. A typical object-based parallel file system mainly includes the client file system software, metadata, and the object storage server software system.

As shown in Figure 47.2, the metadata stored in the background by the metadata server is responsible for the back-end storage system. Metadata server requests arrive via the metadata service processing module. The unified view module is responsible for management of the space within the system storage node; the storage resource management module is responsible for the user's information management, quota management, and distribution of rights. The multiple system metadata server cluster management module (with the MDS) works to establish a mutual connection; the metadata lock module is used to ensure cache coherency, concurrent metadata operations, data consistency, and data caching of different nodes.

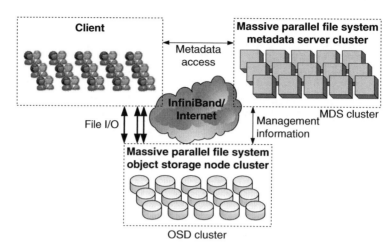

Figure 47.2   *Massive parallel file system architecture*

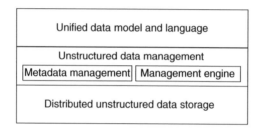

Figure 47.3   *Unstructured data management system architecture*

The object storage device (OSD) (Weber, 2010) consists of persistent objects in the underlying file system. It is responsible for the object's local preservation management. Data object cache file systems built on top of persistent objects are used to improve data access performance. The metadata server, similar to the data-logging module, is used to provide a reliability guarantee and cache coherency. The data object lock module ensures concurrent multiuser data access object data consistency. The cluster management module is used to coordinate multiple object storage device nodes; the object storage device and the metadata server command channel are used for communication.

The object-based parallel file system client and the virtual file-system interface are used to provide users with a POSIX-compliant access interface. The metadata cache and the data object cache module are used to improve access performance. The object lock module is used to provide coherency and consistency in distributed file system.

47.3.3   Data Resource Integration

For data resource integration, an unstructured data management system architecture, shown in Figure 47.3, uses a distributed object file system.

The bottom layer is used to achieve efficient unstructured data storage and fast retrieval. The logical management of unstructured data implements most of the current operation of unstructured data, corresponding to several different operations concerning unstructured data. The upper layer provides a unified operation. The following outlines its data model, user interface language, and data logic management.

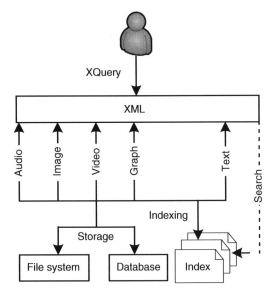

Figure 47.4 *Schematic data model*

47.3.3.1 Data Model

To represent unstructured data there must be a unified data model. One such scalable data model is shown in Figure 47.4.

The model first organizes unstructured data such as audio, images, video, graphics, and text; it creates XML metadata. The metadata includes file name, file type, storage path, index path, and creation time. XML metadata can be stored in the file system (or a database) using unstructured data metasearch. To facilitate a fast search and content-based retrieval of unstructured data, hybrid indexes such as text indexes, graphics indexes, and other indexes must be created. Through this model abstraction, any unstructured data can be stored in the system, and can be processed (based on metadata and content) to add, delete, query, or modify.

Further research and development in unified modeling unstructured data is required in key areas such as unified XML representation of unstructured data and content-based image, audio, video, graphics, and text indexing model.

47.3.3.2 User Interface Language

An unstructured audio data management system's user interface cannot be used to describe a table because of the general characteristics and the special nature of each of the audios. The proposed development can browse, search, and perform processing of audio content, enabling users to describe their query requirements easily and get the appropriate data. Users sometimes do not even know what to look for in audio data; users also do not know how to describe their queries. Hence, unstructured audio data-management systems (for user-interface requirements) not only receive the user's description, but also help users to describe their ideas, find content if required, and show it on the interface.

47.3.3.3 Data Logic Management

Data logic management is responsible for unstructured data management, system logical structure, and characterization management, providing the general and professional unstructured data management engine, and maintaining high data independence.

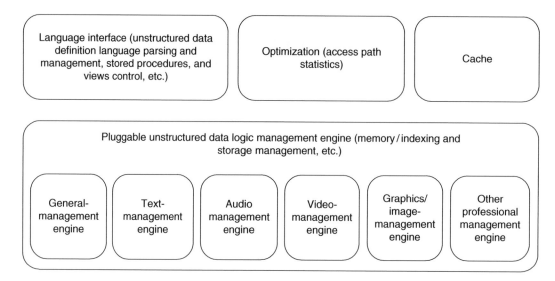

Figure 47.5   *Logical management of unstructured data*

The logical management of unstructured data can be further decomposed into two sublayers, as shown in Figure 47.5. The upper level of the application layer supports unstructured data query language parsing functionality for the application layer to access statistics, performance-optimization features, and system cache policy control functions. The application support layer, depending on the application, sends a request to establish local data structure and characteristics of the logical description, and generates an application associated with the logical view of the data.

The underlying logic of an unstructured data-management engine for the access layer can greatly enhance the data-management system's optimization potential and scalability. For mainstream unstructured data, such as text, audio, video, and graphics / images, it will provide a professional management engine. Different logical management engines may target specific types of unstructured data storage modes and access policies optimization to improve the efficiency of data storage and its access. The internal structure of the management data logic is shown in Figure 47.5. The data-access layer ultimately integrates the management module and returns to the application support layer. Diversity in the unstructured data storage mode is different from the traditional management system; it is an important feature of a structured database.

47.4   Conclusion

We have highlighted current and anticipated challenges in the handling of large data sets in the cloud and have outlined key state-of-the-art data management technologies for extremely large datasets. We outlined the main lessons learned in designing and building data-management solutions, presented the opportunities for deploying CDM. We discussed cloud computing infrastructure for big data storage and computing, services discovery, and content distribution, cross-platform interoperability, query processing and indexing in cloud computing systems, and structured and unstructured data management.

Further advances in research and development of CDM are bound to attract a large and varied pool of applications that use time- and mission-critical data.

References

Anderson, J. C., Lehnardt, J., and Slater, N. (2010) *CouchDB: The Definitive Guide*, O'Reilly Media, Sebastopol, CA.

Borthakur, D. (2013) *HDFS Architecture Guide*, http://hadoop.apache.org/docs/r1.2.1/hdfs_design.html (accessed January 18, 2017).

Chang, F., Dean, J., Ghemawat, S., *et al.* (2006) *Bigtable: A distributed storage system for structured data.* Proceedings of the Seventh USENIX Symposium on Operating Systems Design and Implementation. USENIX/ACM SIGOPS, http://static.googleusercontent.com/media/research.google.com/en//archive/bigtable-osdi06.pdf (accessed January 19, 2016).

Cooper, B. F., Ramakrishnan, R., Srivastava, U., *et al.* (2008) *PNUTS: Yahoo!'s Hosted Data Serving Platform.* Proceedings of the 34th International Conference on Very Large Data Bases (VLDB'08). ACM, http://www.cs.ucsb.edu/~agrawal/fall2009/PNUTS.pdf (accessed January 19, 2016).

DeCandia, G., Hastorun, D., Jampani, M., *et al.* (2007) *Dynamo: Amazon's Highly Available Key-Value Store.* Proceedings of the 21st ACM Symposium on Operating Systems Principles (SOSP'07). ACM, http://www.allthingsdistributed.com/files/amazon-dynamo-sosp2007.pdf (accessed January 19, 2016).

Fielding, R. T. (2000) Architectural Styles and the Design of Network-Based Software Architectures. PhD dissertation. University of California, Berkeley, CA.

Ghemawat, S., Gobioff, H., and Leung, S.-T. (2003) *The Google File System.* Proceedings of the 19th ACM Symposium on Operating Systems Principles (SOSP'03). ACM, http://static.googleusercontent.com/media/research.google.com/en//archive/gfs-sosp2003.pdf (accessed January 19, 2016).

Hirabayashi M. (2010) *Tokyo Cabinet: A Modern Implementation of DBM*, http://fallabs.com/tokyocabinet/ (accessed January 18, 2016).

Membrey, P., Plugge, E., and Hawkins, T. (2010) *The Definitive Guide to MongoDB: The Nosql Database for Cloud and Desktop Computing*, Apress, New York, NY.

Patterson, D., Gibson, G., and Katz, R. (1988) *A Case for Redundant Arrays of Inexpensive Disks (RAID).* Proceedings of the International Conference on Management of Data (SIGMOD), Chicago, IL. ACM, pp. 109–116.

Quinlan, S. and Dorward, S. (2002) Venti: a new approach to archival storage. Proceedings of the First USENIX Conference on File and Storage Technologies (FAST).

Shepler, S., Callaghan, B., Robinson, D., *et al.* (2003) *Network File System (NFS) Version 4 Protocol.* RFC 3530, https://tools.ietf.org/html/rfc3530 (accessed January 19, 2016).

Tolia, N., Kozuch, M., Satyanarayanan, M., *et al.* (2003) Opportunistic use of content addressable storage for distributed file systems. Proceedings of the USENIX Annual Technical Conference (ATC).

Tykhomyrov, O. and Tonkonog, D. (2002) Take command: Starting share files with NFS. *Linux Journal* **93**, 5.

Weber, R. (2010) *SCSI Object-Based Storage Device Commands – 3 (OSD-3)* (Working Draft), T10 Technical Committee Std. Project T10/2128-D, Rev. 02, July 2010, http://www.t10.org/members/w_osd3.htm (accessed January 18, 2016).

Zur Muehlen, M., Nickerson, J. V., Swenson, K. D. (2005) Developing web services choreography standards – the case of REST vs SOAP. *Decision Support Systems* **40**(1), 9–29.

48

Large-Scale Analytics in Clouds

Vladimir Dimitrov

University of Sofia, Bulgaria

48.1   Introduction

The term *large-scale analytics* means advanced analytics techniques applied to big data. This chapter discusses Hadoop as a tool for large-scale analytics.

Information is the data shaped into a form that is useful and meaningful. *Data* consists of streams of raw facts representing events occurring in the organization or in its environment (Laudon and Laudon, 2007). *Data analysis (data analytics)* is the process of shaping data into information. The hierarchy of "data, information, knowledge, and wisdom" (DIKW) defines three levels of data processing: information is about relationships among data; knowledge is information patterns; wisdom is knowledge principles (Baraglia *et al.*, 2010). At every level, connectedness grows with data understanding.

Russom (2011) defines *advanced analytics* as a collection of related techniques and tools used for data analysis. Advanced analytics includes predictive analytics, data mining, statistical analysis, complex SQL, data visualization, artificial intelligence, natural language processing, and database capabilities. Advanced analytics is also *discovery analytics* because it discovers information, knowledge, and wisdom from the data. *Big data analytics* is discovery analytics techniques applied to big data.

48.2   Hadoop and Data Analytics Tools

Apache Hadoop is the driving force behind the big data industry. It is an open-source, Java-based framework. It stores data (Hadoop Distributed File System – HDFS) and executes jobs (MapReduce) on large clusters of commodity servers. Hadoop is very fault tolerant. This framework is very simple but effective for a large class of big data applications. It is scalable from a single server to thousands of servers.

Encyclopedia of Cloud Computing, First Edition. Edited by San Murugesan and Irena Bojanova.
© 2016 John Wiley & Sons, Ltd. Published 2016 by John Wiley & Sons, Ltd.

The two main considerations in big data analytics are big data storage and big data processing. Traditional database systems have limitations on the number of columns, table size, and so forth. They accept only preformatted data for import. Big data does not meet such requirements. A big data file could store many terabytes and has billions of fields. Big data uses a special kind of file storage for data.

Traditional analytics tools use relational data and N-dimensional cubes. The first approach to big data is to extract data from the big data storage and format it for input to traditional analytics tools. The second approach is to develop big data analytic tools directly on big data storage – that is the way of Hadoop. Hadoop combines big data storage and big data analytics. Hadoop Distributed File System is a solution for big data storage and it is a subject of this chapter. Hadoop MapReduce is a solution for big data processing and is a subject of the next chapter.

48.3 Brief History of Hadoop

Doug Cutting (2012) created Hadoop. Mike Cafarella and Doug Cutting started the Apache Nutch project in 2002. Their idea was to create a Web search engine that could index and search one billion Web pages. This engine had to be deployed on half million dollars of hardware and with monthly running costs of $30 000 (Cafarella and Cutting, 2004). The main problem was the index storage. Meanwhile, in 2003, Google engineers published a paper on Google File System (GFS) (Ghemawat *et al.*, 2003). This inspired the Nutch project to create Nutch Distributed Filesystem (NDFS) in 2004.

In 2004, Google engineers Jeffrey Dean and Sanjay Ghemawat published a paper on MapReduce (Dean and Ghemawat, 2004). This influenced the Nutch project and, in 2005, MapReduce was available for Nutch. In 2006, Hadoop was initially a subproject and then, in 2008, it became a top-level project in Apache. The developers realized that the project solutions were useful in a broader area than Web search. Nowadays, many companies such as Yahoo! and Facebook use Hadoop. In 2013, Hadoop is the ultimate leader with 1.42 TB/min sorting.

48.4 Hadoop as Software

Hadoop's framework is not a Web search engine, sort utility, or file system. This chapter and the next try to describe it. Broadly speaking, Hadoop is an open-source implementation of GFS, Google MapReduce, and Google BigTable. Google engineers teach outside the company on Google technologies using Hadoop.

Hadoop is not suitable for grid computing. Grids consist of computing and storage nodes (hosts). Processing transfers data from storage to computing nodes. Grid data processing is long running. Data nodes store huge amounts of data. Data transfers among the nodes are not very frequent. Hadoop does not differentiate hosts in the grid way – one host can run several computing nodes and several data nodes. Computing nodes usually take their inputs from local data nodes (*data locality organization*).

Hadoop is not a stream database system. It does not support unlimited streams. Hadoop operates optimally on big files using data locality. The programmer can partition an unlimited stream into a sequence of files and store it on the Hadoop file system for further processing, but this is not a part of the framework.

Hadoop is open to clouds. It runs on commodity hardware. It can run on a cluster of virtual machines in a cloud. Hadoop is easily scalable for large clusters. It can use underlying cloud infrastructure to achieve better optimization. It is available on the leading clouds, such as Yahoo!, Amazon, and MS Azure.

48.5   Hadoop Components

Hadoop components are:

- *Common* – utilities supporting the other Hadoop modules, components and interfaces.
- *MapReduce (YARN)* – a framework for job scheduling and cluster resource management. It is a programming model and an execution engine running on clusters of commodity servers.
- *Hadoop Distributed File System* – a distributed file system running on clusters of commodity computers.
- *Avro* – a serialization system for efficient, cross-language RPC.
- *Sqoop* – a tool for transfer of data between structured data and HDFS.
- *ZooKeeper* – a distributed, highly available coordination service.
- *Oozie* – a service for running and scheduling workflows of Hadoop jobs.
- *Pig* – a data-flow language and an execution engine for big data.
- *Ambari* – a Web-based tool for provisioning, managing, and monitoring Hadoop clusters.
- *Hive* – a distributed data warehouse.
- *HBase* – a distributed, column-oriented database.
- *Cassandra* – a scalable multimaster database with no single points of failure.
- *Chukwa* – a data collection system for managing large distributed systems.

Common and MapReduce are the Hadoop core. All other components are optional. The HDFS plays a special role in Hadoop: it significantly facilitates the execution of MapReduce jobs.

MapReduce is a framework for parallel processing of large data sets. There are two phases in MapReduce: Map and Reduce. The Map phase applies a user-defined function on a key value pair and generates a list of key value pairs. The Reduce phase sorts all lists of key value pairs in lists, where the values list contains all values generated in Map phase for a specific key. A user-defined function generates a list of values. The next chapter gives more details about MapReduce.

48.6   Hadoop Distributed File System

The Hadoop flagship file system is HDFS (a successor to NDFS). Hadoop can link to any file system that supports its abstract file-system interface. It is an implementation of the abstract file system interface. It is suitable for big data analytics and has been optimized to write very big files once and then to read them many times but it is not suitable for random reads / writes. Hadoop complements traditional database systems – they support random I/O operations.

Hadoop Distributed File System stores very large files (many terabytes and petabytes). It could store tens of millions of files. It can run on hundreds or thousands of commodity servers. A general assumption about HDFS is that hardware failure is a norm – not an exception.

An HDFS file is a sequence of blocks stored in a cluster of multiple servers. Fault tolerance is at block level. HDFS blocks are big ones – 64 MB by default. If a file is smaller than a block, it shares the block with other small files.

Hadoop Distributed File System is designed for applications that access (streaming) data sets successively. It is not suitable for small files or for direct reads and writes. It has a high throughput. Interactive applications on HDFS are not the standard use case. Successful Hadoop applications use the MapReduce framework on HDFS data sets – they are batch applications.

Hadoop moves the computations to the storage nodes. It is the best approach for cases when the computing programs are relatively small and the stored data are big enough. In this case, the network traffic decreases. Hadoop Distributed File System can achieve its maximal parallelism if almost all the cluster servers are involved in the computations. However, the data set has to be big enough and distributed on almost all the

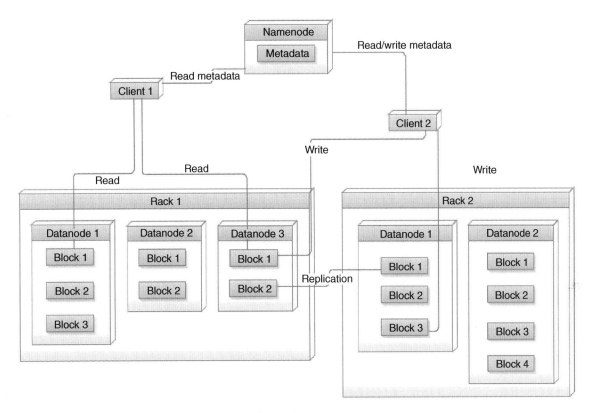

Figure 48.1   *HDFS architecture*

cluster servers. Hadoop Distributed File System is responsible for the file block distribution on the cluster servers. It tries to distribute the file blocks among the cluster servers. However, the file block size must be bigger than the size of the computing programs.

Figure 48.1 describes HDFS architecture. There are *namenode* and *datanode* nodes. The namenode contains metadata for files, directories, file block locations, and so forth. The datanode stores data blocks. The distance between nodes is in this hierarchy: first in the same host, then in the same rack, then in different racks.

The namenode is crucial for the file system's existence – if it fails, then the file system data is lost. The user can start a secondary namenode as an option. The namenode collects all the information needed to restart a new namenode in case of failure. In case of namenode failure, the system recovers in 30 minutes or more. That is an HDFS problem – availability. There are plans for future implementations to resolve this problem with a solution based on ZooKeeper. It automatically switches to the secondary namenode and fences the failed node for all operations that can damage the data.

HDFS Federation scales the cluster with new namenodes. Every namenode manages a different part of the file system – a *namespace volume*. The last one contains metadata information for the namespace and a block pool for the files in that namespace. Namespace volumes do not communicate with each other. One datanode can store blocks from multiple namespace volumes.

The client opens files or directories using the metadata from a namenode. After that, the file datanodes execute the operations. The read operations directly access datanodes in a sequential read access mode. If a read fails, then the datanode uses a block replica. The datanode reports the failed block immediately to its namenode. When HDFS has read all the data from a given block, it chooses the next block among all next block replicas. The replica must be closer to the client. Figure 48.2 describes an example sequence of read operations in HDFS.

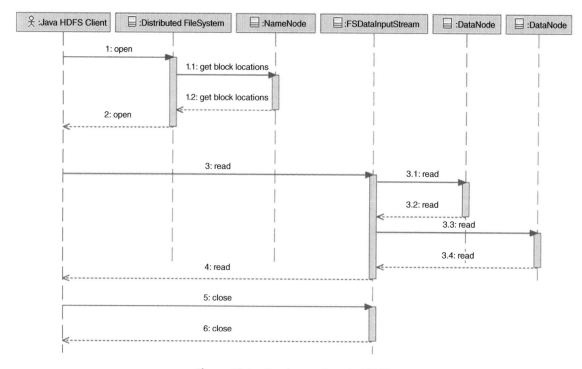

Figure 48.2   *Read operations in HDFS*

Hadoop Distributed File System performs write operations in sequential access mode. The system pipes the new block to the file end through the file datanodes. All write operations are queued in the client. The system pipes a notification backward to the client and removes this operation from the queue. If the write operation fails, then HDFS returns a notification to the client, and performs recovery actions: it removes the failed datanode and redoes all unacknowledged operations. Figure 48.3 describes an example sequence of write operations in HDFS.

HDFS provides some utilities, such as Data Ingest, for moving large stream data into HDFS and Sqoop for importing structured data stores.

Hadoop archives (HAR) package small files. These HAR files are an acceptable input for MapReduce programs, but they do not support compression. Hadoop archives are immutable – they are not updatable and the user must recreate the whole archive if he wants to change a file in it.

Hadoop is a Java implementation. It uses its own serialization format. It is compact and fast, but not open to the other programming languages. Avro is the serialization solution of Hadoop, based on JSON. It is extensible and interoperable.

48.7   Other Hadoop Components

Pig. Complex data analyses apply MapReduce processing many times, so the programmer has to write a MapReduce program for each stage. Pig is a solution for such situations. Pig Latin is a data-flow language for complex processing. It is open and extensible – programmers can replace all kinds of transformations with their own functions. This language is very compact with high expressive power. It dramatically reduces the development cycle. With Pig, the programmer's focus is on big data analyses – not on writing MapReduce programs. The system translates Pig Latin programs into series of MapReduce programs.

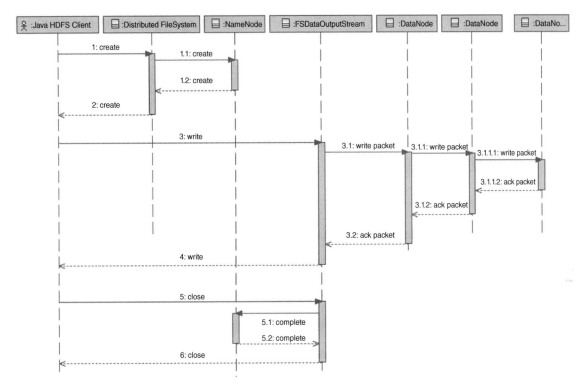

Figure 48.3   *Write operations in HDFS*

A Pig Latin program processes data in three steps. At the first step, it loads data from HDFS. Data sources could be HDFS files or directories but with a user-defined function; any other source is acceptable. At the second step, it transforms the loaded data. Transformations could be row-filtering, joins of two file sets, data grouping for further aggregation, sorting, etc. The third step is to display or store the results. The system can output the results on the screen or can store them in a file for further processing.

Hive is an original Facebook development. Hive query language is HiveQL, which is a SQL-92 dialect. HiveQL does not fully support the SQL-92 standard. Hive is very convenient for programmers with strong SQL skills. Standard analytics tools can use Hive as an interface to big data. The system translates HiveQL queries into MapReduce programs, which execute on the Hadoop cluster. Hive queries are long running and they are not suitable for interactive use. Hive is read oriented and is not suitable for write operations.

ZooKeeper is a Yahoo! development. It is a Java library for development of Hadoop coordination services for distributed applications. The programmer can create a communication system based on highly available loosely coupled interactions with the ZooKeeper. It facilitates development of coordination structures and protocols. ZooKeeper helps to avoid single points of failure. It is a centralized service for configuration support and naming. It provides distributed synchronization and group services suitable for coordination of distributed applications. There are Java and C interfaces with ZooKeeper. Figure 48.4 describes a ZooKeeper configuration.

Sqoop is a tool for the import / export of structured data into / from HDFS. Imported data could be loaded in Hive or HBase tables. Sqoop's import source is a table or an SQL query. Sqoop uses MapReduce for its import / export operations. These operations are maximally parallel and fault tolerant. Sqoop has many connectors to most popular relational databases and data warehouses.

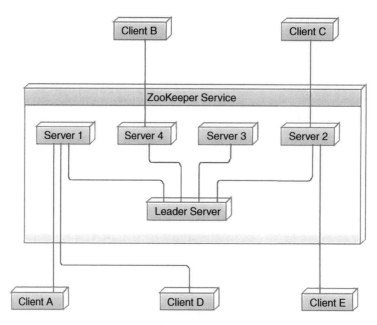

Figure 48.4   *ZooKeeper service*

HBase is a column-oriented database system based on HDFS. It is suitable for sparse data. It is nonrelational and does not support SQL. HBase applications are in Java. It provides Google BigTable capabilities for Hadoop. Figure 48.5 describes HBase architecture.

An HBase database is a set of tables that have columns and rows. Every table has a primary key that is used to access table rows. Every column in the row is an object attribute. A family groups several attributes. Only family attributes are stored together. The user must define the table schema and the families. At every time, the user can add new columns to the family.

HBase supports data compression, in-memory operations and column filtering. HBase tables could be input / output for MapReduce jobs.

Oozie is a workflow scheduler for Hadoop jobs. It is a workflow engine implemented as a Java Web application. Oozie organizes and coordinates the workflow specified as a direct acyclic graph (DAG). The DAG consists of action and control nodes connected to each other. The DAG graph is acyclic. Action nodes are MapReduce job, Pig, Hive, Sqoop application, DistCp task, shell script, SSH, HTTP, Java program or Oozie sub-workflow. Control nodes are start, end, fail, decision, fork, join. The control node accepts control flow and directs it to the other nodes.

Ambari is a framework for provisioning, managing, and monitoring Hadoop clusters. It is a collection of tools and APIs for system administration. Some of its features are wizard-driven installation of Hadoop services; configuration of Hadoop services and components; use of Ganglia for metrics collection and Nagios for system alerts; job diagnostic and troubleshooting tools; RESTful APIs for customization and integration; cluster heat maps. Figure 48.6 describes the Ambari architecture.

Chukwa is a system that collects monitoring data for big distributed systems. It is a toolset for monitoring, analysis, and visualization of collected data. Chukwa is scalable and reliable. Figure 48.7 describes Chukwa architecture. Agents collect data from different application sources, such as application logs, and application metrics. Collectors collect the information delivered from the agents and load it into data sinks. MapReduce components analyze the data and store it permanently in a structured storage. Chukwa detects problems on the cluster and alarms. There are plans for visualization tools for monitored data; however, these are still not available.

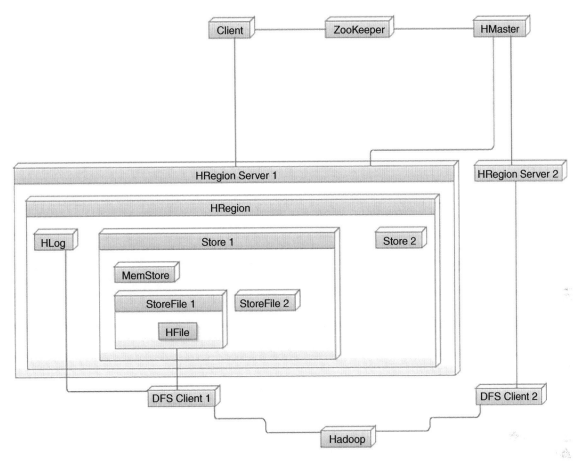

Figure 48.5 *HBase architecture*

48.8 Case Studies

There are many Hadoop case studies in different areas, like politics (Obama's re-election), financial services, health care, human sciences (NextBio), telecoms (China Telecom Guangdong, Nokia), entertainment (Orbitz), logistics (US Xpress), energy (Chevron, OPower), retail (Etsy, Sears), data storage (NetApp), software (SalesForce, Ancestry), imaging (SkyBox, Comcast), and online publishing (Gravity). For more details, see Hadoop Wiki. All Hadoop use cases demonstrate its scalability on large commodity clusters. The simple MapReduce framework is successfully applicable for large classes of big data analytics on different media files, logs, indexing, and so forth.

48.9 Hadoop in Clouds

Hadoop is like an open-source version of Google MapReduce and GFS. Google does not offer direct access to its MapReduce. Google stimulates the usage of Hadoop on its cloud platform. Detailed information about how to use Hadoop on Google cloud is available on the site of the Google Compute Engine, where

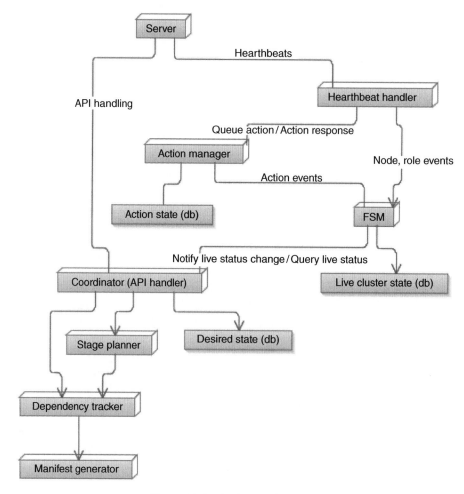

Figure 48.6   *Ambari architecture*

Google offers several scenarios for Hadoop. There is a connector for Hadoop to its cloud storage. Some researchers nominate Hadoop as a better solution than its Google origin.

Amazon Elastic MapReduce (Amazon EMR) is a Web service that offers Hadoop on Amazon EC2. Amazon EMR service starts in minutes. The service is responsible for Hadoop configuration. Data imports / exports from / to Amazon S3 are possible. There are detailed instructions and code templates for application development with Hadoop tools (Hive, Pig), and in different programming languages (Java, Ruby, and Perl). Cluster monitoring is available. Running cluster is easily expandable. When the job finishes, the system automatically removes the cluster.

Microsoft offers Windows Azure HDInsight that is 100% Apache Hadoop. It is a service for the deployment of Hadoop clusters in the Microsoft cloud. An alternative offer is the Hortonworks Data Platform (HDP) on Windows (from Hortonworks) for on-premises cloud deployment. HDInsight integrates with Microsoft Business Intelligence platform, with relational (MS SQL Server) and nonrelational databases via Polybase available with SQL Server 2012 Parallel Data Warehouse, with Tables (NoSQL key-value storage) and Blob (a storage for big objects). The user can create a Hadoop cluster

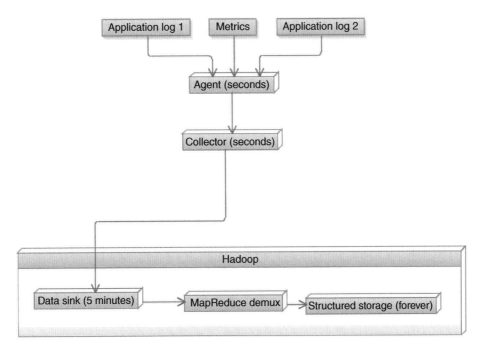

Figure 48.7 *Chukwa pipeline architecture*

in minutes by HDInsight on Microsoft Azure, and he/she can quickly remove the cluster. The Hadoop cluster runs on Linux virtual servers in Windows Azure. There are Hadoop command-line interface and a portal for Hive and Pig.

IBM Platform Symphony with IBM InfoSphere BigInsight is the IBM's grid (cloud) offer. IBM InfoSphere BigInsight is a Hadoop implementation. The company rewrote many Hadoop components for optimization and compatibility. Adaptive MapReduce is faster than the original implementation. Adaptive MapReduce optimizes the system workload using user-defined metrics.

Big SQL is an alternative to HiveQL. It works with big data via MapReduce but also directly queries small data. The General Parallel File System (GPFS) is an alternative to HDFS in this solution. It is a POSIX compatible distributed file system. It is the IBM offer for big-data storage.

A console for monitoring and access control is available in IBM InfoSphere BigInsight. Big Sheets is Web-based tool for analyzing and visualization of big data that is part of the offer. IBM InfoSphere BigInsight integrates InfoSphere Streams (stream processing), InfoSphere Data Explorer (multisource multitype data manager) and Cognos Business Intelligence.

48.10 Packaging

Like many open source projects, Hadoop is freely available but difficult to install – versions, subversions, releases, external libraries, and so forth, require a high level of Hadoop expertise. There are several deployment alternatives, freely offered by commercial vendors such as Cloudera, Microsoft, and Google. They offer a prepackaged version of Hadoop that is easily installable and deployable on the target environment.

Cloudera is a leading promoter of Hadoop. It is advisable to check its offers prior to Hadoop deployment. Cloudera is an initiator and contributor to many open-source projects. Its main expertise is in Hadoop.

Cloudera contributes to HBase, Hive, Pig, Hadoop LZO, HTrace, JCarder, JTrace, Jenkins, MooTools, Record Breaker, and the US FDA Adverse Drug Event System. Cloderians are cofounders of the next Apache open-source projects: Avro, Bigtop, Crunch, Flume, Hadoop, Hive, Lucene, MRUnit, Oozie, Sentry (incubating), Sqoop, Whirr, ZooKeeper; and other open-source projects: Cloudera Development Kit, Crepo, Hue, Impala, Kitten, ML, Seismic Hadoop.

48.11   Hadoop Alternatives

Hadoop alternatives do not use the MapReduce paradigm. Some of them are:

- HPCC Systems, which offers real-time query processing – not only big data batch analytics.
- Twitter / Backtype Storm, which processes unbounded streams.
- Pregel, which executes dynamic graphs of servers and programs.
- Spark, which combines SQL, streaming and complex analytics. It is suitable for machine learning.
- Microsoft Daytona, which is MapReduce rewritten for Windows and uses Azure storage services.
- MongoDB is document-based database management system (DBMS).

HPCC System can have several THOR or several THOR and ROXIE clusters. A THOR cluster (data refinery) is responsible for big data loading, transformation, and indexing. ROXIE cluster offers ad hoc query processing and data warehousing. Both clusters are on top of the distributed file systems with parallel processing. The programmer can use Enterprise Control Language (ECL) to program both clusters. It is a declarative data-flow language. Enterprise Control Language programs compile to C++. Figure 48.8 describes the HPCC architecture. HPCC uses commodity servers. HPCC Community Edition is free. It is the choice of companies such as LexisNexi@, Sandia National Laboratories, and Elsevier.

Twitter/Backtype Storm is a free open-source system for real-time computing of unbounded streams. Storm has a simple programming model that simplifies real-time parallel programming. Its implementation is in

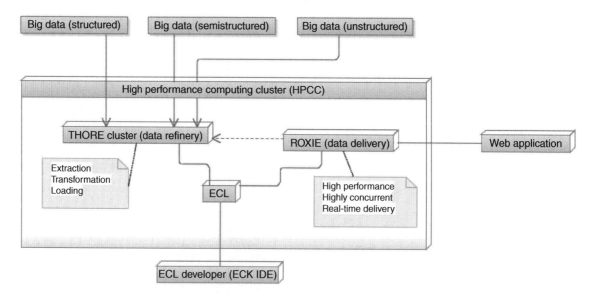

Figure 48.8 *HPCC architecture*

Clojure and it runs under JVM but the system can integrate many different programming languages, queues, and DBMSs in its computations. Storm is fault tolerant and supports horizontal scalability. It can be used for stream processing, continuous computing, distributed RPC, real-time analytics, online machine learning, ETL, and so forth.

Storm runs "topologies." There are master and worker nodes. Topology describes computations with a graph. The nodes of the graph represent processing logic and links among them (data flows). The programmer does not have to write the processing logic only in Clojure. A key abstraction in Storm is the stream. The stream is an unbounded flow of tuples. Storm provides primitives for stream processing. There are two kinds of primitives: "spouts" and "bolts." The spout is a stream source – internally or externally generated. The bolt consummates streams, processes them and eventually generates new streams. It can apply functions, filter tuples, call DBMS, join streams, and so forth. All nodes execute in parallel. The topology is running forever.

Pregel is a fault-tolerant framework for large-scale graph processing in parallel on many servers. Pregel computations are a series of iterations (supersteps) represented as an oriented graph. Every vertex has a unique identifier and is associated with a changeable user defined value. Every directed edge is associated with a source vertex, a user-defined value, and an identifier of its target vertex. During the execution, the graph topology is mutable, i.e. the vertex can change the destination vertexes. The framework calls in parallel the user-defined function for every vertex during the superstep. The function reads messages sent to the vertex in the previous step; sends messages to other vertexes; and changes the vertex state and eventually outgoing edges from the vertex. The programmer focuses only on the local processing in the vertex. The execution stops when all vertexes vote for a halt. When a vertex votes for a halt, the framework does not call it unless someone outside the framework calls the vertex.

The network sends the messages. Supersteps organize message sending in batches. Pregel aggregators organize global communications. Each vertex can send a message to the aggregator, which processes it and can produce as output a message to all vertexes. Pregel is available as C++ API.

Spark is an open-source computing environment. The University of California, Berkeley implemented it in Scala. Spark can integrate Hadoop. It offers a new framework for cluster computing. It is optimal for multiple reuse of read-only datasets from parallel operations. The framework caches the working datasets in the memory to prevent disk read latency. The main term in Spark is the resilient distributed dataset (RDD). It is a collection of read-only objects distributed among a set of nodes. If a part of the RDD fails, the system recreates it automatically. The RDD is a Scala object and Scala applications access it directly. The "drivers" are the Spark applications. The driver implements operations on a single node, or in parallel on multiple nodes. Spark uses a Mesos manager for resource sharing and isolation. There are two types of operations performed by the drivers: actions and transformations. Action processes a dataset and returns a value. Transformation produces a new dataset from an existing one. The user can write Spark applications in Java, Scala and Python.

Microsoft has attempted to deliver more functionality for big data through Windows Azure. There are two directions: storage and computations. Windows Azure data storage has two main solutions: SQL Azure and Azure Table Storage. Microsoft SQL Server is the base of SQL Azure. This is a classical relational database technology. Windows Azure Table Storage stores high volumes of data. It is a fault-tolerant key-value NoSQL storage. The key concept in Azure Table is the table. It is simply a container of rows. Every row has its own schema embedded in it. The row stores pairs of a property name and its typed value. Every row must have three properties: PartitionKey, RowKey, and TimeStamp. These properties control the table performance and its scalability. The PartitionKey is crucial for data distribution. The PartitionKey and RowKey form the table's primary key – the table index. The PartitionKey spreads the data on several servers for workload balancing. A table row size is up to 1MB with up to 255 properties. A table size could be up to 200 TB. Java, PHP, LINQ, and Python programs can query Azure Table. Azure Table supports ADO .NET, Data Series, and REST.

Project Daytona, MapReduce Runtime for Windows Azure, is freely available from Microsoft. It is a MapReduce variant implemented in Microsoft Azure. Daytona automatically deploys interactive MapReduce runtime on virtual servers. It divides data into small portions and distributes them on the servers for parallel Map processing. Then, eventually, it recombines the result (the reduce phase).

Daytona leverages the scalable compute and storage services of Azure. Windows Azure storage is available via input / output streams. The system performs parallel reads and writes via streams from tables, blobs, and queues. Intermediate data are stored in the memory or on the local disks. Daytona offers horizontal scaling and elastic cluster management. The user can easily add and remove servers to / from the cluster. Workers communicate with each other directly. Daytona uses Azure infrastructure services for robustness and dynamic scalability.

Windows Azure mechanisms for data recovery are used and this eliminates the need for a distributed file system. The system supports three replicas by default. Azure storage services support the dynamic data partitioning.

MongoDB is an open source document-oriented DBMS. It is horizontally scalable and fault-tolerant via data replicas. The system automatically recovers the data.

The records in MongoDB are the documents – data structures of field-value pairs. MongoDB documents are JSON-like objects. A field value could be other documents, an array of values or an array of documents. The documents can embed other documents. Indexes (secondary) optimize the access. The documents are stored in collections (groups of related documents with a set of shared indexes). Queries are on the collections, under selection criteria for the documents. A projection operation on document fields is possible – limits, skips, and sort orders are the other options. Data-modification operations are applicable to the collections. Update and delete operations can be done under selection criteria. Write operations are atomic at document level, including embedded documents.

MongoDB provides aggregation operations on single collections. These are pipelines and MapReduce. For MapReduce operations, user-defined JavaScipt operations are used. Alternatively, there can be a finalization phase for MapReduce operations. Aggregations are applicable to the "sharded" collections. These are collections distributed on the cluster servers by a "shard" key.

48.12 Conclusion

Hadoop started from web indexing, document sorting, multimedia processing, and log analysis. It, influenced by MapReduce, entered new areas such as machine learning, and unlimited streams (sensor data, user interactions). This process continues, but the power of this technology raises new ethical problems. It is clear that companies can use this technology to create detailed user profiles, breaching user privacy. However, these problems are beyond this chapter's scope.

Big data analytic tools are evolving in two directions: reuse of currently available tools, and development of new ones dedicated to big data. The first direction benefits from well-established tools for data analysis, but they are not scalable. The second direction is under development.

Undoubtedly big data is here; it contains valuable information and its analysis is a big challenge, requiring new decisions.

References

Baraglia, R., Lucchese, C. and De Francisci Morales, G. (2010) *Large-scale Data Analysis on the Cloud*, http://melmeric. files.wordpress.com/2010/05/large-scale-data-analysis-on-the-cloud-roma-cmg-2010.pdf (accessed January 4, 2016).
Cafarella M. and Cutting, D. (2004) Building Nutch: Open source search. *ACM Queue* **2**(2), http://queue.acm.org/detail. cfm?id=988408 (accessed January 4, 2016).

Cutting, D. (2012) *Intro to Hadoop and MapReduce*, http://www.udacity.com/course/viewer#!/c-ud617/l-306818608/ m-312934728 (accessed January 4, 2016).

Dean J. and Ghemawat, S. (2004) *MapReduce: Simplified Data Processing on Large Clusters*, http://static.googleusercontent. com/media/research.google.com/en//archive/mapreduce-osdi04.pdf (accessed January 4, 2016).

Ghemawat S., Gobioff, H. and Leung, S.-T. (2003) *The Google File System*, http://static.googleusercontent.com/media/ research.google.com/en//archive/gfs-sosp2003.pdf (accessed January 4, 2016).

Laudon K.C. and Laudon, J. P. (2007) *Management Information Systems – Managing the Digital Firm*, 10th edn. Pearson Education, Inc., Upper Saddle River, NJ.

Russom, P. 2011. *Big Data analytics*, TDWI Research, Renton, WA.

49

Cloud Programming Models (MapReduce)

Vladimir Dimitrov

University of Sofia, Bulgaria

49.1   Introduction

There are several cloud programming models: Message Passing Interface (MPI), Directed Acyclic Graph (DAG), Bulk Synchronous Parallel (BSP), and MapReduce.

MPI comes from supercomputing. It is a communication library. Distributed programs based on MPI use many processors that compute in parallel. They exchange messages. MPI strictly orchestrates its computations and communications. Communication time is relatively short in comparison to computation time. The distributed program scales if this relation between communication and computation times is preserved. The supercomputing environment (a supercomputer or cluster of supercomputers) supports that relation but clouds usually use commodity networks, and this relation is not always the case.

Directed acyclic graph (DAG) concept is derived from grid computing. They specify the tasks. Every vertex in such a graph performs some task. Tasks can be actions or controls. Every graph arc describes a control flow. The actions perform computations on local data. The controls define which vertex would be next. The control tasks can dynamically change graph topology. The grid transfers data to the actions. However, it is possible for data to be local but the model does not focus on that. The problem with this model is the big data: many servers have to store the big data and then mapping of vertexes and data distribution on servers is not a trivial job.

Bulk synchronous parallel (BSP) computing defines an abstract computer with many processors, a network that routes messages between pairs of processors, and a synchronization facility for all processors. In this model, *supersteps* divide the computations. In a superstep, the processors compute and exchange data in parallel. When a processor reaches a barrier, it waits for all other processors to reach the barrier too. This model is suitable for iterative graph processing, where data-dependencies across stages are sparse, such as in machine learning. However, it is not suitable for big data analytics.

The MapReduce model is suitable for big data analytics. The idea of MapReduce comes from functional programming. It uses two higher order Lisp functions: *map* and *reduce*. This chapter briefly reinvestigates the original MapReduce, which serves a base for further discussions. However, there are several extensions of MapReduce (beyond the scope of this chapter) that use not only *map* and *reduce* functions, but also *join, cross, match, union, cogroup,* and so forth.

Google (Dean and Ghemawat, 2004) introduced MapReduce as a term. The most popular MapReduce implementation is in the open-source project Hadoop. There are many implementations of MapReduce, and many variants influenced by the original MapReduce paradigm allow inexperienced programmers in distributed computing to write programs that run on large data sets stored on clusters of thousand servers. The programmer can test the program on a single computer and then can easily deploy it on a cluster.

49.2   Map and Reduce in Functional Programming

There are many research areas in functional programming: semantic-based program manipulations; parallel functional programming; hardware specification, synthesis and analyses; large applications with massive parallelism. There are many possibilities for parallel processing, especially in Lisp pure functional implementations. Moreover, a pure functional operating system is available for LispKit. It supports lazy evaluation as a basic mechanism. Lazy evaluation delays function evaluation. It computes the result only on some other function request. Some MapReduce implementations use this mechanism but it is not the leading idea in the computing framework discussed here.

There are very simple data types in functional programming – integers, strings, and so forth – such as the primitive data types in Java. The main data structure is the list. A list elements cannot only be primitive values (atomic values) but can also be a list, for example. In this case, a function accepts lists as arguments and returns a list. In addition, the function description is also a list (lambda expression). Functions can have other functions as arguments. There is a library of standard functions. They are list-processing functions. An example program in LispKit is as follows:

```
(lambda (input_stream)
  (append (quote (Example program))
    (cons newline (append (add_up (until_end input_stream)) (quote
    (Finished))))))
```

This lambda expression processes the list input_stream applying other standard and user-defined functions, and returns a list.

In pure functional programming, functions do not change input lists but they create new lists as a result. This approach opens an area of nearly unlimited parallelism for the functional programs. Potentially, the system can process all list elements in parallel.

Commonly used higher order functions are *map, reduce, filter,* and *close.* Their definitions in LispKit are as follows:

- (map f l) returns the list whose components are the function f applied to components of the list l;
- (reduce f l z) returns the continued application of function f over the list l with zero z, that is (f (head l)) (f (head (tail l)) (f … z) …));
- (filter p l) returns the list of those components c of the list l for which (p c) is true (T);
- (close r l) returns the first value x in the sequence l, (r l), (r (r l)), …, for which x is closure, i.e. (equal x (r x)).

The MapReduce programming model uses the first two functions, *map* and *reduce*. The *map* function has two arguments: a function (argument-function) and a list. The function *map* returns, as a result, a new list containing the consecutive applications of the argument-function on the input list elements. The argument-function may apply independently and in parallel on every element of the list. The *reduce* function has three arguments: an initial value, a function, and a list. The argument-function has two arguments – the first argument is the intermediate result and the second argument is the rest of the argument-list. For the first element, the argument-function applies with the initial value, for the second element it applies with the result returned from the previous step, and so on. The result of the *reduce* function is the result of application of the argument-function to the last element of the argument-list. However, the *reduce* functions cannot execute in parallel, because every execution step depends on the result of the previous one. The MapReduce programming model uses these two functions with some deviations.

49.3   MapReduce Programming Model

The MapReduce programming model presented here follows the original paper (Dean and Ghemawat, 2004) and its implementation in Hadoop (Apache Software Foundation, 2014). MapReduce computation takes on input a set of key / value pairs and produces on output a set of key/value pairs. The computation runs into two phases: Map and Reduce. The programmer has to write two functions: an argument-function for Map and an argument-function for Reduce. The first function name is *map* but, in reality, it is the argument-function of the high-order function *map*. It accepts as a parameter key / value pair and produces a set of intermediate key / value pairs. The MapReduce framework groups all intermediate values associated with the same key and delivers them to the Reduce phase. The second user-defined function is called *reduce,* which again is an argument-function for reduce. It accepts on input an intermediate key and a set of all intermediate values associated with it. This function, eventually, generates a smaller set of values. Usually, the set is empty or contains only one value. An iterator delivers the intermediate values to the *reduce* function. The set of intermediate values can be bigger than the main memory can store. Figure 49.1 presents an example of *map* and *reduce* functions in pseudo code, which the original paper introduced. This example counts the number of occurrences of each word in a large collection of documents.

The phase Map is, in practice, implementation of the Lisp *map* function without the preparations for the next phase. This phase has maximal parallelism. Moreover, the sort operations can run in parallel with *map* applications; the framework could directly deliver every calculated intermediate value to the corresponding intermediate key list, and Reduce can run in lazy-evaluation manner. Strictly speaking, the sort of intermediate

```
map(String key, String value):
// key: document name
// value: document contents
  for each word w in value:
    EmitIntermediate(w, "1");

reduce(String key, Iterator values):
// key: a word
// values: a list of counts
int result = 0;
for each v in values:
  result += ParseInt(v);
Emit(AsString(result));
```

Figure 49.1   *A MapReduce process*

key / values is a part of the Reduce phase. However, the parallelism of the Map phase may be compromised if the user-defined *map* function is not purely functional – if it has side effects.

The phase Reduce is very different from the Lisp function *reduce*. This phase does not produce a single value from a list and has no starting initial value. The reduction applies to the intermediate values lists. The reduction can apply in parallel for every intermediate key. In reality, the Reduction phase is the application of the function reduce on the value lists of every intermediate key. The reduction may run in parallel with Map phase – for every key an instance of *reduce* function can run in parallel, when the Map phase calculates a new value for the key, then the Reduce phase can deliver this value to the iterator of the corresponding *reduce* function instance. This is a very fine parallelism. The output from the *reduce* function usually contains one or zero values. However, there are no limitations – the result may be a list.

In the above presentation, the functional programming was the context of the concepts. For example, the sets of pairs are lists. The element order of the list is not important. However, this programming model in environments, such as C++ or Java, must support the strong typing. In the original paper, user defined functions signatures are:

```
map:    (k1, v1) -> list(k2, v2)
reduce: (k2, list(v2)) -> list(v2)
```

The input key and value types (k1 and v1) are different from the intermediate key and value types (k2 and v2). The intermediate key and value types are the same as the output key and value types. The framework signature is:

```
computation: list(k1, v1) -> list(k2, v2)
```

The framework accepts a list of key / value pairs on input. For each pair, it calls a *map* that generates a list of intermediate key / value pairs. The MapPhase signature is:

```
MapPhase: list(k1, v1) -> list(k2, v2)
```

In the Reduce phase, the framework sorts the lists produced during the previous phase, and the result is a list of intermediate keys / list of values. In this list, each key is associated with all the values generated for it. The preparation signature is:

```
preparation: list(k2, v2) -> list(k2, list(v2))
```

After that, the Reduce phase. For every pair in the list, the framework calls the function *reduce*, which produces a list of values for the key. These values are the same type as the one of the intermediate values. The original paper does not make it very clear what exactly generates the Reduce phase. It points out that the output key and value types are the same as that of intermediate ones. On the other hand, the function *reduce* generates only a list of values. Probably, the framework generates key / value pairs for every call of the function *reduce* using the input key and the generated value list. The reduction signature is:

```
reduction: list(k2, list(v2)) -> list(k2, v2)
```

Finally, the Reduce phase is:

```
ReducePhase = preparation; reduction
```

and its type is:

```
ReducePhase: list(k2, v2) -> list(k2, v2)
```

The MapReduce computations are composition of these two phases:

```
MapReduce = MapPhase; ReducePhase
```

49.4   Google MapReduce Implementation

There are several configurations for the MapReduce programming model: on a single computer with shared memory, on a multiprocessor system, and on a cluster of servers. The Google implementation is a cluster of commodity servers. It is a C++ template library.

The process of model application is as follows: the MapReduce divides the input into several splits of the same size. The first phase, the Map phase, processes the splits in parallel on different servers and delivers the output locally. The framework uses a user-defined parameter M to do this partitioning. The next phase, Reduce, separates and processes the intermediate output / input data on R servers. The user specifies the parameter R. The input may be a single file or a set of files. The user can set the split size from 16 MB to 64 MB. The number of splits, M, defines the number of cluster servers used in the Map phase.

The cluster servers have different roles. One of them has a leading role – it is the master and assigns map and reduce tasks. The other servers are workers (slaves) and they execute jobs assigned to them by the master. Every worker executes the Map phase in parallel on its input data: it reads and parses pairs of its split, calls the *map* function for every pair key / value and buffers *map*-generated results in the main memory. The framework saves the buffered intermediate key / value pairs on the local disks in R lists distributed by their keys. All pairs with the same key are on the same list. The distribution function can be a hash function on the key in the range 0..R-1. The master collects the locations of the local lists for further processing.

In the next Reduce phase, the master assigns reduce tasks to R workers and sends them the locations of their input data. Initially, the reduce worker reads all its data from the remote disks: the first worker reads all first lists, the second worker reads all second lists, and so forth. The workers use remote procedure calls for this reading. After that, the worker sorts its data in the key / values lists. If the data volume is big, the worker uses external sort program. Then, the reduce worker calls the *reduce* function for every element of its sorted list. The function *reduce* returns a list of values for every intermediate key / list of values. Then, the reduce worker generates a list of key / value pairs for every call. Each reduce worker generates an output. The Reduce phase can provide this output lists as files in a single directory.

Figure 49.2 describes a MapReduce process. The Reduce phase processes the intermediate lists in key order and generates the output in the same order.

49.5   Fault Tolerance and Determinism in Google MapReduce

Google MapReduce is a transparent fault-tolerant framework. The master stores information about all workers. This includes the state (idle, in-progress, completed) and identification of every worker server. The master stores the location and the size of respective intermediate lists for each map task. The master polls all the workers from time to time. If a worker does not respond after a fixed amount of time, the master decides that the worker has failed. The failed worker is marked as idle and is available for further processing. If the failed worker has not completed its map tasks, the master reassigns these tasks to another worker for

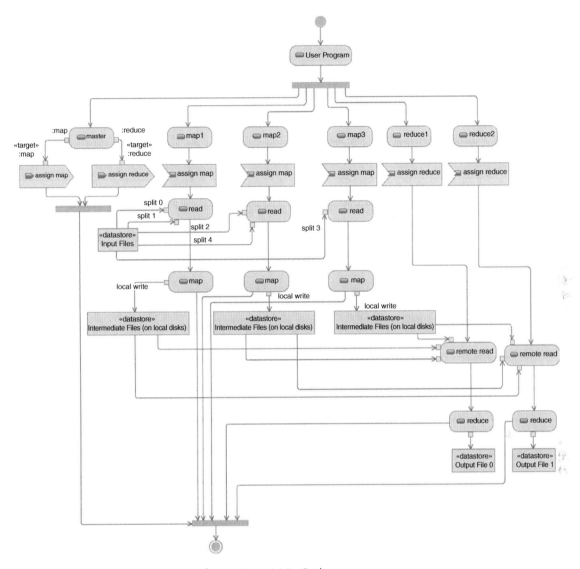

Figure 49.2 *A MapReduce process*

re-execution. The results of the completed tasks are stored on the local disks of the failed worker and therefore they are not accessible by default. When a map task fails, the master informs all reduce tasks that use its results and redirects them to the new worker. The master does not reassign the completed reduce tasks because their results are stored in the distributed file system. If the reduce tasks have not read its input data from the failed worker, they can do that from the new worker when it finishes the execution. If the master fails, the whole job fails. The user must restart its job.

Another important problem of the framework is it determinism – the programmer must pay attention to that. The functions *map* and *reduce* may be deterministic regarding their inputs. This means that the framework guarantees that the program will generate the same output each time – i.e. it will generate the same outputs on a single computer and on a cluster. Moreover, when these user-defined functions are nondeterministic, the framework still guarantees that, for each distributed execution, there exists at least one sequential execution that can produce the same outputs.

Input and output data are stored in the Google File System (GFS). This stores the files in blocks of 64 MB and each block has several replicas (three, by default). The master takes into account the files' locations and tries to execute the map task on the server that stores a replica of the map task input data. If this is impossible, then the master looks for a server that is closest to the input data. Such a nearby server can use the same network switch. This way the framework does not overload the network with unnecessary data transfers.

Sometimes, some servers fail to finish their tasks in a reasonable time. Hardware or software problems prevent them from normal execution. In this case, the overall program execution is delayed. The master prevents these situations from happening by starting "backup" tasks in parallel when a few map or reduce tasks remain at "in-progress" state for long time. When a backup or a normal task finishes, the whole task is marked as completed. This approach allows an acceptable execution time to be achieved for the overall job through a small increase of resources over use for the backup jobs.

49.6   Intermediate Results and Combiners

For optimization purposes, the framework offers the combiners mechanism, which works on the intermediate results. The framework sorts and processes the intermediate key / value pairs in each partition in key order. The system sorts the intermediate lists and the output lists by the intermediate key. Sometimes, the map tasks generate outputs with many repeating keys. If the function *reduce* is commutative and idempotent, a combiner function can be used. An example of such a *reduce* function is the word count *reduce* function.

The *combiner* function can merge the data before sending it over the network. This function runs on the worker after the map task. The framework decides whether or not to use the supplied *combiner* function. Usually, the *combiner* function's code is the same as that of the *reduce* function. The difference is that the *combiner* writes on the local disk, whereas the *reduce* writes on the distributed file system. This approach relieves the network from unnecessary transfers.

49.7   Framework inputs and outputs

The MapReduce library can read the input data in several formats. For example, in text mode, every line is a key / value pair, where the key is the line offset in the file and the value is the line contents. Each input mode knows how to split the input in such a way that a single map processes one split. In the text mode, the system partitions the input on line boundaries. Programmers can add new input modes implementing a simple reader interface. The set of input modes is rich enough and this option is rarely used. It is not obligatory for the MapReduce input to be only from files. It is possible for the reader to read database records.

The output modes work in the same way. The programmer can produce some additional files as output – not only the standard ones. This can be done in map and reduce tasks. These additional files are outside the framework and they may cause side effects. The programmer is responsible for these side effects being atomic and idempotent. These tasks have to be deterministic or they compromise the framework.

49.8   Additional features

The functions *map* or *reduce* may fail in reading or processing some records. That is why the library can install a signal handler in every worker. When the user-defined function fails for a given record, the worker notifies the master. If it happens more than once for a given record, the master instructs worker to skip that

```
#include "mapreduce/mapreduce.h"
// User's map function
class WordCounter : public Mapper {
  public:
    virtual void Map(const MapInput& input) {
      const string& text = input.value();
      const int n = text.size();
      for (int i = 0; i < n; ) {
        // Skip past leading whitespace
        while ((i < n) && isspace(text[i]))
          i++;
        // Find word end
        int start = i;
        while ((i < n) && !isspace(text[i]))
          i++;
        if (start < i)
          Emit(text.substr(start,i-start),"1");
      }
    }
};

REGISTER_MAPPER(WordCounter);
```

Figure 49.3   *Class mapper implementation*

record. This approach is applicable when the program collects some kind of statistics on the data. In that case, several skipped records will not influence the statistics very much.

The MapReduce program runs on a single computer for testing and debugging. In this execution mode, the programmer can use debugging tools.

The counters are another interesting feature of the MapReduce library. Some of them can be defined in the *map* and *reduce* functions, others are automatically embedded in the program. The master periodically receives the information from the counters, aggregates it and sends it back to the workers. Counters are useful for program execution control.

Figures 49.3, 49.4 and 49.5 give an example program using the MapReduce library. It is from the original paper.

49.9   Google Use Cases

Google uses the programming model MapReduce for indexing, information clusterization, machine learning, data extraction for report generation for popular queries, retrieval of properties from Web pages, and graph computations. The model succeeds because it is applied even by programmers who have no experience with parallel and distributed systems because the details on parallelization, fault tolerance, localization optimization, and load balancing are hidden from the programmer. The implementation is easily scalable to clusters of a thousand servers and it effectively uses their resources.

The limitations set on the programming model simplify computing parallelization and distribution, making it fault tolerant. The intensive use of the local disks for intermediate read / write operations optimizes the network bandwidth. Redundant computing allows the impact of the slow servers to be overcome. It protects the program from server faults and loss of data.

```
// User's reduce function
class Adder : public Reducer {
 virtual void Reduce(ReduceInput* input) {
   // Iterate over all entries with the same key and add the values
   int64 value = 0;
   while (!input->done()) {
     value += StringToInt(input->value());
     input->NextValue();
   }
   // Emit sum for input->key()
   Emit(IntToString(value));
 }
};

REGISTER_REDUCER(Adder);
```

Figure 49.4   *Class reducer implementation*

```
int main(int argc, char** argv) {
 ParseCommandLineFlags(argc, argv);
 MapReduceSpecification spec;
 // Store list of input files into "spec"
 for (int i = 1; i < argc; i++) {
   MapReduceInput* input = spec.add_input();
   input->set_format("text");
   input->set_filepattern(argv[i]);
   input->set_mapper_class("WordCounter");
 }
 // Specify the output files:
 // /gfs/test/freq-00000-of-00100
 // /gfs/test/freq-00001-of-00100
 // ...
 MapReduceOutput* out = spec.output();
 out->set_filebase("/gfs/test/freq");
 out->set_num_tasks(100);
 out->set_format("text");
 out->set_reducer_class("Adder");
 // Optional: do partial sums within map tasks to save network bandwidth
 out->set_combiner_class("Adder");
 // Tuning parameters: use at most 2000 machines and 100 MB of memory per task
 spec.set_machines(2000);
 spec.set_map_megabytes(100);
 spec.set_reduce_megabytes(100);
 // Now run it
 MapReduceResult result;
 if (!MapReduce(spec, &result)) abort();
 // Done: 'result' structure contains info about counters, time taken, number of
 // machines used, etc.
 return 0;
}
```

Figure 49.5   *Main program*

49.10   Hadoop MapReduce

The Hadoop implementation of MapReduce follows the original Google implementation. It is a Java package. The programmer has to supply two functions. These functions are Java methods. Their names are *map* in the generic abstract class *Mapper* and *reduce* in the generic abstract class *Reducer.* Hadoop's input list is simply a flat list (file) of pairs. Each pair has a key element and a value element. MapReduce's framework applies the map function to every element of the list and sorts the output list. The result is a list of pairs where the key (Map phase produced key) is unique in the list, but the value is a list of all values for that key produced in the Map phase. The *reduce* function is applied on series of the values with the same key. It has a key as an argument, and all values for that key. This function differs from its functional programming counterpart. MapReduce *reduce* generates lists of key / value pairs, but every pair is a result of the application of the argument-function to all pairs in the list that have the same key.

A Hadoop MapReduce *job* consists of input data, the MapReduce program, and configuration information. The job consists of two types of tasks: *map* tasks and *reduce* tasks. A node *jobtracker* controls the job execution. Several node *tasktrackers* can execute a task. The *tasktracker* reports to its *jobtracker.* If some *tasktracker* fails, then the *jobtracker* runs its task on another *tasktracker.*

The framework divides the input information into *input splits.* They have the same fixed size. For every split, a map task processes all records in the split. Hadoop tries to run the map task on the node where the split resides. This is *data locality optimization.* The splits have to be large enough, otherwise the system will overrun for split management. Hadoop with Hadoop Distributed File Systems (HDFS) divides data into HDFS block-size splits.

If the primary data block node is overloaded, Hadoop schedules the map task on the split block replicas. Following the HDFS replication strategy, Hadoop schedules the map task on another node in the same rack, or in another rack, or in a rack on another datacenter. The map task writes its results on the local disk – not in HDFS. After that, the framework sorts all data from the Map phase and transfers them to the Reduce phase. The reduce tasks do not use locality of data, because their map output lists reside on many severs.

It is possible for more than one reduce task to run simultaneously but it is a subject of configuration. In this case, map tasks must divide their output data in portions – a portion for each reducer task. The framework selects the output by the keys; for example, it transfers the pairs with the same hash value to the same reducer. Every reducer generates its own independent output. It is possible no reduce tasks to run, and then map tasks write their output directly to the HDFS.

Hadoop has so-called *combiner functions* as an option. They run on map task output and combine it for input to reduce tasks. This is useful when map and reduce functions are on different nodes, because data transfer can be minimized. Hadoop decides whether to use or not the supplied combiner function.

The programmer can write the functions *map* and *reduce* in programming languages different from Java. One approach to achieve that is through standard input and output streams and Hadoop commands. The other approach is Hadoop Pipes. It is a library for the different programming languages. The Pipes library uses sockets for communication instead of standard input and output.

Figures 49.6 and 49.7 describe the same example program, introduced above, in Hadoop. The description is from Hadoop system documentation.

49.11   MapReduce Implementations in Clouds

MapReduce is available in clouds as a library, as a deployment package, or as a Web service. The previous chapter discussed these implementations in more detail. Without a solution for the data locality MapReduce does not provide benefits.

```
package org.myorg;
import java.io.*;
import java.util.*;
import org.apache.hadoop.fs.Path;
import org.apache.hadoop.filecache.DistributedCache;
import org.apache.hadoop.conf.*;
import org.apache.hadoop.io.*;
import org.apache.hadoop.mapred.*;
import org.apache.hadoop.util.*;

public class WordCount extends Configured implements Tool {
  public static class Map extends MapReduceBase
    implements Mapper<LongWritable, Text, Text, IntWritable> {
    static enum Counters { INPUT_WORDS }
    private final static IntWritable one = new IntWritable(1);
    private Text word = new Text();
    private boolean caseSensitive = true;
    private Set<String> patternsToSkip = new HashSet<String>();
    private long numRecords = 0;
    private String inputFile;

    public void configure(JobConf job) {
      caseSensitive = job.getBoolean("wordcount.case.sensitive", true);
      inputFile = job.get("map.input.file");
      if (job.getBoolean("wordcount.skip.patterns", false)) {
        Path[] patternsFiles = new Path[0];
        try {
          patternsFiles = DistributedCache.getLocalCacheFiles(job);
        } catch (IOException ioe) {
          System.err.println("Caught exception while getting cached files: " +
            StringUtils.stringifyException(ioe));
        }
        for (Path patternsFile : patternsFiles) {
          parseSkipFile(patternsFile);
        }
      }
    }

    private void parseSkipFile(Path patternsFile) {
      try {
        BufferedReader fis = new BufferedReader(new FileReader(patternsFile.toString()));
        String pattern = null;
        while ((pattern = fis.readLine()) != null) {
          patternsToSkip.add(pattern);
        }
      } catch (IOException ioe) {
        System.err.println("Caught exception while parsing the cached file '" + patternsFile +
          "' : " + StringUtils.stringifyException(ioe));
      }
    }

    public void map(LongWritable key, Text value,
      OutputCollector<Text, IntWritable> output, Reporter reporter) throws IOException {
      String line = (caseSensitive) ? value.toString() : value.toString().toLowerCase();
      for (String pattern : patternsToSkip) {
        line = line.replaceAll(pattern, "");
      }
      StringTokenizer tokenizer = new StringTokenizer(line);
      while (tokenizer.hasMoreTokens()) {
        word.set(tokenizer.nextToken());
        output.collect(word, one);
        reporter.incrCounter(Counters.INPUT_WORDS, 1);
      }
      if ((++numRecords % 100) == 0) {
        reporter.setStatus("Finished processing " + numRecords + " records " +
          "from the input file: " + inputFile);
      }
    }
  }
}
```

Figure 49.6   *Class map*

```
public static class Reduce extends MapReduceBase implements
  Reducer<Text, IntWritable, Text, IntWritable> {

  public void reduce(Text key, Iterator<IntWritable> values,
    OutputCollector<Text, IntWritable> output, Reporter reporter) throws IOException {
    int sum = 0;
    while (values.hasNext()) {
      sum += values.next().get();
    }
    output.collect(key, new IntWritable(sum));
  }
}

public int run(String[] args) throws Exception {
  JobConf conf = new JobConf(getConf(), WordCount.class);
  conf.setJobName("wordcount");
  conf.setOutputKeyClass(Text.class);
  conf.setOutputValueClass(IntWritable.class);
  conf.setMapperClass(Map.class);
  conf.setCombinerClass(Reduce.class);
  conf.setReducerClass(Reduce.class);
  conf.setInputFormat(TextInputFormat.class);
  conf.setOutputFormat(TextOutputFormat.class);
  List<String> other_args = new ArrayList<String>();
  for (int i=0; i < args.length; ++i)
    if ("-skip".equals(args[i])) {
      DistributedCache.addCacheFile(new Path(args[++i]).toUri(), conf);
      conf.setBoolean("wordcount.skip.patterns", true);
    } else {
      other_args.add(args[i]);
    }
  FileInputFormat.setInputPaths(conf, new Path(other_args.get(0)));
  FileOutputFormat.setOutputPath(conf, new Path(other_args.get(1)));
  JobClient.runJob(conf);
  return 0;
}

public static void main(String[] args) throws Exception {
  int res = ToolRunner.run(new Configuration(), new WordCount(), args);
  System.exit(res);
}
}
```

Figure 49.7   *Class Reduce and main program*

49.12 MapReduce in MongoDB

The DBMS MongoDB has a command *mapReduce,* which works on MongoDB collections. Alternatively, a query can preprocess the input. The user can sort or limit the number of documents for command. There are three phases of *mapReduce*: Map, Reduce, and Finalize. The last one is optional. The programmer supplies a JavaScipt function for each phase. The *map* function accepts a document and produces a list of key / value pairs. The result list can be empty. The function *reduce* accepts on input key / list of values produced in the previous phase. The *reduce* function aggregates the list of values into a value (object). For each key, it generates exactly one value. The function *finalize* aggregates the outputs. The user can display the output on the screen or save it in a MongoDB collection.

49.13 MapReduce Use Cases

The classical MapReduce use case is document indexing. For example, a web crawler loads a huge quantity of html documents in HDFS. The programmer writes a function map that accepts on input a document and a list of keywords. The function generates a list of pairs of keywords and the number of their occurrences in the document. The other function that the programmer has to supply is *reduce*. This function accepts on input a keyword and a list of all its occurrences in all documents. It generates a pair of keywords and the sum of all its occurrences. Another example is log files and respective response times; the problem is to calculate the average response time.

Google uses the framework to calculate page ranking. PageRank is the number of documents that point to a given one. The function *map* accepts an URI (document) and URI of the scanned document, and generates a list of URIs (documents) that reference it. The function *reduce* accepts an URI (document) and a list of document URIs that reference it. The function calculates the number of different occurrences in the list and generates a rank for this URI. This is a very simplified version of the real algorithm.

49.14 MapReduce Merits and Limitations

MapReduce is a very simple paradigm: the programmer has to write only two functions that operate locally, while it is responsible for parallelization, fault tolerance, load balancing, and so forth. All attempts to extend the paradigm will make it more complex for programming. MapReduce is useful when data processing is performed at one stage. If the job has several iterations and every stage depends on the data generated in the previous one, then that job uses several MapReduce programs. Some extensions, such as Iterative MapReduce, try to solve the problem in one program. MapReduce is suitable for programs that read all the data only once and generate aggregate results. The processing in *map* and *reduce* functions must be simple. That is the case with big data analytics. MapReduce programs are easily scalable and they utilize all resources in the cluster: more resources mean faster execution. They are suitable for clouds.

49.15 Conclusion

For many years, data programming was a mature technology implemented in DBMS. Now, with big data, it is a challenge to the programmers.

MapReduce uses only two of the higher order functions in Lisp: *map* and *reduce*. This paradigm deviates from the basic concepts of functional programming. Some programming models for big data analytics, such as the pipes in MongoDB, use the third high-order function *filter*. The fourth function is not used in any programming model for big data.

Google developed MapReduce to solve a concrete problem: Web page indexing. It was an initial step to big data analytics. In future, new programming models, which are more consistent and suitable for solving big data problems, may emerge.

References

Apache Software Foundation (2014) *Apache Hadoop NextGen MapReduce (YARN)*, http://hadoop.apache.org/docs/current/hadoop-yarn/hadoop-yarn-site/YARN.html (accessed December 4, 2016).
Dean J. and Ghemawat, S. (2004) *MapReduce: Simplified Data Processing on Large Clusters*, http://static.googleusercontent.com/media/research.google.com/en//archive/mapreduce-osdi04.pdf (accessed January 4, 2016).

50

Developing Elastic Software for the Cloud

Shigeru Imai, Pratik Patel, and Carlos A. Varela

Rensselaer Polytechnic Institute, USA

50.1 Introduction

Developing standalone applications running on a single computer is very different from developing scalable applications running on the cloud, such as data analytics applications that process terabytes of data, Web applications that receive thousands of requests per second, or distributed computing applications where components run simultaneously across many computers. Cloud computing service providers help facilitate the development of these complex applications through their *cloud programming frameworks*. A cloud programming framework is a software platform for developing applications in the cloud that takes care of *nonfunctional concerns,* such as scalability, elasticity, fault tolerance, and load balancing. Using cloud programming frameworks, application developers can focus on the functional aspects of their applications and benefit from the power of cloud computing.

In this chapter, we will show how to use some of the existing cloud programming frameworks in three application domains: data analytics, Web applications, and distributed computing. More specifically, we will explain how to use *MapReduce* (Dean and Ghemawat, 2008) for data analytics, *Google App Engine* (Google, 2014) for Web applications, and *SALSA* (Varela and Agha, 2001) for distributed computing. The rest of the chapter is structured as follows. In section 50.2, we describe nonfunctional concerns supported at different levels of cloud services and go through existing cloud programming frameworks. In section 50.3, we explain MapReduce, Google App Engine, and Simple Actor Language System and Architecture (SALSA). In section 50.4, we illustrate how to use these three programming frameworks by showing example applications. Finally, we conclude the chapter in section 50.5.

Encyclopedia of Cloud Computing, First Edition. Edited by San Murugesan and Irena Bojanova.
© 2016 John Wiley & Sons, Ltd. Published 2016 by John Wiley & Sons, Ltd.

50.2   Programming for the Cloud

50.2.1   Nonfunctional Concerns

Figure 50.1 illustrates how cloud programming frameworks hide nonfunctional concerns from application developers by providing programming languages or application programming interfaces (APIs) to manage application's execution on different cloud service models, such as IaaS and PaaS.

Nonfunctional concerns are not directly related to the main functionality of a system but guarantee important properties such as security or reliability. In the context of cloud computing services, important nonfunctional concerns include the following:

- *Scalability*: the ability to scale up and out computing resources to process more workload or to process it faster as demanded by cloud users.
- *Elasticity*: the ability of an application to adapt in order to scale up *and down* as service demand grows or shrinks.
- *Fault tolerance*: the ability to keep the system working properly even in the event of a computing resource failure.
- *Load balancing*: the ability to balance the workload between heterogeneous networked computing resources.

If not using cloud computing, application developers would have to acquire physical infrastructure (machines, networks, etc.) to support their needs. Using IaaS, developers can create virtual machines (VMs) without up-front costs for hardware; however, they have to install and configure the VM management

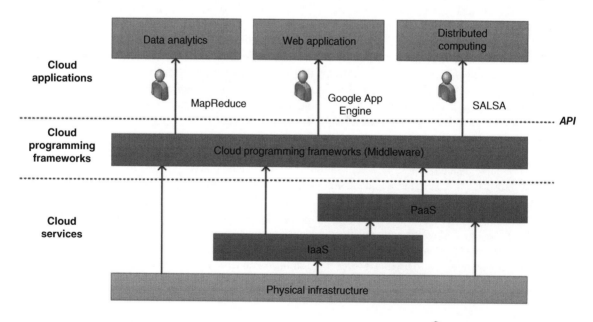

Figure 50.1   *Cloud programming frameworks take advantage of layered services by exposing an Application Programming Interface (API) to developers*

software, networking, operating system, and any additional libraries that their application needs. Finally, different PaaS providers offer from transparent scalability and elasticity to transparent fault tolerance and load balancing services. Typically, these nonfunctional concerns are offered, constraining developers to specific programming models and patterns, such as MapReduce, Web applications, and distributed actor computations.

50.2.2 Overview of Cloud Programming Frameworks

Data analytics. MapReduce is a popular data-processing framework created by Google following a data-parallel programming model based on the *map* and *reduce* higher-order abstractions from functional programming. Hadoop (see http://www.apache.org/, accessed January 5, 2016) is a popular open-source implementation of MapReduce and has a large user and contributor base. There are a number of projects derived from Hadoop. One such project is *Pig* (http://www.apache.org/), which was first developed by Yahoo! Pig offers a high-level language called *Pig Latin* to express data analytics programs. Programs authored in Pig Latin are translated into Java-based Hadoop code and thus Pig users benefit from Hadoop's scalability and fault-tolerance properties as well as Pig Latin's simplicity. Another Hadoop related project is *Hive* (http://www.apache.org/), initially developed by Facebook. Hive is a data warehouse platform built on top of Hadoop. *Mahout* (http://www.apache.org/) is a set of scalable machine learning libraries that also works over Hadoop. *Spark* (http://www.apache.org/) is a cluster-computing framework that focuses on efficient use of data objects to speed up iterative algorithms such as machine learning or graph computation. Whereas Hadoop reads (writes) data from (to) storage repeatedly, Spark caches created data objects in memory so that it can perform up to 100 times faster than Hadoop for a particular class of applications. *Shark* (Xin *et al.*, 2013) is a Structured Query Language (SQL) query engine that has compatibility with Hive and runs on Spark. These frameworks do not support elastic behavior of applications, but have good scalability, load balancing, and fault tolerance.

Web applications. As Web applications have unpredictable and varying demands, they constitute a very good match for cloud computing. There are several PaaS providers that offer a framework for developing and hosting Web applications, such as Google App Engine, Microsoft Azure (Microsoft, 2014), and Heroku (2014). Google App Engine supports Python, Java, PHP, and Go. The runtime environment for Google App Engine is restrictive (e.g., no socket use and no file system access), but you get good scalability with a few lines of code. In contrast, Microsoft Azure offers a more flexible runtime environment; however, it requires you to write more code. Microsoft Azure supports .NET, Java, PHP, Node.js, and Python. Heroku originally supported only Ruby, but now supports Java, Node.js, Scala, Clojure, Python, and PHP. In terms of coding flexibility, Heroku is also as restrictive as Google App Engine; however, it has libraries to support data management, mobile users, analytics, and others.

Distributed computing. Distributed computing systems typically have components that communicate with each other via message passing to solve large or complex problems cooperatively. Erlang (Armstrong *et al.*, 1993) and SALSA are concurrent and distributed programming languages based on the *actor* model (Agha, 1986), in which each actor runs concurrently and exchanges messages asynchronously while not sharing any state with any other actor. Actor systems can therefore be reconfigured dynamically, while transparently preserving message passing semantics, which is very helpful for scalability, elasticity, and load balancing. Erlang is a functional language that supports fault tolerance and hot swapping. SALSA's compiler generates Java code allowing programmers to use the entire Java library collection. It also supports transparent actor migration across the Internet based on a universal naming system. Several research efforts have been made to support nonfunctional concerns for SALSA programs. The *Internet Operating System* (IOS) (El Maghraoui *et al.*, 2006) is a distributed middleware framework that provides support for dynamic reconfiguration of large-scale distributed applications through opportunistic load balancing.

Table 50.1   *Summary of cloud programming frameworks*

Category	Name	Nonfunctional concerns	Description
Data analytics	MapReduce, Hadoop	Scalability, fault tolerance, load balancing	MapReduce exposes simple abstractions: *map* and *reduce*. Hadoop is an open-source implementation of MapReduce.
	Pig, Hive		Pig and Hive are high-level languages for Hadoop.
	Mahout, Spark, Shark		Mahout is a machine-learning library running on top of Hadoop. Spark is a data analytics framework especially for iterative and graph applications. Shark is a Hive-compatible SQL engine running on top of Spark.
Web applications	Google App Engine	Scalability, elasticity, fault tolerance, load balancing	A PaaS from Google. The runtime environment is restrictive, but provides a good scalability. Python, Java, PHP, and Go are supported.
	Microsoft Azure		A PaaS from Microsoft with automatic scaling, automatic patching, and security services.
	Heroku		A PaaS from Heroku with libraries for data management, mobile users, analytics, and others.
Distributed computing	SALSA, Erlang	Scalability, fault tolerance (for Erlang)	General-purpose programming languages based on the actor model.
	IOS, COS	Scalability, elasticity (COS), load balancing	Middleware framework for managing distributed SALSA programs.

In addition to the distributed load-balancing capability, the *Cloud Operating System* (COS) (Imai *et al.*, 2013) further supports elasticity, enabling adaptive virtual machine allocation and de-allocation on hybrid cloud environments, where private clouds connects to public clouds.

Table 50.1 summarizes the cloud programming frameworks mentioned above.

50.3   Cloud Programming Frameworks

In this section, we present existing cloud programming frameworks for data analytics, web applications, and distributed computing, namely, MapReduce, Google App Engine, and the SALSA programming language.

50.3.1   Data Parallelism with MapReduce

MapReduce created by Google is a data-parallel programming model based on the *map* and *reduce* higher order abstractions from functional programming. The implementation of the model is also called MapReduce; it is designed to schedule automatically the parallel processing of large data sets distributed across many computers. MapReduce was designed to have good scalability to achieve high throughput, and fault tolerance to deal with unavoidable hardware failures. These nonfunctional concerns are transparent to application developers; MapReduce provides a simple application programming interface requiring to define only the map and reduce functions that have the format shown in Listing 50.1.

```
map(k1, v1) → list (k2, v2)
reduce(k2, list(v2)) → list (v3)
```

Listing 50.1 *MapReduce abstract application programming interface*

```
map(string key, string value)
    //key: the position of value in the input file
    //value: partial DNA sequence in the input file
    foreach character c in value do
        emit(c, 1);
    end


reduce(string key, string value)
    //key: DNA character (C, G, T, or A)
    //value: a list of counts
    sum = 0;
        foreach count in value do
            sum = sum + count;
        end
    emit(sum);
```

Listing 50.2 *MapReduce pseudo code for DNA sequence analysis*

The computation in MapReduce basically consists of two phases: the "map" phase producing intermediate results, followed by the "reduce" phase processing the intermediate results. Both phases take a (key, value) pair as input and output a list. Keys and values can be arbitrary numbers, strings, or user-defined types. First, the map function gets called by MapReduce and transforms a (k1, v1) pair into an intermediate list of (k2, v2). For instance, in text-mining applications, v1 may be a partial text of an input file. Next, these intermediate lists are aggregated into pairs where each pair has all v2 values associated with the same k2 key. Finally, the reduce function is invoked with a (k2, list(v2)) pair and it outputs a list of v3 values.

Suppose you have a big input file consisting of DNA sequences, which are combinations of {C, G, T, A} characters, one way to write map and reduce functions to count each character occurrence in the input file can be given as shown in Listing 50.2.

The map function emits count 1 as an occurrence of a DNA character. The reduce function takes a list of counts and emits the sum of these counts for a particular DNA character. By calling these simple two functions repeatedly, MapReduce outputs occurrences of each DNA character. An example execution for the DNA sequence counting example is shown in Figure 50.2.

50.3.2 Service-Oriented Programming with Google App Engine

Google App Engine (App Engine hereafter) is a programming framework to develop scalable Web applications running on Google's infrastructure. Developed Web applications are hosted by Google and accessible via the user's own domain name. Google automatically allocates more computing resources when service demand grows and also balances the workload among the computing resources. App Engine is regarded as a PaaS

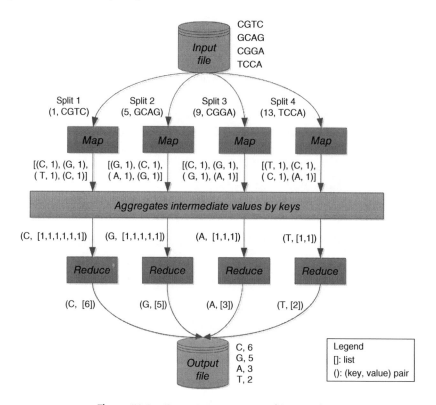

Figure 50.2 *Execution sequence of MapReduce*

implementation and charges costs based on the use of storage, CPU, and bandwidth. App Engine currently supports development in Java, Python, PHP, and Go. App Engine's core features and its programming framework are described in the following sections. We use only Python for brevity.

50.3.2.1 Core Features

The core features of App Engine are as follows:

- *Sandbox.* To prevent harmful operations to the underlying operating system, applications run in an isolated secure environment called *sandbox.* In the sandbox, an application is allowed to only access other computers on the Internet using Uniform Resource Locator (URL) Fetch and Mail services, and application code is able to only run in a response to a Web request. The application is not allowed to write to the file system directly, but instead uses `Datastore` or `Memcache` services.
- *Storing data.* App Engine `Datastore` is a data storing service based on Google's *BigTable* (Chang *et al.,* 2008), which is a distributed storage system that scales up to petabytes of data.
- *Account management.* Applications can be easily integrated with *Google Accounts* for user authentication. With Google Accounts, an application can detect if the current user has signed in. If not, it can redirect the user to a sign-in page.
- *App Engine services.* App Engine provides a variety of services via APIs to applications including fetching Web resources, sending e-mails, using cache and memory instead of secondary storage, and manipulating images.

50.3.2.2   *Programming Framework for Python*

We describe the App Engine's programming framework for Python as visualized in Figure 50.3. When the Web server receives a Hyper-Text Transfer Protocol (HTTP) request from the browser, the Web server passes the request to a framework (a library that helps Web application development) via Web Server Gateway Interface (WSGI). The framework then invokes a handler in a user script (in this example, `main.py`) and the invoked handler processes the request and creates an HTTP response dynamically.

The Python runtime in App Engine uses WSGI as an interface to connect the Web server to the Web applications. WSGI is simple, but reduces programming effort and enables more efficient application development. Here, we give an example of a request handler using the `webapp2` framework that interacts with WSGI as shown in Listing 50.3.

The `application` object – an instance of `WSGIApplication` class of `webapp2` framework – handles the requests. When creating the `application` object, a request handler called `MainPage` is associated with the root URL path (/). The `webapp2` framework invokes the `get` function in the `MainPage` class when it receives an HTTP GET request to the URL /. In the `get` function, it creates an HTTP response with a `Content-Type` header and a body containing a *"Hello, World!"* message.

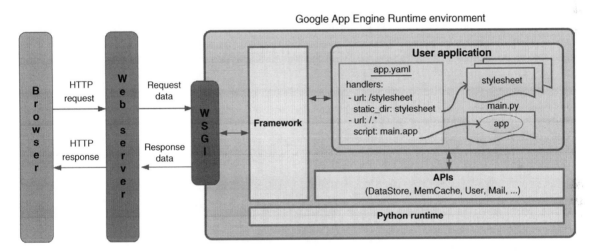

Figure 50.3   *Google App Engine programming framework for Python*

```
import webapp2


class MainPage( webapp2.RequestHandler ):
    def get( self ):
        self.response.headers[ 'Content-Type' ] = 'text/plain'
        self.response.write( 'Hello, World!' )


application = webapp2.WSGIApplication(
                                [('/', MainPage)],
                                debug=True )
```

Listing 50.3   *Application Python script for the Helloworld example*

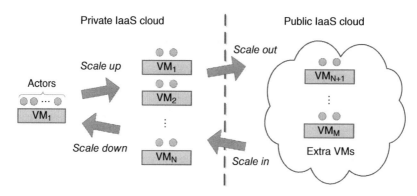

Figure 50.4 *Workload scalability over a hybrid cloud using actor migration. Source: Imai et al. (2013)*

50.3.3   Distributed Actor Systems with SALSA

Simple Actor Language System and Architecture is a concurrent programming language based on the actor model (Agha, 1986). Each actor runs concurrently and exchanges messages asynchronously while encapsulating its state. Since an actor does not share any memory with other actors, it can migrate to another computing host easily. As shown in Figure 50.4, under a hybrid IaaS cloud environment, actors can migrate between the private and public clouds seamlessly with runtime software installed on virtual machines (VMs) on both ends. In this scenario, application developers need either to use COS middleware or support appropriate nonfunctional concerns by themselves as IaaS clouds only provide scalability by means of VM addition / removal. Nevertheless, hybrid clouds can be attractive for those who can access their private computing resources at no additional cost and who need high computing power only occasionally.

In the following subsections, we introduce the actor-oriented programming model followed by a distributed application written in SALSA.

50.3.3.1   *Actor-Oriented Programming*

Actors provide a flexible model of concurrency for open distributed systems. Each actor encapsulates a state and a thread of control that manipulates this state. In response to a message, an actor may perform one of the following actions (see Figure 50.5):

- alter its current state, possibly changing its future behavior;
- send messages to known actors asynchronously;
- create new actors with a specified behavior;
- migrate to another computing host.

Analogous to a class in Java, SALSA programmers can write a *behavior* which includes encapsulated state and message handlers for actor instances:

- New actors are created in SALSA by instantiating particular behaviors with the new keyword. Creating an actor returns its reference. For instance:
  ```
  ExampleActor exActor = new ExampleActor();
  ```
- The message sending operation (<-) is used to send messages to actors; messages contain a name that refers to the message handler for the message and a possibly empty list of arguments. For instance:
  ```
  exActor<-m(1, 2);
  ```
- Actors, once created, process incoming messages, one at a time.

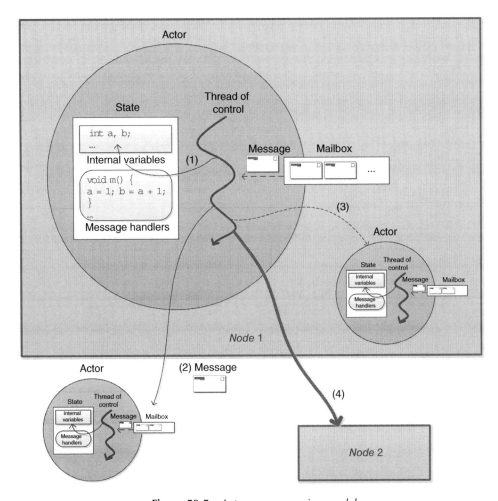

Figure 50.5 *Actors programming model*

50.3.3.2 *Programming Distributed Applications*

A simple distributed SALSA application that migrates an actor with name as specified by an universal actor name (UAN) from a location specified by a universal actor location (UAL) is shown in Listing 50.4. A UAN is an identifier that represents an actor during its lifetime in a location-dependent manner. An actor's UAN is mapped by a naming service into a UAL, which provides access to an actor in a specific location.

Listing 50.4 defines a behavior of the `Migrate` actor. First, the program starts from the `act` message handler and a `Migrate` actor named `migrateActor` is created with a UAN "uan://wcl. cs.rpi.edu:3030/myName" at a UAL "rmsp://host1.cs.rpi.edu:4040/myLoca- tor". Next, the actor receives a `print` message and prints out a string "Migrate actor is here!" in the standard output of `host1`. Right after printing the string, the actor migrates to a new location specified by a UAL "rmsp://host2.cs.rpi.edu:4040/myLocator". Finally, after the migration, the actor prints out the same string at `host2`. Note that SALSA's transparent migration support enables execution of the same print message in two different hosts.

```
behavior Migrate {
    void print() {
        standardoutput<-println("Migrate actor is here!");
    }


     void act(string[] args) {
        UAN uan = new UAN("uan://wcl.cs.rpi.edu:3030/myName");
        UAL ual = new UAL(
            "rmsp://host1.cs.rpi.edu:4040/myLocator");


        Migrate migrateActor = new Migrate() at (uan, ual);
        migrateActor<-print()@
        migrateActor<-migrate(
            "rmsp://host2.cs.rpi.edu:4040/myLocator")@
        migrateActor<-print();
    }
}
```

Listing 50.4   *SALSA migration code example*

50.4   Sample Applications

In this section, we present sample applications for the three programming frameworks. First, we describe a data analytics application that computes the average temperatures of historical weather data using Hadoop. Next, we show a simple Web-based bulletin board application using Google App Engine. Thirdly, we describe a distributed face recognition application using SALSA.

50.4.1   Data Analytics

The application presented in this subsection processes a large amount of data by Hadoop to compute average temperatures for every month using temperature data collected from all over the world.

50.4.1.1   *Global Surface Summary of Day Data (GSOD)*

Global surface summary of day (GSOD) data is a collection of daily weather data produced by the National Climatic Data Center. The weather data has been collected from 1929 to the present by stations all over the world. The elements contained in the daily data include mean temperature, mean sea level pressure, mean visibility, and mean wind speed.

The datasets are provided in ASCII characters. Each row contains weather data for a station on a particular day (see, for example, Listing 50.5). Fields are explained in Table 50.2.

Since we compute average temperatures for every month, the fields we are interested in are YEAR, MODA and TEMP.

```
STN---    YEARMODA    TEMP    DEWP    SLP       STP      ...
030050    19291001    45.3 4  40.0  4 1001.6 4  9999.9 0 ...
```

Listing 50.5 *Example of GSOD data*

Table 50.2 *Field definitions of GSOD data*

Field	Position	Type	Description
STN	1–6	Integer	Station number for the location
YEAR	9–12	Integer	The year
MODA	13–16	Integer	The month and day
TEMP	19–24	Real	Mean temperature for the day in degrees
...	...	...	...

```
map(String key, String value)
    //key: tag, value: number
    pair = (value, 1);
    emit(key, pair);


reduce(String key, String value)
    //key: tag, value: a list of (number, count)pairs
    sum = 0; count = 0;
    foreach pair (n, c) in value do
        sum = sum + n;
        count = count + c;
    end
    average = sum / count;
    emit(average);
```

Listing 50.6 *MapReduce pseudo code for average number calculation*

50.4.1.2 MapReduce Functions for Averaging Numbers

Suppose we are given an input file that consists of (*tag, number*) pairs, where tag is an arbitrary string of a month, year, station, and so on, we can compute an average number for each tag by the map and reduce functions shown in Listing 50.6.

The map function emits an intermediate (*tag, (number, 1)*) pair for every input and then the reduce function computes the average number $\frac{1}{N}\sum_{i=1}^{N} number_i$ for each tag from an aggregated pair (*tag, [(number₁, 1), ..., (numberₙ, 1)]*).

50.4.1.3 GSOD Data Analytics Application

Based on the functions' pseudo code presented in the previous section, we show a Hadoop application that computes average temperatures for every month using GSOD temperature data. The map function is defined as a GsodMapper class implementation that extends the Mapper class as shown in Listing 50.7.

```
public class GsodMapper extends
    Mapper<Object, Text, Text, DoubleIntPair> {
    private Text word = new Text();
    private DoubleIntPair pair = new DoubleIntPair();

    public void map(Object key, Text value, Context context)
        throws IOException, InterruptedException {
        if (value.toString().startsWith( "STN" )) {
            return;
        }
        String[] data = value.split( "+" );
        String yearmon = data[2].substring(0, 6);
        word.set( yearmon );
        pair.set( Double.parseDouble( data[3] ), 1 );
        context.write( word, pair )
    }
}
```

Listing 50.7   *GsodMapper class definition*

The Mapper class is a generic class that takes four type variables – the input key type, the input value type, the output key type, and the output value type. In this example, the input key type is Object, the input key value is Text, the output key type is Text, and the output value type is DoubleIntPair. DoubleIntPair is used to store a pair of double and int values.

The map function repeatedly receives one line of GSOD weather input data in the value variable. If the value starts with an "STN", that means that it is a line for weather field names, therefore we skip it. Otherwise, split the value into a data array in which each element is separated by spaces. Since we want to compute the average temperature for each month, we are interested in the YEARMODA and TEMP fields, which are the third and fourth fields of the weather data. From these two fields, we can extract a year and month by data[2].substring(0,6) in the YYYYMM format and temperature by data[3]. By calling context.write() function, the map function outputs a key value pair in the format (*key*=YYYYMM, *value*=(temperature, 1)).

The GsodReducer class that implements the reduce function is defined in Listing 50.8. The specified input types are Text and DoubleIntPair and output types are Text and DoubleWritable. The reduce function iterates the values, which is a list of a temperature and count, and sums up the temperatures and counts to compute the average temperature.

50.4.2   Bulletin Board Web Application

In this subsection we present a simple bulletin board Web application developed on the Google App Engine framework (see Figure 50.6 for user interface). This application lets users post a message with their usernames to a public bulletin board. The posted messages are stored in the server and shared among the users accessing the application.

```
public class GsodReducer extends
    Reducer<Text, DoubleIntPair, Text, DoubleWritable> {
    private DoubleWritable average = new DoubleWritable();

    public void reduce(Text key,
                    Iterable<DoubleIntPair> values,
                    Context context)
        throws IOException, InterruptedException {
        double sum = 0;
        int count = 0;
        for (DoubleIntPair value: values)
            sum += value.getDouble();
            count += value.getInt();
        }
        average.set(sum / count);
        context.write(key, average)
    }
}
```

Listing 50.8   GsodReducer class definition

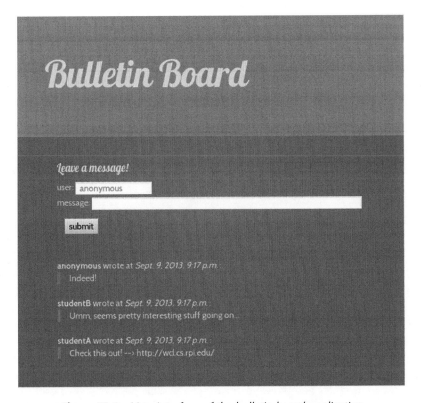

Figure 50.6   User interface of the bulletin board application

```html
<html>
  <head>
    <link type="text/css" rel="stylesheet"
        href="/css/main.css" />
  </head>
  <body>
    <div class="headline">
      <h1>Bulletin Board</h1>
    </div>

    <div class="content">
      <h4>Leave a message!</h4>
      <form action="/postmsg" method="post" class="form-inline">
        <label>user: </label>
        <input type="text" name="user" placeholder="anonymous">
        <label>message: </label>
        <input type="text" name="content">
        <button type="submit">submit</button>
      </form>

      {% for msg in msgs %}
        <b>{{ msg.user }}</b> wrote at <i>{{ msg.date }}</i>:
        <blockquote>{{ msg.content|escape }}</blockquote>
      {% endfor %}
    </div>
  </body>
</html>
```

Listing 50.9   *Jinja2 HTML template for bulletin board application*

50.4.2.1   HTML Template

Unlike the `helloworld` example, which just returns an HTTP response containing a "*Hello, World!*" string to the browser, we can create a more elaborate and dynamically generated response using HTML templates such as *Jinja2* (Ronacher, 2011). Jinja2 is a template engine for Python supported by Google App Engine. For the bulletin board application, we can use a Jinja2 template as shown in Listing 50.9.

In the second `div` section, this HTML file first creates a form to post a username and a message to the Web server. Once these forms are filled by the user and the `submit` button is pressed, an HTTP POST request is sent to the URL specified by `/postmsg`. Then, the message-showing part is expressed using the Jinja2 template. In the Jinja2 template, "`{{ variable }}`" refers to the value of a variable. Similarly, "`{% control logic %}`" refers to control logic such as `for`, `while`, or `if`. In this example, the stored messages are referred to as `msgs` and the username, creation date, and content of each message are displayed by iterating `msgs`. Using HTML templates, we can express dynamic and static HTML contents side by side easily.

```
import os, jinja2, webapp2
from google.appengine.ext import ndb


JINJA_ENVIRONMENT = jinja2.Environment(
    loader=jinja2.FileSystemLoader(os.path.dirname(__file__)),
    extensions=['jinja2.ext.autoescape'])


class Messages(ndb.Model):
    user = ndb.StringProperty()
    content = ndb.StringProperty()
    date = ndb.DateTimeProperty(auto_now_add=True)


class PostMessage(webapp2.RequestHandler):
    def post(self):
        msgs = Messages()
        user = self.request.get('user')
        if user == "":
            msgs.user = "anonymous"
        else:
            msgs.user = user
        msgs.content = self.request.get('content')
        msgs.put()
        self.redirect('/')


class MainPage(webapp2.RequestHandler):
    def get(self):
        msgs = ndb.gql("SELECT * FROM Messages
                        ORDER BY date DESC LIMIT 10")
        template_values = {'msgs': msgs}
        template = JINJA_ENVIRONMENT.get_template('index.html')
        self.response.write(template.render(template_values))


application = webapp2.WSGIApplication(
    [('/', MainPage), ('/postmsg', PostMessage)], debug=True)
```

Listing 50.10   *Application Python script for bulletin board example*

50.4.2.2 *Application Script*

An example script for the bulletin board application is shown in Listing 50.10.

Once the script starts, a global variable JINJA_ENVIRONMENT is instantiated and configured to look for Jinja2 HTML templates from the current directory (specified by "__file__"). This program consists of three classes: Messages class for data model, PostMessage class for handling submitted messages, and MainPage class for handling requests for the main page. Meanwhile, an application object is created by WSGIApplication so that HTTP requests for "/" and "/postmsg" are handled by the MainPage and PostMessage class respectively.

The `Message` class defines a data model that consists of three fields: `user` for the username, `content` for the message content, and `date` for the current time when the model instance is added to the datastore (specified by "`auto_now_add=True`"). The `PostMessage` class defines the behavior when the application receives an HTTP POST message. It extracts the `user` and `content` fields from the request and puts them into a `Messages` data model instance. The `MainPage` class defines the behavior when the application receives an HTTP GET message to the main Web page. Using a SQL-like query called GQL, it retrieves stored messages from the `Messages` datastore up to ten in descending order of date (latest date comes first), and then passes the retrieved `msgs` messages to the template `index.html`. Finally, it creates a complete HTTP response from the template and returns the response to the browser.

50.4.3 Distributed Face Recognition

Face recognition is an application that can be vastly improved by leveraging cloud computing resources. Rather than using a single device, in this case a mobile phone, we can offload parts of the image processing to the cloud (Abolfazli *et al.,* 2014). By using cloud computing, we can save the battery in the mobile device, and also we can consider larger data sets. Using the SALSA programming language and the `FaceRecognizer` API from OpenCV (http://opencv.org/, accessed January 5, 2016), we can design a mobile phone application to recognize a face in a given image using a database of faces (see Figure 50.7).

The face recognition application consists of two stages: the training stage, and the prediction stage. The training stage trains a database of faces using the `FaceRecognizer` method defined in OpenCV. The prediction stage predicts a given face using the desired method with a certain confidence. When the database of faces is small, there is no need to offload computation because the phone can process the faces locally just as fast as it would take to offload to the cloud. As the database grows we run into the limitations mentioned above and offloading to the cloud can be beneficial.

The distributed face recognition model consists of a "farmer actor" that creates *N1* "worker actors" in the cloud. While the farmer and worker actors reside in the cloud, a client on the mobile phone requests the farmer actor to recognize an unknown face. The farmer actor assigns each worker actor a range of faces (*N2/N1* each)

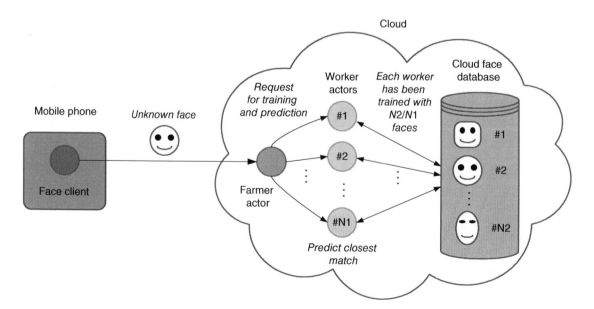

Figure 50.7 *Distributed face recognition using SALSA actors*

to train from the cloud face database containing *N2* faces. The worker actors then predict the closest match to the unknown face based on their assigned database. The farmer actor collects the closest match from each worker actor and calculates the best candidate for the unknown face. Pseudo programs for the client, worker and farmer SALSA programs are shown in Listings 50.11, 50.12, and 50.13 respectively.

In Listing 50.13, note that a `join` block in the `predictAll` message handler is used to synchronize all the worker actors executing predict. After all the workers finish processing `predict`, the join block returns an object array `ImageMetaData[]`, which contains prediction confidence from the workers. Finally, `getBestMatch` takes the array and finds the best matching image with the highest confidence.

50.5 Conclusions

Cloud computing has the potential to bring the benefits of large-scale data analytics and high-performance computing to everyone's fingertips enabling unprecedented societal applications. We have described three programming frameworks for cloud computing: Hadoop's MapReduce, Google App Engine, and SALSA. To illustrate the use of these frameworks, we have shown a weather data analytics application on Hadoop, a

```
behavior FaceClient {
    void act(String[] args) {
        FaceFarmer farmer = (FaceFarmer)
            FaceFarmer.getReferenceByName(
                "uan://nameserver/facefarmer");
        Image unknownImage = new Image(args[0]);
        farmer<-predictAll(unknownImage)@
            displayImage(token);
    }
}
```

Listing 50.11   *SALSA pseudo code for face recognition client actor*

```
behavior FaceWorker {
    FaceRcognizer faceRecognizer;
    void train(Image[] assignedDatabase) {
        faceRecognizer.train(assignedDatabase);
    }


    ImageMetaData preditct (Image testImage) {
        // return the closest match
        return faceRecognizer.predict(testImage);
    }
}
```

Listing 50.12   *SALSA pseudo code for face recognition worker actor*

```
behavior FaceFarmer implements ActorService {
    //CloudFaceDatabase contains N2 faces.
    void act(String[] args) {
        faceWorker[] workers = new faceWorker[N1];
        for (i = 0; i < N1; i++) {
            workers[i] = new faceWorker(new UAN(…));
            workers[i]<-migrate(new UAL(…));
            workers[i]<-train(
                N2/N1 faces from CloudFaceDatabase);


        }
    }
    Image predictAll(Image testImage)
        join {
            for (i = 0; i < N1; i++)
                workers[i]<-predict(testImage);
        }@getBestMatch(token)@currentContinuation;
    }
    Image getBestMatch(ImageMetaData[] closestMatches) {
        // find best match with the highest confidence.
    }
}
```

Listing 50.13 SALSA pseudo code for face recognition farmer actor

simple bulletin board Web application on Google App Engine, and a distributed face recognition application on SALSA offloading computation from a mobile device to the cloud. As we have seen, the target applications for these frameworks are very different from each other, and the support level of nonfunctional concerns is also different between IaaS and PaaS. These frameworks are ideal for the illustrated target application domains, but not necessarily for other application scenarios. Therefore, to get the best from these cloud services, application developers still need to carefully consider the characteristics of their applications and find best matching cloud services in terms of QoS, cost, programmability, vendor lock-in possibility, and others.

Acknowledgment

This work is partially supported by an Amazon AWS in Education Research Grant and a Google Cloud Credits Award.

References

Abolfazli, S., Sanaei, Z., Ahmed, E., Gani, A., and Buyya, R. (2014) Cloud-based augmentation for mobile devices: Motivation, taxonomies, and open challenges. *IEEE Communications Surveys and Tutorials* **16**(1), 337–368.
Agha, G. (1986) *Actors: a Model of Concurrent Computation in Distributed Systems,* MIT Press, Cambridge, MAs.

Armstrong, J., Virding, R., Wikström, C., and Williams, M. (1993) *Concurrent Programming in ERLANG,* Prentice Hall, Englewood Cliffs, NJ.

Chang, F., Dean, J., Ghemawat, S., *et al.* (2008) Bigtable: A distributed storage system for structured data. *ACM Transactions on Computer Systems (TOCS)* **26**(2), 4.

Dean, J. and Ghemawat, S. (2008) MapReduce: Simplified data processing on large clusters. *Communications of the ACM* **51**(1), 107–113.

El Maghraoui, K., Desell, T. J., Szymanski, B. K., and Varela, C. A. (2006) The internet operating system: Middleware for adaptive distributed computing. *International Journal of High Performance Computing Applications* **20**(4), 467–480.

Google (2014) *Google App Engine,* https://cloud.google.com/appengine/docs (accessed January 5, 2016).

Imai, S., Chestna, T., and Varela, C. A. (2013) *Accurate Resource Prediction for Hybrid IaaS Clouds Using Workload-Tailored Elastic Compute Units.* Proceedings of the 2013 IEEE Sixth International Conference on Utility and Cloud Computing (UCC). IEEE.

Heroku. (2014) *Heroku,* http://www.heroku.com/ (accessed January 5, 2016).

Microsoft (2014) *Windows Azure,* http://www.windowsazure.com/ (accessed January 5, 2016).

Ronacher, A. (2011) *Jinja,* http://jinja.pocoo.org/ (accessed January 5, 2016).

Varela, C. A. & Agha, G. (2001) Programming dynamically reconfigurable open systems with SALSA. *SIGPLAN Notices* **36**(12), 20–34.

Varela, C. A. (2013) *Programming Distributed Computing Systems: A Foundational Approach,* MIT, Cambridge, MA.

Xin, R. S., Rosen, J., Zaharia, M., *et al.* (2013) *Shark: SQL and Rich Analytics at Scale.* Proceedings of the 2013 International Conference on Management of Data. ACM, pp. 13–24.

51

Cloud Services for Distributed Knowledge Discovery

Fabrizio Marozzo, Domenico Talia, and Paolo Trunfio

University of Calabria, Italy

51.1   Introduction

The information technology market has been moving from the demand and supply of products towards a service-oriented model in which all resources – processors, memories, data and applications – are provided as services to customers through the Internet. Such convergence between Internet technologies and services, combined with the use of virtualization techniques, has led to the development of the cloud computing paradigm (Mell and Grance, 2011).

In many application areas, knowledge discovery in databases (KDD) techniques are used to extract useful knowledge from large datasets. Very often, distributed KDD approaches must be used because datasets are too large to be analyzed in a single site, or because they are distributed across many locations and cannot be moved to a central site for processing. Several distributed KDD systems have been proposed so far. In most cases, those systems had to face infrastructure-level issues, such as resource allocation, execution management, fault tolerance, and so on (Talia and Trunfio, 2010).

This chapter discusses how cloud computing technologies can be exploited to implement a distributed KDD system without worrying about low-level aspects because they are already addressed by the cloud infrastructure. First, we identify the requirements of a generic distributed KDD system, and discuss how these requirements can be fulfilled by a cloud platform. Then, as a case study, we describe how we used a cloud platform to design and develop the Data Mining Cloud Framework, which supports the distributed execution of KDD applications modelled as workflows.

Encyclopedia of Cloud Computing, First Edition. Edited by San Murugesan and Irena Bojanova.
© 2016 John Wiley & Sons, Ltd. Published 2016 by John Wiley & Sons, Ltd.

51.2   Requirements for a Distributed KDD System

In this section, we identify the main requirements that should be satisfied by a generic distributed KDD system. System requirements are divided into *functional* and *nonfunctional* requirement. The former specify which functionalities the system should provide; the latter include quality criteria mostly related to system performance.

51.2.1   Functional Requirements

The functional requirements that should be satisfied by a generic distributed KDD system can be grouped into two main classes: *resource management* requirement and *application management* requirement. The former refers to requirements related to the management of all the resources (data, tools, results) that may be involved in a knowledge-discovery application; the latter refers to requirements related to the design and execution of the applications themselves.

51.2.1.1   Resource Management

Resources of interests in distributed KDD applications include *data sources*, *knowledge discovery tools*, and *knowledge discovery results*. A distributed knowledge discovery system should therefore deal with the following resource-management requirements:

- *Data management.* Data sources can be in different formats, such as relational databases, plain files, or semistructured documents (e.g., XML files). The system should provide mechanisms to store and access such data sources independently from their specific format. In addition, metadata formalisms should be defined and used to describe the relevant information associated with data sources (e.g., location, format, availability, available views), in order to enable their effective access and manipulation.
- *Tool management.* Knowledge discovery tools include algorithms and services for data selection, preprocessing, transformation, data mining, and results evaluation. The system should provide mechanisms to access and use such tools independently of their specific implementation. Metadata have to be used to describe the most important features of KDD tools (e.g., their function, location, usage).
- *Result management.* The knowledge obtained as the result of a knowledge discovery process is represented by a knowledge (or data-mining) model. The system should provide mechanisms to store and access such models, independently from their structure and format. As for data and tools, data-mining models need to be described by metadata to explain and interpret their content, and to enable their effective retrieval.

51.2.1.2   Application Management

A distributed KDD system must provide effective mechanisms to design KDD applications (*design management*) and control their execution (*execution management*):

- *Design management.* Distributed knowledge discovery applications range from simple data mining tasks to complex data-mining patterns expressed as workflows. From a design perspective, three main classes of knowledge discovery applications can be identified: *single-task applications*, in which a single data mining task such as classification, clustering, or association rules discovery is performed on a given data source; *parameter sweeping applications*, in which a dataset is analyzed using multiple instances of the same data mining algorithm with different parameters; *workflow-based applications*, in which possibly complex knowledge discovery applications are specified as graphs that link together data sources, data-mining algorithms, and visualization tools. A general system should provide environments to design all the classes of KDD applications mentioned above effectively.

- *Execution management.* The system has to provide a distributed execution environment that supports the efficient execution of knowledge discovery applications designed by users. As applications range from single tasks to complex knowledge discovery workflows, the execution environment should cope with such a variety of applications. In particular, the execution environment should provide the following functionalities, which are related to the different phases of application execution: accessing the data sources to be mined; allocating the needed compute resources; running the application based on user specifications, which may be expressed as a workflow; presenting the results to the user. The system should also allow users to monitor the application execution.

51.2.2   Nonfunctional Requirements

Nonfunctional requirements can be defined at three levels: *user*, *architecture*, and *infrastructure*. User requirements specify how the user should interact with the system; architecture requirements specify which principles should inspire the design of the system architecture; finally, infrastructure requirements describe the nonfunctional features of the underlying computational infrastructure.

51.2.2.1   User Requirements

From a user point of view, the following nonfunctional requirements should be satisfied:

- *Usability.* The system should be easy to use by the end users, without the need to undertake any specialized training.
- *Ubiquitous access.* Users should be able to access the system from anywhere using standard network technologies (e.g., web sites), either from a desktop PC or from a mobile device.
- *Data protection.* Data represents a key asset for users; therefore, the system should protect data to be mined and inferred knowledge from both unauthorized access and intentional/incidental losses.

51.2.2.2   Architecture Requirements

The main nonfunctional requirements at the architectural level are:

- *Service-orientation.* The architecture should be designed as a set of network-enabled software components (services) implementing the different operational capabilities of the system, to enable their effective reuse, composition, and interoperability.
- *Openness and extensibility.* The architecture should be open to the integration of new knowledge-discovery tools and services. Moreover, existing services should be open for extension, but closed for modification, according to the open-closed principle.
- *Independence from infrastructure.* The architecture should be designed to be as independent as possible from the underlying infrastructure; in other words, the system services should be able to exploit the basic functionalities provided by different infrastructures.

51.2.2.3   Infrastructure Requirements

Finally, from the infrastructure perspective, the following nonfunctional requirements should be satisfied:

- *Standardized access.* The infrastructure should expose its services using standard technologies (e.g., Web services), to make them usable as building blocks for high-level services or applications.
- *Heterogeneous/distributed data support.* The infrastructure should be able to cope with very large and high-dimensional datasets, stored in different formats in a single datacenter, or geographically distributed across many sites.

- *Availability.* The infrastructure should be in a functioning condition even in the presence of failures that affect a subset of the hardware / software resources. Thus, effective mechanisms (e.g., redundancy) should be implemented to ensure dependable access to sensitive resources such as user data and applications.
- *Scalability.* The infrastructure should be able to handle a growing workload (deriving from larger data to process or heavier algorithms to execute) in an efficient and effective way, by dynamically allocating the needed resources (processors, storage, network). Moreover, as soon as the workload decreases, the infrastructure should release the unneeded resources.
- *Efficiency.* The infrastructure should minimize resource consumption when executing any given task. In the case of parallel / distributed tasks, efficient allocation of processing nodes should be guaranteed. The infrastructure should be used extensively to provide efficient services.
- *Security.* The infrastructure should provide effective security mechanisms to ensure data protection, identity management, and privacy.

51.3 Cloud for Distributed KDD

A key aspect of cloud computing is that end users do not need to have either knowledge or control over the infrastructure that supports their applications. In fact, cloud infrastructures are based on large sets of computing resources, located somewhere "in the Cloud," which are allocated to applications on demand. Cloud resources are provided in highly scalable way, i.e., they are allocated dynamically to applications depending of the current level of requests. Although similar in overall aims to grid systems, clouds are different because they hide the complexity of the underlying infrastructure, providing services ready to use where end users pay only for the resources effectively used (pay-per-use).

Cloud computing vendors classify their services into three categories: *software as a service* (SaaS), where each software or application executed is provided through Internet to customers as ready-to-use services (e.g., Google Calendar, Microsoft Hotmail, Yahoo Maps); *platform as a service* (PaaS), also known as cloud platform services, in which cloud providers offer platform services such as databases, application servers, or environments for building, testing and running custom applications (e.g., Google Apps Engine, Microsoft Azure, Force.com); *infrastructure as a service* (IasS), also known as cloud infrastructure services, which provides computing resources like CPUs, memory, and storage for running virtualized systems over the cloud (e.g., Amazon EC2, RackSpace Cloud).

Clouds can be exploited as effective infrastructures for handling knowledge discovery applications. In particular, KDD services may be implemented within each of the three categories listed above:

- *KDD as SaaS*, where a single well defined data-mining algorithm or a ready-to-use knowledge discovery tool is provided as an Internet service to end users, who may use it directly through a Web browser.
- *KDD as PaaS*, where a supporting platform is provided to developers that have to build their own applications or extend existing ones. Developers can just focus on the definition of their KDD applications without worrying about the underlying infrastructure or distributed computation issues.
- *KDD as IaaS*, where a set of virtualized resources are provided to developers as a computing infrastructure to run their data-mining applications or to implement their KDD systems from scratch.

In all three scenarios listed above, the cloud plays the role of infrastructure provider, even if at the SaaS and PaaS layers the infrastructure can be transparent to the end user. In the following we briefly discuss Windows Azure as an example of a proprietary PaaS environment that can be effectively exploited to implement KDD systems and applications.

51.3.1 An example of PaaS: Windows Azure

Windows Azure is an environment and a set of cloud services that can be used to develop cloud-oriented applications, or to enhance existing applications with cloud-based capabilities. The platform provides on-demand compute and storage resources, exploiting the computational and storage power of the Microsoft datacenters. Azure is designed for supporting high availability and dynamic scaling services that match user needs with a pay-per-use pricing model. The Azure platform can be used to perform the storage of large datasets, execute large volumes of batch computations, and develop SaaS applications targeted towards end users. Windows Azure includes three basic components/services:

- *Compute* is the computational environment to execute cloud applications. Each application is structured into roles: Web role, for Web-based applications; worker role, for batch applications; VM role, for virtual-machine images.
- *Storage* provides scalable storage to manage binary and text data (Blobs), non-relational tables (Tables), queues for asynchronous communication between components (Queues), and NTFS volume (Drives).
- *Fabric controller* whose aim is to build a network of interconnected nodes from the physical machines of a single datacenter. The Compute and Storage services are built on top of this component.

The Windows Azure platform provides standard interfaces that allow developers to interact with its services. Moreover, developers can use IDEs like Microsoft Visual Studio and Eclipse to design and publish Azure applications easily.

Based on our study summarized in Table 51.1, the Azure components and mechanisms can be effectively exploited to fulfill the functional requirements of a generic distributed KDD system that have been introduced in section 51.2. We exploited these components and mechanisms to implement the Data Mining Cloud Framework described in the next section.

51.4 Data Mining Cloud Framework

We worked to design a framework for supporting the scalable execution of knowledge discovery applications on top of cloud platforms. The framework has been designed to be implemented on different cloud systems. However, an implementation of this framework has been carried out using Windows Azure and has been evaluated through a set of data-analysis applications executed on a Microsoft Cloud datacenter.

The framework has been designed to support three classes of knowledge discovery applications: single-task applications, in which a single data-mining task is performed on a given dataset; parameter-sweeping applications, in which a dataset is analyzed by multiple instances of the same data-mining algorithm with different parameters; and workflow-based applications, in which knowledge discovery applications are specified as workflows.

51.4.1 System Architecture

The Data Mining Cloud Framework architecture includes different kinds of components that can be grouped into storage and compute components (see Figure 51.1).

The storage components include:

- A *Data Folder,* which contains data sources and the results of knowledge-discovery processes. Similarly, a *Tool Folder* contains libraries and executable files for data selection, pre-processing, transformation, data mining, and results evaluation.

Table 51.1   *How Azure components fulfill the functional requirements of a distributed KDD system*

KDD system requirements		Azure components
Resource management	Data	*Different data formats*: binary large objects (Blobs); nonrelational tables (Tables); queues for communication data (Queues); relational databases (SQL database). *Metadata support*: tables/SQL databases to store data descriptions; custom description fields can be added to Blobs containing data sources.
	Tools	*Implementation – independent access*: tools can be exposed as Web services. *Metadata support*: Tables/SQL databases to store tools descriptions; custom description fields can be added to Blobs containing binary tools; WSDL descriptions for Web services.
	Results	*Models storing*: Blobs to store results either in textual or visual form. *Metadata support*: Tables/SQL databases to describe models format; custom description fields can be added to Blobs containing data-mining models.
Application management	Design	*Single-task applications*: programming the execution of a single Web service or binary tool on a single Worker role instance. *Parameter sweeping applications*: programming the concurrent execution of a set of Web services or binary tools on a set of Worker role instances. *Workflow-based applications*: programming the coordinated execution of a set of Web services or binary tools on a set of Worker role instances.
	Execution	*Storage resources access*: managed by the Storage layer. *Compute resources allocation*: managed by the Compute layer. *Application execution and monitoring*: Web services / Worker role instances to run single tasks; Tables to store tasks information; Web role instance to present monitoring information. *Results presentation*: Blobs / Tables to store/interpret the inferred models; Web role instance to present results.

- *Data Table, Tool Table* and *Task Table* contain metadata information associated with data sources, tools, and tasks.
- The *Task Queue* contains the tasks ready to be executed.

  The compute components are:

- A pool of *Worker instances*, which are in charge of executing data-mining tasks submitted by users.
- A pool of *Web instances* host the *web site*, by allowing users to submit, monitor the execution, and access the results of their data-mining tasks.

  The *web site* is the user interface to three functionalities: (i) *app. submission*, which allows users to submit single-task, parameter-sweeping, or workflow-based applications; (ii) *app. monitoring*, which is used to monitor the status and access results of the submitted applications; (iii) *data / tool management*, which allows users to manage input / output data and tools.

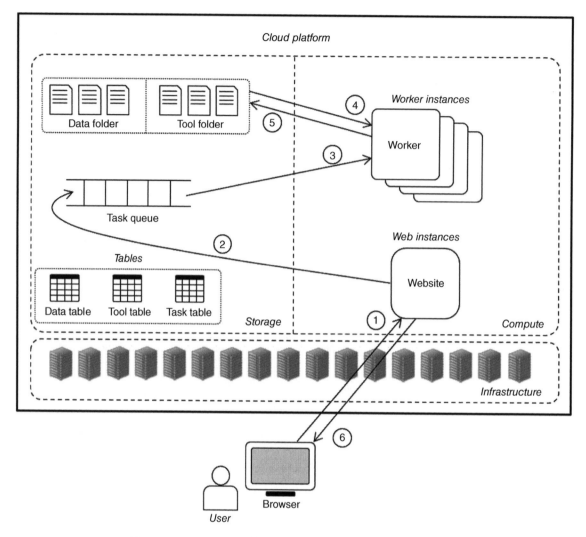

Figure 51.1   *System architecture and application execution steps*

51.4.2   Applications Execution

A user interacts with the system to perform the following steps for designing and executing a knowledge discovery application:

1. The user accesses the web site and designs the application (either single-task, parameter sweeping, or workflow-based) through a Web-based interface.
2. After application submission, the system creates a set of tasks and inserts them into the Task Queue on the basis of the application.
3. Each idle Worker picks a task from the Task Queue, and concurrently executes it.
4. Each Worker gets the input dataset from the location specified by the application. To this end, a file transfer is performed from the Data Folder where the dataset is located, to the local storage of the Worker.

5. After task completion, each Worker puts the result on the Data Folder.
6. The web site notifies the user as soon as her / his task(s) have completed, and allows her / him to access the results.

The set of tasks created in the second step depends on the type of application submitted by the user. In the case of a single-task application, just one data-mining task is inserted into the Task Queue. If the user submits a parameter sweeping application, the tasks corresponding to the combinations of the input parameters values are executed in parallel. In the case of a workflow-based application, the set of tasks created depends on how many data-mining tools are invoked within the workflow; initially, only the workflow tasks without dependencies are inserted into the Task Queue.

The Task Table is dynamically updated whenever the status of a task changes. The web site periodically reads and shows the content of this table, thus allowing users to monitor the status of their tasks.

Input data is temporarily staged on a server for local processing. To reduce the impact of data transfer on the overall execution time, it is important that input data are physically close to the virtual servers where the workers run on.

51.4.3 User Interface

The *App submission* section of the web site is composed of two main parts: one pane for composing and running both single-task and parameter-sweeping applications and another pane for programming and executing workflow-based knowledge discovery applications.

As an example, Figure 51.2 shows a screenshot of the App submission section, taken during the execution of a parameter-sweeping application. An application can be configured by selecting the algorithm to be executed, the dataset to be analyzed, and the relevant parameters for the algorithm. The system submits to the cloud a number of independent tasks that are executed concurrently on a set of virtual servers.

A user can monitor the status of each single task through the App monitoring section, as shown in Figure 51.3. For each task, the current status (submitted, running, done, or failed) and status update time are

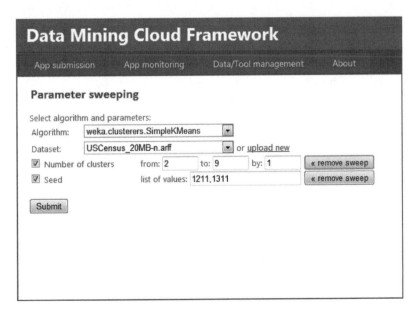

Figure 51.2   *Screenshot of the App submission section*

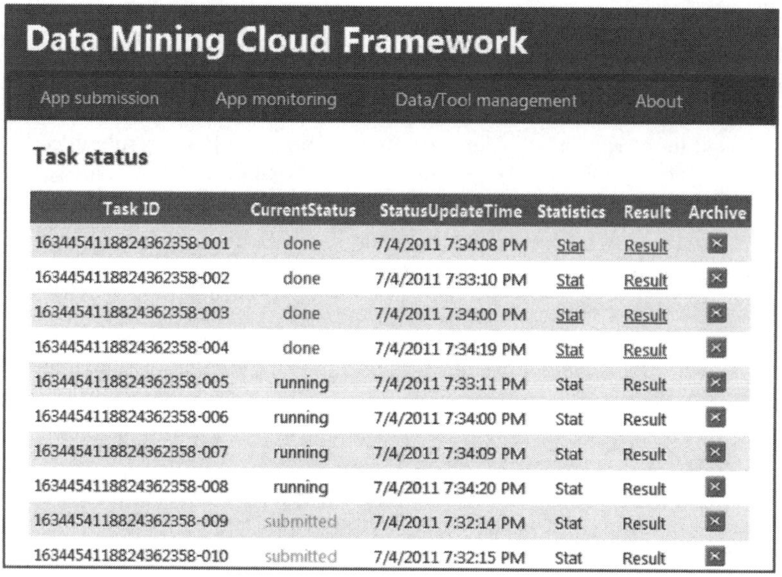

Figure 51.3   *Screenshot of the app monitoring section*

shown. Moreover, for each task that has completed its execution, two links are enabled: the first one (Stat) gives access to a file containing some statistics about the amount of resources consumed by the task; the second one (Result) visualizes the task result.

51.4.4   Workflow Programming

As mentioned above, the framework also includes the programming interface and its services to support the composition and execution of workflow-based knowledge discovery applications. Workflows support research and scientific processes by providing a paradigm that may encompass all the steps of discovery based on the execution of complex algorithms and the access and analysis of scientific data. In data-driven discovery processes, knowledge discovery workflows can produce results that can confirm real experiments or provide insights that cannot be achieved in laboratories.

Visual workflows in our framework are directed acyclic graphs whose nodes represent resources and whose edges represent the dependencies among the resources. Workflows include two types of nodes:

- *Data node*, which represents an input or output data element. Two subtypes exist: Dataset, which represents a data collection, and Model, which represents a model generated by a data analysis tool (e.g., a decision tree).
- *Tool node*, which represents a tool performing any kind of operation that can be applied to a data node (filtering, splitting, data mining, etc.).

The nodes can be connected with each other through direct edges, establishing specific dependency relationships among them. When an edge is being created between two nodes, a label is automatically attached to it representing the kind of relationship between the two nodes.

Data and Tool nodes can be added to the workflow singularly or in array form. A data array is an ordered collection of input / output data elements, while a tool array represents multiple instances of the same tool.

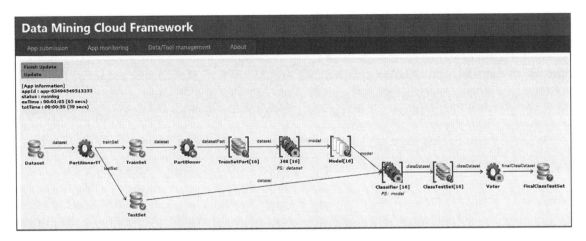

Figure 51.4   *The workflow during its execution*

Table 51.2   *Execution times and speedup of the application using up to 16 virtual machines*

No. of servers	125 MB dataset		250 MB dataset		500 MB dataset	
	Execution time	Speedup	Turnaround time	Speedup	Turnaround time	Speedup
1	00:29:49	1	01:18:07	1	03:16:38	1
2	00:16:03	1.86	00:39:25	1.98	01:44:03	1.89
4	00:08:25	3.54	00:19:54	3.93	00:53:04	3.92
8	00:04:57	6.02	00:12:45	6.13	00:30:47	6.76
16	00:03:05	9.67	00:07:10	10.9	00:19:02	10.9

Figure 51.4 shows a data-mining workflow composed of several sequential and parallel steps as an example for presenting the main features of the visual programming interface of the Data Mining Cloud Framework (Marozzo *et al.,* 2013). The example workflow analyzes a dataset using 16 instances of the J48 classification algorithm provided by the Weka toolkit (Witten and Frank, 2000), which work on 16 partitions of the training set and generate the same number of knowledge models. By using the 16 generated models and the test set, 16 classifiers in parallel produce the same number of classified datasets. In the final step of the workflow, a voter generates the final dataset by assigning a class to each data item, choosing the class predicted by the majority of the models.

Table 51.2 presents execution times and speedup values achieved by using up to 16 virtual machines to execute the workflow on three datasets with size of 125, 250 and 500 MB. They are extracted from the KDD Cup 1999's dataset (Bache and Lichman, 2013).

We can observe that, in this example, the speedup achieved is satisfactory even if it does not increase linearly with the number of servers used because partitioning and voting tools run sequentially in the workflow. On the other hand, there are cases in which the inherent parallelism of applications can be fully exploited, bringing a linear speedup. For example, with a parameter-sweeping data-mining application on large data, discussed in Marozzo *et al.,* 2011), we achieved almost a linear speedup (14.6 on 16 virtual servers). Even when the speedup is not linear, the absolute amount of time saved can be significant when large datasets are analyzed. For instance, in the application whose results are presented Table 51.2, the execution time drops from more than 3 hours using 1 server, to less than 20 minutes using 16 servers. These results show the effectiveness of the proposed approach based on cloud resource exploitation for running data analysis applications.

51.4.5 Other Cloud Solutions for KDD Implementation

In the implementation mentioned before, the following mapping is used between Data Mining Cloud Framework's components and Azure's components: (i) Data Folder and Tool Folder are implemented as Azure's Blob containers; (ii) Data Table, Tool Table, Application Table, Task Table, and Users Table are implemented as Azure's nonrelational tables; (iii) the Task Queue is implemented as an Azure's Queue; (iv) Virtual Compute Servers are implemented as Azure's Worker Role instances; (v) Virtual Web Servers are implemented as Azure's Web Role instances.

Even though the current implementation of framework is based on Azure, it has been designed to abstract from specific Cloud platforms. It can therefore be implemented using other cloud systems, like, for instance, the popular Amazon Web Services (AWS) or the Google App Engine. In particular, AWS offers compute and storage resources in the form of Web services, which comprise compute services, storage services, database services, and app services. Compute services include Elastic Compute Cloud (EC2) for creating and running virtual servers and Amazon Elastic MapReduce for building and executing MapReduce applications (Dean and Ghemawat, 2008). Storage services include Simple Storage Service (S3) for storing and retrieving data via the Internet. Database services include Relational Database Service (RDS) for relational tables and DynamoDB for nonrelational tables. App services include, among others, Simple Queue Service that implements a queue for application-level messages. If AWS is considered as a target Cloud platform, the Data Mining Cloud Framework's components can be implemented on the AWS's components as follows: (i) Data Folder and Tool Folder could be stored on S3; (ii) Data Table, Tool Table, Application Table, Task Table, and Users Table could be implemented as non-relational tables using DynamoDB; (iii) the Task Queue could be implemented using the Simple Queue Service; (iv) Virtual Compute Servers and Virtual Web Servers could be created on top of EC2.

Furthermore, other than on well known public cloud platforms, the framework can be implemented on the top of a private IaaS system by using open source cloud frameworks such as OpenStack. In this case, each component could be implemented using a software library/application deployed on a virtual machine executed by the IaaS system. According to this approach, the Data Mining Cloud Framework components can be implemented as follows: (i) Data Folder and Tool Folder could be implemented as FTP servers (e.g., Filezilla); (ii) Data Table, Tool Table, Application Table, Task Table, and Users Table could be implemented as nonrelational tables (e.g., MongoDB); (iii) Task Queue could be implemented using a message-oriented middleware (e.g., Java Message Service); (iv) Virtual Compute Servers could be implemented as batch applications (e.g., Java applications); (v) Virtual Web Servers could be implemented as Web servers (e.g., Apache / Tomcat).

51.5   Conclusion

In this chapter, we discussed how cloud computing technologies can be exploited to implement a service-oriented distributed KDD system. Starting from the requirements of a generic distributed KDD system, we discussed how these requirements can be fulfilled by a cloud platform. As a case study, we described the Data Mining Cloud Framework, a system that supports the distributed execution of KDD applications. The user interface is very simple and hides the complexity of the cloud infrastructure used to run applications.

The performance of the Data Mining Cloud Framework have been evaluated through the execution of data-mining applications on a pool of virtual servers hosted by a Microsoft Cloud datacenter. The experiments demonstrated the effectiveness of the framework, as well as the scalability that can be achieved through the parallel execution of parameter sweeping data-mining applications on a pool of virtual servers.

Current work is aimed at supporting the design and execution of script-based data analysis workflows on clouds. In Marozzo *et al.* (2015), we introduced a workflow language, named JS4Cloud, that extends JavaScript to support the implementation of cloud-based data analysis tasks and the handling of data on the cloud. We also demonstrated how data analysis workflows programmed through JS4Cloud can be processed by the Data Mining Cloud Framework to make parallelism explicit and to enable their scalable execution on clouds.

References

Bache, K. and Lichman, M. (2013) *UCI Machine Learning Repository*, University of California, School of Information and Computer Science, Irvine, CA, http://archive.ics.uci.edu/ml (accessed January 5, 2016).

Dean, J. and Ghemawat, S. (2008) MapReduce: Simplified data processing on large clusters. *Communications of the ACM* **51**(1), 107–113.

Marozzo, F., Talia, D. and Trunfio, P. (2011) *A Cloud Framework for Parameter Sweeping Data Mining Applications*. Proceedings of the 2011 IEEE Third International Conference on Cloud Computing Technology and Science (CLOUDCOM'11). IEEE Computer Society, Washington, DC, pp. 367–374.

Marozzo, F., Talia, D., and Trunfio, P. (2013) A cloud framework for big data analytics workflows on Azure, in *Clouds, Grids and Big Data* (ed. L. Grandinetti), IOS Press, Amsterdam.

Marozzo, F., Talia, D., and Trunfio, P. (2015) JS4Cloud: Script-based Workflow Programming for Scalable Data Analysis on Cloud Platforms. *Concurrency and Computation: Practice and Experience* **27**(17), 5214–5237. Wiley InterScience.

Mell, P. M. and Grance, T. (2011) *The NIST Definition of Cloud Computing*. Special Publication 800-145. NIST, Gaithersburg, MD, http://www.nist.gov/customcf/get_pdf.cfm?pub_id=909616 (accessed November 25, 2015).

Talia, D. and Trunfio, P. (2010) How distributed data mining tasks can thrive as knowledge services. *Communications of the ACM* **53**(7), 132–137.

Witten, I. H. and Frank, E. (2000) *Data Mining: Practical Machine Learning Tools and Techniques with Java Implementations*, Morgan Kaufmann, San Francisco, CA.

52

Cloud Knowledge Modeling and Management

Pierfrancesco Bellini, Daniele Cenni, and Paolo Nesi

University of Florence, Italy

52.1   Introduction

Almost all relevant infrastructures are using cloud-based approaches to manage their resources, and have set up high-availability solutions addressing different layers such as IaaS, PaaS, and SaaS. Several vendors are covering different aspects and supporting different services natively for cloud solutions. Most of them provide specific products addressing only a limited number of features and services. On the other hand, the availability of a wide range of services is often the basis for selecting cloud solutions. Among the services that are requested is the ability to monitor, change, and move virtual machines and services in the same cloud for resource optimization and among different clouds to increase reliability and for migration purposes. To this end, the modeling and formalization of cloud models and information is becoming more relevant to the formalization of different aspects of a cloud at its different levels – IaaS, PaaS, SaaS – and to specific resources: hosts, virtual machines, networks, memory, storage, processes, services, applications, and so forth, and their relationships.

52.1.1   Modeling Knowledge

In the past, some knowledge representation formalisms were introduced for cloud data modeling. Most of them were rooted in simple data structure with description logic to model more complex relationships. In recent years, with the beginning of the semantic Web, there has been a new interest in knowledge description formalisms. The W3C introduced several recommendations for the description of information on the Web, and information to be interpreted by machines in general. The basis of the standards is the Resource Description Framework (RDF). With RDF (http://www.w3.org/RDF/, accessed January 5, 2016) a fact is

Encyclopedia of Cloud Computing, First Edition. Edited by San Murugesan and Irena Bojanova.
© 2016 John Wiley & Sons, Ltd. Published 2016 by John Wiley & Sons, Ltd.

represented with a "triple," made up of a subject, a predicate and an object or a data value. Moreover the subject, the predicate and the object are represented with a uniform resource identifier (URI). For example, RDF triple http://www.example.com/p.bellini, **http://xmlns.com/foaf/0.1/knows**, http://www.example.com/p.nesi states that the "thing" identified by URI http://www.example.com/p.bellini "knows" the other "thing" identified by http://www.example.com/p.nesi. The "knows" predicate is also defined as an *object property* and it is identified by URI http://xmlns.com/foaf/0.1/knows. This property belongs to the FOAF (Friend Of A Friend) vocabulary defining aspects and characteristics of people and their relations on the Web (http://www.foaf-project.org/, accessed January 5, 2016).

To express it more concisely, a part of the URI can be considered a prefix that identifies the namespace of the thing being described. So, for example, "*ex*" could be the prefix for http://www.example.com/ and "*foaf*" for http://xmlns.com/foaf/0.1/ and thus the same triple is expressed by:

`ex:p.bellini `**`foaf:knows`**` ex:p.nesi`

It is also possible to state that something belongs to a class of things and this fact can be represented by a triple. For example, the following RDF triple states that *ex:p.bellini* identifies something that belongs to the class of people:

`ex:p.bellini http://www.w3.org/1999/02/22-rdf-syntax-ns#type foaf:Person`

There is also the possibility of associating simple data values (strings, numbers, dates, etc.) with a subject URI. In the following example, family name and given name of *ex:p.bellini*, are provided. Furthermore, the *familyName* and *givenName* are called *data properties*:

`ex:p.bellini `**`foaf:familyName`**` "Bellini"`
`ex:p.bellini `**`foaf:givenName`**` "Pierfrancesco"`

When two or more consecutive triples share the same subject URI (as in the previous example), we can write:

`ex:p.bellini `**`foaf:familyName`**` "Bellini"; `**`foaf:givenName`**
`"Pierfrancesco".`

A vocabulary defines the common characteristics of things belonging to classes and their relations. A vocabulary can be also called "an ontology." It is defined using the RDF Schema (RDFS) or Web Ontology Language (OWL). For example, the "*knows*" object property is defined as having a domain and range class *foaf:Person*. When using this information, what can be inferred is that both *ex:p.bellini* and *ex:p.nesi* belong to the class *foaf:Person*. Moreover, the vocabulary states that the class *foaf:Person* is a subclass of a more general class *foaf:Agent*; thus both *ex:p.bellini* and *ex:p.nesi* belong to the class *foaf:Agent*.

Web Ontology Language is a family of three ontology languages: OWL-Lite, OWL-DL, and OWL-Full. The first two languages can be considered syntactic variants of SHIF(D) and SHOIN(D) description logics (DL), respectively, whereas the third language was designed to provide full compatibility with RDF(S) (Bellandi *et al.*, 2012). The OWL version 2 language proposed by W3C is quite powerful; it allows the definition of disjunctive classes, union and intersection of classes, functional properties, symmetric, transitive properties, minimum and maximum cardinality of the associated elements of a property, and other features. OWL 2 is still based on RDF semantics and provides datatype. It has three profiles: OWL2 EL, OWL2 QL and OWL2 RL specifically designed and suitable for reasoning with existential quantifications, query formalization and access, reasoning and formalization of rules, respectively (http://www.w3.org/TR/owl2-profiles/, accessed January 6, 2016).

In order to exploit the information encoded as a set of RDF triples, it can be stored in RDF stores, which are optimized for RDF data management and for the activation of reasoning processes on the collected knowledge. To this end, a specific query language was designed to find RDF store reticular information. Thus, the SPARQL (SPARQL Protocol and RDF Query Language, recursive definition (http://www.w3.org/TR/2004/WD-rdf-sparql-query-20041012/, accessed January 5, 2016)) uses an advanced matching algorithm to match a portion of the RDF graph with a specified template. For example, the following query lists names of people known by a person (identified with his e-mail) directly and indirectly through one or more people:

```
SELECT ?n WHERE {
        ?p1 foaf:mbox <mailto:pbellini@unifi.it>.
        ?p1 rdf:knows+?p2.
        ?p2 foaf:name ?n.
}
```

Moreover, integrated ontologies can be adopted to enforce capabilities in the model – for example, by exploiting vocabularies (ontology segments) to define properties such as the *FOAF* for people and structures, *Dublin Core* for metadata, *wgs84_pos* for latitude and longitude representation, *OWL-Time* or *TimeOnt* for reasoning about time and temporal aspects, *INDL* for infrastructure and network description, and *QoSOnt* to define quality of service aspects.

52.1.2   Exploiting Knowledge

The main motivations for modeling and using a cloud knowledge base are related to its exploitation for semantic computing and thus for reasoning about cloud (Androcec *et al.*, 2012). The modeling can be performed using ontology in OWL and RDF. A cloud ontology and knowledge base consists of an ontology that can be used as model for a big data RDF store, including cloud resource configurations and conditions at the level of IaaS, PaaS, and SaaS, the service level agreements (SLAs) of multitier applications and deployments, monitoring data, supporting reselling, brokerage, and real instance data. In a seminal work, Youseff *et al.* (2008) proposed an approach to the creation of a cloud ontology, decomposing problems into five layers: applications, software environments, software infrastructure, software kernel, and hardware. The work identified the challenges and discussed the ontology. First attempts to model cloud aspects have been grounded on the proposed taxonomical and Simple Knowledge Organization System (SKOS) models (Hoff, 2009; Lairds, 2009).

A more precise understanding of the effective usage of cloud knowledge modeling exploitation can be taken from an analysis of the literature concerning the use of cloud knowledge for: (i) facilitating interoperability among public and private clouds including automated configurations and deployment (e.g., virtual machine, storage and cloning or migration of services); (ii) verification and validation of cloud configuration structures, virtual machine patterns, hosts, and so forth, against available resources and structures; (iii) discovering and brokering services and resources brokering, including SLAs, SLA, analysis and matchmaking; (iv) computing cloud simulation for resource and cost planning, prediction, and optimization; (v) reasoning about cloud workload conditions estimated by monitoring and needed when taking decisions about the use of resources, such as moving virtual machines, changing resource parameters, negotiating different SLA agreements, detecting critical conditions, and so forth; (vi) reasoning about cloud security conditions and evolution.

The Open Grid Forum (OGF) (https://www.ogf.org/ogf/doku.php, accessed January 5, 2016) with its Open Cloud Computing Interface (OGF-OCCI), aims to define interfaces for a unified interface at the level of IaaS. This would allow the creation of an interoperable layer among different vendors, by using a Unified Modeling Language (UML) model. In the mOSAIC EC FP7-ICT project (Moscato *et al.*, 2011), cloud knowledge modeling has been addressed with the aim of creating a common model to cope with the heterogeneity of terms used by different cloud vendors, and with standards referring to cloud systems using different terminology. The major issue for cloud interoperability is the lack of standardized APIs, thus the interaction and migration of VMs is a difficult task. The problem of interoperability has been addressed by IEEE Project, P2301 – Guide for Cloud Portability and Interoperability Profiles (CPIP), and by IEEE P2302 – Draft Standard for Intercloud Interoperability and Federation with the aim of defining common interoperability protocols among federated clouds and defining configuration, functionalities, and management of inter-cloud interoperability (IEEE, 2014).

The problems of discovering services, negotiating them, and their composition for cloud infrastructure based on ontological models have been discussed in Sim (2011). The proposed solution included a reasoner for similarity analysis and compatibility analysis. Dastjerdi *et al.* (2010) proposed an architecture and solution to provide virtual appliances on demand. The idea is mainly derived from the SLA models adopted for grid computing solutions. Some efforts to describe SLAs were made in the past, beginning with WSLA for the definition of SLAs of WebServices (Ludwig *et al.*, 2003).

Regarding cloud simulation, a significant example is CloudSim (Calheiros, *et al.*, 2011), where several layers of a typical cloud stack can be simulated, including IaaS, SLA, and so forth, without using knowledge-based modeling. The solution is suitable for simulating simple cloud solutions but not, for example, problems related to the verification and validation of configurations, smart strategies.

The use of a knowledge base for reasoning about cloud structures and resources, which means automated provisioning and verification of service composition, configuration, optimization and deployment, can be defined as "Smart Cloud." This may consist of a set of semantic modeling and computing tools for cloud status reasoning while considering the cloud status and evolution via the cloud knowledge base. Intelligence on a smart cloud is enforced by means of a set of algorithms to detect and predict critical conditions, verification, and validation of configurations (feasibility in terms of consistency and completeness, while taking into account present and possible future available resources), estimating slack, automated verification of completeness and consistency, verification of the compatibility of the service level agreement (SLA) with available resources, and so forth.

Currently there are several attempts to build smart cloud solutions grounded on ontology and cloud computing (Androcec *et al.*, 2012). Most efforts have been focused on the description of the services available on the cloud, to allow users to search and compare services (Zhang *et al.*, 2012).

The most interesting projects related to this topic are: (i) linked USDL (Unified Service Description Language – http://www.linked-usdl.org/, accessed January 6, 2016) used by the FI-WARE European project, providing a set of vocabularies for the description of the different service aspects (core service description, SLA, security, price and intellectual property rights (IPRs), even though it is focused on service search and discovery; (ii) the mOSAIC project developed a wide ontology covering many aspects from service deployment to service description, and is also focused on cloud service search (Moscato *et al.*, 2011); and (iii) the Icaro Cloud project developed an ontology for the description of both infrastructure and services considering verification and validation of configurations and monitoring information and SLAs (http://www.disit.org/5482, accessed January 6, 2016).

In addition to the above innovative solutions, there are state-of-the-art solutions provided by major vendors such as IBM, VMware, HP, and Microsoft, and some specific additional tools and plugins that enforce intelligence on infrastructure management systems.

52.2   Cloud Knowledge Modeling for Smart Cloud

This section reports the current attempts to model cloud knowledge; focuses on the description of infrastructure, platform, applications, and business processes.

52.2.1 Modeling IaaS Information

The IaaS information contains material related to the physical structure of a datacenter, which is made of host machines connected on one or more networks, while hosts may have virtual machines assigned. What follows is a possible example of a datacenter with 100 hosts, one external storage, and two firewalls described using the vocabulary being developed for the Icaro project (http://www.disit.org/5482, accessed January 6, 2016):

```
ex:datacenter1 rdf:type cld:DataCenter;
        cld:hasName "production data center";
        cld:hasPart ex:host1;
        ...
        cld:hasPart ex:host100;
        cld:hasPart ex:storage1;
        cld:hasPart ex:firewall1;
        cld:hasPart ex:firewall2;
```

Each host machine can have details of the number of CPUs available, the memory size in GB, the disk size in GB, the network adapters and the installed operative system, as in the following example:

```
ex:host1 rdf:type cld:HostMachine;
        cld:hasName "host 1";
        cld:hasCPUCount 16;
        cld:hasCPUSpeed 2.2;
        cld:hasCPUType "Intel Xeon X5660";
        cld:hasMemorySize 16;
        cld:hasDiskSize 300;
        cld:hasLocalStorage ex:host1_disk;
        cld:hasNetworkAdapter ex:host1_net1;
        cld:hasNetworkAdapter ex:host1_net2;
        cld:hasOS cld:vmware_esxi;
        cld:isPartOf ex:datacenter1.
ex:host1_net1 rdf:type cld:NetworkAdapter;
        cld:hasIPAddress "192.168.1.1";
        cld:boundToNetwork ex:network1.
ex:host1_disk rdf:type cld:LocalStorage;
        cld:hasDiskSize 300.
        ...
ex:firewall1 rdf:type cld:Firewall;
        cld:hasName "Firewall 1";
        cld:hasNetworkAdapter ex:firewall1_net1;
        cld:hasNetworkAdapter ex:firewall1_net2.
```

Each host machine contains a number of virtual machines; for example, a virtual machine with 2 CPUs, 1 GB of RAM, 10 GB of disk, one network adapter, with Windows XP professional running on host5, which is described as:

```
ex:vm1 rdf:type cld:VirtualMachine
        cld:hasName "vm 1, windows xp";
        cld:hasCPUCount 2;
```

```
        cld:hasMemorySize 1;
        cld:hasVirtualStorage ex:vm1_disk;
        cld:hasNetworkAdapter ex:vm1_net1;
        cld:hasOS cld:windowsXP_Prof;
        cld:isStoredOn ex:host1_disk;
        cld:isPartOf ex:host1.
ex:vm1_disk rdf:type cld:VirtualStorage;
        cld:hasDiskSize 10.
```

Moreover each element can have associated information needed for monitoring purposes (see section 52.3.4).

Regarding the description of infrastructure resources, ontology like INDL-Infrastructure and Network Description Language may be used (Ghijsen *et al.*, 2012). It defines a generic *Node* that is linked with other nodes through *interfaces* and *links*. *VirtualNode*s are used to represent virtual machines running on nodes. *NodeComponent*s are used to represent the processing, memory, and storage components of a node.

The differences between the two formalizations seem to be mainly related to the description of networking aspects – namely in INDL it is more detailed, although it lacks some details.

The mOSAIC ontology allows some information about the host and the virtual machines to be described (e.g., CPU, memory, storage), but it does not describe how they are connected in the network and how the virtual machines are related to the host machine. Virtual machines are stored and associated with hosts or clusters.

52.2.2   Modeling PaaS Information

This section reports current efforts to model the platform level, while considering the services used to create applications. In the cloud, services are the building blocks used to create more complex and complete applications, which can be used. In general, an application uses services, like a database service, a file system service, a mail service, as well as some Web servers or Web application servers. Generally these services are requested by other specific applications at SaaS level by allocating them on a set of virtual machines. These virtual machines can host and implement more than one service or cooperate with other virtual machines to implement a service (e.g., a DB cluster). Moreover, these virtual machines can either provide services for only one specific customer or can be shared among multiple customers. In the latter case, some kind of authentication and service-sharing mechanism is used.

An application can be modeled using specific constraints (e.g., a maximum of four web servers); the application can be seen as a class containing the specific application instances. A way to represent an application is by defining its relations using the OWL constructs. What is reported below is the general definition of the class of Applications expressed using the OWL2 Manchester syntax (http://www.w3.org/TR/owl2-manchester-syntax, accessed January 6, 2016):

```
Application = Software
        and (hasIdentifier exactly 1 string)
        and (hasName exactly 1 string)
        and (developedBy some Developer) and (developedBy only
        Developer)
        and (createdBy exactly 1 Creator) and (createdBy only Creator)
        and (administeredBy only Administrator)
        and (needs only (Service or Application or ApplicationModule))
        and (hasSLA max 1 ServiceLevelAgreement)
        and (hasSLA only ServiceLevelAgreement)
        and (useVM some VirtualMachine) and (useVM only VirtualMachine)
```

It states that an Application is a Software, which has exactly one identifier and one name, it has been developed by one or more developers (and only by developers!), it has been created (instantiated) by a creator user, it can be administered only by administrator users, it needs only Services, other Applications or ApplicationModules, it has at most one SLA, and it uses some virtual machines. Subclasses of the Service are the services running on a virtual machine. A specific application, for example Joomla, is a subclass of Application with some additional constraints:

```
Joomla SubClassOf Application
  and (needs exactly 1 MySQLServer)
  and (needs exactly 1 HttpBalancer)
  and (needs exactly 1 NFSServer)
  and (needs min 1 (ApacheWebServer and (supportsLanguage value php_5)))
```

The Joomla class is defined as a subclass of the intersection of the Application class with the classes of things that need exactly one MySQL server, one Http balancer, one NFS server, and at least one Apache WebServer supporting PHP 5.

A specific instance of the Joomla application is as follows:

```
ex:Joomla1 rdf:type app:Joomla;
  cld:hasName "Joomla for my business";
  cld:developedBy ex:user;
  cld:createdBy ex:u1;
  cld:needs ex:mysql1, ex:apache1, ex:apache2, ex:httpbalancer1,
  ex:nfsserver1;
  cld:hasSLA ex:sla1;
  ...
ex:mysql1 rdf:type cld:MySQLServer;
  runsOnVM ex:vm1;
  ...
ex:apache1 rdf:type cld:ApacheWebServer;
  cld:runsOnVM ex:vm2;
  cld:supportsLanguage cld:php_5;
  ...
```

52.2.3   Modeling SaaS, and XaaS Information

This section describes current efforts to model the whole service provided by the cloud, by considering the interoperability aspects and the brokerage of services from different clouds, as well as the description of a whole business process.

As shown in section 52.3.2, an application may be described by its parts (the services being used) and it may also have associated pricing information, a description of the provided functionalities, a service-level description, and other aspects. When using this kind of information, a third party can store all such application descriptions, and provide a service making it possible to search for applications having some functionality (e.g., enterprise resource planning – ERP) with some pricing constraints and some other interesting features.

The most interesting project dealing with this kind of information is *Linked USDL*. It allows the description of the pricing information of a service while reusing other popular vocabularies such as GoodRelations (Hepp, 2008), Dublin Core (http://dublincore.org, accessed January 6, 2016) and FOAF. It allows a service to be associated with a *PricePlan* having different *PriceComponents*, which may be based on different *PriceVariables*. It also allows complex dynamic pricing to be modeled. The following example is related to a plan to use a service for 5 euro/month:

```
ex:Joomla rdf:type usdl-core:ServiceOffering;
       usdl-price:hasPricePlan ex:joomlaPriceplan.
       ...
ex:joomaPriceplan rdf:type usdl-price:PricePlan;
       usdl-price:hasPriceComponent ex:ppc;
       ...
ex:ppc rdf:type usdl-price:PriceComponent;
       price:hasPrice [ rdf:type gr:PriceSpecification ;
       gr:hasCurrency "EUR" ;
       gr:hasCurrencyValue "5";
       gr:hasUnitOfMeasurement "MON"
       ] .
```

Another aspect related to SaaS is multitenancy. This approach involves the possibility of exploiting only a part of a shared service and not the entire software application. This portion of the service application is defined as tenant, and it behaves as if it were the full application, whereas the service is shared among all the application tenants. In this case, the shared application has a set of tenants that can have specific SLAs (e.g., they may use a certain amount of storage, a certain amount of network bandwidth, a certain amount of connections, etc.).

When business users create their business processes on cloud, they can decide to use different applications, which can be connected to share information. A BusinessConfiguration can be described as a set of applications or application tenants that may have dependencies on one another – for example, a business configuration with a Joomla instance and a customer relationships management (CRM) tenant.

52.2.4   Modeling SLAs and Monitoring Aspects

Regarding the formalization of SLAs, some efforts have been made beginning with WSLA for the definition of SLAs of WebServices (Ludwig *et al.*, 2003). The SLA is described using XML schema and it is very general, thus allowing service metrics to be defined and composed.

The description defines services (*ServiceDefinition*) on the basis of parameters (*SLAParameter*) that are defined using metrics (*Metric*) and metrics are defined using functions that can use other metrics. An SLA can be associated with some *Obligations* describing the objectives of the service level to be guaranteed.

The WS-Agreement was developed from the Grid Resource Allocation Agreement Protocol Working Group (GRAAP-WG). The *WS-Agreement Specification V1.0*, defined a protocol to specify the agreement between two services, and it was published in May 2007 as an Open Grid Forum Proposed Recommendation. The specification is composed of a schema for the agreement description, a set of "*port type*" and "*operation*" to manage the agreements life-cycle (including creation, termination and agreement state control). An *Agreement* is made of a *Context* and some *Terms* divided into *ServiceTerms* and *GuaranteeTerms*, the latter including the conditions that need to be guaranteed by the service. Conditions can be specified with a target value for the parameter or using expressions, but the syntax to be used to express these conditions is not specified. Any kind of XML or textual representation can be used, thus limiting the description interoperability.

Oldham *et al.* (2006) defined an ontology for matching service requests and offers, beginning with WS-Agreement. This ontology uses *QoSOnt* (Dobson *et al.*, 2005) to define quality of service aspects and *TimeOnt* for temporal aspects. This service (SWAPS) is based on semantic technologies such as IBM SNOWBASE for ontology management and IBM ABLE for the reasoning and for inference rules. It should be noted that WSLA expressions (which can be easily modeled in OWL) are used to define service conditions.

The European project NextGrid, for the definition of a European platform for grid computing, has defined a SLA based mainly on WS-Agreement. Moreover, SLAng (Lamanna *et al.*, 2003) is a different XML schema for SLA definition, which is far less generic than WSLA and WS-Agreement. Moreover, in the context of the FI-WARE project, the SLAware model has been proposed. SLAware defines the formal semantics of the SLA by using Transparent Intentional Logic (a modal temporal logic), while the data model is defined using UML.

In the context of the FI-WARE project, LinkedUSDL has a part related to SLA representation, which is much simpler than SLAware. On the other hand, it seems that the work on SLAware modeling and maintenance has been stopped. In LinkedUSDL-SLA, the service-level profile is associated with a service level that can be a Guaranteed State or a Guaranteed Action. The service level can be associated with a service-level expression, representing, in natural language, the description of the condition to be met, and it is also associated with variables that are taken into account to check whether the condition is fulfilled or not.

In Icaro Cloud, the SLA allows the formalization of a set of conditions based on metric values associated with applications, application tenants, and/or complete business configurations. The Icaro Cloud SLA model follows a simplified WSLA model, composing complex conditions and/or conditions based on comparing metric values to constant values (e.g., defined in the SLA contract). For example, Figure 52.1 depicts a SLA to guarantee a response time less than 5 s for the Apache http server and a database size less than 1 GB. In Figure 52.1 the SLA is represented as an oriented graph where nodes are subject or object URI and arcs are the properties relating them.

In the MOSAIC ontology, the SLA of a service allows it to associate a set of policies. The latter, in turn, can be defined as a set of functional (e.g., monitoring, backup and recovery, replication) and nonfunctional properties

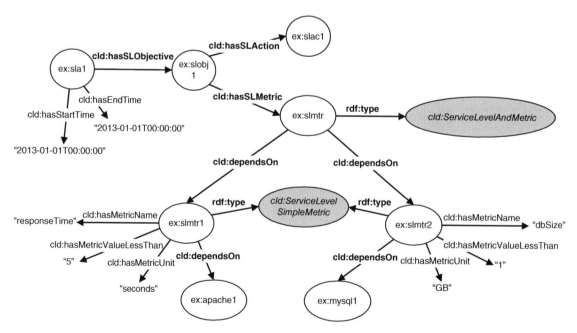

Figure 52.1   *A graph representing a SLA in the framework of the Icaro cloud ontology*

(e.g., CPU speed, network bandwidth, availability). For example, a virtual machine provided as IaaS with x86 CPU architecture, and two CPU cores featuring high replication, can have an SLA represented as follows:

```
ex:vm_sla rdf:type msc:SLA;
        msc:definedForService ex:vm;
        msc:definePolicy ex:vm_policy.
ex:vm rdf:type msc:VirtualMachine;
        msc:hasVirtualizationTechnology msc:Xen;
        ...
ex:vm_policy rdf:type msc:Policy;
        msc:expressRequirement msc:x86;
        msc:expressRequirement [ msc:numberOfCPUCores 2 ];
        msc:expressRequirement msc:HighReplication;
```

The modeling of SLAs can be very different; some approaches are focused on the specification of service metrics bounds and conditions that need to be verified, whereas others, like Mosaic, are more focused on the specification of high-level requirements, which are more difficult to verify. This is due to the fact that some solutions are oriented to SLA verification/checks, whereas others are oriented to allow service search or match. Among solutions focused on SLA checking, there are Linked-USDL and WS-Agreement, which do not have a way to represent the conditions that have to be met. On the other hand, other solutions such as WSLA, SWAPS, and Icaro Cloud, define specific constructs to represent conditions and fit them in their automated computation and reasoning. These latter ones are obviously more suitable for knowledge reasoning.

As far as we know, only Icaro Cloud allows the description of monitoring information associated with host machines, virtual machines, services, and application tenants. The monitoring information may include, for example, the IP address to be used for monitoring a service or specific information regarding the metrics to be monitored – particularly those that are specific to the application and that are used in the SLA. For example, an Apache Web server may have defined a monitor on response time:

```
ex:apache1 rdf:type cld:ApacheWebServer;
        cld:runsOnVM ex:vm2;
        cld:hasMonitorInfo ex:minfo1;
        ...
ex:minfo rdf:type cld:MonitorInfo;
        cld:hasMetricName "responseTime";
        cld:has Arguments "http://..."; #specific arguments to be
        provided to the plugin
        cld:hasWarningValue 1;
        cld:hasCriticalValue 4;
        cld:hasMaxCheckAttempts 3;
        cld:hasCheckInterval 5; #check every 5 min
```

52.3 Smart Cloud vs Industrial Applications

Modeling cloud knowledge can be the basis for enabling a large range of future reasoning applications. Due to the complexity of cloud knowledge models and to the amount of data collected by cloud monitoring systems, cloud reasoning is becoming a big data problem (Bellini *et al.*, 2013). At the industrial level, the closest features to "smart-cloud" reasoning are the so-called resource optimization tools, elasticity, and so forth. Elasticity aims to cope with objectives such as performances, energy consumption, optimization of costs. These approaches could benefit from the presence of cloud knowledge base, and yet in most cases, traditional approaches are used, thus limiting cloud smartness.

As a general rule, elasticity is defined as the ability given to customers to quickly request, receive, and release as many resources as needed. The *elastic* paradigm in cloud computing is strongly related to cloud resource monitoring and prediction, and it should not be confused with scalability, which is the ability of a system to make use of the available increased resources. An elastic application can automatically adapt itself to modify the requested or released resources. Therefore, scalability is defined as a static property, and elasticity as a dynamic one. Elasticity policies are divided into automatic (actions are taken on the basis of rules and settings or SLAs) and manual (the user is responsible for monitoring the cloud environment). GoGrid, Rackspace and Microsoft Azure are notable examples of cloud infrastructures where resources are manually managed with no automatic elasticity policies.

Typically, elastic computing includes three different aspects: replication (i.e., horizontal scale), migration, and resizing (i.e., vertical scale). Replication includes adding/removing resource instances from the cloud environment (e.g., virtual machines, SaaS modules). Migration includes moving a running virtual machine from one physical server to another. Resizing includes adding/removing processing, memory and storage resources from a running resource instance.

Automatic policies are further divided into reactive (based on rules) and predictive. As an example, reactive rules are implemented in Amazon. On the other hand, predictive policies make use of heuristics and mathematical techniques to predict the system behavior, and hence to decide the modality and the amount of scaling. Amazon Web Services includes a replication feature called Auto-Scaling, in the EC2 service (http://aws.amazon.com/ec2, accessed January 6, 2016). This feature makes use of the so-called Auto Scaling Group (ASG) (i.e., a set of instances at the disposal of an application), and it uses an automatic reactive approach where each ASG includes a set of rules defining the number of rules that must be added/removed.

52.4   Conclusions

The cloud knowledge models reviewed and represented in this chapter have been derived from the literature and current attempts at standardization. The state of the art of cloud knowledge is presently in evolution. A major effort is needed to fully cover all the potential capabilities of cloud knowledge base applications. The most widespread applications are in the areas of modeling and reasoning about: (i) cloud interoperability among public and private clouds; (ii) cloud configuration at the different levels of cloud stacks; (iii) cloud service and application discovering and brokering, including SLA matchmaking; (iv) cloud simulation for workload prediction; (v) dynamic analysis to adapt cloud workload conditions, as in the elastic computing paradigm; and (vi) security and security analysis. Most of these application fields need to work on different cloud knowledge models, while a common standard would be needed to make the applications and algorithms interoperable. Modeling cloud knowledge can be the basis for enabling a large range of future reasoning applications. Due to the complexity of cloud knowledge models and the amount of data collected by cloud monitoring systems, cloud reasoning is becoming a big data problem.

References

Androcec, D., Vrcek, N., and Seva, J. (2012) Cloud computing ontologies: A systematic review. Proceedings of The Third International Conference on Models and Ontology-based Design of Protocols, Architectures and Services (MOPAS 2012), Chamonix, France, April 29.

Bellandi, A., Bellini, P., Cappuccio, A., *et al.* Assisted knowledge base generation, management and competence retrieval. *International Journal of Software Engineering and Knowledge Engineering* **32**(8), 1007–1038.

Bellini, P., Di Claudio, M., Nesi, P., and Rauch, N. (2013) Tassonomy and review of big data solutions navigation, in *Big Data Computing* (ed. R. Akerkar), Chapman & Hall/CRC Press, Boca Raton, FL.

Calheiros, R. N., Ranjan, R., Beloglazov, A., *et al.* (2011) CloudSim: a toolkit for modeling and simulation of cloud computing environments and evaluation of resource provisioning algorithms. *Software: Practice and Experience* **41**, 1, 23–50.

Dastjerdi, A. V., Gholam, S., Tabatabaei, H., and Buyya, R. (2010) *An Effective Architecture for Automated Appliance Management System Applying Ontology-Based Cloud Discovery.* Proceedings of the 2010 10th IEEE/ACM International Conference on Cluster, Cloud and Grid Computing (CCGRID '10). IEEE Computer Society, Washington, DC, pp. 104–112.

Dobson, G., Lock, R., and Sommerville, I. (2005) QoSOnt: A QoS ontology for service-centric systems. Thirty-First Euromicro Conference on Software Engineering and Advanced Applications, pp. 80–87.

Ghijsen, M., van der Ham, J., Grosso, P., and de Laat, C. (2012) Towards an infrastructure description language for modeling computing infrastructures. Tenth IEEE International Symposium on Parallel and Distributed Processing with Applications.

Hepp, M. (2008) *GoodRelations: An Ontology for Describing Products and Services Offers on the Web.* Proceedings of the 16th International Conference on Knowledge Engineering and Knowledge Management (EKAW2008), September 29–October 3, Acitrezza, Italy. Springer LNCS, pp. 332–347.

Hoff, C. (2009) *Cloud Taxonomy and Ontology,* http://rationalsecurity.typepad.com/blog/2009/01/cloud-computing-taxonomy-ontology.html.

IEEE P2302 (2014) *Standard for intercloud interoperability and federation (SIIF),* IEEE, Piscataway, NJ.

Lairds, P. (2009) *Cloud Computing Taxonomy.* Proceedings of Interop09, IEEE Computer Society, Las Vegas, May, pp. 201–206.

Lamanna, D. D., Skene, J., and Emmerich, W (2003) SLAng: A language for defining service level agreements. Proceedings of the Ninth IEEE Workshop on Future Trends in Distributed Computing Systems (FTDCS), pp. 100–106.

Ludwig, H., Keller, A., Dan, A., *et al.* (2003) Web Service Level Agreement (WSLA) Language Specification, IBM Corporation, Armonk, NY.

McGuinness, D. L., van Harmelen, F. (2004) *OWL Web Ontology Language Overview. W3C Recommendation,* http://www.w3.org/TR/2004/REC-owl-features-20040210/ (accessed January 6, 2016).

Moscato, F., Aversa, R., Di Martino, B., *et al.* (2011) An analysis of mOSAIC ontology for cloud resources annotation. Proceedings of the Federated Conference on Computer Science and Information Systems (FedCSIS), September 18–21, pp. 973, 980.

Oldham, N., Verma, K., Sheth, A., and Hakimpour, F. (2006) *Semantic WS-Agreement Partner Selection. Proceedings of the 15th International Conference on World Wide Web, (WWW '06),* Edinburgh, May 23–26. ACM Press, New York, NY, pp. 697–706.

Sim, K. M. (2011) Agent-based cloud computing. *IEEE Transactions on Services Computing* doi: 10.1109/TSC. 2011. 52

Youseff, L., Butrico, M., and Da Silva, D. (2008) Towards a unified ontology of cloud computing. Proceedings of the Grid Computing Environments Workshop (GCE '08), November, pp. 1–10.

Zhang, M., Ranjan, R., Haller, A. *et al.* (2012) An ontology-based system for Cloud infrastructure services' discovery. Proceedings of the Eighth International Conference on Collaborative Computing: Networking, Applications and Worksharing (CollaborateCom), 14–17 October, pp. 524, 530.

Part X
Cloud Prospects

53

Impact of the Cloud on IT Professionals and the IT Industry

Cameron Seay,[1] Montressa Washington,[2] and Rudy J. Watson[3]

[1] North Carolina A&T University, USA
[2] IBM, USA
[3] University of Maryland University College, USA

53.1   Introduction

Information workers are familiar with many common cloud applications such as Google, Facebook, iTunes, and Drop Box. These applications access resources that are part of the current information technology (IT) landscape. As consumers, information workers access online banking tools while giving little thought to the implications of these applications for their professions. This chapter will focus on the implications and how cloud computing is impacting the IT industry today and the IT industry of the future.

Perhaps an understanding of the impact on the IT industry depends on the perspective one has of the definition of cloud computing. Different people/institutions define cloud computing with varying degrees of precision and technical specificity. Some take a hardware-centric, services-oriented point of view. It is the infrastructure and the services provided that establish whether or not one is in the realm of cloud computing. Others, like the National Institute of Standards and Technology (NIST), take a much broader view.

On a narrow perspective of cloud computing, the implications for the IT industry are limited to the effect that cloud computing has on the handoffs between technical roles. The focus would be on understanding the overlaps and gaps among various skills required to develop and maintain the infrastructure and provide services. A broader perspective takes into consideration a wider range of skills, including strategic analysis and architecture. Financial implications also come into play.

This chapter takes a broad perspective of cloud computing to look at the effect it has on the overall industry as well as on individual job roles. Cloud computing will impact the overall industry as it relates to where the

Encyclopedia of Cloud Computing, First Edition. Edited by San Murugesan and Irena Bojanova.
© 2016 John Wiley & Sons, Ltd. Published 2016 by John Wiley & Sons, Ltd.

jobs are located. In a traditional structure there is duplication of skills within the service provider and consumer organizations. The nature of cloud computing lends itself to the elimination of this duplication.

The implication is that many of the traditional technical roles will migrate from the consumer organization to the provider of services. As these roles migrate, they will evolve as a result of requiring individuals to possess skills that heretofore have been identified as different disciplines such as networking, database, hardware, software, and security. Cloud computing will blur the lines across these disciplines. This chapter brings into focus the drivers of transformation in the current IT industry, the rationale behind the changes to job roles, and a perspective of the probable future state of the IT industry as a result of the growth of cloud computing.

53.2   Impact on the IT Industry

A perfect storm of hardware cost reduction, a proliferation of Web-related development tools, and a similar proliferation of stable open-source tools have allowed cloud services to mature into the prevailing computing paradigm in IT. Cloud technology allows firms to offload the management of what can be viewed as commodity services – networking, analytics, applications, and infrastructure. Instead of employing a large staff to maintain the corporate infrastructure, many companies are purchasing, as services, functions they used to maintain in-house on a day-to-day basis. Major cloud service providers like Google, Amazon, and IBM are increasing the services they provide, and more customers are using them. As a result, traditional IT service providers, whose offerings in the past have been very similar to the services companies would provide internally, will be challenged to incorporate cloud technologies effectively into their business models. As a consumer of these services, of course, one may not care whether the provider uses cloud technologies or not – one only wants the service that was contracted. But providers will be increasingly compelled to avail themselves of cloud services when these services, as they often do, provide cost or efficiency advantages for the service they provide.

Kim *et al.* (2012) outlines the challenges traditional IT service providers face vis-à-vis areas of outsourcing. They note that maintaining an optimal workforce level for the service provider is difficult due to the volatile nature of IT. For example, in one period, demand for a complex skill like enterprise database management may be high, such as the demand for Oracle database administrators (DBAs), but a year later such demand may dissipate. It should be noted that often these services merely provide flexible staffing for the IT service providers' clients as the contracts provide for on-site expertise. The volatility of the market for IT services makes it hard for service providers to keep adequate staff on hand to match demand precisely. This is compounded by the issue of underutilization of personnel. To keep an adequate level of staff for customers' needs, IT service providers often maintain excess staff, idle for long periods of time. This is of course expensive and inefficient. A cloud-based service model allows IT service providers to maintain skills closer to optimal staffing levels. While specific business-related competencies like DBAs will still be needed for the business, commodity skills like network administration can be provided via cloud services.

The evolving world of IT in various industries has created demands that service providers recognize as opportunities and challenges. Among these are: mobile devices, biometric services, autonomic management of cloud services, healthcare information systems, increased security concerns, and cloud based desktop interfaces. As result, the IT service model is increasingly becoming cloud based. In the past, service providers managed customers' networks, either on site or remotely, using specific protocols for specific customers. As cloud service providers, they must still offer these services but now, with networking as a service (NaaS), the providers' tasks become much more sophisticated and challenging. They must offer multiple platforms to multiple customers, and often, due to virtualization, these protocols will live side by side on the same hardware. Multiple levels of service and completely dynamic provisioning must be provided. Accounts of outages with Amazon's Elastic Community Cloud and Intuit, as well as data access problems at Dropbox and Google, give a clear indication that cloud-based services introduce a new level of sophistication for network service providers (Bouchenak *et al.,* 2013).

Feng *et al.* (2011) propose an approach to manage the increased demands of cloud-based network services. They give an overview of the philosophy behind their approach:

> From the perspective of service, the abstraction of network function and the layer of network protocol stack will be re-organized and divided into three layers: service specification, network capacity, network behavior. In this vision, network service in different abstract forms will regard as middle ground for the continuous resolution of tussles between providers and users.

The approach includes a "cloud-based network" with the ability to deploy network protocols dynamically. This is done using a "protocol service instance," which allows for separate instances of networking protocols as services. A potential major tool for the deployment of networks in the cloud is the open-source tool Open Flow, which is functionality added to commercial switches that facilitate software defined networking. It is a layer of abstraction above the existing infrastructure, which allows easier modification and management. Open Flow, and solutions like it, are seen as the path forward to allowing cloud-based networking to be more dynamic and robust.

Another IT service that is moving rapidly to the cloud is healthcare informatics. As we have seen with the complications of implementing the Affordable Healthcare Act, performance and security are essential issues when it comes to healthcare technology via the Web. The Health Insurance Portability and Accountability Act (HIPPA) is a set of legal guidelines to be followed in protecting patient confidentiality. Managing this data in the cloud provides considerable challenges.

Liu *et al.* (2012) proposed a model for using mobile devices and a virtual integrated medical information system (VIMS). Using such a service, healthcare professionals can share vital patient data via the cloud. The challenge, of course, is to make sure this data is handled securely and efficiently. Liu and her team proposed a cloud-based system to follow both HIPPA guidelines and run in a multiplatform, multidevice environment. Such solutions are absolutely essential if healthcare informatics is to live in the cloud effectively.

Prior to the advent of outsourcing, an organization had to manage all of the components involved in its IT function: hardware, operating systems deployment, backend server deployment including databases, and all involved middleware. Outsourcing allowed the organization to contract services either for cost advantages or to provide expertise the company did not have.

The IT function in organizations has long used outsourcing to extend its capabilities. In the past, IT services were merely third-party solutions for functions normally provided by the IT organization. In 1989, Eastman Kodak was spending $250 million a year on information technology. They entered an agreement with IBM and two other companies for $250 million over 10 years to provide IT services. While not the first or largest company to enter into an outsourcing agreement, this is cited as the first highly visible, successful outsourcing agreement (Marchewka, 2012: 386). It was also the beginning of IT outsourcing as a standard industry practice. By 2000, 54% of IT services purchased in North America were outsourced.

Pre-cloud IT outsourcing did not fundamentally change the nature of the IT services provided. As previously stated, they were for the most part identical in nature to the services companies would provide for themselves internally. The IT function had evolved from the 1950s to 2000 as companies began to see the computer as an essential tool for doing business, and one that needed dedicated resources to manage it.

Around the year 2000, the technical foundation for cloud computing was in place. Both industry and end users began to see the benefit of accessing computing resources remotely. Home users could access the world's information instantly via the World Wide Web, and companies began allowing their information workers to access company LANs remotely. While most of the software used still lived on company desktops and servers, the appearance of SalesForce.com in 1999 marked the beginning of the cloud era (Hayes, 2008). SalesForce offered a suite of programs that users could access, which were housed entirely online. While SalesForce was and is an application and not an IT outsourcing service, it did not take long for the advantages of this model to be understood by the IT industry. As servers grew faster and storage became cheaper, it became more reasonable for applications to be accessed remotely via a cloud model.

The current cloud model encompasses software as a service (SaaS), platform as a service (PaaS), and infrastructure as a service (IaaS), and most recently desktop as a service (Deboosere, 2012). The term "service" implies that an organization needs one of the services mentioned above while engaging in its core competencies such as manufacturing. Under the cloud model, all of these components can be offloaded to a third-party service provider, and the organization pays a fee for access to them. When done correctly, this can result in considerable cost saving to the company. Why? Because the cloud service providers can aggregate the needs of their customers (using hardware virtualization) and enjoy considerable economies of scale, thereby providing services to their customers at a much lower cost. At present the cloud model can be regarded as "mature" as opposed to "emerging." In the near future, we foresee a considerable restructuring of the IT support industry based around the cloud model.

Cloud computing's impact on IT has been dramatic. It is safe to say that the industry is in effect being redefined by it. The core competencies required for a cloud-enable company are the same essential ones required for a company to remain viable, irrespective of its size.

In his book, *The World Is Flat,* Thomas Friedman (2007: 303) outlines for us his view that the global cloud infrastructure has completely flattened the playing field for those firms that know how to use it. Any product or service that can be digitized can be cloud enabled, from a small accounting firm in India doing the routine work for larger firms in the United States and Europe to a fast-food vendor using the Internet to having its order takers located in one state and the actual restaurant in another. This has created a new business model with a new set of competencies required to change an existing business model into one that includes cloud. Due to the abundant new cloud-based services, companies can compete at levels far above where small size has traditionally been a problem. To understand why this is the case, we need to begin with an understanding of the cloud services known as software as a service (SaaS), hardware as a service (HaaS), platform and applications as a service (PaaS), and, peripherally to all three, security as a service. These offerings of cloud computing are well documented, so this chapter does not go into a detailed description. For the uninformed, these services allow users to access computing resources to which a small business would not have access prior to the pervasiveness of the cloud.

For small businesses, SaaS probably has the most immediate impact. A small business can either be a provider of a service, as in the small Indian accounting firm mentioned above, or a consumer of the service. It can also be a startup software developer who has a product for which the firm wants to charge, but instead of installing its code on a customer's system it merely allows customers access to it via cloud enablement. In the following example, a client has a product that automates the billing of a medical procedure. It is a case study of how a small firm can dramatically extend its footprint via the Cloud.

A small five-person software development company wants to focus on developing and refining its product. The product is a healthcare application that automates billing for a particular medical procedure. To deploy this application entails considerable expertise in hardware and network security. The company has neither the staff nor the expertise to maintain its own hardware infrastructure to distribute the product via a cloud environment, nor does it have the staff to address issues like installation problems and maintenance. Providing its product as a cloud service allows it to generate revenue from the product with minimal staff. The company contracts with vendors to host the application, secure it, and provide customer support. The geographic location of the vendors is irrelevant. All that is needed is the right business model for the product or service and the identification of vendors and partners with the necessary expertise to support the business model.

For a medium-sized company (revenues between $50M and $200M) cloud computing can be a mixed blessing. On the one hand, it opens a lot of new opportunities for markets. On the other hand, however, it enables smaller companies to become potential threats due to the opportunities that cloud computing opens to them. In addition, medium-sized companies may have considerable sunk costs in noncloud infrastructure that they may be hesitant to forsake in favor of cloud technologies. For example, a medium-sized furniture retailer (annual revenues of $150m/year), which was trying to expand its customer base via technology, had invested a considerable amount of money in proprietary technology (midrange servers and expensive database product). A decision had to be made to use the existing infrastructure for the expansion or use cloud technologies in lieu of them. The company decided to use its existing infrastructure and experienced some degree of success, but smaller companies can deploy the same functionality (management of branch offices, Web based catalogs and purchases, and other applications) for a fraction of the overall cost. The same cloud computing technology that levels the playing field and allows medium-sized companies to compete with larger global firms, also facilitates smaller companies competing head to head with medium-sized companies.

For large companies, cloud computing is merely the latest step in an evolutionary process of enterprise computing. Large companies have always had a need for systems that allow connectivity throughout a broad geographical area. What the cloud allows, however, is a level of transparency that permits much more flexibility in deployment. Database management systems can be much more lightweight and eclectic (lightweight to the user, that is; for the provider they become more complex). Cloud-friendly technologies like XML allow for seamless integration. So while cloud computing does not represent the same scaling opportunities for large companies that it does for small and medium sized companies, it provides another dimension along the competitive landscape. Large companies have always been challenged to deploy IT strategically; this challenge now includes deploying IT so that it permeates the business. This level of deep integration of IT into the business is increasingly driven by cloud-based technologies.

53.2.1 Changes to Existing IT Roles

Along with the impact of the cloud on the IT function of a firm will come a considerable redefinition of existing roles and new roles will be created. As cloud computing evolves, the long-standing and time-tested concepts that have fueled enterprise computing will increasingly be mapped to the cloud. Enterprise computing is reliable, secure, and high performing: exactly what is needed as the demands of cloud computing continue to grow.

For example, traditionally in IT, often the roles of systems administrator and storage administrator were functionally separate. The systems administrator is the person who managed the day-to-day system's operations; setting up accounts and deploying physical and virtual servers. The storage administrator is the person who configured and managed physical storage. A cloud model requires a merging of these formerly separate functions into a new function, often known as an infrastructure engineer. This type of redefinition of roles is necessitated by the nature of cloud computing. Virtualization, which is the backbone of cloud computing, sees the management of the virtual servers and the storage upon which they reside as transparent. The merging of previously separate functions is influenced by the change in perspective of the IT space that accompanies cloud computing. Storage and systems administration were separate because they were, in fact, two separate functions. Under the cloud model both are merged into an infrastructure service, which is managed as a distinct entity.

Previously, physical servers accessed defined blocks of storage, such as storage area networks. This interaction could be, and perhaps needed to be, managed separately due to the nature of the structure of the in-house IT department. Often multiple departments would access the same blocks of physical storage. A team would be assigned to manage storage across the firm, while departments or business units would have systems administrators for functional units. In a cloud-based world, virtual servers should use a consistent configuration built around specific virtualization products such as VMWare. It is more efficient and less expensive in terms of labor costs for an individual or team to manage the entire cloud infrastructure. This is especially true if multiple firms are using the same cloud. While purchasers of cloud services will often manage their own virtual servers, the infrastructure on which these virtual servers live will be managed by a person or team external to the customer. The service provider (individual or team) will manage both the operating system or hypervisor (the software that controls hardware virtualization) and the storage required by it. This model can be modified to include more or less self-service on the part to the customer as needed.

Similarly, it is anticipated that other roles will also be merged into service entities that will be managed as a whole. Application development and application support will be merged into application services. Database management system support and business intelligence support will be merged into database and analytics services. The enterprise architect and businesses analyst roles will be merged into business process services. These new roles will redefine IT functional groupings for both for corporate IT departments and Cloud based service providers. While it is not cloud technology per se that drives all of this merging of roles, it is the layers of abstraction that cloud computing creates that makes it possible.

For the most part, traditional IT roles like network administrator and DBA will morph into combined roles that focus more on process and service delivery than they do on specific technologies. For example, currently there are senior technical roles for systems administration and storage administration. We will see a general move toward combining these roles into a role that has technical responsibility for both areas. This will be a technical manager, with expertise in both areas, though perhaps in neither to the degree of his or her predecessors when the roles were more separate. This individual will become adept at managing the tools and APIs that emerge as a result of cloud data services. But the new infrastructure engineer will know more about use cases and organizational value than in the past, to allow the decisions they make to be better aligned with the organization's strategic plan. Table 53.1 is an example of the combination of these evolving roles (Belfoure *et al.*, 2013).

From a corporate perspective, a sample IT organization might have the following units: strategic development (overseeing the strategic integration of the other units); application development (not only managing all internal software development but also integrating software development by cloud service providers with the applications developed for and by the firm); business intelligence/analytics development (a combination of the traditional data services units and the business intelligence units); platform management (managing the deployment of applications on the heterogeneous cloud-aware infrastructure); infrastructure design (integrating specific hardware deployments with both the other units and cloud services providers); security

Table 53.1 *Evolving roles*

Systems administrator Storage administrator	Infrastructure engineer
Software developers Support engineers	Development operations engineers
Oracle DBA DB2 DBA Business intelligence administrator	Data engineer
Enterprise architect Business analyst	Business architect

and compliance (managing the increased complexities of IT security internally and in the cloud). In this hypothetical department, each area is managed by a director who reports to the CIO. The overriding difference between these new, cloud-savvy IT departments and their predecessors is that, instead of developing requirements for specific vendors of IT services, the departments will focus on their requirements for specific services, which can be provided via the services markets by a variety of vendors. It could commonly be the case that networking, application development, security and data management will all be provided by different vendors, each with detailed service-level agreements (SLAs).

The total number of full-time employees in corporate IT will almost assuredly decrease. However, this will not cause the absolute number of IT jobs to go down, as many of the positions previously existing within the firm will be pushed to the cloud services provider. The US Bureau of Labor Statistics estimates tremendous job growth for several IT skills in future years.

Since the positions will increasingly become available in cloud service provider organizations, our prediction is that the growth estimates may be conservative. This is especially true for software development and data related jobs. Software development and data-related work are changing at such a rapid pace that it is difficult to predict how the jobs in these areas will look in the near future.

53.2.2   Newly Created IT Roles

Along with the growth in traditional IT job roles, newly created roles will develop. As we enter the era of big data, the role of DBA is evolving to the role of data engineer. It requires the skill to manage the huge data repositories being created across the enterprise and the cloud. This new role will merge the responsibilities of database administration with those of the business intelligence developer. It is also likely that a distinct role will emerge that incorporates the skills of a software developer with the ability to make sense of accessing a vast amount of data.

The roles of the past still need to be carried out but they will reside within different organizations and across different people possessing combined skill sets. For example, the physical infrastructure still needs to be supported but it may be transparent to the organization that uses it. The user organization will have platform engineers responsible for enterprise specific environments required to support its applications. These environments will be used by the application developers and the groups responsible for data analytics. Threading throughout all of these environments is the focus on aligning IT strategy to the corporate strategy and, of extreme importance, security. The need to have a subset of each of these skills residing in a single individual will have a dramatic effect on future roles.

Vast numbers of highly specialized cloud-based enterprise architects and developers will be needed to deploy and maintain the infrastructure that cloud computing will require. Information technology departments of companies will become leaner and more specialized. The internal IT function will focus more on the company's core competencies while offloading more of the commodity functions of database, Web access, and network architecture to cloud providers who specialize in providing such services. While initially IT programs will be reduced in size, the need for professionals with cloud skills will increase considerably. The traditional IT job roles as we have understood them will be restructured to match the evolving needs of supporting the cloud.

53.2.3   The Future IT Industry

It is clear that the future of the IT industry will be tightly coupled with that of cloud computing. The growth of data is exponential and showing no signs of slowing down. As has been the case since the appearance of the personal computer, hardware is becoming faster and cheaper. Datacenters are becoming denser (due to virtualization) and greener at the same time. IT departments are becoming smaller and more efficient but the impact of this increased efficiency does not appear to be reducing the number of IT staff needed in absolute terms. Our current educational structures are not producing, and it appears they will not produce, nearly enough workers with IT skills. New training models will have to emerge to meet the steadily increasing need for skills.

Belfoure *et al.* (2013) say that the IT jobs will move up the stack. What they mean by this is, relative to the architectural layers of the IT environment, the increase in the numbers of jobs will be at the higher levels. Explicitly, this includes the database engineering jobs and the middle-ware engineering jobs. Schadler and McCarthy (2013) says that the CIO will have six major organizations delivering IT services. The organizations will focus on strategy, security, analytics, application development, platform, and infrastructure. They estimate that the workload of the individuals performing various roles in the organizations will decrease due to technology, in essence, allowing companies to do more with less. In the near future there will be a greater need for generalists who have cross-discipline skills and a decreased need for specialists. With this need for generalists, the traditional departmental boundaries will break down. Organizations will become more integrated and matrixed as opposed to functional. This heightens the need for centralized governance systems and metrics that truly measure the IT organization's ability to respond quickly, efficiently, and effectively to changing business requirements.

As the strategic and logistical advantages of Cloud technology become clearer to larger numbers of people, many roles will emerge to facilitate the expansion of cloud-based services. Services like Open Shift, (open source code driven by Red Hat) provide a robust development environment, which allows companies and individuals to completely ignore platform issues and focus on software development. What this entails is a focused group of infrastructure support personnel who work across organizations to provide a standard set of tools. In line with Friedman's view of the world as flat, this horizontal team can stretch as far as it needs to accommodate increasing demand. Some would argue that this will mean fewer jobs overall in IT due do to the ability of a tool like Open Shift to serve a huge customer base. While an argument could be made that in absolute terms we may need fewer database and network administrators, the need for more developers, business analysts, and analytics experts will be ongoing, increasing, and insatiable. There will also be a demand for application owners, vendor relationship managers, and types of expertise we do not yet even fathom. These factors contribute to an extremely healthy outlook for the future of the IT industry, with one provision: that there should develop a clearer collaboration between industry and academia in terms of what IT skills will be needed and how they will be acquired. Traditionally, universities have taught their courses based on an organization of content that had little to do with how industry was organized. Computer science was taught with little regard given to the needs of industry. As the cloud paradigm becomes more pervasive, it is imperative that universities correctly reflect this model in their curricula. There is now a misalignment between industry and academia such that industry, for the most part, has to train new hires in the practical application of the basic IT principles the new hires learn in school. But there is still enough overlap between the two worlds to make this task possible. A relational database is still a relational database, so a student who has taken one or two database courses in college is able to learn the specific implementation used by a given company. But in the cloudcentric world, the nature of how we manage data is always developing new techniques for managing increasingly large and complex data stores. What will be needed are not so much students who are versed in a given technology but students who are trained to manage the complexity and rapid change of the environment. At present, this is not what our universities do.

To conclude, cloud computing has introduced a new way of thinking about the nature of both consumer and business computing. The focus is now more on what is the service that one needs rather than the technology needed. This new way of thinking has opened an exponentially larger universe of possibilities and opportunities that we are only just beginning to uncover.

References

Belfoure, V., Brown, K., Coveyduc, J. L., and Huynh, S. X. (2013) Organizational alignment to achieve integrated system benefits with IBM pure application system. *IBM DeveloperWorks,* http://www.ibm.com/developerworks/cloud/library/cl-ps-aim1305_alignorg/ (accessed January 6, 2016).

Bouchenak, S., Chockler, G., Chockler, H., *et al.* (2013) Verifying cloud services: Present and future. ACM *SIGOPS Operating Systems Review* **47**(2), 6–19.

Deboosere, L., Vankeirsbilck, B., Simoens, P., *et al.* (2012) Cloud-based desktop services for thin clients. *IEEE Internet Computing* **16**(6), 60–67.

Feng, T., Bi, J., Hu, H., and Cao, H. (2011) Networking as a service: A cloud-based network architecture. *Journal of Networks* **6**(7), 1086.

Friedman, T. L. (2007) *The World Is Flat: A Brief History of the Twenty-First Century,* Macmillan, Basingstoke.

Hayes, B. (2008) Cloud computing. *Communications of the ACM* **51**(7), 9–11.

Kim, J. Y., Altinkemer, K., and Bisi, A. (2012) Yield management of workforce for IT service providers. *Decision Support Systems* **53**, 23–33.

Liu, C., Chung, Y., Chiang, T., *et al.* (2012) A mobile agent approach for secure integrated medical information systems. *Journal of Medical Systems* **36**(5), 2731–2741.

Marchewka, J. T. (2012) Information Technology Project Management, 4th edn. John Wiley & Sons, Inc., Hoboken, NJ.

Schadler, T. and McCarthy, C. (2013) Wanted: mobile engagement providers. *Forrester,* http://resources.mobiquityinc.com/rs/mobiquity/images/Forrester_Wanted_Mobile_Engagement_Providers_Schadler%20McCarthy.pdf (accessed January 6, 2016).

54

Cloud Computing in Emerging Markets

Nir Kshetri[1] and Lailani L. Alcantara[2]

[1] *The University of North Carolina at Greensboro, USA*
[2] *Ritsumeikan Asia Pacific University, Japan*

54.1   Introduction

Cloud computing (hereafter: "the cloud"), which is likened to the industrial revolution in terms of technological innovations, structural change, and as a source of economic growth, has started to transform economic activities (Kshetri, 2013a). It has been argued that the cloud reduces infrastructure costs and levels the playing field for small and medium-sized enterprises (SMEs), especially in emerging markets (EMs) (Greengard and Kshetri, 2010; Kshetri, 2010a, b, 2011, 2012; Alcantara and Kshetri, 2013).

Unsurprisingly, EMs are rapidly embracing the cloud-based economy as evidenced by more rapid growth in cloud deployment in these markets compared to more developed markets (Kshetri, 2013b). For instance, according to Cisco's third annual Global Cloud Index, during 2012–17, the Middle East and Africa region will have the highest cloud workload (which is amount of work run on the cloud) growth rate (CAGR: 45%), followed by Asia Pacific (CAGR: 40%) and then by Central and Eastern Europe (CAGR: 31%) (http://tinyurl. com/nmdqn7q, accessed January 7, 2016). Likewise, in 2012, Tata reported that emerging economies are the most aggressive adopters of the cloud.

Some early signs indicate a number of cloud-led economic and social transformations affecting society. The cloud may provide an opportunity to leapfrog and overcome barriers related to information and communications technology (ICT) infrastructures. A comparison of how the cloud is being used in EMs and in industrialized economies, however, reveals basic similarities as well as striking differences. For instance, for a small segment of the population in EMs the use of the cloud closely resembles that in industrialized countries. Cloud-based services related to entertainment, media, social networking, and communications such as e-mail, social media, YouTube and cloud-based TV are becoming as popular in EMs as in industrialized countries. At the same time, a number of interesting and creative techniques involving the cloud have been developed and deployed in the EMs, which are unique to these markets.

Due to the attractiveness of EMs, a number of global cloud providers have entered these markets. IBM has built cloud centers in a number of EMs such as China, India, Kenya, Vietnam, and Brazil. Microsoft, VMware, Hewlett-Packard, Salesforce, Parallels, and other high-profile global cloud providers are also active in EMs. Similarly, firms based in the developing world have jumped on the cloud bandwagon. Cloud-related venture capital investments are flowing in these markets. Global and local companies have made creative adaptation of the cloud to the local market to suit modern and traditional economic sectors.

Cloud diffusion patterns and associated factors among EMs are highly heterogeneous. For instance, South Korea had over 2700 secure servers per 1 million people compared with less than 2 in a number of least developed countries (LDCs) such as Afghanistan, Bangladesh, Myanmar, and Nepal (UNCTAD, 2013). The patterns also exhibit a wide rural-urban gap on various indicators as well as differences across various sectors of the economy. This chapter will analyze the current status of the cloud industry in EMs and examine the fundamental forces driving the use and deployment of the cloud in these countries, as well as major constraints facing them. It also provides a comparison of cloud deployment in traditional and modern sectors of EMs, and discusses some examples of cloud deployment in EMs that can become a role model.

54.2 Diffusion of Cloud Computing in Emerging Markets: A Survey

As Table 54.1 illustrates, the cloud has been deployed in EMs in diverse activities, which involve various levels of sophistication.

The diffusion pattern of the cloud and the associated environmental factors vary widely across EMs. In this section, we focus on the top three countries on the list of the top 20 EMs released by *Bloomberg Businessweek* in January 2013: China, South Korea, and Thailand. As shown in Figure 54.1, these three economies differ drastically in terms of development attributes and risks related to the cloud. This figure is based on the Cloud Readiness Index developed by the Asia Cloud Computing Association for 14 countries in Asia. The Index represents three types of environmental factors: information regulations (intellectual property (IP) protection, freedom of access to information, data sovereignty, and data privacy), information infrastructure (broadband quality, datacenter risk, power grid and green policy, and international connectivity), and business and government environment (business sophistication and government online services, and ICT prioritization). Figure 54.1 shows that South Korea performs better than China and Thailand in all the dimensions. The performances of China and Thailand are relatively lower in most dimensions, particularly in terms of information regulations and infrastructure.

54.2.1 Environmental Factors Related to the Cloud in China

Cloud computing is considered by the Chinese government to be one of the so-called strategic emerging industries (http://tinyurl.com/ax9ae4g, accessed January 7, 2016). In fact, cloud computing is part of the Chinese government's twelfth Five-Year Plan, which aims to transform the economy into a steadier and more stable trajectory. The government has promoted five pilot cloud-computing cities in Beijing, Shanghai, Shenzen, Hangzhou, and Wuxi (http://tinyurl.com/7ycljl5, accessed January 7, 2016).

As reported in Kshetri (2010a, b, 2011), the Chinese government-run IBM Cloud Center in Wuxi City's Science and Education Industrial Park in Jiangsu province illustrates how clouds could help SMEs. IBM provides technology, including system x and system p servers on a secure virtual local area network. It is worth noting that, on average, software firms in China have 25 employees, whereas in India, the average is

Table 54.1   *Some representative examples of cloud deployment in EMs*

Area of cloud deployment	Examples of organizations / consumers
R&D	Yahoo! collaborates with the Indian Institute of Technology Madras (IIT-Madras) and establishes Grid Computing Lab, which allows researchers to access and conduct research on big data and cloud computing.
Education	The University of Information Technology, part of Vietnam National University, is using IBM PureFlex System, IBM Tivoli Service Delivery Manager, and IBM Workload Deployer to build a Smarter Computing IT infrastructure that hosts the university's virtual campus and deploys virtual education services.
	The South Korean Ministry of Education, Science and Technology is implementing a program that will turn the nation's classrooms paperless by 2015. This program will provide each student a tablet and an access to textbooks and other educational materials from a cloud computing system.
Healthcare	Capacity Kenya, a USAID funded project, in collaboration with the Kenya Medical Training College and Africa Medical and Research foundation, has hosted a local cloud environment that is used to locate and map healthcare specialists and develop virtual learning platforms for medical students (http://tinyurl.com/pahlah8, accessed January 7, 2016).
	The Medical Informatics Group of the Centre for Development of Advanced Computing in India has made the Mercury Nimbus Suite available for healthcare service providers. This suite offers telemedicine services to people in remote areas through public and private cloud.
Agriculture and farming	Haiti's Ministry of Agriculture has strengthened its knowledge management system through its adoption of cloud technology. This project is funded by the World Bank and makes Haiti's agricultural infrastructure, which was destroyed by earthquake in 2010, resilient to natural disasters (World Bank, 2013)
E-business/e-commerce	KT Corporation, a large telecommunication service provider in Korea, offers a cloud computing service, Olleh U-cloud, which allows end consumers to watch videos and photos on TV via cloud for a monthly fixed fee.
	Alibaba Group Holding Ltd., the largest e-commerce company in China, offers a cloud service, known as Ju Baopen. It allows bank and securities firms to provide online payment services (http://tinyurl.com/nlgvvyh, accessed January 8, 2016).
Business process and IT outsourcing	Turkey's Türk Telekom has implemented IBM Smart Cloud Entry, which allows the company to shorten provisioning time, delegate server administration to internal IT users, and provide a metering and billing system of cloud services to end users.
Banking and finance	Twelve Mexican financial institutions including Sofol Tepeyac, Grupo Agrifin, Findeca, Soficam and C. Capital Global use Temenos' T24, the most technically advanced banking system that runs in the cloud (http://tinyurl.com/q8t4ppn, accessed January 8, 2016).
	Fourteen commercial banks in China's Shandong Province have formed an alliance to deploy cloud service platforms, providing them with a unified IT system and various settlement services.
Environmental monitoring and protection	Hadley Center for Climate Prediction and Research has sponsored a researcher in West Africa for free access to cloud services (Kshetri, 2013b).
	Burkina Faso's University of Ouagadougou has modeled the movement of pollutants in the Sourou River drainage basin (Kshetri, 2013b).

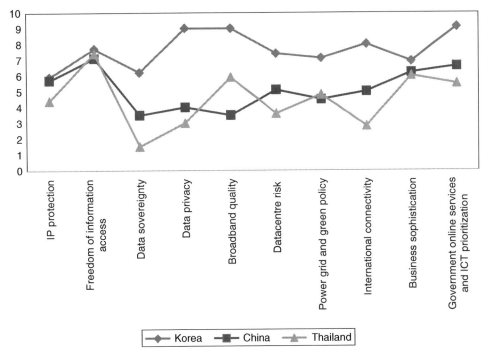

Figure 54.1   *Development of the cloud industry and market: A comparison of China, South Korea and Thailand*
Source: Asia Cloud Computing Association (2012)

174. Park tenants (mostly start-up software and chip-making companies) have access to an entire IT infrastructure. They do not need to buy servers, applications, or tools, and pay only for the services they use. Using virtualized resources lowers upfront investment and product development costs. The industrial park plans to support "several hundred thousand developers across hundreds of companies." In July 2009, IBM and Wuxi Park launched the PangooSky SaaS platform, which targets SMEs. Within a few weeks, 21 enterprises had signed agreements for the platform.

The significant involvement of the private sector in the diffusion of cloud is also reported by Accenture in 2010. A subsidiary of the Alibaba Group founded in 2007, Alisoft, announced in 2009 its $146 million worth 3-year promotion of its new free "Shopkeeper" accounting and financial management SaaS services to SMEs. It was reported that a 2-week trial run of "Shopkeeper" attracted over 135 000 users (Wei, 2009). Together with Suzhou Industrial Park and Fengyun Network services, Microsoft has also started to provide SaaS services to SMEs. In May 2012, Microsoft launched its Microsoft Cloud Accelerator (MCA) for Windows Azure, which provides a cloud computing platform for startup businesses in China, with the hope of accelerating cloud-based innovations in the country. In November 2012, Microsoft announced the expansion of its premier commercial cloud services, Office 365 and Windows Azure, into China under a strategic partnership with the Shanghai Municipal Government and 21Vianet Group, Inc. Moreover, aimed at promoting cloud education, Microsoft has partnered with the Qingdao Municipal Government and Weidong Real Estate to develop a cloud computing center in the Jiaodong Peninsula.

Although there are cases indicating the diffusion of cloud among SMEs, the cloud-computing in China is still at the early stage of development. There is a lack of knowledge about the cloud. Chinese officials and business leaders are also concerned about losing highly sensitive data about the economy, military, government, and commercial secrets. The security concern is largely attributed to the country's lack of comprehensive data privacy laws and regulations.

To move to a different issue, the Chinese government's cyber-control measures have led to an inability of businesses and consumers to realize the cloud's potential. For instance, China's filtering system makes it difficult or impossible to access cloud services provided by foreign vendors such as Google (specifically Google Docs), and Dropbox and causes significant connectivity speed and capacity reduction (Getting to grips with cloud in China, 2011).

Cybercontrol measures have also discouraged foreign investments. Google's 2009 report indicated that it had discovered an attack on its infrastructures that originated in China. In 2008, Google's CEO said that his company would work with Chinese universities, starting with Tsinghua University, on cloud-related academic programs. The country's unfavorable environment from the security standpoint, however, led to Google's withdrawal from China.

54.2.2 Environmental Factors Related to the Cloud in South Korea

There is a huge market for cloud-based IT businesses in South Korea. According to the OECD, South Korea ranked sixth for total number of fixed broadband subscriptions and third for wireless broadband subscriptions in 2012. When measured per 100 inhabitants, South Korea had more wireless broadband subscriptions than inhabitants (104.2). Moreover, South Korea had the highest average Internet connection speed of 14.7 Mbps (http://tinyurl.com/osbshhf, accessed January 8, 2016). It also leads the world in terms of "fiber to the home" connectivity with 62% of the homes connected to fiber as of 2012 (UNCTAD, 2013). This provides a healthy infrastructure for the cloud.

The South Korean government has taken initiatives to improve its IT environment and promote the adoption of the cloud. According to the National Computing and Information Agency (NCIA), since the beginning of the 2000s, the government has tackled challenges facing its e-government system such as the lack of systematic strategies to respond to security attacks and natural disasters, high demand for stable and consistent services, poor quality management of various agencies' administrative services, and high cost yet inefficient computing systems. As a part of its solution, the government constructed the world's first Governmental Internet Data Center (GIDC), which integrates and operates the entire information system of its 40 central administrative organizations. The GIDC was completed in December 2007 under the supervision of the NCIA, and is now recognized as the leading e-government solution in the world. It has developed a comprehensive defense system called e-ANSI (Advanced National Security Infrastructure) that has dramatically enhanced the security of the e-government system (Kim and Kim, 2011). In 2012, with the continuing effort of the NCIA to increase the stability and security of the GIDC, South Korea ranked as the world's most e-ready government, up from the 13th place in 2003 by the United Nations.

In addition, in 2011, the NCIA deployed a government cloud computing platform, the Government Cloud Computing Service (G Cloud), which has improved the services and reduced the costs of GIDC and supported the ICT industry. The NCIA's president Woo-Han Kim noted: "[NCIA] is implementing various policies in order to shape a market environment to support the growth of small and medium-sized enterprises. NCIA has not only strengthened its capabilities by going into partnership with local companies but also in actively introducing new trends and technologies and spreading them across all sectors of the government" (as quoted in Kim and Kim, 2011). In fact, 50% of NCIA yearly projects are participated by SMEs (Kim and Kim, 2011).

The private sector has also played a key role in expanding the cloud market in South Korea. In 2011, the South Korean telecom provider, KT, launched a joint cloud computing venture with Japan's Softbank, named kt-SB data service, in Gimhae City, with initial funding of about US$37 million (http://tinyurl.com/okb8vpl, accessed January 8, 2016). This venture targets Japanese companies seeking reliable data storage safe from natural disasters and power outages.

The collaboration between the local government and private sector has also been a key driver in the diffusion of cloud in South Korea. According to a Groupe Speciale Mobile Association (GSMA) report in 2012, with a total investment of US$320 million, the local government of Busan together with Cisco and KT built Busan green ubiquitous city (u-city). The u-city uses a cloud-based infrastructure based on the blueprint and business architecture developed by Busan metropolitan government and Cisco. In the first phase of the blueprint, in 2010, Busan Mobile Application Center (BMAC) was set up to provide a platform as a service (PaaS) for SMEs developing online apps and services. During the first year of operation, BMAC supported the launch of 13 new companies and development of 70 new apps by SMEs, generating revenues of US$2.2 million and online sales revenue of US$42 000. The second phase of the project was planned to deliver SaaS cloud services, while in the third phase, the cloud-based services are made accessible to all citizens.

54.2.3   Environmental Factors Related to the Cloud in Thailand

As shown in Figure 54.1, much improvement is needed in the cloud infrastructure, regulations, and business environment in Thailand. In fact, Thailand, along with Vietnam, had the lowest cloud preparedness among the 14 countries included in the index. Nonetheless, efforts to promote and diffuse cloud services in the country are noticeable.

In May 2010, the True Internet Data Center, the leading Internet datacenter provider in Thailand, introduced the first cloud computing service in the country through a collaboration with Trend Micro, a global Internet content security company founded in the United States and headquartered in Japan. A month later, NIIT Technologies and Hitachi Asia (Thailand) collaborated to launch cloud services. NIIT was responsible for the design and architecture of the cloud system, while Hitachi implemented and maintained the system. In 2011, Japan's NEC launched its cloud computing solutions to provide low-cost offerings to SMEs. In addition, CAT Telecom Plc (CAT), a state-owned telecom service company, signed an agreement with IBM in April 2013 to provide cloud-based services ranging from basic IT infrastructure to specialized business applications, to midsized businesses in the country.

Despite the low score of Thailand in the Cloud Preparedness Index, the Asia Cloud Computing Association reported in March 2013 that the cloud market in Thailand will increase by 22%, amounting to US$77 million, particularly among medium size companies. This increase will be driven by further developments in cloud infrastructure such as the introduction of 3G wireless broadband and big data analytic services.

54.2.4   Rural-Urban and Intersectoral Differences in the Use and Deployment of Cloud Computing

Many EMs also exhibit a rural-urban gap on various cloud-related indicators as well as differences across various sectors of the economy. In sub-Saharan Africa (SSA), for instance, while 53% of the urban population has access to electricity, only 8% of the rural population has such access. Moreover, rural-income gaps in income and other factors translate to such gaps in the diffusion of the cloud.

The traditional and modern economic sectors in EMs also exhibit wide variations in cloud deployment patterns. The deployment of the cloud in modern economic sectors highlights the growing demand for more sophisticated cloud-based applications in. For instance, South Africa's Nedbank uses the IBM Cognos platform in budgeting and forecasting solutions (http://tinyurl.com/o9onzax, accessed January 8, 2016). In order to enhance security, Nedbank and South Africa's Entersect Technologies have developed a system that uses digital certificates placed on the phone and push notifications. Push notifications, which allow messages to be sent directly to a mobile device, resemble text messages. This means that the end user's experience is similar to receiving a text (http://tinyurl.com/pzn3c3o, accessed January 8, 2016). In this way the system adds security without sacrificing the ease of use. A user enters a PIN to approve or deny transactions such as online purchases, wire transfers, and ATM withdrawals when prompted on the mobile phone. The system takes advantage of AWS, which can handle massive amounts of mobile phones connecting to its infrastructure. In this way, the bank can cope with a large number of users connecting to its system.

Financial institutions such as Nedbank are likely to face the fluctuation of demand within a year and within a month. For instance, the ends of the months are high transaction periods because employees are paid. Likewise, the end of the year and the holiday and shopping seasons are characterized by a comparatively high demand for services. The cloud is ideal to deal with a high degree of fluctuation in demand. During high transaction periods, the system scales up and, in the slower periods, the infrastructure scales down.

As an example of cloud deployment in the traditional economic sector, one can consider the Apps4Africa award-winning app, iCow developed by Green Dreams, which is based on a mobile application, and helps small-scale dairy farmers track and manage their cows' fertility cycles. The app informs farmers about important days of the cow gestation period, collects and stores milk and breeding records, and sends farmers best practices. It also helps a farmer find the nearest vet and other service providers. Green Dreams has also formed a simple system involving Google Docs.

Compared to iCow, Nedbank's e-banking application is somewhat more costly for the end consumers. For instance, digital certificates need to be installed on the user's phone. The system looks for this certificate to authenticate transactions.

Fraud-related risks are increasing in banks in EMs. Compared to the users of iCow, Nedbank's e-banking customers require a high degree of cybersecurity. In this regard, the Entersekt system sends fully encrypted data from the bank, which passes through AWS.

Finally, local firms developing cloud-based applications for the modern sectors are likely to be of higher caliber. For instance, while both Green Dreams (the developer of iCow) and Entersect's systems are SSA-based firms, the latter is more globalized. For instance, Entersect's system is being sold in the United States by Transecq (http://tinyurl.com/pzn3c3o, accessed January 8, 2016).

54.3 Factors Driving the Use and Deployment of Cloud Computing in Emerging Markets

In this section, we discuss the key factors driving the cloud industry and market in EMs.

54.3.1 The Roles of Local Cloud Providers

The current global market is dominated by a few US-based cloud providers. Nonetheless EM-based firms have also jumped on the cloud bandwagon. Especially there are a number of local cloud players in big EMs such as China, India, and South Africa. CRL and AdventNet, for instance, are among high-profile Indian cloud providers. TCS and Wipro have also entered the cloud market. TCS started cloud pilot projects in 2009 and had 130 clients by February 2011. TCS claims that its cloud can make savings of up to 30–40% for SMEs (http://tinyurl.com/o2b7q24, accessed January 8, 2016). Infosys has dedicated 175 engineers to identify potential areas attractive for the cloud (http://tinyurl.com/363c8ht, accessed January 8, 2016), while Hyderabad-based Pressmart provides SaaS based e-publishing and digitization services to the print industry. The Pressmart solution can help firms deliver content across multiple platforms such as the Web, mobiles, Really Simple Syndication (RSS), podcasts, blogs, social networking sites, articles directories and search engines.

In China local players account for the bulk of cloud investments. The cloud has also been a critical component at Huawei and a number of other Chinese companies.

In Africa, for instance, South Africa's Integr8 and MTN, and Zimbabwe's Twenty Third Century Systems, have launched cloud offerings. MTN MyOffice supports accounting, human resource, customer relationship management, e-mail and videoconferencing, storage and back-up for SMEs in manufacturing, hospitality, microfinance, and advertising. In 2011, Kenya's Safaricom launched the Safaricom Cloud, hosting platforms for government agencies and corporations and other offerings. By 2011, it had invested US$150 million and announced plans to invest another US$200 million (http://tinyurl.com/ko5gfq6, accessed January 8, 2016).

A number of EM-based cloud firms have collaborated with foreign companies in order to enhance their cloud offerings. Safaricom teamed up with Cisco (storage), EMC (security), and Seven Seas Technology (overseeing and training). Some EM-based cloud players have also acquired foreign firms. For example, in early 2012, Bangalore-based Aditi Technologies acquired Cumulux, a US-based cloud startup.

54.3.2   Multinationals' Entry into the Cloud Sector

The development of the cloud industry and market in EMs is associated with and facilitated by the entry of global multinationals into this sector. Consider India, for instance. In 2008, IBM opened a cloud center in Bangalore for mid-market vendors, universities, government bodies, and microfinance and telecommunications companies. Then, in early 2012, it helped the India-based Tulip Telecom construct a 900 000 square foot datacenter – the largest in the country – which will provide cloud infrastructure services. Parallels has also been operating in India since late 2008. In 2009, Microsoft started offering productivity apps on the cloud for approximately $2 per month, including e-mail, collaboration, and conferencing services. Also in 2009, VMware opened a cloud center in Pune. Salesforce.com started operations in 2005, and in September 2011, it acquired a social customer-service SaaS startup, Assistly, for US$50 million. Salesforce.com has many high-profile clients in India, including Bharti, eBay India, SIFY Technology, Polaris, and the National Research Development Corporation (Kshetri, 2012).

Similarly, in 2008, IBM opened a cloud center in Beijing. Its Shanghai R&D facility also has the cloud as a primary area. In 2009 it opened a Healthcare Industry Solution Lab for hospitals and rural medical co-operatives. IBM has a partnership with China's Range Technology to construct a 6.2 million square foot state-of-the-art center in Langfang city, which will be comparable to the Pentagon, with 646 000 square feet devoted to the datacenter (Wilson, 2011).

Similar developments are taking place in other EMs (Kshetri, 2013b). Consider, for instance, Africa. In addition to its Johannesburg datacenter, IBM built its 41st global innovation center in Kenya in 2013. Other global providers, such as HP and VMware, have a significant presence in sub-Saharan Africa. Amazon's customer service center in South Africa opened in 2011 and employs 1400 people, and in 2012, it launched the AWS Developer Support office in Cape Town. Specialized providers have also entered sub-Saharan Africa. In 2012, Switzerland-based Sofgen launched the core banking platform, Temenos T24. It targets banks, microfinance institutions, and savings and credit cooperatives and has built-in fraud detection capabilities.

54.3.3   The Roles of International Agencies

International agencies' roles in facilitating the cloud industries in EMs deserve mention. For instance, in 2006, the World Bank (WB) and the African Development Bank financed a US$280 million cable project serving 23 countries. Likewise, in 2007, the International Finance Corporation invested US$32.5 million in EASSy, connecting 21 countries (http://tinyurl.com/pq57roq, accessed January 8, 2016). Similarly, Sierra Leone received US$31 million to connect to ACE cable through the WB-funded West Africa Regional

Communications Infrastructure Program. The WB also provided bandwidth subsidies to many universities (http://tinyurl.com/ko5gfq6).

54.3.4   Philanthropic Causes

Philanthropic and charitable causes have also been a factor in stimulating the cloud industry in EMs. For instance, Worldreader, which describes its mission as to "make digital books available to children and their families in the developing world, so millions of people can improve their lives," uses AWS to download books to computers in Africa (http://tinyurl.com/dyf9yqo, accessed January 8, 2016). Likewise, the UK-based Indigo Trust funded over £65 000 (about $100 000) to develop iCow, which helped to cover the core costs, legal fees and customer care supports (http://tinyurl.com/plx4mpb, accessed January 8, 2016).

54.4   Factors Constraining the Diffusion of Cloud Computing in Emerging Markets

A number of barriers are hindering the cloud's use and deployment in most EMs. We discuss some of the most important hindrances in this section.

54.4.1   Economic and Infrastructural Constraints

The most important barrier to cloud diffusion in many EMs relates to the lack of availability of basic infrastructures. For instance, regular power outages in some EMs disrupt communications between cloud users and providers as well as among various cloud providers (UNCTAD, 2013). We contrast the gap between the availability of datacenters in the least developed countries (LDCs) and industrialized countries to illustrate how the lack of various types of infrastructures could hinder the growth of the cloud in EMs. In 2013, about 85% of the world's datacenters that offered co-location services were in developed economies. UNCTAD (2013) refers to this gap as the "datacenter divide" and notes that high-income economies had about 1000 times more secure data servers per capita than in LDCs in 2013.

Among the important barriers are also high prices of cloud offerings. In EMs economies, bandwidth, electricity and other traditional IT infrastructures are costly in absolute as well as relative terms. For instance, electricity in Kenya costs $0.20/KWH, which is 50% higher than in the United States. Excluding cooling and management, a server in Kenya costs more than $1800 a year to run (http://tinyurl.com/pzpotor, accessed January 8, 2016). There are prohibitively high costs in SSA economies relying on satellite. According to the World Bank, bandwidth in Sierra Leone costs ten times as much as in east Africa and 25 times the price in the United States. Some embassies, banks, and mining companies in Sierra Leone have reportedly secured 2 MBPS satellite-based connection known as very small aperture terminals (VSAT), which cost US$8000 a month plus a very high installation fee. NATCOM estimated that there were 100–150 VSATs in the country in 2011. One level below is 512 kbps connections provided by ISPs, which disburse leased bandwidth through wireless networks. Such connections typically cost US$570 a month for a household plus US$445 for the modem. Services offered by mobile phone companies, which charged US$70 a month, distribute satellite bandwidth via GPRS or Edge modems packaged as USB sticks. These services are among the cheapest but have extremely slow speed and totally unusable at busy times (http://tinyurl.com/3eg9v7j, accessed January 8, 2016). Even in more internationally connected economies, such as Ghana and Nigeria, prices are higher compared with international standards.

Due to the lack of availability and high prices of local bandwidth, some EM-based cloud users have found the offerings of providers in foreign locations more attractive. For instance, in India, many firms use cloud offering of providers that are located in Singapore due to cheaper bandwidth and more developed infrastructure (UNCTAD, 2013).

54.4.2   Regulatory Constraints

Regulatory and human resources issues also represent a barrier to the development of the cloud industry and market in EMs. While there have been regulatory demands for long-term retention of data in EMs, underdeveloped regulatory institutions have hindered the growth of the cloud industry in most EMs. For instance, according to the Business Software Alliance's (BSA) evaluation of national laws and regulations in seven policy areas related to cloud computing, South Africa, which has one of the most developed cloud industries in Africa, ranked 20th out of 24 economies considered in 2013.

54.4.3   Human Resources Constraints

The development of the cloud industry in EMs is also hampered by the lack of sufficient human resources and expertise (ITU, 2012). Due to the lack of the availability of IT-skilled teachers and other related factors, many EMs have failed to incorporate curriculums related to IT skills development in their education systems at various levels (UNCTAD, 2013).

54.4.4   Concerns Related to Security

Research has suggested that a significant gap remains between vendors' claims and regulators' and users' views of the cloud's security, privacy and transparency (Kshetri, 2013a). Issues such as security, privacy, and availability are among the topmost concerns in organizations' cloud adoption decisions rather than the total cost of ownership (McCreary, 2008). This concern is equally, if not more, applicable to EMs. For instance, security concern among CIOs and IT managers in Africa has been a key barrier to a wider adoption of the cloud in the continent (http://tinyurl.com/oo8v43e, accessed January 8, 2016). Likewise, in a survey conducted in 2010 among Indian enterprises, 72% of the respondents cited privacy and data security as among extremely significant concerns for the cloud (http://tinyurl.com/nj42n4y, accessed January 8, 2016). Another survey found that Indian companies were concerned about maturity and the capability of cloud vendors and 86% believed that external certification would increase their trust in the vendors (http://tinyurl.com/qgr3nwd, accessed December 8, 2016).

Potential cloud users have also been concerned about the possibility that third parties may access sensitive data stored in cloud. One fear has been foreign governments' access to sensitive data (UNCTAD, 2013). For instance, Brazil's IT policy secretary Virgilio Almeida discussed the possibility that the Brazilian government may store its sensitive data locally rather than in the cloud (http://tinyurl.com/pnzoyo5, accessed December 8, 2016). Brazil's president Dilma Rousseff also asked Brazil's Congress to introduce regulations that may require foreign technology companies such as Facebook and Google to store data generated by Brazilians on servers physically located in Brazil (Brooks and Bajak, 2013).

54.5   Discussion and Implications

While it took many years and large investments for the industrialized economies to acquire infrastructure, datacenters and customized applications, the cloud has made it possible for the EMs to access them easily. In this regard, the chapter identified certain key issues associated with the diffusion and adoption of the cloud in EMs and explored a number of factors that facilitate or hinder the development of the cloud industry in these markets.

As noted above, underdeveloped regulatory institutions have hindered the cloud's growth in most EMs. That said, few encouraging signs have emerged in recent years to suggest that governments in EMs are becoming more serious about introducing regulatory and policy measures to drive the cloud industry. In order to develop a government cloud and a cloud strategy, for instance, the Indian Government established an Empowered Committee, which consisted of representatives from government departments. It also established a special task force with private-sector participation (UNCTAD, 2013). Likewise, the cloud industry in Vietnam is being driven by the government's initiatives to build a skilled workforce. The cloud is used to link government agencies, universities, private-sector research, start-ups and other organizations (Cleverley, 2009). Vietnam started collaboration with IBM in 2007.

Another barrier hindering EMs' embracement of the cloud economy centers around poor infrastructure development required for the cloud economy. For instance, while South Korea has the highest household fiber penetration, as mentioned earlier, it is also worth noting that less than three dozen economies in the world had a household fiber penetration of over 1% in 2012 (UNCTAD, 2013). This means that most EMs perform poorly in terms of fiber penetration to capitalize on the cloud's potential.

Foreign and local firms can take a wide variety of measures intended to help the development of the cloud industry and market in EMs. Foreign companies, for instance, could benefit by collaborating with local cloud providers, characterized by lean cost structures and experience in developing low-cost products. For EM-based cloud providers, on the other hand, their ability to deliver value for money in the domestic and regional markets could give them a competitive advantage in foreign markets, especially if they are in a position to reconfigure their resources to operate effectively in other EMs.

54.6 Concluding Remarks

Cloud computing is in the infant stage of development in EMs and currently cloud usage has been shallow, narrow, and vanishingly small in these markets. Cloud-based innovations and business models are, as yet, far from inclusive of SMEs in EMs, especially in the least developed small nations. However, as economic and regulatory factors improve in EMs, the cloud certainly holds promise for bridging the digital divide. The EMs must thus exploit the opportunities afforded by the cloud while minimizing the associated risks to allow access to advanced IT infrastructure, datacenters and applications, and to protect sensitive information.

References

Alcantara, L. and Kshetri, N. (2013) Diffusion of cloud computing among SMEs in emerging markets. Proceedings of the International Conference on Small and Medium Sized Enterprises in a Globalized World, 25–28 September, Cluj-Napoca, Romania.

Asia Cloud Computing Association (2012) *Cloud Readiness Index,* http://www.asiacloudcomputing.org/research/cri2012 (accessed January 8, 2016).

Brooks, B. and Bajak, F. (2013) *Brazil looks to break from US-centric Internet,* https://www.yahoo.com/news/brazil-looks-break-us-centric-internet-040702309.html (accessed December 2, 2013).

Cleverley, M. (2009) Viewpoints: emerging markets – how ICT advances might help developing nations. *Communications of the ACM* **52**(9), 30–32.

Greengard, S., and Kshetri, N. (2010) Cloud computing and developing nations. *Communications of the ACM* **53**(5), 18–20.

ITU (2012) *Cloud computing in Africa: Situation and Perspectives,* http://www.itu.int/ITU-D/treg/publications/ Cloud_Computing_Afrique-e.pdf (accessed January 8, 2016).

Kim, J., and Kim, S. (2011) Korea's Government Integrated Data Center Sets a New Benchmark on e-Government IT. *Korea IT Times*, http://www.koreaittimes.com/story/19156/koreas-government-integrated-data-center-sets-new-benchmark-e-government-it (accessed January 8, 2016).

Kshetri, N. (2010a) Cloud computing in developing economies. *IEEE Computer* **43**(10), 47–55.

Kshetri, N. (2010b) Cloud computing in developing economies: Drivers, effects and policy measures. Proceedings of the Pacific Telecommunications Council's (PTC) Annual Conference, Honolulu, HI, January 16–20.

Kshetri, N. (2011) Cloud computing in the Global South: Drivers, effects and policy measures. *Third World Quarterly* **32**(6), 995–1012.

Kshetri, N. (2012) Cloud computing in India. *IEEE IT Professional* **14**(5), 5–8.

Kshetri, N. (2013a) Privacy and security issues in cloud computing: The role of institutions and institutional evolution. *Telecommunications Policy* **37**(4–5), 372–386.

Kshetri, N. (2013b) Cloud computing in Sub-Saharan Africa. *IEEE IT Professional* **15**(6), 64–67.

McCreary, L. (2008) What was privacy? *Harvard Business Review* **86**(10), 123–131.

UNCTAD (2013) *Information Economy Report 2013: The Cloud Economy and Developing Countries.* United Nations Conference on Trade and Development. United Nations, Geneva, Switzerland.

Wei, M. (2009) Alisoft to spend $146 mln to market its software. *Reuters*, http://www.reuters.com/article/2009/03/31/alibaba-china-idUSPEK18477820090331 (accessed January 8, 2016).

Wilson, D. (2011) *China Readies Cloud Computing City: IBM and 6.2 Million Square Feet,* http://www.techeye.net/hardware/china-readies-cloud-computing-city (accessed January 8, 2016).

World Bank (2013) *Agriculture in Haiti: a Quiet Revolution Under the Cloud,* http://www.worldbank.org/en/news/feature/2013/03/08/agriculture-in-haiti-a-quiet-revolution-under-the-cloud (accessed January 8, 2016).

55

Research Topics in Cloud Computing

Anand Kumar, B. Vijayakumar, and R. K. Mittal

Birla Institute of Technology and Science, United Arab Emirates

55.1   Introduction

This chapter considers research needs, indicators, and trends from the Extreme Science and Engineering Discovery Environment (XSEDE) together with industry trends of some of the major cloud players, which were discussed in research papers published in *IEEE Transactions on Cloud Computing*, at the ACM Symposium on Cloud Computing (SOCC), the Cloud Computing Conference (CLOUDCOM), and at the International Conference on Cloud Computing Technologies, Applications and Management (ICCCTAM).

55.2   Virtualization

On the applied research front, virtualization technologies and products can be evaluated for their performance with regard to latency (ACM Symposium on Cloud Computing, 2013). Efficient techniques for replication of virtualization are also of interest (ACM SOCC, 2013) as well as related availability performance measures. Methods to monitor and diagnose virtual networks are essential to determine the performance of virtualization (ACM SOCC, 2013; see also http://blog.zhaw.ch/icclab/category/projects/the-init-cloud-computing-lab/, accessed January 9, 2016). The impact of various virtualization technologies on parameters like computability and network bandwidth, bandwidth guarantees, computability and network latency, memory size and bandwidth, throughput and I/O has been researched but needs further investigation.

One of IBM's white papers examined the role of virtualization in disaster recovery, when data must be backed up and key processes need to continue even if the organization's datacenter is disabled due to a disaster (IBM Global Technology Services, 2013). Migration in the context of virtualization and associated throughput optimization must be researched (http://2013.cloudcom.org/, accessed January 9, 2016). *Live migration and co-migration of virtual machines is important.* Virtualization's impact on cloud datacenter design is also an important area for investigation, in particular to ensure optimal use of energy, optimal availability of bandwidth, and optimal scalability (http://2013.cloudcom.org/, http://www.cloudbus.org/research_probes.html, accessed January 9, 2016).

The interplay between virtualization and architecture is worth examining more closely to simultaneously support resource awareness and scalability, adaptive memory management, and critical infrastructure services (http://2013.cloudcom.org/). Virtual resource representation, provisioning, and management are important (http://2013.cloudcom.org/, http://www.cloudbus.org/research_probes.html). Research must be conducted to identify robustness metrics for virtualization-based cloud-computing datacenters. Formal analysis, modeling, and verification of VM-based cloud platforms is needed (see the first issue of *IEEE Transactions on Cloud Computing* in 2013).

55.3 Provisioning

Microsoft has shown interest in infrastructure provisioning. Universities and some budget-constrained organizations that explore high-performance computing (HPC) or the "Internet of things" (IoT) would be interested in provisioning that considers cost as a constraining factor in the optimal delivery of service. Service provisioning, in general, and for mobile phone users, is an applied research area worth exploring. Some industry and university labs are researching such areas (http://www.cloudbus.org/research_probes.html; http://www.hpl.hp.com/research/, accessed January 9, 2015).

55.4 Monitoring

Research into monitoring design, methods, and techniques is important for infrastructure provisioning (Microsoft System Center, 2012), security, performance, and quality of service measurement. Monitoring network resources to alleviate congestion and in dynamic routing provides scope for research. Such monitoring would assist in future system design. Furthermore real-time and dynamic monitoring would include traffic monitoring and perhaps support network reconfiguration. Monitoring is a key element in fault and configuration management (http://2013.cloudcom.org/, accessed January 9, 2016). Research in monitoring supports and affects big data, cloud architecture, and grid monitoring, implying specialized research might be required for each context rather than generalized research (http://2013.cloudcom.org/, http://www.cloudbus.org/research_probes.html). Methods to monitor and analyze elasticity of cloud services are essential. For telephony applications supported on the cloud, latency monitoring is also essential.

55.5 Cloud Design

In terms of cloud design, the following associated areas require research: network planning; data storage design; design of an efficient multiuser environment; network design incorporating high QoS, low costs, and low CO_2 emission; design of scalable replicable protocols; distributed memory caching; security design; open cloud design; and disruptive memory technologies. New algorithms and software architectures are needed that support energy efficiency of the cloud datacenters. Research into modeling network and computational components to achieve load balancing and simultaneously reduce carbon emissions is relevant. Research into scalability and elasticity that span multiple clouds is necessary. Quality-of-service aware data-replication algorithms are a very much needed outcome from research. Highly available and reliable cloud design must be also researched. Holistic resource management frameworks that connect geographically distributed resources with a backbone network but meet regulatory and legal requirements on data must be researched and designed. The legal issues of distributed data must be studied in conjunction with risks associated with cross-border services. Classical robustness metrics may not be appropriate for cloud datacenters, and hence new metrics are needed. Programming models are needed to convert normal programs into elastic cloud applications.

Some services, like sensor network as a service, perhaps should be designed and deployed with a service-level agreement (SLA) focus. Efficient search algorithms for big data in cloud context are essential, as is dynamic exception handling in federated clouds. Brokering algorithms for optimizing the availability and cost of cloud storage services is essential. The impact of market on cloud platforms cannot be underestimated. Metrics for service performance must be standardized and used by a majority of vendors. Cognition has a major role in mobile cloud gaming, and this must be researched.

Given the volume of applications and their varied nature, applied research to develop suitable programming models for each application is necessary and important (Lifka *et al.*, 2013; see also the Proceedings of ICCCTAM-12, which are available on IEEE XPlore).

As major cloud benefits identified pertain to reduced costs for the user, it is necessary to research and identify cloud-licensing mechanisms for tools like MATLAB, which will bring about a win-win situation for the end user, the cloud service provider, and the tool vendor. Identification of such cloud-licensing mechanisms will be essential in most software as a service (SaaS) offerings.

Furthermore, the charging / billing mechanisms for cloud usage should be researched to provide a win-win situation for the end user and cloud service provider (Lifka *et al.*, 2013). The charging / billing mechanisms should also provide maximum flexibility for end users in making choices on how to mix and match their requirements on CPU, memory, data storage and speed of execution. Cloud design must support data and application migration as well as application execution across federated clouds.

55.6   Application Deployment and Architecture

There are a variety of areas of applied research that can be researched. Some of these areas are identified here. Architectures must be proposed to support computing that is data intensive and scalable. Systems must be architected to support dynamic storage management. Federated architectures are worth looking into. The impact of dynamic resource composition, multicloud environments, scaling, large in-memory computation, fast deployment, cloning services, service-level agreements (SLAs), SLA performance, and dynamic optimization of SLA-based services on the architecture are worth exploring. Architecture can be fine tuned to support automatic fault diagnosis, cloud performance prediction, monitoring, and reconfiguration of cloud resources. Other areas of interest would be self-organizing architectures and architectures to evaluate distributed application deployments in multiclouds.

Architectures will be required to support networking across distributed cloud datacenters. Scaling that is time aware, resource aware and energy aware has to be considered in the architecture. Decision support for migration must be built into the architecture. The impact of Hadoop, MapReduce, and similar techniques on the architecture must be evaluated. Support for dynamic and flexible provisioning on the architecture can be investigated. Frameworks supporting self-healing and self-adaptation should be explored.

Hybrid clouds must be investigated further to enable a mix of private clouds that protect intellectual property and steady-state workloads and public/community/national clouds for computational scaling and burst modes (Lifka *et al.*, 2013).

55.7   Security and Privacy

Real and perceived security concerns are major barriers for cloud computing adoption, although it has many advantages such as large-scale computation and data storage, virtualization, high expansibility, high reliability, and reduced up-front commitment, lower long-term cost, and green friendliness. A holistic view of cloud computing security spans the issues and vulnerabilities connected with virtualization infrastructure, software platform, identity management, and access control, data integrity, confidentiality and privacy, physical and process security, and legal compliance. Before people fully embrace cloud services, they must feel assured that the services are addressing these legitimate concerns.

In this section, we analyze cloud computing security problems, highlight some recent trends, and outline a number of critical issues that researchers and practitioners should consider as they develop future solutions for data security in the cloud. The cloud security issues can prevent the rapid development of cloud computing and hence need to be addressed from the point of view of cloud service providers (CSPs), cloud consumers, and third parties such as governments.

Conditions invariably exist that can result in security of data stored in cloud being compromised. The general security concerns emanate from:

- a shared cloud environment;
- a lack of client control;
- potential system failure;
- service provider problems.

To address these legitimate concerns, we have tools like data auditing and encryption, which require a continuous upgrade.

55.7.1   Data Auditing

Data auditing techniques provide continued verification of the accuracy and security of remotely stored information for both the cloud service provider and the client. To ensure effective and efficient data auditing, the cloud data auditing protocol (CDAP) requirements can be enumerated as follows:

- Cloud data, being stored in encrypted format for obvious reasons, requires auditing protocols capable of verifying integrity of encrypted information while maintaining privacy.
- Data auditing protocols should timeously detect any corruption or destruction of user data for whatever reason – system failure or tampering.
- To resolve possible disputes between clients and CSPs, the protocol should provide for third-party verification without compromising the confidentiality of information and involving as little metadata as possible.
- The cloud data auditing protocol should provide for batch processing support as multiple clients might need to verify data at the same time.
- Adaptability of the CDAP is required to adapt to rapidly changing technology and the dynamic fluidity of virtual machines.
- Cost is always a concern and the CDAP should have low overheads – in terms of both communication and computation costs.

No single traditional auditing protocol meets all these requirements. To accomplish the requirements of the clients and CSPs, a full auditing protocol for encrypted data that maintains cloud data integrity and privacy remains a challenge for future research. Some of the keywords for these are: secure dynamic auditing, static and dynamic provable data, proof of retrievability, and privacy-preserving auditing.

55.7.2   Data Encryption

Traditional data encryption requires decryption and invites huge data-processing overheads. Newer cryptographic schemes allow computing to be performed on the encrypted data. An ideal encryption algorithm for cloud data should offer both security and flexibility while providing a cost-effective, searchable, and computable (structured) encryption that is homomorphic and order preserving.

We believe that some combination of traditional encryption techniques with newer schema can ultimately achieve these goals. Ideally, the encryption process should be amenable to CDAP and enhance overall performance of cloud computing. We see versatile data-auditing protocols and encryption schemas as steps towards fully protecting cloud users' data from cyberattacks while assuring high availability.

55.8   Big Data and Analytics

Further research is required to determine, tune, and optimize the performance of compute engines like Google Compute Engine for Scientific Computing and for methods like MapReduce. New, accurate, and perhaps distributed algorithms are needed to handle large-scale graph processing, cluster-size scaling, and large datasets. Real-time requirements of network monitoring data need to be researched for data-intensive frameworks. The quality of data is a performance parameter and must be investigated for cloud data store replication.

Analytics and defined, repeatable processes are an important dimension of the virtualization and consolidation process in building a cloud or migrating to a cloud (IBM Global Technology Services, 2012). Research into new platforms, distributed and secure analytics, and workloads at cloud scale is needed (http://www.hpl.hp.com/research/, accessed January 9, 2016). Enabling and porting current data analytics on to the cloud and making business intelligence (BI) affordable for organizations is an area that must be further explored.

55.9   Government and Cloud Computing Research

Much government engagement in cloud computing research comes from exploring cost-effective and energy-efficient computing paradigms for scientists to accelerate discoveries in a variety of disciplines, including analysis of scientific data sets in biology, climate change, and physics. There is also an interest in creating an inexpensive cloud platform quickly and securely for military and associated personnel.

In the civilian sector, there is interest from some governments in increasing operational efficiencies, optimizing common services and solutions across organizational boundaries, and enabling transparent, collaborative, and participatory government.

NASA's research project, Matsu, aims to create a cloud of satellite imagery data and make it available to interested users and for disaster assistance; this came in the aftermath of the Haiti earthquake in 2010.

Japan has looked to cloud computing to boost its economy. One of the emphases of the Canadian government is to leverage the country's vast landscape and cold climate for the construction of large, energy-efficient cloud computing datacenters.

55.10   Conclusions

The survey from XSEDE brings out the benefits of using cloud computing for research and education. Scalable computing, storage, and services enable research to be pushed into advanced frontiers that would greatly benefit mankind. Most research funding comes from government sources, which implies that government initiatives must be in place. Governments venturing into cloud computing are motivated by improved operational efficiency, the principle of open government, a wish to provide digital equality for the common man, to achieve excellence in government services, to boost the country's economy, and to leverage local climate conditions, among other factors.

The performance of clouds with respect to dedicated servers, and their associated ease of use, are important parameters for clouds to be more extensively used in research and education. Security did not seem to be much of a concern for the research and education communities but must be a major research parameter for industry to leverage cloud computing more extensively. Portability and interoperability between clouds and cloud service providers are essential for this progress.

Much research in industry is focusing on particular companies' current strengths and adapting them to the cloud scenario, be they hardware, operating systems, databases, search engines, or other applications. The major thrust in academic research is big data and analytics.

The main areas that are likely to be the focus of research are virtualization, provisioning, monitoring, cloud design, architecture, security, privacy, and associated performance and QoS measures.

References

ACM Symposium on Cloud Computing (2013) http://www.socc2013.org/home/program (accessed January 9, 2016).

IBM Global Technology Services (2012) *Accelerate Server Virtualization to Lay the Foundation for Cloud: Analytics Combine with Standard, Repeatable Processes for Quicker Transformation and Improved ROI*. Thought Leadership White Paper, https://www.google.co.uk/url?sa=t&rct=j&q=&esrc=s&source=web&cd=1&cad=rja&uact=8&ved=0ahUKEwiCwOusxZ3KAhUGqxoKHVu9DmkQFggpMAA&url=http%3A%2F%2Fwww.inteligencija.com%2Findex.php%2Fen%2Fcomponent%2Fphocadownload%2Fcategory%2F29-ibm-article%3Fdownload%3D194%3Aanalysis-and-reporting&usg=AFQjCNGhwToeJj9QrXT8emKARryz2QTwaw&bvm=bv.111396085,d.d2s (accessed January 9, 2016).

IBM Global Technology Services (2013) *Virtualizing Disaster Recovery using Cloud Computing: Protect your Applications Quickly with a Resilient Cloud*. Thought Leadership White Paper, https://www.hashdoc.com/documents/18641/virtualizing-disaster-recovery-using-cloud-computing (accessed January 9, 2016).

Lifka, D., Foster, I., and Mehringer, S. (2013) *XSEDE Cloud Survey Report*, www.cac.cornell.edu/technologies/XSEDECloudSurveyReport.pdf (accessed January 9, 2016).

Microsoft System Center (2012) *Microsoft Private Cloud – Making it Real*, https://www.google.co.uk/url?sa=t&rct=j&q=&esrc=s&source=web&cd=1&cad=rja&uact=8&ved=0ahUKEwihxoWXppzKAhVHMhoKHTcfChQQFgg9MAA&url=http%3A%2F%2Fdownload.microsoft.com%2Fdownload%2F1%2F0%2F7%2F107D3951-9732-421D-8B57-AC19530F24D1%2FPrivate%2520Cloud%2520Making%2520It%2520Real.pdf&usg=AFQjCNHaLd5QpGQCJ-PpY7-LHrVRYZiztw&bvm=bv.111396085,d.d2s (accessed January 9, 2016).

56

Cloud Outlook: The Future of the Clouds

San Murugesan[1] and Irena Bojanova[2],*

[1] *BRITE Professional Services and Western Sydney University, Australia*
[2] *National Institute of Standards and Technology (NIST), USA*

56.1   Introduction

Cloud computing's transformational potential is huge and is yet to be fully embraced. Driven by several converging and complementary factors, it is advancing as an IT service-delivery model at a staggering pace and is causing a paradigm shift in the way we deliver and use IT services and applications. Cloud computing is also helping to close the digital (information) divide.

In order to embrace the cloud successfully and harness its power for traditional and new kinds of applications, we must recognize the features and promises of one or more of the three foundational cloud services – software as a service (SaaS), platform as a service (PaaS), and infrastructure as a service (IaaS). We must also understand and properly address several other aspects such as security, privacy, access management, compliance requirements, availability, and functional continuity in case of cloud failure. Furthermore, adopters need to learn how to architect cloud-based systems that meet their specific requirements. We may have to use cloud services from more than one service provider, aggregate those services, and integrate them on premises' legacy systems or applications.

To assist cloud users in their transition to the cloud, a broader cloud ecosystem is emerging that aims to offer a spectrum of new cloud support services to augment, complement, or assist the foundational SaaS, IaaS, and PaaS offerings. Examples of such services are security as a service, identity management as a service, and data as a service. Investors, corporations, and startups are eagerly investing in promising cloud computing technologies and services in developed and developing countries. Many startups and established companies

* This work was completed by Irena Bojanova and accepted for publication prior to her joining NIST.

continue to enter into the cloud arena offering a variety of cloud products and services, and individuals and businesses around the world are increasingly adopting cloud-based applications. Governments are promoting cloud adoption, particularly among micro, small, and medium enterprises. Thus, a new larger cloud ecosystem is emerging.

56.2 Bright Prospects

The prospects for the cloud are bright. Several converging and complementary factors are driving the further ascension of the cloud. The increasing maturity of cloud technologies and cloud service offerings by cloud service providers (CSPs) coupled with users' greater awareness of the cloud's potential benefits (and limitations) is accelerating its adoption. Better Internet connectivity, intense competition among CSPs, and digitalization of micro-, small-, and medium-sized enterprises are increasing the cloud's use. Changing attitudes and mindsets toward the cloud among users now accustomed to the growing ubiquity of mobile devices and applications is also greatly improving the cloud's adoption. Cloud technologies offer appealing responses to the growing demand from emerging markets for computing services at an affordable cost, and government support and initiatives are propelling clouds. Cloud computing is not just an IT paradigm change, as some perceive. It is redefining not only the IT and communication industry but also enterprise IT in all industry and business sectors. It is also helping to close the digital (information) divide, and facilitating the deployment of new applications that would otherwise not be feasible.

A cloud ecosystem has begun to evolve to provide an array of services that support the deployment of cloud-based solutions for applications across several domains. As the market for cloud computing continues to grow there will be more offerings of cloud services, and competition among CSPs will further intensify. New cloud-deployment types, value-added cloud services, and innovative costing and business models will emerge to serve varied client needs. Both private clouds and hybrid clouds will increase in number and capacity, as more medium and large enterprises start embracing them. Open-source cloud software will become popular and may be used widely.

Of course, there are a few major concerns about, and limitations of, the cloud that remain barriers to exploiting its fuller potential, and they need to be addressed satisfactorily. Several initiatives are under way that focus on standards for cloud security and interoperability, data virtualization through advanced analytics and parallel-processing optimization, and special services for development and deployment of mobile applications. Another key area being addressed is interoperability among clouds, which would let users scale a service across disparate providers, while maintaining the appearance of a single offering. Cloud federation – the interconnection of cloud services from different providers and networks – is a promising approach that lets providers wholesale or rent computing resources to other providers to balance workloads and handle spikes in demand.

There are also several ongoing developments, outlined in the next section, which are aimed at further advancing the cloud and its widespread adoption, addressing its limitations and concerns about it.

56.3 Advances and the Way Forward

The cloud computing field is fertile and cloud researchers, developers, cloud vendors, IT industry, regulatory agencies, and government must work together to advance the cloud further, addressing its limitations and concerns about it. Several new developments and initiatives are under way, which will further aid cloud computing. They span a range of areas, including software-defined systems, community clouds, cloud standards, cloud interoperability and portability, fog computing, cloudlets, cloud aggregation, cloud gaming, streamlined and standardized service-level agreements (SLAs), and cloud regulations. Major drivers of cloud adoption are big data analytics, the Internet of Things, ubiquitous mobile applications, and the increasing demand for IT services by individuals and businesses in emerging markets that are yet to fully embrace the power of IT.

56.3.1 Software-Defined Datacenters

The software-defined datacenter (SDDC), also known as a virtual datacenter, is an emerging next step in the evolution of virtualization and cloud computing. A traditional cloud computing environment has a flexible arrangement for resource allocation, utilization, and management gained through virtualization. However, its hardware environment is still "stiff" and hard to modify and adapt in an integrated fashion, and it poses considerable challenges regarding the flexibility, dependability, and security that next-generation systems will require. For example, networking topologies, many aspects of the user control over IaaS, PaaS or SaaS layers, construction of XaaS services, and others have limited flexibility and management options. To address these limitations, software-defined systems (SDS) that add software components to help abstract actual IT equipment and other layers, are gaining interest. Examples of SDS are: software-defined network (SDN), software-defined storage (DSD), software-defined servers (virtualization), software-defined security (SDSec), SDDC, and software-defined cloud (SDCloud).

In a SDDC, all infrastructures are virtualized and the control of the datacenter is fully automated by software – that is, the hardware configuration is managed by intelligent software system. This is in contrast to traditional datacenters, where the infrastructure is typically defined by hardware and devices. Ongoing research and development is poised to advance these areas and drive the next generation of clouds.

56.3.2 Smaller Regional Clouds, Community Clouds

Besides major global cloud vendors such as Amazon, Microsoft, Google, Rackspace, and IBM, which have huge cloud infrastructures to cater for a large number of clients, many small regional vendors have begun to offer cloud services that are intended to cater for the needs of local clients, providing local technical and application support. Such offerings also address the problem of legal jurisdiction, data residence, compliance requirements, if any, and pricing commensurate with local/regional conditions. New local cloud vendors are setting up datacenter facilities in their region, which would also help address concerns of cloud users related to global vendors.

Further, more community clouds, also known as vertical clouds or industry clouds, which are optimized and specially deployed for use by a particular industry sector or a group of users meeting specific requirements that are crucial to them, are emerging.

56.3.3 Innovative Applications Based on Combination of Cloud Services

Several cloud-based services will be combined in previously unimagined ways to offer entirely new applications and services. For example, let us consider an application such as Uber, which uses a few different cloud services. It utilizes geolocation to find both users and nearest cars, analytics to price on demand, communication to negotiate and connect with a traveler, and finally payment services to close the transaction (Abrosimova, 2014). Companies like Onshape, Proto Labs and MarkForged are offering new services that facilitate design, modeling, testing and collaboration to support manufacturing using 3D printing and other technologies. Many such new applications, which innovatively combine several services, will emerge in the near future.

56.3.4   Cloud Service Brokers and Aggregators

Enterprises will use multiple cloud environments delivered on heterogeneous platforms. This will require specialist skills and capabilities to manage the different cloud environments with a consistent management framework. Rather than obtaining in-house capabilities, which many SMEs cannot afford, users may prefer to utilize services of intermediaries such as cloud brokers and cloud aggregators who can provide value-added services that help cloud users. For example, a cloud broker can help a client to find a suitable cloud service, review its SLA, and negotiate and manage a deal on behalf of that client. A cloud aggregator can offer tailor-made cloud services / applications by drawing on a few different services from different CSPs, and integrate and manage the services offered to its client. As there will be increased demand for such value-added cloud services by intermediaries, new players will emerge to offer cloud brokering and aggregation services.

56.3.5   Cloud in Emerging Markets

Cloud computing is, and will continue to be, a catalyst for change in emerging markets (EMs) — nations in the process of rapid growth, industrialization, and socioeconomic development, which represent two-thirds of the global population. Although these markets have historically lagged behind advanced economies in adopting and innovatively leveraging IT to address their problems, they are now harnessing cloud computing in novel ways in a range of areas including business, education, socioeconomic development, health care, and governance. Emerging markets are fostering a new level of transformation, facilitated by the cloud. The cloud, combined with pervasive mobile phones, is transforming developing countries by facilitating improved productivity and economic and social progress. The potential of the cloud in developing countries is enormous. Hence, multinationals, local companies, policy makers, and IT professionals in emerging markets pay greater interest to develop and deploy the cloud and foster adoption of cloud computing particularly by micro, small, and medium enterprises (MSMEs), and by individuals. A number of supporting factors are contributing to the rise of clouds in EMs. Multinational IT businesses' interest in EMs, emergence of local CSPs, government and international organizational initiatives, collaborative initiatives and growing interest in and need for using IT for socioeconomic development, education, governance, banking and finance, and other factors, will raise the cloud offerings and adoption in EM.

56.3.6   Internet of Things

The Internet of things (IoT), in which things such as cars, TVs, machinery, and electrical appliances will be connected to the Internet, will create a whole range of new applications and services – smart grids, smart homes, smart cities, healthcare applications, and more. The large volume of data collected from vast array of sensors will be stored and analyzed at one or more clouds, and accessed from anywhere by mobile phones and other devices. Widespread adoption of the IoT in the coming years will drive demand for traditional and special cloud services, such as data analytics and location-based services, and will be a key driver for cloud's accession.

56.3.7   Fog Computing

Fog computing, also known as edge computing, is a new incarnation of the cloud. Instead of using computing and storage resources at a centralized location, as in a traditional centralized cloud, the data is stored and processed where it is generated or needed by placing some processes and resources at the network ends. Fog computing aims to reduce communication bandwidth needed by aggregating data at certain access points and then transmitting to a central cloud, rather than sending every bit of data. It is closely linked with the

IoT and sensor networks and is primarily for applications that require very low and predictable latency. Large distributed-control applications and some mobile and IoT applications have embraced fog computing as it facilitates simple processing and decisions where data is generated or aggregated near to the source, minimizing long-distance data transmission. It is also a way of bypassing the wider Internet, whose speed can vary considerably and largely depend on carriers.

56.4 Cloud Computing: The New Normal

Cloud computing is pervasive and is the new normal. However, while hailing the features of current and new cloud services that help users adopt and tailor the services they use according to their needs, it is important to recognize that the new interlinked cloud ecosystem still presents a few challenges and concerns. Such concerns are those relating to interoperability, the quality of service of the entire cloud chain, compliance with regulatory requirements and standards, and security and privacy of data, access control and management, and service failures and their impact. All these issues need to be addressed innovatively, and this calls for collaboration among various players in the cloud ecosystem. Good news is that investors, established corporations, and startups are eagerly investing in promising cloud computing technologies and services, and are willing to collaborate (to an extent) to raise the clouds to newer heights. We can hope for a brighter, bigger, more collaborative cloud ecosystem that benefits all of its stakeholders and society at large. Cloud service providers, the IT industry, professional and industry associations, governments, and IT professionals all have a role to play in shaping, fostering, and harnessing the full potential of the emerging cloud ecosystem.

In this encyclopedia, we offered a holistic and comprehensive view of the cloud from different perspectives. Innovations in technology, service delivery, and business models are needed to make further inroads and embrace the cloud ecosystem's untapped potential. We hope this encyclopedia kindles your thinking and helps to make the grand vision of an all-encompassing, interoperable, collaborative cloud ecosystem that benefits the society at large a reality in the near future.

We welcome your feedback on this encyclopedia and your thoughts on the emerging cloud ecosystem at cloudcomputingencyclopedia@gmail.com.

Reference

Kate Abrosimova (2014) *Uber Underlying Technologies and How it Actually Works*, https://medium.com/yalantis-mobile/uber-underlying-technologies-and-how-it-actually-works-526f55b37c6f#.nvaqkjpxr (accessed January 10, 2016).

Index

Encyclopedia of Cloud Computing, First Edition. Edited by San Murugesan and Irena Bojanova.
© 2016 John Wiley & Sons, Ltd. Published 2016 by John Wiley & Sons, Ltd.